T3-AKC-487

DISCARDED
THE UNIVERSITY OF WINNIPEG
PORTAGE & BALMORAL
WINNIPEG, MAN. R3B 2E9
CANADA

Quantitative geography: a British view

G
70
.Q36
1981

Quantitative geography: a British view

Edited by

N. Wrigley

and

R.J. Bennett

Routledge & Kegan Paul
London, Boston and Henley

First published in 1981
by Routledge & Kegan Paul Ltd
39 Store Street, London WC1E 7DD
9 Park Street, Boston, Mass. 02108, USA, and
Broadway House, Newtown Road, Henley-on-Thames, Oxon RG9 1EN
Set in 9pt Press Roman by Elephant Productions, London SE15
and printed in Great Britain by
Lowe & Brydone Printers Ltd, Leeds
© Routledge & Kegan Paul 1981
No part of this book may be reproduced in
any form without permission from the
publisher, except for the quotation of brief
passages in criticism

British Library Cataloguing in Publication Data

Quantitative geography.

1. Geography – Methodology
I. Wrigley, N. *II. Bennett, R.J.*
910'01'8 *G70*

ISBN 0-7100-0731-0

Contents

Part 1

Retrospect and prospect on British quantitative geography

Chapter 1

Introduction

R.J. Bennett and N. Wrigley

Rationale of the volume

Any text, particularly one as large as this, requires a justification for imposing itself upon an already overburdened academic community. In our case the justification is clear-cut and we hope it will satisfy our readers. This book is to be published during the eighteenth year of operation of the Institute of British Geographers Quantitative Methods Study Group (QMSG) and is intended to mark the Group's 'coming of age'. At any time, a coming of age is a suitable opportunity to take stock and to speculate on the delights to come, but in the case of the QMSG it also coincides with the beginnings of a new decade, the Fiftieth Anniversary of the parent body, the Institute of British Geographers, and the development of a truly international dimension to the activities of the QMSG. Several of these developments, particularly the Fiftieth Anniversary of the IBG, were felt to require a response by the QMSG, and the idea, therefore, of a retrospective and prospective review of the work of British quantitative geographers began to take shape. This volume is the realization of that idea.

The aim of this volume is to offer as broad as possible a view of recent developments in quantitative geography in Britain. From this retrospect it is intended to draw out major lines of future development in quantitative geography as a prospect on the areas in which British quantitative geographers can be expected to make significant contributions. Against this background, it has been our major purpose to include as large a range of contributors as possible in order to obtain a synoptic set of views. As such the volume is an updating and major extension of previous reviews of quantitative geography in Britain.

The aim of centering attention on quantitative geography in Britain does not, we hope, derive from a limited parochialism. Certainly it is not our intention to exclude, or in any way diminish, the very significant contributions to the subject made in other countries where, especially in North America, such contributions have been a major stimulus to quantitative research. The aim of emphasizing British work derives from three main stimuli. First, it is hoped to display the distinctive contribution that the British geographers have made to quantitative geography as, in some senses at least, a national caucus. Second, it is hoped to facilitate international contacts between British and other groups of geographers by displaying the current areas and emphases of research together with the likely lines of future development. Third, it is a major aim of this volume to project the Study Group in Quantitative Methods as a major forum for discussion at both national and international levels. In so doing, it is hoped that this volume will be useful to both British and other geographers in obtaining financial support to improve research contacts between the British and other national groups of quantitative and theoretical geographers. Certainly any retrospect and prospect of quantitative geography in Britain must acknowledge the considerable impact on the QMSG in the development of the subject and section 1.2 below gives a short review of the role of this group.

The rationale for the present volume provides a framework for setting the contributions into a particular context as retrospect and prospect. As such, however, it is not the editors' intention to detract from the merits of the individual contributions, which stand alone as significant comments on the particular areas of work covered. Clearly, it is also impossible, from a range of thirty-eight contributions from forty-two authors, to draw very rigorously defined lines of retrospect or prospect for the subject. Although the range of contributions is both large and wide in its interests, the pressure of space has prevented all possible contributions and ranges of interest being included. Similarly, in our comments in this introductory chapter, and in the linking chapters to each part of this volume, we have had to be selective as to which points to emphasize and isolate for particular attention. The individual chapters are presented for the reader to use in constructing his own view of the development and prospect for the subject. The journals which quantitative geographers use extensively for publication (especially *Environment and Planning*, and *Geographical Analysis*) are available to the reader to determine how far this volume as a whole is a fair representation of the subject. Hence, we choose as editors unashamedly to emphasize, in the introductory and linking chapters, our own views, as gleaned by compiling this volume. We hope, therefore, that the contributors will accept this comment as an apologia should they disagree markedly with our views.

The volume has been divided into five main parts which emphasize broad features of the development of the subject. Following the present introductory chapter, Part 2 addresses specific issues of information collection, data handling, and the effects of data sampling and structure on the resulting analyses that can be undertaken and the conclusions that can be drawn. Part 3 consists of a large range of papers on statistical methods and models which draw together themes that have particularly characterized British work. Some of these themes, such as factor analysis, multidimensional scaling and point pattern analysis, have been extensively developed outside of Britain. Other themes, especially spatial and temporal analyses, exploratory data and categorical data analyses, have been developed very largely by British workers. In Part 4 of the volume mathematical models are discussed both within major areas of application, such as urban and programming models, but also with emphasis on major lines of development that are required: especially responding to the stimuli of mathematical topology, stochastic process theory, control theory and Q-analysis. Part 5 of the volume addresses lines of development of quantitative geography as they have affected other branches of the discipline in both physical and human geography. Finally, in Part 6 of the volume, the issue of teaching quantitative geography is raised, with treatment given to the problems of presenting material at both school and university level.

British quantitative geography and the IBG Quantitative Methods Study Group

For eighteen years the IBG Quantitative Methods Study Group has played a major and often underestimated role in maintaining the momentum of quantitative geography in Britain. The origins of the group have been summarized by S. Gregory (1976), its first chairman. The QMSG was formed in January 1964, and for its first five years operated outside the formal research group structure of the IBG. These were the early years of the 'quantitative revolution' in geography and the Group acted as a forum for the few British geographers who were either using quantitative techniques in research or pioneering the introduction of quantitative methods into undergraduate degree courses in geography. In addition the group acted as a catalyst of change in the nature of British geography. Committee members in this period included Chorley, Haggett, Gregory, Garner and Cole: leaders of the movement which resulted in a 'paradigm shift' in British geography towards the approaches of model building and spatial science.

In 1968 the Group became incorporated as a Study Group of the Institute. By that time quantitative methods had become an established part of geographical research and teaching and membership of the Group was increasing rapidly (223 members by 1969). The next few years saw the development of the now established pattern of three formal meetings per year, and control of the Group passed to a younger set of individuals, notably Robson and Goddard as Chairman and Secretary. Meetings in the period 1969-71 (on sampling problems, trend surface analysis, scaling techniques, network analysis, the teaching of quantitative methods, computer simulation, space-time budgets, interaction models, the analysis of spatial series, classification and regionalization) reflected the widening interests and increasing activity of quantitative geography (see the QMSG annual reports in *Area* 1, p. 46; 2, p. 61; 3, p. 60). In retrospect, this was the period in which the Group can be seen to have established itself as an important constituent within the structure of the IBG, and in which the framework of the Group's organization was laid down.

The period 1972-5 saw the next major change in the Group's activities. Since its foundation, one of the Group's major objectives had been to develop links between statisticians, mathematicians and geographers. In its early years the Group had achieved this through the participation of individual statisticians and mathematicians in its meetings (e.g. C.W.J. Granger and J.K. Ord), and through informal contacts. In the period 1972-5 formal contacts were established, and the first joint conferences and publications took place. These included: a joint conference with the Applied Stochastic Processes Study Group in 1975 (ASPSG, 1976) which atttacted a grant from the Statistics Committee of the Social Science Research Council; the review paper by Cliff and Ord (1975) to the Royal Statistical Society Research Section; and a special issue of *The Statistician* (1974). Membership of the Group continued to rise throughout the period, reaching 400 by the end of 1975, and again control of the QMSG rested with a younger group, passing to Taylor and Unwin as Chairman and Secretary. Meetings in the period considered topics such as: cartographic automation and geographic analysis: Markov models in geography; analysis; Markov models in geography; measurement and analysis techniques in behavioural geography; multivariate analysis in geography; computer applications in geographical education; whither spatial analysis; applied spatial analysis; spatial interpolation techniques; mathematical and theoretical geography; problems and potentials in survey research; geographic information systems; and patterns and processes in the plane (see the QMSG annual reports in *Area* 4, pp. 72-3; 5, p. 77; 6, p. 76; 7, pp. 132-3). Residential meetings were established as part of the Group's annual programme, the first of the Group's publications was launched, and S. Gregory, the Group's first chairman, became President of the IBG - the *bête noire* of the 1960s had truly become integrated into the establishment.

After the rapid growth of the late 1960s and early 1970s, the period 1976-80 saw a stabilization in the membership of the Group at around 400. This was brought about by two linked factors. First, the rapid growth in other study groups within the IBG (from 6 in 1970 to 15 in 1980), and second, the growing disenchantment of many younger geographers with the spatial science model-based paradigm. This disenchantment generated a search for alternative philosophical and methodological positions, and in human geography hermeneutic and critical science perspectives replaced empirical analytic approaches as the fashionable areas of interest.

Despite this changing environment outside the QMSG, however, the late 1970s were years of considerable success for the Group. The policy of joint meetings and publications with research bodies outside the geographical fraternity was continued and meetings were held with the Royal Statistical Society, the Operations Research Society, the Remote Sensing Society, the Institute of Mathematics and its Applications, and the Regional Science Association. In addition, many eminent guest speakers from the fields of statistics, mathematics and computing were invited to QMSG meetings and the links with these disciplines were progressively strengthened. Extending this policy of developing external links, a policy of fostering international contacts between sister groups of the QMSG was then developed, and this period saw an Anglo-Swedish symposium in 1977, the First Anglo/Franco/German Colloquium in Quantitative and Theoretical Geography in 1978 (Unwin, 1979), the Second Anglo/Franco/German Colloquium in Quantitative and Theoretical Geography (which also included members of Dutch and Italian quantitative geography groups) in 1980 (Bennett *et al.* 1981), and the development of strong relationships with the Association of American Geographers Speciality Group in Mathematical Models and Quantitative Methods (founded in 1979 as the AAG equivalent of the QMSG). In the field of publication, the Group's CATMOG (Concepts and Techniques in Modern Geography) mongraph series launched in 1975 proved to be a considerable commercial and academic success and, under the editorships of P.J. Taylor, R.J. Johnston and D.J. Unwin, more than twenty-eight mongraphs had been produced by 1980 and were in wide use throughout the world. In addition, books of collected conference papers from the Group's joint meetings with other research bodies were published (Martin, Thrift and Bennett, 1978; Wrigley, 1979). Finally, the Group was involved in the first national research training courses to be run for postgraduate geographers in Britain, sponsored by the Social Science Research Council, and by 1980 the First and Second SSRC Research Training Courses in Data Collection and Analysis for Geographical Research had been held at the University of Bristol (Hepple and Wrigley, 1979; Unwin and Beaumont, 1980).

Hence, although the late 1970s will be remembered by some as a period in which criticism of the quantitative, empirical analytic paradigm was widespread in Britain, it was nevertheless a period in which the QMSG reached a new peak of activity. In this period, control of the Group once again passed to a younger group, first Unwin and Wrigley, and then Wrigley and Bennett as Chairman and Secretary, and more meetings than ever before were held on topics such as: cartography in the 1980s; dynamic models and spatial representation in geography and regional science; the use and analysis of census data; statistical applications in the spatial sciences; stochastic process models in physical geography; the teaching of quantitative geography; data collection and analysis in less-developed countries; research in quantitative geography; aspects of exploratory data analysis; data handling and remote sensing; spatial applications in operational research; the analysis of categorical data; developments in modelling geographical systems; and quantitative geography and public policy (see the QMSG Annual Reports in *Area* 8, pp. 159-60; 9, p. 131; 10, pp. 72-3; 11, p. 25; 12, p. 91; 13, p. 91).

Up to this point, therefore, the history of the QMSG has been one of a progressively increasing range of activities. Publications, research training, a vigorous policy emphasizing contacts with research groups in other disciplines, and growing liaison with sister groups in other countries have been grafted onto the Group's core functions. The prospect for the 1980s in Britain is a period in which the higher education sector will be static, or contracting. As a result the number of recruits to both the subject and to the QMSG will also be static or may contract and, hence, it is likely that the average age of the Group's members will increase. In such circumstances, the policy of external links is likely to take on an even more important role in maintaining the Group's vitality, and, as noted above, it is hoped that this volume will facilitate this policy by emphasizing the QMSG's role as a major forum for the discussion of quantitative and theoretical geography at both the national and international levels.

Retrospect and prospect

Quantitative geography in the 1960s was in the vanguard of research in the subject as a whole, both in Britain and elsewhere: to the extent in fact that it was dubbed a 'revolution' in the subject, and indeed formed the basis of a 'new' geography. In contrast, the 1970s have seen a period of maturity and 'coming of age' of quantitative geography in Britain, and elsewhere. Whereas the 1960s had often been characterized by early exploratory use of technique, the testing of many previously untried tools, and, inevitably, the committing of a number of statistical and methodological blunders, the 1970s have seen the emergence of a maturity, a deeper understanding, and an accepted commitment to quantitative work. To such an extent has the quantitative side of the subject matured and come to pervade the whole discipline, that quantitative geography in Britain, as in North America, has been seen increasingly as part of the 'establishment' of the subject. And as part of the establishment it has been subject to attack by other 'new' geographies, especially those advocating a greater degree of 'relevance', 'humanism' or 'social commitment'.

It is possible here only to highlight a few of the more important features of the shifts of the past which we consider to be important indicators of prospects for future development: four interrelated points deserve particular attention:

1. there has been a paradigm shift in geography as a whole away from spatial geometry;
2. there has been a reappraisal of the nature and appropriateness of statistical inference within the methodology of quantitative geography;
3. applied social, political and other frame disciplines have provided increasingly imporant stimuli for quantitative work;
4. there has been a growing criticism of quantitative geography as part of the critique of positivism.

Geography and the paradigm of geometry

A convenient distinction in discussing the nature of a subject is that between 'core' and 'frame' disciplines. Core disciplines may differ at different time periods, but at any one time are those areas of thinking which are providing new systems of terminology, forming new concepts, and developing new explanatory paradigms. Over periods of time, such core disciplines have profound effects on other subjects as the influences of their new concepts and paradigms diffuse. Seventeenth-century astronomy and Newtonian mechanics, nineteenth-century Darwinism, twentieth-century quantum mechanics, statistical mechanics, probability theory, control engineering and management science, have each successively formed cores of new concepts and paradigms which have diffused and have had profound influences on other subjects often highly unrelated to the original objects of study.

The nature of the geographical core disciplines which define the methods of observation and analysis in geography, and the form of geographical objects from which derives the basis of a specific geographical nomenclature has always been a hotly disputed question. Such a core discipline has been sought in the 1960s and 1970s within the spatial geometric view. In this view the geographical object is one of point, line, area or other spatial relationship; and from this view derive concepts of central place systems, network analysis, and so forth.

Frame disciplines, on the other hand, are those which derive in methodology, object of study, and terminology, from other external subjects: for geography such frame disciplines are, for example, climatology, hydrology, geomorphology, social geography, economic geography, demography, etc. Lichtenberger (1978) has addressed major importance to this distinction between such core and frame disciplines as far as it characterizes the different research approaches of geographers, their careers, the institutions they work in, and the teaching structures which these institutions operate.

In the early days of quantitative research in geography there was a growing momentum which supported a specifically spatial quantitative focus as the core. The University of Michigan's group of mathematical geographers, the Regional Science Research Institute, and the work of Christaller, Lösch, Isard, Hägerstrand, Dacey, Berry, Curry, Olsson, Kansky, Bunge, Haggett, Chorley, and many others, seemed to point to the emergence of a specifically spatial set of quantitative techniques and models and a distinctive spatial methodology. More recently, the work of Cliff and Ord, *Spatial Autocorrelation* (1973), the Haggett, Cliff and Frey *Locational Analysis in Human Geography* (2nd edn. 1977), and Bennett's *Spatial Time Series* (1979), also seem to point towards a specifically spatial focus for geographical techniques. But nowhere is this approach better exemplified than in the two volumes (1962, 1966) by W. Bunge on *Theoretical Geography,* and in P. Haggett's *Locational Analysis in Human Geography* (1965). The geographical objects of study are those of point, area, region or surface which represent the spatial relations between phenomena. Geographical phenomena are accepted as the legitimate objects of interest by other disciplines, but geography is distinguished from other disciplines by its basis in the study of the spatial relations between phenomena. Hence, the paradigm within which geography was coming increasingly to work was one which characterized the subject by the distinctiveness of its spatial point of view. It is not surprising, therefore, that concepts such as networks and topological structures should increasingly occupy attention. And not surprising also that quantitative geography should adapt very readily to the geometric point of view: in fact one was the stimulus to the other and it is not realistic in retrospect to divide components which at the time, for many researchers, seemed inseparable.

Increasingly in recent years, however, the paradigm of spatial geometry has been weakened. Indeed many researchers would now reject that a specifically spatial view can be used as a means of defining the core of the discipline. At one extreme, for human geography, the spatial view has been characterized as a special form of fetishism which obscures more fundamental social issues. Thus, for example, Castells (1976) states that there is no specifically spatial theory which is not part of a more general social theory, i.e. that human geography is no more than a special branch of social science. At another extreme, for physical geography, the methodology of other subjects is increasingly seen as overriding any rather weak spatial view that geographical analysts may adopt. Indeed, the criteria of intellectual legitimacy and acceptance within the subject are increasingly ones which dictate publication of research findings in journals outside of geography which often provide higher quality referees and where the research will find a wider and more interested audience.

Hence, many of the questions deriving from the methodology of spatial geometry have been left behind. In their place is, in part, a methodological vacuum as far as providing a paradigm for the subject as a whole. In part also the spatial geometry view has been replaced by a more specific emphasis on spatial processes. Where geometry emphasized the relations between phenomena and often led to purely descriptive models, an emphasis on spatial process provides a more fundamental understanding of dynamics, response, rate of change and *functional* relationships, and this offers a much more satisfactory explanatory power. This is partly related to other shifts, to be discussed below, of a greater concern with deductive over inductive thinking.

The major shift which has affected quantitative geography, however, has been towards a view that places quantitative geography largely as a frame discipline no longer at the centre of development of the geographic core. With the decline in acceptability of the spatial geometric paradigm, the specifically spatial emphasis of technique development has been very largely eroded. This leads to the important consequence that quantitative geographers have increasingly sought recognition, criticism and involvement with other disciplines. In Britain, this has characterized the increasing formal and informal contacts of the QMSG with other groups, for example the Royal Statistical Society, Operations Research Society, Institute of Mathematics and its Applications, Remote Sensing Society, etc.

A further important question, however, concerns the relationship of quantitative geography to the rest of geography. Here again, recent years have seen an erosion of the number of people who would research exclusively on, or would term themselves specialists in, quantitative geography. This is in contrast to the growing numbers of geographers who would accept other assignations but who would make extensive use of quantitative techniques; for example urban geographers, social geographers, biogeographers, hydrologists, oceanographers, glaciologists, etc. Geography has become more specialized, and at the same time, the sub-disciplines have taken on board a quantitative methodology such that most quantitative geographers are now also specialists in other fields of enquiry. This certainly increasingly characterizes the members of the British QMSG.

As a consequence of these shifts, a major feature of the 'coming of age' of quantitative geography is that the crusade for the methodology of quantitative work needs no longer to be waged. The battles for acceptance which characterized early years of the subject (see Gregory, 1976; Taylor, 1976) have resulted in a fairly pervasive use of quantitative methods throughout the discipline. Whilst mistakes and blunders in the use of such methods continue and will continue to be made, most of the major sub-disciplines within geography make and will continue to make considerable use of quantitative methods. As a consequence the research focus for concern in quantitative techniques has shifted to a considerable extent into these specific areas. This is reflected in the papers of this volume, particularly in chapters 25 to 36 which provide a synoptic discussion of major areas of applications of quantitative methods. A small number of specialist quantitative geographers will continue the development of specifically spatial statistical techniques, often in close association with statisticians and other specialists outside the subject. Indeed major research fields in data handling, and spatial and temporal analysis are clearly identified in this volume. In addition, important developments will continue to be made in mathematical models and theory, and again this volume identifies mathematical topology, stochastic process theory, network theory (including Q-analysis) and control theory as areas where important developments can be expected. However, in the future, it seems likely that the majority of developments will be made by geographers in specialist frame sub-disciplines where the theory, models and techniques are dictated by, and derived from, the rigours and specific characteristics of applied research problems.

Anuchin (1970a; b) has suggested that periods of increasing generalization within disciplines are often succeeded by periods of increasing fragmentation, which are then absorbed into a period of further generalization, and so on in a cyclic fashion. Geography as a whole in the 1960s underwent a period in which general concepts (derived from viewing the subject as spatial science) absorbed many of the more particular aspects of the subject. In the

1970s this generality has gradually given way to the emergence of an increased emphasis on particularity and specialization within the subject. The contributors to this volume in many ways reflect this fragmentation. It seems likely that in the 1980s there will be a period in which fragmentation will increase yet further before geography is again ripe for the development of a new integrating paradigm such as that which characterized the discipline in the 1960s.

The reappraisal of statistical inference

Associated with shifts in emphasis of the subject as a whole, a major change in the approach of quantitative geographers has been an overall decline in the use of modes of analysis based on the theories of inferential statistics and a more sophisticated understanding of their appropriateness. Statistical inference was seized upon in the 1960s as a panacea for geographical methodology. The supposed objectivity of the Neyman-Pearson theory, the seeming rigour of methods of repeated sampling and trials, the convenience of the critical region as a criterion of choice, and in many cases the sheer intellectual beauty of the methodology itself, made the methods of inferential statistics a very attractive area for geographers. One had only to grasp the necessary tools and the subject could attain all the characteristics of an applied science, and could stand equal comparison with other social sciences, especially economics, which were also rapidly incorporating these methods into their sub-disciplines.

In the 1970s, however, many geographers have come increasingly to question the role and appropriateness of the methodology of statistical inference in the subject, and Gould (1970), for example, has gone so far as to ask whether *statistix inferens* is the name of a geographical wild goose! In part, this has resulted from a growing realization of the difficulties involved in using classical inferential procedures which assume independent observations with geographical data which typically exhibit systematic ordering over space and time, and British quantitative geographers such as Cliff, Haggett, Hepple and Bennett have made a notable contribution in demonstrating and overcoming these difficulties. However, it also reflects wider issues relating to the developing critique of positivism in geography, and the search for appropriate modes of quantitative enquiry in the social and environmental sciences. Leaving aside, for the moment, the question of the critique of positivism, to be discussed later, an internal debate within the corpus of quantitative geographers has focused on the nature of appropriateness of statistical inference in geographical research. Three aspects of this debate are discussed below.

First, it has been increasingly suggested that modes of statistical inference evolved for agricultural trials and controlled experiments are inappropriate for the social sciences. These latter are distinguished from other sciences most particularly by the large quantity of prior knowledge which is available, both from internal self-appraisal and phenomenological interpretation, and from the characteristic that we can ask components of the system why they respond in the way they do. Hence Neyman-Pearson theory and the associated concept of the null hypothesis based on pure poisson, Binomial and normal distributions are quite inappropriate. Instead a method of approach allowing the incorporation of variable degrees of prior information should be pursued. This leads us naturally to the construction of likelihood functions and Bayesian modes of inference, in contrast to the Neyman-Pearson theory.

Second, there has been an increasing concern with the limitations of the data available in many areas of geographical research. A growing band of British geographers has suggested that in the past quantitative geographers have asked too much of or given too little consideration to the nature of their data, have been prepared to accept too cavalier criteria as to the assumptions which the data satisfy, or have sought too high a degree of sophistication in the nature of the conclusions which we can expect to draw in either the social or environmental sciences. This in turn has led workers such as Openshaw and Taylor to consider the relationship between statistical inference and zonal data, and to argue that the fundamental importance of the modifiable areal unit problem casts doubts on many of the previous uses of inferential procedures which have employed zonal data. It has also led researchers such as Wrigley to argue that geographers should learn to deal successfully with the low-order categorical data characteristic of much social science research where data is often derived from social surveys. In both these cases a more sophisticated form of statistical inference is seen as being required to deal with more difficult data. In contrast, another group of British geographers has argued that there is also a role for much less sophisticated procedures. Workers such as Cox and Jones have followed the approach of John Tukey and argued that geographers should place emphasis on exploratory rather than confirmatory modes of statistical analysis. Instead of adopting the restrictive apparatus of statistical inference, they argue that geographers should show more willingness to engage in uninhibited exploration of their data, and that they should often be content with simple but robust numerical measures and imaginative graphical displays.

Third, there has been a shift in emphasis in the subject of geography as a whole from primarily inductive to increasingly deductive methodology. Quantitative geography was never as inductive in emphasis as the critics of positivism (see e.g. Gregory, 1978) would like to pretend. However, it was characterized in its very early days in North America by an inductive emphasis, and it is probably true to say that in Britain the emphasis on induction has been maintained for a much longer period. North American workers such as Les Curry, Waldo Tobler and Georgiou Papageorgiou have often commented (at least in private) on the relative absence of deductive theory on the east side of the Atlantic! In the 1970s, and certainly in the 1980s, the emphasis of British work has shifted towards a more coherent use of models and equations derived with considerable foreknowledge from well-developed prior theory. Instead of inference, increasing concerns have been, and will continue to be, with the estimation, fitting and calibration of models specified largely *a priori* (using existing theory and empirical results), and with the appropriate design of spatial representations to give the best models to display and use known relationships. What is the point in setting up a null hypothesis to test whether a relationship does exist when the relationship is an established part of prior theory and existing empirical knowledge? Despite this obvious statement, many of the papers in the field of quantitative geography in the 1960s were characterized by such a methodology, and a shift away from this false emphasis is related to the growing concern with applied research discussed below.

For the future, then, we must expect that statistical inference will continue to be used, often in a much more sophisticated way, but overall to a much less important extent than in the past. In addition three lines of development will increasingly emerge: one based on the more extensive use of prior theory and applied research and concentrating attention on estimation, calibration and representation; a second deriving from the use of prior theory to define likelihood and Bayesian probability functions; and a third which places increasing emphasis on exploratory rather than confirmatory modes of statistical analysis.

The stimulus of applied and policy research

Underpinning the shift away from both the emphasis on statistical inference and from spatial geometry as a core of the subject has been the increasing emphasis in geography as a whole on its applied or 'frame' disciplines. Few commentators now worry very greatly about the 'Nature of Geography', and many of those who do set geography in the field of social science (e.g. Harvey, 1973, Gregory, 1978, 1982) or some other applied science. This trend may not continue through the whole of the 1980s, but in the immediate future the emphasis of the subject has shifted to its applied areas. This is evidenced by the proliferation of IBG study groups and AAG paper sessions and interest groups, by the growing body of geographers who publish in the journals of other disciplines, by the body of other researchers publishing in geographical journals, and by the rapid expansion of joint meetings and contacts outside of the subject.

Two themes which arise from the stimulus of this applied emphasis merit particular attention. First, applied research has lead to a rebirth of subjectivity in the subject. Statistical methods and mathematical models were never objective tools, although it has often been claimed that they are; they can only be applied within a framework of assumptions, premises, and hypotheses, which reflect prior theory; and prior theory has an important subjective as well as an objective dimension. Subjectivity enters especially from the experience of the analyst working with data and problems he knows well. The expert in a field applies not only his objective critical and logical facilities, but also his subjective appraisal of the importance of facts, his judgment of the significance of errors and inaccuracy and, in the social sciences, his empathetic understanding of the social situation. As quantitative geography has 'come of age', this subjectivity has increased and the expertise of geographers in specialist fields has become combined with a specific set of techniques and methods deriving from quantitative geography.

A second theme deriving from applied research results from the increased presence of pressures for 'relevance', and to questions of social and political importance. This has resulted in a whole suite of implications which are discussed more fully in chapters 34 to 36 on political geography, public policy and the geography of elections. Here it is only necessary to highlight two main features. First, the 'new' geography of the 1960s has been supplemented by a new 'new' geography of the 1970s concerned with radicalism, social relevance, and social responsibility. Like the first 'quantitative revolution', this second 'revolution' has derived from North America. As far as quantitative geography is concerned, this new 'revolution' has resulted in quantitative techniques being developed for applications to new questions of social distribution. Moreover, it is now clear that many of the important questions of Marxist and related analyses can be answered only by very meticulous quantitative analysis. The second consequence of this new applied emphasis is the emergence of a new normative emphasis in geography as a whole which has affected quantitative geography in important ways. Most important amongst the impacts of normative thinking has been a new emphasis on the goals of society. Deriving from this have been the twin interests of, on the one hand, determining the likely progress and goals of society unfettered by policy, and on the other hand, of determining the impact of public policy on the social and economic geography of the State. This leads to a natural concern with social goals, policy instruments, policy assessment and policy recommendations, all of which require approaches which are, in part, formulated quantitatively.

In the 1980s, therefore, we believe that quantitative geography will shift its emphasis towards subjective questions including goals and norms, and will find increasing application to areas of social policy. This will naturally lead to a greater emphasis on deductive thinking and model building, and to more exploratory analysis of data. The criteria applied to the acceptability and significance of this work will be less those of statistical or mathematical elegance, and more those of the meaningfulness of the results (a) to social and political questions, and (b) to the fundamental research questions of the sub-disciplines involved.

Quantitative geography and the critique of positivism

It would be inappropriate to present a volume such as this without some comment on the critique of positivism within geography which has emerged in the 1970s. The comment contained in the following paragraphs, we feel, states the position at the opening of the 1980s and one which will increasingly characterize the subject over the next decades. It is our contention that the critique of positivism in geography, whilst interesting in itself, is largely *irrelevant* to quantitative geography. We take note of the criticisms made, but contend that these are directed at an abstract view of quantitative geography as positivism which attributes methods, views and conclusions to quantitative geographers which most have never held, or if they did ever hold, have since abandoned, or if they hold in some form, hold only in part alongside wider views which make the critique of positivism at best a misrepresentative irrelevance, and at worst a fatuous distraction.

The critique of positivism has emerged through a series of writings deriving from Marxist, neo-Marxist, humanist, phenomenological, 'structuralist' and other viewpoints. Major examples are those of Harvey (1973), Gregory (1978, 1982) and Sayer (1976) amongst many other contributions. It is the contention of these writers, in various degrees of emphasis, that quantitative geography is 'evil' and 'alienating'. First, it creates a false sense of objectivity by attempting to remove the observer from the observed, which in turn lends an ability to control and manipulate. Second, its use of machines and techniques subsumes man under a paradigm of mathematics and machinery, reifying man to an atomistic level by losing the elements of soul and humanistic concerns. Third, it is descriptive of existing behaviour and hence supports the *status quo*, especially of social distribution, within society. Fourth, it allows no consideration of values and hence of the norms by which systems, and hence society, *should* be organized. Fifth, it attempts to construct models and theories of universal generality by a mode of inductive logic moving from the particular to the general.

These contentions of the critics of positivism each carry some small element of truth. Indeed, we have already recognized that much quantitative geography in the 1960s was concerned, to too great an extent, with techniques *per se*, with creating an artificial sense of objectivity, and with induction and inference rather than use of prior theory. However, the best quantitative work of that period, and the majority of quantitative work in more recent years, has been far more catholic than its critics would allow. Moreover, as the preceding paragraphs have highlighted, modern quantitative geography has already incorporated many of the elements the critics see as lacking: for example, the use of less abstract approaches which allow 'soft' and perceptual information to be incorporated; the assessment of normative and ontological goals for models of *becoming* as well as of the *status quo* of being; the treatment of questions of social distribution; the mix of deductive and inductive theory; and the use of prior information.

Indeed, in the chapters of this volume, many of the characteristics the critics would wish to develop are already clearly evident: an 'applicable' quantitative geography is already at hand and is being used to answer the fundamental questions of social norms, social distribution, policy impacts and humanistic concerns which the critics rightly emphasize. Moreover, the burden of the social critique of quantitative methods is one largely characteristic of the English-speaking countries only. In West European, East European, Chinese, and other literature, techniques have been accorded a primary place in answering socially and politically relevant questions. Perhaps tangentially to this point, it is worthy of note that an extensive part of Marxist-Leninist philosophy is concerned with 'political arithmetic' such that in many respects Lenin can be seen as an important originator of the input-output model together with Leontief and Isard.

Hence, in concluding this introductory chapter, we would contend that the critique of positivism is largely irrelevant to quantitative geography in the future. Quantitative geography has never been truly positivist, as the critics contend, and the questions which the critics would seek to emphasize are already being successfully tackled by quantitative geographers. We would conclude, with Hay (1979) that the critics of positivism are victims of their own misunderstandings and misrepresentations. In setting up a polarized and unrealistic view of the 'quantitative geographer', and the structure of his methodology and constructs, the critics have constructed a straw man, or what Wilson (1978) terms an 'Aunt Sally': a windmill to tilt at quixotically. Moreover, much of the critique, or support for it, has come from those who seek an easy way out. It is, after all, easy to resort to rhetoric or polemic as a mode of reasoning rather than undertake detailed quantitative empirical analysis to assess the validity of cherished assumptions. As a result, we would conclude, with Haggett (1965, p. 310) that 'In the long run the quality of geography in this century will be judged less by its sophisticated techniques or its exhaustive detail, than by the strength of its logical reasoning.'

References

Anuchin, V.A. (1970a) 'Mathematization and the geographical method', *Soviet Geography Review and Translation,* 11, 71-81.

Anuchin, V.A. (1970b) 'On the problems of geography and the task of popularizing geographical knowledge', *Soviet Geography Review and Translation,* 11, 82-112.

ASPSG (1976) 'Patterns and processes in the plane', Conference Report by the Applied Stochastic Processes Study Group, in *Advances in Applied Probability,* 8, 651-8.

Bennett, R.J. (1979) *Spatial Time Series: Analysis, Forecasting and Control,* Pion: London.

Bennett, R.J. *et al.* (1981) 'The Second European Colloquium in quantitative and theoretical geography', *Area,* 13.

Bunge, W. (1962) *Theoretical Geography,* 1st edn, Lund: London.

Bunge, W. (1966) *Theoretical Geography,* 2nd edn, Lund: London.

Castells, M. (1976) *The Urban Question,* Arnold: London.

Cliff, A.D., and Ord, J.K. (1973) *Spatial Autocorrelation,* Pion: London.

Cliff, A.D., and Ord, J.K. (1975) 'Model building and the analysis of spatial pattern in human geography', *Journal of the Royal Statistical Society,* B 37, 297-348.

Gould, P.R. (1970) 'Is *Statistics inferens* the geographical name for a wild goose?', *Economic Geography (Supplement),* 46, 439-48.

Gregory, D.E. (1978) *Ideology, Science and Human Geography,* Hutchinson: London.

Gregory, D.E. (1982) *Social Theory and Spatial Structure,* Hutchinson: London.

Gregory, S. (1976) 'On geographical myths and statistical fables', *Transactions Institute of British Geographers,* New Series 1, 385-400.

Haggett, P. (1965) *Locational Analysis in Human Geography,* 1st edn, Arnold: London.

Haggett, P., Cliff, A.D., and Frey, A. (1977) *Locational Analysis in Human Geography,* 2nd edn, Arnold: London.

Harvey, D. (1973) *Social Justice in the City,* Arnold: London.

Hay, A. (1979) 'Positivism in Human Geography: response to critics' in D.T. Herbert and R.J. Johnston (eds), *Geography and the Urban Environment: Progress in Research and Applications,* Vol. II, 1-26, Wiley: Chichester.

Hepple, L.W., and Wrigley, N. (1979) 'SSRC research training course', *Area,* 11, 261-2.

Lichtenberger, E. (1978) 'Quantitative geography in the German-speaking countries', *Tijdschrift voor Economische en Sociale Geografie,* 69, 362-73.

Martin, R.L., Thrift, N.J., and Bennett, R.J. (eds) (1978) *Towards the Dynamic Analysis of Spatial Systems,* Pion: London.

Sayer, R.A.F. (1976) 'A critique of urban modelling from regional science to urban and regional political economy', *Progress in Planning,* 6, 189-254.

The Statistician (1974) Special issues, 23, Nos. 3 and 4.

Taylor, P.J. (1976) 'An interpretation of the quantification debate in British geography', *Transactions Institute of British Geography,* N.S. 1, 129-42.

Unwin, D.J. (1979) 'Theoretical and quantitative geography in North-West Europe', *Area,* 11, 164-6.

Unwin, D.J., and Beaumont, J.R. (1980) 'Data collection and analysis for geographical research: a postgraduate training course', *Journal of Geography in Higher Education,* 4, 72-80.

Wilson, A.G. (1978) Review of Sayer (1976), *Environment and Planning,* A 10, 1085-6.

Wrigley, N. (ed) (1979) *Statistical Applications in the Spatial Sciences,* Pion: London.

Part 2

Data collection management and processing

Methods of data collection and organization are fundamental to quantitative geography, and since the formation of the QMSG in 1964 significant changes have occurred in these fields. First, accompanying technological progress, new methods of data collection have become available, for example, remote sensing of environmental data from satellites. Second, shifts in research orientation within geography have created a demand for different types of data. For example, in human geography a concern with problems of spatial behaviour and choice has resulted in a growing demand for microscale social survey data of various types, and in physical geography a similar demand for more microscale data has accompanied the shift in orientation from studies of form to studies of process. Third, and most important, the burgeoning power and rapidly reducing real costs of computer hardware have taken geography, along with the rest of the academic community, and society at large, out of the pre-computer age of the 1950s into the silicon-chip, home-computer age of the 1980s. In so doing the new technology has dramatically transformed the geographer's information handling and analysis capabilities. The chapters in this section review some of these major changes. In addition, they show how computing power has helped to extend the geographer's ability to handle traditional sources of data, such as census data, and to deepen his understanding of traditional data collection and analysis problems; for example, the 'modifiable areal unit problem' which results from much geographical data being collected or reported in areal units of arbitrary and varying size.

The first chapter, by Rhind, addresses the central issue of how computer power has changed the data organization and handling capabilities of British geographers. In the 1960s and early 1970s quantitative geographers were by no means slow in using computers in research and teaching. Indeed SSRC and QMSG surveys in 1973 (Unwin, 1974) confirmed that geographers, together with economists and psychologists, were the major users of computers amongst social scientists, that most undergraduate geographers were provided with courses in computer use, that geography departments often had a relatively high hardware investment, and were major computer users in many British universities. Much of this use was related to statistical analysis, to the development of mathematical models for use in urban and transport planning, or to the routine use and development of computer mapping programs. In terms of the potential of computer systems for the organization and manipulation of geographical data, however, Rhind characterizes this work as being rather limited and essentially of a 'one-off' type. Most geographical studies were concerned with the manipulation and mapping of individual data files. The use of computers to interrelate geographical data sets drawn from different sources and pertaining to different variables and/or different moments in time (i.e. to develop 'geographical information systems') was extremely limited. Since the mid 1970s there has been some increase in the amount of research in this area and operational geographical information systems of varying degrees of complexity have been developed at Durham, Manchester, Edinburgh, University College London, the LSE and Cambridge. Nevertheless, the interest in geographical information systems amongst British geographers has remained comparatively low. In this respect, Rhind sees the 1980s as presenting a major challenge for quantitative geographers. He believes that it is essential for geographers to come to terms with three issues: first, to acquire the capacity to analyse the torrent of typically extremely large data files continuously being created by governments and international monitoring and data collection agencies; second, to create computer systems which can interrelate data sets collected on different geographical bases at different moments in time but for the same geographical area, and which can provide rapid data display; and third, to use such systems as the means by which quantitative techniques can be applied to *non-trivial* problems. Only in this way does he see geographers being able to influence government data collection, analysis and dissemination practices, and perhaps ultimately policy.

Many of the issues which Rhind raises are illustrated in more detail in the chapters by Harris and Evans. Both geo-coded national population censuses, and remote sensing Earth-resource satellites create geographical data files of immense size and create associated problems of computerized data management and processing. In both cases, the sheer size of data files has necessitated research on methods of data compression, and substantial degrees of compression have often been achieved. In such circumstances the organization of data files becomes a crucial issue, particularly if comparisons between data sets are to be made. Once again considerable improvements in file access times have been achieved by use of organization techniques. In addition, both chapters provide illustration of how automated cartography can be used to provide rapid and often high-quality data display, and how this forms an essential part of a geographical data processing system.

The final two chapters in this section form a bridge to the issues of statistical analysis discussed in Part 3 of the volume. Evans discusses the statistical problems confronted in the analysis of the geo-coded 1 km grid square data from the 1971 population census of Great Britain, and illustrates approaches which link this chapter to the exploratory data analysis and closed number system data chapters of Part 3. He also raises an issue of central importance in geographical data collection and analysis - the problems caused by the collection and/or reporting of much geographically relevant information in the form of zonal data (i.e. relating to a mesh of areal units of arbitrary, and perhaps varying, sizes). Evans draws attention to the problems caused by the continuing experimentation with the spatial basis of the British census, and raises the issue of how changing the scale of the analysis affects the statistical results which can be derived from it. However, as Openshaw and Taylor point out, the *scale problem,* whereby results vary as areal units are combined to form new larger units at a new scale, is only one aspect of the problems caused by the zonal nature of much geographical data. Another issue is the *aggregation problem* whereby results vary due to the large number of zoning systems which can be employed at any given scale (i.e. due to the number of ways of positioning the mesh of areal units, even when the number of units remains constant). Together, the scale and aggregation problems comprise the *modifiable areal unit problem.* Openshaw and Taylor consider this problem to be of fundamental importance in geographical analysis, and believe that it casts doubts on many (if not the majority) of applications of quantitative techniques to zonal data. This has important implications for some of the approaches discussed in Part 3, and Openshaw and Taylor argue that it is essential for geographers to become more aware of the zoning systems they study and to seek for an essentially geographical solution to the modifiable areal unit problem.

Geographical data sets, of course, are by no means all zonal in nature. Spatially referenced micro-scale data collected by a variety of methods and agencies are widely used in all branches of geographical study. Although the chapters in this section do not explicitly consider such data, it should be noted that the QMSG has paid considerable attention to its collection and analysis, and many chapters in Parts 3 and 5 assume such data as the norm. Moreover, the development of large-scale integrated geographical information systems, and the closer contact with government organizations which this is likely to involve, will bring with it the promise, and also the frustrations, of gaining (or failing to gain) access to government-collected micro-scale data. As Rhind suggests, there are major challenges in this area for the quantitative geographer in the 1980s.

Reference

Unwin, D.J. (1974) 'Hardware provision for quantitative geography in the United Kingdom' *Area,* 6, 200-4.

Chapter 2

Geographical information systems in Britain

D.W. Rhind

Introduction

'Geographical Information Systems', usually abbreviated to GIS, is a term which first appeared in general use in the late 1960s. It is normally used to describe general-purpose and extensible computer facilities which handle data pertaining to areas of ground or to individuals or groups of people who can be defined as living or working in specific geographical locations. In this chapter use of the term is restricted to those computer systems which have the capability to interrelate data sets pertaining to different variables and/or to different moments in time. Thus facilities solely for the manipulations or mapping of individual files are not here considered as geographical information systems. Rhind (1976) argued that 'information systems' was, in any case, a misleading title and proposed that 'data processing systems' was a more apposite description of the systems in current use (e.g. see Tomlinson, Calkins and Marble, 1976). Whatever the terminology, the concept and prospect of facilitating all of the quantitative procedures laboriously utilized in an *ad hoc* manner by geographers since the 'quantitative revolution', of using the torrent of government-produced data (CSO, 1978) with negligible effort and, beyond this, of interrelating data sets - hitherto separate and unrelatable through resource limitations or inadequate referencing - are attractive. They are made even more attractive as the costs of computer power decrease: the Computer Board Working Party on Micro-Computers estimated in 1979 that, over the last fifteen years, these costs had diminished by an order of magnitude every five years and would continue to do so for the foreseeable future. Partly because of this diminution in cost and of their rapidly increasing storage capacities, computer facilities are becoming more ubiquitous; it is also significant that increasing amounts of data are being captured directly in digital form, thus further reducing costs and minimizing effort. Direct data capture is particularly feasible, of course, in the environmental sciences and, to some extent, in physical geography.

Despite these obvious incentives, the interest in geographical information systems amongst British geographers has remained comparatively low especially if computer mapping *per se* is excluded (see Rhind, 1977). We may speculate that this is related to the considerable level of personal commitment and long lead times before significant results are obtainable. Alternatively, it may be because both building and using a comprehensive GIS demands team, rather than individual, effort. It may even be because some of the functional requirements of geographers are complex and, on occasions, may be impossible with present resources (see below). It is proper, however, to relegate technical problems to a later stage and begin by enquiring why we should or should not be involved in such work. We do this by a brief, personalized interpretation of recent British quantitative geography which differs from that of Unwin (1978) in being specifically concerned with those geographers involved in empirical work using a computer.

Information systems, geographers and government

The academic approach

British academic geographers, like their North American counterparts, have been enthusiastic and even prolific so far as applying and evaluating statistical procedures is concerned. In the main, these procedures have been applied to secondary data, i.e. those collected by other workers. Only certain classes of geographical problems have been addressed by these studies. Over time, almost all of the procedures have come to be carried out on a computer and, increasingly, standard computer packages such as SPSS, OSIRIS and CLUSTAN have been employed. The relative advantages and dangers of such a pre-programmed approach are not the subject of this paper: Carmer (1975), Francis (1973a;b), Janaceck and Negus (1974), Slysz (1974), Velleman and Francis (1975) and others have catalogued the comparative merits of different packages and the methodological problems which are often ignored (e.g. see in Johnston, 1978). Here we merely point out that the view of geography usually taken is a simplistic one: the archetypical data organization in such studies has been the rectangular data matrix - n cases, one for each geographical individual or entity (described usually as a point, line or area) by m variables or attributes pertaining to each individual. In essence, then, these analyses are based upon a selection of or from one slice or two parallel and adjacent time slices in Berry's (1964) geographical data cube. Where parallel slices are used, this has almost invariably been at a level of spatial resolution (such as the English county) where no changes in form of 'collecting unit' or 'geographical individual' have taken place between the time slices. Such studies have often encoded some abstraction of geography but usually one so crude - such as a zone centroid - as to limit severely the spatial aspects of the possible analysis.

These 'area profiles', 'factorial ecologies' and 'regional taxonomies', carried out mainly over the two decades from 1960 onwards, have arguably made good use of existing resources, notably population censuses from many different countries. In Britain, analyses such as those by Knox (1974), PRAG (1975), Coulter (1978) and many others have been reviewed by Hakim (1977) and 'officially approved' in that the government census agency has itself published the results of national factorial ecologies based on their data (Webber and Craig, 1978). The methodological bases of the ordination and classifications most commonly used in these studies were reviewed by Spence and Taylor (1970). Increasing amounts of criticism, however, have been levelled at such 'handle turning' exercises. Some such studies seem to have been obscurantist, as in the study of Liverpool in which separate data sets were individually factor analysed and the resulting factor scores combined into one file and factor analysed again (PRAG, 1975); Armstrong (1967), of course, parodied (and also anticipated) such work and demonstrated some of its shortcomings. Freestone (1979), in reviewing the Comparative Atlas of America's Great Cities (Adams, Abler and Ki-Suk Lee, 1976) - which was substantially based on census data - described urban geographers of the early 1970s as 'data fetishists'. It is certainly the case that most such studies are *ad hoc* and entirely descriptive: even more seriously, as Moore (1979) - following Coombs (1964) and many others - has shown, such studies are not 'objective' in any sense other than 'replicable'. In the first place, the possible analyses are inevitably constrained by the available data: longitudinal data are rarely available and aggregate cross-sectional data are the norm, the time between the cross-section varying from one to ten years. While fixed five- or ten-year periods between surveys simplifies technical aspects of organizing data, they can lead either to aliasing or to total loss in detected events because of the coarse temporal resolution (although most area profile studies are even more limited in pertaining largely to one moment in time, providing only a static description of geography). Even allowing for the limitations of the available data, the selection of variables to analyse, the data groupings and the assignment of class intervals, as well as the selection of algorithms for more sophisticated operations like factor analysis, reflect - at best and perhaps unconsciously - implicit theories of behaviour. At worst, such decisions are unconscious ones through sloth, conceptual inertia or incomprehension of the technicalities involved.

The more computationally accomplished geographers of the last decade have, by and large, been concerned with one form or another of modelling (Unwin, 1978). Even here, considerable criticism has been levelled at the proponents (see Batty, 1978). One of the more serious accusations to be levelled is that many model specifications make quite unreasonable demands upon existing data. But, to this author at least, there is also a direct parallel between 'modellers' and 'factorial ecologists' in that both groups have commonly seen their work as 'one-off'. The emphasis in modelling has been on the estimation method and the algorithm, rather than on modularity, ability to be re-configured easily, place within a system or ease of use, understanding and application. A promising exception to this generalization was the work of Baxter and co-workers in the early and mid 1970s (Baxter, 1974, 1976a).

The government approach

Simultaneously with much of the statistical analysis of census-type data and modelling by academics, a number of organizations and individuals in different branches of government in the United Kingdom were planning, implementing, repairing and updating various geographically disaggregated data bases and creating related computer programs. It is convenient, if a simplification, to detect two threads to these developments - the building of data bases for administrative purposes and the experimental use of data and programs for scientific or quasi-scientific purposes, such as mapping. In the former instance, two separate types of system grew up. The first was an aggregate-data statistical reporting system, similar to some of those in government research laboratories (below) or in universities - LINMAP (Gaits, 1969, Gaits and Hackman, 1976) was a good example of the *genre*, having embryonic data manipulation facilities as well as crude mapping capabilities. University-trained geographers frequently had considerable influence in their formulation and use. The second development involved far greater resources and pertained in the main to micro-data: few, if any, geographers were involved. One of the two most significant developments of this type in Britain was the LAMIS system (Harrison, 1978) - originally set up in Leeds and now run by six local authorities - which is primed with data, including boundaries, pertaining to all individual properties plus some data on households. The second significant development was the Joint Information System in Tyne and Wear. The latter was jointly sponsored by DoE and the Tyne and Wear County Council in its putative stage (when it was known as the National Gazetteer Pilot Study). Spicer *et al.* (1979) and Rhind and Hudson (1980) have described the salient features of this system, which contains over 500,000 point-referenced records for individual properties in the County, each of which is updated when change in land use status is detected through the Development Control System or other monitoring systems in local government. Additional data may be linked to the central filing system in both LAMIS and JIS via a unique identifier constructed from the street address.

Because of the scale of investment devoted to the building of these and other information systems with geographical components and because there is a real prospect that their use could diffuse through local government, considerable need was seen for coordinating the miscellaneous developments; the Department of the Environment, for instance, published a manual of advice on methods of point referencing properties (DoE, 1973), as well as being associated with a number of other published reports on both technical and non-technical aspects of the management of information pertaining to land and people (HMSO, 1972; DoE, 1974; DoE, 1975; ICL, 1976). These developments in local government, however, were often outside the remit of DoE: the MOSS system, for example, was developed by a consortium of local authority Highways Departments (Craine, 1975). As a consequence, coordination had variable success. Latterly (from c. 1975), other government agencies such as the Ordnance Survey and the Army have become involved in building computer systems to collect geographically disaggregated data for reasons other than replicating an existing mapping purpose: digital terrain models (DTMs), for instance, have become necessary for flight training simulators and, presumably, for targeting Cruise missiles. The development of GIS within these organizations is still, however, in its infancy (compare this with the situation in the USA where between 100 and 200 DTMs per month are purchased from the USGS (Doyle, 1978a) and where numerous uses of these data have sprung up (PERS, 1978).

The second thread of government development, that of the experimental creation of GIS, was rather different. The most relevant - but hardly typical - example of an organization which falls within its definition was the Experimental Cartography Unit, funded largely by the Natural Environment Research Council and concerned primarily, but not exclusively, with making maps by computer. Throughout its lifetime, the large majority of employees in the Unit have been geographers and computer scientists. As yet, no overall history of the work of the Unit exists although various papers such as Rhind (1971) describe this at various stages. Margerison (1977) has given a partial summary of the published papers emanating from the Unit, claiming that 200 scientist man-years had been devoted to building up geographical data bases and computer systems to interrelate and map them. Quite apart from the ECU, however, other organizations such as the Institute of Terrestrial Ecology, the Institute of Oceanographic Sciences, the Institute of Geological Sciences (Jeffrey, Gill and Henley, 1974), the Harwell Laboratory (Jackson *et al.*, 1980) and the Transport and Road Research Laboratory (Perrett, 1976) were busily creating programs, systems and data for research purposes from the early 1970s. Some of these were subsequently pressed into uses which were never anticipated: the use of Perrett's TRAMS system and its structured street network by the West Midland Police to route policemen to scenes of crimes is one such example.

Towards an academic/government tapestry?

Thus, in direct government-financed organizations, the 1970s saw rapid expansions in the number and size of data bases with implicit or explicit geographical references, together with investment in associated computer programs and systems. Most of the few important new methods - as opposed to methods made feasible by newer, faster technology - were devised by computer scientists (e.g. Aldred, 1975; Stocker, 1975). On the basis of their publications, academic geographers as a computer user group were largely interested in substantive or, alternatively, rather esoteric ends and tailored their research to what was easy and/or specific. A small number of academics, however, were actively embroiled in more complex conceptual and technical problems of representing, encoding, cross-referencing and analysing data with a geographical basis: though no formally titled and institutionally-based group such as the Geographic Information Systems Laboratory at Buffalo has appeared in Britain to work specifically on GIS, the contributions of British academics are significant and are described later in this paper. At the time of writing, however, it is fair both to say that few of the fruits of their work have fed into general usage in Britain and only a few have actually been involved in collaboration with government institutions - though most have willingly used data from such sources (cf. Adams and Rhind, 1981; Rhind and

Hudson, 1980). Only one group in academe (see Rhind and Hudson, 1980) is known to have produced major data bases from the survey stage onwards and only a few academics (e.g. Baxter, 1974; Duffield and Coppock, 1975; Rhind, 1980) have set up a variety of linked data bases.

Yet collaboration between government and academica is, at least at first sight, of mutual benefit: academics gain access to data collected on a scale far beyond that of individual or university resources whilst government gets - or has a right to expect - helpful criticism and improved theories and methodologies from academics. Wood (1977), in pointing out the comparative lack of interest of British geographers in information systems, itemized some of the consequences for the discipline. He argued that, without academic inputs, new approaches to the collection of spatial data in Britain would be justified largely on the requirements of the current planning machinery or on local considerations. It can readily be demonstrated that data needed for local authority purposes can be unsuitable for many research purposes: Rhind and Hudson (1980), for instance, have described the very considerable difficulty experienced in comparing land use data because of different classifications, each focused on specific and partially incompatible local government requirements.

To all those who have been involved in substantive, quantitatively-based research, the biggest difficulties have always stemmed from the nature of the data available. Rarely, for example, do aggregate geographical data meet closely the stringent preconditions for many statistical tests (e.g. see Johnston, 1978, p. 252 onwards). In an ideal world, human geographers need longitudinal profiles of each human being in their study - where each person lived and worked at different times, their health history and much else related to the individual such as the life history of industrial plants - if meaningful causal studies are to be carried out. Some micro-data - a poor substitute for such profiles but a step forward from area aggregates alone - have been made available on a sampled basis from the US Census of Population: a discussion paper on its provision in the UK has been published (Hakim, 1978) but, at the time of writing, no prospect of making such data available to researchers was held out. A further data shortcoming is the concentration on (sometimes repeated) static inventorying: even government data sets which contain migration data, for instance, very rarely have this disaggregated by origin and destination. Improvement in both or either of these areas would have implications for techniques employed in GIS, such as larger data files and more complex cross-references; they would necessitate more sophisticated use of computers in government departments and, most serious of all, they would arise at a time of increasing concern over data banks containing personal information (HMSO, 1975; HMSO, 1978; Bulmer, 1979a, b; JRSS, 1979) and the role of government data collection agencies.

More, perhaps, than in any other topic of interest to geographers, work involving GIS benefits from collaborating closely with government organizations. Without this, the academic working with geographical information systems - as opposed to 'one-off' studies from data purchased 'off the shelf', often long after it is available to 'insiders' - is reduced to working merely on untestable technical solutions to assumed or postulated problems. In addition, the desirable access to highly disaggregated data such as establishment data from the Annual Census of Employment is rarely possible without formal collaboration with government (usually consolidated by the signing of the Official Secrets Act). Yet, set against these benefits is the inevitable restriction - in Britain at least (cf. Sweden: see Rystedt, 1977) close collaboration with local and national agencies invariably restricts academic freedom. Working on and publishing results derived from certain readily available data sets - such as census-derived aggregate urban housing statistics - is highly unlikely to lead to opposition or withdrawal of support unless the results are directly opposed to entrenched policy of the sponsoring agency. But, as Moore (1979) has pointed out, such data severely limit the value of many analyses. To obtain access to micro-data (that pertaining to individuals, often on a sampled basis, or to individual establishments) relating to such topics as crime, illness, wealth, manufacturing production, employment characteristics and the like is a very different matter: to obtain it is difficult and to publish on the basis of it involves extremely detailed scrutiny of papers by sponsors prior to submission of papers. Closeness to government does not, of course, guarantee access to such micro-data. In one major study in Britain involving the creation of a geographical information system sponsored by a government department, for instance, the academic researchers were unable to obtain detailed employment statistics from a closely allied government department, even though the figures were instantly available to research workers in local government and to the sponsors!

The technical problems

These may be subdivided into those resulting from the non-availability of local skills or computer resources, those which relate to the most appropriate organization of the data and system facilities to meet defined needs, and those which involve fundamental problems, either computational or geographical. The first set is ignored here because of its local nature, even though it may be substantial constraint on the development of a GIS. It is, for example, singularly difficult to make maps which even approximate to manually produced products if no pen-plotter is linked into the local computer - as still occurs at a number of academic computer centres in Britain. In looking ahead to the 1980s, such a physical resource problem can be expected to diminish, not least because of the growing availability of computer networks. What is rather more likely to remain a constraint is the availability of easily used and reliable software (see below).

Data organization and system design

Data structure General principles may readily be promulgated for the manner in which geographical data are organized and stored (see Martin, 1975; Pfaltz, 1977, and numerous other texts, plus Peuker and Chrisman, 1975; Rhind, 1976). The two most general principles are that the organization should reflect the structure in the data themselves and also should be orientated towards the primary uses of the data. In the frequent absence of data-base management systems on many university machines, this is rarely a trivial exercise. Even prior to this, however, the receipt and conversion of data from government sources is, at present, often less than straightforward when it is in bulk machine-readable form; apparently trivial problems can absorb much of the researcher's time and energy. Two examples will be given. The first concerns the supply of the 1971 Population Census Small Area Statistics by 1 km grid squares to the Census Research Unit (CRU) at Durham. This consisted almost entirely of a single 'time slice' relating to 150,000 areas, for each of which there were 1,571 variables (Rhind *et al.*, 1980; Visvalingam, 1976). The data were supplied nearly three years later than originally planned and this caused obvious problems in research grant scheduling. More technically serious, however, was that, when the whole data set was received, transferred to a different method of encoding and reorganized, blank areas were found in resulting maps. The investigation of the reasons for these blanks and for other inconsistencies in the data, together with the supply of the missing data, resulted in further unplanned effort and delay. The second example relates to a pilot information system built for the Manpower Services Commission to handle employment, unemployment and demographic data; even though both the employment and unemployment data sets pertained to the same areas of ground (the Employment Office Areas) and were produced by the same organization (the Department of Employment), the identifiers on the identical area were different in the two data sets! While readily solved by constructing a look-up table, these experiences emphasized that bulk transfer from certain government departments may still be troublesome. Examples of the converse situation also exist, e.g. 500 Ordnance Survey digital map sheets were transferred without significant trouble to a university computer.

It has been argued (Rhind, 1976) that various elements of the geographical individual can be distinguished conveniently so far as GIS are concerned. Only three will be considered here. The first of these is the geometry - encoded usually on a dimensional basis (see Waugh, 1974) as points, lines or areas. The measurement scale(s) used can be either nominal (such as street name), ordinal (normal house numbering within a street) or the quantitative scales. Where quantitative measurement scales are utilized, the coordinate systems may be planar or spherical and a variety of units have been utilized - between which there is some approximate means of conversion, such as the formulae for conversion of latitudes and longitudes to plotting coordinates. The second element is the set of scalar attributes related to the geographical individual, which may be on any measurement scale: Webster (1977), for instance, describes the use of a simple geographical information system of soil data in which the attributes are binary presence-or-absence in form. In other situations, different variables are on different measurement scales and some form of data dictionary is needed to obviate blatant misuse of the data. The third element, which is sometimes not included in geographical data bases, is the relationships between geographical individuals or their attributes. Those most frequently encoded are equivalence or proximity in geographical space or in property space. Hence, topology is encoded via a table of links from each line segment to one or more others; it may be complemented by a file of all the segments emanating from each node. Examples of all three elements are shown in Figure 2.1. Topological encoding is used as a data quality control and as an aid to building polygons from bounding segments in the TRAMS systems (e.g. Perrett, 1976), having first been used in a major way by the US Bureau of Census in their DIME (Dual Independent Map Encoding) system (see Loomis, 1965; Cooke and Maxfield, 1967; Corbett, 1975). Alternatively, topology can - in most instances and often laboriously - be generated from the geometry by hunting for adjacent end points of line segments. Some systems are built entirely upon this presumption (Cook, 1978; Thompson, 1978). Such systems make major demands upon the fidelity and completeness of the geometric elements in the data base: missing segments or sections of them can create substantial difficulties. Another example of relationships sometimes encoded is that of the links between the nearest stored examples of certain types of features (such as outcrops of stratigraphic units); in this case 'nearest' may be nearest in terms of the file organization, rather than in geographical space. Yet another example of relationships which may be inbuilt (although usually outside the raw data file) is that of hierarchies, e.g. of all districts forming one county and all counties forming one region and Figure 2.2 illustrates some examples of such encoding. It will be obvious that many of these

Figure 2.1 Examples of organization of geographically disaggregated data, to record explicit topology and geometry in vector 'notation'.

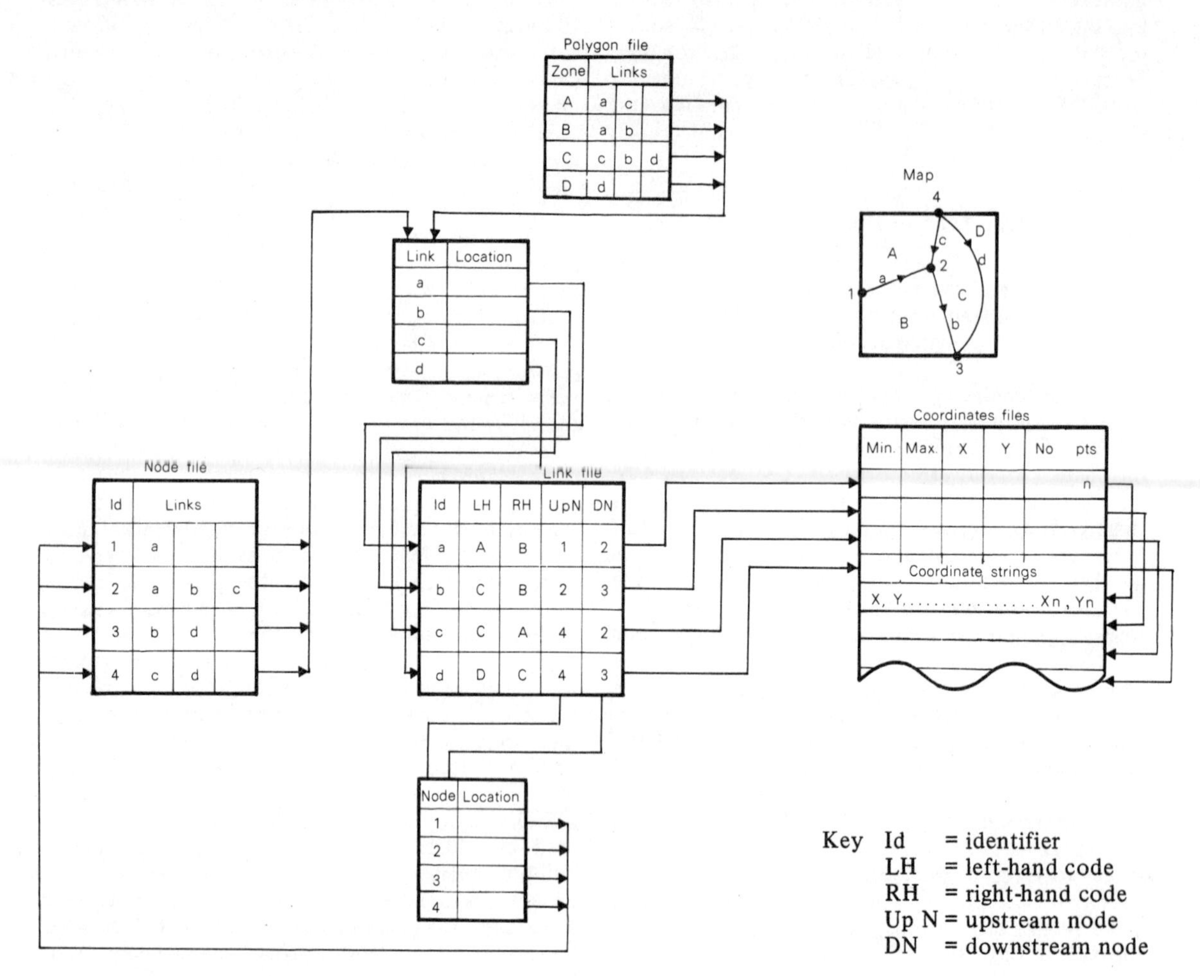

relationships may be deduced from the data base itself, provided enough detail by way of attributes or sufficient geometric accuracy is provided in the files in the first instance. Baxter (1976b) took the opposite approach in encoding details of an administrative hierarchy via the attributes of boundary segments in his file of district, county and regional data for Great Britain.

The above description hints at the practical complexities which may result in building data files with explicit geography (or, as stated above, geometry). The situation is complicated still further by three other considerations: the 'notation' (Rhind, 1975b) in which the data are encoded, the current and potential size of geographical data bases, and the likely uses of the data base.

Spatial 'notation' of the data Almost all of the discussion thus far has implied that the data model to be used for storing geographical individuals is a string-based one where vectors of X and Y coordinates define linear features or boundaries of zones and where, in the degenerate case, one coordinate pair defines the perimeter of an area of zero size, i.e. a point. This is not the only 'notation' used - in many cases, especially where data are obtained from scanning radiometers or linear arrays of sensors in remote sensing (see Doyle, 1978b; L.L. Thompson, 1979), the data are organized in a raster scan format with the X and Y coordinates and some topology implicit in the sequencing of the Z values. Such notation is generally, but not invariably, of constant spatial resolution; the inevitable consequence is that the geographical units thus described are artificial ones (see Harvey, 1969) and are usually mixed (see Rhind and Hudson, 1980). Switzer (1975) discussed the accuracy of encoding polygonal data by such cells. Other methods of defining the geometry (e.g. Pfaltz and Rosenfeld, 1967) have not found general use. In essence, the main reason for the great and growing popularity of raster encoding and storage is convenience: various remote sensing

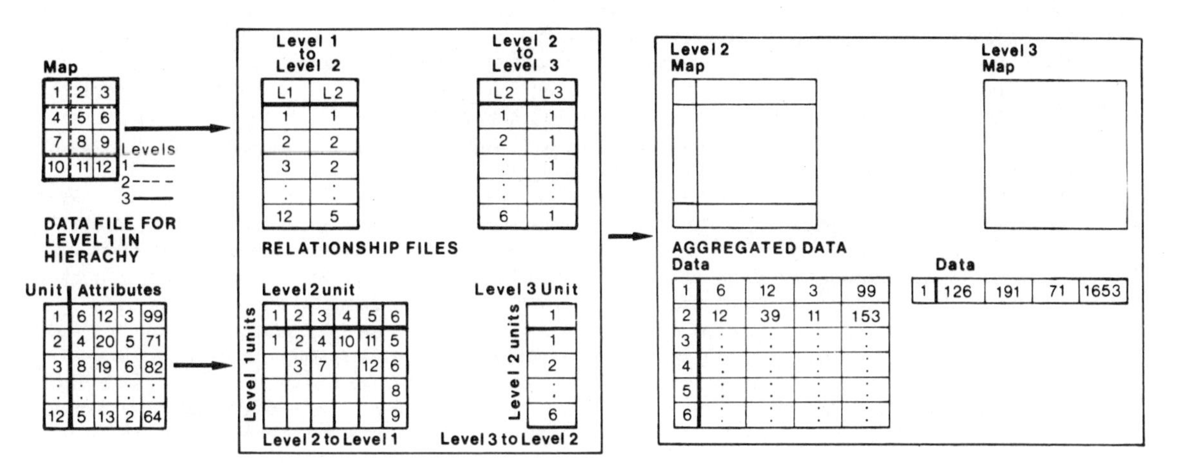

Figure 2.2 Encoding of hierarchical relationships of geographical individuals where the individual areas are nested

devices collect the data in this 'notation' and certain operations such as overlaying two data sets for the same area (Tomlinson, 1974) are greatly simplified using it. Other operations, however, pertaining to linear or boundary features - such as calculating the length of boundaries of individual zones of the same land use type - are rather easier and more accurate with suitably organized vector data; considerably less data may also be needed to store distributions using strings and that 'notation' is inherently one which permits variable spatial resolution to be recorded to meet local situations. It seems inevitable, therefore, that both methods of organizing data will continue to be used and appropriate conversion facilities from one to another will be required (see below) in geographical information systems. Proceeding from vector to raster organization is a well-defined operation, if sometimes lengthy. The reverse operation is extremely time-consuming and constitutes a major algorithmic challenge (Williams, 1979).

Size characteristics of geographical data bases Most geographical data bases created other than by academics are large (Rhind and Adams, 1981) while the operators applied to them are algorithmically simple. It has been estimated, for instance, that to store the whole of the Ordnance Survey basic scale topographic data for Great Britain would require of the order of 550 magnetic tapes using existing OS methods of storage (Adams and Rhind, 1981). A regular grid of ground altitudes covering the whole of the United States is stored on some 2,000 magnetic tapes. Currently the most useful set of international boundaries, water features and transport links in digital form is the CIA-produced World Data Bank II: even though this was encoded from maps at scales as small as 1:1 million to 1:4 million (Anderson, Angel and Gorny, 1978), some 5.6 million coordinate pairs are utilized. In all of these cases and in many attribute files used by geographers, substantial degrees of data compression - on occasions by factors of 5 to 10 (Freeman, 1974; Visvalingam and Rhind, 1977) - can be achieved and redundancy can sometimes be eliminated safely. Even so, we are seeing only the beginnings of geographical data files - census data files, for instance, will continually accumulate. Satellite imagery has also added a new dimension to geographical information systems. Digital storage of each 185 km square 'picture' from the current version of Landsat satellite produces approximately 60 x 10^6 bits (binary digits, 0 or 1) per waveband; each picture is imaged in four wavebands and is repeated at least every 18 days (NASA, 1976). The present resolution of Landsat is about 80 metres but forthcoming earth resource satellites will work to resolutions of 30 metres or better, multiplying the data volumes by at least another factor of seven.

Doyle (1978a) has pointed out that to encode ground altitudes alone at 1 m resolution all over the earth's surface would require 10^{14} bits - to load this into the core of a large contemporary computer would require of the order of three years!

There is no suggestion here that all geographical information systems will need to be primed with all available geographical data or to have to deal with satellite data - but only that direct data capture devices are becoming commonplace and that encoding data locked up in maps to acceptable levels of resolution leads to substantial data sets. With data sets of this size and frequency, some degree of organization is certainly essential if inter-data set comparisons are to be made, since sequential access is unacceptably slow: improvements in access time to areal sub-sets of individual CRU population census data sets by factors of more than 100, for instance, have been achieved when the data are organized in a known fashion, as compared to selection based on sequential searching through the original file.

Defining user needs We have already suggested that some geographical problems - notably those dependent on cross-tabulation and analysis of statistical associations - can not be tackled ideally without micro-data. Apart from this specific difficulty, most of the technical problems discussed thus far have arisen from internal shortcomings in data quality or from the need to carry out operations which were not anticipated by the originators of the data. Such problems are inevitable since omniscience is an unreasonable characteristic to demand of data suppliers; the range of operations which users may wish to carry out is also, in practice, almost infinite. Despite this, however, some form of specification of user needs is a prerequisite before any geographical information system can be designed and constructed. Here we discuss it after other technical considerations only because definitions of terminology and technical constraints provide essential background.

At one extreme are users such as 'factorial ecologists' who simply wish to use data to describe a situation: at the other is the user like Tobler (1979) who suspects that evolution of form might permit one to deduce process or, better still, the 'rules of the geographical chess game'. Such uses differ greatly in their demands on systems and data organization: the degree of detail, for instance, needed to record explicit topology, etc. varies greatly with the application. Anticipating even some of those applications and structuring data appropriately has turned out to be extremely expensive if done at the encoding stage (see Southard, 1978, for one such approach). The situation is complicated still further by the gestation periods over which the systems are built and the data collected and the consequent need to plan on the basis of future, rather than present, requirements. A recent government committee set up to define the role of Ordnance Survey (HMSO, 1979) over the next twenty years took more than one million words of evidence in over 500 submissions from interested parties: despite this, extreme difficulty was encountered in defining what digital topographic data would be required by users and when - it is already clear from recent experience that the existence of systems and data generate new demands and data uses.

Perhaps *the* major problem is that while it is easy to define data and system needs in a 'low-level' administrative function, such as collecting local taxes (e.g. rates), defining these for essentially idiosyncratic 'high-level' functions such as geographical research is a very different matter. In the face of such difficulty, the development of information systems for 'high-level' tasks has thus far begun by focusing on a single, comparatively well-defined requirement (e.g. Duffield and Coppock, 1975; Rhind, Evans and Visvalingam, 1980; see also Note 1). These 'bottom up' systems grow by accretion as new demands are encountered. Figure 2.3 illustrates the contemporary state of one such system (Rhind, 1980). The alternative, 'top down', approaches have mostly been confined to mapping systems (e.g. see Waugh and Taylor, 1976). An alternative viewpoint to that of defining or cultivating systems is that of defining primitive or basic functions and then providing these, usually in sub-program form. Rhind (1976), GAG (1977)

Figure 2.3 A conceptual representation of the system set up in Durham. The different lines indicate some of the differing routeways used to combine and compare data thus far: most data have also been analysed and/or mapped prior to combination. Details of some of the individual projects and contents of the data files are given in Rhind (1980)

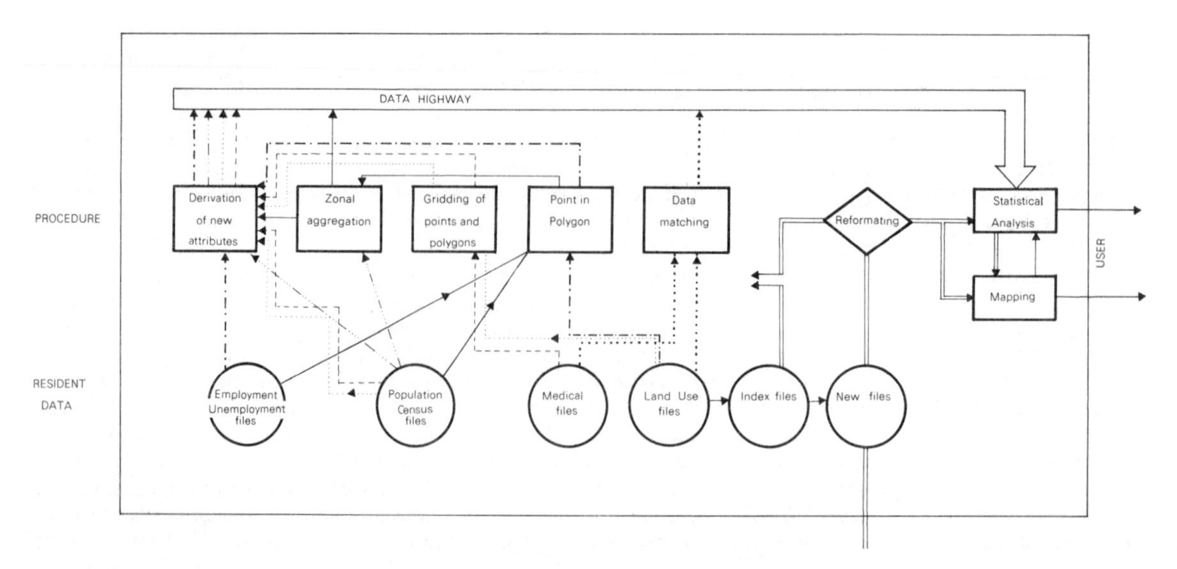

and Rhind and Hudson (1980) have attempted to define a totality of these functions: a list of them is given in Table 2.1.

Basic technical problems

A number of detailed technical problems have already been identified. In this section we discuss some of the more basic and intransigent ones, the perception of which is predicated upon the list of basic functions given in Table 2.1. Only the computationally simpler and the more general requirements are discussed: other functions which will be of particular importance to some users such as pattern recog-

Table 2.1 Functions in a comprehensive geographical information system

Stage	*Functions*
1 Input	The ability to accept data with different forms of spatial reference (point, line, area) in a number of standard formats, including data on stocks and flows.
	Performing consistency checks upon these data, e.g. that all data are within map sheet limits, that lines with the same attributes join at edges of map sheets.
	The ability to generate data descriptors from the data e.g. minimum and maximum X and Y coordinates of each string.
	Graphical output at convenient (e.g. input map) scale to permit visual checking of input data.
	Editing functions to correct any errors detected. The editing functions required are move, insert or delete specified points, lines and areas and fuse lines, plus change, insert or delete attribute values.
2 Manipulation	To be capable of retrieving data for any geographical area(s) specified by name or coordinates and on any combination of attributes of the geographical individual.
	Internal decision rules to assess which operations are feasible given the data available to the system (e.g. to check for coverage of specified geographical area, and to present arithmetic operations on nominal scale variables).
	Facilities to transform data from one coordinate system to another on an empirical basis using control points (e.g. from digitizing table to National Grid coordinates).
	Facilities to transform data on the basis of global expressions (e.g. from coordinates in latitude and longitude to plotting coordinates on a specified map projection).
	Facilities to update records for given geographical individuals automatically by matching new and old records on the basis of a given, unique identifier and creating file linkages.
	Facilities to convert data in string 'notation' into raster 'notation' and vice versa.
	Facilities to build polygons from supplied line segments. The capability is to be possible both where left-hand and right-hand area attributes are present and where no information other than segment end points is present.
	Facilities to aggregate geographical individuals, their geometric representation and their attributes to larger units, on the basis of some in-built hierarchy and on the basis of an externally-supplied list - irrespective of the measurement scale(s) of attributes.
	Facilities to disaggregate geographical individuals and their attributes, on the basis of the overlay of other individuals and specified rules for the allocation of attributes over space.
	Facilities either to carry out simple statistical sampling and summation (e.g. measurement of lengths of lines, frequency of different attributes in which missing data may occur) or to re-format data appropriately for a standard statistical or modelling package.
	To have clearly documented interfaces to allow import and export of data to/from

	the system at any stage.
3 Presentation to user	To be able to generate tabular reports on data in formats specified by the user.
	To be able to graph the data, e.g. as histograms and bivariate scatter plots.
	To be able to map the data at scales and with symbolism specified by the user, in dot distribution, point symbol, choropleth, isarithmic or cartogram form.
4 All	To maintain a log of operations, specifically of commands to the system and programs, plus details of the user identifier and file access for all operations.
	To build up a composite 'picture' of the user so that default values for parameters are progressively and automatically adjusted closer to a personal norm as he or she makes more use of the system.

In addition to these functions, access to an interactive computer system with a graphics computer terminal is pre-supposed.

---

nition (e.g. see Rosenfeld and Kak, 1976), present major problems which cannot be covered in this brief review. We should note again that some constraints are geographically specific - there are, for instance, now many machines available for encoding data from graphic documents (on this subject see Rogers and Dawson, 1979) but the essential machine and software combination is not yet available in many British geography departments (Rhind, 1979).

The first and most extreme real difficulty is that certain operations carried out by the best known existing methods are computationally impossible or, rather, would take longer than the age of the earth on the largest computers extant. Knowledge of members of this set of 'NP complete' problems is growing (Shamos and Bentley, 1978) and inevitably includes some geographical concerns: it is, for example, impossible to guarantee to optimize plotting of a network so as to minimize total distance travelled by a plotter pen, although good heuristic approximations can usually be made to this solution. The problem is a variant of the 'travelling salesman' one (at the time of writing a possible solution to this by Khachian has evoked considerable discussion).

The great majority of other specific problems in this field hinge on devising improved algorithms to make practicable operations which are presently too costly to perform on a routine basis or to reduce errors. A good example of the former is the Green and Sibson (1978) algorithm for defining Thiessen polygons from X, Y point data: this can cope with between one and two orders of magnitude more points than its predecessors (e.g. Brassel, 1975) and computes the polygons several orders of magnitude more quickly. It is of considerable potential value in such tasks as conversion of point- to area-based data (see Rhind, 1973) and in defining median lines between coasts or boundaries considered as connected point sets. An important example of work providing both cost and accuracy benefits is that of Openshaw and Ramsey (1978) in automatic address matching. Given a gazetteer of names describing geographical individuals (such as a house address) and some higher scale geographical description (e.g. centroid grid reference), address matching procedures permit any other files containing only the gazetteer entry to be tagged with the other descriptor; alternatively, updates can be made automatically on a file. Address matching schemes are thus a basic shorthand in providing geographical descriptors on many files created through the administrative responsibilities of local authorities, especially health records and the like. They are not trivial to construct, largely because of errors embedded in such files: Perrett (pers. comm., 1977) encountered 49 spellings or abbreviations of Hampshire in an early Annual Census of Employment file. As a consequence of these errors, probability matching procedures have to be incorporated in address matching routines.

Perhaps the most widely quoted problem (see Tomlinson, 1974) in geographical information systems is that of overlay. Given the anarchic procedures for defining aggregate data collection areas in Britain and the lack of moves to rationalize these (Rhind, 1975a), almost any comparison of different data sets has to be carried out through the property of space shared by data collection areas - the geographical individuals - in the two data sets. The mechanism by which geographically-based comparisons of data sets is carried out is 'overlay'. Important assumptions concerning the distribution within the geographical individual are inherent in any such analysis even though, on some occasions, logical considerations can minimize the importance of the assumptions: Rhind and Trewman (1975), for example, used land-use data to reallocate census data to residential areas within Enumeration Districts, rather than to the whole of the District, and a similar approach was adopted in the census atlas of Newcastle (Taylor, *et al.*, 1976).

The ease of implementation and reliability of the overlay procedure depend substantially upon the data 'notation': where the data are in raster format, the results are imprecise (Switzer, 1975) but the procedure simple and rapid. Where strings define the boundaries of geographical individuals, the procedure is time-consuming and may generate spurious 'sliver polygons' (Goodchild, 1978). Such overlay of strings is, however, a vital component in the use of Ordnance Survey digital data (Thompson, 1978) and a major contract costing £360,000 (HMSO, 1979) placed with a software house (then containing negligible geographical expertise) has resulted in transportable software for breakdown of digitized features into 'atomic, homogeneous' units and their reconstruction into attribute-defined 'minimum polygons' (Figure 2.4); other reconstruction rules are to be programmed to provide networks.

Perhaps the most intractable and hence longest-term problem of all is that of the language to instruct the system and, along with it, the creation of some elements of real intelligence in the so-called

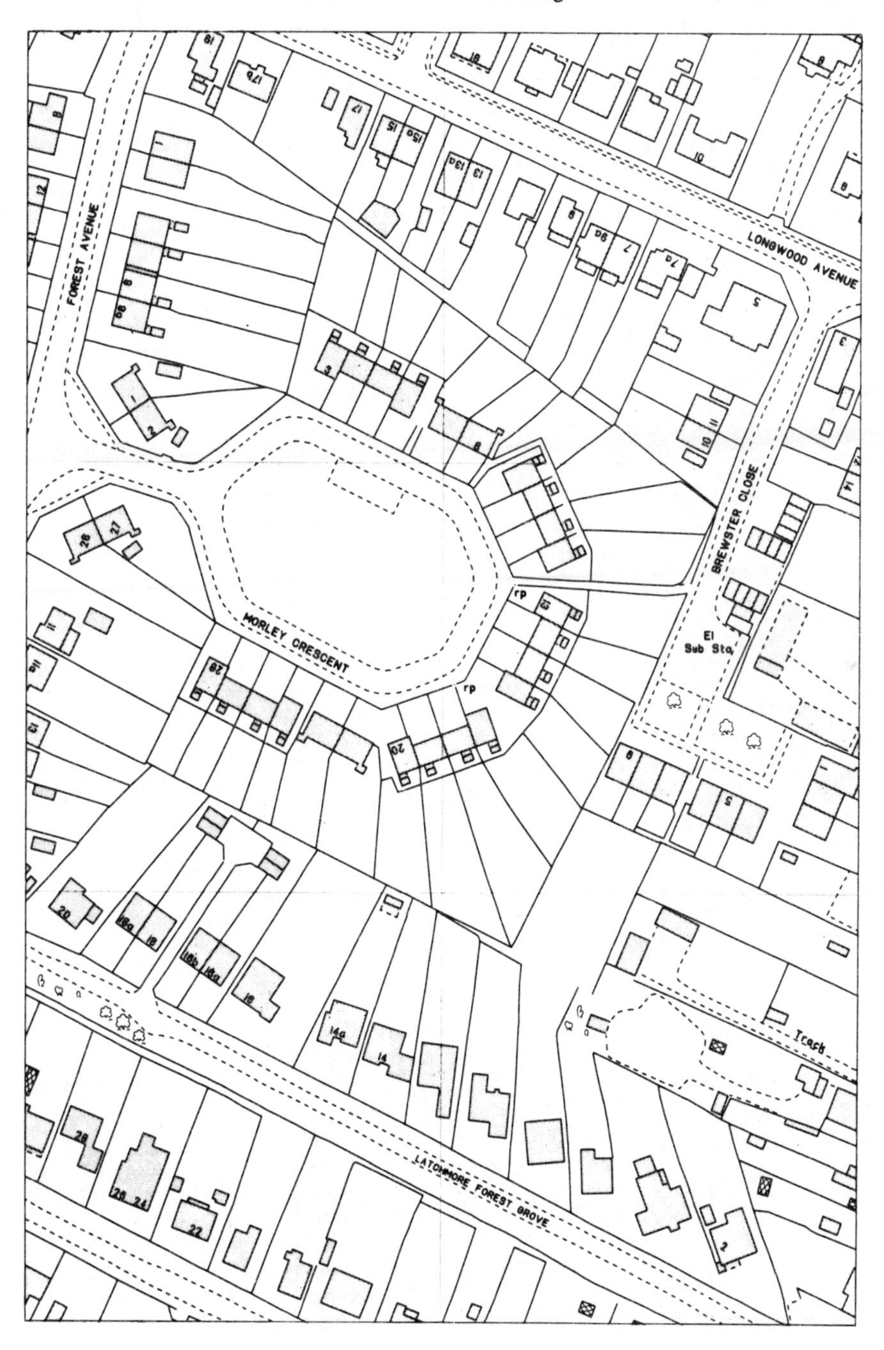

Figure 2.4 The breakdown of Ordnance Survey digitized topographic data into 'atomic units' and its reorganization into property zones: (a) published plan; (b) restructuring-links and nodes; (c) automatic parcel formation (Crown Copyright reserved; from Thompson, 1978)

information systems. Rhind (1976) suggested the need for a specific geographical language and Cox, Aldred and Rhind (1980) have simplified certain aspects of the implementation of such a language by their proposal for a geographical data type. Russell (1979, p. 340), while arguing the need - in a highly speculative essay - for spatial languages in geography generally, implies these earlier approaches are minimalist by saying 'this is not a question of the two-dimensional generalization of context-free transformational grammar, nor of finessing varieties of cellular automata, nor of perfecting geometry theorem machines, nor of borrowing arbitrary pattern or picture languages'. Unfortunately, he gives few clues to the solution. Beyond such obvious matters as ensuring that the user cannot make gross logic errors as in comparing data sets on different measurement scales (Rhind, 1976), the incorporation of intelligence into systems is a much more complex matter. Waugh (forthcoming) has

Figure 2.4b

suggested some preliminary ways in which this might be introduced via a 'thinking', computer-based map which constantly senses its own physical location within the area described by its stored data and adds to this data through encountering unpredicted events. Though most of the basic technology is available, building such systems is far beyond the conceptual capacity and expertise of geographical research workers at present: it would appear to contain elements in advance of the achievements of those working in machine intelligence.

British contributions

Contributions by British geographers to the development of GIS have been made in two ways. The first is by those geographers such as Clarke, Lawton and Wise who have expended time and effort on government committees in the hope of persuading the

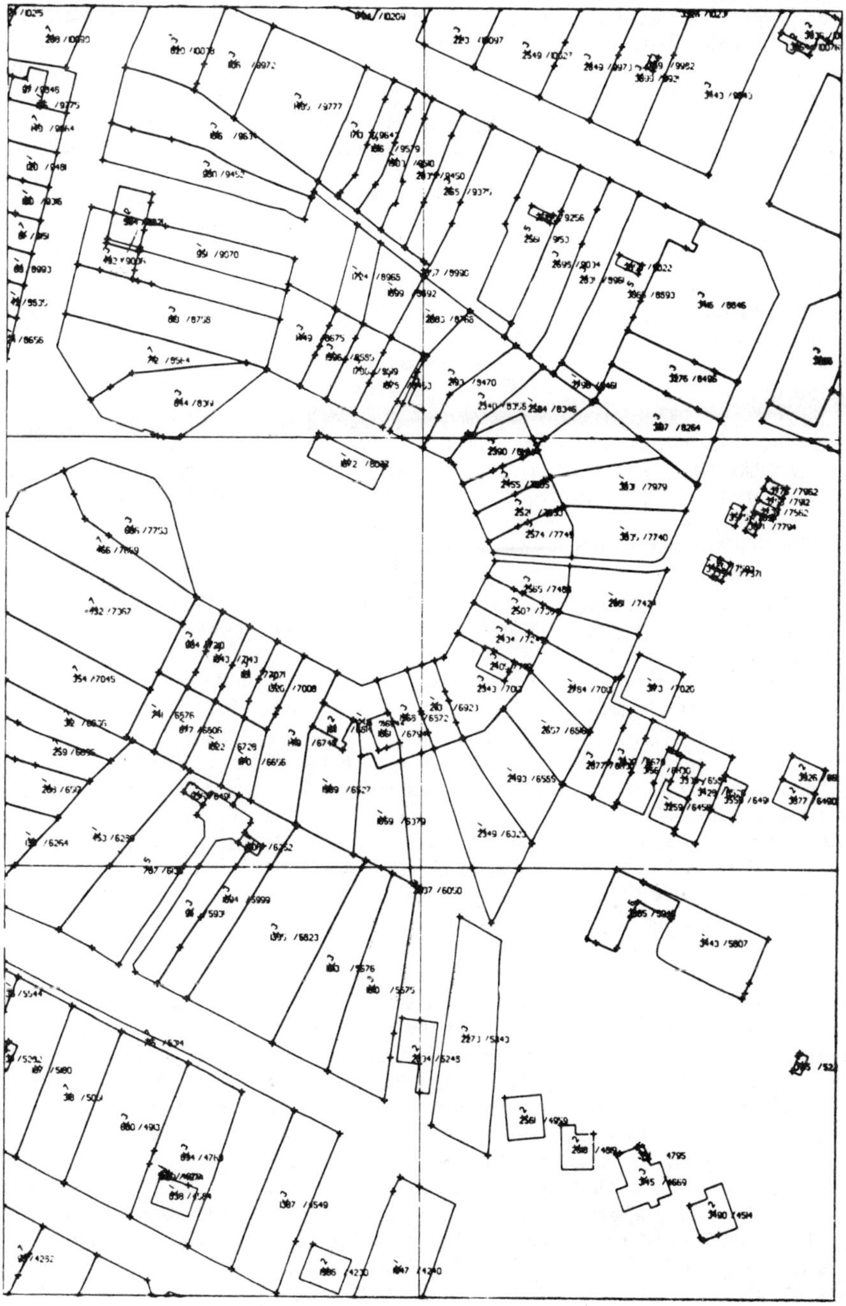

Figure 2.4c

'establishment' to collect and distribute geographical data and to allocate research grants on a more widespread basis. Such contributions are largely unheralded; they go almost unrecorded in the literature. The most recent instance (and one of the most important) is the contribution of J.T. Coppock to the Ordnance Survey Review Committee. These contributions are necessary, but not sufficient, conditions for much research: without adequate data, made available within acceptable time periods of collection and with minimal restrictions as to use, there is little point in developing geographical information systems or using those which exist.

On the second, and more technical, direction, a number of signal academic contributions by British geographers to research and development in geographical information systems have already been itemized in passing. The work of Baxter (1976a, b), Campbell *et al.* (GAG, 1977), Coppock (1974, 1975) and Waugh (1974, 1979, forthcoming), as well as various workers at the University of Durham, have already been mentioned. Other relevant work includes that of geographers in the North West Region Industrial Research Unit, who have set up a simple but operational GIS incorporating detailed industrial statistics for this large fraction of Britain, providing interactive mapping and some analytical capabilities (Blakemore and Lloyd, 1979; Blakemore and Reeve, 1977). It is arguable also that Waugh's GIMMS system (Waugh and Taylor, 1976) is a GIS though it is primarily orientated towards mapping; certainly it is becoming the most widely used second generation thematic mapping system in the world, being run overseas in Australia, the Philippines, France, the USA and Canada (where it has been adopted by Statistics Canada as their main mapping program) and is partially incorporated in other systems such as the Harvard Computer Graphics Laboratory ODYSSEY system and the TRAMS (Perrett, 1976).

A number of British contributions have also been made which, in a piecemeal fashion, move towards or encourage the development of GIS. Virtually the first attempt to implement a definition of functional requirements by creating a well-documented library of primitive geographical functions - building-blocks written to defined standards from which others might make their own programs - has arisen from the Geography Algorithms Group (GAG, 1977; Campbell, 1977). This approach, modelled on the extremely successful Numerical Algorithms Group, has also been followed in the algorithms section of 'Environment and Planning' although the stringent quality demanded of the supplied code (see Baxter, 1977) has apparently discouraged potential contributors. The approach differs totally from that based on 'stand alone' programs of the Geographical Program Exchange (GPE), based initially in Michigan and then in Hawaii and Buffalo, with a European offshoot in Zurich. A GPE-style approach has been adopted by the Geographical Algorithms and Program Exchange group (GAPE; Shepherd, 1976) based in Loughborough University. Their *modus vivendi* has been to distribute well-tested and documented, simple 'stand alone' programs to secondary school, colleges and universities. In passing, we should note the involvement of various geography classes in selected schools in the use of computer-processed and - analysed data (Robinson *et al.*, 1978): based in Birmingham University, the Computer Assisted Learning in Upper School Geography (or CALUSG) project formed part of the National Development Programme in Computer-Assisted Learning. While it did not lead to developments in geographical information systems as such, it seems certain to have stimulated student awareness of this approach in geography. Similarly, the growing use of personal micro-computers in universities such as Liverpool and Nottingham is leading students towards acceptance of a machine-related research culture. At a more advanced educational level, comparatively few postgraduate theses have yet been produced on aspects of geographical information systems (but see Tomlinson, 1974; Waugh, 1974; Thomas, 1976; Adams, 1979).

With the possible exception of work in Sweden (see Wastesson *et al.*, 1977 and, in particular, Rystedt, 1977) and of a few developments in North America (see Tomlinson, Calkins and Marble, 1976), the most unique contribution by the British thus far has been the definition of the results of practical experience in setting up systems and in their use for comparatively low-level functions such as monitoring change. Scarcely surprisingly, these have not impinged on academic work (but see Visvalingam, 1976; Rhind and Hudson, 1980). Thus Spicer *et al.*, (1979) have produced possibly the most complete and valuable - as opposed to speculative - description of the pitfalls and uses of a machine-based gazetteer. The LAMIS system (Harrison, 1978) is one of the very few operational management information systems using geographical disaggregation of data and combining micro-scale and meso-scale applications, while the MOSS system, used primarily for highway design, is in use in over forty local authorities. The Ordnance Survey (Thompson, 1978; 1979) has encoded and distributed more detailed and more voluminous topographic data (apart from digital terrain models) than any other survey organization in the world; they have also partially implemented the most general approach to data set cross-referencing - through shared geographical space. The collaboration between the Harwell Image Analysis Group and the Department of the Environment has resulted in procedures within a GIS which are at least as successful as anywhere else for monitoring mid-latitude urban change from satellites (Jackson *et al.,* forthcoming). In addition, the utility of certain British hardware and software is outstanding in building or running geographical information systems - notably the Laserscan HRD-1 display and the FASTRAK automatic digitizer (Howman and Woodford, 1978; Thompson and Woodford, 1979), the ICL Content Addressible File Store and the Distributed Array Processor.

Conclusions

Given the present commitments of resources and plans, it is scarcely conceivable that developments in geographical data handling *qua* information systems will not continue. It is particularly likely that, in Britain, this will continue and even expand within government organizations although cutbacks in government expenditure have, at the time of writing, led to a reduction in funding for some programmes. If this medium-term view is correct, the consequences for most academic geographers at least may be an unintentional safeguarding of their independence of thought and right to criticize: the corollary will be a relative diminution in their ability to make use of government-produced statistical data, especially for small areas. We have already argued that much of government data are unsatisfactory so far as quality and appropriateness in meeting research needs are concerned; but to remain remote from government is to lose an opportunity to influence which data are collected, how this is carried out and the manner in which they are presented, plus - ultimately the only important consideration - policy matters. Though we are only likely to get piecemeal improvements in our data collection and dissemination procedures, to this author, these are still worth the effort (see above).

Most of the immediate research problems faced in designing and building geographical information systems are, in reality, 'development' rather than research *sensu stricto*. There are, however, formidable theoretical issues in comparing data sets on different geographical bases, if only to provide error bounds to our conclusions. Likewise, there are major theoretical issues in the devising of languages to communicate with the machine and to describe our results. Without substantial work in these areas and, I suspect, team work to draw on different skills and experience, we shall not have the routine ability to analyse, compare and display geographically disaggregated data in the foreseeable future. This would be catastrophic for, in the longer term, 'quantitative geography' only has significance if the approaches it embraces and the techniques it espouses are ones which can be widely, easily and successfully applied to non-trivial examples; to be esoteric is not enough. Without the creation of computer systems - as opposed to *ad hoc* computer programs of uncheckable reliability - no material progress is likely and cynical attitudes to quantification, such as those in Taylor (1976), may well become even more prevalent.

Acknowledgments

Writing a general overview such as this has been immensely simplified by numerous friendly and totally frank discussions with others in the field. In Britain, Dick Baxter, Bill Campbell, Ken Perrett, Stan Openshaw, Mahes Visvalingam and Tom Waugh have been unfailing sources of stimulation over the years: from disagreement arose progress and understanding. Dick Baxter made numerous helpful suggestions for improving this paper. Latterly many others - such as Chris Thompson - have further contributed to my education. In a topic in which there are world-wide developments, however, it is also certain that I have gained much benefit from discussions with others in different countries: I wish to acknowledge my thanks to those numerous individuals in Australia, Canada, France, Mexico, the Netherlands, Sweden, West Germany and the USA with whom I have had fruitful discussions or working relationships in the last few years. Rowena Barker patiently and efficiently typed drafts of my appalling manuscripts into readable form whilst I was on sabbatical leave at the Australian National University.

Note

1 This refers to Coppock and Barrett's study of the implementation of information systems in local authorities for the Scottish Development Department, Waugh and Rhind's survey of the present and future role of automated cartography in DoE and Davey, Harris and Preston's review of automated digitizing hardware and software, both for the Department of the Environment.

References

Adams, J., Abler, R., and Ki-Suk Lee (1976) *The comparative atlas of America's great cities*, Ballinger: Cambridge, Mass.

Adams, T. (1979) 'Characteristics of a national topographical digital data base', Unpublished MSc dissertation, University of Durham.

Adams, T., and Rhind, D.W. (1981) 'The projected characteristics of a national digital topographic data bank', *Geographical Journal.*

Aldred, B.K. (1975) 'The implementation of a locational data type', *IBM UK Scientific Centre*, UKSC Report 0065 (see also 'The case for a location data type', UKSC Report 0054, 1974).

Anderson, D.E., Angel, J.L., and Gorny, A.J. (1978)

'World Data Bank II: Content, Structure and Application', in G. Dutton (ed) op. cit., Vol. 2.
Armstrong, J.S. (1967) 'Derivation of theory by means of factor analysis, or Tom Swift and his electric factor analysis machine', *American Statistician*, 21, 17-21.
Batty, M. (1978) 'Ten years of urban modelling in Britain', *Area*, 10, 11-15.
Baxter, R.S. (1974) 'The management of data and the presentation of results from a computerised information system', in Perraton and Baxter (eds), op. cit., 211-30.
Baxter, R.S. (1976a) *Computer and statistical techniques for planners*, Methuen: London.
Baxter, R.S. (1976b) 'A parsed data set of British Local Authority boundaries', *Building Research Establishment Note* 138/76.
Baxter, R.S. (1977) (ed) 'Notes for authors on the preparation and description of computer algorithms', *Environment and Planning A*, 9, 223-8.
Berry, B.J.L. (1964) 'Approaches to regional analysis: a synthesis', *Annals Associations American Geography*, 54, 2-11.
Blakemore, M.J., and Lloyd, P.E. (1979) 'Interactive digitising and display procedures for a computer-based industrial information system' in *Harvard Library of Computer Graphics, Mapping Collection Vol. 2: Mapping Software and Cartographic Data Bases*, Harvard, Mass.
Blakemore, M.J., and Reeve, D.E. (1977) 'CHORO - A users' guide to choropleth mapping by computer', Working Paper 3, North West Industry Research Unit, University of Manchester.
Brassel, K. (1975) 'Neighbourhood computations for large sets of data points', in *Proceedings of the International Symposium on Computer-assisted Cartography: Auto-Carto II*, US Department of Commerce, Reston, Va.
Bulmer, M. (1979a) 'Data protection: Social scientists need to be vigilant', *Survey Archive Bulletin*, 14, 7-8.
Bulmer, M. (ed) (1979b) *Censuses, Surveys and Privacy*, Macmillan: London.
Campbell, W.J. (1977) 'Computer algorithms for spatial data', *Area*, 9-2, 106-8.
Carmer, S.G. (1975) 'One Statistician's view of consumer evaluation of statistical software', *Proceedings of Computer Science and Statistics 8th Annual symposium*, 149.
Cook, B.G. (1978) 'The structure and algorithmic basis of a geographic data base', in G. Dutton, op. cit., Vol. 4.
Cooke, D., and Maxfield, W. (1967) 'The Development of a Geographic Base File and its uses for mapping', Paper from the 5th Annual Conference of the Urban and Regional Information Systems Association, 207-18.
Coombs, C.H. (1964) *A theory of data*, Wiley: New York.
Coppock, J.T. (1974) 'Geography and public policy: challenges, opportunities and implications', *Transactions Institute British Geographers*, 63, 1-16.
Coppock, J.T. (1975) 'Maps by line printer' in Davis, J.C, and McCullagh, J.J. (eds), *Display and Analysis of Spatial Data*, 137-54, Wiley: London.
Corbett, J.P. (1975) 'Topological principles in cartography', *Proceedings of the International Symposium on Computer-assisted Cartography, Auto-Carto II*, U.S. Department of Commerce, Reston, Va.
Coulter, J. (1978) 'Grid Square Census data as a source of the study of deprivation in British conurbations', Census Research Unit Working Paper 13, Department of Geography, University of Durham, 105 pp.
Cox, N.J., Aldred, B.K. and Rhind, D.W. (1980) 'Towards a Geographical Data Type', *Geoprocessing*, 1, 217-29.
Craine, G. (1975) 'MOSS: a modelling system for highway design and related disciplines' in B.K. Aldred (ed), *Proceedings of the IBM UK Scientific Centre Seminar on Geographic Data Processing*, UKSC Report 0073, 87-93.
CSO (1978) *Guide to Official Statistics*, Central Statistical Office: London (2nd edn).
Dear, M., and Clark, G. (1978) 'The state and geographic process', *Environment and Planning A*, 10, 2, 173-83.
DoE (1973) *Manual of point-referencing*, Department of the Environment: London.
DoE (1974) 'Point in polygon project, stage 1', *Research Report 2*, Department of the Environment: London.
DoE (1975) 'General review of local authority management information systems', *Research Report 1*, Department of the Environment: London.
Doyle, F.J. (1978a) 'Contribution to panel discussions on digital terrain models', *Photogrammetric Engineering and Remote Sensing*, 44, 12, 1490-3.
Doyle, F.J. (1978b) 'The next decade of satellite remote sensing', *Photogrammetric Engineering and Remote Sensing*, 44, 2, 155-64.
Duffield, B.S., and Coppock, J.T. (1975) 'The delineation of recreational landscapes: the role of a computer-based information system', *Transactions Institute British Geographers*, 66, 141-8.
Dutton, G. (ed) (1978) 'Proceedings of the First International Advanced Study Symposium on Topological Data Structures', *Harvard Papers on Geographic Information Systems*, Eight volumes, Harvard. Mass.
Francis, I. (1973a) 'An evaluation of factor analysis programs', *Bulletin International Statistical Institute*, 45, 423.
Francis, I. (1973b) 'Comparison of analysis of

variance programs', *Journal American Statistical Association*, 68, 860.

Freeman, H. (1974) 'Computer processing of line drawing images', *Computing Surveys*, 6, 57-97.

Freestone, M. (1979) Review of Adams, Abler and Ki-Suk Lee (1976) *Cartography*, 11, 2, 115-7.

GAG (1977) *Feasibility and design study for a computer algorithms library*, report to Social Science Research Council by the Geography Algorithms Group. Available from the Department of Geography, University College, London.

Gaits, G.M. (1969) 'Thematic mapping by computer', *Cartographic Journal*, 6, 1, 50-68.

Gaits, G.M., and Hackman, G. (1976) 'The LINMAP computer mapping system', *Management Services in Government*, 31, 3, 1-13.

Gale, S., and Olsson, G. (eds) (1979) *Philosophy in Geography*, Reidel: Dordrecht.

Goodchild, M.F. (1978) 'Statistical aspects of the polygon overlay problem', in G. Dutton, op. cit., Vol. 6.

Green, P.J., and Sibson, R. (1978) 'Computing Dirichlet tessellations in the plane', *Computer Journal*, 21, 2, 168-73.

Hakim, C. (1977) *Census-based Area Profiles: a Review*, Office of Population Censuses and Surveys Census Division Occasional Paper 2: London.

Hakim, C. (1978) *Census confidentiality, microdata and census analysis*, Office of Population Censuses and Survey Census Division Occasional Paper 3: London.

Harrison, J.C. (1978) 'The LAMIS system', paper presented to Royal Institute of Chartered Surveyors, Land Surveyors Division and North East London Polytechnic conference on 'Data banks and digital mapping', November 1978.

Harvey, D.W. (1969) *Explanation in Geography*, Arnold: London.

Harvey, D.W. (1974) 'What kind of geography for what kind of public policy', *Transactions Institute British Geographers*, 63, 18-24.

HMSO (1972) *General Information Systems for Planning*, report of the Joint Study Team of Local Authorities, the Scottish Development Department and the Department of the Environment: London.

HMSO (1975) *Computers and Privacy*, Cmnd 6353.

HMSO (1978) *Report of the Committee on Data Protection* (Chairman: Sir Norman Lindop), Cmnd 7341.

HMSO (1979) *The Report of the Ordnance Survey Review Committee.*

Howman, C., and Woodford, P.A. (1978) 'The Laserscan FASTRAK automatic digitising system', paper presented to the 9th International Conference on Cartography, University of Maryland , USA.

ICL (1976) 'Spatial retrieval for point referenced data - a system specification study', report to the Department of the Environment.

Jackson, M.J., Carter, P., Gardner, W.G., and Smith, T.F. (1980) 'Urban land use mapping from remotely sensed data', *Photogrammetric Engineering and Remote Sensing*, 46, 1041-50.

Janacek, S., and Negus, B. (1974) 'Some observations on design and accuracy of the Biomedical computer programs package', *Bulletin Institute Mathematics and its Applications*, 10, 166-71.

Jeffrey, K.G., Gill, E.M., and Henley, S. (1974) *G-EXEC system: user's manual.* Atlas Computer Lab/Institute of Geological Sciences, London.

Johnston, R.J. (1978) *Multivariate statistical analysis in geography*, Longman: London.

JRSS (1979) 'Proceedings of a Symposium on the Data Protection Committee Report', *Journal Royal Statistical Society*, A, 142, 3.

Knox, P. (1974) 'Spatial variations in level of living in England and Wales in 1961', *Transactions Institute of British Geographers*, 61, 1-24.

Loomis, R.G. (1965) 'Boundary networks', *Communications of the Association of Computing Machinery*, 8, 44-8.

Margerison, T.A. (1977) *Computers and the renaissance of cartography*, Natural Environment Research Council: London.

Martin, J. (1975) *Computer data base organisation*, Prentice Hall: Englewood Cliffs, NJ.

Moore, E.G. (1979) 'Beyond the census: data needs and urban policy analysis', in S. Gale and G. Olsson, op. cit., 269-86.

NASA (1976) *Landsat data users handbook*, National Aeronautics and Space Administration Document No. 76SDS 4258, Goddard Space Flight Centre: Greenbelt, Md.

Openshaw, S., and Ramsay, J.B. (1978) 'An automatic address matching and survey coding program (ACP)', *Bulletin of the Urban and Regional Information Systems Association*, 33, 7.

Perraton, J., and Baxter, R.S. (1974) (eds) *Models, Evaluation and Information Systems for Planners*, Land Use and Built Form Studies Conference Proceedings No. 1, Cambridge.

Perrett, K.E. (1976) 'Transport Referencing and Mapping System', Transport and Road Research Laboratory, Crowthorne, Berkshire.

PERS (1978) 'Proceedings of the ASP Digital Terrain Model Symposium', *Photogrammetric Engineering and Remote Sensing*, 44, 12, 1435-586.

Peuker, T.K., and Chrisman, N. (1975) 'Cartographic data structures', *American Cartographer*, 2, 55-69.

Pfaltz, J.L. (1977) *Computer data structures*, McGraw Hill: New York.

Pfaltz, J.L., and Rosenfeld, A. (1967) 'Computer representation of planar regions by their skeletons', *Communications of the Association of Computing Machinery*, 10, 119-25.

PRAG (1975) 'Liverpool Social area study, 1971 data: final report', TP 14, Planning Research Applications Group, Centre for Environmental Studies: London.

Rhind, D.W. (1971) 'Towards instant and efficient maps: the work of the Experimental Cartography Unit', *Rev de Geographie de Montreal,* 24 (4) 391-8.

Rhind. D.W. (1973) 'Computer mapping of drift lithology from borehole records', *Report of the Institute of Geological Sciences,* 73/6, 12 pp.

Rhind, D.W. (1975a) 'The reform of areal units', *Area,* 7, 1, 1-3.

Rhind, D.W. (1975b) 'The "state of the art" in geographic data processing' in B.K. Aldred (ed) *Proceedings of the IBM UK Scientific Centre Seminar on geographic data processing,* IBM UKSC Report 0073, 4-41.

Rhind, D.W. (1976) 'Towards universal, intelligent and usable automated cartographic systems', *ITC Journal* 1976, 4, 515-45.

Rhind, D.W. (1977) 'Computer aided cartography', *Transactions Institute British Geographers,* NS 2, 1, 71-97.

Rhind, D.W. (1979) 'Why digitize?', *Area,* 11, 211-13.

Rhind, D.W. (1980) 'A multi-temporal, multi-purpose, multi-user, multi-resolution geographical information system', in *Urpis 7, Proceedings of the Australian Urban and Regional Information Systems Association annual conference,* Newcastle, NSW.

Rhind, D.W., and Adams, T. (1981) 'Coordinate data bases: availability and characteristics', *Proceedings Harvard Computer Graphics Week,* Laboratory for Computer Graphics, Harvard University.

Rhind, D.W., and Hudson, R. (1980) *Land Use,* Methuen: London.

Rhind, D.W., and Trewman, T. (1975) 'Automatic cartography and urban data banks - some lessons from the UK', *International Yearbook of Cartography,* 15, 143-57.

Rhind, D.W., Evans, I.S., and Visvalingam, M. (1980) 'Making a national atlas of population by computer', *Cartographic Journal,* 17, 3-11.

Robinson, R., Boardman, D., Fenner, J., and Blackburn, J. (1978) *Data in Geography: Drainage Basin Units,* Longman: London.

Rogers, A., and Dawson, J.A. (1979) 'Which digitiser?', *Area,* 11, 1, 69-73.

Rosenfeld, A., and Kak, A.C. (1976) *Digital picture processing,* Academic Press: New York.

Russell, D. (1979) 'An open letter on the dematerialisation of the geographic object', in S. Gale and G. Olsson, op. cit., 329-44.

Rystedt, B. (1977) 'The Swedish Land Data Bank - a multipurpose information system' in O. Wastesson *et al.,* op. cit., 19-48.

Shamos, M.I., and Bentley, J.P. (1978) 'Optimal algorithms for structuring geographic data' in G. Dutton, op. cit., Vol. VI.

Shepherd, I.D.H. (1976) 'Bridge that gap with GAPE', *Area,* 8, 3, 173-4.

Slysz, W.D. (1974) 'An evaluation of statistical software in the social sciences', *Communications of the Association of Computing Machinery,* 17, 3-6.

Southard, R.B. (1978) 'Development of a digital cartographic capability in the National Mapping Program', paper given to the 9th International Conference on Cartography, University of Maryland, USA.

Spence, N.A., and Taylor, P.J. (1970) 'Quantitative methods in regional taxonomy', *Progress in Geography,* 2, 1-64.

Spicer, J., and members of the Joint Information System team (1979) *Property Information Systems for Government,* Research Report 30, Department of the Environment: London.

Stocker, P.M. (1975) 'Research study in geographic data bases', in B.K. Aldred, (ed) *Proceedings of the Scientific Centre Seminar on Geographic Data Processing,* Report UKSC 0073, 94-9.

Switzer, P. (1975) 'Sampling of planar surfaces', in J.C. Davis and M. McCullagh (eds), *Display and Analysis of Spatial Data,* Wiley: New York.

Taylor, P.J. (1976) 'An interpretation of the quantification debate in British geography', *Transactions Institute British Geographers,* NS 1, 129-42.

Taylor, P.J., Kirby, A.M., Harrop, K.J., and Gudgin, G. (1976) *Atlas of Tyne and Wear,* Department of Geography, University of Newcastle Research Paper 11.

Thomas, A.N. (1976) 'Spatial models in computer-based information systems', Unpublished PhD thesis, University of Edinburgh.

Thompson, C.N. (1978) 'Digital mapping in the Ordnance Survey 1968-1978', paper given to ISP Commission IV Inter-Congress Symposium, Ottawa, October 1978.

Thompson, C.N. (1979) 'The need for a large-scale topographic data base', paper given to Commonwealth Survey Officers Conference, Cambridge, July 1979.

Thompson, C.N., and Woodford, P.A. (1979) 'The Ordnance Survey topographic data base concept for the 1980s', paper given to the Second UN Regional Cartographic Conference for the Americas, Mexico City, September 1979.

Thompson, L.L. (1979) 'Remote sensing using solid-state array technology', *Photogrammetric Engineering and Remote Sensing,* 45, 1, 47-55.

Tobler, W. (1979) 'Cellular geography', in S. Gale and G. Olsson, op. cit., 379-86.

Tomlinson, R.F. (1974) 'The application of electronic computing methods to the storage, compilation and assessment of mapped data', unpublished PhD thesis, University of London.

Tomlinson, R.F., Calkins, H.W., and Marble, D.F. (1976) *Computer handling of geographical data,* UNESCO, Natural Resources Research Series, No. XIII: Paris.

Unwin, D.J. (1978) 'Quantitative and theoretical geography in the United Kingdom', *Area,* 10, 5, 337-44.

Velleman, P.F., and Francis, I. (1975) 'Measuring statistical accuracy of regression programs', *Proceedings of Computer Science and Statistics 8th annual symposium,* 122.

Visvalingam, M. (1976) 'Storage of the 1971 UK Census data: some technical considerations', Census Research Unit Working Paper 4, Department of Geography, University of Durham.

Visvalingam, M., and Rhind, D.W. (1977) 'Compaction of the 1971 UK census data', *Computer Applications,* 3, 3/4, 499-511.

Wastesson, O., Rystedt, B., and Taylor, D.R.F. (1977) (eds) 'Computer Cartography in Sweden', *Cartographica Monograph,* 20, 114 pp.

Waugh, T.C. (1974) 'Geographic data in a system environment', Unpublished M.Phil. thesis, University of Edinburgh.

Waugh, T.C. (1979) *GIMMS Reference Manual,* Inter-University/Research Council Series Report 30, Program Library Unit, University of Edinburgh.

Waugh, T.C. (forthcoming) 'The intelligent map', *Canadian Cartographer.*

Waugh, T.C., and Taylor, D.R.F. (1976) 'GIMMS/An example of an operational system for computer cartography', *Canadian Cartographer,* 13, 2, 158-66.

Webber, R.J., and Craig, J. (1978) 'A socio-economic classification of Local Authorities in Great Britain', *OPCS Studies in Medical and Population Subjects,* 35, HMSO, 117 pp.

Webster, R. (1977) *Quantitative and numerical methods in soil classification and survey,* Clarendon Press: Oxford.

Williams, C.M. (1979) 'Scanning digitising: automating the input of drawings', *Computer Aided Design,* 11, 4, 227-30.

Wood, P.A. (1977) 'Information for geography - cause for alarm?', *Area,* 9, 2, 109-13.

Chapter 3

Remote sensing

R. Harris

Introduction

The subject of remote sensing is difficult to summarize in a short paper. In some ways it is representative of the structure of quantitative methodology, but so far has stood largely outside the interests of quantitative geographers in Britain. Much remote sensing is appropriate to quantitative geographic work because it is *a fortiori* spatial (see Heuseler, 1976). The many problems posed by quantitative remote sensing can be addressed by techniques of interest to geographers (Swain and Davis, 1978), problems such as digital data handling, sampling, spatial interpolation, classification and discrimination. Remote sensing specialists often come from the 'hard' scientific backgrounds of physics, mathematics and electrical engineering, where there is much interest in satellite orbits, sensor design, data transmission and data display (Harris, 1978). However, where remote sensing is most useful is in the *application* of the remotely sensed data to understanding and answering questions and problems set by the real world, to use satellite technology in a problem-solving rather than data-collection role. Following Barrett and Curtis (1976), Lintz and Simonett (1976a) and Reeves (1975), the components of remote sensing may be identified as follows:

1. data collection from aircraft, rockets and satellites;
2. preprocessing the data to clean up noise, check for validity, and massage the data into a suitable form for further analysis;
3. corroboratory data collection from ground truth sites;
4. processing and analysis;
5. display;
6. field checking and interpretation of the results of the previous five steps.

Many of the components of this structure can be attacked by quantitative methods, and an awareness of quantitative methodology is vital for the proper use of remote sensing data.

Quantitative work in remote sensing developed principally outside Geography, and outside Britain. The launch of the Tiros 1 weather satellite on 1 April 1960 began an era of environmental data provision which continued to expand throughout the 1960s and 1970s. The Tiros satellite series was followed in 1966 by the polar orbiting Essa series which gave better quality data over a wider area of the globe (Barrett, 1974). Further improvements came in the 1970s with polar orbiting Noaa and Tiros-N weather satellites, and geostationary satellites such as Meteosat (Honvault, 1978). While the weather satellite programme was developing, the US National Aeronautics and Space Administration (NASA) also investigated the provision of a satellite system to investigate Earth resources (Lintz and Simonett, 1976b), and this investigation led to the Landsat programme (NASA, 1976), and the successful launch of Landsat 1 in 1972. For most parts of the Earth there now exists data coverage by Landsat imagery (World Bank, 1976), and although this presents invaluable opportunities for mapping and data analysis, it also presents problems of handling the data mountains (Thomas, 1977) produced by the Landsat satellites. This has led to work on data compression, following models of information theory such as those developed by Shannon and Weaver (1949), and also to the use of forms of automated cartography for rapid data display and dissemination.

Remote sensing work in Britain has drawn on both US data and US experience, but is now developing within a European context, with the establishment of the European Space Agency (ESA), the EARTHNET data transfer system (Proca, 1978), and various European remote sensing projects such as AGRESTE (Berg *et al.*, 1978) and EURASEP (Sturm *et al.*,

1978). Development will continue within this organizational framework, particularly with the establishment of a UK National Remote Sensing Centre. Within a thematic framework, much quantitative work will follow the general lines of image processing and pattern recognition, owing much to the attempts at modelling vision and acquisition (Overington, 1976; Rosenfeld, 1967), and the comparative studies of human and machine interpretation (Jagoe and Paton, 1978).

Data collection and pre-processing

Digital remote sensing data come from two principal sources: either from scanning film transparencies with a microdensitometer or similar instrument, or by using the pixel density data provided by data dissemination agencies such as the EROS Data Center in the USA (Harris, 1979). Microdensitometers allow monochrome and colour films from aircraft and spacecraft to be converted into digital density arrays for further processing. The film is mounted on a drum, or a flat bed, and is density digitized by a light sensitive scanner with a small scanning spot, typically of the order of 50 to 100 μm (Bryant, 1975). Below this size confusion with the grain size of the film is often a problem. The equivalent ground resolution of a microdensitometer system is a function of the scale of the photography, the size of the film transparency, and the size of the scanning spot. Owen-Jones (1977) used microdensitometric data collection techniques for crop studies in Argentina from Skylark imagery and for terrain mapping in Austrailia from aerial photography, while Carter and Gardner (1977) described the use of a flying spot scanner as part of the Harwell Image Processing System which has been used for land use mapping.

Scanning radiometers on board satellites provide pixel density data direct. Normally the data are collected in analogue mode, then sampled to give digital pixel values which are transmitted to ground receiving stations on radio carrier frequencies. The most common sets of readily available digital data are probably those from the Landsat satellites, with data provided on computer compatible tapes (CCTs). The ground resolution of the multispectral scanner on Landsats 1, 2 and 3 is 79 m, so the digital pixel values are the amounts of reflected energy for each 79 m x 79 m area of the Earth's surface (or clouds) in the four wavebands of the scanner (NASA, 1976). Landsat 3 has been successful in providing thermal infrared data, with a ground resolution of 237 m, and high resolution return beam vidicon data with a pixel size of approximately 40 m (Justice and Townshend, 1979). Figure 3.1 shows the format of EROS Data Center Landsat CCT sets prior to 1979 (Harris, 1979). The pixels are interleaved in pairs for the four wavebands, and the image area is split into

Figure 3.1 One of the formats of the EROS Data Center CCT sets for Landsat data

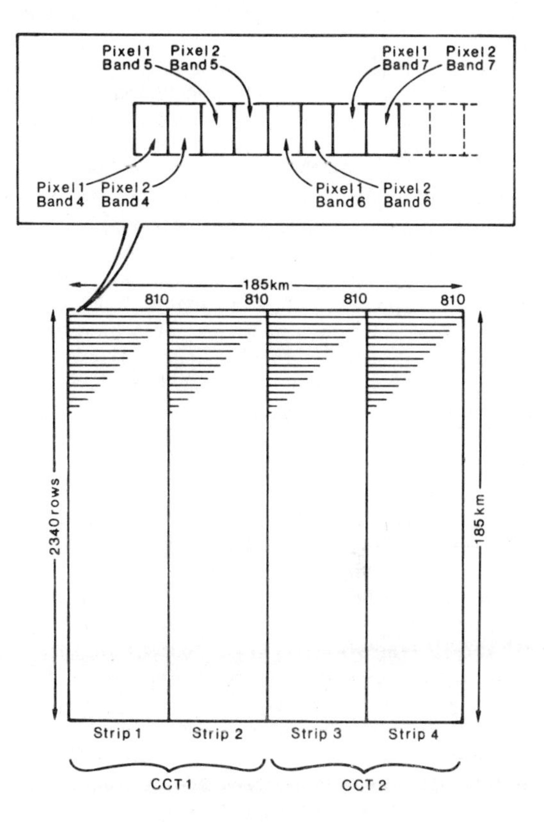

four strips (Thomas, 1975). With 2,340 rows of data, 3,240 pixels per row, and 4 wavebands per pixel, each Landsat frame has over 30 million pixel density values: obviously computer processing is vital. Since January 1979 the EROS Data Center has instituted a new pipeline processing system whereby Landsat CCTs are available in either band sequential or band interleaved by line format (Holkenbrink, 1978). The latter format is that normally used by the Landsat receiving station at Fucino near Rome, which receives Landsat data for the European area. In addition to the receiving stations in Europe and the USA, there are also Landsat receiving stations in Canada, Brazil, Argentina, Japan, Australia, India and Zaire, and each has its own local tape formating procedures.

Digital pixel data are also available from other satellites, such as Meteosat, Noaa or Tiros-N. In the early 1960s weather satellite pictures were received by the Automatic Picture Transmission (APT) system, which gave very poor quality imagery, but was still being used by the Meteorological Office at Bracknell in 1974. Better quality data are now received at the Dundee University satellite receiving station which archives data in CCT form for a number of weather satellites. The ground resolution of these data vary, but currently the polar orbiting

Noaa and Tiros-N satellites have highest ground resolutions of 0.9 km and 1.1 km respectively, and Meteosat has a highest ground resolution of approximately 4 km.

Radar imaging systems appear to offer great possibilities for remote sensing, particularly the synthetic aperture radar which offers both a high ground resolution and the ability to sense through clouds (Corless, 1977; van Genderen, 1975). An active imaging radar was carried on the Seasat satellite, and with a ground resolution of 25 m gave excellent results after preprocessing. The advantages of radar were shown in the K-band radar mapping project of Darien Province in Panama (Viksne *et al.*, 1969). Reasonable quality maps could be drawn of the area for virtually the first time, as previously cloud cover had prevented aerial survey, and difficult terrain had limited ground survey (Sabins, 1976).

So far, data collection has been discussed in relation to imagery, but there is also a branch of remote sensing which uses non-imaging sensors. Precise information about the vertical temperature structure of the atmosphere is of great value to meteorologists (Houghton, 1979), and a group at Oxford has worked extensively on data from sensors such as the Selctive Chopper Radiometer on Nimbus III and IV (Harwood, 1977). The data from sensors such as these can be used for temperature retrieval at various levels in the atmosphere by inversion methods (Fymat and Zuev, 1978; Twomey, 1977) and by comparison with other temperature sensing systems (Barnett *et al.*, 1975), and Peckham (1978) gives a review of atmospheric temperature remote sounding from satellites.

Digital data in their various forms normally need some form of preprocessing before they can be used correctly in remote sensing. Aerial photographs often suffer from vignetting, whereby the lens darkens an image towards its outside edges (Bristor *et al.*, 1966). Aerial photography can be scanned with a microdensitometer and the vignetting effect removed by digital processing before rewriting the digital data to a film with the vignetting effect absent. Satellite data suffer from a range of data quality problems which necessitate preprocessing (Fortuna and Hambrick, 1974), and these can be divided into two categories: radiometric correction and geometric correction (Simonett, 1974; Ullman, 1973).

Radiometric correction

Scan calibration A scanning radiometer on board a satellite needs to be continually calibrated to prevent the sensor response from drifting. This is achieved by allowing the scanner to scan an in-house spacecraft reference and outer space before each scan of the Earth below.

Limb darkening As the zenith look angle of a sensor increases there is an increasing depth of atmosphere to distort the reflected or transmitted energy from the ground to the sensor. This results in a darkening of the edges or limbs of each scan, which must be corrected by brightness normalization.

Blackbody temperature Each thermal infrared sensor is calibrated before launch and has a calibration curve of sensor response with temperature. The temperature recorded is an equivalent blackbody temperature which assumes uniform temperature of the sensed object. This assumption may not be fulfilled when a pixel has contributions from various different types of surface conditions.

Geometric correction

Distortions arise in the geometry of an image if a spacecraft changes its orientation in orbit.

Pitch variation occurs when the spacecraft dips on its along-flight axis, pointing the scanner backwards or forwards.

Yaw variation occurs when the spacecraft points in a different direction, thereby modifying the across track pixel orientation.

Roll variation is the result of the spacecraft body rolling in orbit, giving curved edges to the image area.

Altitude changes result in image pixel area changes.

Spacecraft velocity variation results in the distance between scan lines changing.

In addition to these variations, there are also those specific to certain satellites. For example, in Landsat satellites the scanner is a multiple scan instrument which scans six lines over the Earth's surface at once. A problem known as sixth line banding has resulted from this scanner operation and the images are often striped horizontally because there are groups of six scan lines together.

All of these preprocessing tasks must be tackled, but most CCTs have calibration data to assist with radiometric correction, and orientation data to assist with geometric correction (Thomas, 1975).

Data processing

Much quantitative remote sensing work falls into the larger class of image processing, and in image processing the principal methodology is pattern recognition. Pattern recognition can be defined as 'the categorization of input data into identifiable classes via the extraction of significant features or attributes of the data from a background of irrele-

THE
UNIVERSITY OF WINNIPEG
PORTAGE & BALMORAL
WINNIPEG, MAN. R3B 2E9
CANADA

vant detail' (Tou and Gonzalez, 1974), and is obviously concerned with broader issues than just image processing (Fu and Rosenfeld, 1976). Remotely sensed data are often used in picture format, before or after one of the data collection stages, and pictures have certain characteristics (Rosenfeld, 1969):

1 a picture function is a real valued function of two real variables ($f(x,y)$);
2 the function is nonzero within the bounded region of the physical extent of the picture;
3 for any possible picture function f, then $0 \leqslant f(x,y) \leqslant M$ for all (x,y), where M is an upper bound to the possible image brightness.

Pratt (1978) suggests further that a picture function is also bounded by time (t) and wavelength (λ), and defines a picture function as $f(x,y,t,\lambda)$. Following the brightness response of the human eye, Pratt (1978) defines an image light function in a multispectral imaging system in which the *i*th spectral image field is given by

$$F_i(x,y,t) = \int_0^\infty f(x,y,t,\lambda)\, S_i(\lambda)\, d\lambda$$

where $S_i(\lambda)$ is the spectral response of the *i*th sensor.

A continuous picture function is not measurable by image scanning, but the digital picture function resulting can be used as a representation of the original image (Harris, 1977b), and Rosenfeld (1969) gives the characteristics of a digital picture function:

1 a digital matrix representing a picture function defines a piecewise constant digital picture function;
2 any picture function is indistinguishable from a digital picture function;
3 any picture function is indistinguishable from a quantized picture function.

It is assumed that digital density pixel arrays of raw data also have the same characteristics as digital picture functions. The elements of a digital picture are normally termed picture points; and the term pixel normally refers to scanner data, but for the present the term pixel is used to cover both types of data.

Analysing digital pixel arrays can be as straightforward as single band density slicing (Carter, 1974). For each pixel a decision is made concerning the

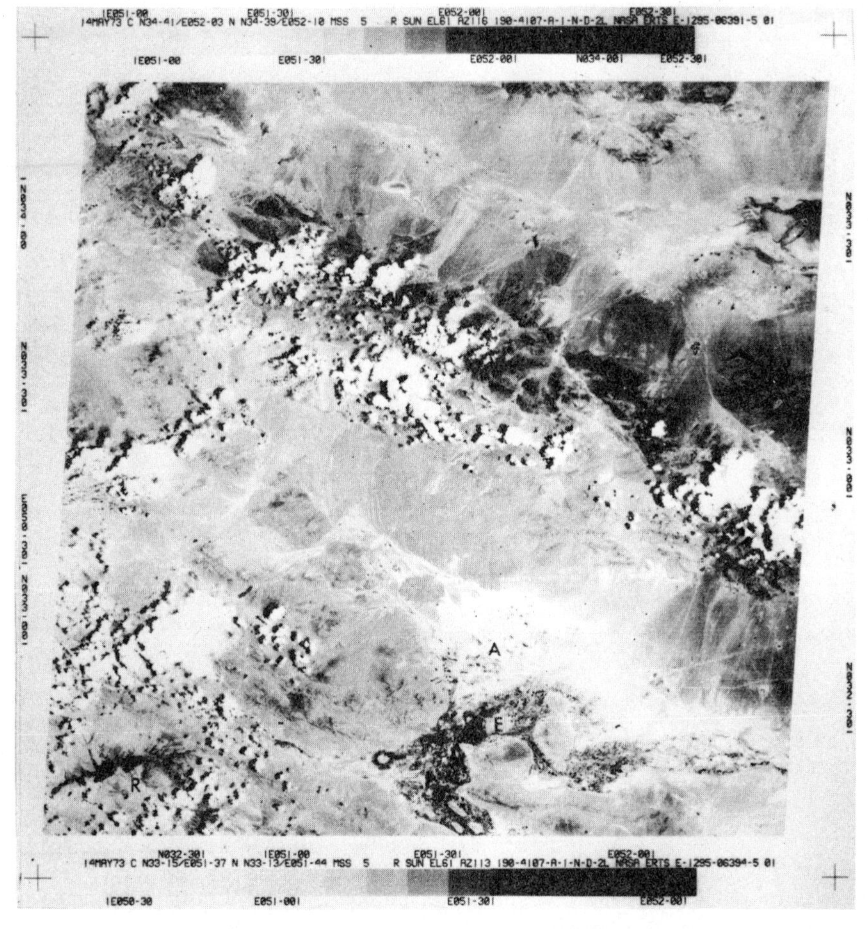

Figure 3.2 (a) Landsat 1 scene of the Esfahan region, central Iran, 14 May 1973. A, E and R mark the areas for which there are spectral signatures in (b) and (c)

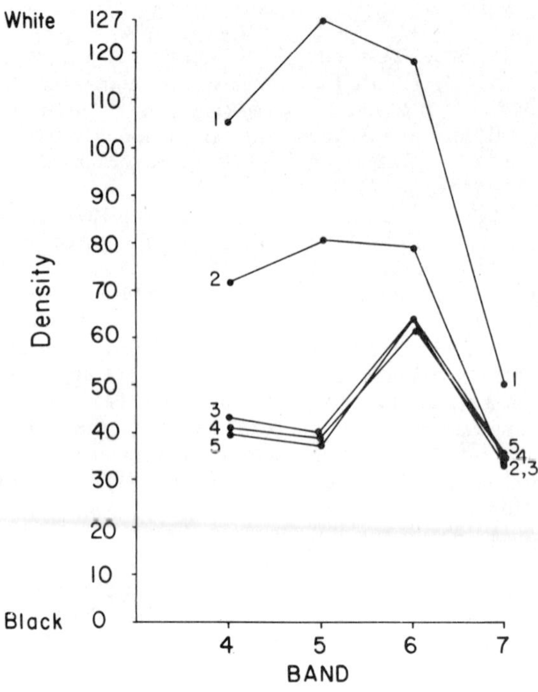

(b) Spectral signatures for terrain cover types on (a) for Esfahan city (E) and selected agricultural areas (A)

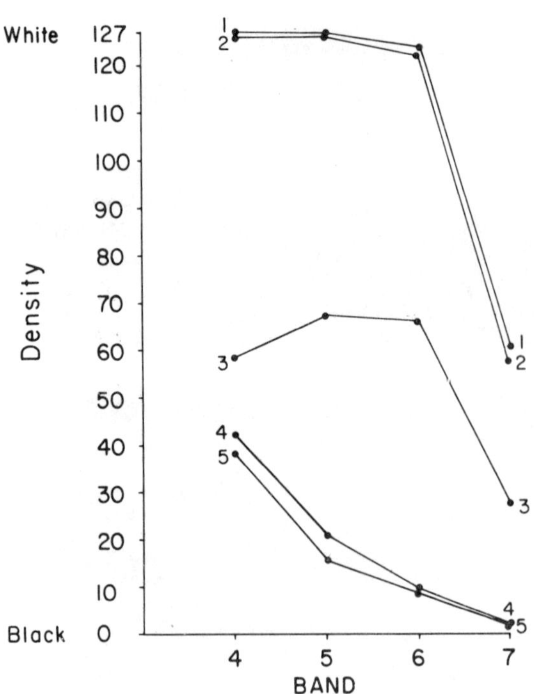

(c) Spectral signatures for clouds, water and terrain on (a) for the Shah Abbas reservoir region (R), west of Esfahan

pixel's density. A picture matrix Z can be transformed into a binary picture matrix X by

$$x_{ij} = 1 \quad \text{if} \quad z_{ij} > w$$
$$x_{ij} = 0 \quad \text{if} \quad z_{ij} \leqslant w$$

where w is a threshold density. The picture matrix X then shows areas above the threshold density, and the simple assumption is that the areas so identified are directly related to a terrain type, and thus a terrain type classification, albeit binary, has been performed. Pixel density values can also be manipulated algebraically with multispectral data when pixel density values are added, multiplied etc. A common technique is band ratioing (Berg *et al.*,

1978) in which a new picture matrix X is produced from the ratio of two picture matrices Y and Z, the digital arrays of bands n and m of a multispectral scanner. Again, terrain type classification can be produced by examining the ratio map.

Multispectral image analysis is a common tool in remote sensing, and became particularly important after the launch of Landsat 1, with its multispectral scanner. Spectral signatures can be constructed for different terrain cover types, and Figure 3.2 shows an example of spectral signatures of sample areas from a Landsat 1 scene of central Iran. Multispectral data can also be used for terrain classification, and the two general classes of model used are unsupervised and supervised classification (Armstrong and Clayton, 1977; Owen-Jones, 1975; Preston, 1975). Unsupervised classification is a means of clustering the digital data, commonly using sequential or K-means clustering (Preston, 1975), while supervised classification uses sample data from training areas to establish a set of group boundaries by discriminant analysis (Andrews, 1972). Of the two methods, Steiner and Salerno (1975) express a preference for supervised classification because classification accuracy is generally higher. Maximum likelihood models have been used by Carter (1974) and Kettig and Landgrebe (1975) to classify remote sensing image data, and Beakley and Tuteur (1972) have suggested a Neyman-Pearson model for discrimination which is a distribution-free measure. This model does not appear to have been used in remote sensing data analysis in Britain, but it has certain advantages which make it worthy of attention.

Multispectral data have been used in such classification schemes for the mapping of terrain cover (Biehl and Silva, 1976; Carter, 1974; Carter *et al.*, 1976; Gardner *et al.*, 1977; Kumar and Silva, 1977; Owen-Jones, 1975; Tarnocai and Kristof, 1976), for geological investigations (Anderson and Schubert, 1976; Lawrence and Herzog, 1975), and for cloud analysis (Barrett and Grant, 1978; Shenk *et al.*, 1976). However, classification accuracy using classifications of mutispectral pixel data has often been found to be less than was desirable, so the use of measures based on spatial features has recently been investigated. Such features use density information surrounding each pixel as well as the density value of the pixel itself to characterize portions of an image, and have been termed texture features. Thomas and Davey (1977) give some examples of work in Britain on texture analysis.

Descriptions of image texture are normally either statistical or structural (Lipkin and Rosenfeld, 1970), and it is the statistical group descriptions which are easier to model. Small sub-images or window areas of a picture are examined to determine their texture (Harris, 1977b), and the analysis of the texture of these window areas provides an input to the type of classification models discussed above. Three main orders of texture analysis have been recognized (Harris, 1977a). The first order measures are summaries of the density variation within a digital sub-image, normally of between 3 x 3 and 16 x 16 pixels in size. Edge counting, variance measures, information theory, and autocorrelation functions are some of the first order measures which have been used to characterize image texture (Harris, 1977b; Harris and Barrett, 1978; Hawkins, 1970; Sutton and Hall, 1972; Triendl, 1972; Viglione, 1970). The second order texture measures have largely been proposed by Haralick (1974), who suggests that the texture information of a sub-image is adequately specified by an angular nearest neighbour grey tone spatial dependence matrix (**P**), which is the matrix of relative frequencies with which two pixels, one with density *i*, the other with density *j*, and separated by distance *d*, occur on the sub-image. Figure 3.3 shows a digital sub-image of 4 x 4 pixels and the subsequent production of four **P** matrices for eachof the angular relationships 0°, 45°, 90°, 135°. Haralick (1974) then lists fourteen statistical summaries of these derived matrices, some of which are similar to first order grey tone texture measures, and these are then input into a

Figure 3.3 Generation of angular nearest neighbour grey tone spatial dependence matrices
A = Original density array of 4 x 4 pixels
B = General form of the spatial dependence matrix (**P**)
C = Horizontal relationship ($\mathbf{P}_h$)
D = Right diagonal relationship ($\mathbf{P}_{rd}$)
E = Vertical relationship ($\mathbf{P}_v$)
F = Left diagonal relationship ($\mathbf{P}_{ld}$)

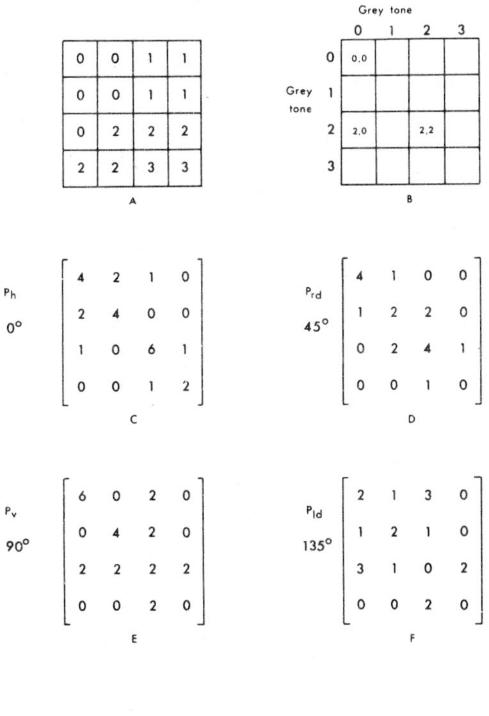

A

0	0	1	1
0	0	1	1
0	2	2	2
2	2	3	3

B

Grey tone	0	1	2	3
0	0,0			
1				
2	2,0		2,2	
3				

C

P_h 0° $\begin{bmatrix} 4 & 2 & 1 & 0 \\ 2 & 4 & 0 & 0 \\ 1 & 0 & 6 & 1 \\ 0 & 0 & 1 & 2 \end{bmatrix}$

D

P_{rd} 45° $\begin{bmatrix} 4 & 1 & 0 & 0 \\ 1 & 2 & 2 & 0 \\ 0 & 2 & 4 & 1 \\ 0 & 0 & 1 & 0 \end{bmatrix}$

E

P_v 90° $\begin{bmatrix} 6 & 0 & 2 & 0 \\ 0 & 4 & 2 & 0 \\ 2 & 2 & 2 & 2 \\ 0 & 0 & 2 & 0 \end{bmatrix}$

F

P_{ld} 135° $\begin{bmatrix} 2 & 1 & 3 & 0 \\ 1 & 2 & 1 & 0 \\ 3 & 1 & 0 & 2 \\ 0 & 0 & 2 & 0 \end{bmatrix}$

classification algorithm (Haralick and Shanmugan, 1974). Third order texture measures are concerned with transforms of image data. The information on spacing and orientation given by Fourier transforms have been used as texture indicators in both optical (Barnett and Harnett, 1977) and digital (Dyer and Rosenfeld, 1976) systems, and have provided useful summary information on the texture of sub-images.

Weszka *et al.* (1976) compared the three orders of texture measures and concluded that third order measures performed least successfully, while first and second order measures performed equally well. As most of the first order texture measures are simpler, easier to program, and faster to compute than the second order measures, then it seems appropriate to prefer the group of first order measures in the automatic analysis of remote sensing image data.

Post-processing

While texture analysis produces useful improvements in classification accuracy, it does not yet provide the avenue to completely accurate image analysis. One route currently being investigated is the use of context information in image classification. Human beings see most objects (houses, buses, paintings) in context, so, in attempting to emulate some of the better human image interpretation abilities, incorporating the spatial context of pixels or groups of pixels into classification algorithms seems a route likely to lead to success. Context information has been used by Carter and Stow (1979) in the identification of urban change in the Reading area from Landsat imagery. They used distance, size and shape criteria to decide whether to accept or reject a pixel as new urban development. The importance of context analysis has been expressed by several authors, but so far in Britain there have been few explorers in this line of research.

Once analysis of the remotely sensed data has been performed using the type of procedures outlined above, then the results need to be displayed in a convenient fashion. Monochrome TV displays have now given way to colour monitors which can use multiband data in an effective way, and show both original image data and classified maps in colour. Such display devices are incorporated into dedicated image analysers such as the Quantimet (Wignall, 1977) and the Plessey IDP 3000 (Allan *et al.*, 1979; Balston, 1977), a version of which is currently available for remote sensing users in Britain at the Royal Aircraft Establishment in Farnborough. Colour hard copy devices include the photowriters (Bryant, 1975) and laser plotters, which operate on the reverse principle to a microdensitometer, writing light spots to an unexposed piece of film rather than reading spot densities. There is a link here with automated cartography, where with an integrated system it would be possible to accomplish rapid map revision by entering into a geocoded data base newly acquired remote sensing data.

In conjunction with remote sensing data analysis, calibration information needs to be collected, and this is normally termed ground truth. Williams (1977) and Williams and Curtis (1977) discuss the important considerations in ground truth monitoring, and in comparing ground truth data with imagery and with maps from classification schemes. Hay (1979) and van Genderen *et al.* (1978) address the problem of testing the accuracy of thematic maps produced from remote sensing data, but fewer workers have seen far beyond the output products to the limitations placed upon those products by the quality of the ground truth data.

Future progress

Progress in satellite sensor technology will probably be the single most important development in the early part of the 1980s. Landsat-D, currently planned for launch in mid 1982, will carry a thematic mapper with six or seven short wavelength bands and a ground resolution of approximately 30 m, and a thermal infrared band with a ground resolution of 120 m (Doyle, 1978). This will be followed by Landsat-D' around 1983 and together they are planned to provide an operational, rather than experimental, Earth resources monitoring system. Already 40 m resolution has been achieved with the return beam vidicon camera on Landsat 3 (Justice and Townshend, 1979), and proposals appear from time to time about a 10 m ground resolution. Such resolution may be possible with the so-called *pushbroom* scanners as replacement for the multispectral scanners (Colvocoresses, 1979a; Thompson, 1979). These new scanners are solid state devices with a linear arrangement of sensing cells which are scanned electronically rather than by a moving mirror. Direct mapping from space might be possible in future with Stereosat (Doyle, 1978), or a satellite such as the Mapsat proposed by Colvocoresses (1979b). Stereosat, which has yet to be approved, would consist of two 600 mm focal length telephoto lenses focused on a 1,872 element linear array and would have a possible ground resolution of 19.3 m x 16.6 m. However, in the design so far, the spacecraft has limited attitude control and no spatial positioning capability (Doyle, 1978), so its use in accurate topographic mapping would seem rather limited. One new satellite of particular interest is the Space Shuttle, which will be equipped with a variety of sensors (Plevin, 1977), for example, Orbital Test Flight 2 will carry a carbon monoxide monitor, a ten channel near infrared non-imaging sensor, an ocean colour scanner, a movie camera to study lightning, and an L-band synthetic aperture radar for terrain studies (Doyle, 1978). Experiment packages such as the metric camera will enable topographic mapping to be performed with a spot height accuracy of up to 13 m (Petrie, 1979).

New satellite systems such as these will greatly increase our ability to monitor the environment. However, before such monitoring is possible it is necessary to transmit, record, preprocess, analyse and display the vast quantities of information likely to be received from the new generation of sensor systems. In addition, the data should be readily available to the user community without a long ordering delay. Future developments in remote sensing are likely to be concerned with:

1 efficient data collection and transmission;
2 more robust multispectral, texture and context analysis procedures to refine classification accuracy;
3 classification algorithms to deal with the large data amounts and the complex nature of boundaries between classes in the feature space;
4 interactive data analysis systems;
5 integration of remote sensing data with conventional ground surveys to aid in map revision and the monitoring of environmental change.

Such developments will help in the greater use of remote sensing data on the one hand, and on the other the integration of remote sensing within geography by the use of precise quantitative methods.

References

Allan, J.A., Custance, N.D.E., and Latham, J.S. (1979) 'New era for Landsat', *Geographical Magazine*, 51, 428-31.

Allan, J.A., and Harris, R. (eds) (1979) *Remote sensing and national mapping*, Remote Sensing Society, Reading.

Anderson, A.T., and Schubert, J. (1976) 'ERTS-1 data applied to strip mining', *Photogrammetric Engineering and Remote Sensing*, 42, 211-19.

Andrews, H.C. (1972) *Introduction to mathematical techniques in pattern recognition*, Wiley: New York.

Armstrong, A.C., and Clayton, K.M. (1977) 'Objective generalization of Landsat images', in E.C. Barrett and L.F. Curtis (eds), op. cit., 163-89.

Balston, D.M. (1977) 'An interactive image processing system', in R.F. Peel *et al.*, op. cit., 80-90.

Barnett, J.J., Harwood, R.S., Houghton, J.J., Morgan, C.G., Rogers, C.D., and Williamson, E.J. (1975) 'Comparison between radiosonde, rocketsonde, and satellite observations of atmospheric temperatures', *Quarterly Journal Royal Meteorological Society*, 101, 423-36.

Barnett, M.E., and Harnett, P.R. (1977) 'Optical processing as an aid in analysing remote sensing imagery', in E.C. Barrett and L.F. Curtis, (eds), op. cit., 125-42.

Barrett, E.C. (1974) *Climatology from satellites*, Methuen: London.

Barrett, E.C., and Curtis, L.F. (1976) *Introduction to environmental remote sensing*, Chapman & Hall: London.

Barrett, E.C., and Curtis, L.F. (eds) (1977) *Environmental remote sensing 2*, Edward Arnold: London.

Barrett, E.C., and Grant, C.K. (1978) 'An appraisal of Landsat 2 cloud imagery and its implications for the design of future meteorological observing systems', *Journal of the British Interplanetary Society*, 31, 3-10.

Beakley, G.W., and Tuteur, F.B. (1972) 'Distribution-free pattern verification using statistically equivalent blocks', *IEEE Transactions on Computers*, C-21, 1337-47.

Berg, A., Flouzat, G., and Galli de Paratesi, S. (1978) *AGRESTE Project, Final Report*, Commission of the European Communities: Ispra, Italy.

Biehl, L.L., and Silva, L.F. (1976) 'Machine aided multispectral analysis utilizing Skylab thermal data for land use mapping', *IEEE Transactions on Geoscience Electronics*, GE-14, 49-54.

Bristor, C.L., Callicot, W.M., and Bradford, R.E. (1966) 'Operational processing of satellite cloud pictures by computer', *Monthly Weather Review*, 94, 515-27.

Bryant, M. (1975) *Digital image processing: the answer to tomorrow's information needs*, Optronics: Mass.

Carter, P. (1974) 'Application of digital image processing to multispectral aerial and satellite photography', *Oxford conference on computer scanning*, 515-31.

Carter, P., and Gardner, W.E. (1977) 'An image-processing system applied to earth-resource imagery' in E.C. Barrett and L.F. Curtis (eds), op. cit., 143-62.

Carter, P., and Stow, B. (1979) 'Clean-up of digital thematic maps of urban growth extracted from Landsat imagery' in J.A. Allan and R. Harris (eds), op. cit., 27-40.

Carter, P., Gardner, W.E., and Smith, T.F. (1976) 'The use of Landsat imagery for the automated recognition of urban development' in J.L. van Genderen and W.G. Collins (eds), *Land use studies by remote sensing*, Remote Sensing Society: Reading. 54-88.

Colvocoresses, A.P. (1979a) 'Multispectral linear arrays as an alternative to Landsat D', *Photogrammetric Engineering and Remote Sensing*, 45, 67-9.

Colvocoresses, A.P. (1979b) 'Proposed parameters for Mapsat', *Photogrammetric Engineering and Remote Sensing,* 45, 501-6.

Corless, K.G. (1977) 'Remote sensing by radar' in R.F. Peel *et al.,* op. cit., 38-53.

Doyle, F.J. (1978) 'The next decade of satellite remote sensing', *Photogrammetric Engineering and Remote Sensing,* 44, 155-64.

Dyer, C.R., and Rosenfeld, A. (1976) 'Fourier texture features: suppression of aperture effects', *IEEE Transactions on Systems, Man and Cybernetics* SMC-6, 703-5.

Fortuna, J.J., and Hambrick, L.N. (1974) *The operation of the Noaa polar orbiting satellite system,* NOAA Technical Memorandum NESS 60, Washington, DC.

Fu, K.S., and Rosenfeld, A. (1976) 'Pattern recognition and image processing', *IEEE Transactions on Computers* C-25, 1336-46.

Fymat, A.L., and Zuev, V.E. (1978) *Remote sensing of the atmosphere: inversion methods and applications,* Developments in atmospheric science 9, Elsevier: Amsterdam.

Gardner, W.E., Carter, P., and Smith, T.F. (1977) 'Digital analysis of multispectral aerial and Landsat data for land use planning in Britain' in R.F. Peel *et al.,* op. cit., 96-107.

Haralick, R.M. (1974) *Texture-tone study with application to digitized imagery,* Remote Sensing Laboratory Technical Report 182-6, University of Kansas Center for Research.

Haralick, R.M., and Shanmugan, K.S. (1974) 'Combined spectral and spatial processing for ERTS imagery data', *Remote Sensing of Environment* 3, 3-13.

Harris, R. (1977a) 'Automatic analysis of meteorological satellite imagery', in J.O. Thomas and P.G. Davey (eds) *Texture analysis,* 45-72, University Physics Photographic Unit: Oxford.

Harris, R. (1977b) *Procedures for satellite nephanalysis,* PhD Thesis, University of Bristol.

Harris, R. (1978) *Processing satellite imagery from Landsat satellites,* Conference on Contemporary problems in quantitative and theoretical Geography, 28-30 September 1978, Strasbourg.

Harris, R. (1979) 'Access to Landsat data', *Area,* 11, 63-6.

Harris, R., and Barrett, E.C. (1978) 'Toward an objective nephanalysis', *Journal of Applied Meteorology,* 17, 1258-66.

Harwood, R.C. (1977) 'Some recent investigations of the upper atmosphere by remote sounding satellites' in R.F. Peel *et al.,* op. cit., 111-25.

Hawkins, J.K. (1970) 'Textural properties for pattern recognition' in B.S. Lipkin and A. Rosenfeld (eds), op. cit., 347-70.

Hay, A.M. (1979) 'Sampling designs to test land use accuracy', *Photogrammetric Engineering and Remote Sensing,* 45, 529-33.

Heuseler, H. (1976) *Die Erde aus dem All,* Deutsche Verlags-Anstalt and Georg Westermann: Stuttgart.

Holkenbrink, P.F. (1978) *Manual on characteristics of computer-compatible tapes produced by the EROS Data Center Digital Image Processing System,* version 0.0, United States Geological Survey.

Honvault, C. (1978) 'The in-orbit performance of, and early results from, Meteosat', *ESA Bulletin,* 13, 17-20.

Houghton, J.T. (1979) 'The future role of observations from meteorological satellites', *Quarterly Journal Royal Meteorological Society,* 105, 1-23.

Jagoe, R., and Paton, K. (1978) 'Generalized counting in digital pictures', *Computer graphics and image processing,* 7, 52-66.

Justice, C.O., and Townshend, J. (1979) 'Clear images from space', *Geographical Magazine,* 51, 828-31.

Kettig, R.L., and Landgrebe, D.A. (1975) 'Classification of multispectral image data by extraction and classification of homogeneous objects', *Symposium on Machine Processing of Remotely Sensed Data,* Laboratory for Applications of Remote Sensing, 75-CH-1009-0-C, Purdue: Indiana.

Kumar, R., and Silva, L.F. (1977) 'Separability of agricultural cover types by remote sensing in the visible and infrared wavelength regions', *IEEE Transactions on Geoscience Electronics* GE-15, 42-9.

Lawrence, R.D., and Herzog, H.H. (1975) 'Geology and forestry classification from ERTS-1 digital data', *Photogrammetric Engineering and Remote Sensing,* 41, 1241-51.

Lintz, J., and Simonett, D.S. (eds) (1976a) *Remote sensing of environment,* Addison-Wesley: Reading, Mass.

Lintz, J., and Simonett, D.S. (1976b) 'Sensors for spacecraft' in J. Lintz and D.S. Simonett (eds), op. cit., 323-48.

Lipkin, B.S., and Rosenfeld, A. (1970) *Picture processing and psychopictorics,* Academic Press: New York.

NASA (1976) *Landsat data users handbook.* Document no. 76SDS4258, Goddard Space Flight Center: Greenbelt, Md.

Overington, I. (1976) *Vision and acquisition,* Pentech Press: London.

Owen-Jones, E.S. (1975) 'Pattern classification of agricultural and non-agricultural areas' in J.L. van Genderen and W.G. Collins (eds), op. cit., 73-95.

Owen-Jones, E.S. (1977) 'Densitometric methods of processing remote sensing data, with special reference to crop-type and terrain studies' in E.C. Barrett and L.F. Curtis (eds) op. cit., 101-24.

Peckham, G.E. (1978) 'Remote measurements of atmospheric properties from satellites' in T. Lund, *Surveillance of environmental pollution and resources by electromagnetic waves,* 95-110,

D. Reidel: Dordrecht.

Peel, R.F., Curtis, L.F., and Barrett, E.C. (eds) (1977) *Remote sensing of the terrestrial environment*, Butterworth: London.

Petrie, G. (1979) 'The status of topographic mapping from space imagery' in J.A. Allan and R. Harris (eds), op. cit., 1-16.

Plevin, J. (1977) 'Remote sensing from Spacelab: a case for international cooperation' in E.C. Barrett and L.F. Curtis (eds), op. cit., 48-71.

Pratt, W.K. (1978) *Digital image processing*, John Wiley: New York.

Preston, G. (1975) 'Automatic data processing for non-mathematicians' in J.L. van Genderen and W.G. Collins (eds), op. cit., 53-72.

Proca, G.A. (1978) 'LEDA: une banque de données de l'ESA consacrée aux images de la Terre vue de l'Espace', *ESA Bulletin*, 13, 47-51.

Reeves, R.G. (ed) (1975) *Manual of remote sensing*, Keuffel & Esser: Virginia, USA.

Rosenfeld, A. (1967) 'On models for the perception of visual texture' in W.W. Dunn (ed.) *Models for the perception of visual form*, 219-23, MIT Press: Cambridge, Mass.

Rosenfeld, A. (1969) *Picture processing by computer*, Academic Press: New York.

Sabins, F.F. (1976) 'Geologic applications of remote sensing' in J. Lintz and D.S. Simonett (eds), op. cit., 508-71.

Shannon, C.E., and Weaver, W. (1949) *The mathematical theory of communication*, University of Illinois Press: Urbana.

Shenk, W.E., Holub, R.J., and Neff, R.A. (1976) 'A multispectral cloud type identification method developed for tropical ocean areas with Nimbus-3 MRIR measurements', *Monthly Weather Review*, 104, 284-91.

Simonett, D.S. (1974) 'Quantitative data extraction and analysis of remote sensor images' in J.E. Estes and L.W. Senger (eds), *Remote sensing: techniques for environmental analysis*, 51-81, Hamilton Publ. Co.: California.

Steiner, D., and Salerno, A.E. (1975) 'Remote sensor data systems, processing, and management' in R.G. Reeves (ed.) op. cit., 611-803.

Sturm, B., Lundgren, B., Maracci, G., and Mehl, W. (1978) *Some results of the EURASEP OCS experiment in 1977*, Commission of the European Communities: Ispra, Italy.

Sutton, R.N., and Hall, E.L. (1972) 'Texture measures for automatic classification of pulmonary disease', *IEEE Transactions on Computers* C-21, 667-76.

Swain, P.H., and Davis, S.M. (1978) *Remote sensing: the quantitative approach*, McGraw-Hill: New York.

Tarnocai, C., and Kristof, S.J. (1976) 'Computer-aided classification of land and water bodies using Landsat data, MacKenzie delta area, NWT, Canada', *Arctic and Alpine Research*, 8, 151-9.

Thomas, J.O. (1977) 'Texture analysis in imagery processing' in J.O. Thomas and P. Davey (eds), op. cit., 1-43.

Thomas, J.O., and Davey, P. (1977) *Texture analysis*, University Physics Photographic Unit: Oxford.

Thomas, V.L. (1975) *Generation and physical characteristics of Landsat 1 and 2 MSS computer compatible tapes*, Goddard Space Flight Center, NASA X-563-75-233: Greenbelt, Md.

Thompson, L.L. (1979) 'Remote sensing using solid-state array technology', *Photogrammetric Engineering and Remote Sensing*, 45, 47-55.

Tou, J.T., and Gonzalez, R.C. (1974) *Pattern recognition principles*, Addison-Wesley: Reading, Mass.

Triendl, E.E. (1972) 'Automatic terrain mapping by texture recognition', *Proceedings of the 8th International Symposium on Remote Sensing of the Environment*, 771-6, Ann Arbor, Michigan.

Twomey, S. (1977) *Introduction to the mathematics of inversion in remote sensing and indirect measurements*, Elsevier: Amsterdam.

Ullman, J.R. (1973) *Pattern recognition techniques*, Butterworth: London.

Van Genderen, J.L. (1975) 'Visual interpretation of remote sensing data and electronic image enhancement techniques' in J.L. van Genderen and W.G. Collins (eds), op. cit., 19-52.

Van Genderen, J.L., and Collins, W.G. (eds) (1975) *Remote sensing data processing*, Remote Sensing Society: Reading.

Van Genderen, J.L., Lock, B.F., and Vass, P.A. (1978) 'Remote sensing: statistical testing of thematic map accuracy', *Remote sensing of environment*, 7, 3-14.

Viglione, S.S. (1970) 'Applications of pattern recognition technology' in J.M. Mendel and K.S. Fu (eds), *Adaptive learning and pattern recognition systems*, 115-62, Academic Press: New York.

Viksne, A., Liston, T.C., and Sapp, C.D. (1969) 'SLR reconnaisance of Panama', *Geophysics*, 34.

Weszka, J.S., Dyer, C.R., and Rosenfeld, A. (1976) 'A comparative study of texture measures for terrain classification', *IEEE Transactions on Systems, Man and Cybernetics* SMC-6, 269-85.

Wignall, B. (1977) 'A critical review of the Quantimet 720 Image Analyser in remote sensing' in R.F. Peel *et al.*, op. cit., 71-9.

Williams, T.H.L. (1977) 'The role of ground truth data and an approach to its collection' in F. Shahroki (ed.), *Remote sensing of Earth resources*, vol. 6, 39-49, University of Tennessee.

Williams, T.H.L., and Curtis, L.F. (1977) 'Development of a ground observation system in relation to air photography', *Photogrammetric Record*, 9, 55-70.

World Bank (1976) *Landsat index atlas of the developing countries of the world*, Washington, DC.

Chapter 4

Census data handling

I.S. Evans

Introduction

Census data continue to be a major source for British quantitative studies in geography. This chapter contains a discussion of their advantages and disadvantages, of the mechanics of their use including how variables are defined, and of their areal basis and the effect of areal scale. Finally, some recent uses of census data for mapping and statistical analysis are reviewed, especially for innovations concerning large data sets. This chapter is concerned mainly with the Census of Population in Great Britain, but the initial discussion applies to census data in general. Emphasis is placed upon the Small Area Statistics (SAS).

Advantages and disadvantages of census data

Those who use census data are striking a Faustian bargain. They gain a large data set without expending the man-years of effort involved in generating it. The hidden cost may be the loss of conventional scientific method; the freedom of researchers to define their own variables and their own experimental design in terms of the questions asked, the definition of individuals, the areal divisions used and the timing. Obviously, then, there are many hypotheses which cannot be tested with census data. Most human geographers nevertheless find it necessary, at some stage, to use census data: why?

A national survey would involve great expenditure and the training of numerous questioners or enumerators to produce consistent results; the researcher is soon up against constraints comparable to those of the census. Indeed, if he needs national data his best course of action is to attempt, in co-operation with other users, to influence the design of the census. Here he will discover how the limitations of the census arise, as compromises are reached between many different requirements. How could a compulsory, national census be otherwise? The 1981 census is estimated to cost £62 million for Great Britain, i.e. about £3 per household spread over a five- or ten-year period. In practice, local and central government departments are the largest users of census data, and the researcher would do well to win their support for his proposals.

The great advantages of census data are in the size of the data set, the response rate and the national uniformity. A census is a very thorough survey, designed by experts, with pre-tests, accuracy checks and a lengthy examination of the questions. Questions in the survey of an individual scientist will not have been probed in such depth, and are more likely to be misunderstood and inaccurately answered; unforeseen problems are more likely to arise during the survey.

Data set size is extremely important. The universal teaching of inferential statistics to University geographers during recent years does not seem to have inculcated this fact: as in psychology some years ago (Cohen, 1962), many research papers are based upon samples too small to establish the effects hypothesized. The complete coverage of a census gives huge 'samples', unless very small populations are studied.

The population census coverage in Great Britain is at least 99 per cent of households. This is achieved only by the backing of legal sanctions for non-response. The public are told that census results are essential to government and local authority administration. They are also given a categorical assurance of the confidentiality of their individual responses; even other Government departments cannot see them (Hakim, 1978a; and in Bulmer, 1979; Durbin, 1979). Clearly such conditions cannot apply to questionnaire surveys by individual scientists, which are therefore likely to provide poorer data with response rates as low as 60 per cent quite common.

The uniformity of the census within Great Britain, from Kent to the Outer Hebrides, is a great advantage. This constraint is, up to a point, desirable in that it makes studies in different areas more comparable than they would otherwise be. Increasing agreement on social indicators derived from the census (Hakim, 1978b) should be welcomed by geographers: if different indicators are used in studies of, say, deprivation in Glasgow and in Liverpool, how much of the difference in results reflects this? Real differences may be masked by differences of definition. Use of census data facilitates comparative studies with identical variables and procedures (Coulter, 1978). Measures are in hand to promote greater international comparability (Brown, 1978).

Census data, then, have these three major advantages: their large 'sample' size, minimal non-response, and national uniformity; in addition, their disadvantages may be less than at first appears. The introduction of new variables which require new questions is clearly slowed down by the need for consultation, trial and compromise in the census: but the 1,571 items of information in (1971) Small Area Statistics allow a huge number of variables to be derived by aggregation and ratio formation. The broad range of topics included facilitates interrelation.

Data handling

Data from British population censuses are available in an increasing variety of forms, and through an increasing number of agencies: Hakim (1978d) has provided a thorough account of data dissemination. This includes workplace and migration data for local authorities, which will not be discussed here because of my concentration on Small Area Statistics (SAS). Output from the 1971 census comprised 37,618 published and 1,589,492 unpublished 'pages' of tables: this compares with 8,168 pages (all published) for 1951. SAS are now the most used type of census data. This trend is in line with developments in the USA and in Canada (Stinson, 1977). Although printed SAS and published county and national volumes remain available, an increasing part in the dissemination of census data is played by machine-readable material: punched cards and (ICL) magnetic tapes.

Conversion of ICL tapes for reading on IBM and other machines is not undertaken by OPCS (The Office of Population Censuses and Surveys), and has caused months of delay in some organizations. Since 1978, some researchers have been able to avoid this by obtaining data for Enumeration Districts (EDs), administrative areas or constituencies from the Social Science Research Council (SSRC) Social Survey Archive at Essex University (see their Bulletins and Data Catalogue). The Census Research Unit (Durham) supplied grid square data on IBM tapes during 1977 and 1978. For very large data sets magnetic tape is unavoidable and, if this must be obtained from OPCS, conversion of the mixed binary and character-coded data on the ICL tapes to a form compatible with the local computer installation is a barrier which non-ICL users must overcome (Visvalingam, Norman and Sheehan, 1976). The extension of networking may reduce this barrier for 1981 census data.

Processing of a large data set - the SAS for every 1 km square in Britain (Rhind, 1975) - for use on an IBM 370/168 computer was described by Visvalingam (1976). Tenfold compaction was achieved by 'trimming' incomplete (suppressed) records, using bit-map representation or a form of run length encoding, as appropriate (Visvalingam, 1976; Visvalingam and Rhind, 1977) to condense the large numbers of zero values, and storing non-summary data in 16-bit integer mode (maximum 2^{15} -1 = 32,767), compacted to 8-bit wherever possible. Checks were made (manually, graphically and by program) to avoid duplication or omission in the various copying stages (Visvalingam and Perry, 1976). The compressed files were compared against the original files by program.

SAS for 1971 (OPCS, 1976) are grouped into 28 tables, most of which are two-way classifications, e.g. number of rooms by number of persons for private households, a 7 x 6 table (Table 19, which also contains row and column totals for both households and rooms); or sex by birthplace, a 2 x 16 table (Table 8). In these and most other tables, counts (numbers of people or households) are given; but some summary tables give only ratios per thousand, for example Table 11 with several age-groups, married females in three age-groups, and students, for persons present in private households only.

The 1971 SAS were split into four computer records: 100 per cent population data, 100 per cent household data, and two records based on 10 per cent sample data (dealing with employment, qualifications, migration and travel to work). After previous difficulties with sampling by enumerators in the field, the 1971 10 per cent sample was drawn in the office during processing, as 10 per cent of households within each ED. This meant that the proportion sampled per grid square varied, especially in thinly populated areas.

Two measures taken by the Census Offices in order to preserve the confidentiality of information for identifiable individuals (Hakim, 1978a) can have drastic effects from the point of view of SAS users. The (100 per cent) population record for an areal division is suppressed if fewer than 25 people are present; only number of males, number of females and total are then given. The (100 per cent) households record is suppressed if fewer than 8 households are present, and only number of households is given: this is a roughly equivalent criterion since the average

1971 household contained 2.9 people. The 10 per cent sample records are suppressed if only one household is sampled: this is a more stringent criterion since, if the corresponding number of households (10) were found in the 100 per cent data, they would not usually be suppressed.

Except for studies of sex ratios or population or household totals, users are concerned mainly with the unsuppressed areas, which contain the great bulk of the data. There are 67,546 one-km squares with unsuppressed 100 per cent population data, 68,412 with unsuppressed household data, 54,464 with unsuppressed data for the first 10 per cent record and 54,153 for the second 10 per cent record (Visvalingam, 1977, p. 4). Although a large proportion of 1-km squares suffer suppression, about 98.54 per cent of the total population live in squares with unsuppressed 100 per cent data, and about 97.31 per cent in squares with unsuppressed 10 per cent data. Only 800 out of some 125,000 EDs in Great Britain in 1971 were affected by suppression.

So that there can be no certainty that information relates to any identifiable person or household, the Census Offices add 1, 0 or -1 to each constituent count in the 100 per cent SAS, unless a negative value would result. This adjustment is done randomly, but systematized so as to reduce cumulative errors obtained by adding counts (Newman, 1978). Nevertheless, errors do cumulate: the least populous square with full 100 per cent population data has 17 people, and has been adjusted down from 25 by subtraction of 1 from each of eight constituent counts.

Analyses based on ratios usually require two further processing operations. Most ratio denominators correlate closely with population; ratios are therefore more reliable for more populous squares. In correlation, regression and related analyses, it is best to weight each square by its population: thus a square with 1,000 people receives the same weight as ten squares with 100 people each.

The frequency distributions of many ratios (especially for minorities) are highly skewed, both for EDs (Holtermann, 1975) and for grid squares. Where authors do not give skewness, its presence may be suspected wherever standard deviations exceed half the mean (on a ratio scale); where they exceed the mean itself, skewness is likely to be troublesome. For closed ratios with means in excess of 50 per cent (100 per cent – mean) should be taken for this comparison, since variables with a mean of 95 per cent are likely to be negatively skewed for the same reasons as those with a mean of 5 per cent are positively skewed.

Skewed variables are likely to violate assumptions of the General Linear Model, especially linearity of relationships, additivity of effects, and equality of error variances (i.e. conditional distributions of regression residuals should approximate zero mean and equal variance). Hence transformations are usually required: angular, probit or logit for closed ratios (Chapter 12), power series (including reciprocal and logarithmic) for open ratios (Chapter 12; Mosteller and Tukey, 1977; Kruskal, 1968). Transformation should be viewed as a necessary modification of General Linear Model techniques to suit them to much real data. For 1-km grid square data, Evans (1979) has discussed the effects of weighting and transformation.

Definition of variables

A number of figures in the Small Area Statistics (SAS), especially the ratio lines, are used directly as statistical variables or social indicators. Most of those used by analysts, however, are aggregates of similar subsets, or row or column totals which collapse a cross-classification to suit the purpose in hand. There is a huge number of ways in which this can be done, and Hakim (1978b) has very carefully tabulated the more important or commonly used definitions of 'social indicators' from the 1961, 1966 and 1971 censuses under the headings demographic (25 indicators), housing (73), household and family (36), employment (72), country of birth (37) and migration (17). She also noted 26 different aggregations of socio-economic groups which could be permuted with many of these 260 indicators.

For a particular analysis, a consistent set of variables should be defined on the basis of certain principles:

1. overlap should be avoided;
2. denominators should be carefully chosen, for example, working married females are expressed more usefully in proportion to married females (in the same age group) than to working females;
3. only subsets which are very similar, or highly intercorrelated, should be aggregated, for example, 15-44 years is a poor age grouping because the 15-24 and 25-44 age groups correlate negatively (Evans, 1979, p. 175);
4. where several variables are subsets of the same total, they should be on a comparable level of subdivision;
5. as far as is reasonable, the same denominator should be used for subsets of a single classification;
6. closure problems can be reduced by using finer subsets rather than broad ones. In multivariate analysis of proportions of the same total, one subset should be omitted to avoid excessive multicollinearity and consequent problems in eigenvalue analysis (see the conclusion to Chapter 12);
7. for the sake of comparability, previously used definitions are preferred unless there is good reason for change;

8 low variability throughout the study area limits the value of a variable;
9 ratios with denominators which are small in many areal divisions are statistically unstable; they also take on only a limited number of values (e.g. 0/3, 1/3, 2/3 or 3/3), so the scale of measurement becomes discrete rather than continuous - their omission, or choice of a different denominator, should be considered;
10 variables which are zero for most areal divisions may be an embarrassment; their omission or further aggregation should be considered. The degree of disaggregation should be appropriate to the populations involved, for example, coarse age divisions for EDs, 5-year groups for wards, and 1-year groups for local authorities;
11 instead of dividing a frequency distribution into many subsets, it can be summarized in terms of an average measure (such as persons per room) and two extreme groups (in this case, households with under 0.5, or with over 1.5, persons per room). In this way the number of variables can be economized while the extremes, which are usually of greater interest, are retained.

The technical problems of deriving new variables were discussed by Rhind, Evans and Dewdney (1977), who presented the (Durham) Census Research Unit's list of 102 ratio variables together with a listing of the computer program, parts of which can be applied in other contexts. This list of variables is a revision of two previous lists, following experience in mapping and analysing data for County Durham (Dewdney and Rhind, 1975). Again, the full set of 102 variables would not be used in a single multivariate analysis; divisions of age and of socio-economic class at two levels of detail were included so that choice between them could be made at a later date.

Some improvement will be made for the 1981 census; some unreliable variables such as '5-year migrants' and some amenity variables will be dropped. A 'cheap package' of ratio variables may encourage standardization, since its use will be easier than defining new variables. Comparability with previous censuses will be (marginally) reduced, however, by the switch to 'normally resident' (*de jure*) rather than 'present' (*de facto*) population as the basis for most counts. In addition to extensive data relating to the whole of Great Britain, there will be some special tables for Scotland, Wales, and England and Wales to meet local needs. Although provision of an anonymous 'public use' sample of individual data is being considered, there is no prospect of any British Government being prepared to pay the political cost of releasing data on identifiable individuals as suggested by Moore (1979). Despite good intentions, it seems unlikely that results will be delivered much sooner, unless there is a major injection of labour into the Census Offices.

Areal basis of the Census

It could be argued that the most fundamental problem in quantitative geography is the definition of 'geographic individuals', areas with meaningful and easily-agreed boundaries which are treated as homogeneous units either by design or because more detailed data are unavailable. Precise measurement requires precise definition, but meaningful analysis requires either non-arbitrary definition or insensitivity of results to variations in definition. Yet whenever variations in areal definition have been explored, it has been found that differing results are produced (Openshaw and Taylor, 1979). Since results depend upon the boundaries of the areas used (not just upon the size of areas), meaningful results are obtained only for meaningful boundaries, related to the problem in question. Constituencies are meaningful areal individuals for (Westminster) electoral analyses, and local authorities are meaningful individuals when the causes or effects of their taxes or expenditures are under study, but for most sociogeographic studies these areal divisions are arbitrary, that is, they could be modified without affecting the reasoning involved. Their boundaries are crossed by large numbers of people on the way to work, shop, play or meet, and the boundaries often cross homogeneous areas while quite different types of area are included inside the same division. There is a problem, then, of *modifiable area divisions*.

Openshaw and Taylor (1979) have made a powerful case for much fuller investigation of the effects of areal divisions in geographic analyses. Faced with the modifiable division problem, it is advisable for geographers to (i) start from the smallest divisions available, or the smallest they can process; (ii) aggregate these in a fashion relevant to their investigation, and (iii) assess the repeatability of their results for several different aggregations at the same areal scale. In some cases it will be important also to (iv) assess the sensitivity of results to varying scale. A major problem with non-census data is the coarseness of areal resolution and the varied divisions for which data are given; see the 'Catalogue of UK statistical data potentially available for mapping' (Experimental Cartography Unit, 1971, Appendix 1).

The UK Census of Population has a statutory duty to provide statistics for administrative areas and for constituencies. Published county reports also provide limited data for wards and parishes. Special tabulations of data for these and smaller areas have become, under pressure of user demand, an increasingly important aspect of the census: for 1961, 1966 and 1971, data were provided for enumeration districts (i.e. subdivisions of wards and parishes) and for 1971 comparable statistics for 1 km

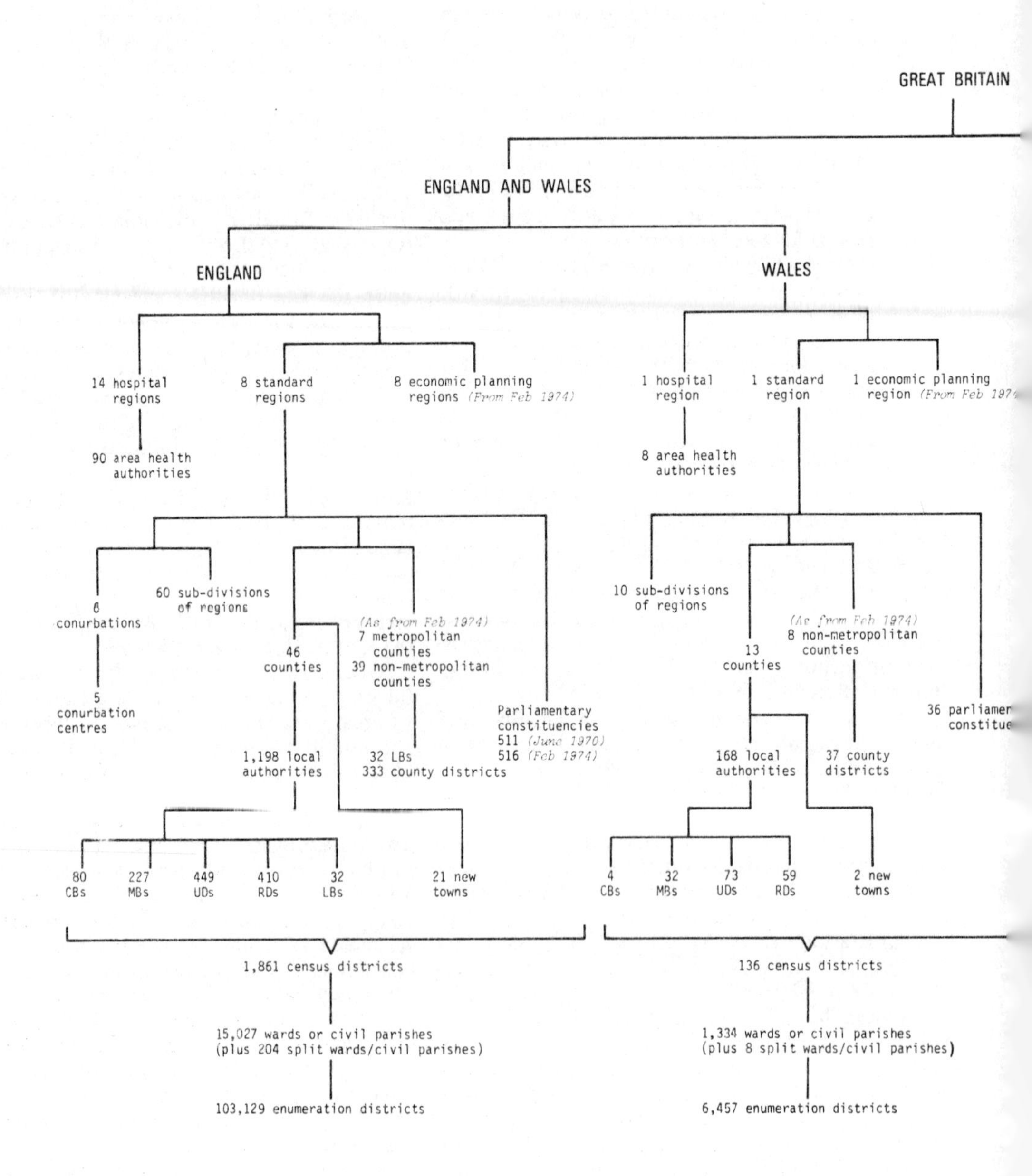
GREAT BRITAIN
ENGLAND AND WALES
ENGLAND
WALES
14 hospital regions
8 standard regions
8 economic planning regions (From Feb 1974)
90 area health authorities
6 conurbations
60 sub-divisions of regions
5 conurbation centres
46 counties
(As from Feb 1974) 7 metropolitan counties 39 non-metropolitan counties
Parliamentary constituencies 511 (June 1970) 516 (Feb 1974)
1,198 local authorities
32 LBs 333 county districts
80 CBs
227 MBs
449 UDs
410 RDs
32 LBs
21 new towns
1,861 census districts
15,027 wards or civil parishes (plus 204 split wards/civil parishes)
103,129 enumeration districts
1 hospital region
1 standard region
1 economic planning region (From Feb 197
8 area health authorities
10 sub-divisions of regions
13 counties
(As from Feb 1974) 8 non-metropolitan counties
36 parliamen constitue
168 local authorities
37 county districts
4 CBs
32 MBs
73 UDs
59 RDs
2 new towns
136 census districts
1,334 wards or civil parishes (plus 8 split wards/civil parishes)
6,457 enumeration districts

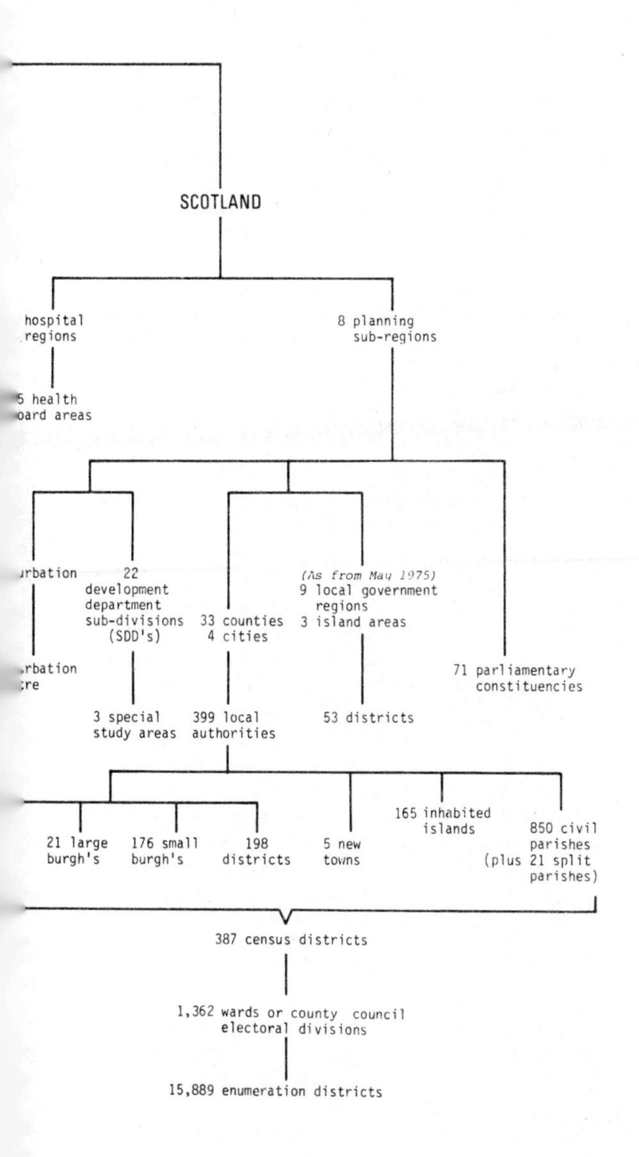

Figure 4.1 Number and types of areas for which 1971 census data have been produced (Crown copyright reserved, reproduced from Hakim, 1978d, by kind permission of the Office of Population Censuses and Surveys)

and for some 100 m grid squares were eventually produced. Turnock (1978) discussed a further source of population-only data, for 6,321 'localities' in Scotland. Census data are thus available for a broad range of areas (Figure 4.1), and experimentation is continuing with postcodes and functional regions for 1981: the Census is far from providing even medium-term stability of areal divisions, let alone long-term stability. Denham (1980) has discussed the 1971 and 1981 frameworks for England and Wales.

Chisholm (1975) discussed the effects of 1974 reorganization of local authorities, including the reduction in relative variability of District populations. At present, data from the Office of Population Censuses and Surveys are available for areas smaller than Districts only for the Census of Population, but shortly there will be the potential to produce birth and death statistics for wards in England and Wales. The situation was summarized by Population Statistics Division (1979), who related current local authority areas to pre-1974 areas and to health authority areas. Except that area health authorities are based on counties and metropolitan districts, there is often no direct correspondence by aggregation one way or the other. Figure 4.1 shows the numbers of divisions at each hierarchical level.

Districts, health areas and parliamentary constituencies are all aggregates of wards. For smaller areas, three rival modes of division are in use: enumeration districts, postcodes, and grid squares.

Enumeration districts (EDs) are defined for the administration of the census: each must be a manageable workload. For 1981, each should cover 125-220 households in urban areas or 55-135 in rural areas. In practice, some EDs turn out to have more households, and some are found to be unpopulated (e.g. following demolition; or unpopulated civil parishes); numbers of households therefore varied in 1971 from 0 to 300. The subset of urban EDs (excluding those with fewer than 50 people) studied by Holtermann (1975) had a mean population of 470 and standard deviation of 159.

ED boundaries do not cross administrative boundaries such as wards, civil parishes and Area Health Districts. They should follow recognizable features such as rivers, railways and streets (Denham, 1980). (Although combination of both sides of a street might provide a better social unit, boundaries between backyards are sometimes more difficult to check, and more houses might be missed.) Within the above constraints, local authorities are invited to influence the delimitation of boundaries, e.g. to fit traffic zones or 'General Improvement Areas': hence the quality of EDs as geographic individuals (e.g. their compactness and homogeneity) varies according to how earnestly this possibility has been taken up. Maps of EDs (which have to be specially purchased) show that some have quite contorted shapes (Rhind and Hudson, 1980, Figure 3.4) or include several varied parts without direct road connection. Nevertheless, many census users have requested stability of ED definition as a prime consideration. OPCS have agreed to take this view into account for 1981, but they must at least subdivide or amalgamate to provide equitable workloads for enumerators; they expect that 30 to 40 per cent of EDs will have boundaries different from those of 1971, although stability at the level of groups of EDs will be sought (Denham, 1980).

Problems of analysing data for irregular areas were discussed by the Experimental Cartography Unit (1971, pp. 28-32), and some of Coppock's (1960) points about agricultural data for parishes and the need for careful aggregation apply also to EDs. Hunt (1952) showed the difficulty of mapping data for irregular areas, and Robertson (1969) demonstrated the problems which result from the instability and irregularity of ED boundaries. EDs are designed as working units, not as homogeneous statistical units; demanding that they serve as the most detailed areal division for census statistics may be asking too much. Occasionally EDs consisting of two contrasted areas are split, but this increases the weight of paperwork: in 1971, there were 125,000 EDs in Great Britain, for 97,000 enumerators.

Postcodes (Experimental Cartography Unit, 1971) are an innovation which have been adopted for Small Area Statistics for the 1981 Census in Scotland and which may be used in England and Wales as a stage in geocoding. At their highest resolution they cover 1-70 addresses, with an average of 15. It is hoped that because of their small size they can be aggregated to a variety of larger areas, such as grid squares or districts, without too much overlap or underlap. In Scotland some statistics may be published for medium levels of the postcode hierarchy; maps of these divisions show that they often form dumbbell or even fragmented shapes, and hence are poor geographic individuals unless aggregated specifically for this purpose. At very broad scales, postal codes may approximate functional regions, at least in the USA (Abler, 1970).

Postcodes may also be used for other OPCS data such as births and deaths, but a considerable clerical effort is involved in adding missing postcodes. Use of postcodes by the public is partial because letters arrive just as quickly without them; introduction of sorting machines using the codes is known to have encountered problems and delays. Although postcodes are used for a number of data-gathering operations, e.g. in market research, their continuation stands or falls by their value for the postal system itself, and the Post Office will revise codes as necessary.

Grid squares appear 'unnatural' and their boundaries have an arbitrary relation to the real world. This means that if 'real' irregular areas are of interest grid squares should be relatively small. But they do have a number of major advantages: (i) their

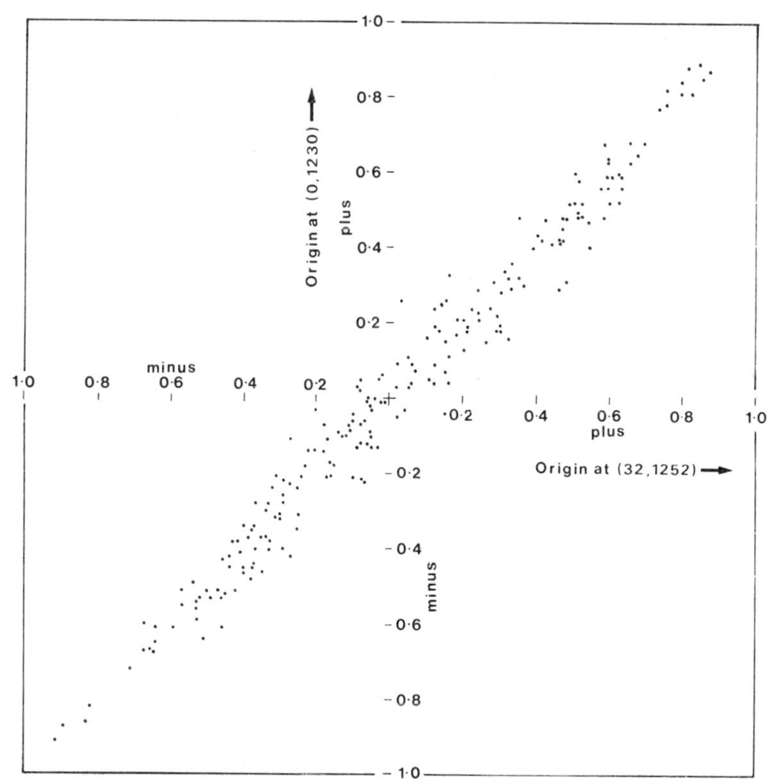

Figure 4.2 Correlations between twenty-one 100% variables for 64 x 64 km squares with (x) and without (y) displacement of the grid by 32 km in both directions. Two hundred and ten pairs of product-moment correlation coefficients are plotted against each other. Variables were transformed to reduce skewness, and correlations are weighted by population per square

area is constant; (ii) their shape is constant; (iii) they are constant over time; (iv) those used in the British Census of Population are marked on Ordnance Survey maps: no special maps are required; (v) the same grid squares are used for a number of other data-gathering operations, including some of the 'natural' environment, and they form a usefully neutral basis for interrelating different distributions. The first three points justify the use of the term areal *units* for grid squares: they are constant units, like metres or kilograms. Enumeration districts and postcode areas are too variable to justify the term 'units', and only the broader term 'areal divisions' should be used for them.

Densities calculated for grid squares are mutually comparable, whereas those calculated for areas varying widely in size are not (variability in density decreases as area increases). Grid square data are easy to map since each data unit can be represented by a single symbol, for which the area available is known beforehand (J.I. Clarke *et al.,* 1980). The areal basis is thus clear and direct, whereas in choropleth maps for irregular areas a large area with an intricate outline may be shaded for a value which relates to only a small part of the area (e.g. Shrimpton, 1977). The use of grid squares is so attractive for many forms of spatial analysis that authors such as Craig (1975) have generated 'pseudo-grid-square data' by allocating data for irregular divisions to grid squares; Robertson (1972) used topographic maps for detailed guidance. If, however, the population of an irregular division is allocated to a centroid, the grid superimposed must be much coarser than the initial divisions, otherwise whether a square is empty or not depends not on reality, but on the positions of the centroids: this is one of the problems with many maps produced from ward and parish data using the program LINMAP (Gaits, 1969).

The production of 1 km and (partial) 100 m grid square data for the 1971 Census was a bold innovation, which provided the largest multivariate spatial series in human geography. Unfortunately, the experiment had low priority and the data became available only between 1974 and 1976. This greatly reduced the rate of use, since workers with deadlines to meet had to switch to the ward, parish and ED data which were available earlier. It would be tragic if, in consequence, high quality grid square data are not produced for 1981; if so, their great advantage for studies over time would be lost, the experiment would be incomplete and the long-term value of the investment would not be realized. Being compact units, grid squares are desirable whenever density, proximity or spatial interaction (especially on foot) are major concerns. Use of a regular grid for measurement of commuting, for example, is the only way of avoiding the problems of comparison described by Chisholm (1960). Unfortunately, there are no plans to use grid squares for defining travel to work data. Willis (1972) is one of the few authors to have analysed the effect which the size and shape of areal divisions have upon apparent spatial interaction

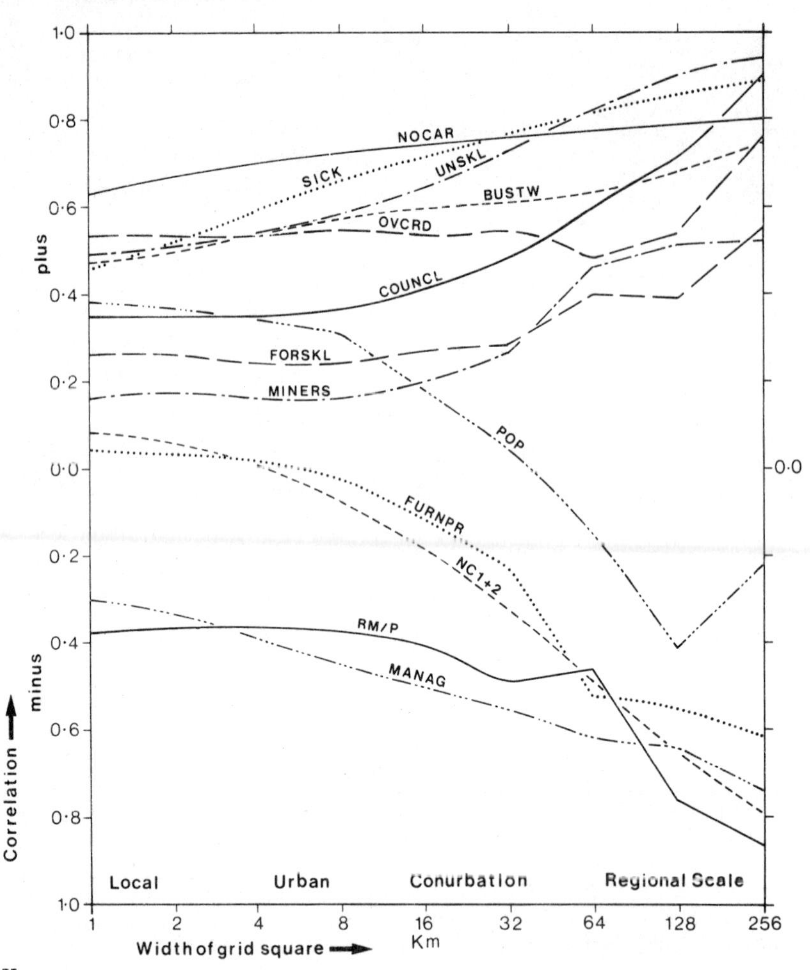

Figure 4.3 Variation in correlations of male unemployment rate with areal scale

Key
NOCAR = households with no car
SICK = sick, economically active persons
UNSKL = unskilled workers
BUSTW = travelling to work by bus
OVCRD = overcrowded households (over 1.5 persons per room)
COUNCL = council-rented households
FORSKL = foremen and skilled workers
MINERS = miners
POP = population per square
FURNPR = furnished private-rented households
NC1 + 2 = both parents, or one parent and self, born in New Commonwealth
RM/P = rooms per person
MANAG = employers and managers (Socio-Economic Groups (SEGs) 1, 2 and 3)

(migration rates).

Because grid squares are constant in area, they are far from constant in population. Although less variable in population, enumeration districts and postcodes do vary considerably in this as well as in area and shape (and over time), and it is desirable to hold at least one of these constant. The greater population variation for grid squares makes it essential to adopt analytical techniques which allow for this, e.g. weighting each grid square by its population in a correlation or multivariate analysis, or calculating chi-square values instead of ratios (Visvalingam, 1979a). Such techniques are desirable also, however, for districts (and postcodes), where the effects of variation in population have all too often been swept under the carpet. Visvalingam (1979b) showed that the effects of variation of population size between local authorities can be very important.

Since so many variables used in human geography

are ratios with population (or a closely related count) as the denominator (see Chapter 12 below), it would be very convenient if data could be provided for *areas equal in population*. It is quite challenging to attempt definition of these: an example was given by Robertson (1972, Figure 5). Geographers should experiment with the use of such areal divisions for statistical analysis.

Finally, it will be possible for statistics from the 1981 Census to be produced for *human functional regions* (Coombes *et al.*, 1979, 1980) at two hierarchical levels. Both levels would be at a much coarser spatial scale than considered above. A much better delineation of functional regions will be possible when 1981 travel to work data for wards become available.

Areal repeatability and scale effects

The demonstration by Openshaw and Taylor (1979) that a very wide range of correlations for the same pair of variables can be obtained from different areal aggregations (zoning systems) may be depressing, until it is realized that many of the areal aggregates have extreme shapes. If the requirement for contiguity is strengthened by one for compactness, more repeatable results are intuitively expected. With grid square data, it is possible to test the repeatability of results for aggregates by displacement of the origin.

A limited test of this nature was applied to 64 x 64 km squares, of which there are 108 whether the origin is at (0.1230) or at (32.1252) km. Figure 4.2, where the correlations observed with one grid incidence are plotted against correlations with the other grid incidence, shows that the relationship is close. Similar results are obtained whether the pair of variables correlated are both from 100 per cent data (as here, for 21 major variables), both from 10 per cent sample data, or one from each (Evans and Coulter, forthcoming).

It is most encouraging that even at the broad scale of 64 x 64 km correlations are highly repeatable. I conclude that the aggregation effect of Openshaw and Taylor (1979) is a much less important source of variance in correlations when the zones are grid squares than when they are irregular. As these authors demonstrated, aggregation variance is less important also when larger numbers of zones are involved.

While we hope that correlations will be insensitive to changes in the definition of the arbitrary zones which are treated as statistical individuals, we expect that correlations will differ when these individuals are quite different, for example, when they are grid squares of different sizes. Just as aggregate correlations for neighbourhoods may be expected to differ from correlations based on individual households, so regional correlations may differ from neighbourhood correlations.

Again, these expectations can readily be tested with aggregations of 1 km grid squares. This was done for 31 important variables for a series of grid squares with widths arranged in the doubling geometric series 1, 2, 4, 8, 16, 32, 64, 128, 256 km. Area increases fourfold at each step. The numbers of squares involved (unsuppressed for 100 per cent population data) are 67,546; 30,640; 11,023; 3,458; 1,072; 329; 106; 38; 13.

Only a small part of these results can be presented here; others are given in Evans (forthcoming). Figure 4.3 shows areal scale variations in 12 of the variables most strongly correlated with male unemployment rate. The dominant impression is of correlation strengthening (whether positive or negative) toward broader scales: this is in line with the pioneer observations of Gehlke and Biehl (1934). Since the number of cases involved at 32 or 64 km scale is quite large, the increased strength is not due to small sample size; most relationships really are stronger, and prediction can be more accurate, at the broader scales. The trend of change in correlation is usually consistent across the range of scales considered here, even for 256 km scale (except for population and overcrowding).

The general consistency of the trend for each correlated pair in Figure 4.3 is further evidence for the repeatability of the results, except sometimes at 256 km scale. It seems that, while areal scale is an important consideration in geographic studies based on correlation (Clarke, 1975), differences between adjacent scales in this series - between 1 and 2 km, or between 16 and 32 km - are fairly small. Larger contrasts are found between different or differently-defined study areas (Coulter, 1978).

Data uses: mapping and classification

The uses of census SAS data range from the simple presentation of summaries for one or more areas to provide 'area profiles', through mapping, to multivariate analysis. In each case, summarization and selection of SAS data are involved. Because of the size of the SAS data set, and its multipurpose nature, no one uses it directly in its entirety.

Recent papers by Hakim (1978b, c) give much information about census use and remove the need for a thorough account, so I will concentrate here on some relatively innovative uses involving large data sets from the SAS. The 1961 census was mapped for Ireland and the UK at 1/2,400,000 scale by a group of members of the Institute of British Geographers (Hunt, 1968). The maps were conventional choropleth maps for administrative districts and boroughs, varying greatly both in area and in population. The class interval system for each map was (reluctantly) based on octiles, each containing one eighth of the districts - but not one eighth of the area, nor one eighth of the population.

For 1966 SAS data, computer mapping by line printer came into use, notably in *Character of a conurbation* (Rosing and Wood, 1971). The 1971 census SAS were again portrayed in a number of county atlases produced via the line printer. A useful innovation in the Atlas of Tyne and Wear (Taylor *et al.*, 1976) was the restriction of shading to built-up residential areas. This not only provided an extra base map element, but also moderated the effect of choropleth maps in giving excessive weight to extensive but thinly populated areas. Though not applicable to largely rural counties, this device can be recommended for further conurbation maps.

By this time better output devices were coming into use. The census atlas of Northern Ireland (Compton, 1978) mapped 332 areal divisions at 1/750,000 by choropleth shading produced by computer-controlled graph plotter. Most maps used five classes, o.8 standard deviations wide (on a transformed scale if necessary), with the extreme classes open-ended (Evans, 1977).

The Atlas of South Yorkshire (Coates, 1974) was originally designed for line printer mapping, and each ED was encoded as a set of rectangular printer positions. It was possible to switch to laser plotting (Rhind, 1974) for high quality shading, and multi-colour printing was used, but the stepped ED outlines remained. Where adjacent EDs fell into the same mapping class the boundary was suppressed; this made the base map provided by boundaries incomplete and hid the areal basis of the statistics mapped.

The first atlas to be based on 1971 grid square data was *People in Durham* (Dewdney and Rhind, 1975). Half the maps used shades of red produced on a laser plotter; the other half used squares of different sizes produced on a graph plotter. Monochrome shading produced in this way was not as clear as multicolour mapping (Rhind *et al.*, 1976). The pattern of variation over the Durham coalfield is complex (unlike the 'bullseye' patterns around some conurbations), and it was found that whereas the pattern of familiar parishes or administrative districts on conventional choropleth maps provided a base map of its own, the unfamiliar outlines of grid squares needed support from a more complete base map. The atlas was essentially experimental and innovative. The more fundamental problem with these maps is their basis in ratios, and this applies also to all the previously noted atlases (Chapter 12). The chi-square technique (Visvalingam, 1979a) was adopted for most of the maps in *People in Britain* (Clarke *et al.*, 1980), where multicolour printing was available and the laser plotting technique permitted very high resolution. The maps of Great Britain at almost 1/4 million consist of coloured dots 0.254 mm across, each representing a 1 km square (Rhind, Evans and Visvalingam, 1980). A wall chart of the same title is available from the Customer Information Unit, OPCS, and includes a 1/1,500,000 population map.

Decisions on which variables to map were based in part upon population-weighted correlations between transformed ratios, and on measures of dissimilarity in the distribution of counts per grid square (Evans, 1979). Nationwide small-area-based correlations are usually weaker than those for more restricted study areas such as conurbations (Coulter, 1978), but provide a useful basis for comparison. The ratio variables do not fall into clearly separated factors, but loose clusters involving (a) age and household size, (b) crowding and major tenures, (c) deprivation, (d) immigrants plus shared accommodation, and (e) housing quality and unfurnished private renting can be recognized.

Another nationwide study was Webber's (1978, 1979) classification of EDs, following on his previous work for Districts, wards and constituencies. Similarities between a sample of 4,000 EDs were measured over 40 variables, standardized to have equal weight. It was found convenient to recognize 60 clusters of EDs, using an iterative relocation procedure based upon minimum error sum of squares. These clusters were grouped progressively until 8 broader families were produced, so that the classification could be applied at two levels of detail. The remaining 116,000 EDs were then allocated to the clusters to which they were most similar, but cluster definition was not further modified.

Several difficulties with this procedure have been pointed out by Openshaw, Cullingford and Gillard (1980). They showed that quite different classifications may be produced from the same variables in a study for a particular conurbation, and that locally important types of neighbourhood may have been missed by the 1 in 30 sampling procedure.

Despite their arbitrary nature, and the validity of these criticisms, the classifications produced by Webber have undoubtedly been useful to local authorities. Each classification must be interpreted, however, as just one of many possible classifications, not in any sense an optimal or natural classification. On the assumption that areal divisions essentially form a continuum without clearly separated clusters, it may be preferable to divide property space into arbitrary but simple divisions, as in Visvalingam's (1979b) application of the signed chi-square technique. Here, eight types of household composition were mapped and analysed, on the basis of having a notable excess of one type of household. Holtermann's (1975) study of deprivation was similarly based on arbitrary thresholds, this time for the percentage of people or households suffering from each 'problem'. By examining the overlap in the worst areas on each criterion, she was able to conclude that for unemployment and poor housing 'the spatial coincidence of these problems is far from complete; and, further, that although there is some degree of spatial concentration of deprived people into the "worst" areas, there are large numbers of deprived people who live outside them' (Holterman, 1975, p. 16).

Finally, a number of studies have applied census

data to spatial searching and allocation problems. Clarke *et al.*, (1976) have used ward and parish data to plot population as a weighted cumulated function of distance from nuclear power stations, while Openshaw (1980) used grid square data. Rhind, Stanness and Evans (1977) plotted population and various groups as functions of distance from city centres, while Evans and Rhind (unpublished) have provided centrographic measures and mapped population potential. Robertson (1972, 1974, 1978) has proposed locations of facilities to optimize access by population.

Conclusion

Census data - and their analysis - have strong political overtones. Some radical writers charge that social scientists analysing these 'objective' data and producing research reports for academic, administrative and/or business elites are simply perpetuating the *status quo* with all its injustices, or helping these elites to maintain power by making adjustments which avoid or delay radical change. Whatever degree of sympathy this view receives, it is difficult to argue that reduced availability of information will help the underprivileged; rather, this would increase the relative advantage in information of powerful groups. Knowledge gives power; ignorance facilitates oppression. What should be done is for the Census Offices and for researchers to put more effort into the dissemination of their information and research results to the public.

In some fields the Government fails to provide the public with information at a reasonable cost; the high prices of Royal Commission reports are hardly compatible with a commitment to open government. *Social Trends, 1980*, supposedly a popular summary of government-compiled statistics, cost £12.90; the man or woman in the street will therefore see only brief and biased summaries in newspapers, if that. The Census Offices have a relatively good record, in that published and unpublished tables are available to all users on an equal basis and the cost of collecting and coding data are not passed on to users. The prices charged for published volumes reflect printing and distribution costs. One might argue, however, that these must be printed because of demand from Government departments, and the Government should therefore cover the cost of setting up the print run, leaving other purchasers to cover the cost of printing an extra copy, plus the distribution cost.

Unpublished output such as the SAS are inevitably more expensive, since the cost of preparation in this form must be covered. Again, though, there is a political commitment to produce SAS for the whole country, and local and central Government departments are major users. It would be reasonable for other purchasers to cover only the cost of duplication and distribution. Any public library *can* hold SAS; a decision should be taken for central libraries in every district to hold SAS for that and neighbouring districts, and to facilitate copying by the public for the cost of duplication. Although the cost of SAS is low per areal division, it is relatively large when a few variables are needed for many divisions - for example, for mapping. OPCS have been co-operative in making special arrangements with the SSRC data archive and with the Institute of British Geographers for distribution of 1971 SAS, and it is hoped that an enlightened attitude to copyright will continue. Subsets of the national areas x variables matrix should be made available on a more flexible basis.

Researchers should also play their part by participating in the census consultation process: Universities were poorly represented at the open discussion meetings for planning the 1981 SAS. Researchers must not only publish their results, but also publicize them and express them in simple language without omitting vital qualifications.

Unless a more open policy to the dissemination of information is pursued, there may be a popular reaction which, through decreased truthfulness and increased refusal rates, decreases the accuracy of census data and of all other social science data. This goes far beyond the question of individual privacy (Hakim, 1978a) and could reduce the value of present uses of the census as well as preventing desirable extensions.

The 1976 census was cancelled in an economy drive. Up-to-date information on the 1981 census, which will differ from that of 1971, can be obtained from

The Census Information Unit,
OPCS,
St Catherine's House,
10 Kingsway,
London, WC2B 6JP,
England

Their Census Monitors series is valuable, especially CEN 80/8. After economies and the results of trials, the 1981 census will *not* include questions on ethnic origin, nationality, year of entry, parents' country of birth, school-level qualifications, hours worked, five-year migration or number of dwellings.

Acknowledgments

This account forms part of the work of the Census Research Unit, University of Durham, funded by SSRC and using data kindly supplied by OPCS. I am indebted to Chris Denham, Peter Norris, David Rhind and Mahes Visvalingam for comments which have led to considerable improvements in this section, and to C. Hakim, N.J. Cox, J.C. Dewdney and all my Census Research Unit colleagues. Responsibility for the sometimes contentious views expressed here is entirely my own. A considerably extended version of this chapter is available as a Census Research Unit Working Paper.

References

Abler, R.F. (1970) 'Zip-code areas as statistical regions', *Professional Geographer*, 22, 270-4.

Brown, A. (1978) 'Towards a world census', *Population Trends*, 14, 17-19.

Bulmer, M. (ed.) (1979) *Censuses, surveys and privacy*, Macmillan: London.

Chisholm, M.D.I. (1960) 'The geography of commuting', *Annals, Association of American Geographers*, 50, 187-8 and 491-2.

Chisholm, M.D.I. (1975) 'The reformation of local government in England' in R. Peel, M.D.I. Chisholm and P. Haggett (eds), *Processes in Physical and Human Geography: Bristol essays*, 305-18, Heinemann: London.

Clarke, J.I. (1975) 'Population and scale', Census Research Unit, University of Durham, *Working Paper 1*, 19 pp.

Clarke, J.I., Denham, C., Dewdney, J.C., Evans, I.S., Rhind, D.W., and Visvalingam, M. (1980) *People in Britain: a census atlas*, HMSO.

Clarke, R.H., Fitzpatrick, J., Goddard, A.J.H., and Henning, M.J. (1976) 'The use of census data in predicting spatial distributions of collective dose', *Journal of the British Nuclear Energy Society*, 15, 297-303.

Coates, B. (ed.) (1974) *Census atlas of South Yorkshire*, Department of Geography, University of Sheffield.

Cohen, J. (1962) 'The statistical power of abnormal-social psychological research', *Journal of Abnormal and Social Psychology*, 65, 145-53.

Compton, P.A. (1978) *Northern Ireland: a census atlas*, Gill and Macmillan: Dublin, 169 pp.

Coombes, M.G., Dixon, J.S., Goddard, J.B., Openshaw, S., and Taylor, P.J. (1979) 'Daily urban systems in Britain: from theory to practice', *Environment and Planning A*, 11, 565-74.

Coombes, M.G., Dixon, J.S., Goddard, J.B., Taylor, P.J. and Openshaw, S. (1980) 'Functional regions for the 1981 Census of Britain: A user's guide to the CURDS definitions of functional regions for the 1981 census of Britain', Centre for Urban and Regional Development Studies, Newcastle University, *Discussion Paper*, 30.

Coppock, J.T. (1960) 'The parish as a geographical-statistical unit', *Tijdschrift voor Economische en Sociale Geografie*, 51, 317-26.

Coulter, J.M.M. (1978) 'Grid square census data as a source for the study of deprivation in British conurbations', Census Research Unit, University of Durham, *Working Paper*, 13.

Craig, J. (1975) 'Population density and concentration in Great Britain 1931, 1951 and 1961', *OPCS Studies on Medical and Population Subjects*, No. 30, HMSO, 148 pp.

Denham, C. (1980) 'The geography of the census, 1971 and 1981', *Population Trends*, 19, 6-12.

Dewdney, J.C., and Rhind, D.W. (eds) (1975) *People in Durham - a census atlas*, Department of Geography, University of Durham, 146 pp.

Durbin, J. (1979) 'Statistics and the Report of the Data Protection Committee', *Journal, Royal Statistical Society, A*, 142, 299-328.

Evans, I.S. (1977) 'The selection of class intervals', *Transactions, Institute of British Geographers*, NS 2, 98-124.

Evans, I.S. (1979) 'Relationships between Great Britain census variables at the 1 km aggregate level' in N. Wrigley (ed.), *Statistical Applications in the Spatial Sciences*, 145-88, Pion: London.

Evans, I.S. (forthcoming) 'The effect of areal scale on relationships between census variables'.

Evans, I.S., and Coulter, J.M.M. (forthcoming) 'The areal repeatability of correlations between census variables'.

Evans, I.S., and Rhind, D.W. (unpublished) 'Macrogeography of population in Great Britain, 1971'.

Experimental Cartography Unit (1971) *Automatic Cartography and Planning*, Architectural Press: London. 232 pp.

Gaits, G.M. (1969) 'Thematic mapping by computer', *Cartographic Journal*, 6, 50-68.

Gehlke, C.E., and Biehl, K. (1934) 'Certain effects of grouping upon the size of the correlation coefficient in census tract material', *Journal of the American Statistical Association*, Supplement, 29, 169-70.

Hakim, C. (1978a) 'Census confidentiality, microdata and census analysis', Office of Population Censuses and Surveys (OPCS) *Occasional Paper 3*, 18 pp. London.

Hakim, C. (1978b) 'Social and community indicators from the census', OPCS *Occasional Paper 5*.

Hakim, C. (1978c) 'Census data and analysis: a selected bibliography', OPCS *Occasional Paper 6*, 30 pp.

Hakim, C. (1978d) 'Data dissemination for the population census: a data user's guide', OPCS *Occasional Paper 11*, 62 pp.

Holtermann, S. (1975) *Census indicators of urban deprivation*. WN-6, ECUR Division, Department of the Environment (see also *Social Trends*, (1975) 6, 33-47).

Hunt, A.J. (1952) 'Urban population maps', *Town Planning Review*, 23, 238-48.

Hunt, A.J. (ed.) (1968) special number on population maps of the British Isles, 1961,

Transactions Institute of British Geographers, 43.

Kruskal, J.B. (1968) 'Transformations of data' in D.L. Sills (ed.), *International Encyclopaedia of the Social Sciences,* 15, 182-93, Macmillan: New York.

Moore, E.G. (1979) 'Beyond the census: data needs and urban policy analysis' in S. Gale and G. Olsson (eds), *Philosophy in Geography,* 269-86, D. Reidel: Dordrecht.

Mosteller, F., and Tukey, J.W. (1977) *Data analysis and regression,* Chapters 4 and 5, Addison-Wesley: Reading, Mass.

Newman, D. (1978) 'Techniques for ensuring the confidentiality of census information in Great Britain', OPCS *Occasional Paper 4,* 20-30.

OPCS (1976) *1971 Census: Standard Small Area Statistics (Ward Library) Explanatory Notes.*

Openshaw, S., Cullingford, D., and Gillard, A. (1980) 'A critique of the national classification of OPCS/PRAG', *Town Planning Review.*

Openshaw, S. (1980) 'A geographical appraisal of nuclear reactor sites', *Area,* 12 (4), 287-90.

Openshaw, S., and Taylor, P.J. (1979) 'A million or so correlation coefficients: three experiments on the modifiable areal unit problem' in N. Wrigley (ed.) *Statistical Applications in the Spatial Sciences,* 127-44, Pion: London.

Population Statistics Division, OPCS (1979) 'Geographical area units', *Population Trends,* 15, 16-21.

Rhind, D.W. (1974) 'High-speed maps by laser beam', *Geographical Magazine,* 46 (8), May, 393-4.

Rhind, D.W. (1975) 'Geographical analysis and mapping of the 1971 UK census data', Census Research Unit, University of Durham *Working Paper 3,* 20 pp.

Rhind, D.W., and Hudson, R. (1980) *Land Use,* Methuen: London.

Rhind, D.W., Evans, I.S., and Dewdney, J.C. (1977) 'The derivation of new variables from population census data', Census Research Unit, University of Durham, *Working Paper 9,* 84 pp.

Rhind, D.W., Evans, I.S., and Visvalingam, M. (1980) 'Making a national atlas of population by computer', *Cartographic Journal,* 17, 3-11.

Rhind, D.W. Stanness, K., and Evans, I.S. (1977) 'Population distribution in and around selected British cities', Census Research Unit, University of Durham, *Working Paper 11,* 47 pp.

Rhind, D.W., Visvalingam, M., Perry, B., and Evans, I. (1976) 'People mapped by laser beam', *Geographical Magazine* 49 (3), December, 148-52.

Robertson, I.M.L. (1969) 'The census and research: ideals and realities', *Transactions, Institute of British Geographers,* 48, 173-87.

Robertson, I.M.L. (1972) 'Population distribution and location problems: an approach by grid squares in central Scotland', *Regional Studies,* 6, 237-45.

Robertson, I.M.L. (1974) 'Scottish population distribution: implications for locational decisions', *Transactions Institute of British Geographers,* 63, 111-24.

Robertson, I.M.L. (1978) 'Planning the location of recreation centres in an urban area: a case study of Glasgow', *Regional Studies,* 12, 419-27.

Rosing, K.E., and Wood, P.A. (1971) *Character of a conurbation: a computer atlas of Birmingham and the Black Country,* University of London Press, 126 pp.

Shrimpton, M. (1977) *Census atlas of Newfoundland, 1971,* Institute for Social and Economic Research, Memorial University of Newfoundland.

Stinson, J.G. (1977) 'Improving the accessibility of data from the census of Canada', *Demography,* 14, 121-9.

Taylor, P.J., Kirby, A.M., Harrop, K.J. and Gudgin, G. (1976) *Atlas of Tyne and Wear,* University of Newcastle upon Tyne, Department of Geography, *Research Paper 11,* 37 pp.

Turnock, D. (1978) 'A new source for the study of Scotland's settlement pattern', *Area,* 10 (2) 89-94.

Visvalingam, M. (1976) 'Indexing with coded deltas - a data compaction technique', *Software - practice and experience,* 6, 397-403.

Visvalingam, M. (1977) 'A locational index for the 1971 Kilometre-square population census data for Great Britain', Census Research Unit, University of Durham, *Working Paper 12,* 25 pp.

Visvalingam, M. (1979a) 'The signed chi-square measure for mapping', *Cartographic Journal,* 16, 93-8.

Visvalingam, M. (1979b) 'The identification of demographic types: a preliminary report on methodology', Census Research Unit, University of Durham, *Working Paper 14,* 55 pp.

Visvalingam, M., and Perry, B. (1976) 'Storage of the grid-square based 1971 GB census data; checking procedures', Census Research Unit, University of Durham, *Working Paper 7,* 21 pp.

Visvalingam, M., and Rhind, D.W. (1977) 'Compaction of the 1971 UK census data', *Computer Applications* (Nottingham) 3 (3-4), 499-511.

Visvalingam, M., Norman, M.J., and Sheehan, R. (1976) 'Data interchange on industry-compatible tapes', Census Research Unit, University of Durham, *Working Paper 6,* 18 pp.

Webber, R. (1978) 'Which residential neighbourhoods are alike?', *Population Trends,* 11, 21-6.

Webber, R. (1979) 'Census enumeration districts: a socio-economic classification', OPCS *Occasional Paper 14,* 51 pp.

Willis, K. (1972) 'The influence of spatial structure and socio-economic factors on migration rates. A case study: Tyneside 1961-66', *Regional Studies,* 6, 69-82.

Chapter 5

The modifiable areal unit problem

S. Openshaw and P.J. Taylor

Introduction

The validity of nearly all applications of quantitative techniques to zonal data depend on the assumption that the spatial units being studied are *a priori* given (see Bartels 1979). Whilst this may be a tolerable statistical assumption for people who do not know any better, it is doubtful whether this is a satisfactory geographical assumption on which to base the developments and applications of quantitative techniques in geography. This is the essence of the problem that this chapter is concerned with.

If the nature of zoning systems were such that they can be regarded as divinely given so that their constituent zones are uniquely meaningful fixed entities, there would be no problem. However, zoning systems are not divinely given, they are not fixed or unique, and the same individual data can be aggregated to provide different areal representations which yield data with little resemblance to the 'real' data that existed before any spatial aggregations occurred. While this may conjure up the ecological fallacy type of problem (Robinson, 1950), this is but a minor difficulty in comparison with the uncertainties associated with the choice of zoning systems and the ramifications this holds for geographical analysis.

At present geographers differ greatly in their perceptions of the importance of this modifiable areal unit problem. Many believe, or hope, that it is of only localized significance to geographical studies of data for zonal units. The chapter starts, therefore, by considering the nature of the modifiable areal unit problem. Subsequently, a survey is made of the prevailing ambivalent attitudes that geographers display and the general absence of any sense of verisimilitude is emphasized. Finally, a critical review is made of several alternative approaches to handling the problem.

The modifiable areal unit problem

The obvious questions that need answering are as follows. What is the nature of the modifiable areal unit problem? Why is it so important to geographical analysis that it cannot be ignored? The problem exists because geographers have long been interested in studying data for zoning systems which are invariably arbitrary and modifiable. If there are no standard zoning systems then the results of spatial study will depend to some extent on the units that are used.

The problem can be avoided simply by not studying spatially aggregated data, although this is not a practical proposition. Indeed there is a widespread use of zoning systems as spatial data reporting frameworks precisely because they avoid many of the problems associated with the use of individual data; for example, confidentiality. In addition there are considerable computational advantages of studying data in an aggregated form; for instance, spatial aggregation greatly reduces the size of data matrices and by converting categorical measurement scales into more continuous ones avoids many of the problems of analysing data with discrete scales of measurement. In addition it is recognized that geography has a long tradition of studying data for 'places' or 'regions' for which zonal approximations can be found. Thus it is hardly surprising that the vast majority of applications of quantitative techniques have used zonal data. If the problem of studying data for modifiable units cannot be avoided, then it is essential that the consequential limitations of such studies are clearly understood.

Since a study area over which data are collected is continuous, it follows that there will be a tremendously large number of different ways by which it can be divided into non-overlapping areal units for the purpose of reporting spatial aggregations of individual data. Viewed as a combinatorial problem, the number of different zoning systems of m-zones that data for n-individuals can be aggregated to, becomes unimaginably large for even modest values of n (Cliff and Haggett, 1970; Keane, 1975). The problem also repeats, so that there is another combinatorial explosion when a zoning system of n-zones is aggregated to a zoning system with fewer zones. Indeed, it is noted that most zonal data available for analysis have been aggregated a number of times.

Although there is virtually an infinite number of ways by which a study area can be areally divided, data are normally only presented or analysed for one particular set of units. Furthermore, it is highly likely that the units which are used are not representative of the set of alternative zoning systems from which they were selected.

It is argued that this uncertainty regarding the nature or zonal entities is inherently a geographical problem. The zones in a zoning system constitute the objects, or geographical individuals, that are the basic units for the observation and measurement of spatial phenomena. Usually in a scientific experiment the definition of the objects of study precede any attempts to measure their attributes. This is not the case with areal data where the spatial objects only exist after data collected for one set of entities are subjected to an arbitrary aggregation to produce another set. This would be quite reasonable if the aggregation was performed according to fixed rules or internationally accepted conventions, but there are no rules and there are no standards. As Taylor and Goddard (1974, p. 153) put it 'It is unfortunate that there is no spatial equivalent to the day, month, or year'. Quite simply the definition of the zonal objects used for many geographical studies are modifiable and vary greatly according to the whims and fancies of whoever is doing the aggregating.

Whereas data collected by the census relates to carefully defined natural, indivisible, and fixed objects; such as the household or person; the data made available for analysis relates not to these carefully defined objects but to arbitrary aggregations of them on an areal basis. If the nature of zones as areal objects are arbitrary then the value of work based on them must be thrown into doubt. In particular it is necessary to consider the following questions: are the units meaningful objects given the purpose of the exercise, are they comparable in the sense of being the product of the consistent application of a fixed set of rules, and how sensitive are the results that are obtained to the areal definitions that are used. What is surprising about the modifiable areal unit problem is the fact that many geographers know about its existence but deliberately choose to ignore its implications for geographical analysis.

The existence of the problem has been known for a long time. In electoral geography there is a long history of attempts to use the freedom inherent in the choice of zoning systems to deliverately distort voting intentions by the careful manipulation of boundaries; for instance the famous 1810 gerrymander (Taylor and Johnston, 1979, pp. 371-4). Many of the problems discussed here have already been investigated as redistricting problems for electoral purposes (Weaver and Hess, 1963; Weaver, 1970; Littschwager, 1973). Gradually the magnitude of the problem has been recognized. Neutral districting procedures could not be produced (Taylor, 1973; Taylor and Gudgin, 1976). Instead zoning systems which are deliberately biased to be 'fair' have been produced; for example, the bipartisan approach of Nagel (1965, 1972), and this has parallels with what is later referred to as the optimal zoning approach. Unfortunately, the more general implications of the electoral geographer's gerrymandering problems for spatial analysis have not been recognized.

There are also some early indications that the modifiable areal unit problem has an effect on statistical analysis (Gehlke and Biehl, 1934). Certainly Kendall and Yule (1950) clearly recognized the implications when they wrote that the results 'measure, as it were, not only the variation of the quantities under consideration, but the properties of the unit mesh which we have imposed on the system in order to measure it' (p. 312). Many geographers would accept this statement whilst querying how much of the variation can be attributed to the zoning system. Anyway, despite early beginnings the entire subject has been largely ignored, partly because it has proved very difficult to investigate by means of theoretical study at a time when such approaches were fashionable, and partly because of doubts about the significance of zoning system effects.

Looking back at how the subject was handled in the standard texts of the 1960s, such as Haggett (1965), Chorley and Haggett (1967), Harvey, (1969), it is apparent how little understood the problem was. The situation has not improved much in the 1970s as such matters receive scant attention in the important books by Wilson (1970, 1974), Batty, (1976b), Haggett, Cliff and Frey (1977), and Bennett (1979). Indeed it is strange that so many geographers completely fail to even understand the problem. Almost by tradition, they have considered only the scale problem, whereby aggregation reduces the number of areal units and thus increases the scale of the study. To confuse things further, the aggregation problem is often only viewed as a scale problem; see for example, Clark and Avery (1976). This is not right, since in addition to deciding the scale of a study, interpreted as selecting an appropriate number of zones, it is also necessary to select an appropriate zoning system from the set of alternatives with the required number of zones. For example, any attempt to identify scale effects by considering what happens when the number of zones are reduced from, for instance, 99 to 66 to

44 to 22, also involves deciding which aggregation of 66, 44, and 22 zones will be used. Indeed it should be obvious that there is far more freedom of choice in the selection of any aggregation of 99 zones to say 66 than there is in the choice of 66 zones.

The exaggerated importance afforded to scale effects seems to be a reflection of the greater ease in which scale problems can be studied; see Moellering and Tobler (1972). It is only recently that systematic methods for identifying alternative zoning systems at a fixed scale have been developed, mainly as a spin-off from redistricting algorithms (Taylor, 1973; Openshaw, 1977a). Thus it is not until the late 1970s that the first empirical studies begin to provide a detailed picture of the effects that the choice of zoning systems have on the results of spatial study and of the interaction between scale and aggregation effects that are the primary components of the modifiable areal unit problem. In a series of papers Openshaw (1977b, 1977c, 1978b, 1979) presents results which demonstrate both the severity and the universality of zoning system influences on geographical analysis, something that many have long feared but preferred to ignore. The principal development that made this possible was the creation of an efficient heuristic procedure which can approximately optimize any general function by systematically manipulating the zoning system: see Openshaw (1977b, 1978a) for details. For the first time it was possible to directly identify the magnitude of zoning system effects on any method of quantitative analysis.

The fact that the correlation coefficient can be made to vary between quite wide limits is perhaps not too remarkable. Nevertheless, for a small data set of 99 Iowa counties, 12 zone aggregations could be engineered that produce correlations between -.97 and +.99 a regression slope coefficient of between -12.7 and +12.2, and a level of fit which could be nearly perfect or incredibly poor, and a robust regression coefficient that varied from -14.6 to +16.2: see Openshaw (1978b, 1979). Similar pronounced zoning system effects have also been observed for various multivariate statistical techniques and spatial interaction models. Indeed there are no longer grounds for believing that methods exist which are not affected, unless there is proof. It cannot be assumed as a matter of faith.

The only known exceptions are not what they may at first sight appear to be. For example, both Bearwood (1972), Bearwood and Kirby (1975) and Rogers (1969, 1972) describe rules which allow for consistency of results after aggregation. This does not mean that there are methods which are aggregation insensitive, only that by various means a degree of consistency can be achieved. Consideration may also be given to the fact that in some areas of geographical study broadly similar results have been obtained despite the modifiable areal unit problem. The best example is the numerous factorial ecological studies which yield similar results, although this may be explained as due to the use of the same set of highly intercorrelated variables which totally dominates the factor analyses. More generally, the effects of the problem have been reduced by the tendency of geographers not to engage in comparative studies. In any case it must be emphasized that the level of zoning system effects present in a study varies with circumstances. It is not possible to obtain any result merely by manipulating the zoning system as the data being aggregated has an effect. Nevertheless, on many occasions the magnitude of the zoning effect that has been observed is far too large to be ignored.

Comments on the way in which the problem has been treated

Thus the modifiable areal unit problem is viewed as being endemic to all spatially aggregated data. It consists of two interrelated aspects. First, there is uncertainty about what constitutes the objects of spatial study, involving the scale and aggregation problem; and second, there are the implications this holds for the methods of analysis commonly applied to zonal data and for the continued use of a normal science paradigm which can neither cope nor admit to its existence.

The first of these problems can also be viewed as a spatial representation problem. Indeed the zoning system is the principal way in which spatial structure is incorporated into geographical analysis. It is useful, therefore, to review some of the different types of areal units that have been employed for purposes of spatial representation (see Table 5.1).

As has been noted previously, nearly all the areal units that are listed are totally unrelated to virtually all geographical purposes. They are termed *a priori* given units in Table 5.1 and are considered to be geographically irrelevant. Such units tend to be 'officially' defined on the basis of criteria which have no relevance to any conceivable geographical research purpose. They are used because of their convenience and availability and because it is assumed that the choice of units is unimportant. The effects that their design criteria have on the results of analysis have not been thought through; consider for example, the implications of using post-code sectors for analysing the 1981 census.

The reaction of some geographers to this representation problem has been to provide their own areal units. These are referred to as '*a posteriori* units' in Table 5.1. The 'simplest' of these involves a geometric reaction whereby purely arbitrary lattices are located upon the data. The initial examples involve the use of hexagon (see Robinson *et al.*, 1961), but latterly grid squares have become fashionable. Thus the principal research advance

Table 5.1 The variety of areal units

Type of unit	*Purpose of unit*	*Delimitation criteria*
A priori *given units*		
local government areas	provide local services	population and political control
standard regions and sub-regions	policy analysis by government	mixture of formal and functional structure
census enumeration districts	organization of census data collection	population and visibility on the ground
postal code units	computerize addresses	contiguity and postman's walk
constituencies and wards	political representation	population, local government boundaries, and rural-urban distinctions
A posteriori *units*		
geometric	objectivity: defined as unrelated to any data	pure location
formal regions	to produce regions	minimum intra-regional variance
functional regions	to produce functional entities	minimize inter-regional flows
model associated	integrated as part of modelling analysis	to best operate a model of spatial analysis

seems to have been a move from an analogue of a Loschian landscape to Michael Bentine's 'It's a Square World'. The purpose is to produce neutral aggregations although it is not clear why locationally arbitrary areal units should be of any interest to geography. A more common development has been what can be termed a regionalization reaction. This involves the use of clustering algorithms to define regions, although most early applications of this approach degenerated into defining regions for their own sake. This is not the case, however, with the definition of functional regions and model-associated units. The latter are an attempt to identify purposeful units by deliberately reaggregating data in order to compensate for prior failings in their definitions. This approach is discussed in more detail later.

The second of the problems follows directly from this uncertainty concerning the choice of zonal units. If the empirical analyses referred to previously had shown that current methods of analysis were insensitive to the choice of zoning systems so that broadly the same results could be obtained regardless of the zoning systems being used, then the problem could be ignored. However, this is not the case. Different areal arrangements of the same data produce different results, so we cannot claim that the results of spatial studies are independent of the units being used and the task of obtaining valid generalizations or of comparable results becomes extraordinarily difficult. Despite the potential havoc that this may well cause to all manner of geographical studies, it is apparent that geographers have been very slow at recognizing the implications. They have been equally reticent about making clear their assumptions with respect to this problem. Indeed, the standard practice is to simply ignore it altogether in the development and application of methods of analysis that completely fail to take it into account. What is even worse, many geographers see no need to even try.

As a result there is an overwhelming presumption that the modifiable areal unit problem in particular, and the spatial representation problem in general, is of little consequence. Consider, for example, the development of model building methods for zonal data that explicitly exclude any consideration of the effects or nature of space on the data used; for example, entropy-maximizing methods (Wilson, 1970) and even the so-called spatial entropy methods of Batty (1974) only take into account zone size. Consider also the attempts that have been made to handle the effects of spatially dependent observations on classical statistical methods; see Cliff and Ord (1975) and Martin *et al.*, (1978). These studies conveniently ignore the fact that spatial autocorrelation is but one reflection of the zoning systems being used. It may be increased or reduced by the use of alternative zoning systems. Yet the expected moments of the standard measures of spatial autocorrelation have been calculated under two sets of assumptions (Cliff and Ord, 1973), both of which assume that zonal data are fixed.

There is no denying that these and many other papers concerned with related themes are elegant solutions to the specified statistical problems, but doubts exist about whether the problems were correctly formulated in the first place. Thus the specification of dependency structures and the redefinition of existing statistical methods to take this into account (Haining, 1978; Streitberg, 1979) may well solve the immediate statistical problem but it gives no consideration to the fact that the data being studied are modifiable and not fixed. Claims that zoning system effects are only ignored to allow the initial development of theory so that some progress can be made with existing tools, needs to be treated with caution if this initial assumption can never be relaxed, as is usually the case. It is emphasized that the statistical and geographical problems are not independent nor is the latter subordinate to the former: indeed, as geographers we should be putting it the other way. Likewise, methods of studying zonal data which start with the aim of removing what are considered to be the irrelevant and undesirable effects of either space or the zoning system so as to expose the 'real' relationships to analysis (for example, Tobler, 1963; Batty and March, 1976) can only ever be of limited geographical value.

It would appear that geographers are doing their best to build what can only be described as intrinsically non-geographical models of geographical phenomena based on unreasonable assumptions regarding the nature of zonal data. This clearly demonstrates the modern geographer's almost complete failure to understand the geography, rather than the statistics or mathematics, of what they are doing. The inescapable conclusion is that at present the modifiable areal unit problem is viewed as being irrelevant, or if not irrelevant as being something that is too difficult to handle given existing methods so that it is simply ignored altogether. It is suggested that this is not a sensible attitude for geographers to adopt towards something as fundamental to geographical analysis as this is.

The most visible example of this total disregard for the geography of the modifiable areal unit problem can be seen in the production of computer atlases: see Rhind (1977) for a review. Computer-based cartographers it seems are quite content to produce maps of data for virtually any set of base units, ranging from enumeration districts to 1 km grid squares. Yet it is readily apparent that very different map patterns can be produced for the same data merely by utilizing a different set of areal units. It is of little consequence whether you use grid squares or administrative units, they are both arbitrary and have little meaning to the data being mapped. They are neither homogeneous, nor comparable except sometimes in terms of their geometry, nor neutral. The use of 1 km grid squares for mapping census data is a good example (Dewdney and Rhind, 1976). It is difficult to imagine a worse set of geographical units for presenting information about people.

A review of alternative approaches

Despite the fact that few geographers show any explicit concern at all about the zoning systems they use, it is possible to recognize a number of different approaches. The really fundamental distinction is between those who hold the view that the definition of zoning systems must be independent of the purposes for which they are used and those who are willing to allow at least a degree of dependency.

One justification for the total independence view rests on the strict adoption of a normal science paradigm which holds that the formulation and testing of hypotheses must be independent of the identification and measurement of the objects being studied. This requires that the results of studying zonal data be independent of the definition of the zones that are being studied, and it has already been argued that this is an unreasonable assumption. Two corollaries of this approach are; first, the belief that substantially the same results would be obtained with virtually any set of zones, and second, that a general theory of the modifiable areal unit problem will be forthcoming to enable existing methods to continue being used with any zoning system. One example of this approach is the various attempts at weighting values in correlation analysis in order to eliminate 'differences in parameters which may be attributed to differences in the size of areal units from those differences which are owing "truly different" relationships' (Thomas and Anderson, 1965, p. 432). This of course does not work and it is very doubtful whether this sort of approach is a viable proposition.

One variation on this theme is to accept a degree of dependency between the selection of zoning systems and the purpose of study, but without suggesting that the strength of the relationship is sufficient to invalidate the normal science assumption of independence. Usually this involves no more than attempts to design zoning systems for the purposes of either data analysis or comparability across different data sources; see, for example, Kirby and Taylor (1976). Even then, in order to proceed with the analysis, it is blithely assumed that the results are not particularly affected by these manipulations. Consider also the use of functional regions as a basis for examining changes in the national urban system

(Drewett *et al.*, 1976). The problem here is that the results depend very strongly on the areal definitions that are used and are not capable of supporting zone-free generalizations, although this is precisely what happens. For example, the pattern of increasing suburbanization that the Drewett *et al.* (1976) study identifies may be partly, or even largely, the result of a mismatch between the patterns picked up by the zoning system that is used and the patterns that a more disaggregate level of analysis might have revealed.

The general philosophy of this type of approach is neatly put by Batty and Sammons (1978, p. 748) when they declare that what is needed is a means of designing zoning systems to provide an efficient data organization prior to spatial analysis under the assumption that 'general principles exist which enable the analyst to organize his data set in an optimal way, quite independent of the context of analysis'.

An alternative to this approach is to follow Openshaw (1977b, 1978a) who argues that a context-free approach is contrary to geographical common sense irrespective of what other virtues it may have. The design of zoning systems is viewed as an important geographical component in spatial analysis, one that should be used rather than ignored. Consider an example. The values assigned to the undetermined parameters in a model of some kind for zonal data are dependent on the zoning system that is used. The current practice of seeking what are considered to be efficient statistical estimates without considering the zonal system at all is regarded as being as inappropriate to a geographer as the analogous process of fitting a model with completely arbitrary parameters solely by manipulating the zoning system would be to a statistician. With zonal data the processes of parameter estimation and zone design are completely interdependent. More generally, the results obtained from spatial study depend on the data and the zoning system that is used. It is not sensible to leave the choice of zoning system undetermined, instead it must reflect either the results that are desired or the purpose of the study. This dependency might be considered a drawback to many methods of analysis which cannot cope with them, but it also provides a new way of approaching the problem of analysing zonal data which is described later.

In between these extremes there are numerous variations, six of which are itemized for further discussion. They are: filtering, arbitrary design criteria, information theoretic, statistical, traditional, and optimal zoning.

Filtering

Curry (1966) and Casetti (1966) both recognized that the aggregation of space series filters out harmonics whose wavelengths are smaller than the size of the zones in a zoning system and that this provides a basis for handling aggregation problems. Following Tobler (1969, 1975) it is possible to apply spatial filtering methods to try to identify trends or real patterns by removing the noise caused by aggregation effects. It can be argued that the resulting smoothed map generalizations may prove to be relatively insensitive to aggregation effects. The problem is of course to prove it. In any case the underlying assumption that there is a real pattern that can be identified from data for virtually any zoning system may not be appropriate. What for example does this 'real' pattern relate to? Is it that which would have been recognized if disaggregated data has been available, or is it something that reflects the zoning system to which the data has been aggregated before analysis begins, or is it something that is created by interaction between data, zoning system, and filtering model?

Arbitrary design criteria

For certain purposes it has been recognized that special types of zoning systems might be better than others. For example, spatial interaction models can only describe inter-zonal flows so it seems sensible to design zoning systems for these models that have a minimum of intra-zonal interaction (Broadbent, 1969). A number of other zone design criteria have also appeared. Some are model specific; see, for example, Williams (1976); whilst others are more general purpose; see, for example, the maximally homogeneous aggregations of Cliff *et al.* (1975, pp. 17-19), or the approximately equal population and compact zoning criterion of Sammons (1976). Far more elaborate multiple criteria zone design sequences have been suggested by Masser, Batey, and Brown (1975, 1978).

In practice this is a half-hearted approach which largely ignores the likely effects that arbitrary design criteria have on the results that are obtained. Indeed, there are as yet few studies of the performance of these arbitrary zone design criteria and thus no measure of their usefulness as 'rules of thumb'. It seems likely that different criteria merely produce different results (see Openshaw, 1978b). A major problem with this approach has been the failure of model builders to specify what criteria the zoning systems should satisfy before they can be used with their models.

Information theoretic

To make up for the absence of any theory, Batty (1978) suggests an approach using information theory. This is based on the observation that scale and aggregation affects the amount of information contained in zonal data. Batty mainly concentrates on the zone number or scale problem, whilst trying to hold constant zone configuration or the aggregation component. To this end he suggests a series of spatial entropy measures (Batty, 1974; 1976a), the aim being to determine how many zones are needed to describe or model a given phenomenon. While he recognizes that there is a relationship between the variation in the phenomenon being studied and the zoning system used to measure it (Batty, 1978, p. 115), he remains firmly wedded to the normal science view that the zoning system must be independent of the phenomenon being studied (Batty, 1978, p. 146). A major problem is, therefore, the specification of an appropriate information function that best measures the meaningfulness of the information. Surely this must depend on the proposed method of analysis and cannot be general purpose. So far the initial results have not been too promising (Batty and Sammons, 1978).

Statistical

A far more convenient approach is to try to develop a statistical theory of the modifiable areal unit problem. Two possibilities have emerged. The first of these is the sampling analogy, see Cliff (1973) and Williams (1976, 1979). This implies that the procedures of sampling and zoning are analogous. The scale problem can be viewed as analogous to the sample size problem and the aggregation problem to sampling error. If it can be proven this analogy would be exceptionally convenient because it appears to provide a ready-made statistical basis for treating the variability in the results of spatial study due to the use of different zoning systems. These effects could be estimated using standard formulae for calculating sampling errors under the assumption of simple random sampling.

Unfortunately, the analogy is less than satisfactory. Consider a sample survey of households in a study region. Suppose these data are aggregated to a zoning system. From a geographical point of view the zonal data now represents a population since the zones completely cover the study region. From a statistical point of view it still contains sampling errors because it is based on survey data. Suppose that these data are subsequently reaggregated and a statistic calculated from it. If this aggregation process could be repeated a sufficiently large number of times and if it is a random one, then it is possible that the parameters of the distribution of the statistic might be approximated by sampling theory.

There are a number of problems. The zonal data contain sampling errors as well as aggregational variability so that sampling theory is required somehow to take both into account. Additionally, the question arises as to what the zonal estimates for the statistic concerned are estimates of. Another problem is the fact that the observed correspondence between empirically derived zoning and sampling distributions is not too good (Openshaw, 1978b; Openshaw and Taylor, 1979). Furthermore, the analogy only holds if the choice of zoning system is random and this assumption is virtually never satisfied in practice.

An alternative approach towards a statistical theory is based on a Blalockian conjecture (see Taylor, 1977, p. 222). This fails because it completely ignores the existence of aggregation effects and because the basic idea is far too simple for such a horribly complex problem (see Openshaw and Taylor, 1979).

Traditional

This constitutes the first of two intrinsically geographical solutions. The modifiable areal unit problem only exists because of uncertainty as to what are the objects of spatial study. Remove that uncertainty and the problem disappears. All that is required is for geographers to try to identify zones as meaningful objects to study in an explicit albeit subjective fashion. The resulting zones will still be arbitrary but they would at least have a geographical rationale and thus would be sensible units for a particular purpose. This does not mean that it is only necessary to declare that, for example, all administrative boundaries define meaningful entities, instead it must involve a careful process of definition. An example of this approach are the attempts by Coombes *et al.* (1978, 1979) to identify a set of functional urban definitions as a reporting framework for national census data. These units are justified solely in geographical terms; that is they have a theoretical basis, they appear to make sense, and they are the result of the rigorously consistent application of a fixed set of rules. Undoubtedly they are very non-random and geometrically irregular but these are no longer relevant criteria to judge them against once it is decided that the units are uniquely meaningful entities. However, such units will not be suitable for all purposes and are probably best suited to the simple presentation of data.

Optimal zoning

A final approach is radically different from the previous ones. If it is accepted that the results of studying zonal data depend on the zoning systems used, then the choice of zones must reflect the results that are desired. This is a complete reversal of the normal approach yet the alternative is to naively accept virtually any result which in part reflects the arbitrary zoning system used. The use of this or that zone design criterion does not help in the slightest. The problem is how to use the freedom inherent in the choice of zoning systems to develop a new general paradigm for geographical analysis.

One possible approach is as follows. First, a speculation is made concerning the expected result for a given model or method of analysis. Second, an attempt is made to achieve this target by seeking an appropriate zoning system using the heuristic procedure described in Openshaw (1977b, 1978a). If the desired result cannot be obtained then the associated hypothesis should be rejected. If it can be obtained, then it is necessary to consider the range of different zoning systems that also produce the target result. It is also possible to introduce various statistical and geographical constraints on both the nature of the data produced by the zoning systems and on the configuration of the zones themselves.

This paradigm is really a methodology for formulating and testing spatial hypotheses by deliberately and purposefully exploiting the uncertainty that exists in zonal data because of the modifiable areal unit problem. Further discussion is given in Openshaw (1977b, 1978a, 1979). It is noted that this approach is messy, computer-dependent, and at present lacks any theoretical basis; but it seems to work, it is intensely geographical, and it provides perhaps the only real solution so far to an immensely complicated problem in geographical analysis.

Conclusions

There are three justifications for ignoring the modifiable areal unit problem. First that it is insoluble. Second that it is of trivial importance. Third that to acknowledge its existence would cast doubts on the applicability of nearly all applications of quantitative techniques to zonal data. It has been argued that the first two of these assumptions are incorrect, the third is quite correct. After all, if the measurement of observations cannot be independent of the hypotheses being tested and, further, if it is doubtful whether there really is a single 'real' result that can be revealed by analysis, then the manner in which existing methods are being applied cannot reasonably continue. Problems of this magnitude cannot simply be ignored. A poor analogy would be the study of survey data without estimating error bounds due to sampling errors. The principal difference is that whereas estimates can usually be made of sampling error, the analogous variability in zonal data due to the choice of zoning system is a largely unknown quantity.

Obviously total abandonment of all the available methods of analysis is not a useful way to proceed, if there is nothing better to replace them. Instead it is necessary for geographers to be more aware of the geography of the methods they employ and the zoning systems they study. In the longer term it would be nice if the possibilities opened up by the optimal zoning approach for a more geographical form of geographical analysis could be realized. Indeed it is particularly important that the modifiable areal unit problem is given a geographical solution, since it seems very unlikely that there will ever be either a purely statistical or mathematical solution. In fact there appears to be something seriously wrong with the work of those geographers who have become so blinkered in their view of quantitative geography that they can neither see the basic geographical problems nor conceive of geographical solutions to them. Their works may well qualify as good statistics or good mathematics but hardly as good geography.

References

Bartels, C.A.P. (1979) 'Operational statistical methods for analysing spatial data' in C.A.P. Bartels and R.H. Ketellapper (eds), op. cit., 5-50.

Bartels, C.A.P., and Ketellapper, R.H. (eds) (1979) *Exploratory and explanatory statistical analysis of spatial data,* Maitinus Nijhoff: Leiden.

Batty, M. (1974) 'Spatial entropy', *Geographical Analysis,* 6, 1-31.

Batty, M. (1976a) 'Entropy in spatial aggregation', *Geographical Analysis,* 8, 1-21.

Batty, M. (1976b) *Urban modelling: algorithms, calibrations, predictions,* Cambridge University Press.

Batty, M. (1978) 'Speculations on an information theoretic approach to spatial representation' in I. Masser and P.J.B. Brown (eds), op. cit., 115-47.

Batty, M., and March, L. (1976) 'The method of residues in urban modelling', *Environment and Planning A,* 8, 189-214.

Batty, M., and Sammons, R. (1978) 'On searching for the most informative spatial pattern', *Environment and Planning A,* 10, 747-79.

Bearwood, J.E. (1972) 'The space-averaging of deterrent functions in gravity model trip distribution calculations', TRRL Report LR 462, Crowthorne, Berkshire.

Bearwood, J.E., and Kirby, H.R. (1975) 'Zone definition and the gravity model: the separability, excludability, and compressibility properties', *Transportation Research,* 9, 363-9.

Bennett, R.J. (1979) *Spatial time series: analysis-forecasting-control,* Pion: London.

Broadbent, T.A. (1969) 'Zone size and singly constrained interaction models', CES-WN-132, Centre for Environment Studies, London.

Casetti, E. (1966) 'The analysis of spatial association by trigonometric polynomials', *The Canadian Geographer,* 10, 199-204.

Chorley, R.J., and Haggett, P. (1967) *Models in geography,* Methuen: London.

Clark, W.A.V., and Avery, K.L. (1976) 'The effects of data aggregation in statistical analysis', *Geographical Analysis* 8, 428-38.

Cliff, A.D. (1973) 'A note on statistical hypothesis

testing', *Area*, 5, 240.
Cliff, A.D., and Haggett, P. (1970) 'On the efficiency of alternative aggregations in region building problems', *Environment and Planning* 2, 285-94.
Cliff, A.D., and Ord, J.K. (1973) *Spatial autocorrelation*, Pion: London.
Cliff, A.D., and Ord, J.K. (1975) 'Model building and the analysis of spatial pattern in human geography', *Journal of the Royal Statistical Society Ser. B*, 37, 297-348.
Cliff, A.D., Haggett, P., Ord, J.K., Bassett, K.A., and Davies, R.B. (1975) *Elements of spatial structure: a quantitative approach*, Cambridge University Press.
Coombes, M.G., Dixon, J.S., Goddard, J.B., Openshaw, S., and Taylor, P.J. (1978) 'Towards a more rational consideration of census areal units: daily urban systems in Britain', *Environment and Planning A*, 10, 1179-85.
Coombes, M.G., Dixon, J.S., Goddard, J.B., Openshaw, S., and Taylor, P.J. (1979) 'Daily urban systems in Britain: from theory to practice', *Environment and Planning A*, 11, 565-74.
Curry, L. (1966) 'A note on spatial association', *The Professional Geographer*, 18, 97-9.
Dewdney, J.C., and Rhind, D.W. (1976) *People in Durham: a census atlas*, Census Research Unit: University of Durham.
Drewett, R., Goddard, J.B., and Spence, N. (1976) 'British cities: urban population and employment trends, 1951-1971', *Research Report* 10, Department of the Environment: London.
Gehlke, C.E., and Biehl, K. (1934) 'Certain effects of grouping upon the size of the correlation coefficient in census tract material', *Journal of the American Statistical Association*, 29, 169-70 (supplement).
Haggett, P. (1965) *Locational Analysis in Human Geography*, Arnold: London.
Haggett, P., Cliff, A.D., and Frey, A. (1977) *Locational analysis in human geography*, 2nd Edition, Arnold: London.
Haining, R.P. (1978) 'Specification and estimation problems in models of spatial dependence', *Northwestern University Studies in Geography* no. 24, Northwestern University Press, Evanston, Ill.
Harvey, D.W. (1969) *Explanation in Geography*, Arnold: London.
Keane, M. (1975) 'The size of the region-building problem', *Environment and Planning A*, 7, 575-7.
Kendall, M.G., and Yule, G.U. (1950) *An introduction to the theory of statistics*, Griffin: London.
Kirby, A.M., and Taylor, P.J. (1976) 'A geographical analysis of the voting pattern in the EEC referendum, 5 June 1975', *Regional Studies*, 10, 183-92.
Littschwager, J.M. (1973) 'The Iowa redistricting system', *Annals New York Academy of Science*, 219, 221-35.
Martin, R.L., Thrift, N.J., and Bennett, R.J. (1978) *Towards the dynamic analysis of spatial systems*, Pion: London.
Masser, I., and Brown, P.J.B. (eds) (1978) *Spatial representation and spatial interaction*, Martinus Nijhoff: Leiden.
Masser, I., Batey, P.W.J., and Brown, P.J.B. (1975) 'The design of zoning systems for spatial interaction models', in E.L. Cripps (ed.) *Regional science - new concepts and old problems*, 168-87, Pion: London.
Masser, I., Batey, P.W.J., and Brown, P.J.B. (1978) 'Sequential treatment of the multi-criteria aggregation problem: a case-study of zoning system design' in I. Masser and P.J.B. Brown (eds), op. cit., 27-50.
Moellering, H., and Tobler, W.R. (1972) 'Geographical variances', *Geographical Analysis*, 4, 34-50.
Nagel, S.S. (1965) 'Simplified bi-partisan computer re-districting', *Stanford Law Review*, 17, 863-99.
Nagel, S.S. (1972) 'Computers and the law and politics of redistricting', *Polity*, 5, 77-93.
Openshaw, S. (1977a) 'Algorithm 3: a procedure to generate pseudo-random aggregations of N zones into M zones, where M is less than N', *Environment and Planning A*, 9, 1423-8.
Openshaw, S. (1977b) 'A geographical solution to scale and aggregation problems in region-building, partitioning, and spatial modelling', *Transactions of the Institute of British Geographers, NS* 2, 459-72.
Openshaw, S. (1977c) 'Optimal zoning systems for spatial interaction models', *Environment and Planning A*, 9, 169-84.
Openshaw, S. (1978a) 'An optimal zoning approach to the study of spatially aggregated data', in I. Masser and P.J.B. Brown (eds), op. cit., 93-113.
Openshaw, S. (1978b) 'An empirical study of some zone-design criteria', *Environment and Planning A*, 10, 781-94.
Openshaw, S. (1979) 'Appropriate methods for testing hypotheses derived from studies of spatially aggregated data', Paper presented at IBG Annual Conference, Manchester (mimeo).
Openshaw, S., and Taylor, P.J. (1979) 'A million or so correlation coefficients: three experiments on the modifiable areal unit problem' in N. Wrigley, (ed), *Statistical Applications in the Spatial Sciences*, Pion: London.
Rhind, D. (1977) 'Computer aided cartography', *Transactions of the Institute of British Geographers, NS* 2, 71-97.
Robinson, A.H., Linberg, J.B. and Brinkman, L.W. (1961) 'A correlation and regression analysis applied to rural farm population densities in the Great Plains', *Annals Association of American*

Geographers, 51, 211-21.
Robinson, W.S. (1950) 'Ecological correlations and the behaviour of individuals', *American Sociological Review*, 15, 351-7.
Rogers, A. (1969) 'On perfect aggregation in the matrix cohort survival model of inter regional population growth', *Journal of Regional Science*, 9, 417-24.
Rogers, A. (1972) *Matrix methods in regional analysis*, Holden Day: San Francisco.
Sammons, R. (1976) 'Zoning systems for spatial models', *Reading Geographical Papers* 52.
Streitberg, B. (1979) 'Multivariate models of dependent spatial data' in C.A.P. Bartels and R.H. Ketellapper (eds), op. cit., 139-78.
Taylor, P.J. (1973) 'Some implications of the spatial organisations of elections', *Transactions of the Institute of British Geographers*, 60, 121-36.
Taylor, P.J. (1977) *Quantitative methods in geography*, Houghton Mifflin: Boston.
Taylor, P.J., and Goddard, J.B. (1974) 'Geography and statistics: an introduction' *The Statistician*, 13, 149-55.
Taylor, P.J., and Gudgin, G. (1976) 'The statistical basis of decision making in electoral districting', *Environment and Planning A*, 8, 43-58.
Taylor, P.J., and Johnston, R.J. (1979) *Geography of elections*, Penguin: Harmondsworth.
Thomas, E.N., and Anderson, D.L. (1965) 'Additional comments on weighting values in correlation analysis of areal data', *Annals of the Association of American Geographers*, 55, 492-505.
Tobler, W.R. (1963) 'Geographic area and map projections', *Geographical Review*, 53, 59-78.
Tobler, W.R. (1969) 'Geographical filters and their inverses', *Geographical Analysis*, 1, 234-53.
Tobler, W.R. (1975) 'Linear operators applied to areal data' in J.C. Davis and M.J. McCullagh (eds), *Display and analysis of spatial data*, 14-37, Wiley: London.
Weaver, J.B. (1970) *Fair and equal districts*, National Municipal League: New York.
Weaver, J.B., and Hess, S.W. (1963) 'A procedure for nonpartisan redistricting', *Yale Law Journal*, 73, 288-308.
Williams, I.N. (1976) 'Optimistic theory validation from spatially grouped regression: theoretical aspects', *Transactions of the Martin Centre*, 1, 113-45.
Williams, I.N. (1978) 'Some implications of the use of spatially grouped data' in R.L. Martin, R.J. Bennett and N.J. Thrift (eds), op. cit., 53-64.
Wilson, A.G. (1970) *Entropy in urban and regional modelling*, Pion: London.
Wilson, A.G. (1974) *Urban and regional models in geography and planning*, Wiley: London.

Part 3

Statistical methods and models

Statistical methods, and the underlying deductive and inductive models from which they derive, have been a central concern of geography in the 1960s and 1970s, and they must remain a major centre of attention into the foreseeable future. Whilst not all geographical questions are amenable to answers which can be found using statistical technique, a great number of economic, social and political questions do rely for answers upon the methodology of statistical inference, deductive statistical model building, or exploratory statistical analysis.

The chapters presented in this section review some of the major species of statistical methods with which British geographers have been particularly associated. These chapters demonstrate that a continued concern has been with developments and applications of well-tried existing techniques familiar to most geographers; for example, the general linear model encountered in regression analysis, discussed in the chapter by Silk, factor analysis, discussed in the chapter by Mather, and point pattern analysis, discussed in the chapter by Thomas. However, a strong emphasis of specifically British work has been concerned with spatial and temporal analysis. Particularly important are the developments by Cliff and Ord of techniques concerned with spatial autocorrelation and by these workers together with Haggett, Hepple, Haining, Bennett, and others concerned with time series analysis. An additional and new thrust of British work, however, has been with the less sophisticated tools of exploratory data analysis employing the modes of approach developed by Tukey. The chapter by Cox and Jones discusses this field. In addition categorical data analysis and the analysis of closed number system data, both long neglected as systematic areas by geographers, have been given considerable attention in Britain in recent years and these topics are discussed in the chapters by Wrigley, Evans and Jones. The work by Wrigley has been particularly important in stimulating a concern with categorical data analysis, and he has drawn the attention of geographers to the publications of Goodman, Fienberg, McFadden and Koch. Again, multidimensional scaling, developed to a high level in psychology and related disciplines, has recently received considerable attention in geography, and this work is reviewed by Gatrell.

The quantitative revolution in geography in the 1960s in Britain, as elsewhere, must be seen now, in retrospect, as largely concerned with the application of many of the statistical techniques in this section of the present volume. It may well now appear, however, that this early period is characterized more by the application of techniques than the development of methodology, and the 1970s have seen a critical reappraisal of much of the earlier work. From this line of critical reappraisal, perhaps five main thrusts can be identified as of central relevance for the development of statistical techniques in geography in the future.

First, the more recent developments have seen the continued use of many well-tried techniques, but a much more critical understanding of their underlying assumptions. As a result, considerably better applications of techniques have emerged, and the methodological blunders undermining many early applications have not been committed. Nowhere is this better understanding more evident than in the use of factor and principal component analyses. For example, Mather in Chapter 14 of this book emphasizes that such techniques cannot be used outside of the particular framework of objective measurement, but that the methods of observation determine the choice of technique adopted, the forms of rotation employed, the interpretation of factor scores, the methods of communality estimation adopted, and other issues.

A second thrust of concern, related to more critical understanding of underlying assumptions, has been the exploration of detailed alternative methods for attaining given statistical conclusions. Considerable attention, for example, has been directed to alternative methods of estimating regression equations and this has encompassed maximum likelihood, weighted least squares, instrumental variable, logit, ridge regression, simultaneous equation, and reduced form estimates. Again, in time series and spatial time series studies, there has been extensive discussion of the merits of various alternative forms of stochastic process and transfer function process representation, and with a wide variety of alternative non-stationary estimation methods.

As a third thrust of concern, and deriving from a greater awareness of both the limiting assumptions of techniques and the availability of alternatives, there has been a growing understanding of the limitations of data itself. This thrust has led to a concern to construct simple rather than complex models. It is certainly true that the 1960s and 1970s saw the use of over-sophisticated techniques for the tackling of many problems: 'the use of a sledgehammer to crack a walnut'. For example it is doubtful if any of the factorial ecology papers based on factor and principal component analyses advanced us, in an understanding of the city, beyond the early work of Shevky and Bell in 1956 who used prior definition of factors. Again, it is doubtful if either the space-time autoregressive integrated moving-average or the spatial Markov field models developed in the late 1970s really contribute a great deal to our understanding of spatial diffusion and related processes.

This new concern with simple rather than complex models has not derived from an interest in parsimony of the model itself, but instead from a concern with the limitations of the data. Thus the late 1970s have seen new thrusts of concern with exploratory and categorical data analysis and multidimensional scaling. On the one hand, the exploratory data analysts, such as Cox and Jones in their chapter in this book, have directed our attention to uninhibited exploration of data, making extensive use of prior knowledge, and refraining from procedures of statistical inference which make rigorous and highly optimistic assumptions with regard to the quality of available data. On the other hand, categorical data analysis, as demonstrated in Wrigley's chapter, has shown the potential of parametric analysis with nominal and ordinal data, or with mixtures of nominal and ordinal with interval and ratio scale data. Again, multidimensional scaling has shown the potential for application of statistical methods to the rather intractable problems of assessing personal preference functions.

The fourth new thrust of concern in the 1970s, which perhaps underlies each of the three shifting emphases discussed above, is the use of statistical techniques not as ends in themselves, but increasingly as applied tools to answer specific problems. Thus, if quantitative geography has 'come of age' in the 1970s it has been through the use of a particular technique as a natural consequence of the mode of analysis adopted, the questions asked, and the specific characteristics of the problem in hand. As such, then, the concern has shifted in quantitative geography from statistical methods and models *per se* to applied problems requiring particular types of solution.

Deriving from each of the thrusts identified above there has been the general 'paradigm shift' in quantitative geography noted in the introduction to this volume: that the emphasis on spatial geometry so characteristic of the 1960s has been replaced by an emphasis on spatial processes. Many important research problems still remain with spatial phenomena, and the chapter by Thomas raises the many issues surrounding the analysis of point patterns. But in making a shift away from an emphasis on spatial geometry, the subject has left behind many other unsolved problems in that area. Instead it has thrust new problems to attention concerned with change, dynamics, evolution and policy involvement. This change in emphasis particularly characterizes the chapters concerned with spatial and temporal analysis in this volume; for example, Haining emphasizes the need for more *a priori* and deductive model formulation, Bennett stresses the need for credible adaptable spatial time series models which are applicable to policy, Hepple reviews the importance of frequency as well as time-domain specification of models, and Cliff and Ord stress the role of tests of spatial autocorrelation as a diagnostic checking stage in model building.

Chapter 6

The general linear model

J. Silk

Introduction

Our discussion of the General Linear Model (GLM) falls into two major sections. The first is concerned with classical use of the GLM in geography, so called because the techniques are employed in a manner scarcely differing from that found in disciplines not sharing the geographer's interest in spatial relationships. However, adjustments made to allow for the special nature of spatial data are also described. In the second section, attention focuses on special applications of the GLM in which spatial and temporal components are explicitly incorporated, and we will call these space-time models. The rest of this introduction briefly describes the GLM.

A variety of techniques and models used by geographers which may not appear to have many features in common are in fact versions of the GLM, whose structure may be expressed as

$$Y_i = \theta_1 X_{i1} + \theta_2 X_{i2} + \ldots + \theta_K X_{iK} + \epsilon_i \quad (1)$$
$$(i = 1, 2, \ldots, N)$$

where Y_i is an observation of a response or dependent variable, the θ_i are fixed but unknown parameters, the X_{ij} $(j = 1, \ldots, K)$ fixed values of observed independent variables measured without error, and ϵ_i an unobservable random variable known as the error or disturbance term. The requirement of an additive or linear relationship as shown in (1) is less restrictive than it first appears, as a variety of apparently nonlinear but intrinsically linear relationships may be transformed to comply with it (Draper and Smith, 1966, pp. 132-4). Furthermore, none of the independent variables should be perfect, or near-perfect, linear function of any or all of the remaining independent variables, i.e. the no multicollinearity assumption. The ϵ_i are assumed to be random variables each with mean zero and unknown variance σ^2, and to be mutually uncorrelated. For statistical inferences to be made, it is further assumed that the ϵ_i follow a normal distribution which also makes them mutually independent.

The type of GLM most familiar to the geographer is the *linear regression model,* typically written as

$$Y_i = \beta_1 X_{i1} + \beta_2 X_{i2} + \ldots + \beta_K X_{iK} + \epsilon_i \quad (2)$$

where $X_{i1} = 1$ for all i, so that $\beta 1$ represents the intercept or constant term of the regression equation and the remaining β_i are slope coefficients. The flexibility of this particular model enables it to incorporate aspects of space and time as dependent or independent variables so that temporal and spatial series can be investigated. Thus, each independent variable in (2) may represent a term in the polynomial product of the spatial coordinates of some phenomenon of interest, as in *trend surface analysis,* or instead involve trigonometric functions in which case we obtain the model used in *Fourier analysis* of time series and spatial data (Krumbein, 1967, p.40). If the $X_{i\cdot}$ represent variables whose values are lagged over time alone, we obtain a model of dependencies over time, if over space alone, of dependencies over space. Space-time models incorporating dependencies in both modes have also been developed (Haggett *et al.*, 1977, Ch. 16).

There is also a form of the linear regression model which assumes a multivariate normal distribution of all the variables involved, as well as normality of the error term, and these assumptions apply also to the *correlation model.*

The *Analysis of Variance model* may, with suitable coding, be written so that it has the structure of (1), and also combined with the linear regression model to permit comparison of regression lines and specification of the Analysis of Covariance model (Silk, 1976, 1979a; Horton, 1978). More complex techniques such as factor and components analysis, canonical correlation and discriminant analysis can be regarded as multivariate analogs of the GLM in that there are several response or dependent variables rather than just one.

Estimates of the θ_i in (1) or the β_i in (2) are obtained from sample observations by the method of ordinary least squares (OLS) which minimizes the sum of the squared deviations about the fitted surface.

Classical use of the general linear model in geography

Early use of the GLM in both human and physical geography strongly reflected the overall reliance in the discipline upon classical aspatial statistical techniques. As Hepple (1974, pp. 96-7) remarks, the relevant theory on statistically non-independent data was scattered in the literature, and no texts were available. Geographers relied on standard statistical texts, classical and experimental in treatment, and assuming spatial independence of error terms in their models. Indeed, Unwin (1977) argues that the popularity of morphological systems analysis in geomorphology from the mid 1960s to the early 1970s, in which a 'timeless' equilibrium state is assumed and spatial variation not explicitly incorporated, was largely due to its symbiotic relationship with the classical techniques in which temporal and spatial variations cannot be included in a straightforward manner.

Regression

A number of features - not necessarily mutually exclusive - have characterized regression (and correlation) studies in geography. We note that both techniques tended to be regarded as interchangeable in early studies because hypothesis-testing rather than estimation was emphasized.

In human geography particularly, a striking feature of many studies was the description of, and testing of hypotheses about, relationships between multivariate spatial data sets, i.e. the observations on each variable were located at various points or in different regions (Cliff and Ord, 1975a). Although distributed over space, space *per se* did not explicitly enter into the relationships specified. Until the 1970s, the spatial input to such studies came from 'eyeballing' the pattern of residuals about the regression line (Thomas, 1968). These studies of 'areal association' have continued throughout the last decade (e.g. Keeble and Hauser, 1972; Johnston, 1976; Pooley, 1977). Many were carried out within the framework of the positivist scientific method of sequential testing and refinement of hypotheses initially proposed by McCarty (1956), refined by Haggett (1965, pp. 278-80) and generalized by Harvey (1969) and Johnston (1978).

Another feature has been the identification of empirical regularities, first, to aid this process of hypothesis-testing and, second, to provide some basis for forecasting and control. Examples of the former are cited in Chorley (1966, pp. 340-57) and later studies show this tradition to be still healthy (e.g. Brunsden, 1971; Gregory and Walling, 1973; Richards, 1977). The latter is evident in hydrology and climatology (e.g. Nash and Shaw, 1966; Stocking and Elwell, 1976) and also in the study of spatial interaction, both in geography (e.g. Olsson, 1965; Taylor, 1975) and cognate fields such as transport planning (e.g. Domencich and McFadden, 1975; Stopher and Meyburg, 1975).

In the late 1960s and early 1970s an uneasiness about the use of quantitative techniques in geography gave rise to a feeling that the assumptions underlying use of the linear regression model were poorly understood and also that spatial data, and the processes giving rise to them, posed statistical problems that geographers had barely considered. The statement that '[in geography] everything's related to everything else, but near things are more related than distant things' (Tobler, 1969, p. 7) and Gould's (1970) contention that the assumption of spatial independence in regression and correlation analysis was totally at variance with the research aims and findings of geographers, seemed to crystallize this mood. In this context, it was hardly surprising that the assumption of spatial independence underlying use of the GLM took on an almost exaggerated importance in the early 1970s, once it was shown that certain spatial relationships could be conceived of statistically in terms of spatial autocorrelation (Cliff and Ord, 1973).

Extensions of aspatial techniques adopted by British geographers during the 1970s include those of simultaneous-equation regression analysis, enabling the researcher to deal explicitly with two-way relationships between variables whether dependent or independent (Todd, 1979), and others for dealing with dichotomous and polychotomous dependent variables (Wrigley, 1975, 1976). It is noteworthy that two of the three references cited are to monographs in the CATMOG series.

Correlation

Simple correlation analysis has been routinely employed in most branches of the discipline since the mid 1960s, but more sophisticated applications are relatively rare in the geographic literature, British or otherwise.

Path analysis was first used to estimate 'coefficients of inbreeding' and other genetic correlations (for references, see Wright, 1960). In sociology, Duncan (1966) suggested using this methodology to tackle the reverse problem of estimating the paths which may account for an observed set of correlations, and also pointed out that *causal modelling* (Blalock, 1964) can be regarded as a special case of path analysis. Both techniques are particularly valuable in requiring the researcher to make the interrelationships between all variables, not just between the dependent and independent variables, explicit and so expose the 'inner workings' of the model (Duncan, 1975). In this sense they resemble simultaneous-equation regression analysis. There appear to be only three studies employing path analysis, two from Britain (Davidson, 1976; Bonnar, 1979) and one from North America (Golledge and Spector, 1978). For studies based on causal modelling, see Cox (1968), Moore (1969) and Mercer (1975) and also the exchange between Taylor (1969) and Cox (1969). The reason for such meagre interest is partly that such techniques have not been covered in quantitative geographic texts until very recently (Ferguson, 1978, pp. 36-41; Johnston, 1978, pp. 93-8) - although see Harvey, 1969, pp. 392-400). A further reason may be that the investigator cannot avoid a complex specification task before plugging in to a computer program. However, there is considerable potential for use of these techniques, particularly in areas of human geography which make use of survey research data.

The Analysis of Variance

Use of the Analysis of Variance (ANOVA) by geographers appears to accord with Draper and Smith's (1966, p. 243) comment that if the researcher is asked 'what model are you considering?' the reply is frequently 'I am not considering one - I am using analysis of variance'. A formal statement of the ANOVA model and its assumptions is almost impossible to find in any geographic text except Norcliffe (1977, p. 159), although a detailed treatment is now available in Silk (1979a: 1981). This drawback becomes critical when more than one factor, or continuous variables, are introduced. It is only recently that the links between the linear regression model and the ANOVA model as versions of the GLM have been made explicit in the geographic literature, initially in the context of dummy variable analysis (Silk, 1976; Johnston, 1978) and then in discussion of the Analysis of Covariance and comparison of regression lines (Silk, 1979a).

In consequence, technical problems such as those associated with *a posteriori* significance testing in ANOVA or the use of various stepwise regression procedures are poorly known (Hauser, 1974; Mather and Openshaw, 1974; Silk, 1981), and there has been virtually no mention of the difference between fixed and random effects models. Furthermore, there has been almost no recent treatment of the principles of experimental design in the geographic literature. Early contributions were made by Haggett (1964; 1965, pp. 299-303) and Chorley (1966, pp. 335-40), and extended in Haggett *et al.* (1977, pp. 270-9), but concepts such as those of the control group, randomization or confounding are nowhere to be found. However, there is a clear need for such knowledge in behavioural geography and much of (1976, pp. 87-8), and also in physical geography if the observation that 'it has become more experimental in approach' (Unwin, 1977, p. 207) is correct. It is also surprising that some of the formal similarities between regions and classes have not made ANOVA popular (e.g. Murdie, 1969), or led to imitation of Zobler's (1958) early work on assigning areas to given regional classes.

Finally, we note here that resemblances between a far wider class of modelling techniques than those based on regression and ANOVA procedures have been discussed under the heading of the *generalized linear model* (Nelder and Wedderburn, 1972), and are particularly concerned with relationships between categorical variables - see the discussion in Wrigley's article in Chapter 11.

Extensions of the GLM

A number of techniques extend the logic of the GLM by relating one or more sets of given or derived variables while simultanously maximizing the value of some criterion. The first three sets of techniques to be described were all initially employed in geographic research in North America before diffusing outwards, and made their first appearance in a geographic text in King (1969).

Factor and principal components analysis began to be heavily employed in Britain in the late 1960s and a number of British texts provide detailed coverage (Daultrey, 1976; Goddard and Kirby, 1976; Mather, 1976; Taylor, 1977, Chapter 6; Johnston, 1978, Chapter 5). For useful reviews see Clark *et al.* (1974), Unwin (1977, pp. 192-4) and Mather's contribution to this volume (Chapter 14).

Canonical correlation, used to relate sets of criteria (dependent) and predictor (independent) variables so that their overall correlation is maximized, appeared early in North America (Berry, 1966; Gauthier, 1968). The technique is explained in the CATMOG by Clark (1975), and Johnston (1978, Chapter 6) provides a three-fold classification of studies - relating groups of interrelated variables, e.g. Corsi and Harvey (1975); integration of formal and functional regions, e.g. Clark (1973); comparison of two equivalent data sets, e.g. Lankford (1974).

Discriminant analysis first appeared in geographic studies by Casetti (1964, 1966) and King (1967). In the British literature Spence and Taylor (1970, pp. 45-8) provide a short review, but no suitable text (or CATMOG) appeared until Johnston (1978) although Mather (1976) and Unwin (1977) provide reviews in physical geography, the former rather technical. The technique is designed to discriminate between overlapping populations or classifications - including regional classifications - so as to assign new observations to established classes with the minimum probability of error. Johnston (1978, Chapter 8) gives an example of classifying new observations, but although it has been used in this way by Mather and Doornkamp (1970), most studies involve either testing and generation of hypotheses (thereby identifying the most important discriminators between groups), e.g. Norris and Barkham (1970), or evaluation of prior classifications, e.g. Ahmed (1965), Barker (1976). Discriminant functions may be rotated (Norris and Barkham, 1970), and an iterative procedure used to deal with binary variables (Mather, 1976, pp. 453-9). We also note that multivariate analysis of variance (MANOVA) may be employed to test the hypothesis that multivariate means of several groups are equal, either as a check before using discriminant analysis, or for hypothesis testing (Mather, 1976, pp. 432-7).

Apart from factor and components analysis, none of these complex techniques has been heavily employed by British geographers at any stage, partly because the quantitative surge has subsided, and also because the cutting edge of quantitative research lies in spatial and temporal analysis, and to a lesser extent in the introduction of new aspatial methods such as exploratory data analysis (Chapter 13) and categorical data analysis (Chapter 11).

Technical difficulties

Mather and Openshaw (1974, p. 296) classified technical difficulties facing the geographer trying to fit some version of the GLM under four headings which we use here, together with a fifth concerning the special nature of spatial data.

Choice of functional form in regression analysis is, as noted earlier, wider than might first appear as any model linear in the parameters may be fitted by OLS, including step functions as specified by the classical ANOVA model or by the inclusion of dummy variables in multiple regression analysis (Silk, 1976, 1979a). However, if we wish to fit an intrinsically nonlinear model then numerical approximation techniques, as described by Mather (1976, pp. 198-211) and in a wide-ranging review by Batty (1976) may be used. For an up-to-date evaluation in the context of calibrating spatial interaction and activity allocation models, see Batty's contribution to this volume (Chapter 17).

Independent variable selection is not adequately trêated in any geographic text, and researchers usually refer to Draper and Smith (1966) or Daniel and Wood (1971). For students, the answer may lie in more imaginative teaching methods, based on interactive use of minicomputers, which make it clear that such a difficulty is one of many encountered in the complex sequential decision-making process of data analysis.

Fulfilment of the assumptions of the model received considerable attention as a result of the unease of the late 1960s described earlier (Colenutt, 1968; Poole and O'Farrell, 1971). Relevant monographs in the CATMOG series (Wrigley, 1976; Ferguson, 1978), the detailed treatment in Mather (1976, Chapter 2), the useful survey in Haggett *et al.* (1977, Chapter 10) and better introductory undergraduate tests have all helped improve the situation. Three developments deserve more detailed comment. First, techniques which minimize the sum of squares of distances normal to the fitted surface, rather than vertically, may be used if values of the independent variables cannot be measured without error (Sturgal and Aiken, 1970). Where the problem arises because dependent and independent variables cannot be clearly identified the reduced major axis may be fitted (Till, 1973). However, more fundamental issues here have been raised by Mark and Peucker (1978), who suggest that functional analysis is often more appropriate than regression analysis (see also Cook and Falchi, 1979; Sheppard, 1979; Mark and Peucker 1979). Second, the problem of multicollinearity is well-known to geographers (King, 1969; Johnston, 1978). Recently, the technique of *ridge regression* (Hoerl and Kennard, 1970a, b) has been cautiously recommended to geographers by Mather and Openshaw (1974), Mather (1976) and Unwin (1977), as it allows the investigator to trade off a reduction in standard error against an increase in bias in the estimated regression coefficients, but the only examples of its use are Moriarty (1973) and Pinch (1978). Finally, important work by British geographers on the effect of spatial dependence in the error term will be discussed under the special nature of spatial data.

Computational problems are frequently overlooked, but Mather and Openshaw (1974) and Mather (1976, pp. 52-73, 95-101) warn us to bear them in mind, especially if data matrices are thought to be poorly conditioned.

The special nature of spatial data was dimly realized by geographers in the mid 1950s and led to consideration of the modifiable units problem, i.e. the effect of varying the size and shape of areal units to which data refer (Robinson, 1956). Other North American work by geographers has appeared on this topic, and on the associated one of interpreting ecological correlations (Thomas and Anderson, 1965; Curry, 1966; Clark and Avery, 1976; Smit, 1978). In Britain, simulation techniques have been used to show that estimated regression coefficients are not independent of the choice of zoning system (Openshaw, 1977a, b; 1978a, b) and are discussed in the detailed survey by Openshaw and Taylor in Chapter 5 of this volume.

A major research effort has been made by British geographers in the study of spatial dependence in general and its influence on use of the GLM in particular. Since a basic assumption of the regression model is that the error terms must be uncorrelated, or independent, according to intention, it follows that for spatially distributed data the errors should not be spatially dependent. Failure to meet this assumption gives rise to upward or downward bias in estimates of sampling variances of the regression coefficients and of the residual error, according to the form of spatial autocorrelation present (Haggett *et al.*, 1977, pp. 334-6). Methods for coping with spatial dependence are identified in Cliff and Ord (1973, 1975a) and detailed procedures described in Cliff and Ord (1973). Modifications of the test to allow for spatial dependence are discussed in Cliff and Ord (1975b), and Griffith (1978) has generalized their treatment of regression models to the one-way ANOVA fixed effects model. Sen and Sööt (1977) provide a rank test for spatial autocorrelation which is not sensitive to the normality assumption as are the tests generally recommended. For a detailed review see the contribution by Cliff and Ord, Chapter 10.

It has also been realized that in studies based on the unconstrained gravity model, effects of the function of distance and the spatial pattern of points or areas generating and terminating flows may be confounded. Olsson (1970) initially raised the issue, and speculations by Curry (1972) and Johnston (1973) were followed by a lively exchange between British and North American 'teams' in Regional Studies (Cliff *et al.*, 1974; Johnston, 1975; Curry *et al.*, 1975; Cliff *et al.*, 1975; Sheppard *et al.*, 1976; Cliff *et al.*, 1976).

Applications of the general linear model to the study of spatial and temporal series

The first group of applications consists of map generalization techniques (Bassett, 1972) which explicitly incorporate a spatial component alone, and the second of space-time models used chiefly for prediction and forecasting.

Map generalization techniques

The most popular technique has been *trend surface analysis,* first introduced in the geographic literature by Chorley and Haggett (1965). Both its popularity and early introduction can be attributed to previous widespread use in geology and availability in texts (Miller and Kahn, 1962; Krumbein and Graybill, 1965), availability of computer routines, and its ability to produce isopleth maps. The polynomial regression terms in combinations of 'eastings' and 'northings' of a Cartesian coordinate system are taken to represent large-scale or regional trend components, and the error term small-scale or local components, in map patterns of the dependent variable (Krumbein, 1956). However, Robinson (1970) pointed out that systematic local covariation and purely random variation at a point may be confounded in the error term. A further problem is that although geological processes could be credibly invoked to account for spatial variations at both scales, this proved difficult for socioeconomic data where 'with a commingling of processes at many scales, a distinction between large-scale and local processes may be particularly arbitrary . . .' (Bassett, 1972, p. 228). Despite these reservations, the technique has been used by human geographers, chiefly as a descriptive device, and includes comparison of map patterns in different regions (Bassett, 1972; Cliff, Haggett *et al.*, 1975). More examples are to be found in physical geography and for reviews and references at various stages see Chorley and Haggett (1965), Doornkamp (1972) and Unwin (1977).

In geology the technique has been extended to generate canonical surfaces so that simultaneous variation in several dependent variables may be studied (Lee, 1969), and in geography to handle dichotomous and polychotomous dependent variables allowing construction of probability surface maps (Wrigley, 1977a, b).

Many logical, statistical and computational problems thrown up by applications of trend surface analysis have been explored during the 1970s (Unwin, 1977, p. 191). The arbitrary nature of many fitted surfaces is shown by the high R^2 values sometimes obtained for purely random data values when there are few data points (Howarth, 1967; Norcliffe, 1969; Unwin, 1970), and by marked differences between surfaces roughly equivalent with respect to the number of terms and R^2 values (Miesch and Connor, 1968). Effects of the spatial pattern of data points, and at the boundary of the study area, have also attracted attention (Miesch and Connor, 1968; Doveton and Parsley, 1970; Davis, 1973; Unwin, 1975a) and recent work has shown how coefficients may be derived which are invariant under translation of origin and rotation of the coordinate grid (Cliff and Kelly, 1977), facilitating intraregional comparisons (over time) and interregional comparison. Cliff, Haggett *et al.* (1975, Chapter 4) also show that over 2,000 different surfaces may be generated by setting combinations of the coefficients of a cubic polynomial equation to zero. In human geography, it is recognized that trend interpolation is hampered because of the modifiable units problem and consequent arbitrary location of control points (Johnston, 1978, p. 93). As the order of the specified surface increases, more accurate parameter estimates may be obtained using orthogonal polynomials (Whitten, 1970, 1972), but multicollinearity is still not avoided and Mather (1976, p. 142) suggests experiments with ridge regression in this area.

Other spatial surface-fitting techniques based on the GLM may be considered under the general heading of *spectral analysis,* used where it is thought that the surface is characterized by saptial periodicities whereas trend surface analysis describes smoothly varying surfaces without regular changes in direction and shape. Rayner (1971) provides an introduction with an excellent classification reference section, and there are reviews in Bassett (1972), Cliff, *et al.* (1975), Haggett *et al.* (1977).

Other techniques are also available. *Kriging* (Matheron, 1970; Watson, 1971; Delfiner and Delhomme, 1975) is a spatial interpolation technique with a statistical base and has been used in physical geography (Hutchinson, 1974; McCullagh, 1975). For cartographic applications of this and other techniques see Rhind (1975).

Space-time modelling

Spatial, and temporal, dependencies could be regarded either as nuisances when applying classical statistical techniques to geographic data, or as worth modelling in their own right. The 'nuisance aspect' has already been covered, and British geographers have adapted a number of formal modelling techniques developed in other disciplines to the study of space-time processes (Cliff, 1977), or what Curry (1970, p. 241) called 'forecasting future maps'. This work is described in Cliff *et al.* (1975), Haggett *et al.* (1977) and for a planning audience in Openshaw (1978c). Work on time series and spatial time series is reviewed by Hepple (Chapter 8) and Bennett (Chapter 9) respectively elsewhere in this volume. These developments appear to have been encouraged by traditional interests in spatial variation as expressed in simulation modelling of spatial diffusion processes begun by Hägerstrand (1952) and by the desire to use more elegant analytic techniques, so increasing the discipline's academic respectability. A powerful stimulus has also come from the geographer's desire to contribute to public policy formulation (Chisholm *et al.*, 1971; Coppock, 1974; Harvey, 1974). Thus, much of the work is concerned with forecasting (Haggett *et al.*, 1977, Chapter 16), with implications for our understanding and possible control of phenomena which have only recently been given much attention in geography (Olsson, 1975; Bennett and Chorley, 1978; Silk, 1979b). This is in addition to the need for sound theory on the structure of the space-time processes involved if these approaches are not to lapse into curve-fitting exercises (Cliff, 1977, p. 495).

Conclusions

Classical uses of the GLM, as defined above, have been strongly in evidence in British geography since the late 1960s. The unglamorous but essential task of providing technical back-up is being carried out by an increasing number of (mainly British) elementary and advanced texts, and by contributions to the CATMOG series sponsored by the Quantitative Methods Study Group of the IBG. At the time of writing, contributions on path analysis and causal modelling, on experimental design, and on the generalized linear model - perhaps in the CATMOG series - would be very welcome. The evidence of this volume alone suggests that there are still enough quantitative *cognoscenti* in British geography to provide the major route by which developments, whether based on the GLM or other modelling techniques, are made known to all workers in the discipline. However, because those at the research frontier have, I would guess, less time or inclination to write 'nuts and bolts' texts, and the supply of new quantifiers may have declined (Taylor, 1973), it may not be easy to find individuals to write such books and papers. A further difficulty arises at a higher conceptual level. Elementary texts - if they deal with the question at all - tend to recommend testing of prior hypotheses, whereas the picture becomes far less clear-cut at more advanced levels. For example Johnston (1978, p. 252) implies that regression and ANOVA models are best used in this manner, but Ferguson (1978, p. 30) argues that most geographic applications of regression analysis have been exploratory. Similarly contrasting views have been expressed with respect to trend surface analysis (Norcliffe, 1969, pp. 340-1; Bassett, 1972, p. 232). There is certainly no easy answer to this problem, but revival and extension of the discussion in Harvey (1969) would be highly desirable.

Texts are available describing applications of the GLM to spatial and/or temporal data (Cliff *et al.*, 1975; Haggett *et al.*, 1977), but neither provides a treatment at the 'nuts and bolts' level, except the CATMOGs on trend surface analysis by Unwin (1975b) and Wrigley (1977b). It is perhaps too much to expect this particular gap to be filled yet, as most developments began in the 1970s and are still under way, but this remains a vital task for the 1980s. An important feature of many of the space-time models is that, apart from their technical complexity, careful specification of their structure is required. This requirement applies as much, if not more, to techniques such as path and causal analysis, simultaneous-equation regression analysis, complex Analysis of Variance and Covariance designs and to procedures available for fitting the generalized linear model. A poorly thought-out, poorly-structured analysis is far more difficult to carry out when such techniques are used. An increase in their use during the 1980s would be most welcome on these grounds, and should compensate for fears expressed by Cooke and Robson (1976, p. 84) that there is as much lack of comprehension between first- and second-generation quantifiers as ever there was between quantifiers and non-quantifiers of the 1960s.

The leading role of human geographers in introducing or developing new techniques for classical or space-time analysis in the subject seems likely to continue, although there appear to be ample opportunities for their application in physical geography (Unwin, 1977, pp. 202-5).

References

Ahmed, Q. (1965) *Indian cities: characteristics and correlates*, University of Chicago, Department of Geography Research Paper 102.

Barker, D. (1976) 'Hierarchic and non-hierarchic grouping methods: an empirical analysis of two techniques', *Geographiska Annaler* B, 42-58.

Bassett, K. (1972) 'Numerical methods for map analysis' in C. Board, R.J. Chorley, P. Haggett and D.R. Stoddart (eds), *Progress in Geography*, 4, 217-54, Arnold: London.

Batty, M. (1976) *Urban modelling*, Cambridge University Press.

Bennett, R.J., and Chorley, R.J. (1978) *Environmental systems: philosophy, analysis and control*, Methuen: London.

Berry, B.J.L. (1966) *Essays on commodity flows and the spatial structure of the Indian economy*, University of Chicago, Department of Geography Research Paper 111.

Blalock, Jr H.M. (1964) *Causal inferences in non experimental research*, University of North Carolina Press: Chapel Hill.

Bonnar, D. (1979) 'Life cycle factors and housing choice in the private sector: a temporal economic study using path analysis' in I.G. Cullen (ed.) *Analysis and decision in regional policy*, 64-84, Pion: London.

Brunsden, D. (ed.) (1971) *Slopes: form and process*, special publication No. 3, Institute of British Geographers.

Casetti, E. (1964) 'Multiple discriminant functions', *Technical Report No. 11*, Computer applications in Earth Sciences Project, Department of Geography, Northwestern University: Evanston, Ill.

Casetti, E. (1966) 'Analysis of spatial association by trigonometric polynomials', *The Canadian Geographer*, 10, 199-204.

Chisholm, M.D.I., Frey, A.E., and Haggett, P. (eds) (1971) *Regional forecasting*, Butterworth: London.

Chorley, R.J. (1966) 'The application of statistical methods to geomorphology' in G.H. Dury (ed.) *Essays in Geomorphology*, 275-387, Heinemann: London.

Chorley, R.J., and Haggett, P. (1965) 'Trend-surface mapping in geographical research', *Transactions and Papers of the Institute of British Geographers*, 37, 47-67.

Clark, D. (1973) 'Urban linkage and regional structure in Wales: an analysis of change', *Transactions of the Institute of British Geographers*, 58, 41-58.

Clark, D. (1975) *Understanding canonical correlation analysis. Concepts and Techniques in Modern Geography* 3, Geoabstracts: Norwich.

Clark, D., Davies, W.K.D., and Johnston, R.J. (1974) 'The application of factor analysis in human geography', *The Statistician*, 23 (3/4), 259-82.

Clark, W.A.V., and Avery, K.L. (1976) 'The effects of data aggregation in statistical analysis', *Geographical Analysis*, 8, 428-38.

Cliff, A.D. (1977) 'Quantitative methods: time series methods for modelling and forecasting', *Progress in Human Geography* 1 (3), 492-502.

Cliff, A.D., and Kelly, F.P. (1977) 'Regional taxonomy using trend-surface coefficients and invariants', *Environment and Planning*, 9, 945-55.

Cliff, A.D., and Ord, J.K. (1973) *Spatial autocorrelation*, Pion: London.

Cliff, A.D., and Ord, J.K. (1975a) 'Model building and the analysis of spatial pattern in human geography', *Journal of the Royal Statistical Society* B, 37, 297-348.

Cliff, A.D., and Ord, J.K. (1975b) 'The comparison of means when samples consist of spatially autocorrelated observations', *Environment and Planning* A, 7, 725-34.

Cliff, A.D., Martin, R.L., and Ord, J.K. (1974) 'Evaluating the friction of distance parameter in gravity models', *Regional Studies* 8, 281-6.

Cliff, A.D., Martin, R.L., and Ord, J.K. (1975) 'Map pattern and friction of distance parameters: reply to comments by R.J. Johnston, and by L. Curry, D.A. Griffith and E.S. Sheppard', *Regional Studies*, 9, 285-8.

Cliff, A.D., Martin, R.L., and Ord, J.K. (1976) 'A reply to the final comment', *Regional Studies*, 10, 341-2.

Cliff, A.D., Haggett, P., Ord, J.K. Bassett, K.A., and Davies, R.B. (1975) *Elements of Spatial Structure*, Cambridge University Press.

Colenutt, R.J. (1968) 'Building linear predictive models for urban planning', *Regional Studies*, 2, 139-43.

Cook, T., and Falchi, P. (1979) 'Regression analysis and geographic models: a comment', *Canadian Geographer*, 23, 71-5.

Cooke, R.U., and Robson, B.T. (1976) 'Geography in the United Kingdom 1972-76', *Geographical Journal*, 142, 3-22.

Coppock, J.T. (1974) 'Geography and public policy: challenges, opportunities and implications', *Transactions of the Institute of British Geographers*, 63, 1-16.

Corsi, T., and Harvey, M.E. (1975) 'The socio-economic determinants of crime in the city of Cleveland; the application of canonical scores to geographical processes', *Tijdschrift voor Economische en Sociale Geographie*, 66, 323-36.

Cox, K.R. (1968) 'Suburbia and voting behaviour in the London metropolitan area', *Annals of the Association of American Geographers*, 58, 111-27.

Cox, K.R. (1969) 'Comments in reply to Kasperson and Taylor', *Annals of the Association of American Geographers*, 59, 411-15.

Curry, L. (1966) 'A note on spatial association', *The*

Professional Geographer, 18, 97-9.
Curry, L. (1970) 'Univariate spatial forecasting', *Economic Geographer*, 46, 241-58.
Curry, L. (1972) 'A spatial analysis of gravity flows', *Regional Studies* 6, 131-47.
Curry, L., Griffith, D.A., and Sheppard, E.S. (1975) 'Those gravity parameters again', *Regional Studies*, 9, 289-96.
Daniel, C., and Wood, F.S. (1971) *Fitting equations to data*, Wiley-Interscience: New York.
Daultrey, S. (1976) *Principal components analysis. Concepts and techniques in modern geography* 8, Geoabstracts: Norwich.
Davidson, N. (1976) *Causal inferences from dichotomous variables. Concepts and techniques in modern geography* 9, Geoabstracts: Norwich.
Davis, J.C. (1973) *Statistics and data analysis in geology*, Wiley: New York.
Delfiner, P., and Delhomme, J.P. (1975) 'Optimum interpolation by Kriging', in J.C. Davis and M.J. McCullagh (eds), *Display and Analysis of Spatial Data*, Wiley: London.
Domencich, T.A., and McFadden, D.L. (1975) *Urban travel demand: a behavioural analysis*, North-Holland: Amsterdam.
Doornkamp, J.C. (1972) 'Trend surface analysis of plantation surfaces, with an East African case study' in R.J. Chorley (ed.) *Spatial analysis in geomorphology*, Methuen: London.
Doveton, J.H., and Parsley, A.J. (1970) 'Experimental evaluation of trend surface distortions produced by inadequate data point distributions', *Transactions of Institute of Mining and Metallurgy*, B197-B208.
Draper, N.R., and Smith, H. (1966) *Applied regression analysis*, Wiley: New York.
Duncan, O.D. (1966) 'Path analysis: sociological examples', *American Journal of Sociology*, 72, 1-16.
Duncan, O.D. (1975) *An introduction to structural equations*, Academic Press: New York.
Ferguson, R. (1978) *Linear regression in geography. Concepts and techniques in modern geography* 15. Geoabstracts: Norwich.
Gauthier, H.L. (1968) 'Transportation and the growth of the Sao Paulo economy', *Journal of Regional Science*, 8, 77-94.
Goddard, J.B., and Kirby, A.M. (1976) *An introduction to factor analytical techniques. Concepts and techniques in modern geography* 7. Geoabstract: Norwich.
Golledge, R.G., and Spector, A.N. (1978) 'Comprehending the urban environment: theory and practice', *Geographical Analysis*, 10, 386-402.
Gould, P.R. (1970) 'Is *statistix inferens* the geographical name for a wild goose?', *Economic Geography*, 46 (Supplement), 439-48.
Gregory, K.J., and Walling, D.E. (1973) *Drainage basin form and process: a geomorphological approach*, Arnold: London.
Griffith, D.A. (1978) 'A spatially adjusted ANOVA model', *Geographical Analysis*, 10, 296-301.
Hägerstrand, T. (1952) 'The propagation of innovation waves', *Land Studies in Geography, B, Human Geography*, 4, 3-19.
Haggett, P. (1964) 'Regional and local components in the distribution of forested areas in south east Brazil: a multivariate approach', *Geographical Journal*, 130, 365-80.
Haggett, P. (1965) *Locational analysis in human geography*, Arnold: London.
Haggett, P., Cliff, A.D., and Frey, A.E. (1977) *Locational Methods*, Vol. 2 of *Locational analysis in human geography*, (2nd edn), Arnold: London.
Harvey, D.W. (1969) *Explanation in geography*, Arnold: London.
Harvey, D.W. (1974) 'What kind of geography for what kind of public policy?', *Transactions of the Institute of British Geographers*, 63, 18-21.
Hauser, D.P. (1974) 'Some problems in the use of stepwise regression techniques in geographical research', *Canadian Geographer*, 18, 148-58.
Hepple, L.W. (1974) 'The impact of stochastic process theory upon spatial analysis in human geography' in C. Board, R.J. Chorley, P. Haggett, D.R. Stoddart (eds), *Progress in Geography*, 6, 89-142.
Hoerl, A.E., and Kennard, R.W. (1970a) 'Ridge regression: biased estimation for nonorthogonal problems', *Technometrics*, 12, 55-67.
Hoerl, A.E., and Kennard, R.W. (1970b) 'Ridge regression: applications to nonorthogonal problems', *Technometrics*, 12, 69-82.
Horton, R.L. (1978) *The general linear model*, McGraw-Hill: New York.
Howarth, R.J. (1967) 'Trend surface fitting to random data - an experimental test', *American Journal of Science*, 265, 619-25.
Howarth, R.J. (1973) 'Preliminary assessment of a non-linear mapping algorithm in a geological context', *Mathematical Geology*, 5, 39-57.
Hutchinson, P. (1974) 'Progress in the use of the method of optimum interpolation for the design of the Zambian raingauge network', *Geoforum*, 20, 49-62.
Johnston, R.J. (1973) 'On frictions of distance and regression coefficients', *Area*, 5, 187-91.
Johnston, R.J. (1975) 'Map pattern and friction of distance parameters', *Regional Studies*, 9, 281-3.
Johnston, R.J. (1976) 'Political behaviour and the residential mosaic', in D.T. Herbert and R.J. Johnston (eds) *Social areas in cities Vol. 2* John Wiley: Chichester.
Johnston, R.J. (1978) *Multivariate statistical analysis in geography*, Longman: London.
Keeble, D.E., and Hauser, D.P. (1972) 'Spatial analysis of manufacturing growth in outer South-East England 1960-1967. II Method and results', *Regional Studies*, 6, 11-36.
King, L.J. (1967) 'Discriminatory analysis of urban

growth patterns in Ontario and Quebec, 1951-1961', *Annals of the Association of American Geographers*, 57, 566-78.

King, L.J. (1969) *Statistical analysis in geography*, Prentice-Hall: Englewood Cliffs, NJ.

Krumbein, W.C. (1956) 'Regional and local components of facies maps', *Bulletin of the American Association of Petroleum Geologists*, 40, 2163-94.

Krumbein, W.C. (1967) 'The general linear model in map preparation and analysis' in D.F. Merriam and W.C. Cocke (eds), *Colloquium on trend analysis*, University of Kansas, State Geological Survey, Computer Contribution 12.

Krumbein, W.C., and Graybill, F.A. (1965) *An introduction to statistical models in geology*, McGray-Hill: New York.

Lankford, P.M. (1974) 'Testing simulation models', *Geographical Analysis* 6, 294-302.

Lee, P.J. (1969) 'The theory and application of canonical trend surfaces', *Journal of Geology*, 77, 303-18.

McCarty, H.H. (1956) 'Use of certain statistical procedures in geographical analysis', *Annals of the Association of American Geographers*, 46, 263.

McCullagh, M.J. (1975) 'Estimation by Kriging of the proposed Trent telemetry network', *Computer Applications*, 2, 357-74.

Mark, D.M., and Peucker, T.K. (1978) 'Regression analysis and geographic models', *Canadian Geographer*, 22, 51-64.

Mark, D.M., and Peucker, T.K. (1979) 'Regression analysis and geographic models: reply', *Canadian Geographer*, 23, 79-81.

Mather, P.M. (1976) *Computational methods of multivariate analysis in physical geography*, Wiley: London.

Mather, P.M., and Doornkamp, J.C. (1970) 'Multivariate analysis in geography with particular reference to drainage basin morphology', *Transactions of the Institute of British Geographers*, 51, 163-87.

Mather, P.M., and Openshaw, S. (1974) 'Multivariate methods and geographical data', *The Statistician*, 23, 283-308.

Matheron, G. (1970) 'Random fluctuations and their application in geology' in D.F. Merriam (ed.), *Geostatistics*, Plenum: New York.

Mercer, J. (1975) 'Metropolitan housing quality and an application of causal modelling', *Geographical Analysis*, 7, 295-302.

Miesch, A.T., and Connor, J.J. (1968) 'Stepwise regression and nonpolynomial models in trend analysis', University of Kansas, State Geological Survey, Computer Contribution 27.

Miller, R.L., and Kahn, J.S. (1962) *Statistical analysis in the geological Sciences*, Wiley: New York.

Moore, E.G. (1969) 'The structure of intra-urban movement rates: an ecological model', *Urban Studies*, 6, 17-23.

Moriarty, B.M. (1973) 'Causal inferences and the problem of nonorthogonal variables', *Geographical Analysis*, 5, 55-61.

Murdie, R.A. (1969) *Factorial ecology of metropolitan Toronto 1951-1961: An essay on the social geography of the city*, University of Chicago, Department of Geography Research Paper 116.

Nash, J.E., and Shaw, B.L. (1966) 'Flood frequency as a function of catchment characteristics', *Institute of Civil Engineers, Symposium on River Flood Hydrology*, 115-36, London.

Nelder, H.A., and Wedderburn, R.W.M. (1972) 'Generalized linear models', *Journal of the Royal Statistical Society* A, 135, 370-84.

Norcliffe, G.B. (1969) 'On the uses and limitations of trend surface models', *Canadian Geographer*, 13, 338-48.

Norcliffe, G.B. (1977) *Inferential statistics for geographers*, Hutchinson: London.

Norris, J.M., and Barkham, J.P. (1970) 'A comparison of some Cotswold beechwoods using multiple-discriminant analysis', *Journal of Ecology*, 58, 603-19.

Olsson, G. (1965) 'Distance and human interaction: a review and bibliography', *Regional Science Research Institute, Bibliography Series* 2.

Olsson, G. (1970) 'Exploration prediction, and meaning variance: an assessment of distance interaction models', *Economic Geography*, 46 (Supplement), 223-31.

Olsson, G. (1975) *Birds in Egg*, Department of Geography, University of Michigan: Michigan Geographical Publication 15, Ann Arbor.

Openshaw, S. (1977a) 'A geographical solution to scale and aggregation problems in region-building, partitioning and spatial modelling', *Transactions of the Institute of British Geographers*, NS 2, 459-72.

Openshaw, S. (1977b) 'Optimal zoning systems for spatial interaction models', *Environment and Planning* A, 9, 169-84.

Openshaw, S. (1978a) 'An optimal zoning approach to the study of spatially aggregated data' in I. Masser and P.J.B. Brown (eds), *Spatial representation and spatial interaction*, 95-113, Nijhoff: Leiden.

Openshaw, S. (1978b) 'An empirical study of some zone-design criteria', *Environment and Planning* A, 10, 781-94.

Openshaw, S. (1978c) *Using models in planning*, Retail and Planning Associates: Corbridge.

Pinch, S.P. (1978) 'Patterns of local authority housing allocation in Greater London between 1966 and 1973: an inter-borough analysis', *Transactions of the Institute of British Geographers*, NS 3, 35-54.

Poole, M.A., and O'Farrell, P.N. (1971) 'The assumptions of the linear regression model',

Transactions of the Institute of British Geographers, 52, 145-58.

Pooley, C.G. (1977) 'The residential segregation of migrant communities in mid-Victorian Liverpool', *Transactions of the Institute of British Geographers,* NS 2, 364-82.

Rayner, J.N. (1971) *An introduction to spectral analysis,* Pion: London.

Rhind, D. (1975) 'A skeletal overview of spatial interpolation techniques', *Computer Applications* 2, 293-309.

Richards, K.S. (1977) 'Channel and flow geometry: a geomorphological perspective', *Progress in Physical Geography* 1, 65-102.

Robinson, A.H. (1956) 'The necessity of weighting values in correlation of areal data', *Annals of the Association of American Geographers,* 46, 233-36.

Robinson, G. (1970) 'Some comments on trend-surface analysis', *Area,* 3, 31-6.

Sen, A.K., and Sööt, S. (1977) 'Rank tests for spatial correlation', *Environment and Planning* A, 897-903.

Sheppard, E.S. (1979) 'Regression and functional analysis', *Canadian Geographer,* 23, 75-9.

Sheppard, E.S., Griffith, D.A., and Curry, L. (1976) 'A final comment on mis-specification and auto-correlation in those gravity parameters', *Regional Studies,* 10, 337-43.

Silk, J.A. (1976) 'A comparison of regression lines using dummy variable analysis', Geographical Paper 44, Department of Geography, University of Reading.

Silk, J.A. (1979a) *Analysis of covariance and comparison of regression lines. Concepts and techniques in modern geography* 20, Geoabstracts: Norwich.

Silk, J.A. (1979b) 'Quantitative developments in geography and planning in a practical reasoning framework' in B.G. Goodall and A.M. Kirby (eds), *Resources and Planning,* 139-53, Pergamon: Oxford.

Silk, J.A. (1981) *The Analysis of Variance. Concepts and techniques in modern geography.* Geoabstracts: Norwich.

Smit, B. (1978) 'On aggregation and statistical analysis', *Geographical Analysis,* 10, 190-3.

Spence, N.A., and Taylor, P.J. (1970) 'Quantitative methods in regional taxonomy', *Progress in Geography,* 2, 1-64.

Stocking, M.A., and Elwell, H.A. (1976) 'Rainfall erosivity over Rhodesia' *Transactions Institute of British Geographers,* NS 1, 231-45.

Stopher, P.R., and Meyburg, A.H. (1975) *Urban transportation modelling and planning,* D.C. Heath: Lexington.

Sturgal, J.R., and Aiken, C. (1970) 'The best plane through data', *Mathematical Geology,* 2, 325-31.

Taylor, P.J. (1969) 'Causal models in geographic research', *Annals of the Association of American Geographers,* 59, 402-4.

Taylor, P.J. (1973) 'Whither spatial analysis?', *Area,* 5, 312-15.

Taylor, P.J. (1975) *Distance decay in spatial interactions. Concepts and methods in modern geography* 2, Geoabstracts: Norwich.

Taylor, P.J. (1977) *Quantitative methods in geography: an introduction to spatial analysis,* Houghton Mifflin: Boston.

Thomas, E.N. (1968) 'Maps of residuals from regression' in B.J.L. Berry and D.F. Marble (eds), *Spatial analysis,* Prentice Hall: Englewood Cliffs, New Jersey. 326-52.

Thomas, E.N., and Anderson, D.L. (1965) 'Additional comments on weighting values in correlation analysis of areal data', *Annals of the Association of American Geographers,* 55, 492-505.

Till, R. (1973) 'The use of linear regression in geomorphology', *Area,* 5, 303-8.

Tobler, W.R. (1969) 'A computer movie simulating urban growth in the Detroit region', Paper prepared for meeting of the IGU Commission on Quantitative Methods, Ann Arbor.

Todd, D. (1979) *An introduction to the use of simultaneous equation regression analysis in geography. Concepts and techniques in modern geography* 21, Geoabstracts: Norwich.

Unwin, D.J. (1970) 'Percentage RSS in trend surface analysis', *Area,* 2, 25-8.

Unwin, D.J. (1975a) 'Numerical error in a familiar technique: a case study of polynomial trend surface analysis', *Geographical Analysis,* 7, 197-203.

Unwin, D.J. (1975b) *An introduction to trend surface analysis. Concepts and techniques in modern geography* 5, Geoabstracts: Norwich.

Unwin, D.J. (1977) 'Statistical methods in physical geography', *Progress in Physical Geography,* 1, 185-221

Watson, R.S. (1971) 'Trend surface analysis', *Mathematical Geology,* 3, 215-26.

Whitten, E.H.T. (1970) 'Orthogonal polynomial trend surfaces for irregularly spaced data', *Mathematical Geology,* 2, 141-52.

Whitten, E.H.T. (1972) 'More on "irregularly spaced data and orthogonal polynomial trend surfaces"', *Mathematical Geology,* 4, 83.

Wright, S. (1960) 'Path coefficients and path regressions: alternative or complementary concepts?', *Biometrics,* 16, 189-202.

Wrigley, N. (1975) 'Analyzing multiple alternative dependent variables', *Geographical Analysis,* 7, 187-95.

Wrigley, N. (1976) *Introduction to the use of logit models in geography. Concepts and techniques in modern geography* 10, Geoabstracts: Norwich.

Wrigley, N. (1977a) 'Probability surface mapping: a new approach to trend surface mapping', *Transactions of the Institute of British Geographers,* NS 2, 129-40.

Wrigley, N. (1977b) *Probability surface mapping. An introduction with examples and FORTRAN programmes. Concepts and techniques in modern geography* 16, Geoabstracts: Norwich.

Zobler, L. (1958) 'Decision making in regional construction', *Annals of the Association of American Geographers,* 48, 140-8.

Chapter 7

Spatial and temporal analysis: spatial modelling

R.P. Haining

The setting

A fundamental concern in geography is with the description and explanation of the distribution of objects and events in space. From a spatial viewpoint this encompasses at least two distinctive research frameworks. On the one hand studies in the 'areal differentiation' of objects and events on the earth's surface (stressing problems of regionalization, comparative and multivariate analysis) and on the other hand studies in 'spatial relationships' between objects and events (stressing concern for the nature of spatial interdependencies between one or more variables across collections of sites often defined on uniform surfaces).

The so-called quantitative revolution was not allied exclusively to either of these research frameworks and indeed both have their origins prior to this revolution (Taaffe, 1974). What the revolution did further, however, was more rigorous and more objective description and explanation and indeed introduced a range of techniques and methods that gave particular (but by no means exclusive) impetus to the study of spatial relationships which had hitherto been the weaker of the two traditions.

The study of spatial relationships is concerned with the identification of order and regularity in a spatial distribution (the description of spatial form) and also with the explanation of that form in terms of those processes that are responsible for the observed regularity. It is an important assumption of the movement that spatial form reflects the nature of the underlying generating processes and that these processes are themselves 'spatial'.

The acquisition of a scientific outlook allied to this meant that such description and explanation was to be carried out at the highest level of generality. The search was for general methods of description that, inasmuch as they produced quantitative measures of form, permitted broad comparative analyses. Such a programme also included the search for spatial laws and spatial principles with which to explain spatial form.

The description and explanation of spatial form, using an extended range of quantitative techniques and with the acquisition of general scientific principles, characterized what might be called the 'new' geometric tradition of the early 1960s. It was 'new' inasmuch as the analysis of form involved not just the geometric properties of size, length and area (etc.) but the statistical properties of mean, variance and distribution. Similarly the study of process involved not just the elaboration of deterministic models (such as central place models) but also stochastic models (such as stochastic extensions of central place models, diffusion processes, etc.).

The empirical achievements of this project, that had been given post-war methodological expression by Schaefer (1953), were first summarized in Haggett (1965). That book, a landmark in early British quantitative geography, reflected the strength of the new geometric tradition in the United States and, indirectly, the lack of similar developments in Britain.

The geometric tradition embraces the study of point, line and area patterns as well as the underlying processes. Point pattern studies are reviewed elsewhere in this book. This chapter considers only those spatial models relating to a fixed system of counties or cells (or alternatively system of lattice points) that exhaust a region. Associated with each of these counties is a value and the collection of values is assumed to be the realization of some process. It is one aspect of this research tradition to be concerned with describing the spatial variation in values and offering explanations for that variation.

In the next section we shall discuss the history of developments since the mid 1960s and in the final section we will consider the prospects of further research.

A survey of developments since 1965

Acquisition of techniques

The acquisition of new techniques of data analysis was perhaps the most obvious change taking place in the geometric tradition during the early years of the quantitative revolution. The inductive route to theorizing and spatial modelling starts with the collection of accurate data and proceeds through the objective analysis of that data to the identification of structure. New techniques often permit more incisive map analysis and the identification of interpretable structure not apparent with the use of existing techniques. Much of this early work was pursued in a spirit of exploration - seeing how the new concepts fared in the realms of spatial analysis (Hepple, 1974).

The early history of the geometric tradition reflected strongly the pioneers' belief that spatial form was so laden with meaning and insights into the processes responsible for the pattern that data analysis offered the most appropriate route to theory formulation. It was not surprising therefore that acquiring new techniques and learning how to use them would figure prominently. The scope of this acquisition has been summarized in Bassett (1972). It had two main thrusts - the development of techniques for identifying firstly deterministic attributes and secondly stochastic attributes.

One of the earliest techniques used for separating deterministic surface attributes was trend surface analysis (Chorley and Haggett, 1965). Not only could it be used to describe regional surfaces and make comparisons between different regions (Haggett, 1968; Haggett and Bassett 1969) it also offered a method of testing spatial theories and suggesting the nature of underlying causal processes (Chorley, 1969). The configuration of the regional surface was assumed to be an intimate response to the underlying spatial processes and the fitting of linear, quadratic, cubic or higher order surfaces offered one way of identifying these relationships. Eventually included in this range of techniques was harmonic analysis which identified periodic components in spatial data (see for example Harbaugh and Preston, 1968).

These techniques might be broadly classified as identifying deterministic elements in the spatial surface. New techniques including spectral and autocorrelation analysis permitted the analysis of surface properties that were assumed to be the outcome of a stochastic process - such techniques had obvious appeal in a discipline where outcomes, even if they were not always seen as being caused by truly probabilistic events, were seen as the result of a highly complex and ultimately unspecifiable multitude of causal factors.

In Britain this stage of development was led from Bristol and also by those more widely scattered researchers who had received their early training in North America.

Development of process studies

In the 1960s, process studies within the geometric (process/form) tradition were largely of two kinds. On the one hand it was hoped that by describing the form of the spatial surface, using techniques of the kind discussed above, the underlying generating processes might become clearer. On the other hand much research was also carried out on fitting different probability distributions particularly to count data. It soon emerged however that neither provided an unambiguous route to theory construction. Some of the problems were reviewed in a series of papers by Harvey (1966, 1968a, b) and by Dacey (1968) with specific reference to count data obtained from quadrat analysis. Some of the more important problems were to provide research foci in the following years. Firstly there was the problem of choosing a model to fit to the data - often more than one model would do an adequate job in describing the frequency distribution of cell counts so which one should be used? Secondly, if a satisfactory fit could be obtained for any model not only was this fit often dependent on the scale of the sampling grid but also there was no simple way of relating that model to any specific process since in many cases a single probability distribution could be derived from several different processes. Different processes could yield, statistically, the same surface. Thirdly, cell counts were not always independent, so it was not enough simply to model the aspatial frequency distibution of counts, it was also necessary in many cases to account for the spatial arrangement of values across the study region.

The first problem emphasized that the inductivist route to theory construction was highly suspect in the absence of a rigorous set of model specification criteria, a point that was never fully appreciated in the early geographical literature on scientific method. The second problem argued for the development of a more deductivist approach to spatial modelling - a point of view that Harvey (1969) considered at length as a basis for all geographical explanation. The third problem suggested the need for the development of spatial models that specifically included arrangement properties of maps. We will now consider how these three themes have been developed.

Process studies continued to be largely inductive, that is they continued to apply different techniques and fit different probability models to spatial data in order to try and identify the nature of the underlying spatial processes rather than testing process theories from which map pattern properties have been deduced. However, interest did tend to focus more on the analysis of the dependence properties between variates at different sites - a reflection of the importance attached to diffusion and interaction processes in moulding spatial form.

Before discussing examples it is useful to conceptualize map analysis in terms of a three-tiered structure. Map form or map pattern is the surface tier which we observe and on which we base empirical enquiry. This is assumed to be a realization of some probability distribution (the second tier) which is itself assumed to be induced by some spatial process (the first tier). The inductivist route therefore moves from statements about the surface tier of map form to inferences about the nature of the spatial process. (Not all map analysis is of this type, many of the autocorrelation tests for example are tests of map pattern organization rather than tests aimed at making inferences about spatial processes - the relationship between the two is by no means simple.) The deductivist route, by contrast, moves from prior assumptions concerning the nature of the spatial process and deductions concerning the properties of the induced map surface to tests on the map pattern as a basis for confirming or refuting these process assumptions.

Techniques, such as spectral and autocorrelation analysis, already existed as we have seen for investigating dependence in spatial data. In the United States, Tobler (1969) and Rayner and Golledge (1972, 1973) analysed population distribution and settlement patterns using both of the above techniques. In Britain, Cliff *et al.* (1975) used spatial correlograms to suggest the nature of the process that produced recurrent measles outbreaks in the South West of England. The study depended on a visual interpretation of the correlogram to suggest a 'central place plus wave diffusion' mechanism of spatial spread. Other techniques have been developed both here and in the United States for treating further aspects of this problem, notably papers by Batty (1974) and Moellering and Tobler (1972).

A different avenue of research was to try to fit probability distributions to mapped data sets (discrete and real valued variables) that showed evidence of spatial dependence. The rationale for this programme was similar to earlier studies involving quadrat count data. The structure of the model might offer insights into the nature of the underlying spatial interactions. In the early 1970s only the statistical models for real valued variables discussed by Whittle (1954) were referenced in the geographical literature (Granger, 1969; Bassett, 1972) though at this time a number of research papers were appearing by Bartlett and Besag on a class of conditional spatial interaction models that culminated in Besag's paper (1974) to the Royal Statistical Society.

These models were specifically designed by statisticians as tools for identifying structure in spatial data - an area formerly much neglected by statisticians. However, if they were to be useful in an inductive context and not lead to the first type of problem noted above then not only had the range of available models to be defined but also their specification and estimation properties (as well as other properties such as their space-time generators). Such results would be vital if the worst sorts of inferential ambiguities were to be avoided. In the geographical literature Haining (1977, 1978a, b, c, 1979) has considered the appropriateness of these models for analysing different types of geographical maps and has also developed specification and estimation properties for certain models that seem of particular interest. However we should record at this stage the scepticism evident in certain reviews, notably by Granger (1969) concerning the utility of these models for geographical scale processes. His objections are important and we shall return to them in the final section.

There have been few examples for geographers taking a more deductive approach to spatial modelling. In the 1960s, Dacey (1964, 1966) analysed maps of urban settlements in terms of various location models, but then as now most geographers have had a preference for data analysis rather than rigorous model formulation through prior specification of the underlying process. In Britain this tendency parallels the growing interest in problems of regional forecasting (Curry, 1970). The emergence of this interest in the 1970s is in part the result of the discipline's new quest for 'relevance' at a policy level. As a research goal it elevates the methods of data analysis over those of rigorous model formulation through the need to provide answers to difficult and often inherently messy problems. Only the simplest spatial processes are capable at the present time of being given a rigorous formulation and there is a tendency for them to seem trivial and unrealistic when set against the expansive problems of predicting regional unemployment levels and forecasting the space-time evolution of epidemics.

We turn now to the prospects for future research.

Prospects

Many of the problems that emerged in the 1960s and which stimulated later research still await further development. We consider some of these now.

Perhaps one of the most disquieting features of much inductive spatial modelling relates to the inconsistency often evident between the spatial structure of the model and the structure of the space to which it is applied. Little can be gained in an inductive sense from fitting models, ostensibly in order to identify structure in a data set, if the assumptions of the model are at odds with what we know about the processes under study. All the models gathered from the statistical literature assume a stationary process (defined on a homogeneous collection of sites). Implicit in this is the fact that interactions between sites are uniform depending only on lag or distance separation. However in most interesting geographical situations these assumptions are clearly not tenable. On a central place lattice, inter-site associations are a function of site characteristics which are of course differentiated by order. In other cases the structure of a regional space is organized around a small number of dominant nodes which strongly influence the pattern of regional flows and the nature of inter-site variation. The structured nature of geographical space may also undermine the value of some of the techniques mentioned above (such as autocorrelation and spectral analysis) that compute average measures of association based on all pairs of sites separated by a fixed distance.

It was difficulties like this that Granger (1969) referred to when he cast doubt on the applicability of models like those in Whittle (1954). However the case is perhaps more fairly stated by saying that the onus is on those who would use such models to recognize the importance of these assumptions. There are geographical situations where spatial homogeneity and process stationarity are reasonable assumptions. In other cases where the processes are non stationary then it may still be possible to utilize these models in a modified context. Non-stationary means may be extracted by trend surface (Haining, 1978e) or regression (Hepple, 1976) methods leaving the residual to be adequately described by a stationary model. These non-stationary processes are spatially homogeneous. In the case of non-homogeneous lattices such as central place lattices where inter-site associations are determined by order characteristics, modified versions of the above models may be applicable (Haining, 1978d). The important feature about non-homogeneous lattices is the fact that not only may the mean or deterministic component vary spatially but also the interaction or covariance components are not uniform so that different parameters may be required for different sub-collections of sites.

This distinction between homogeneous and non-homogeneous models has important implications for all aspects of spatial modelling. Consider just one area. Specification rules based on averaging techniques like autocorrelation and likelihood ratio tests assume process stationarity. They are important, as has been observed, because only with detailed specification rules can we hope to avoid the more sterile aspects of 'curve fitting'. It is to be hoped that further research will develop criteria particularly for small sample situations. However, what is to be done if the process is non-stationary and the space is non-homogeneous, that is the site interactions are more structured? Averaging techniques are now inappropriate for not only do they compute inter-site properties that are not even useful in a general sense (indeed they may well be meaningless) but they are referencing a class of models that are inconsistent with the nature of the problem. The need is clearly for different types of specification rules and here is where one hopes to see a stronger relationship developing between geographers working in this field and the wider aspects of the discipline. There is scope, for example, for using interaction data in order to specify at least the skeletal structure of inter-site dependencies as a preliminary to fitting specific models. The study of nodal regions in urban geography has proceeded using a variety of different data sets including shopping trip data (Berry, 1967), traffic flow data (Brush, 1953), and telephone call data (Nystuen and Dacey, 1961). Different techniques have been used to identify the principal aspects of the linkage structure - Nystuen and Dacey (1961) for example used graph theoretic methods. Haining (1980) has used findings from Brush (1953), who identified spatial linkages in a system of small urban places in south-western Wisconsin, to model spatial variations in population distribution. It is important for the future development of this area of spatial modelling for there to be a closer integration between model selection and substantive studies of those real world processes that are moulding the structure of the space and in particular the nature of inter-site associations.

Finally it is to be hoped that there will be more interest in the properties of theoretical spatial processes as a basis for making more studied interpretation of correlograms, spectral output and the results of fitting probability distributions. There is always a need for theory to provide justification for the use of techniques and to aid in the interpretation of results, as for example in Curry (1967) on the relationship between spectral analysis and central place theory.

Some processes such as Hägerstrand-type diffusion models are capable of quite deep analysis which may provide a quite clear set of hypotheses for empirical testing (Mollison, 1977). In other cases the nature of the empirical space, or the process itself, may be too complex to make such analysis worthwhile. But there is no disgrace in resorting to simulation methods in order to try to identify the shapes of different diagnostic functions, given different process assumptions, as a preliminary yardstick by which to judge empirical results. Finally, even 'informal' deductive methods, where certain expectations are posited on the basis of informal process analysis, at least help to provide some structure to data analysis (Haining 1978c). In all these ways one may hope to see further progress in the area of spatial modelling.

References

Bassett, K. (1972) 'Numerical methods for map analysis' in C. Board, R.J. Chorley, P. Haggett, D.R. Stoddart (eds), *Progress in Geography*, 4, Arnold: London.

Batty, M. (1974) 'Spatial entropy', *Geographical*

Analysis, 6, 1-31.

Berry, B.J.L. (1967) *The geography of market centres and retail distributions*, Prentice-Hall: Englewood Cliffs, NJ.

Besag, J. (1974) 'Spatial interaction and the statistical analysis of lattice systems', *Journal of the Royal Statistical Society* Ser. B, 36, 192-236.

Brush, J.E. (1953) 'The hierarchy of central places in south-western Wisconsin', *Geographical Review*, 43, 380-402.

Chorley, R.J. (1969) 'The elevation of the lower greensand ridge, south-west England', *Geological Magazine*, 106, 231-308.

Chorley, R.J., and Haggett, P. (1965) 'Trend-surface mapping in geographical research', *Transactions and Papers. Institute of British Geographers*, 37, 47-67.

Cliff, A.D., Haggett, P., Ord, J.K., Bassett, K.A. and Davies, R.B. (1975) *Elements of spatial structure: a quantitative approach*, Cambridge University Press.

Curry, L. (1967) 'Central places in the random space economy', *Journal of Regional Science*, 7 (supplement), 217-38.

Curry, L. (1970) 'Univariate spatial forecasting', *Economic Geography*, 46 (supplement) 241-58.

Dacey, M.F. (1964) 'Modified Poisson probability law for point pattern more regular than random', *Annals of the Association of American Geographers*, 54, 559-65.

Dacey, M.F. (1966) 'A county seat model for the areal pattern of an urban system', *Geographical Review*, 56, 527-42.

Dacey, M.F. (1968) 'An empirical study of the areal distribution of houses in Puerto Rico', *Transactions and Papers. Institute of British Geographers*, 45, 51-69.

Granger, C.W.J. (1969) 'Spatial data and time series analysis' in A.J. Scott (ed.) *London papers in Regional Science, Vol. 1, Studies in Regional Science*, 1-24, Pion: London.

Haggett, P. (1965) *Locational analysis in human geography*, Arnold: London.

Haggett, P. (1968) 'Trend surface mapping in the inter-regional comparison of intra-regional structures', *Papers and Proceedings of the Regional Science Association*, 20, 19-28.

Haggett, P., and Bassett, K.A. (1969) 'The use of trend surface parameters in inter-urban comparisons', *Environment and Planning*, 2, 225-37.

Haining, R.P. (1977) 'Model specification in stationary random fields', *Geographical Analysis*, 9, 107-29.

Haining, R.P. (1978a) *Specification and estimation problems in models of spatial dependence*, Northwestern University Studies in Geography 24, Northwestern University Press: Evanston, Ill.

Haining, R.P. (1978b) 'The moving average model for spatial interaction', *Transactions and Papers: Institute of British Geographers*, NS 3, 202-25.

Haining, R.P. (1978c) 'Estimating spatial interaction models', *Environment and Planning, A*, 10, 305-20.

Haining, R.P. (1978d) 'Interaction modelling on central place lattices', *Journal of Regional Science*, 18, 217-28.

Haining, R.P. (1978e) 'A spatial model for High Plains agriculture', *Annals of the Association of American Geographers*, 68, 493-504.

Haining, R.P. (1979) 'Statistical tests and process generators for random field models', *Geographical Analysis*, 11, 45-64.

Haining, R.P. (1980) 'Intra-regional estimation of central place population parameters', *Journal of Regional Science* 20, 365-75.

Harbaugh, J.W., and Preston, F.W. (1968) 'Fourier series analysis in geology' in B.J.L. Berry and D.F. Marble (eds), *Spatial Analysis: a reader in statistical geography*, Prentice Hall: Englewood Cliffs, NJ.

Harvey, D.W. (1966) 'Geographical processes and point patterns: testing models of diffusion by quadrat sampling', *Transactions and Papers: Institute of British Geographers*, 40, 81-95.

Harvey, D.W. (1968a) 'Some methodological problems in the use of the Neyman Type A and negative binomial probability distributions in the analysis of spatial series', *Transactions and Papers: Institute of British Geographers*, 43, 85-95.

Harvey, D.W. (1968b) 'Pattern, process and the scale problem in geographical research', *Transactions and Papers: Institute of British Geographers*, 45, 71-8.

Harvey, D.W. (1969) *Explanation in Geography*, Arnold: London.

Hepple, L. (1974) 'The impact of stochastic process theory upon spatial analysis in human geography', *Progress in Geography*, 6, 89-142.

Hepple, L. (1976) 'A maximum likelihood model for econometric estimation with spatial series' in T. Masser (ed.), *Theory and Practice in Regional Science*, 90-104, Pion: London.

Moellering, H., and Tobler, W.R. (1972) 'Geographical variances', *Geographical Analysis*, 4, 34-50.

Mollison, D. (1977) 'Spatial contact models for ecological and epidemic spread', *Journal of the Royal Statistical Society Ser. B*, 39, 283-326.

Nystuen, J.D., and Dacey, M.F. (1961) 'A graph theory interpretation of nodal regions', *Papers and Proceedings of the Regional Science Association*, 7, 29-42.

Rayner, J.N., and Golledge, R.G. (1972) 'Spectral analysis of settlement patterns in diverse physical and economic environments', *Environment and Planning*, 4, 347-71.

Rayner, J.N., and Golledge, R.G. (1973) 'The spectrum of US route 40 re-examined', *Geographical Analysis*, 4, 338-50.

Schaefer, F.K. (1953) 'Exceptionalism in geography: a methodological examination', *Annals of the Association of American Geographers*, 43, 226-49.

Taaffe, E.J. (1974) 'The spatial view in context', *Annals of the Association of American Geographers*, 64, 1-16.

Tobler, W.R. (1969) 'The spectrum of US 40', *Papers and Proceedings of the Regional Science Association*, 23, 45-52.

Whittle, P. (1954) 'On stationary processes in the plane', *Biometrika*, 41, 434-49.

Chapter 8

Spatial and temporal analysis: time series analysis

L.W. Hepple

Introduction

In the 1960s time series analysis was not yet a theme in quantitative geography. It was not a topic in Haggett (1965), and was only a possibility for the future in King (1969). During the last decade this situation has changed markedly, and this chapter examines applications in human geography. Parallel developments have taken place in physical geography (notably in hydrology and climatology), and some of these are discussed in Chapter 25.

In human geography, time series analysis has been very closely associated with research on regional economic fluctuation and short-run regional change, and particularly with work on the regional and sub-regional time series data for employment and unemployment that is available for many Western countries. Both Christaller and Lösch discussed the geography of economic fluctuations, and it was a topic in the early post-war years of regional analysis (reviewed in Isard, 1960).

However, it was neglected in quantitative geography during the 1960s (when the emphasis was on static spatial analysis or long-term spatial shifts), until a pioneering paper by the economist Brechling (1967) awakened geographers' interest in time series and regional fluctuations (King, Casetti and Jeffrey, 1969; Bassett and Haggett, 1971).

Time series analysis has a long history, going back to the earliest years of mathematical statistics at the end of the nineteenth century, but it is worth noting that at the very time geographers were becoming interested in the subject, time series analysis was having a new flowering, with rapid developments in several different (and often disparate) directions: new techniques in econometrics, Kalman filters in control engineering, the fast Fourier transform developing spectral methods, and the work of Box and Jenkins (Box and Jenkins, 1970).

Time domain techniques

Most time series studies in quantitative human geography have employed time-domain regression and correlation techniques. Regression methods have been used to model regional and local responses to national economic fluctuations, and to estimate relations between variables within a region (e.g. inflation-unemployment curves (Martin, 1979), and migration-unemployment relations (Bennett, 1975). Correlation-based methods have been used to study the linkages and lags between cities and regions in their reactions to economic fluctuations.

Time series analysis has a very large and sophisticated literature, but it differs from standard statistical analysis in two central respects: its recognition of the problem of autocorrelation in time series and in residuals from time series models and the effect of this on standard inference, and its development of alternative specifications that allow dynamic (lagged) relationships between time series. The earlier time series applications in the geographical literature made only limited use of these techniques, but one of the most notable features about time series applications in geography has been the way researchers have rapidly absorbed the large literature on time series and econometrics, and applied the techniques.

The basic regional response model may be written:

$$y_t = \sum_{k=-m}^{+m} \beta_k x_{t-k} + \epsilon_t \qquad (1)$$

where y_t, t = 1 . . . T, is the time series for employment or unemployment in a town or region, x_t, t = 1 . . . T, is the national series, and ϵ_t is a random disturbance term. The vector of β_k coefficient traces out the regional responses at each lead and lag.

This type of regional response model has been applied in a wide variety of studies, e.g. fluctuations in American cities (King, Casetti and Jeffrey, 1972), Canadian regions (King and Clark, 1978), local areas of South West England (Cliff *et al.*, 1975) and North East England (Hepple, 1975). The recognition of leading or lagging responses has been central to this work (though substantive results differ on how significant lags are), but most studies have assumed a discrete lead or lag, and identified this lead or lag either by trial and error or by peaks in cross-correlation functions. A few studies have used time series techniques for distributed lags. Bennett (1974) and Clark (1979) use Box-Jenkins identification procedures to specify a set of terms in the β_k-vector, while Hepple (1975) uses spectral regression techniques.

Autocorrelation is frequently detected in the residuals from these models, and this can seriously affect the β_k estimates and test-statistics, and increasingly appropriate time series techniques are used to incorporate this autocorrelation. Clark (1978b) and Bennett (1979) use the first-order autoregressive Cochrane-Orcutt transformation, Hepple (1975) uses spectral generalized least squares, while Clark (1979) uses Box-Jenkins methods to identify and fit general autoregressive-moving average (ARMA) models of the form

$$\epsilon_t = \sum_{k=1}^{m} \alpha_k \epsilon_{t-k} + \sum_{j=0}^{n} \delta_j \zeta_t \qquad (2)$$

where ζ_t is a genuinely random disturbance term.

Correlation-type studies of linkages between time series for different cities and regions and the leads-lags between them have used cross-correlation techniques to identify the lags and measure the correlations (King, Casetti and Jeffrey, 1969; Bassett and Haggett, 1971), but such identification is notoriously difficult (because of the autocorrelation structure of the series and the presence of distributed lags), and spectral methods have provided an alternative approach (see next section).

A frequently declared goal of these time series applications is regional forecasting, and this was the title of the major Colston research conference in Bristol in 1970 (Chisholm, Frey and Haggett, 1971). However, very few of the studies have actually produced forecasts or evaluated them. Exceptions are the studies by Bennett (1975) of annual forecasts of unemployment, employment and population in North West England, and studies by Cliff *et al.* (1975) and Hepple (1978) of monthly regional unemployment forecasting. These studies showed that useful short-term forecasts can be produced, though more recently interest has shifted towards monitoring change and control rather than pure forecasting as a goal.

Frequency (spectral) methods

An important perspective in the time series literature has been the viewing of time series as different periodic or quasi-periodic components of different wavelengths (or frequencies). Spectral analysis uses Fourier transform techniques to decompose the variance of a time series into these frequency components, and during the 1960s estimation techniques, notably the fast Fourier transform, were being developed and computational procedures becoming available (Jenkins and Watts, 1968). For economic time series, spectral methods allow measurement of the relative importance of long-wave business cycle fluctuations, shorter sub-cycles, seasonal frequencies, and rapid short-wave fluctuations. Geographical applications have used the spectral analysis of local unemployment series to assess the significance of various frequency-bands in local and regional economic fluctuations (Bartels, 1977; Bassett and Haggett, 1971; Hepple, 1975); for example, Cliff *et al.* (1975) use spectral analysis to measure and map the seasonality of local economies in South West England.

The extension of spectral methods to study relations between time series (cross-spectral analysis) has been important in geographical applications. For example, in regional response modelling, the timing and magnitude of the local response to long-wave (business cycle) national fluctuations may differ from the response to short-wave fluctuations or seasonal cycles, and cross-spectral analysis allows the explicit differentiation of these responses; differentiation that may be obscured in time-domain regression models. For each selected frequency-band, cross-spectral analysis provides estimates of coherence (correlation), gain (regression slope) and phase (time-lag) between the time series. Geographical applications have revealed the importance of such frequency variation, e.g. Hepple (1975) found that rural and market town labour markets in North East England had lower coherence with national cycles at the low frequencies, whereas the major urban and industrial centres had their highest coherences at these frequencies.

Cross-spectral analysis has also been employed to examine the interaction between cities and towns in terms of correlation-links and relative lags at different frequencies and the identification of clusters (Bartels, 1977; Bassett and Haggett, 1971). The development of multivariate matrix spectral analysis (Brillinger, 1975) has led to interregional coherence (correlation) matrices for important frequency-bands (Bartels, 1977; Hepple, 1975), and to the use of spectral principal components analysis and factor analysis to identify clusters at specific frequencies and regional 'reference cycles' (factors).

Frequency-domain models are Fourier transformations of time-domain models, and a recent theme in econometrics and statistics has been to develop closer links between the two perspectives, notably with the development of spectral regression methods (Harvey, 1978, 1980). Bennett (1974, 1979) examined the spectra that corresponded to particular time-domain specifications and then used empirical spectra to identify time-domain models, and Hepple (1975) used spectral techniques to estimate the general distributed lag model with general autocorrelated errors.

Instability and time varying models

The time series models and techniques discussed above assume that, after specific filtering and detrending procedures, the basic model structure (e.g. cyclical responsiveness) is a constant and stable one throughout the sample period. This assumption underlies most economic time series modelling, but the economic upheavals since the late 1960s make such stability a less reasonable presumption than during the comparative macroeconomic stability of the 1950s and 1960s. Recent regional analysis has used three types of approach to incorporate such instability. First, discrete shifts or changes of regime in a relationship can be dealt with by dummy variables and related techniques, and these have been widely used to incorporate shifts in the British labour market in 1966-7. This is a fairly restrictive case, however. The general testing and modelling of parameter evolution and instability is more difficult. In control engineering and econometrics, the last decade has seen the development of recursive least-squares, maximum likelihood and Kalman filter techniques to model directly the changing statistical relationships in time series. The β-vector in the general linear model is made time-dependent.

$$y_t = X_t \beta_t + \epsilon_t$$

and evolution of β_t is governed by an additional statistical relationship, such as the random-walk model

$$\beta_t = \beta_{t-1} + \zeta_t$$

where ζ_t is a K x 1 vector of independent disturbances.

These methods are discussed by Martin (1978) and Bennett (1979). Recursive least-squares tests for stability are applied to regional unemployment responses in the UK by Hepple (1979) and to regional inflation models by Martin (1979), and regional time-varying parameter models are used by Bennett (1975) and Hepple (1979).

Whereas this work leads to a refinement and generalization of the linear statistical model, the third approach tends to reject formal statistical tools. This third approach is based on the National Bureau of Economic Research (NBER) methods of business cycle analysis. The time series are seasonally filtered and then smoothed (by moving averages) to reveal underlying cycles. For each national cycle, local turning points, peaks, troughs, lags and amplitudes are identified, and these may vary from cycle to cycle. This descriptive approach therefore enables one to track changing regional impacts of successive recessions. Recent regional work in this vein includes Clark's study of Canadian cycles (Clark, 1978a), the Rand study of cycles in the USA (Vaughan, 1976) and parts of Cliff *et al.* (1975) for South West England. The NBER approach has long been unfashionable in economics as 'measurement without theory', but it is worth noting that this type of work is now becoming more popular again in macroeconomics (Sims, 1980). However, it is known that moving average and seasonal filtering can seriously distort leads and lags between time series (Nerlove, 1964), and there is also a real danger in NBER methods of overinterpreting the smoothed series and treating all differences as 'significant'. There is now a need to relate the NBER approach to the statistical time-varying parameter tests and models, and to examine their different assumptions.

Prospects

Turning now to the major promising directions for future time series research in quantitative geography, these can be discussed under two main themes: (a) further econometric and technical advances, and (b) applications, and particularly policy applications.

First, in terms of econometric and time series techniques, it is vital to note that this is still a very active and fast developing research field, and regional time series work can benefit greatly from these current developments. A notable feature of recent developments has been the convergence and bringing together of different strands in time series modelling, an integration that is in great contrast to the divergent trends in the late 1960s. This convergence should lead to a more unified approach in geographical applications. Particularly relevant is the recent work that has related the engineering-oriented time-varying parameter techniques to standard econometric methods and tests (Harvey, 1980), and the integration of the general linear model with frequency-domain (spectral) techniques in terms of spectral regression, band-limited regression and diagnostic tests (Harvey, 1978).

The other major avenue of research is likely to be policy-oriented applications. Much of the initial stimulus for time series analysis in human geography came from a concern with applied questions - much more so than in most areas of quantitative geography - but the policy links are likely to be strongly reinforced. The regional and spatial dimensions of economic policy become of greater importance each year, and regional time series analysis can provide valuable studies of both the dynamic problems and the evaluation of policy impacts. Despite the applied origins of regional time series analysis, the main effort during the 1970s decade has gone into evaluating alternative ways of modelling the available data and developing expertise in time series modelling techniques. This developed experience now gives a solid basis for useful applied work. 'Policy relevant' applications will, of course, mean different things to different groups of geographers: already several politically-based standpoints can be seen, such as the radical economics implicit in Clark (1978a), the liberal position (e.g. Vaughan, 1976), and the monetarist perspectives in studies of regional monetary time series in the USA.

Relevant to all three positions is the potential use of time series techniques for 'spatial monitoring' of the economy and of the impact of policies. In the formulation and evaluation of spatial policy it is extremely unlikely that new data will emerge rapidly, and so it is vital to make full use of the rich time series data already available, even though they are limited to labour market variables. The regional time series work gives a basis for such spatial monitoring, not only through structural and forecasting models, but also (and equally important) through monitoring the 'non-fit' or deviation models as they are updated and so detecting the emergence of new economic black spots and problem regions. One of the powers of regional time series analysis is the ability to detect and discriminate very specific features as well as broad structural responses in the space-economy, to detect shifts in responses and impacts in very small local economies as well as modelling large-scale regional responses. In developing these policy-oriented applications it will be important to consolidate links with areas such as manpower economics and regional econometric modelling (Glickman, 1977).

These contemporary economic applications are likely to be the main geographical focus of time series studies in the 1980s, but other areas of application should not be forgotten. Although data series are limited, there is valuable scope for time series applications in historical geography, to studies of regional linkage and the evolution of spatial integration in the economy using employment and price series (Hepple, 1975), and the spatial diffusion of epidemics is a solidly-founded area (Cliff and Haggett, 1979).

References

Bartels, C.P.A. (1977) 'The structure of regional unemployment in the Netherlands. An exploratory statistical analysis', *Regional Science and Urban Economics,* 7, 103-35.

Bassett, K.A., and Haggett, P. (1971) 'Towards short-term forecasting for cyclic behaviour in a regional system of cities' in M. Chisholm, A.E. Frey and P. Haggett (eds), *Regional forecasting,* 389-413, Butterworth: London.

Bennett, R.J. (1974) 'Process identification for time series modelling in urban and regional planning', *Regional Studies,* 8, 157-74.

Bennett, R.J. (1975) 'Dynamic systems modelling of the North-west region: 4. adaptive spatio-temporal forecasts', *Environment and Planning A* 7, 887-98.

Bennett, R.J. (1979) *Spatial Time Series,* Pion: London.

Box, G.E.P., and Jenkins, G.M. (1970) *Time Series Analysis, Forecasting and Control,* Holden-Day: San Francisco.

Brechling, F.P.R. (1967) 'Trends and cycles in British regional unemployment', *Oxford Economic Papers,* 19, 1-21.

Brillinger, D.R. (1975) *Times series: data analysis and theory,* Holt, Rinehart & Winston: New York.

Chisholm, M., Frey, A.E., and Haggett, P. (eds) (1971) *Regional forecasting,* Butterworth: London.

Clark, G.L. (1978a) 'The political business cycle and the distribution of regional unemployment', *Tijdschrift voor Economische en Sociale Geografie,* 69, 154-64.

Clark, G.L. (1978b) 'Regional labour supply and national fluctuations: Canadian evidence for 1969-1975', *Environment and Planning A* 10, 621-32.

Clark, G.L. (1979) 'Predicting the regional impact of a full employment policy in Canada: a Box-Jenkins approach', *Economic Geography,* 55, 213-26.

Cliff, A.D., and Haggett, P. (1979) 'Geographical aspects of epidemic diffusion in closed communities' in N. Wrigley, op. cit., 5-44.

Cliff, A.D., Haggett, P., Ord, J.K., Bassett, K., and Davies, R. (1975) *Elements of spatial structure,* Cambridge University Press.

Glickman, N.J. (1977) *Econometric analysis of regional systems,* Academic Press: New York.

Haggett, P. (1965) *Locational analysis in human geography,* Edward Arnold: London.

Harvey, A.C. (1978) 'Linear regression in the frequency domain', *International Economic Review,* 19, 507-12.

Harvey, A.C. (1980) *Time series analysis and econometrics,* Philip Allan: Dedington, Oxford.

Hepple, L.W. (1975) 'Spectral techniques and the study of interregional economic cycles' in R. Peel, M. Chisholm and P. Haggett (eds), *Process in physical and human geography: Bristol essays,* 392-408, Heinemann: London.

Hepple, L.W. (1978) 'Forecasting the economic recession in Britain's depressed regions' in R.L. Martin, N.J. Thrift and R.J. Bennett (eds),

Towards the dynamic analysis of spatial systems, 172-90, Pion: London.

Hepple, L.W. (1979) 'Regional dynamics in British unemployment, and the impact of structural change' in N. Wrigley op. cit., 45-63.

Isard, W. (1960) *Methods of regional analysis: an introduction to regional science,* MIT Press: Cambridge, Mass.

Jenkins, G.M., and Watts, D.G. (1968) *Spectral analysis and its applications.* Holden-Day: San Francisco.

King, L.J. (1969) *Statistical analysis in geography,* Prentice-Hall: Englewood Cliffs, NJ.

King, L.J., and Clark, G.L. (1978) 'Regional unemployment patterns and the spatial dimensions of macro-economic policy: the Canadian experience 1966-1975', *Regional Studies,* 12, 283-96.

King, L.J., Casetti, E., and Jeffrey, D. (1969) 'Economic impulses in a regional system of cities: a study of spatial interaction', *Regional Studies,* 3, 213-18.

King, L.J., Casetti, E., and Jeffrey, D. (1972) 'Cyclical fluctuations in umemployment levels in US metropolitan areas', *Tijdschrift voor Economische en Sociale Geografie,* 63, 345-52.

Martin, R.L. (1978) 'Kalman filter modelling of time-varying processes in urban and regional analysis' in R.L. Martin, N.J. Thrift and R.J. Bennett (eds), *Towards the dynamic analysis of spatial systems,* 104-26, Pion: London.

Martin, R.L. (1979) 'Subregional Phillips curves, inflationary expectations, and the intermarket relative wage structure: substance and methodology' in N. Wrigley, op. cit., 64-110.

Nerlove, M. (1964) 'Spectral analysis of seasonal adjustment procedures', *Econometric,* 32, 241-86.

Sims, C.A. (1980) 'Macroeconomics and reality', *Econometrica,* 48, 1-48.

Vaughan, R.J. (1976) 'Public works as a countercyclical device: a review of the issue', *Rand Corporation, Santa Monica, Report R-1990-EDA.*

Wrigley, N. (ed.) (1979) *Statistical applications in the spatial sciences,* Pion: London.

Chapter 9

Spatial and temporal analysis: spatial time series

R.J. Bennett

Introduction

Spatial time series concerns the analysis of dynamic relationships which are distributed across a set of spatial regions or lattice points. Since the late 1960s a central thrust of British quantitative research has been concerned with such relationships, and the emphasis of this work has differed considerably from early static, geometric views of spatial relationships so ably reviewed in Haggett's *Locational Analysis in Human Geography* (1965). Instead of geometry, emphasis has been placed on stable, quasi-stable and slowly changing spatial relationships; and more recently the additional features of sharp discontinuities, catastrophes and shifts from one quasi-stable state to another have occupied increasing attention.

The vigour of this new work has tended to leave behind many of the problems of the spatial geometers unsolved, but in so doing it has often demonstrated that the solutions were in any case not required; instead it was the philosophy and representation of the problem which was at fault. For example, problems such as spatial stationarity, the interpretation of purely spatial processes on the plane (such as spatial autoregressions), the difficulties of spatial Markovity, have all been rather bypassed. These problems remain for the purely spatial modeller, but the new thrust of concern within spatial time series modelling, in emphasizing the dynamics of spatial relationships, has avoided the philosophical traps of the geometers. The lacunae of geometry and topology have been replaced with the dynamics of change, action, response, feedback, simultaneity, and intervention.

Hence models have moved from the ivory towers of beautiful mathematical structures to the dirty, messy and inconvenient real world of unstable structures, simultaneous spatial and multivariate interdependence, interaction with human values, and questions of political and ideological underpinnings. Model builders have increasingly asked the question: models for what? and in answering have tended to find the neat geometrical approach of the 1960s irrelevant. The new emphasis has been instead with real processes, with rates of change and adaptation, and has shown itself more directed to applied thinking, to simultaneity and feedback rather than linear causal structures, and in so doing has created the necessary conditions for a take-off into new areas of concern which are of direct relevance not only to the model builder and quantitative geographer, but also to other branches of the subject and to the policy maker. However, in this brief review it will not be possible to do more than sketch the nature of these new emphases and then to indicate the development prospects for the next few years.

Models of spatial time series

In the simplest terms it is possible to distinguish four main approaches to spatial time series analysis: systems models, regional leading indicators, the weights matrix approach, and purely spatial models. Each is briefly reviewed below.

Spatial systems models

The simplest approach to spatial time series in conceptual terms is that based upon representing the structure of spatial interactions over time as a system structure. This has been suggested by Hordijk and Nijkamp (1977), Bennett and Chorley (1978), Bennett (1979), and Bartels (1979), and allows the simultaneous equation estimation of spatial models. The structure of the resulting spatial time series model is as follows:

$$\begin{bmatrix} Y^1_{t1} \\ \vdots \\ Y^{m_1}_{t1} \\ \vdots \\ Y^1_{tN} \\ \vdots \\ Y^{m_N}_{tN} \end{bmatrix} = \begin{bmatrix} S^{11}_{11} & S^{12}_{11} & \cdots & S^{1n_1}_{11} & \cdots & S^{11}_{1N} & S^{12}_{1N} & \cdots & S^{1n_N}_{1N} \\ \vdots & \vdots & \ddots & \vdots & & \vdots & \vdots & \ddots & \vdots \\ S^{m_1 1}_{11} & S^{m_1 2}_{11} & \cdots & S^{m_1 n_1}_{11} & & S^{m_1 1}_{1N} & S^{m_1 2}_{1N} & \cdots & S^{m_1 n_N}_{1N} \\ \vdots & \vdots & & \vdots & \ddots & \vdots & \vdots & & \vdots \\ S^{11}_{N1} & S^{12}_{N1} & \cdots & S^{1n_N}_{N1} & & S^{11}_{NN} & S^{12}_{NN} & \cdots & S^{1n_N}_{NN} \\ \vdots & \vdots & \ddots & \vdots & & \vdots & \vdots & \ddots & \vdots \\ S^{m_N 1}_{N1} & S^{m_N 1}_{N1} & \cdots & S^{m_N n_N}_{N1} & \cdots & S^{m_N 1}_{NN} & S^{m_N 2}_{NN} & \cdots & S^{m_N n_N}_{NN} \end{bmatrix} \begin{bmatrix} X^1_{t1} \\ \vdots \\ X^{n_1}_{t1} \\ \vdots \\ X^1_{tN} \\ \vdots \\ X^{n_N}_{tN} \end{bmatrix} \qquad (1)$$

or $\mathbf{Y}_t = \mathbf{SX}_t$

where,

$$\underset{(n_1 + \ldots + n_N) \times 1}{\mathbf{X}_t} = \begin{bmatrix} X^1_{t1} \\ X^2_{t1} \\ \vdots \\ X^{n_1}_{t1} \\ \vdots \\ X^1_{tN} \\ \vdots \\ X^{n_N}_{tN} \end{bmatrix} \quad \underset{;\ (m_1 + \ldots + m_N) \times 1}{\mathbf{Y}_t} = \begin{bmatrix} Y^1_{t1} \\ Y^2_{t1} \\ \vdots \\ Y^m_{t1} \\ \vdots \\ Y^1_{tN} \\ \vdots \\ Y^{m_N}_{tN} \end{bmatrix} \qquad (2)$$

and where S^{kl}_{ij} is the set of regression parameters relating endogenous and exogenous variables. These terms represent the interaction between exogenous k and endogenous variable l in any region i. The diagonal elements give the intraregional interaction, and the off-diagonal elements give the interregional interactions between region i and region j. There are assumed to be N regions. Each region has m_j exogenous variables and n_i endogenous variables, and there may be different numbers of exogenous and endogenous variables in each region.

The major advantage of the system structure given by equation (1) is that it is identical to the multivariate structure used for simultaneous equation estimation problems. Hence these pre-established estimation methods can also be applied to spatial time series problems and this offers the major advantage that no new statistical sampling theory is required. However, a major disadvantage of representations such as (1) is the large dimension of the resulting estimation problem and the resultant loss of degrees of freedom, or requirement for very large data sets. This may cause the analyst to eliminate some of the interactions and reduce the dimension of the estimation problem if sufficient prior information is available. Alternatively, analysis can proceed by use of the weights matrix approach discussed below. Finally, a principal component solution can be sought. This has been suggested by Arora and Brown (1977), Tan (1977), Bartels (1979), Bennett (1979), and others, and is an attractive prospect. The method is particularly useful when combined with instrumental variables. Instrumental variables are used to reconstruct a set of lagged endogenous variables which are independent of the noise sequences, then the principal components solution is used to resolve the multicollinearity problem between the independent variables (and the lagged endogenous variables). Problems in interpretation of the principal components can usually be overcome by translation from the components back to the original variables. The method, as yet little used in spatial time series applications, offers considerable potential.

Regional leading indicators

A simplified form of spatial system model is that using regional leading indicators. This has been especially developed for forecasting and consists of a pairwise analysis of the relationships in equation (1). Changes in one region are used as leading indicators of changes in other regions, hence reducing the spatial pattern to a bivariate time series. This type of model has been applied first by King *et al.* (1969) and Haggett (1971) to regional unemployment levels, and there are now a large number of other applications available. These are summarized in Hepple (1975), Haggett *et al.* (1977) and Bennett (1979). A major drawback to the approach is its very simplicity; in ignoring the complex set of multivariate and spatial interactions in the full matrix structure of equation (1) considerable identification problems can occur, and it is likely that the final model will lack full behavioural validity or uniqueness of estimates.

The weights matrix approach

This is a development of the leading indicator methodology, but instead of using one single region as a leading indicator to forecast changes in other regions, a weighted sum of changes in a whole set of regions is used to forecast in the region of interest. This methodology has been developed most especially by Cliff and Ord (1975; 1980), Cliff *et al.* (1975), and Haggett *et al.* (1977) to which the reader is referred for more detailed reviews. The resulting model structure is given by:

$$Y_{t+1,\, i} = \sum_{k=1}^{p} \sum_{s=o}^{R} a_{ks} L^s Y_{t-k+1,\, i} + \sum_{k=o}^{q} \sum_{s=o}^{R} b_{ks} L^s U_{t-k+1\,,i} \qquad (3)$$

for the spatial regression model. For the spatial stochastic model (the so-called space time autoregressive integrated moving-average model (STARIMA)) the analogous equation is:

$$Y_{t+1,\, i} = \sum_{k=1}^{p} \sum_{s=o}^{R} a_{ks} L^s Y_{t-k+1,i} + \sum_{k=1}^{v} \sum_{s=o}^{R} c_{ks} L^s e_{t-k+1,\, i} + e_{t+1,\, i} \qquad (4)$$

The b_{ks} terms are the regression parameters, and the a_{ks} and c_{ks} terms are the respective autoregressive and moving-average parameters. The important change from previous models is the introduction of the weights matrix operator L^s for weights matrix with elements w_{ij} defined by:

$$L^s Y_{ti} = \sum_{j=i}^{k} w_{ij} Y_{tj} \quad s > o \qquad (5)$$

$$L^s Y_{ti} = \sum_{j=1}^{k} w_{ij} Y_{tj} \quad s > 0$$

where k is defined as the number of regions j which interact with region i, and s is the number of regions that are s spatial lags, or other metrics, away from zone i. The weights matrix w_{ij} incorporates the structure of spatial dependence so that spatial lags can be defined as the order of nearest neighbour, as distances, or some other definition. In addition the sum of all the weights terms equals unity.

A major problem of the weights matrix approach is the arbitrariness which often enters into the definition of terms. When the weighting terms have been defined, essentially all the spatial interactions in the model are determined, except to the level of empirical constants. Moreover, there are usually no very satisfactory criteria available for choosing one weights matrix rather than another, and many applications have been forced to make rather adventurous assumptions.

Purely spatial forecasts

In some special circumstances purely spatial forecasts may be required, for example, if it is desired to forecast from one part of a map to another. Most commonly this requirement arises with data interpolation (e.g. the 'Kriging' methods used in geology). In addition, specialized Markov field models may be relevant to some empirical cases. Such cases are fairly limited in practice and are mainly applicable to *equilibrium* spatial fields, situations where equilibrium *seems* to exist (due to long temporal sampling intervals and very short reaction times), or cases in which model fitting rather than explanation is required. These methods are complex and the reader is referred for further discussion to Besag (1974), Bennett (1979, pp. 499-531), Cliff and Ord (1980), and to the previous chapter by Haining.

Prospects

A central concern of time series models has been dynamics. This has permitted representation of variable reaction, relaxation, lead, lag, and frequency response components of geographical relationships. As such, this emphasis has redirected the analyst away from problems of pattern to ones of process. Hence, the prospects for development in the coming years using these methods result from this new and more direct process concern.

Of course, the resulting dynamic models can be just as descriptive as those models resulting from the geometric approach, and many of the applications of 'Box-Jenkins' and other time series techniques have been open to this criticism. But spatial time series models contain an added potential for linkage with prior theory formulations, and thus for more relevant application and more adequate explanation. A major thrust of this development has been linked with systems analysis (see Bennett and Chorley, 1978). Certainly the dynamics of process interrelationships have lent themselves to ready interpretation as input-output or stimulus-response. However, the major prospects for future developments do not derive so much from systems theory as such, but more from adaptability and potential of the mathematical representation which the systems analytic framework has stimulated. The spatial systems representation summarized in equation (1) has a flexibility and potential which other representations can achieve only by much more tortuous logic, and it is from this heuristic stimulus that many of the prospects for fruitful developments of spatial time series methods are likely to derive.

One major characteristic of the spatial time series representation has already been noted: the ready applicability of the statistical sampling and distribution theory of multivariate estimators. The reader is referred to Cliff and Ord (1975), Ord (1975) and Bennett (1979) for further discussion of such applications. However there are three major barriers to the development of these techniques. First, the models are particularly sensitive to their boundary conditions. These accord system closure, but are usually ignored totally in most analyses. Certainly increased concern and treatment of models on the boundary of spatial systems is required. A second barrier derives from the lack of adequate small sample theory for multivariate and spatial system estimators. This applies to each of the spatial representations (1) to (5) and to all spatial autocorrelation test statistics. It is unlikely that much will be gained in this area from statistical theory, since small sample properties are very heavily dependent upon the unique structure of each estimation problem. Instead, the most fruitful prospects for development will probably lie with the use of Monte Carlo, and other simulation studies, of estimation properties in typical practical situations. A third area of difficulty relates to the specificity of each practical situation: the spatial arrangement, size, shape and organization of the regional cells or lattice for which data is derived will affect the structure of the process identified and the parameters estimated. Hence, model structure is not independent of the data base adopted: the so-called 'Curry effect' (see Curry, 1972; Sheppard, 1978). This problem is not important if the data base is stable, and it is not sought to make comparative studies. However, the frequent occurrence of non-compatible data bases requiring interpolation, and the urgent need for more comparative quantitative studies, creates a strong imperative for the development of understanding of the relation of model estimates to the data base chosen. Preliminary estimates of such structures have been given by Openshaw (1977) and Williams (1978), but a general theory of such dependence is still awaited. From each of these difficulties flow prospects for future research foci, and it looks likely that each will be overcome in the next decade. However, the major prospects for the future of spatial time series models leads to the consideration of six main areas as particularly fruitful for future development: the modelling of (1) structural change, (2) simultaneity, (3) feedback, (4) forecasting, (5) control, and (6) the inclusion of political and value criteria.

The modelling of structural change is a particularly important area for future concern since most economic and environmental processes are not stable over their temporal and spatial ranges. A considerable battery of both theoretical discussion and statistical technique now exists which is concerned with structural change, and an important prospect for future research will evolve from meshing these two. On the one hand, there now exists a growing array of theories of change in physical processes such as climate, downstream river morphology and ecosystems, or in economic processes resulting from learning, adaptation and evolution. On the other hand, mathematical topology and bifurcation theory and adaptive parameter statistical estimators, such as the Kalman filter, allow the incorporation of structural changes into differential equations, and statistical estimation of the pattern of instability. The results in empirical applications within geography of bifurcation theory (Amson, 1972, 1974; Wilson, 1976, 1978; Wagstaff, 1977) and Kalman filtering (Bennett, 1975, 1976, 1979) have so far been very encouraging, and it is likely that, as these techniques become more widely known, they will stimulate further research into more complex adaptive models, and thus create the prospect of considerably more realistic models which can switch with spatial or temporal location, incorporate nonlinearities and non-stationarity as part of the model, and predict threshold and switching behaviour.

Simultaneity has always been a major concern of theoretical geography. In one sense a curse of complexity, but in another sense simultaneity represents the very *raison d'être* of the discipline: the phenomenon of spatial interdependence, or 'everything is related to everything else'. Spatial time series, especially when based on analysis with the systems equation (1), rather than compressing such simultaneity, has thrown new emphasis on this phenomenon and it is likely that, as the difficulties of small samples and boundary conditions are overcome, considerable progress will be made. However, as noted above, it may be that due to the loss of degrees of freedom and the complexity of intercorrelation between variables, this approach requires principal components or some other form of canonical solution.

Feedback processes are related to simultaneity. They arise when endogenous variables, or lags of endogenous variables, enter into the right-hand side of equation (1). The effect is to create both spatial and temporal autocorrelation through autoregressive structures. In addition, when policy control has been responsible for feedback, the resulting geographical relationships may become self-cancelling with the result that functional structures and input-output patterns become unidentifiable. The spatial and temporal autocorrelation problem has been extensively discussed in the past (see Dacey, 1966; Cliff and Ord, 1973, 1980) but successful estimators or test statistics still rely on asymptotic sampling theory. As in the case of full simultaneous equation estimation, small sample distribution theory is still required. In the case of policy feedback, considerable progress has beeen made in overcoming the problem of under-identification and the major range of techniques available has been reviewed by Bennett (1978). A major improvement in many economic models will result from the more widespread use of these techniques.

Forecasting models have found increasing currency in geography (see Haggett, 1971, 1973; Haggett *et al.*, 1977; and Bennett, 1979, Chapter 8 for recent reviews). However, two central dilemmas have surrounded the use of these models. First, too often forecasts have been made for their own sake, rather than as part of a problem requiring foreknowledge, e.g. for the implementation of a countervailing policy. As Haggett (1973) comments 'models are to use'. But where geographical forecasting models have been used, they have often given rise to a second dilemma: that the forecasts have been based upon models which have no meaningful behavioural interpretation. Haggett (1973, p. 234) and others have been happy to proceed using the pragmatic tenet that 'models are to use, not to believe', but an increasing body of criticism has been directed at such a position. It is difficult to use models, however accurate they may be, if those people who must suffer the adverse consequences of planning decisions based upon their use cannot understand their assumptions or structure, or if the model builder himself cannot justify that structure by other than statistical fit or forecasting error. Such models undermine the potential for effective participation in the planning processes and divorce the model builder from the political process (see for example the discussions of road and air traffic forecasts by Bennett, 1978). One of the most important prospects for future developments of geographical forecasting must therefore be with micro-economic and behavioural models rooted in accepted and well-understood theories. Considerable progress can be expected here from combining the micro-simulation studies of the Leeds group (Williams, Clarke and Kees, 1979) with the general spatial time series models outlined above.

Control, like forecasting, is also rooted in the pragmatics of 'models to use'. This generates two impetuses. On the one hand, to reshape existing models, especially those so extensively used in urban and regional planning such as the Lowry-Wilson approach, so that their essentially static structure can encompass both dynamic relationships, but also allow more direct policy appraisal. Few models as yet have explicitly included policy instruments of even descriptively assessed policy. An important exception is the US-based econometric work of Glickman (1977), but British work in this area has been lacking. On the other hand, the policy stimulus is likely to lead to more explicit control models. These are discussed in detail in a later chapter and it is sufficient here to note that these will place new impetus on the search for policy measurement, the development of more complex feedback and feedforward model structures directly oriented to monitoring, anticipating and counteracting changes, the development of goal and target specifications, the more flexible definition of objective functions, and a more explicit analysis of optimization and measurement of policy-effectiveness. However, it is likely that this approach will draw more on *ad hoc* rather than optimal solutions to policy questions, and is also likely to shift attention away from the static layout of facilities towards the more recurrent and shorter-term issues of resource allocation, and social redistribution and public finance; issues which are discussed in more detail in Chapters 34 and 36.

The development of control models in turn leads naturally to consideration of political and ideological underpinnings. The overwhelming emphasis of quantitative geography up to the late 1970s in Britain and North America has been that of the scientific methodology as evolved in the physical sciences; an epitome of this approach is given by Harvey (1969). As such it has been largely positivist in stance and has sought to objectify spatial and temporal processes, explicitly to remove human value judgments, and to keep political questions remote from the analysis. This emphasis has not accorded with the continental European countries, nor is it the emphasis which must necessarily characterize the study of spatial time series. As Lichtenberger (1976) cogently notes, the German-speaking countries, for example, have been concerned with a much more explicitly political element in theory development and model building; whilst the French geographers have emphasized a more holistic approach encompassing values, human judgment and explicit ideological stances. These differences have also been evident in exchanges between the English, French and German quantitative geography groups (see Unwin, 1978). It is all the more surprising, therefore, that radical critics should direct so much attention at quantitative geography as a *positive* science since it is neither *necessarily* positive, nor does it have this emphasis outside of the English-speaking countries (a good discussion of this paradox is given by Hay, 1979). Certainly the major prospects for the development and application of spatial time series models will derive from recognizing these differences, and through the use of control and policy involvement, to permit value and ideological criteria to enter the model building process: into the choice of objective functions, the choice of variables, and the choice of problems studied.

The resulting developments and prospects of time series models can be summarized as, first, the development of better and explicit behavioural and theoretical bases; second, the inclusion of a holistic grasp of the wider value and ideological underpinnings of the models adopted; third, the inclusion of explicit value and political choice into manipulable policy instruments; and finally, the development of models which are largely pragmatic, particular, and frequently unique in structure constrained by the specificity of each (policy) application. From this development it can be hoped, then, that it will be possible to obtain models which can not only be used, but also believed.

References

Amson, J.C. (1972) 'Equilibrium models of cities: an axiomatic theory', *Environment and Planning,* A 4, 429-44.

Amson, J.C. (1974) 'Equilibrium and catastrophic models of urban growth' in A.G. Wilson (ed.), London Papers in Regional Science 4, *Space-time concepts in urban and regional models,* Pion: London.

Arora, S.S., and Brown, M. (1977) 'Alternative approaches to spatial autocorrelation: an improvement over current practice', *International Regional Science Review,* 2, 67-78.

Bartels, C.A.P. (1979) 'Operational statistical methods for analysing spatial data' in C.A.P. Bartels and R.H. Ketellapper (eds), *Exploratory and Explanatory Statistical Analysis of Spatial Data,* Nijhoff: Leiden.

Bennett, R.J. (1975) 'Dynamic systems modelling of the north-west region; four parts', *Environment and Planning,* A 7, 525-38, 539-66, 617-36, 887-98.

Bennett, R.J. (1976) 'Nonstationary parameter estimation for small sample situations', *International Journal of Systems Science,* 7, 257-75.

Bennett, R.J. (1978) 'Forecasting in urban and regional planning closed loops: the examples of road and air traffic forecasts', *Environment and Planning* A 10, 145-62.

Bennett, R.J. (1979) *Spatial Time Series: Analysis, forecasting and control,* Pion: London.

Bennett, R.J., and Chorley, R.J. (1978) *Environmental Systems: Philosophy, Analysis and Control,* Methuen: London, and Princeton University Press: NJ.

Besag, J.E. (1974) 'Spatial interaction and the statistical analysis of lattice systems', *Journal of the Royal Statistical Society,* B 36, 192-236.

Cliff, A.D., and Ord, J.K. (1973) *Spatial Autocorrelation,* Pion: London.

Cliff, S.D., and Ord, J.K. (1975) 'Model building and the analysis of spatial pattern in human geography', *Journal of the Royal Statistical Society,* B 37, 297-348.

Cliff, A.D., and Ord, J.K. (1980) *Spatial Processes,* Pion: London.

Cliff, A.D., Haggett, P., Ord, J.K., Bassett, K., and Davies, R. (1975) *Elements of Spatial Structure,* Cambridge University Press.

Curry, L. (1972) 'A spatial analysis of gravity flows', *Regional Studies,* 6, 131-47.

Dacey, M.C. (1966) 'A county-seat model for the areal pattern of an urban system', *Geographical Review,* 56, 527-42.

Glickman, N.J. (1977) *Econometric analysis of regional systems: explorations in model building and policy analysis,* Academic Press: New York.

Haggett, P. (1965) *Locational Analysis in Human Geography, First Edition,* Arnold: London.

Haggett, P. (1971) 'Leads and lags in inter-regional systems: a study of cyclic fluctuations in the south west economy' in M.D.I. Chisholm and G. Manners (eds), *Spatial Policy Problems of the British Economy,* Cambridge University Press.

Haggett, P. (1973) 'Forecasting alternative spatial, ecological and regional futures' in R.J. Chorley (ed.), *Directions in Geography,* Methuen: London.

Haggett, P., Cliff, A.D., and Frey, A. (1977) *Locational Analysis in Human Geography,* 2nd edn, Arnold: London.

Harvey, D. (1969) *Explanation in Geography,* Arnold: London.

Hay, A. (1979) 'Positivism in human geography: response to critics' in D.T. Herbert and R.J. Johnston (eds), *Geography and the Urban Environment,* Vol. 2, Wiley: Chichester.

Hepple, L.W. (1975) 'Spectral techniques and the study of interregional economic cycles' in R. Peel, M. Chisholm and P. Haggett (eds), *Processes in Physical and Human Geography: Bristol Essays,* Heinemann: London.

Hordijk, L., and Nijkamp, P. (1977) 'Dynamic models of spatial autocorrelation', *Environment and Planning* A 9, 505-19.

King, L.J., Casetti, E., and Jeffrey, D. (1969) 'Economic impulses in a regional system of cities: a study of spatial interaction', *Regional Studies,* 3, 213-18.

Lichtenberger, E. (1976) 'Quantitative geography in the German-speaking countries', *Tijdschift voor Economische en sociale Geografie,* 69, 362-273.

Openshaw, S. (1977) 'A geographical solution to scale and aggregation problems in region-building, partitioning and spatial modelling', *Transactions Institute of British Geographers,* NS 2, 459-72.

Ord, J.K. (1975) 'Estimation methods for models of spatial interaction', *Journal American Statistical Association,* 70, 120-6.

Sheppard, E. (1978) 'Gravity Parameter estimation', *Geographical Analysis,* 11, 120-32.

Tan, K.C. (1977) *The estimation of spatial systems,* University College London Department of Geography, unpublished mimeo.

Unwin, D. (1978) 'Quantitative and theoretical geography in the United Kingdom', *Area,* 10, 337-44.

Wagstaff, J.M. (1977) 'A possible interpretation of settlement pattern development in terms of

"Catastrophe theory"', *Transactions Institute of British Geographers*, NS 3, 165-78.

Williams, H.C.L., Clarke, M., and Kees, P. (1979) 'Household dynamics and economic forecasting: a micro-simulation approach', presented at Regional Science Association meeting, Leeds.

Williams, I.N. (1978) 'Some implications of using spatial grouped data' in R.L. Martin, N.J. Thrift and R.J. Bennett (eds), *Towards the dynamic analysis of spatial systems*, Pion: London.

Wilson, A.G. (1976) 'Catastrophe theory and urban modelling: an application to model choice', *Environment and Planning*, A 8, 351-6.

Wilson, A.G. (1978) 'Towards models of evolution and genesis of urban structure' in R.L. Martin, N. Thrift and R.J. Bennett (eds), *Towards the dynamic analysis of spatial systems*, Pion: London.

Chapter 10

Spatial and temporal analysis: autocorrelation in space and time

A.D. Cliff and J.K. Ord

Introduction

It is often necessary to consider the spatial distribution of some phenomenon, whether in the 'counties' or 'states' of a 'country', in the 'quadrats' of a study area, or as a map of (point) locations of occurrences. Two basic questions may then be asked: (1) is the spatial pattern displayed by the phenomenon significant in some sense and therefore worth interpreting? If it is, (2) can we obtain any information on the processes which have produced the observed pattern from an analysis of the mapped distribution of the phenomenon? Question (1) implies that we are scanning maps for spatial pattern in the variation of values of some variate across the map (cross-sectional analysis). If there is systematic spatial variation in the variate, the phenomenon being studied is said to exhibit *spatial autocorrelation;* it is natural to try to establish the nature of the processes which have produced this (question (2), above) by looking for systematic *temporal* variation in a series of maps showing the distribution of the phenomenon at different points in time - that is, by searching for temporal autocorrelation or time-structure in the process. Our task is to unravel the complex patterns of autocorrelation in both time and space to gain some insight into the functional dependencies between areas implied by the presence of autocorrelation. Ability to do this will depend upon the problem in hand. Although the statistical methods provide no hard and fast answers, inferences based upon the circumstantial evidence of mapped patterns form a basis of inductive theory building in geography. In this paper, we review some measures of autocorrelation in time-space and ways of modelling the processes identified.

Detecting time-space autocorrelation

A space-time index (STI)

Before any test of the level of space-time interaction among counties can be performed, it is necessary to specify on *a priori* grounds which counties are believed to influence each other. This is usually done by defining a weighting matrix, **W**, whose typical element w_{ij}, reflects the strength of the link (join) or interaction between counties i and j; $w_{ij} = 0$ if counties i and j are believed to be unrelated (cf. Bennett in this volume, Chapter 9). An appropriate measure of space-time *inter*dependence is then

$$STI = \sum_{i}\sum_{j \atop i \neq j} w_{ij}\, y_{ij}; \tag{1}$$

the $\{w_{ij}\}$ are the spatial weights given above, and the $\{y_{ij}\}$ measure closeness in time for example, if the ith event occurred at time t_i, we may define

$$y_{ij} = 1 \text{ if } |t_i - t_j| \leqslant u$$

and

$$y_{ij} = 0, \text{ otherwise,}$$

for a suitable choice of u.

The space-time interaction (STI) coefficient given in (1) was first proposed by Knox (1964) and generalized by Mantel (1967). It was used by these authors to examine the proximity in time and space of successive outbreaks of a disease. For a recent application to a cholera outbreak, see Cliff and Ord (1980, section 1.7.3).

The form of the STI coefficient is very flexible. By redefining the $\{y_{ij}\}$ or the $\{w_{ij}\}$ we may include all the standard temporal and spatial autocorrelation measures within the framework. Thus, suppose we set $x_i = 1$ if an event has occurred in the ith county (equivalent to colour coding black, B) and $x_i = 0$ if an event has not occurred (equivalent to colour coding white, W). It follows that all the so-called *join count statistics* used to search for spatial autocorrelation in colour maps are but special cases of (1) for appropriate choices of the $\{y_{ij}\}$. Thus

$$y_{ij} = x_i x_j \quad (2)$$

yields the BB join count statistic, while

$$y_{ij} = (x_i - x_j)^2 \quad (3)$$

yields the BW join count statistic (see Cliff and Ord, 1980, Chapter 1 for details). If we now relax the condition that the $\{y_{ij}\}$ are binary we can use STI as a very general interaction (autocorrelation) coefficient in which w_{ij} measures the 'closeness' of locations i and j in space and y_{ij} measures their 'closeness' with respect to some other dimension (time, income or whatever). We always take $y_{ii} = 0$ for all i. This approach permits development of the I statistic proposed by Moran (1950) and Cliff and Ord (1973) to test for spatial autocorrelation among interval scaled variate values on a map, namely

$$I = (n \sum_{i}\sum_{j \atop i \neq j} w_{ij} z_i z_j) / S_o \sum_{i=1}^{n} z_i^2), \quad (4)$$

where $S_o = \sum_{i}\sum_{j \atop i\neq j} w_{ij}$, $z_i = x_i - \bar{x}$, x_i is the variate value and $n\bar{x} = \sum_{i=1}^{n} x_i$, in addition to earlier notation. Hubert (1978) has used this method to develop an I statistic for ranked data. The c statistic of Geary (1954) follows with $y_{ij} = (x_i - x_j)^2$.

The kth temporal autocorrelation coefficient, r_k, is obtained by setting $w_{ij} = 1$ if i and j are k units apart and $w_{ij} = 0$ otherwise, so that for $y_{ij} = z_i z_j$ we arrive at $r_k = \sum_{i=1}^{n-k} z_i z_{i+k} / \sum_{i=1}^{n} z_i^2$ where $z_i = x_i - \bar{x}$ as before. For both I and r_k, we have introduced the denominator term Σz_i^2; this will be constant under random permutations of the $\{z_i\}$, so that the distribution theory of the next section is not affected.

Distribution theory

In order to determine whether any calculated value of STI is statistically significant or not in the Neyman-Pearson sense, we need to determine the sampling distribution of STI under the null hypothesis of no association (or no autocorrelation). Under H_o, we may regard all the assignments of the $\{w_{ij}\}$ as fixed and then consider every possible permutation of the rows (and corresponding permutations of the columns) of the n x n matrix, $\mathbf{Y} = \{y_{ij}\}$. This procedure is known as randomization, and permits evaluation of the moments of STI whatever the underlying distribution of the population(s) generating the data. Following Mantel (1967) and Cliff and Ord (1980, Chapter 2), we have

$$E(STI) = \frac{S_o T_o}{n(n-1)} \quad (5)$$

where $T_o = \sum_{i}\sum_{j \atop i\neq j} y_{ij}$.

To specify the variance, we set

$$T_1 = \tfrac{1}{2}\sum_{i}\sum_{j \atop i\neq j} (y_{ij} + y_{ji})^2 \text{ and } T_2 = \sum_{i=1}^{n} (y_{i.} + y_{.i})^2.$$

Here, $y_{i.} = \sum_j y_{ij}$ and $y_{.i} = \sum_j y_{ji}$. In the same way, put

$$S_1 = \tfrac{1}{2}\sum_{i}\sum_{j \atop i\neq j} (w_{ij} + w_{ji})^2, \; S_2 = \sum_{i=1}^{n} (w_{i.} + w_{.i})^2 \text{ and}$$

$w_{i.} = \sum_j w_{ij}$. Mantel's results for the variance were derived without restrictions upon the $\{y_{ij}\}$, but we shall assume that $y_{ij} = y_{ji}$ for i and j. Then the variance may be written as

$$\text{var}(STI) = \frac{S_1 T_1}{2n^{(2)}} + \frac{(S_2 - 2S_1)(T_2 - 2T_1)}{4n^{(3)}} + \frac{(S_0^2 + S_1 - S_2)(T_0^2 + T_1 - T_2)}{n^{(4)}} - [E(STI)]^2. \quad (6)$$

Whenever STI is reduced to a measure such as BB, BW or I, it can be demonstrated that the sampling distribution of STI is approximately normal for reasonable n (say $\geqslant 25$), and a test of significance is provided by using (5) and (6) to evaluate STI as a standardized normal deviate. If STI is left in its general form, we need to have information on the structure of $\{y_{ij}\}$ to make statements about the shape of the sampling distribution. However, when both the $\{w_{ij}\}$ and the $\{y_{ij}\}$ are binary and the number of non-zero values for each is small (specifically, both S_0/n and T_0/n should stay finite as n goes to infinity) then the distribution of ½STI is approximately Poisson with mean $\tfrac{1}{2}S_0 T_0/n(n-1)$; the factor of ½ is included to avoid double counting of the joins. This result is demonstrated in Cliff and Ord (1980), section 2.2.4). The alternative to randomization is to assume a particular functional form for the probability distribution of the $\{y_{ij}\}$ or special cases thereof. Thus, for the join count statistics we may consider a binomial scheme for 'free' sampling (Clifford and Ord, 1980, sections 2.2 and 2.3), while for the I statistic we may assume that the $\{X_i\}$ are normally distributed. For a further discussion, see Cliff and Ord (1980, section 2.3.1).

Correlograms

While the interaction between sites may be greatest between immediate neighbours, its strength will often vary in a complex way with 'distance' in both time and space. Thus measles epidemics usually spread among contiguous areas from one time period to the next, but may jump from one major urban area to another, leaving intervening rural areas initially unaffected (a hierarchical effect). In the time domain, epidemic cycles create harmonics of autocorrelation corresponding to the recurrence interval of the waves. The use of correlograms to detect such variations in time series is well known (Kendall, 1976, p. 70). Here we develop the correlogram to examine the order of spatial processes.

Consider a system of n sites with random variables $X_1, \ldots, X_n$ and let the sites i and j be g^{th} order *neighbours* (or g spatial steps apart). Various definitions of neighbourliness are possible. Thus, two sites i and j may be g steps apart if

(a) the shortest path from i to j on the graph connecting adjacent sites has g edges; that is, the path passes through (g-1) intervening sites ($g \leqslant D$, where D is known as the *diameter* of the graph).
Then the set C(g) contains all pairs of subscripts whose sites are separated by (g-1) intervening sites, or

$$C(g) = \{(i, j) \text{ such that i and j are separated by } (g\text{-}1) \text{ intervening sites}\}.$$

(b) the distance, d_{ij}, between sites i and j falls in the gth distance class, $a_{g-1} < d_{ij} \leqslant a_g$, so that the set C(g) contains all pairs of subscripts for which d_{ij} is within these limits, or

$$C(g) = \{(i, j) \text{ such that } a_{g-1} < d_{ij} \leqslant a_g\}$$
$$g = 1, \ldots, D$$

where $a_0 = 0$, and $a_D \geqslant \max_{(i,j)} d_{ij}$.

Given the classification of each pair of sites into one of the sets, C(g), we may define the weighting matrices **W**(g) with elements

$$w_{ij}(g) = 1 \text{ if } (i, j) \in C(g)$$
$$= 0 \text{ otherwise.}$$

Then the gth order sample spatial autocorrelation is

$$I(g) = [n/S_0(g)]\ \mathbf{z}'\mathbf{W}(g)\mathbf{z}/\mathbf{z}'\mathbf{z} \tag{7}$$

where $\mathbf{z}' = (z_1, \ldots, z_n)$. $z_i = x_i - \bar{x}$, $i = 1, \ldots, n$, and $S_0(g) = \sum_{i}\sum_{j,\, i \neq j} w_{ij}(g)$. Alternatively, this may be written as

$$I(g) = [n/S_0(g)] \sum_{(i\,j) \in C(g)} w_{ij}(g) z_i z_j / \sum_{i=1}^{n} z_i^2.$$

We observe that the symmetric form of **W** in (7) means that each term appears twice in the summation. The means and variances for these measures follow directly from the spatial component of equations (5) and (6). The plot of I(g) against g yields the *spatial correlogram,* and enables us to determine the manner in which the autocorrelation function varies with distance.

Models of spatial processes

So far, we have looked at ways of determining the level of autocorrelation between regions separated in time and space. However, while hypothesis testing about autocorrelation has been the main focus of applications in the past, as Bennett and Haining have indicated in earlier chapters in this book, the detection of pattern is not an end in itself. More and more interest is being focused upon the development of spatial and space-time models, and this is a trend which we expect to continue. Chapter 9 by Bennett, like his book (1979), stresses multi-region time series models, and so as a contrast we focus upon the development of purely spatial models using the autoregressive-moving average approach. Estimation details are given in Haining (1977, 1978b), Bennett (1979, Chapter 7) and Cliff and Ord (1980, Chapter 6).

Autoregressive models

Here, primary interest is upon the covariance structure among observations, either directly or through an autoregressive framework. To see how we may develop such a scheme, it is useful to consider the first order autoregressive model for the time-series X_t ($t = \ldots, -1, 0, 1, \ldots$), which may be described in three equivalent ways (putting $|\rho| < 1$ to ensure stationarity):

(1) $$X_t = \rho X_{t-1} + \epsilon_t \tag{8}$$

with $E(X_t) = 0$, $E(\epsilon_t) = 0$ $\text{var}(\epsilon_t) = \sigma^2$ and $\text{cov}(\epsilon_t, \epsilon_{t-s}) = 0$, $\text{cov}(X_{t-s}, \epsilon_t) = 0$ for $s > 0$;

(2) $$E(X_t | x_{t-1}, x_{t-2}, \ldots) = \rho\, x_{t-1} \tag{9}$$

and $$\text{var}(X_t | x_{t-1}, x_{t-2}, \ldots) = \sigma^2; \tag{10}$$

(3) $$E(X_t) = 0 \text{ and } \text{cov}(X_t, X_{t-s}) = \sigma_X^2 \rho^{|s|} \tag{11}$$

where $\sigma_X^2 = \sigma^2/(1-\rho^2)$.

In this case the three formulations, which we refer to, respectively, as

(1) *simultaneous* autoregressive (Whittle model)
(2) *conditional* autoregressive, and
(3) *covariance*

are determined readily one from another and each has an intuitive appeal. However, in the spatial case, the *multilateral* dependence of the observations has two important repercussions:

(i) the specification of linear conditional means and constant conditional variances will usually imply multivariate normality (see Kendall and Stuart, 1973, p. 367);

(ii) a simple covariance structure does not possess a convenient autoregressive form, or vice-versa. Bearing in mind these rather sobering results, we now look at spatial models defined by analogy with (1) - (3) above.

Whittle's model. The seminal paper on spatial models is that of Whittle (1954), who proposed a *simultaneous* autoregressive scheme of the form

$$\mathbf{X} = \mathbf{GX} + \boldsymbol{\varepsilon} \tag{12}$$

where $\mathbf{X}' = (X_1, \ldots, X_n)$, $\boldsymbol{\varepsilon}' = (\epsilon_1, \ldots, \epsilon_n)$ and $\mathbf{G} = \{g_{ij}\}$ is an n x n matrix. We may then write

$$E(\boldsymbol{\varepsilon}) = \mathbf{O} \text{ and var } (\boldsymbol{\varepsilon}) = \boldsymbol{\Sigma} = \begin{pmatrix} \sigma_1^2 & & \mathbf{O} \\ & \ddots & \\ \mathbf{O} & & \sigma_n^2 \end{pmatrix}$$

so that $\boldsymbol{\Sigma}$ is a diagonal matrix. From (12)

$$\mathbf{X} = (\mathbf{I} - \mathbf{G})^{-1} \boldsymbol{\varepsilon}$$

so that, since $E(\mathbf{X}) = \mathbf{O}$,

$$\begin{aligned} \text{var } (\mathbf{X}) = \mathbf{V} &= E(\mathbf{XX}') \\ &= (\mathbf{I} - \mathbf{G})^{-1} E(\boldsymbol{\varepsilon\varepsilon}') (\mathbf{I} - \mathbf{G}')^{-1} \\ &= (\mathbf{I} - \mathbf{G})^{-1} \boldsymbol{\Sigma} (\mathbf{I} - \mathbf{G}')^{-1}. \end{aligned} \tag{13}$$

No distributional assumptions are necessary to arrive at (13), but if we do assume normality for ϵ, it follows that $\mathbf{X}$ is multivariate normal with covariance matrix $\mathbf{V}$, or

$$\mathbf{X} \sim \text{MVN}(\mathbf{O},\mathbf{V}).$$

If $\mathbf{X}^* = \mathbf{X} + \boldsymbol{\mu}$, then

$$\mathbf{X}^* \sim \text{MVN}(\boldsymbol{\mu}, \mathbf{V})$$

so that non-zero means are introduced readily into the original model (12).

Example 1 Suppose that the n sites lie on a regular grid. The subscripts i, j are used to describe the location in row i and column j of the grid. An isotropic first order simultaneous autoregressive scheme would be

$$X_{ij} = \rho(X_{i-1,j} + X_{i+1,j} + X_{i,j-1} + X_{i,j+1}) + \epsilon_{ij},$$

with modifications at the edges of the study area where necessary and with $|\rho| < 1/4$ to ensure stationarity. The matrix $\mathbf{G}$ could be written in the form $\rho\mathbf{W}$, where the elements of $\mathbf{W}$ would be one or zero depending upon whether the two sites concerned were one step apart on the grid or not.

The conditional approach

Following Bartlett (1971) and Besag (1974) we may specify the conditional autoregressive model as follows. We let $\mathbf{x}_i^* = \{x_j, j \neq i\}$ that is $\mathbf{x}_i^*$ denotes $\mathbf{x}$ after deletion of x_i; as before we take the $\{X_i\}$ to have zero means. The model is

$$E(X_i | \mathbf{x}_i^*) = \sum_{j \neq i} g_{ij} x_j, \tag{14}$$

$$\text{and} \quad \text{var}(X_i | \mathbf{x}_i^*) = \sigma_i^2 \tag{15}$$

for $i = 1, 2, \ldots, n$. When the conditional distributions are each normal, the joint distribution is

$$\mathbf{X} \sim \text{MVN}(\mathbf{O},\mathbf{V})$$

$$\text{where } \mathbf{V}^{-1} = \mathbf{D}(\mathbf{I} - \mathbf{G}), \mathbf{D}^{-1} = \boldsymbol{\Sigma} = \begin{pmatrix} \sigma_1^2 & & \mathbf{O} \\ & \ddots & \\ \mathbf{O} & & \sigma_n^2 \end{pmatrix}$$

and $\mathbf{G} = \{g_{ij}\}$. Clearly $\mathbf{DG}$ must be symmetric, so that equations (14) and (15) must be specified subject to $\sigma_j^2 g_{ij} = \sigma_i^2 g_{ji}$ for all i. See Cliff and Ord (1980, Chapter 6) for details.

Example 2 As in example 1, we consider a regular grid. If we let the conditional mean for X_{ij} depend only and equally upon its four immediate neighbours, and $\mathbf{x}_{ij}^*$ denotes $\mathbf{x}$ after delection of x_{ij} then we have

$$E(X_{ij} | \mathbf{x}_{ij}^*) = \rho(x_{i-1,j} + x_{i+1,j} + x_{i,j-1} + x_{i,j+1});$$

$|\rho| < 1/4$ to ensure stationarity.

The models in examples 1 and 2 seem analogous to the *simultaneous* and *conditional* schemes given for time-series in equations (8)-(10). However, they are *different* as we now show.

For simplicity, let $\sigma_i^2 = \sigma^2$ for all i. Then the *simultaneous* scheme has the covariance matrix

$$\mathbf{V} = \sigma^2 (\mathbf{I} - \mathbf{G})^{-1} (\mathbf{I} - \mathbf{G}')^{-1} \tag{16}$$

where $\mathbf{G}$ need not be symmetric, but the *conditional* scheme has

$$\mathbf{V} = \sigma^2 (\mathbf{I} - \mathbf{G})^{-1} \tag{17}$$

where $\mathbf{G}$ must be symmetric to ensure that $\mathbf{V}$ is also symmetric. The merits and drawbacks of the simultaneous and conditional schemes are considered at length in Cliff and Ord (1980, Chapter 6). Here, it is worth noting that, in the normal case, any *simultaneous* scheme with defining matrix $\mathbf{G}_s$ may be expressed as a *conditional* scheme with defining matrix

$$\mathbf{G}_c = \mathbf{G}_s + \mathbf{G}_s' - \mathbf{G}_s\mathbf{G}_s'.$$

Another discussion of these alternatives and their applications in spatial modelling is provided by Haining (1979).

Moving-average models

By analogy with time series (Box and Jenkins, 1970), it is possible to specify moving average spatial models as follows. We take $E(X_i) = 0$ without loss of generality and let

$$X_i = \epsilon_i + \sum_{j \neq i} g_{ij}\epsilon_j \qquad (18)$$

where $E(\epsilon_i) = 0$ and $E(\epsilon_i^2) = \sigma_i^2$ and $E(\epsilon_i\epsilon_j) = 0$ if $i \neq j$. Equation (18) can be rewritten in matrix form as

$$\mathbf{X} = (\mathbf{I} + \mathbf{G})\,\boldsymbol{\varepsilon}$$

whence $\quad \text{var}(\mathbf{X}) = (\mathbf{I} + \mathbf{G})\,\boldsymbol{\Sigma}\,(\mathbf{I} + \mathbf{G}'), \qquad (19)$

taking $\boldsymbol{\Sigma} = \begin{pmatrix} \sigma_1^2 & & 0 \\ & \ddots & \\ 0 & & \sigma_n^2 \end{pmatrix}$ as before. The process is spatially stationary provided only that

$$\sum_{j \neq i} g_{ij}^2 < \infty,$$

and invertible (representable as an autoregressive scheme) if and only if

the largest eigenvalue of $\mathbf{G}$ is less than one in absolute value.

These properties are given by Haining (1978a) who has provided a comprehensive development of the schemes. When $\sigma_i^2 = \sigma^2$ (for all i), the elements of (19) may be written as

$$\text{var}(\mathbf{X}) = \sigma^2 (\mathbf{I} + \mathbf{G})(\mathbf{I} + \mathbf{G}') \qquad (20)$$

or $\text{cov}(X_i, X_j) = \sigma^2 \left[\sum_k g_{ik}g_{jk} + g_{ij} + g_{ji}\right], \; i \neq j.$

Models of covariance structure

The various processes examined so far all consider a finite set of locations. This emphasis is justified because, in many geographical applications, we will be interested in a set of sites (such as cities) or spatial aggregates (counties). However, in other fields, for example, geology and meteorology, it is more plausible to assume that the process is defined continuously over space. The n locations are then sampling points in space rather than sites of particular interest. In these circumstances, it is better to specify the covariance structure directly rather than to attempt to formulate a linear dependence model as was done above. This is because, if a linear scheme is justified at one spatial scale, we would need a scheme that was non-linear in ρ at any different spatial scale.

If we consider the process $X(\mathbf{u})$ defined at location $\mathbf{u} = (u_1, u_2)'$, the mean and covariance structure for a stationary process may be specified as

$$E[X(\mathbf{u})] = \mu$$

and $\quad \text{cov}[X(\mathbf{u}), X(\mathbf{v})] = \sigma^2 c(\mathbf{u}, \mathbf{v})$

where $c(\mathbf{u}, \mathbf{u}) = 1$ for all $\mathbf{u}$. Various forms have been suggested for the covariance function, usually relying upon the distance $d = \left\{(u_1 - v_1)^2 + (u_2 - v_2)^2\right\}^{1/2}$ between locations $\mathbf{u}$ and $\mathbf{v}$. Thus rewriting $c(\mathbf{u}, \mathbf{v})$ as $c(d)$ and keeping $c(0) = 1$, we might consider

(i) $c(d) = e^{-\alpha d}, \; \alpha > 0,$
(ii) $c(d) = (d + \beta)^{-\phi}, \; \beta > 0, \phi > 0,$
or (iii) $c(d) = \begin{cases} 1 + \theta_1 d + \theta_2 d^2 + \theta_3 d^3, & d \leq d_0 \\ 0 & d > d_0, \end{cases}$

where $\alpha, \beta, \phi, \theta_1, \theta_2$ and θ_3 are parameters and the θ_i must be restricted to ensure that $c(d)$ is a valid covariance function [for example, $c(d) \leq 1$ for all d, $c(d_0) = 0$]. A further alternative, suggested by Kooijman (1976), is to take

(iv) $c(d) = c_k$ if $D_{k-1} < d \leq D_k$,

where $D_0 = 0$ and the intervals (D_i, D_{i+1}) form a suitable partition of the interval $(0, D_{max})$. Here, D_{max} is the largest distance between locations in the study area.

The advantage of scheme (iv) is that it suggests a straightforward (if inefficient) estimation procedure, namely to let

$$\hat{c}_k = \frac{1}{n_k} \sum x(\mathbf{u})\, x(\mathbf{v}) - \hat{\mu}^2$$

for all $\mathbf{u}, \mathbf{v}$ such that $D_{k-1} < d \leq D_k$; there are n_k such pairs. By making the successive intervals in (iv) narrower, one can approximate any form of covariance function but, in so doing, the estimators become very inaccurate. Further research is needed to establish an appropriate trade off between interval width (bias) and accuracy. The methods of spectral estimation in time series analysis should offer some useful guidelines. The estimation of schemes such as (iii) has been considered by Matheron (1971), while (i) and (ii) require either an iterative search over the likelihood function or some *ad hoc* procedure (cf. Cliff and Ord, 1980, section 6.3.5).

A further extension of the covariance approach is the use of *regionalized variables* (Matheron, 1971). To weaken the assumption of spatial stationarity, Matheron considered

$$E[\{X(\mathbf{u}) - X(\mathbf{v})\}^2] = \gamma(d);$$

$\gamma(d)$ is known as the variogram. The crucial feature of the variogram is that $\gamma(d)$ does not approach zero as $d \to 0$ because of the inherent local variation of the process. Another way of expressing this would be to let

$$X(\mathbf{u}) = Z(\mathbf{u}) + \delta(\mathbf{u}),$$

where $\quad \text{cov}[Z(\mathbf{u}), Z(\mathbf{v})] = \sigma_Z^2\, c(d),$

but $\delta(\mathbf{u})$ is a *white noise process* with

$$\text{var}[\delta(\mathbf{u})] = \sigma^2$$

and $\text{cov}[\delta(\mathbf{u}), \delta(\mathbf{v})] = 0$ for all $\mathbf{u} \neq \mathbf{v}$.

Although a continuous white noise process is not physically realizable, given that any sample occupies a finite area, however small, the formulation is mathematically convenient yet still effectively realizable. Given this structure (a spatial Kalman filter model), it follows that

$$\gamma(0) = 0$$

and $\gamma(d) = 2\sigma^2 + 2\sigma_Z^2\,[1 - c(d)]$, $d > 0$.

As $c(d) \to 0$, $\gamma(d)$ approaches a maximum of $2\sigma^2 + 2\sigma_Z^2$. With this scheme, any of the models (i)-(iv) for $c(d)$ could be used for $\gamma(d)$; the available estimation procedures are as before although σ^2 needs to be extrapolated from $\gamma(d)$ as $d \to 0$. See Matheron (1971), Delfiner and Delhomme (1975) and Huijbrechts (1975) for further details.

Model identification

Before any of the above models can be estimated, we need to select an appropriate form (autoregressive or moving average for the problem in hand and also to specify the spatial (and temporal) order of lagged terms to be included. This constitutes the *identification problem;* identification is usually carried out using the correlogram and partial correlogram. Details are given in Bennett (1975) and Martin and Oeppen (1975). To show the relationship of the material presented in the section on correlograms above to the discussion of autoregressive and moving average models, we concentrate here upon autoregressive processes.

The partial correlogram We defined the spatial correlogram in equation (7) as

$$I(g) = [n/S_o(g)]\ \mathbf{z}'\,\mathbf{W}(g)\mathbf{z}/\mathbf{z}'\mathbf{z}$$

for $g = 1, 2, \ldots$, and it would appear that partial correlograms could be defined in terms of the $I(g)$. Indeed, if we are interested in the *covariance* structure of the process, then this definition is appropriate. However, if we wish to identify a model, then we must take account of the number of sites g steps away from a given site. To make the discussion more concrete, consider a series of conditional autoregressive schemes of order $g = 1, 2, \ldots, D$. That is, for the ith site we specify the conditional mean and variance of the gth order scheme as

$$E(X_i|X_j = x_j, j \neq i) = \rho_1 \sum_{j\epsilon C(1)} x_j + \rho_2 \sum_{j\epsilon C(2)} x_j + \ldots + \rho_g \sum_{j\epsilon C(g)} x_j, \qquad (21)$$

$$\text{var}\,(X_i|X_j = x_j, j \neq i) = \sigma_g^2,$$

for $g = 1, 2, \ldots$, where the $\{\rho_j\}$ are unknown parameters and the sets $C(j)$ are as defined for equation (7). That is, we group all the sites one, two, ..., g steps away from site i, up to the maximum of D steps. If the appropriate form of (21) involves non-zero coefficients ρ_j only up to g_o ($\rho_j = 0$ for $j > g_o$) then we say that the correct model is a g_o^{th} order autoregressive scheme. In such circumstances, we would expect estimates of the ρ_j to be close to zero for $j > g_o$.

The least squares estimators are consistent for the conditional autoregressive scheme, so that we could fit autoregressive schemes of order g for $g = 1, 2, \ldots$, by this means. If we let

b_{jg} denote the estimate of ρ_j in the gth order autoregressive scheme, $j = 1, \ldots, g$ and $g = 1, 2, \ldots D$,

then the partial correlogram is the plot of

b_{gg} against $g (g = 1, 2, \ldots, D)$.

To evaluate successive b_{gg}, we do not need to fit the next order model (21) each stage, but may proceed directly by Durbin's method, which was developed originally for time series (Durbin, 1960).

Let r_g be the gth modified spatial autocorrelation, defined as

$$r_g = \frac{\sum_{i=1}^{n} z_i z_i(g)}{\left\{\sum_{i=1}^{n} z_i^2 \sum_{i=1}^{n} [z_i(g)]^2\right\}^{1/2}} \qquad (22)$$

where $z_i = x_i - \bar{x}$, and the $\{z_i(g)\}$ are given by the vector $\mathbf{z}(g) = \mathbf{W}(g)\mathbf{z}$. That is

$$r_g = \frac{S_o(g)\ I(g)}{n} \left[\frac{\Sigma z_i^2}{\Sigma z_i^2(g)}\right]^{1/2};$$

see equation (7). Given the $\{r_g\}$, the $\{b_{gg}\}$ are evaluated successively as follows:

$$b_{11} = r_1$$

$$b_{g+1,g+1} = (r_{g+1} - \sum_{j=1}^{g} r_{g+1-j}\, b_{jg})/(1 - \sum_{j=1}^{g} r_j b_{jg})$$

$$b_{j,g+1} = b_{jg} - b_{g+1-j,g}\, b_{g+1,g+1}$$

for $j = 1, \ldots, g$ and $g = 1, 2, \ldots D$.

When the covariance structure is of primary interest, the same computational procedure is followed, but with $I(g)$ in place of r_g. In the time series case, $x_i(g) = x_{i-g}$ so that $I(g)$ and r_g are usually indistinguishable unless n is very small; for computational convenience $I(g)$ is commonly used. In spatial studies, there will be several terms in each summation in (21) so that, typically,

$$\Sigma z_i^2(g) < \Sigma z_i^2;$$

this extra damping in the correlogram formed from the $I(g)$ distorts the resulting partial correlogram when used for model identification.

Conclusion

In this paper we have defined a test for space-time autocorrelation and shown that the conventional tests of spatial autocorrelation can be derived as special cases of it. We have suggested that whereas the emphasis in the past in this field has been upon hypothesis testing, future developments are more likely to focus upon models for spatial and space-time processes. To complement the earlier chapters, we have outlined some of the main alternatives for purely spatial models currently in use, and have shown how an appropriate form can be identified through correlogram methods. In Chapter 9 Bennett stresses time rather than space; here we have reversed the emphasis. Integrating the two approaches is likely to be a major future problem, particularly when we bear in mind, as discussed in Cliff and Ord (1975), that space-time autoregressive-moving average models converge to spatial equilibrium models with a different lag structure.

References

Bartlett, M.S. (1971) 'Physical nearest-neighbour models and non-linear time series', *Journal of Applied Probability*, 8, 222-32.

Bennett, R.J. (1975) 'The representation and identification of spacio-temporal systems: an example of population diffusion in north-west England', *Transactions Institute of British Geographers*, 66, 73-94.

Bennett, R.J. (1979) *Spatial Time Series* Pion: London.

Besag, J.E. (1974) 'Spatial interaction and the statistical analysis of lattice systems' (with discussion), *Journal of the Royal Statistical Society*, B 36, 192-236.

Box, G.E.P., and Jenkins, G.M. (1970) *Time Series Analysis, Forecasting and Control*, Holden-Day: San Francisco.

Cliff, A.D., and Ord, J.K. (1973) *Spatial Autocorrelation*, Pion: London.

Cliff, A.D., and Ord, J.K. (1975) 'Model building and the analysis of spatial pattern in human geography' (with discussion), *Journal of the Royal Statistical Society*, B 37, 297-348.

Cliff, A.D., and Ord, J.K. (1980) *Spatial Processes: Models and Applications*, Pion: London.

Delfiner, P., and Delhomme, J.P. (1975) 'Optimum interpolation by Kriging' in J.C. Davis and M.J. McCullagh (eds), *Display and Analysis of Spatial Data*, 96-114, Wiley: London.

Durbin, J. (1960) 'Estimation of parameters in time-series regression models', *Journal of the Royal Statistical Society*, B 22, 139-53.

Geary, R.C. (1954) 'The contiguity ratio and statistical mapping', *The Incorporated Statistician*, 5, 115-45.

Haining, R.P. (1977) 'Model specification in stationary random fields', *Geographical Analysis*, 9, 107-29.

Haining, R.P. (1978a) 'The moving average model for spatial interaction', *Transactions of the Institute of British Geographers*, NS 3, 202-25.

Haining, R.P. (1978b) 'Estimating spatial interaction models', *Environment and Planning*, A 10, 305-20.

Haining, R.P. (1979) 'Statistical tests and process generators for random field models', *Geographical Analysis*, 11, 45-64.

Hubert, L. (1978) 'Nonparametric tests for patterns in geographic variation: possible generalizations', *Geographical Analysis*, 10, 86-8.

Huijbrechts, C. (1975) 'Regionalized variables and quantitative analysis of spatial data' in J.C. Davis and M.J. McCullagh (eds), *Display and Analysis of Spatial Data*, 38-53, Wiley: London.

Kendall, M.G. (1976) *Time-Series*, 2nd edn, Griffin: London.

Kendall, M.G., and Stuart, A. (1973) *The Advanced Theory of Statistics*, Vol. II, 3rd edn, Griffin: London.

Knox, E.G. (1964) 'The detection of space-time interactions', *Applied Statistics*, 13, 25-9.

Mantel, N. (1967) 'The detection of disease clustering and a generalized regression approach', *Cancer Research*, 27, 209-20.

Martin, R.L., and Oeppen, J.E. (1975) 'The identification of regional forecasting models using space-time correlation functions', *Transactions of the Institute of British Geographers*, 66, 95-118.

Matheron, G. (1971) *The Theory of Regionalised Variables*, Centre de Morphologie Mathematique: Fontainebleau.

Moran, P.A.P. (1950) 'Notes on continuous stochastic phenomena', *Biometrika*, 37, 17-23.

Whittle, P. (1954) 'On stationary processes in the plane', *Biometrika*, 41, 434-49.

Chapter 11

Categorical data analysis

N. Wrigley

Introduction

Had this book been prepared ten or even five years ago, it is unlikely that a separate chapter would have been devoted to the analysis of categorical data. Quantitative geographers were certainly aware of the need for special methods to handle the low order, nominal or ordinal scale data with which they were often confronted, but the prevailing opinion was that, when required, suitable non-parametric statistical methods could always be found which would adequately meet the need. In the late 1960s and early 1970s, therefore, when rapid developments in the use of statistical methods in geography were being made, based upon a much closer relationship with the approach of econometrics and drawing on the integrated structure provided by the general linear model, the analysis of low order, non-metric or categorical data remained to a large extent outside the mainstream of development in quantitative geography for it was equated with non-parametric methods which in comparison appeared to be a non-developing, non-integrated collection of *ad hoc* techniques. Ironically, during this same period a 'major advance in the analysis of categorical data, an area where statistical methodology had previously been weak' (Coxon, 1977) was beginning to take shape elsewhere. At the turn of the decade several influential statements on the analysis of categorical data were published (Cox, 1970; Goodman, 1970, 1972; Grizzle, Starmer and Koch, 1969; Theil, 1969, 1970) which succeeded in reaching and influencing a wide audience, and at the same time doctoral dissertations or early research papers were completed by a group of scholars who were to go on to produce equally influential books and papers on the topic in the mid 1970s (Bishop, Fienberg and Holland, 1975; Fienberg, 1977; McFadden, 1974; Domencich and McFadden, 1975; Haberman, 1974a, 1978).

In the late 1970s it is now clear that the work of these and countless other scholars has succeeded in creating a long-awaited unified approach to the analysis of categorical data, and that this unified approach has significant advantages over traditional methods. Amongst the most important of these advantages are that it links the analysis of categorical data to the general linear model, provides a comprehensive and unified scheme for the analysis of multidimensional contingency tables, allows the development of computer programs comparable in generality to those available for the analysis of linear models, demonstrates how many statistical procedures seen and taught as distinct entities can be viewed as part of an integrated and powerful general methodology, and provides the basis of the significant progress which has been made in developing traditional consumer theory to encompass choice among discrete alternatives.

In geography, awareness and exploitation of the techniques which compose this unified approach has developed only slowly, and although there have now been reviews (Wrigley, 1979), introductory accounts of various aspects of the approach (Wrigley, 1976a; Lewis, 1977, pp. 128-47; Upton and Fingleton, 1979), specialized applications (Wrigley, 1977a, b, 1980a, d; Whitney and Boots, 1978; Baxter, 1979; Odland and Blazer, 1979; Upton, 1981) and a symposium of the IBG Quantitative Methods Study Group on the topic (Wrigley, 1980c), the methods have made only the briefest incursion into general statistical texts for geographers (Johnston, 1978, pp. 264-7) and have yet to become the standard research tools which they now are in many social science disciplines. Given the increasing availability of appropriate computer software and suitable introductory accounts of these methods, however, it is likely that the early 1980s will see a major expansion in the application of these methods in geographical research. This expansion will occur first in human geography, but there is evidence of sufficient interest in the methods in related environmental sciences (Chung, 1978; Fienberg, 1970b) to suggest that the adoption of the methods in physical geography cannot be long delayed.

A family of statistical models

At the heart of the unified approach to categorical data lies the interrelated family of logistic, linear logit and log-linear models. Although in a review of this length there is clearly not sufficient space to provide details of these models, it will be useful to outline their general structure and the class of statistical problems for which each type of model is appropriate (see Wrigley, 1979, for further details).

A useful framework for this purpose is that outlined in Table 11.1 which shows a classification of statistical problems on the basis of the type of response and explanatory variables involved. Table 11.1 is organized in such a way that as we move from cell (a) to cell (f) and cell (g) so the categorical data problem becomes more pervasive and the level of expertise of most geographers declines. Whereas most geographers are familiar with the regression models appropriate for the problems of cell (a) to (c), few are familiar with the logistic, linear logit and log-linear models which provide us with methods of handling the problems in cells (d) to (f) and, in addition, allow us to handle the problems of cell (g) in which all variables are categorical and there is no division of variables into response and explanatory (i.e. in which all variables are response variables).

Table 11.1 Classes of statistical problems

Response variables	*Explanatory variables*			
	Continuous	Mixed	Categorical	None
Continuous	(a)	(b)	(c)	
Categorical	(d)	(e)	(f)	(g)

Categorical response variable, continuous or mixed explanatory variables

Although there are a number of methods of handling the problems of cells (d) and (e) of Table 11.1 (see Domencich and McFadden, 1975, pp. 102-8; Pindyck and Rubinfeld, 1976, pp. 238-49; Wrigley, 1976a, pp. 9-11), those which have become the most widely used over the past ten years are based on the logistic and linear logit models. In the case of a dichotomous (i.e. 2-category) response variable, the logistic model has the form

$$P_i = \frac{e^{f(X_i)}}{1 + e^{f(X_i)}} \qquad (1)$$

where,

$$f(X_i) = \alpha + \sum_{k=1}^{K} \beta_k X_{ik} \qquad (2)$$

and where P_i represents the probability that the first category will be selected by the ith individual or at ith locality, given the values of the K explanatory variables. The logistic model can be rewritten and shown to be equivalent to a linear model of the form

$$\log_e \frac{P_i}{1 - P_i} = f(X_i) \qquad (3)$$

The left-hand side of this model is a transformation of P_i known as the logit transformation, and the model is known as the linear logit model. In both these models, categorical explanatory variables can be included in the function (2) by adopting the well-known principles used in the extension of the conventional regression models of cell (a) of Table 11.1 to the 'dummy' variable regression models of cell (b).

In the case of polychotomous or 'polytomous' (i.e. more than 2-category) response variables, the dichotomous models (1) and (3) generalize in a straightforward manner (see Mantel, 1966; Cox, 1970, p. 105; Mantel and Brown, 1973, p. 654; Wrigley, 1975, p. 191; Schmidt and Strauss, 1975a, p. 485). In the case of a J-category response variable, the logistic model takes the form

$$P_{r|i} = \frac{e^{f_r(X_i)}}{\sum_{s=1}^{J} e^{f_s(X_i)}} \qquad r = 1, \ldots J \qquad (4)$$

where,

$$f_r(X_i) = \alpha_r + \sum_{k=1}^{K} \beta_{kr} X_{ik} \qquad (5)$$

and where $P_{r|i}$ represents the probability that the rth category will be selected by the ith individual or at the ith locality, given the values of the K explanatory variables. The linear logit model for the same case generalizes (see Theil, 1970, pp. 117-20; Pindyck and Rubinfeld, 1976, pp. 256-9; Wrigley, 1976a, pp. 22-4) to the form

$$\log_e \frac{P_{r|i}}{P_{J|i}} = f_r(X_i) \qquad r = 1, \ldots J\text{-}1 \qquad (6)$$

In principle, there are two methods which can be used to estimate the parameters of these models; a non-iterative weighted least squares procedure or an iterative maximum likelihood procedure. The weighted least squares procedure is appropriate when the sample under investigation includes a sufficient number of repeated observations for each combination of values of the explanatory variables (see Wrigley, 1979, pp. 320-2). Although this is normally the case in cell (f), this condition is rarely satisfied for problems in cells (d) and (e) and, instead, an iterative maximum likelihood procedure which has the advantage of not requiring any repetitions of values of the explanatory variables (i.e. it can use 'ungrouped' data) is used. Details of the method are to be found in Cox (1970, p. 87), Mantel and Brown (1973, pp. 654-5), Wrigley (1975, pp. 191-3), Domencich and McFadden (1975, pp. 110-12), and Schmidt and Strauss (1975a, pp. 484-5).

Categorical response variable, categorical explanatory variables

In the case of cell (f) of Table 11.1, all variables are categorical and the sample data are most logically displayed in the form of a contingency table. The problems of cell (f) lie in a 'zone of transition' between the domains of logistic and linear logit models on the one hand, and log-linear models on the other, and, as a result, two types of models are now widely used to handle the problem of this cell.

The first type are linear logit models which in the polychotomous case take the form

$$\log_e \frac{P_{r|j}}{P_{J|j}} = E_A \left[\log_e \frac{f_{r|j}}{f_{J|j}} \right] = \qquad (7)$$

$$\alpha_r + \sum_{k=1}^{K} \beta_{kr} X_{jk} \quad r = 1, \ldots J\text{-}1$$

The subscript j replaces the i in (3) and (6) and indexes a set of groups or sub-populations formed from the cross-classification of the categorical explanatory variables (e.g. if there are two explanatory variables having 3 and 4 categories respectively, then there are 12 sub-populations and $j = 1, \ldots 12$). $f_{r|j}$ denotes the observed proportion of the sample drawn from the jth sub-population who select response category r, E_A denotes the fact that the logit in the probabilities is the 'asymptotic expectation' of the logit in the observed proportions, and the explanatory variables are all dichotomous 'dummy' variables. Given that the sample sizes for each of the sub-populations j are sufficiently large, the parameters of these models can be estimated using the non-iterative weighted least squares procedure described by Grizzle, Starmer and Koch (1969) and Theil (1970).

The second type of models used are log-linear models. These are models which are linear in the logarithms of the expected cell frequencies of the contingency tables in which the sample data are displayed. In the case of three-dimensional contingency tables, with typical element n_{ijk}, for example, the most general log-linear model takes the form

$$\log_e m_{ijk} = E(n_{ijk}) = u + u_{1(i)} + u_{2(j)} + u_{3(k)} + u_{12(ij)} + u_{13(ik)} + u_{23(jk)} + u_{123(ijk)} \qquad (8)$$

and is known as the saturated model. By setting different u terms, the parameters of the log-linear model, to zero, a range of different models describing the relationship between the response and explanatory variables can be specified, and the task is to select the most acceptable of these on the basis of its goodness-of-fit, parsimony, and substantive meaning. The parameters of these models, and the expected cell frequencies, are estimated using either the iterative proportional fitting procedure described by Fienberg (1970a, 1977, pp. 33-6) and Bishop *et al.* (1975, pp. 83-97), or the iterative weighted least squares procedure described by Nelder (1974) and Payne (1977). In using such models for the problems of cell (f) the marginal totals of the contingency tables corresponding to the explanatory variables are treated as fixed (i.e. we condition on the marginal totals involving the explanatory variables). As a result, the log-linear model fitted must always include certain of the u terms to ensure that the marginal totals of the estimated frequency table equal the observed marginal totals.

Although the logit and log-linear models appropriate for cell (f) look considerably different, it is simple to demonstrate that they are formally equivalent (see Fienberg, 1977, p. 78; Wrigley, 1979, p. 348). In other words, linear logit and log-linear models are linked together, and thus logistic, linear logit and log-linear models can be seen to be members of an interrelated family.

All variables categorical but no division into response and explanatory

In the case of cell (g), problems similar to those of cell (f) are being considered (i.e. all variables are categorical) but either all variables are 'natural' response variables, or the investigator chooses to treat them as such and not to make use of any response-explanatory distinction which exists. In this case, log-linear models are the appropriate models to use (see Fienberg, 1977; Haberman, 1978; Payne, 1977; Upton, 1978, for introductory accounts). Contingency tables of this type, with no division of variables into response and explanatory are sometimes termed 'symmetrical' tables, whilst those in which a division is recognized are termed 'asymmetrical' tables. The aim in the analysis of 'symmetrical' tables is not to assess the effects of explanatory variables on a response variable, but to describe the structural relationships among the variables corresponding to the dimensions of the table.

One of the simplest 'symmetrical' tables is a two-dimensional I x J table. The most general log-linear model which can be fitted to such a table is the saturated model

$$\log_e m_{ij} = \log_e E(n_{ij}) = u + u_{1(i)} + u_{2(j)} + u_{12(ij)} \quad (9)$$

where u is the overall mean of the logarithm of expected frequencies, $u_{1(i)}$ represents the main effect of being at level i of variable 1, $u_{2(j)}$ represents the main effect of being at level j of variable 2, and $u_{12(ij)}$ represents the two-way or 'interaction' effect between the two variables. Since the $u_{1(i)}$ and $u_{2(j)}$ terms represent deviations from the overall mean, the constraints

$$\sum_{i=1}^{I} u_{1(i)} = \sum_{j=1}^{J} u_{2(j)} = 0 \quad (10)$$

hold, and in addition

$$\sum_{i=1}^{I} u_{12(ij)} = \sum_{j=1}^{J} u_{12(ij)} = 0 \quad (11)$$

By setting different u terms in the saturated model to zero, a range of alternative models of the structural relationships among the variables can be specified. In practice, only a partial set of the possible specifications is considered. This set is known as a hierarchical set and has the property that higher order terms can only be included in a model if all related lower order terms are also included (i.e. u_{12} can only appear in a model if both u_1 and u_2 also appear). Unlike the log-linear models used for the problems of cell (f), the log-linear models fitted to 'symmetrical' tables do not normally require conditioning upon fixed marginal totals. As a result, there are no u terms which must always be included in the models, and all terms in the saturated model can potentially be set to zero as long as the resulting model is a member of the hierarchical set.

The saturated model for a three-dimensional contingency table is shown in equation (8) and there are a set of associated constraints which are extensions of those in (10) and (11). The set of hierarchical models considered in this case is given by Wrigley (1979, p. 344) and full discussion of these models is to be found in Bishop *et al.* (1975, pp. 37-8), Haberman (1978, pp. 198-207), Payne (1977, pp. 119-20). Extension of the saturated log-linear model (8) to a form suitable for a higher-dimensional table is straightforward and involves the same principles as employed in extending the two-dimensional saturated model to the three-dimensional form.

The range of application

The framework used in the previous section to discuss the interrelated family of statistical models can be used with advantage once again to classify the types of application of these methods which have occurred in geography and related disciplines, or which potentially offer the greatest promise for the future.

Categorical response variable, continuous or mixed explanatory variables

A simple example of a dichotomous logistic/linear logit model in which the parameters have been estimated by an iterative maximum likelihood procedure can be seen in (12). This shows a model fitted by the author to a data set relating to 124 oil wells drilled in Stafford County, south-central Kansas (Doveton, 1973; Harbaugh, Doveton and Davis, 1977, pp. 192-219). The model attempts to account for the probability that a well penetrating the 'B' division of the Mississippian Osage Series at locality i will be a 'producing' rather than dry well, on the basis of three explanatory variables: X_{1i} - thickness of the Mississippian B at i; X_{2i} - local geological structure at i (as measured by the residual from a first order trend surface of the top of the Mississippian B); and X_{3i} - the shale content of the Mississippian B at i (an indicator of permeability, measured by the mean digitized gamma ray log reading).

$$\hat{P}_i = \frac{e^{f(X_i)}}{1 + e^{f(X_i)}} \quad \text{or} \quad \log_e \frac{\hat{P}_i}{1 - \hat{P}_i} = f(X_i) \quad (12)$$

where,

$$f(X_i) = \underset{(0.5578)}{-1.1265} + \underset{(0.0213)}{0.0614X_{1i}} + \underset{(0.0061)}{0.0115X_{2i}} - \underset{(0.4278)}{8.1628X_{3i}}$$

$\hat{P}_i$ = predicted probability of well i being a 'producing' well. Standard errors in brackets.

This simple model shows Mississippian B production to be most likely in localities where the B zone is thick, where the shale content is low, and where there are local positive structures.

The form of a simple polychotomous model is shown in (13), which is taken from an exploratory study of the noise expectations of migrants into the area surrounding Luton Airport (Wrigley, 1976b). The model attempts to explain, on the basis of certain environmental and socio-economic variables, whether migrants found aircraft noise 'worse', 'the same', or 'better' than they had expected prior to moving into the area, and it includes categorical 'dummy' explanatory variables entered in the simplest form.

$$\log_e \frac{P_{r|i}}{P_{3|i}} = \alpha_r + \beta_{1r}D_i + \beta_{2r}N_i + \beta_{3r}L_i + \beta_{4r}NMP_i + \beta_{5r}P_i \qquad r = 1,2 \qquad (13)$$

where the response categories are: r = 1 = 'noise better than expected'; r = 2 = 'noise worse than expected'; r = 3 = 'noise same as expected.'

$P_{r|i}/P_{3|i}$ = the odds that household i will give the rth response rather than response r = 3 ('noise same as expected').

D_i = straight line distance from airport to location of household i.

N_i = Noise and Number Index contour value at locality of household i.

L_i = length of residence of household i in study area.

NMP_i = dummy variable indicating socio-economic group of head of household (1 if non-manual and non-professional/managerial, 0 otherwise).

P_i = dummy variable indicating socio-economic group of head of household (1 if professional/managerial. 0 otherwise).

Although dichotomous and polychotomous logistic and linear logit models of this type have only occasionally been used by geographers, they are now widely used in transportation science and economics to handle problems as diverse as mode choice on work and shopping trips (e.g. de Donnea, 1971; Watson, 1974; Domencich and McFadden, 1975; Train, 1978), destination choice on work and shopping trips (Richards and Ben-Akiva, 1974, 1975; Domencich and McFadden, 1975; Koppelman and Hauser, 1977), trip timing and frequency choices (Domencich and McFadden, 1975; Adler and Ben-Akiva, 1976), occupational choice and attainment (Schmidt and Strauss, 1975a, b; Boskin, 1974), car ownership (Cragg and Uhler, 1970; Lerman and Ben-Akiva, 1976), choice of residential location, housing type and tenure (Quigley, 1976; Lerman, 1977, Hensher, 1978b), selection of routes for urban freeways (McFadden, 1976b), residential choice of students (Kohn, Manski and Mundel, 1976), retailer's choice of location (Miller and Lerman, 1979), choice of finance by corporations (Baxter and Cragg, 1970), and the structuring of the asset portfolios of households (Uhler and Cragg, 1971). Of particular interest to geographers are those applications dealing with spatial choice problems in the widest sense, ranging from short-term destination choices to long-term location decisions, and the attempts being made to model the interrelationship of location decisions and destination choices (Lerman and Adler, 1976; Ben-Akiva, 1977). Often in the polychotomous versions (4) or (6) of these models the function defined in equation (5) is altered to take account of the fact that the response categories or alternatives from which each individual in the sample can choose are not necessarily the same in number or kind. (In the transportation science literature this is known as an 'unranked' set of alternatives.) In such cases, the explanatory variables in the function (5) must be of what is termed a 'generic' variable type (see Richards and Ben-Akiva, 1975, pp. 30-3; Domencich and McFadden, 1975, pp. 117-19; Wrigley, 1979, p. 324).

There are as many potential extensions of these dichotomous and polychotomous logistic and linear logit models as there are distinct specifications of the functions (2) and (5). Two special cases of interest to geographers can be derived by specifying categorical variable analogues to the traditional trend surface and space-time forecasting models.

(i) The categorical variable analogue to the traditional polynomial trend surface model is known as the polynomial probability surface model (see Wrigley, 1977a, b; Bielawski and Waters, 1979). In the dichotomous case, it is simply a model of the form (1) or (3) in which $f(X_i)$ is replaced by

$$f(U_i,V_i) = \alpha + \beta_1 U_i + \beta_2 V_i + \beta_3 U_i^2 + \beta_4 U_i V_i + \beta_5 V_i^2 + \ldots \qquad (14)$$

and in which U_i and V_i are the geographical coordinates of locality i.

(ii) The categorical variable analogue to a traditional space-time forecasting model (see Haggett *et al.*, 1977, pp. 517-40; Cliff, 1977) is simply (in the dichotomous case) a model of the form (1) or (3) with P_i replaced by $P_{i,t}$ and $f(X_i)$ replaced by

$$f(X_{i,t-1}) = \alpha + \beta X_{i,t-1} + \gamma \sum_j w_{ij} X_{j,t-1} \qquad (15)$$

Models of this form postulate that the probability with which the individual at locality i selects the first category at time t, depends upon the value of the explanatory variable X at locality i in the previous time period t-1, and also upon a weighted sum of values of X in time period t-1 in the j localities surrounding (and thus influencing) events at locality i at time t. The w_{ij} terms are pre-specified non-negative structural weights with w_{ij} proportional to the influence of j on i (see Wrigley, 1979; Odland and Blazer, 1979, for further discussion and application).

Finally, it should be noted that logistic and linear logit models of this type are not confined to single-equation specifications. Multi-equation systems of such models have also been used. For example, Schmidt and Strauss (1975b) have discussed the form of a simultaneous equation system of logit models, provided the likelihood functions necessary for parameter estimation, and applied the models to the problem of predicting the occupation and industry of employment of a sample of individuals drawn from the US Department of Labor data files.

Categorical response variable, categorical explanatory variables

A simple example of the linear logit models used to handle problems of this type can be seen in (16). This shows the basic additive form of a model of home ownership fitted by Li (1977) to household data for the Boston and Baltimore Standard Metropolitan Statistical Areas (SMSAs) extracted from the 1970 US Census of Metropolitan Housing Characteristics.

$$\log_e \frac{P_j}{1-P_j} = E_A\left[\log_e \frac{f_j}{1-f_j}\right] = \alpha + \sum_{k=1}^{4} \beta_k X_{jk} + \sum_{k=5}^{7} \beta_k X_{jk} + \sum_{k=8}^{10} \beta_k X_{jk} + \beta_{11} X_{j11} \quad (16)$$

In this model $f_j/1\text{-}f_j$ denotes the odds that household type j (where household types are defined on the basis of the cross-classification of the explanatory variables) is a home owner rather than renter. Explanatory variables 1 to 4 are a set of four dummy variables denoting five age of head of household categories. Explanatory variables 5 to 7 are a set of three dummy variables denoting four income classes. Explanatory variables 8 to 10 are a set of three dummy variables denoting four family size classes, and explanatory variable 11 is a dummy variable denoting the race of the head of the household (X_{j11} = 1 if non-black, 0 if black). The parameters of this model are estimated using weighted least squares, and Li extends the model to take account of interactions between the categorical explanatory variables.

In practice, logit models of this type are usually handled and displayed in matrix form. If $\log_e f_j/1\text{-}f_j$ in (16) is denoted $\bar{L}_j$, the general matrix formulation of these models takes the form

$$\mathbf{E_A}\,[\bar{\mathbf{L}}] = \mathbf{X}\beta \quad (17)$$

In the Grizzle, Starmer and Koch (1969) approach to models of this type, logit vectors such as that on the left-hand side of (17) are normally generated by using an extremely flexible and computationally simple sequence of matrix operations, in the form

$$\bar{\mathbf{L}} = \mathbf{K}\,(\log_e [\mathbf{Af}]) \quad (18)$$

where **f** is a vector of observed proportions, and **A** and **K** are matrices whose forms are determined by the structure of the data under investigation and the model to be fitted. Using this general matrix formulation it is conceptually and computationally simple to extend dichotomous and polychotomous logit models in which all the explanatory variables are categorical to handle multivariate response variable problems (i.e. cases in which there are two or more, possibly polychotomous, response variables, e.g. Lehnen and Koch, 1974a, pp. 300-09) and also to handle a wide range of seemingly unrelated problems. Of particular interest in this context is the work of Koch and his associates, and the extensions of the models they have proposed to handle what are termed 'repeated measurement research designs' (Koch *et al.*, 1976, 1977; Lehnen and Koch, 1974b; Imrey, Johnston and Koch, 1976; Landis and Koch, (1977). Many of the measurements collected by geographers are generated from such designs and can thus be handled in a unified way using these extended models (see Wrigley, 1980a, d).

Some simple geographical examples of the log-linear models used to handle problems in which there is a division of variables into response and explanatory are presented by Upton and Fingleton (1979). In one of these, three explanatory variables, age (variable 1), income (variable 2), and car ownership (variable 3) are used to account for the responses of a sample of 407 consumers to a question on shopping centre usage. The authors condition on the marginal totals involving the explanatory variables, and as a result fit a hierarchical set of models all of which include the terms u_1, u_2, u_3, u_{12}, u_{13}, u_{23} and u_{123}. The model selected as the most acceptable from the hierarchical set takes the form

$$\log_e m_{ijk\ell} = u + u_1(i) + u_2(j) + u_3(k) + u_{12}(ij) + u_{13}(ik) + u_{23}(jk) + u_{123}(ijk) + u_4(\ell) + u_{34}(k\ell) \quad (19)$$

where variable 4 is the response variable. The interpretation of this model is that for this sample of consumers there is no significant relationship between shopping centre usage and age or income, but there is a relationship (u_{34}) between car ownership and shopping centre usage. As noted above, log-linear models of this type have an equivalent linear logit model form. Wrigley (1980b) shows how (19) can be expressed in logit model form.

Log-linear models of this type are now used in a wide range of social and biomedical studies, many of which are of direct interest to geographers. Some of these studies involve multidimensional contingency tables with a considerable number of cells and associated large sample sizes. For example, Darroch and Ornstein (1980) have undertaken a national level study of the relationship between ethnicity and occupational structure in nineteenth-century Canada which uses a random sample of 10,000 households drawn from the manuscript records of the 1871 census. As part of this study, they consider the extent of urban-rural and regional variations in the relationship between ethnicity and occupational structure, and fit log-linear models to a four-dimensional 6 x 2 x 4 x 8 contingency table (3 explanatory variables, 1 response variable) which thus has 384 cells.

As in the case of the linear logit models discussed above, many extensions of the simple log-linear models are possible. Recently, for example, attention has been directed to methods for handling contingency tables with ordered categories (Haberman, 1974b; Fienberg, 1977, pp. 52-8), incomplete contingency tables (Fienberg, 1977, pp. 108-27; Fienberg and Larntz, 1976) and the relationship of log-linear models to latent structure models (Goodman, 1978; Haberman, 1979).

All variables categorical but no division into response and explanatory

Many examples of log-linear models used to handle problems of this kind are presented in Bishop *et al.* (1975), Fienberg (1977) and Haberman (1978). A simple example in the geographical literature is the investigation by Lewis (1977, pp. 141-6) of data taken from a study of the relationships between building age (variable 1), building decay (variable 2), and building use (variable 3) in a part of north-east London. 1407 buildings in the area were cross-classified into a 5 x 5 x 3 contingency table and a hierarchical set of log-linear models was fitted to the contingency table data. The most acceptable model was found to have the form

$$\log_e m_{ijk} = u + u_1(i) + u_2(j) + u_3(k) + u_{12}(ij) + u_{13}(ik) + u_{23}(jk) \qquad (20)$$

This model is a pairwise association model, which implies that each pair of variables is associated and each two-variable effect is unaffected by the level of the third variable. There is no three-variable effect in the model, i.e. $u_{123} = 0$.

In the case of higher-dimensional contingency tables, there are a large number of possible hierarchical log-linear models which can be fitted. The many biomedical, social and political applications of these models are distinguished by the variety of model selection strategies (Fienberg, 1977, pp. 62-76; Brown, 1976; Benedetti and Brown, 1978; Aitkin, 1979) adopted. Another distinguishing feature is that whereas most North American applications have used an iterative proportional fitting estimation procedure, the widespread availability of the GLIM computer package (Baker and Nelder, 1978) in Britain has resulted in wide use of the iterative weighted least squares estimation procedure (Payne, 1977; Whittaker and Aitkin, 1978).

Discrete choice models

For many social scientists, much of the utility of the statistical models discussed above derives from the fact that they are intimately connected to a developing theory of consumer behaviour in the face of discrete choices (sometimes called quantal or qualitative choice theory). In traditional economic analysis of consumer behaviour, commodities are assumed to be finely divisible, and the arguments entering into the individual's utility function are the quantities of the various commodities consumed. Traditional theory is thus not appropriate when the commodities demanded are discrete (i.e. when choices must be made from a finite set of discrete alternatives) but many decisions which a consumer must take are clearly of this type and do involve selection from a finite set of alternatives. Examples include: choice of a car, house, occupation, the mode of travel on a work trip or shopping trip, the destination of a shopping trip, the brand of commodity required, and so on. Over the past ten years significant progress has been made in developing traditional consumer theory to encompass choice among discrete alternatives. In the work of McFadden (1974, 1976a, 1979), Domencich and McFadden (1975), Ben-Akiva (1973), Richards and Ben-Akiva (1975), Manski (1973, 1977) Hensher (1978a, 1979) and others, the links between theories of individual choice behaviour and population choice behaviour have been explored, and it has been shown that a logically consistent qualitative choice theory can be developed which, moreover, is intrinsically connected to computationally tractable statistical models of the type discussed above.

In this approach, each individual decision-maker i is assumed to be faced with a set of J available choice alternatives

$$A_i = \{A_{1i}, \ldots A_{ri}, \ldots A_{Ji}\} \qquad (21)$$

and each individual is assumed to choose the alternative which yields the highest utility (utility maximization). This means that if the utility of alternative r to individual i is denoted U_{ri}, alternative r will be selected if and only if,

$$U_{ri} > U_{gi} \quad \text{for } g \neq r, g = 1, \ldots J \qquad (22)$$

Since the utility values are stochastic, the event that the condition in equation (22) holds will occur with some probability which can be denoted

$$P_{r|i} = \text{Prob}\left[U_{ri} > U_{gi} \text{ for } g \neq r, g = 1, \ldots J\right] \qquad (23)$$

where $P_{r|i}$ denotes the probability that individual i faced with choice set A_i will select alternative r. With complete generality, each utility value can then be partitioned into two components as shown in (24); a systematic component V which reflects the 'representative' tastes of the population, and a random component η which reflects the idiosyncratic tastes of individual i and/or the unobserved attributes of the choice-alternative.

$$U_{ri} = V(Z_{ri}, S_i) + \eta_{ri} \qquad (24)$$

where,

Z_{ri} = a vector of attributes of the choice-alternative r faced by individual i,
S_i = a vector of socio-economic characteristics of individual i.

Given this partitioning of U_{ri}, equation (23) can then be rewritten as

$$P_{r|i} = \text{Prob}\left[\eta_{gi} - \eta_{ri} < V(Z_{ri}, S_i) - V(Z_{gi}, S_i) \text{ for } g \neq r, g = 1, \ldots J\right] \qquad (25)$$

and this is known as the random utility model.

A probabilistic choice model which is computationally tractable can be derived from (25) by assuming a specific joint distribution of the random components ($\eta_{1i}, \ldots \eta_{ri}, \ldots \eta_{Ji}$). Assuming that the random components are independently and identically distributed with the double exponential distribution produces (see McFadden, 1974; Domencich and McFadden, 1975, pp. 49-69 for proofs) the familiar form of a polychotomous logistic model

$$P_{r|i} = \frac{e^{V(Z_{ri},S_i)}}{\sum_{g=1}^{J} e^{V(Z_{gi}, S_i)}} \qquad (26)$$

or its linear logit equivalent, and as a result the intrinsic connection between the statistical models discussed above and discrete choice models can be seen.

Although the logistic/logit model (26) is by far the most computationally tractable, probabilistic multiple-choice model which can be derived in this way, it has a property known as 'independence from irrelevant alternatives' (IIA) which in some circumstances can be restrictive and undesirable (Domencich and McFadden, 1975; Stopher and Meyburg, 1976, pp. 5-8; Hensher, 1978a). As a result, a number of attempts have recently been made to develop tests of when the IIA property can be accepted or rejected in empirical studies (McFadden *et al.*, 1976; Hensher, 1978a; Hensher and Stopher, 1979), and to develop alternative choice model formulations which circumvent this problem. One of these alternative formulations is the dogit model (Gaudry and Dagenais, 1979; Gaudry and Wills, 1979) which in certain cases collapses to the logistic/logit specification but in general is unconstrained by the IIA property. A feature of the model is that it is flexible enough to allow the choice among specific pairs of alternatives to be consistent with the IIA property, but to allow simultaneously the choice among other pairs to be inconsistent with the IIA property. Other models which circumvent the IIA problem are the nested logit and generalized extreme value models. As in the case of the dogit model, the basic logistic/logit model can be shown to be a special case of both these. Finally, a more general model which circumvents the IIA problem is the multinomial probit model. The useful properties of this model for discrete choice analysis have been known for a considerable time but in the early 1970s it was felt to be a computationally intractable model. More recently, however, it has been the object of renewed interest directed particularly at the construction of efficient estimation algorithms (Daganzo *et al.*, 1977; Albright *et al.*, 1977; Lerman and Manski, 1977; Hausman and Wise, 1978).

Computer programs for categorical data

The gradually increasing availability of appropriate computer software has played a significant role in encouraging the wider use of the interrelated family of logistic, logit and log-linear models. From a position in 1970 where only a few first generation programs were circulating amongst the *cognoscenti*, the situation has been transformed and in the late 1970s many social scientists are now aware of the major categorical data analysis programs.

Perhaps the most widely known of the programs are those used to fit log-linear models. ECTA (Everyman's Contingency Table Analysis) developed by L.A. Goodman at the University of Chicago (see Fay and Goodman, 1975) is widely available and used in North America and Europe, whilst GLIM (General Linear Interactive Modelling) developed by Nelder and his group at Rothamsted under the sponsorship of the Royal Statistical Society (see Baker and Nelder, 1978) is available on the majority of British university computer systems. ECTA uses an iterative proportional fitting procedure which is essentially a cyclic ascent optimization procedure that converges very quickly and handles higher-dimensional contingency tables efficiently. (An algorithm and Fortran listing for this procedure is given by Haberman, 1972). GLIM uses an iterative weighted least squares procedure. This procedure was suggested by Nelder and Wedderburn (1972) to obtain maximum likelihood estimates in a wide range of so-called 'generalized linear models'; models which have error distributions belonging to the exponential family. Log-linear models are merely a special case of such models in which the errors are Poisson distributed. GLIM allows estimation of log-linear models as one option in a wide range of possibilities, including classical multiple regression models, dichotomous logit models, probit models, and models with Gamma error functions.

In the case of linear logit models appropriate for cell (f) of Table 11.1, there is an equally general, widely available, but possibly slightly less well-known program, GENCAT, developed by Landis *et al.* (1976) and available from the Department of Biostatistics, University of Michigan. It uses a non-iterative weighted least squares estimation procedure and has been developed from a series of earlier more restricted versions (e.g. Forthofer *et al.*, 1971). GENCAT is an extremely flexible program capable of handling the linear models for the analysis of categorical data suggested by Grizzle, Starmer and Koch (1969) and the many extensions of the basic models suggested by Koch and his associates in recent years. Its very flexibility, however, requires the user to have a reasonably sophisticated knowledge of the class of problems it is designed to handle. (Example listings for geographical problems are provided in Wrigley, 1980a, d).

In the case of logistic/logit models appropriate

for cells (d) and (e) of Table 11.1, whose parameters are estimated by a direct iterative maximum likelihood procedure, there are now a considerable number of programs available, many of which can be traced back to a common parentage in the set of programs written in the late 1960s by the economist J.G. Cragg. Three of the most widely used are MLOGPRO developed by Hensher (1978c) at Macquarie University, Australia, QUAIL developed by McFadden and his associates at Berkeley (Berkman, Brownstone and associates, 1979), and a program developed by Manski (1974) at Massachusetts Institute of Technology (MIT) and used in work such as that by Richards and Ben-Akiva (1975). Other programs include those by Nerlove and Press (1973) for the RAND corporation, some British programs, including a set written by the author, which use algorithms from the AERE Harwell Subroutine Library, and that by Chung (1978) for the Canadian Geological Survey. With all of these programs, the user must take care to ensure that the likelihood function being maximized in the program is appropriate for the problem he wishes to study. This is particularly true in the case of polychotomous models because of the occurrence of the 'unranked' sets of alternatives and 'generic' variables referred to above. Programs designed solely for 'ranked' sets of alternatives will tend to differ in their structure from those which can handle 'unranked' sets of alternatives.

[For a more recent review of computer software, see O'Brien and Wrigley, 1980.]

Conclusion

The major advances which have occurred in categorical data analysis and related discrete choice modelling in the 1970s have transformed research practice in a number of disciplines, and as yet there is no evidence of a slackening of the pace of progress in this quickly growing area of knowledge. The technical challenges of the early 1980s lie in issues such as sample design (particularly in discrete choice analysis, e.g. Lerman and Manski, 1979), the investigation of missing data and errors in measurement problems, the design of model selection strategies, the investigation of inferential problems including the power of test statistics and their small sample properties, the development or improvement of models for simultaneous equation systems and special structure contingency tables, the use of experiments in discrete choice modelling, and the development of discrete choice models which attempt to incorporate dynamic aspects of behaviour.

Although geographers have been somewhat slow in recognizing the importance of the developments in the analysis of categorical data, there are signs that the 1980s will see major expansion in the application of these methods in geographical research. As applications multiply, it is essential that an attempt be made to forge a link with the methods which quantitative geographers have been developing over the past decade to analyse the spatial or spatial-temporal dependencies which are typically exhibited in geographical data (see Cliff and Ord, Haining, Bennett, and Hepple, this volume). This is likely to be a task of major proportions but it is a field in which quantitative geographers could make a definitive contribution. It is also essential that geographers resist the temptation to become myopically involved with just one element of the unified approach to categorical data analysis, and that they develop a broad and integrated view of these developments which avoids the partial perspectives which are characteristic of some of the recent literature in other disciplines. In this respect, coming late to these developments may prove to be an advantage.

References

Adler, T.J., and Ben-Akiva, M.E. (1976) 'Joint choice model for frequency, destination and travel mode for shopping trips', *Transportation Research Record* 569, Transportation Research Board: Washington DC.

Aitkin, M. (1979) 'A simultaneous test procedure for contingency table models', *Applied Statistics* 28, 233-42.

Albright, R.L., Lerman, S.R., and Manski, C.F. (1977) *Report on the development of an estimation program for the multinomial probit model,* final report prepared by Cambridge Systematics Inc., (Boston) for US Federal Highway Administration.

Baker, R.J., and Nelder, J.A. (1978) *The GLIM system: release 3,* Numerical Algorithms Group: Oxford.

Baxter, M.J. (1979) 'The application of logit regression analysis to production constrained gravity models', *Journal of Regional Science,* 19, 171-7.

Baxter, N.D., and Cragg, J.G. (1970) 'Corporate choice among long term financing instruments', *Review of Economics and Statistics,* 52, 225-35.

Ben-Akiva, M.E. (1973) 'Structure of passenger travel demand models', PhD dissertation, Department of Civil Engineering, MIT.

Ben-Akiva, M.E. (1977) 'Passenger travel demand forecasting: applications of disaggregate models and directions for research' in E.J. Visser (ed.) *Transport decisions in an age of uncertainty,* Martinus Nijhoff: The Hague.

Benedetti, J.K., and Brown, M.B. (1978) 'Strategies for the selection of log-linear models', *Biometrics*

34, 680-6.

Berkman, J., Brownstone, D., and associates (1979) *QUAIL 4.0 user's and programmer's manuals,* Department of Economics, University of California, Berkeley.

Bielawski, E., and Waters, N.M. (1979) 'The use of probability mapping in predicting cultural affiliation from site locations in Aston Bay, Somerset Island, Northwest Territories' mimeo, Dept. of Geography, University of Calgary.

Bishop, Y.M.M., Fienberg, S.E., and Holland, P.W. (1975) *Discrete multivariate analysis,* MIT Press: Cambridge, Mass.

Boskin, M.J. (1974) 'A conditional logit model of occupational choice', *Journal of Political Economy,* 82, 389-97.

Brown, M.B. (1976) 'Screening effects in multidimensional contingency tables', *Applied Statistics,* 25, 37-46.

Chung, C.F. (1978) *Computer program for the logistic model to estimate the probability of occurrence of discrete events,* Geological Survey of Canada, Paper 78-11, Ottawa, Canada.

Cliff, A.D. (1977) 'Quantitative methods: time series methods for modelling and forecasting', *Progress in Human Geography* 1 (3), 492-502.

Cox, D.R. (1970) *The analysis of binary data,* Methuen: London.

Coxon, A.P.M. (1977) 'Recent developments in social science software', *SSRC Newsletter,* 33, 6-9.

Cragg, J.G., and Uhler, R.S. (1970) 'The demand for automobiles', *Canadian Journal of Economics,* 3, 386-406.

Daganzo, C.F., Bouthelier, F., and Sheffi, Y. (1977) 'Multinomial probit and qualitative choice: a computationally efficient algorithm', *Transportation Science,* 11, 338-58.

Darroch, A.G., and Ornstein, M.D. (1980) 'Ethnicity and occupational structure in Canada in 1871: the vertical mosaic in historical perspective', *Canadian Historical Review,* 61.

de Donnea, F.X. (1971) *The determinants of transport mode choice in Dutch cities,* Rotterdam University Press.

Domencich, T.A., and McFadden, D.L. (1975) *Urban travel demand: a behavioural analysis,* North-Holland: Amsterdam.

Doveton, J.H. (1973) 'Numerical analysis relating location of hydrocarbon traps to structure and stratigraphy of the Mississippian 'B' of Stafford County, South-Central Kansas', Technical Report, KOX Project, Geologic Research Section, Kansas Geological Survey.

Fay, R.E., and Goodman, L.A. (1975) *ECTA program: description for users,* Department of Statistics, University of Chicago.

Fienberg, S.E. (1970a) 'An iterative procedure for estimation in contingency tables', *Annals of Mathematical Statistics,* 41, 907-17.

Fienberg, S.E. (1970b) 'The analysis of multidimensional contingency tables', *Ecology,* 51, 419-33.

Fienberg, S.E. (1977) *The analysis of cross-classified categorical data,* MIT Press: Cambridge, Mass.

Fienberg, S.E., and Larntz, K. (1976) 'Log-linear representation for paired and multiple comparisons models', *Biometrika,* 63, 245-54.

Forthofer, R.N., Starmer, C.F., and Grizzle, J.E. (1971) 'A program for the analysis of categorical data by linear models', *Journal of Biomedical Systems,* 2, 3-48.

Gaudry, M.J.I., and Dagenais, M.G. (1979) 'The dogit model', *Transportation Research B,* 13B, 105-11.

Gaudry, M.J.I., and Wills, M.J. (1979) 'Testing the dogit model with aggregate time-series and cross-sectional travel data', *Transportation Research B,* 13B, 155-66.

Goodman, L.A. (1970) 'The multivariate analysis of qualitative data: interactions among multiple classifications', *Journal of the American Statistical Association,* 65, 226-56.

Goodman, L.A. (1972) 'A general model for the analysis of surveys', *American Journal of Sociology,* 77, 1035-86.

Goodman, L.A. (1978) *Analyzing qualitative/categorical data: log-linear models and latent structure analysis,* Abt Books: Cambridge, Mass.

Grizzle, J.E., Starmer, C.F., and Koch, G.G. (1969) 'Analysis of categorical data by linear models', *Biometrics,* 25, 489-504.

Haberman, S.J. (1972) 'Log-linear fit for contingency tables', *Applied Statistics,* 21, Algorithm AS 51, 218-25.

Haberman, S.J. (1974a) *The analysis of frequency data,* University of Chicago Press.

Haberman, S.J. (1974b) 'Log-linear models for frequency tables with ordered classifications', *Biometrics,* 30, 589-600.

Haberman, S.J. (1978) *Analysis of qualitative data. Volume 1, Introductory topics,* Academic Press: New York.

Haberman, S.J. (1979) *Analysis of qualitative data. Volume 2, New developments,* Academic Press: New York.

Haggett, P., Cliff, A.D., and Frey, A. (1977) *Locational analysis in human geography,* 2nd edn, Edward Arnold: London.

Harbaugh, J.W., Doveton, J.H., and Davis, J.C. (1977) *Probability methods in oil exploration,* John Wiley: New York.

Hausman, J.A., and Wise, D.A. (1978) 'A conditional probit model for qualitative choice: discrete decisions recognising interdependence and heterogeneous preferences', *Econometrica,* 46, 403-26.

Hensher, D.A. (1978a) *A review of individual choice modelling,* Report prepared for Australian Department of Environment,

Housing and Community Development, Housing Research Grant AHR – Project 73.

Hensher, D.A. (1978b) 'The demand for location and accommodation - a qualitative choice approach in a policy formulating environment', School of Economic and Financial Studies, Macquarie University, Australia.

Hensher, D.A. (1978c) *MLOGPRO 78, a statistical estimation technique for qualitative choice behaviour - binary logit, binary probit, multiple logit,* School of Economic and Financial Studies, Macquarie University, Australia.

Hensher, D.A. (1979) 'Individual choice modelling with discrete commodities: theory, and application to the Tasman Bridge re-opening', *Economic Record,* 50, 243-61.

Hensher, D.A., and Stopher, P.R. (eds) (1979) *Behavioural travel modelling,* Croom Helm; London.

Imrey, P.B., Johnson, W.D., and Koch, G.G. (1976) 'An incomplete contingency table approach to paired-comparison experiments', *Journal of the American Statistical Association,* 71, 614-23.

Johnston, R.J. (1978) *Multivariate statistical analysis in geography,* Longman: London.

Koch, G.G., Freeman, J.L., and Lehnen, R.G. (1976) 'A general methodology for the analysis of ranked policy preference data', *International Statistical Review,* 44, 1-28

Koch, G.G., Landis, J.R., Freeman, J.L., Freeman, D.H., and Lehnen, R.G. (1977) 'A general methodology for the analysis of experiments with repeated measurement of categorical data', *Biometrics,* 33, 133-58.

Kohn, M.G., Manski, C.F., and Mundel, D.S. (1976) 'An empirical investigation of factors which influence college-going behaviour', *Annals of Economic and Social Measurement,* 5, 391-419.

Koppelman, F.S., and Hauser, J.R. (1977) 'Consumer travel choice behaviour: an empirical analysis of destination choice for non-grocery shopping trips', WP-414-09 Transportation Centre, Northwestern University, Evanston, Ill.

Landis, J.R., and Koch, G.G. (1977) 'The measurement of observer agreement for categorical data', *Biometrics,* 33, 159-74.

Landis, J.R., Stanish, W.M., Freeman, J.L., and Koch, G.G. (1976) 'A computer program for the generalized chi-square analysis of categorical data using weighted least squares (GENCAT)', *Computer Programs in Biomedicine,* 6, 196-231.

Lehnen, R.G., and Koch, G.G. (1974a) 'A general linear approach to the analysis of nonmetric data: applications for political science', *American Journal of Political Science,* 18, 283-313.

Lehnen, R.G., and Koch, G.G. (1974b) 'The analysis of categorical data from repeated measurement research designs', *Political Methodology,* 1, 103-23.

Lerman, S.R. (1977) 'Location, housing, automobile ownership, and mode to work: a joint choice model', *Transportation Research Record* 610, Transportation Research Board: Washington DC.

Lerman, S.R., and Adler, T.J. (1976) 'Development of disaggregate trip-distribution models' in P.R. Stopher and A.H. Meyburg (eds) op. cit.

Lerman, S.R., and Ben-Akiva, M.E. (1976) 'A disaggregate behavioural model of auto ownership', *Transportation Research Record* 569, Transportation Research Board: Washington DC.

Lerman, S.R., and Manski, C.F. (1977) 'An estimator for the generalized multinomial probit choice model', *Transportation Research Record* 623, Transportation Research Board: Washington DC.

Lerman, S.R., and Manski, C.F. (1979) 'Sample design for discrete choice analysis of travel behaviour: the state of the art', *Transportation Research A,* 13A, 29-44.

Lewis, P. (1977) *Maps and statistics,* Methuen: London.

Li, M.M. (1977) 'A logit model of homeownership', *Econometrica,* 45, 1081-97.

Manski, C.F. (1973) 'Qualitative choice analysis', PhD dissertation, Department of Economics, MIT, Cambridge, Mass.

Manski, C.F. (1974) *The conditional/polytomous logit program: instructions for use,* Working Paper, Carnegie-Mellon University, Pittsburg, Pa.

Manski, C.F. (1977) 'The structure of random utility models', *Theory and Decision,* 8, 229-54.

Mantel, N. (1966) 'Models for complex contingency tables and polychotomous dosage response curves', *Biometrics,* 22, 83-95.

Mantel, N., and Brown, C. (1973) 'A logistic re-analysis of Ashford and Sowden's data on respiratory symptoms in British coal miners', *Biometrics,* 29, 649-65.

McFadden, D. (1974) 'Conditional logit analysis of qualitative choice behaviour' in P. Zarembka (ed.) *Frontiers in econometrics,* Academic Press: New York.

McFadden, D. (1976a) 'Quantal choice analysis: a survey', *Annals of Economic and Social Measurement,* 5, 363-90.

McFadden, D. (1976b) 'The revealed preferences of a government bureaucracy: empirical evidence', *The Bell Journal of Economics,* 7, 55-72.

McFadden, D. (1979) 'Quantitative methods for analyzing travel behaviour: some recent developments' in D.A. Hensher and P.R. Stopher, op. cit.

McFadden, D., Tye, W., and Train, K. (1976) 'Diagnostic tests for the independence from irrelevant alternatives property of the multinomial logit model', Working Paper No. 7616, Urban Travel Demand Forecasting Project, University of California, Berkeley.

Miller, E.J., and Lerman, S.R. (1979) 'A model of

retail location, scale, and intensity', *Environment and Planning A*, 11, 177-192.

Nelder, J.A. (1974) 'Log-linear models for contingency tables: a generalization of classical least squares', *Applied Statistics*, 23, 323-9.

Nelder, J.A., and Wedderburn, R.W.M. (1972) 'Generalized linear models', *Journal of the Royal Statistical Society*, Series A, 135, 370-84.

Nerlove, M., and Press, S.J. (1973) *Univariate and multivariate log-linear and logistic models*, RAND Corporation Report R-1306-EDA/NIH, Santa Monica, California.

O'Brien, L.G. and Wrigley, N. (1980) 'Computer programs for the analysis of categorical data', *Area*, 12, 263-8.

Odland, J., and Blazer, B. (1979) 'Localized externalities, contagious processes and the deterioration of urban housing: an empirical analysis', *Socio-Economic Planning Sciences* 13, 87-93.

Payne, C. (1977) 'The log-linear model for contingency tables' in C.A. O'Muircheartaigh and C. Payne (eds), *The analysis of survey data: volume 2*, John Wiley: London.

Pindyck, R.S., and Rubinfeld, D.L. (1976) *Econometric models and economic forecasts*, McGraw-Hill: New York.

Quigley, J.M. (1976) 'Housing demand in the short run: an analysis of polytomous choice', *Explorations in Economic Research*, 2, 76-102.

Richards, M.G., and Ben-Akiva, M.E. (1974) 'A simultaneous destination and mode choice model for shopping trips', *Transportation*, 3, 343-56.

Richards, M.G., and Ben-Akiva, M.E. (1975) *A disaggregate travel demand model*, Saxon House: Farnborough.

Schmidt, P., and Strauss, R.P. (1975a) 'The prediction of occupation using multiple logit models', *International Economic Review*, 16, 471-86.

Schmidt, P., and Strauss, R.P. (1975b) 'Estimation of models with jointly dependent qualitative variables: a simultaneous logit approach', *Econometrica*, 43, 745-55.

Stopher, P.R., and Meyburg, A.H. (eds) (1976) *Behavioural travel-demand models*, D.C. Heath: Lexington.

Theil, H. (1969) 'A multinomial extension of the linear logit model', *International Economic Review*, 10, 251-9.

Theil, H. (1970) 'On the estimation of relationships involving qualitative variables', *American Journal of Sociology*, 76, 103-54.

Train, K.E. (1978) 'The sensitivity of parameter estimates to data specification in mode choice models', *Transportation*, 7, 301-9.

Uhler, R.S., and Cragg, J.G. (1971) 'The structure of the asset portfolios of households', *Review of Economic Studies*, 38, 341-57.

Upton, G.J.S. (1978) *The analysis of cross-tabulated data*, John Wiley: Chichester.

Upton, G.J.G. (1981) 'Log-linear models, screening, and regional industrial surveys', *Regional Studies*, 15.

Upton, G.J.G., and Fingleton, B. (1979) 'Log-linear models in geography', *Transactions of the Institute of British Geographers* NS 4, 103-15.

Watson, P.L. (1974) *The value of time, behavioural models of modal choice*, D.C. Heath: Lexington.

Whitney, J.B., and Boots, B.N. (1978) 'The examination of residential mobility through the use of the log-linear model: 1, theory', *Regional Science and Urban Economics*, 8, 153-73.

Whittaker, J., and Aitkin, M. (1978) 'A flexible strategy for fitting complex log-linear models', *Biometrics*, 34, 487-95.

Wrigley, N. (1975) 'Analyzing multiple alternative dependent variables', *Geographical Analysis*, 7, 187-95.

Wrigley, N. (1976a) *An introduction to the use of logit models in geography*, Concepts and techniques in modern geography, 10, Geo. Abstracts: Norwich.

Wrigley, N. (1976b) 'An analysis of the aircraft noise expectations of migrants into an area around Luton Airport', Institute of Sound and Vibration Research Memorandum 560, University of Southampton.

Wrigley, N. (1977a) 'Probability surface mapping: a new approach to trend surface mapping', *Transactions of the Institute of British Geographers* NS 2, 129-40.

Wrigley, N. (1977b) *Probability surface mapping: an introduction with examples and Fortran programs*, Concepts and techniques in modern geography, 16, Geo. Abstracts; Norwich.

Wrigley, N. (1979) 'Developments in the statistical analysis of categorical data', *Progress in Human Geography*, 3, 315-55.

Wrigley, N. (1980a) 'Paired comparison experiments and logit models: a review and illustration of some recent developments', *Environment and Planning A*, 12, 21-40.

Wrigley, N. (1980b) 'Log-linear models in geography: comments on the recent article by Upton and Fingleton', *Transactions of the Institute of British Geographers* NS 5, 113-17.

Wrigley, N. (1980c) 'Analysis of categorical data', *Area*, 12, 65-66.

Wrigley, N. (1980d) 'Categorical data, repeated measurement research designs, and regional industrial surveys', *Regional Studies*, 14, 455-71.

Chapter 12

Ratios and closed number systems

I.S. Evans and K. Jones

Introduction

Haynes (1978) has shown that among a sample of quantitative human geography papers, 76 per cent used proportions and 24 per cent densities, compared with 69 per cent using distance and 68 per cent using counts. Some papers used many variables expressed as proportions. The 105 papers covered in his study are not atypical of recent work: ratios, including proportions, are dominant in human geography and very important in physical geography. Clearly geographers require training in the analysis of ratios: yet the available texts pay scant attention to the special characteristics of ratios and the special techniques which these necessitate. Not surprisingly, preconditions and qualifications mentioned in the more fundamental statistical texts are blithely disregarded. Most geographers - British and otherwise - analyse ratios as if they were simple numbers.

Ratios are formed when one number (a count or measurement) is divided by another. In other words, two numbers are necessarily involved: a numerator and a denominator. (Occasionally, as in pH measurement by litmus paper, ratios are observed directly and further measurement is required to establish the denominator.) Ratio formation is an act of data compression; two numbers are summarized in one, which is considered to be more important. Below, however, we contend that it is foolish to neglect the second number: the bivariate nature of ratios must be accepted explicitly.

Ratios may be expressed as proportions, percentages, per thousands, or in other ways; the transformation involved in shifting the decimal point is trivial, but sometimes causes confusion. It is more important to recognize that there are two types of ratios: closed ratios, which are bounded by proportions of 0 and 1, and open ratios, which have no upper bound but often have a lower bound of zero. Closed ratios arise in the special case where the numerator is a subset of the denominator, and therefore cannot exceed it. They occur when a total count is divided into mutually exclusive subsets, e.g. people are classified by birthplace or by age, pollen by taxa, or pebbles by lithology. Commonly, each subset is expressed as a percentage of the total, and the resulting ratios vary between 0 per cent and 100 per cent. They also occur when a total quantity is subdivided, e.g. per cent of Gross National Product arising from manufacturing industry or per cent (by weight) of a soil sample which is clay.

Open ratios include those such as population density and drainage density, which have dimensions of length^{-2} and length^{-1} respectively, and those such as children per mother, persons per room, and height/length which are dimensionless. Each of these must be non-negative, but has no upper bound. Percentage change has a lower bound of -100 per cent, and no upper bound: it too is an open ratio. The fact that changes, or children per mother, are commonly expressed as 'percentages' or 'per thousand' does not mean that they are closed ratios; they are open, because ratios in excess of 100 per cent are possible and the denominator does not include the numerator. A special case is provided by the ratio of width to length where width is defined to be no greater than length. Although the ratio is constrained to vary from 0 to 1, it is not a closed ratio because the denominator does not include the numerator.

Ratios should be calculated only for numbers (counts, measurements) expressed on a ratio scale, i.e. one which has a true zero. Note that not only the resulting ratios, but also the absolute numbers (numerators and denominators) are on a ratio scale.

When absolute numbers are available, why do researchers convert them into ratio form prior to further analysis? It may be to create a new variable which is considered more interesting, e.g. gradient rather than height and length considered separately. More often it is to control for a master variable, such as population, which has a dominating influence; thus number of deaths is divided by (initial) number of people to give death rate. The main assumption here is that the first variable is linearly proportional to the master variable; this is sometimes but not always justified.

Choice of an appropriate denominator is extremely important. Phillips (1973) gave an example where the sign of a correlation involving 'business burglary and robbery rate' was reversed, depending on whether 'population' or 'business employment' was taken as its denominator.

Displaying the denominator

The first point to emphasize is that it is misleading to suppress information about the second dimension involved in a ratio; the ratio may seem more important than the original numbers, but it is dangerous to overlook these. If the numbers are counts, the significance of a ratio in both the ordinary and the statistical sense depends on the counts: is it 30 per cent of 10 or 30 per cent of 10,000? To assess the ratio, we need to know. If the numbers are measurements, this consideration may seem unimportant, but absolute size may still be highly relevant: an 80° slope 1,000 m high requires more explanation than one 2 m high. Mean gradient increases as we measure it over shorter distance (Evans, 1979a) and shape may also be size-dependent (Mosimann, 1970). Segregation indices are highly scale-specific (Jones and McEvoy, 1978). Hence the possibility of size (scale) effects necessitates attention to absolute numbers as well as ratios: ratios so affected may be compared only if their denominators are approximately constant. More generally, a reader cannot properly interpret a set of ratios unless he is given their denominators (or numerators).

This has important implications in graphics. The commonest form of thematic map in human geography, easily produced either manually or - in large series - automatically, is the choropleth map in which administrative areas (varying greatly in size and in shape) are shaded in relation to some ratio. This 'lazy man's map' is acceptable if the denominator of the ratio is geographic area, since those ratios with larger denominators then have greater visual weight. If, however, the denominator is number of people, or some other variable without a very high correlation with geographic area, the weighting is quite misleading (Williams, 1976). Attention is concentrated on large areas, which often have low populations: urban areas, where much of the information is concentrated, are made to appear unimportant. This invalidates visual generalization of many maps.

The simplest solution is to display the denominator by the size of a proportional symbol (usually a circle), and the ratio by the shading of the symbol, or better still by the angular extent of a shaded sector. Good examples can be found in Cole's (1964) text on Italy. Below county scale, especially in Britain, the congestion of symbols in conurbation areas may become so great that displacement or overlap is no longer feasible, and space itself must be distorted to give a cartogram (Howe, 1970). Alternatively, the ratio scale may be forsaken in favour of a scale such as 'signed chi-square' (Visvalingam, 1978) which combines absolute and relative deviation from a norm: this is used for most of the maps in 'People in Britain' (Clarke *et al.*, 1980) and for one of the drought atlas (Clark, 1980). Another possibility is to map the probability of a ratio deviating from expectation to the extent observed, on the assumption of independence and of a frequency distribution such as Poisson (Choynowski, 1959, and White, 1972; summarized in Haggett, Cliff and Frey, 1977, pp. 306-8) or Gaussian (Norcliffe, 1972).

Interrelationships

Geographers are often interested in a relationship between two ratios as they vary between places. Unfortunately, the correlation coefficient between two ratios may be interpreted in the usual way only if the counts or measurements involved in each are different and have no inbuilt relation to each other. By 'inbuilt' we mean, for example, when a denominator is the sum of several numerators; or when one denominator is 'married women' and the other is 'working married women'.

When one of the counts or measurements occurs in both ratios, it is unrealistic to compare the observed correlation with an expected value of zero. If the denominators are identical then, whether the numerators are related or not, an element of positive correlation is expected. Likewise, if the numerator of one ratio is the denominator of the other, a tendency towards negative correlation is expected. The actual magnitude of this tendency depends on the means and variances of the numerators and denominators (Pearson, 1897: Chayes, 1971).

Recent debates both in biometrics (Atchley, Gaskins and Anderson, 1976, and discussion) and in geography (Pethick, 1975; Richards, 1978; Ferguson, 1978) show that the ramifications of interpreting correlations between open ratios can be vexatious. We shall start by considering the more clear-cut case of correlation between closed ratios, which is also the most important case for human geography.

Closure correlation

We are often concerned with the correlation between a pair of occupational groups, age-groups or ethnic groups as they vary from place to place. Within a given, mutually exclusive classification, these groups are usually expressed as proportions of the total, e.g. miners and quarrymen as a proportion of all civilian workers. Often, multivariate analyses (based on correlations) include the three 'variables' (i.e. categories of one variable), per cent workers in manufacturing, in service and in 'primary' (farming, forestry, fishing and mining) employment. The latter category was only 4.4 per cent of the British workforce in 1971; manufacturing was 34.5 per cent and the remainder were in services. Hence, where a greater proportion is employed in manufacturing, this is very likely to be at the expense of a smaller proportion being in services: the correlation between the two has a negative bias. If there are 70 per cent in manufacturing, there must be 30 per cent or less in services, and vice versa. A Cartesian scatter plot of one closed ratio against another is confined to a triangular area such that $(x + y) \leqslant 100$ per cent (Figure 12.1); hence the need for a ternary plot (Figure 12.2). Here there are three axes, at 60° to each other: the third is, by implication, the remainder of the closed data set.

On the other hand, we may wish to correlate the proportion who travel to work by motor cycle with the proportion who travel by pedal cycle. These average 1.5 per cent and 4.2 per cent nationally, and so each can vary with little constraint from the other. The closure effect of inbuilt neg-

Figure 12.1 A Cartesian plot of percentage (of total harvested cropland) corn v. percentage hay for 88 counties, Ohio, 1940 (King, 1969, pp. 175-6)

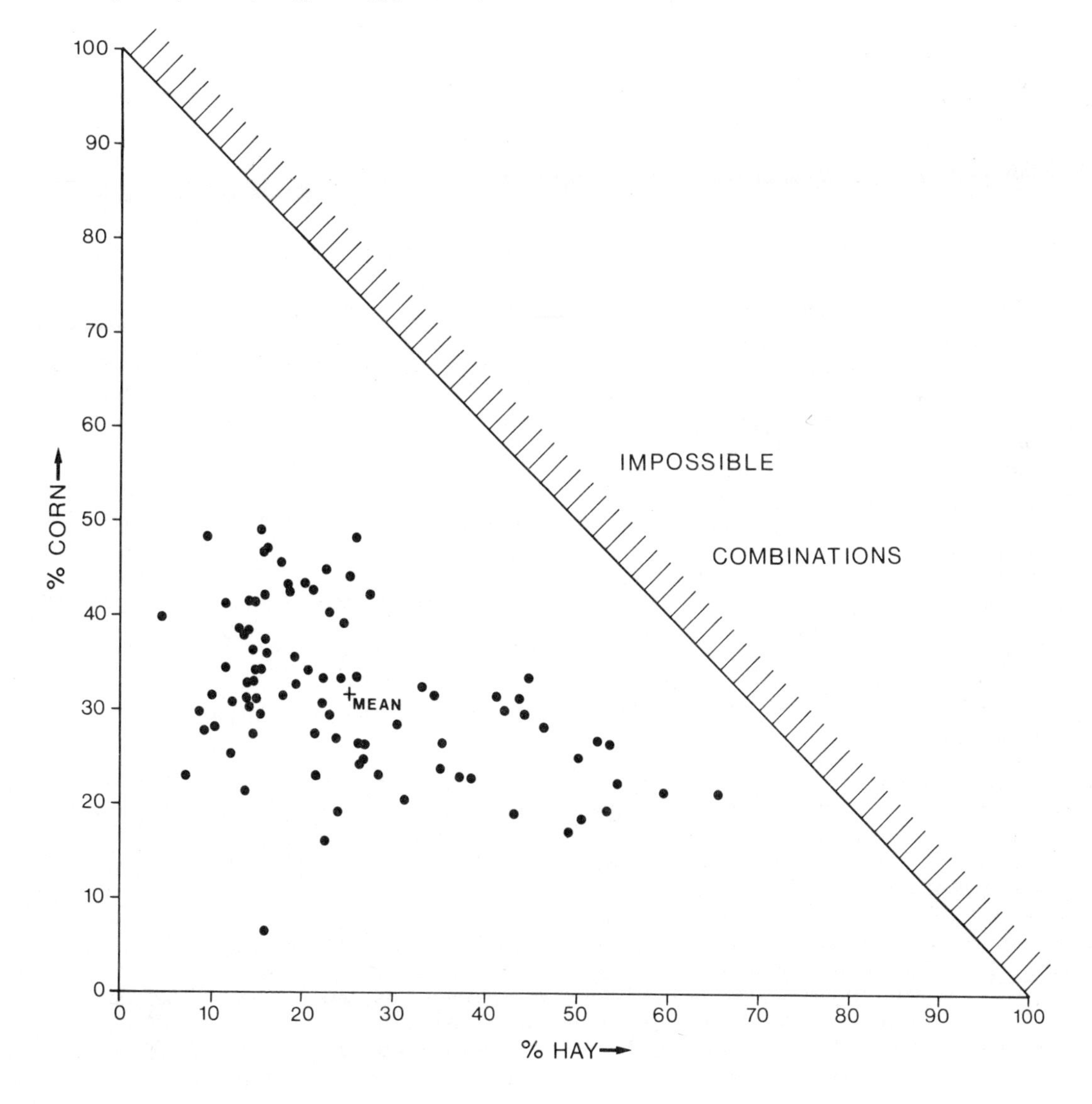

Figure 12.2 A ternary plot of the same data as Figure 12.1

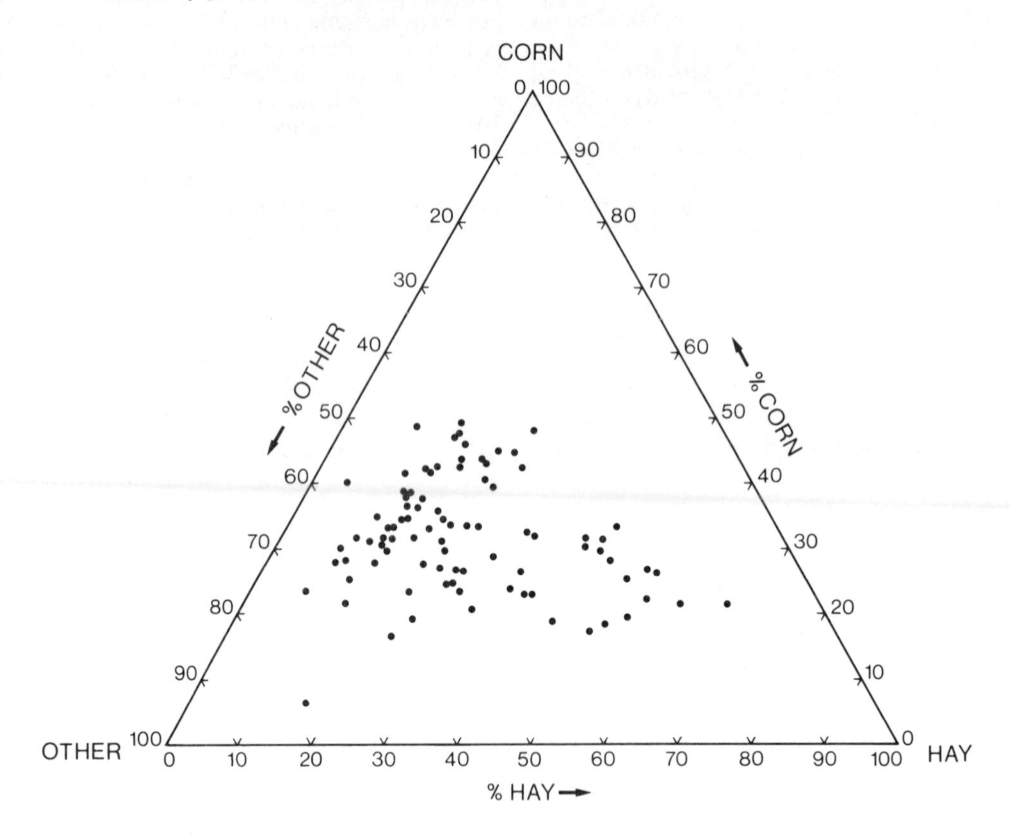

ative correlation is still present, but very small. We can say, then, that this 'distortion' (of the conventional interpretation of correlation) is greatest for large proportions and hence is more likely when the total is divided into only a few categories. The distortion can be reduced by subdividing the categories until each has a small mean proportion, say less than 0.1.

If variance is low, the constraint shown by the diagonal on Figure 12.1 is less likely to be felt than if it is high (in relation to the mean). Correlations between more variable closed groups are more constrained, and hence more difficult to interpret. This effect is less important than that of mean proportion, since within a single classification variance does tend to a close relation with the mean.

One obvious way to avoid the closure constraint is by a 'remaining space' transformation, redefining one of the ratios. Hence manufacturing/total would be related to services/(total-manufacturing). However, in a series of such ratios, some categories would have different denominators for different correlations (Johnston, 1978, p. 270); a correlation matrix could not be constructed in this way. Instead, then, a 'null model' is sought, to give expected correlations 'due solely to the closure constraint'. Two null models are available: those of Mosimann (1962) and Chayes and Kruskal (1966). (A further model elaborated by Darroch (1969) differs from Mosimann's in its derivation, but gives identical 'null correlations' in the bivariate case which is considered here: see also Darroch and Ratcliffe, 1978.)

Each model is based on certain assumptions. Mosimann assumes multinomial frequency distibutions, such that variance is proportional to mean for each numerator. This holds, very approximately, within particular classifications. The null correlation between variables i and j, where p_i and p_j are the corresponding overall proportions, is given simply by:

$$-\sqrt{\frac{p_i\, p_j}{(1-p_i)\,(1-p_j)}}$$

This value lies between 0 and -1: it is larger as the overall proportions are larger. Hence it reflects our observation concerning the effect of the means; it is not influenced by variability since this is assumed to be proportional to mean. (A similar formula was derived independently by Sarmanov, 1961.) If means (and, by implication, variances) are equal the formula reduces to: $-1/(m-1)$ where m is the number of variables. Table 12.1 gives null correlations for various combinations of overall proportions.

Does this make any difference to interpretations

made by geographers who ignore characteristics of ratios? Since null correlations may approach -1.0, it certainly does. Mosimann null correlations were calculated for the Ohio cropland example given by King (1969, p. 177), omitting two crops with 0.14 and 0.16 per cent of the cropland (King had already omitted 6.03 per cent of the total). King noted the possibility of closure effects, but Table 12.2 shows that the null expectations are considerable, and require modified interpretation of at least the negative correlations observed.

The correlation between hay and corn is essentially accounted for by the closure effect: where there is more hay there is less room for corn. Closure also accounts for much of the correlations between hay and oats, hay and wheat, and corn and oats. Conversely, the positive association between wheat and corn has been weakened by the closure effect and is really quite strong. Soybeans are positively related to oats, and 'avoid' hay to a greater extent than does wheat. The effect of closure on King's principal components analysis is unknown, but it seems inadvisable to base this on the observed correlations. King's (pp. 179-82) rather complex interpretation of the results seems less natural than a resolution into three components: hay; wheat + corn; and soybeans + oats.

While Mosimann's model is straightforward and should be included in every introductory statistics text for social or environmental scientists, that of Chayes and Kruskal (1966) is more complex and should be considered only at graduate level: fortunately, the values calculated often relate quite closely to the Mosimann values. Starting from Pearson's (1897) pioneer work on ratio correlation, Chayes and Kruskal considered the effect not only of mean proportions, but also of variances. Their null model is thus less restrictive than Mosimann's, but this relaxation brings problems which seem insuperable for its widespread application.

First, they invoke an hypothetical set of uncorrelated open variables (which, on closure through division by case totals, yields the observed closed variables); this does not always exist. Attempts to calculate it by the Chayes-Kruskal procedure may lead to 'negative variances', a warning sign which indicates that there must be a 'real' association somewhere in the matrix, but we cannot state which categories are involved. Second, the formulae used by Chayes and Kruskal are approximations in which terms in deviations squared, and higher powers, are ignored. Large errors arise when the coefficient of variation (ratio of standard deviation to mean) of a numerator or denominator exceeds 0.15 (Chayes, 1975), as it does for almost all census variables for small areas (Chapter 4), and probably for the absolute numbers underlying Table 12.2 (Ohio counties vary little in total crop area). Coefficients of variation of census counts for 2 km grid squares in Great Britain, 1971, are commonly around 3.0: they range from 1.4 (for number employed in agriculture) to 11.1 (for those born in the UK with parents born in the New Commonwealth). Skewness varies from 3.3 to 26.9. (We should note in passing that the coefficient of variation, being itself a ratio, suffers from problems of instability as the mean approaches zero, and its use assumes that the standard deviation is closely and linearly related to the mean; Cox, 1978.)

A third and possibly fatal problem is that the open matrix of uncorrelated variables invoked by Chayes and Kruskal is but one of many open matrices which, on closure, would yield the observed matrix. Kork (1977) has provided a mathematical demonstration that the related covariance matrices vary considerably, and it is difficult even to establish what properties they have in common. It is dangerous, then, to place reliance on just one matrix, which may be atypical. Was the observed matrix closed from one with uncorrelated categories, or one of the others with quite different 'real' correlations: what are the probabilities? The Chayes-Kruskal procedure is useful only 'to determine whether it is possible that the theoretical open population variables were all uncorrelated' (Kork, 1977, p. 552), not to indicate which correlations might be 'real'. Since in geography we wish to assess the effect of closure on specific relationships, the Chayes-Kruskal procedure is not relevant. Instead, Mosimann's null model should be applied despite its

p_j / p_i	*0.05*	*0.10*	*0.20*	*0.33*	*0.50*	*0.70*	*0.90*
0.05	−0.05	−0.08	−0.11	−0.16	−0.23	−0.35	−0.69
0.10	−0.08	−0.11	−0.16	−0.24	−0.33	−0.51	−1.0
0.20	−0.11	−0.16	−0.25	−0.35	−0.50	−0.76	
0.33	−0.16	−0.24	−0.35	−0.50	−0.71		
0.50	−0.23	−0.33	−0.50	−0.71	−1.0		
0.70	−0.35	−0.51	−0.76			IMPOSSIBLE	
0.90	−0.69	−1.0				COMBINATIONS	

Table 12.1 Null correlations (those due solely to the closure constraint) between variables i and j expected according to the Mosimann (1962) model, for various combinations of overall proportions, p_i and p_j. Correlations of −1.0 are inevitable whenever p_i plus p_j equals 1.0.

Table 12.2 Product-moment correlation coefficients between proportions of cropland under different crops in 1940, over the 88 counties of Ohio. The corresponding Mosimann null correlations are given in brackets; p is the mean proportion and SD its standard deviation over the 88 counties.

(1-p)	*p*	*SD*	*CROP*	CORN	WHEAT	OATS	SOYBEANS	HAY
0.6822	0.3178	0.0867	CORN					
0.7984	0.2016	0.0568	WHEAT	+0.46 (-0.34)				
0.8961	0.1039	0.0646	OATS	-0.33 (-0.23)	+0.08 (-0.17)			
0.9403	0.0597	0.0673	SOYBEANS	+0.01 (-0.17)	+0.13 (-0.13)	+0.38 (-0.09)		
0.7463	0.2537	0.1378	HAY	-0.45 (-0.40)	-0.64 (-0.29)	-0.32 (-0.20)	-0.60 (-0.15)	

apparently restrictive assumptions.

However impossible it is to put into operation, the Chayes-Kruskal model cannot simply be ignored. It expresses more than our simple intuitive notion of closure as competition for the finite space of 100 per cent. Chayes (1971) stated that 'remaining space' plots do not solve the closure problem because both (services) and (total-manufacturing) are components of (total); there is a ratio correlation, for which Chayes derives a 'null value'. Hence his concept of closure includes ratio correlation in addition to Mosimann's and Darroch's concept of closure.

Chayes-Kruskal null correlations can be positive: this contradicts our intuitive notion of closure. It is, however, a perfectly reasonable outcome when variances are allowed to vary other than in proportion to means. The closure effect is greater for the more variable categories, which have more negative null correlations than in the Mosimann model. To balance this, the less variable categories must have less negative null correlations, and given certain combinations of means and variances (Chayes, 1971, p. 39) some of these null correlations are positive. In Table 12.2, for example, hay is the most variable category: its Chayes-Kruskal null correlations are more negative than the Mosimann values. Johnston (1977) also demonstrated the effect of variability.

In a real or hypothetical closed matrix, the sum of deviations for a variable must be zero, hence the variable's sum of covariances, including its variance, must be zero: since the variance must be positive, the sum of the other covariances must be negative by a corresponding amount. The Chayes-Kruskal procedure attempts to apportion the negative covariance occasioned by the observed variances: the Mosimann procedure apportions the hypothetical negative covariance which would result if the categories followed a multinomial distribution with the observed mean proportions. In the Chayes-Kruskal model, if one correlation in a closed matrix is significantly different from the null expectation, the null expectations for the others are shifted in the opposite direction. In the Mosimann model, because the variances are hypothetical, the others are unaffected.

An important question is whether observed variances differ from multinomially expected ones only because of real processes such as segregation: if so, the 'ratio correlation' component of the Chayes-Kruskal model cannot be regarded as an inbuilt or 'spurious' effect. Since Darroch and Ratcliffe (1978) have shown that only the Dirichlet distribution (for which Mosimann's formula applies) is compatible with two-way neutrality between a pair of variables, the Chayes-Kruskal deviations from this imply that each variable follows a distribution of a different family.

Alternatives to correlating closed ratios

Those who correlate closed ratios should, at least, apply the Mosimann procedure as an aid to interpreting the results. This is not ideal, because if a single measure of relationship is required, Δr (the difference between observed and Mosimann values) is inadequate because its statistical properties are unknown. Fortunately, there are a number of alternatives to the correlation of closed ratios. First, transforms such as angular, logit or probit may be correlated: these change the closure effect, but do not eliminate it since some combinations of the two variables are impossible (the shape of the 'impossible' area in Figure 12.1 is simply distorted).

Second, partial correlations between counts or measurements may be calculated, allowing for the effect of the total. This is promising, but transformations (such as log or square root) are usually required first, to reduce skewness of the absolute numbers. Third, chi-square values may be correlated, so long as the sign of (observed-expected) is retained; M. Visvalingam is investigating this approach.

Fourth, the dissimilarity index (D) may be calculated (Duncan and Duncan, 1955): this is based on absolute numbers. Evans (1979b) has compared D matrices with r matrices and suggested that the former give the more meaningful results. These alternatives are discussed in more detail by Evans (forthcoming).

Ratio correlation and its alternatives

While the effect of closure is intuitively obvious, the effects of the other types of ratio correlation discussed by Chayes (1971) are more controversial: the realism of hypothetical situations invoked to show these effects may more easily be challenged. Pearson's (1897) null model of uncorrelated absolute numbers was challenged by Yule (1910), who pointed out that one might equally invoke uncorrelated ratios: absolute numbers derived from these would then have 'spurious' relationships! Both Albrecht and Hills, in their comments on Atchley, Gaskins and Anderson (1976), emphasized that the whole point of calculating ratios was the existence of correlation between numerator and denominator. Atchley, Gaskins and Anderson did in fact consider the case of correlated absolute numbers both in their original paper and in their reply; consideration of the uncorrelated case is simply a necessary step in the exposition. They showed that with a common denominator, inbuilt ratio correlations occur even where the three absolute numbers are highly correlated, unless both the correlations and the coefficients of variation are identical for all three: the effect is very sensitive to variations in these relationships. Both they and Albrecht (in his comment) suggested that ratios are best avoided since it may be more troublesome to validate the use of ratios for a particular problem, than to use better, more elaborate procedures such as covariance analysis.

The 'null' ratio correlations calculated by Atchley, Gaskins and Anderson (1976) are nevertheless inapplicable to many geographic data sets, for a different reason. As noted above, the approximations involved in (Pearson's) estimates of null correlations break down with absolute numbers which have coefficients of variation above 0.15; these include almost all the skewed distributions of absolute numbers encountered in geography. The present situation is that we cannot estimate null values for correlations between ratios based on these variables. We cannot demonstrate that the inbuilt effect is important, or that it is unimportant. We suspect that it is quite important in many cases studied by geographers, even where numerators and denominators are highly correlated.

The basic point is that the correlations between ratios with a common element, or between a ratio and one of its elements, are constrained by the means and variances of the elements. A large part of an observed correlation may be accounted for by such constraints, so that substantive interpretation of the correlation may be misleading. In geomorphology, authors such as Blong (1975) have placed excessive faith in the ability of ratios to provide, for example, 'length-independent' measures: it is difficult to disentangle the substantive from the inbuilt parts of the resulting correlations. Benson (1965) gave hydrologic examples; Riggs (1963, pp. 81-4) gave a useful discussion with physiological examples. Here Mosimann's (1970) theoretical conclusions make sobering reading. Sometimes it is difficult to conceive of a variable (e.g. population density, Jones, 1978; or death rate, Yule, 1910) except in ratio form, but it is often preferable to work instead with absolute numbers (Fuguitt and Lieberson, 1974). This leads to the use of regression, and the analysis of residuals from regression in preference to ratios (Jones, 1978).

A misleading example of ratio correlation was provided by a scatter plot on p. 59 of Haggett, Cliff and Frey (1977), retained from Haggett's first edition. (The fact that this was a visual portrayal of correlation and no coefficient was given is irrelevant.) For each of 100 counties of Santa Catarina State, Brasil, Haggett had data for a number of contiguous counties (N), for area (A) and for population (Q). He plotted 'contact index' (=N/A) against population density (=Q/A) to show that densely populated counties had relatively more neighbours: the result (on log. - log. paper) was a strong positive relationship. It is likely, however, that N, which ranges from 2 to 14, is much less variable than A and Q (the ratios have roughly thousandfold ranges). In other words, N is relatively constant, and 1/A is being plotted against Q/A; a strong relationship is therefore unavoidable, and of no substantive interest. It would be more interesting to correlate (the logarithms of) N, Q and A and thence calculate the partial correlations, especially that between N and Q allowing for A. N increases with A (this is a geometric necessity in a simple random model), and it would be interesting to know whether Q strengthens this relationship. Alternatively, N could be plotted against Q/A, as the hypothesis stated by Haggett implies.

The first alternative to correlating open ratios, then, is to reframe our hypotheses in terms of absolute numbers and to correlate them, with or without moving on to partial correlations. If ratios are of greater interest, or if we are uncertain, we may also calculate partial correlations for ratios (Brown, Greenwood and Wood, 1914). With a common denominator, it is appropriate to allow for its inverse (Fleiss and Tanur, 1971). In any case, it is essential to analyse the frequency distributions of and relations between the absolute numbers on which the ratios are based.

Secondly, a logarithmic transformation may be useful: the situation for open ratios is different from that for closed. Since log (Q/A) is equal to (log Q — log A), the relation between a ratio and its elements is simpler in logarithmic terms (Hills, 1978). This usually reduces but does not remove inbuilt correlation, which is a function of the standard deviations of the logarithms: given these and any one correlation, the other correlations are fixed (Craig and Haskey, 1978, Fig. 1: these authors show the relationship between regressions of log (Q/A) on Q and log A on log Q). Pethick (1975) and Ferguson (1978) regressed log (L/A) on log A where L is total length of streams; the all-important controls of this relationship are the variations in (L/A) and in A (Richards, 1978), not just the covariances. One might also note the need to use functional relationships rather than regressions in each of these examples (Mark and Peucker, 1978; Mark and Church, 1977; Jones, 1979). A third and better alternative is to hold the denominator constant. For density variables, this means using grid squares: for ratios related to population, units equal in population are required. Though difficult to set up, such analyses may provide the only answer where intractable controversies concerning interpretation have arisen (Ferguson, 1978). Note, however, that they do not remove the effect of closure. The excellent review by Church and Mark (1980) has finally resolved this area/drainage density problem.

Closed ratio dependent variables: logit regression

Moving on to consider regression, the most obvious problems arise when a dependent variable is a closed ratio. Although such regression models are frequently estimated by ordinary least squares (OLS) it is not commonly realized that this quite generally leads to nonsensical predictions of the dependent variable, and to the regression coefficients of such a model being no longer the 'best' (minimum variance) estimates. Many geographers (for example, Davis and Casetti, 1978, p. 205; Norton and Rees, 1979, p. 144; Pooley, 1977, p. 371) persist in using such OLS estimates despite the warnings contained in the expository papers of Wrigley (1973, 1976) and the consideration of some aspects of the problem in undergraduate texts (Johnston, 1978, pp. 263-7; Norcliffe, 1977, p. 236).

The first aspect of the problem, that of nonsensical values, can be demonstrated by an example. For each of the counties of England and Wales let X_i be a measure of drought in 1976 and let p_i be the observed proportion of houses with foundation collapse (Clark, 1980). If the linear model

$$P_i = E(p_i) = \beta_1 + \beta_2 X_i \qquad (1)$$

is estimated by OLS then it is possible that some of the predicted proportions could be less than zero or greater than one. But it is meaningless to suggest that less than no houses had foundation collapse or that more than all houses had foundation collapse. Even if the fitted values of the dependent variable do not fall within the range 0 to 1 it is usually possible to choose plausible values of the independent variable that result in impossible predictions of the dependent variable. We require some function to relate the dependent to the independent variable so that the predicted values are constrained to lie between 0 and 1. There are an infinite number of such functions but, in practice, two functions (based on the logistic or on the cumulative Gaussian distribution) have proved to be the most popular; a relationship based on the angular transformation (arcsine square root of the proportion) would also be feasible.

The logistic model is given by

$$P_i = 1/(1 + \exp(-\beta_1 - \beta_2 X_i)) \qquad (2)$$

and has the shape shown in Figure 12.3. This model is obviously non-linear but a transformation of the dependent variable

$$L_i = \log_e \frac{P_i}{1 - P_i} \qquad (3)$$

enables equation (2) to be written in a familiar linear form

$$L_i = \beta_1 + \beta_2 X_i \qquad (4)$$

The ratio P_i / (1 - P_i) represents the odds of an individual house in a county collapsing from foundation failure; it ranges from 0 to infinity. The natural logarithm of this value (the so-called 'logit' L_i) will range from minus infinity to plus infinity as P_i goes from 0 to 1. Thus, the logit transformation achieves the result of modifying the dependent variable so that the values predicted are constrained to lie between 0 and 1.

The second commonly used transformation is called the 'probit' and is based on the cumulative Gaussian distribution. The normal and logistic functions, however, are very similar (differences only show in the tails, with the Gaussian curve approaching the asymptote more quickly than the logistic curve - Figure 12.3); the choice between the transformations is often one of convenience. In the following discussion the logit transformation will be considered because of its simpler mathematical properties.

To understand the estimation problems posed by a dependent variable being a closed proportion, we must examine the composition of that variable in more detail. An observed proportion can be written as

$$p_i = \frac{r_i}{n_i} \qquad (5)$$

where r_i is the number of houses in a county with foundation collapse and n_i is the number of houses

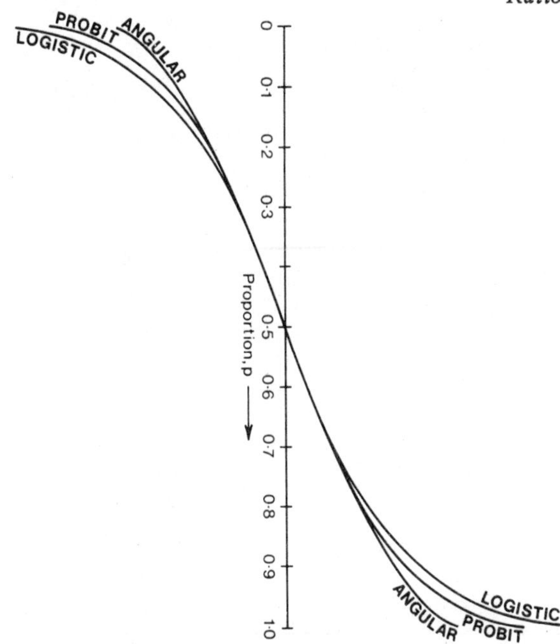

Figure 12.3 Two-bend transformations of proportions: the logistic (logit), probit and angular (Kruskal, 1968)

in a county. All the individual houses in a county have been classified as to whether they have suffered foundation collapse (r_i) or not (n_i - r_i). Such a classification is termed binomial and the resultant counts might be expected to follow a binomial distribution.

When a binomially distributed variable is used as a dependent variable and the model is estimated by OLS the problem of heteroscedasticity arises, that is, the error variance of the model will vary with differing values of the independent variable (Chatterjee and Price, 1977, p. 38). Such non-constancy of variance will lead to the OLS regression coefficients remaining unbiased but having a large variance; in any particular application they may be very different from the true coefficients. Moreover, the commonly used methods of evaluating a model (the t and F tests and R^2, the coefficient of determination) are also likely to be incorrect.

Fortunately, this problem can be overcome by using our knowledge of the form of the non-constancy of the error term to derive an improved estimator. Whereas in OLS estimation a model is fitted that minimizes the sum of squared residuals

$$\Sigma e_i^2$$

models with closed dependent variables require that the fitted model minimizes

$$\Sigma w_i e_i^2$$

where the weights (w_i) are derived from our knowledge of the error variance; for the logit model

$$w_i = n_i p_i (1 - p_i) \quad (6)$$

As the total number of houses in a county (n_i) increases, so the associated proportion for that county has an increased influence on the overall estimation of the model. Furthermore, for a given value of n_i, less weight is attached to a particular observation when the observed proportion approaches 0 or 1.

Such weighted least squares estimates of the logit-transformed model offer a considerable improvement over OLS estimates, but three important points must be considered. Firstly, the logit model is non-linear and presumes that moving P_i from 0.49 to 0.50 requires a smaller change in X_i than a change from 0 to 0.01 or 0.99 to 1.00. While this may be realistic for many applications (Abler *et al.*, 1972, pp. 144-7), an analyst should consider the appropriateness of the model for the particular task in hand. Secondly, when the observed proportions are exactly 0 or 1 the logit is undefined and the weights ensure that such observations are effectively excluded from the estimation of the model. Berkson (1953) regards this as an unwarranted loss of information and suggests that 0 and 1 be replaced by nearby working values. An exploratory approach (Cox and Jones, this volume) would be to estimate the model with and without such working values and, if the results differ considerably, the nature of the extreme data points should be considered more carefully. Finally, for each county it has been assumed that every individual house has the same chance of suffering foundation collapse. Similarly, there is only one measure of drought for each county and thus all parts of the county are assumed to be equally affected. For counties having exactly the same drought value their r_i values should be added, their n_i values should be added and a new proportion calculated.

The message is clear: regression models with a

dependent variable which is a closed proportion should not be estimated by ordinary least squares. It is recommended that the proportions are transformed and the model estimated by weighted least squares; this task can easily be accomplished by using statistical packages such as GENSTAT or GLIM.

Conclusions and recommendations

Over the last twenty years, British geographers have progressed from the incautious application of standard statistical techniques which are aspatial and assume independent measurements (or independent stochastic errors), to more selective applications and the development of techniques more appropriate to their data and hypotheses. There has been occasional recognition of the special and problematic properties of ratios, mainly from those who have noted the important (but inconclusive) work in geology, biometrics, econometrics and sociology. Yet this recognition has usually come as an afterthought, failing to permeate the body of a work: for example, after many chapters using ratios and closed percentages as the main examples, Johnston (1978) concluded with a section suggesting that there may be problems with closed percentages.

Clearly the problems which we note here for correlation coefficients carry over into analyses based on correlation matrices, such as R-mode component analysis (Butler, 1978; Johnston, 1977; Chayes and Trochimczyk, 1978), factor analysis and classification. Other problems occur in calculating means, since $\Sigma\ (x/y)$ differs from $\Sigma x/\Sigma y$ Unbiased estimation of a population ratio from sample values is in fact quite complex, and Flueck and Holland (1976) reanalysed an important example where the use of biased estimators had misleadingly suggested that cloud seeding increased precipitation significantly; the experiments quoted were essentially inconclusive. We have concentrated here on spatial data sets which lead to ratios: the essentially aspatial ratio techniques discussed in books such as Fleiss (1973) do not seem directly applicable.

It is to be hoped that future work will be based on a more thorough understanding of the special properties of ratios and especially of the limitations of ratio correlation. Careful consideration of absolute numbers is required, together with use of transformations and partial correlations. This may be associated with a move towards use of path analysis (Schuessler, 1973) in the search for deeper understanding than factorial ecology has provided. The dissimilarity index (and the related index of segregation) has proved useful for closed data sets; this and the more experimental approaches mentioned above should expand at the expense of ratio correlations. If ratios can reasonably be avoided, they should be: this may save embarrassment.

Acknowledgments

We are very grateful to Nick Cox for his comments on this paper, to Florence Blackett for typing and to David Hume for drawing the figures.

References

Abler, R., Adams, J.S., and Gould, P.R. (1972) *Spatial organisation*, Prentice-Hall: Englewood Cliffs, NJ.

Atchley, W.R., Gaskins, C.T., and Anderson, D. (1976) 'Statistical properties of ratios. I – empirical results', *Systematic Zoology*, 25, 137-48. Discussions 1977: 26, 211-14; 1978: 27, 61-83.

Benson, M.A. (1965) 'Spurious correlation in hydraulics and hydrology', *Journal of Hydraulics Division, American Society of Civil Engineers*, 91 (HY4), 35-42.

Berkson, J. (1953) 'A statistically precise and relatively simple method of estimating the bio-assay with quantal response based on the logistic function', *Journal of the American Statistical Association*, 50, 130-62.

Blong, R.J. (1975) 'Hillslope morphometry and classification: a New Zealand example', *Zeitschrift für Geomorphologie*, NF 19, 405-29.

Brown, J.W., Greenwood, M., and Wood, F. (1914) 'A study of index correlations', *Journal of the Royal Statistical Society*, 77, 317-46.

Butler, J.C. (1978) 'Visual bias in R-mode dendrograms due to the effect of the closure', *Journal of Mathematical Geology*, 10, 243-52.

Chatterjee, S., and Price, B. (1977) *Regression analysis by example*, John Wiley: New York.

Chayes, F. (1971) *Ratio correlation. A manual for students of geochemistry and petrology*, University of Chicago Press.

Chayes, F. (1975) '*A priori* and experimental approximation of simple ratio correlations' in R.B. McCammon (ed.) *Concepts in Geostatistics*, 106-37, Springer Verlag: Berlin.

Chayes, F., and Kruskal, W. (1966) 'An approximate statistical test for correlations between proportions', *Journal of Geology*, 74, 692-702. Corrected ibid. 1970: 78, 380.

Chayes, F., and Trochimczyk, S. (1978) 'An effect of closure on the structure of principal components', *Journal of Mathematical Geology*, 10, 323-33.

Choynowski, M. (1959) 'Maps based on probabilities', *Journal of the American Statistical Association*, 54, 385-8. Reprinted in B.J.L. Berry and D.F. Marble (eds) (1968) *Spatial Analysis*, 180-3, Prentice-Hall: Englewood Cliffs, NJ.

Church, M., and Mark, D.M. (1980) 'On size and scale in Geomorphology', *Progress in Physical Geography*, 4 (3), 342-90.

Clark, M.J. (1980) 'Property damage by foundation failure' in J.C. Doornkamp and K.J. Gregory (eds.) *Atlas of drought in Britain 1975-6* 63-4, Institute of British Geographers Occasional Publication, London.

Clarke, J.I., Dewdney, J.C., Evans, I.S., Rhind, D.W., Visvalingam, M., and Denham, C. (1980) *People in Britain: a census atlas,* HMSO: London for OPCS.

Cole, J.P. (1964) *Italy,* Chatto & Windus: London.

Cox, N.J. (1978) 'Hillslope profile analysis', *Area,* 10, 131-3.

Craig, J., and Haskey, J. (1978) 'The relationships between the population, area, and density of urban areas', *Urban Studies,* 15, 101-7.

Darroch, J.N. (1969) 'Null correlation for proportions', *Journal of Mathematical Geology,* 1, 221-7.

Darroch, J.N., and Ratcliffe, D. (1978) 'No-association of proportions', *Journal of Mathematical Geology,* 10, 361-8.

Davis, D., and Casetti, E. (1978) 'Do black students wish to live in integrated socially-homogeneous neighbourhoods? A questionnaire analysis', *Economic Geography,* 54, 197-209.

Duncan, O.D., and Duncan, B. (1955) 'A methodological analysis of segregation indices', *American Sociological Review,* 20, 210-17.

Evans, I.S. (1979a) *An integrated system of terrain analysis and slope mapping,* Final Report on Project DA-ERO 891-73-G0040 Department of Geography, University of Durham.

Evans, I.S. (1979b) 'Relationships between Great Britain census variables at the 1 km aggregate level' in N. Wrigley (ed.) *Statistical applications in the spatial sciences,* 141-84, Pion: London.

Evans, I.S. (forthcoming) 'Strategies for coping with closure and ratio correlations', mimeo, Dept. of Geography, University of Durham.

Ferguson, R.I. (1978) 'Drainage density - basin area relationship (comment)', *Area,* 10, 350-2.

Fleiss, J.L. (1973) *Statistical methods for rates and proportions,* John Wiley: New York.

Fleiss, J.L., and Tanur, J.M. (1971) 'A note on the partial correlation coefficient', *The American Statistician,* 25, 43-5.

Flueck, J.A., and Holland, B.S. (1976) 'Ratio estimators and some inherent problems in their utilisation', *Journal of Applied Meteorology,* 15, 535-42.

Fuguitt, G.V., and Lieberson, S. (1974) 'Correlation of ratios or difference scores having common terms' in H.L. Costner (ed.), *Sociological Methodology 1973-74,* 128-44, Jossey-Bass: San Francisco.

Haggett, P., Cliff, A.D., and Frey, A. (1977) *Locational Analysis in Human Geography,* 2nd edn, Edward Arnold: London.

Haynes, R.M. (1978) 'A note on dimensions and relationships in human geography', *Geographical Analysis,* 10, 288-91.

Hills, M. (1978) In discussion on Atchley *et al.* 1976.

Howe, G.M. (1970) *National atlas of disease mortality in the United Kingdom,* Nelson: London.

Johnston, R.J. (1977) 'Principal components analysis and factor analysis in geographical research: some problems and issues', *South African Geographical Journal,* 59, 30-44.

Johnston, R.J. (1978) *Multivariate statistical analysis in geography,* Longman: London.

Jones, K. (1978) 'Percentages, ratios and inbuilt relationships in geographical research: an overview and bibliography', *Department of Geography,* University of Southampton, Discussion Paper, 2. ISSN 0140 9875.

Jones, T.A. (1979) 'Fitting straight lines when both variables are subject to error. I: maximum likelihood and least squares estimation', *Journal of Mathematical Geology,* 11, 1-25.

Jones, T.P., and McEvoy, D. (1978) 'Race and space in cloud-cuckoo land', *Area,* 10, 162-6. Discussion, 10, 365-7; 1979: 11, 84-5 and 221-3.

King, L.J. (1969) *Statistical analysis in geography,* Prentice-Hall: Englewood Cliffs, NJ.

Kork, J.O. (1977) 'Examination of the Chayes-Kruskal procedure for testing correlations between proportions', *Journal of Mathematical Geology,* 9, 543-62.

Kruskal, J.B. (1968) 'Statistical analysis, special problems of II. Transformations of data' in D.L. Sills (ed.), *International Encyclopedia of the Social Sciences,* Vol. 15, 182-93, Macmillan: New York.

Mark, D.M., and Church, M. (1977) 'On the misuse of regression in earth science', *Journal of Mathematical Geology,* 9, 63-75.

Mark, D.M., and Peucker, T.K. (1978) 'Regression analysis and geographic models', *Canadian Geographer,* 22, 51-64.

Mosimann, J. (1962) 'On the compound multinomial distribution, the multivariate beta distribution, and correlation among proportions', *Biometrika,* 49, 65-82.

Mosimann, J. (1970) 'Size allometry: size and shape variables with characterizations of the log-normal and generalized gamma distributions', *Journal of the American Statistical Association,* 65, 930-45.

Norcliffe, G.B. (1972) 'Probability mapping of growth processes', *Economic Geography,* 48, 428-38.

Norcliffe, G.B. (1977) *Inferential statistics for geographers,* Hutchinson: London.

Norton, R.D., and Rees, J. (1979) 'The product cycle and the spatial decentralisation of American manufacturing', *Regional Studies,* 13, 141-51.

Pearson, K. (1897) 'Mathematical contributions to the theory of evolution - On a form of spurious correlation which may arise when indices are used in measurement of organs', *Proceedings of the Royal Society, A,* 60, 489-502.

Pethick, J.S. (1975) 'A note on the drainage density-basin area relationship', *Area,* 7, 217-22.

Phillips, P.D. (1973) 'Risk-related crime rates and crime patterns', *Proceedings of the Association of American Geographers,* 5, 221-4.

Pooley, C.G. (1977) 'The residential segregation of migrant communities in Mid-Victorian Liverpool', *Transactions of the Institute of British Geographers,* NS 2, 364-82.

Richards, K.S. (1978) 'Yet more notes on the drainage density-basin area relationship', *Area,* 10, 344-8.

Riggs, D.A. (1963) *The mathematical approach to physiological problems: a critical primer,* Williams & Wilkins: Baltimore.

Sarmanov, O.V. (1961) 'False correlations between random variables' (in Russian), *Trudy MIAN SSSR,* 64, 173-84.

Schuessler, K. (1973) 'Ratio variables and path models' in A.S. Goldberger and O.D. Duncan (eds), *Structural equation models in the social sciences,* 201-28, Seminar Press: New York.

Visvalingam, M. (1978) 'The signed chi-square measure for mapping', *Cartographic Journal,* 15, 93-8.

White, R.R. (1972) 'Probability maps of leukaemia mortalities in England and Wales' in N.D. McGlashan (ed.), *Medical Geography: techniques and field studies,* 173-85, Methuen: London.

Williams, R.L. (1976) 'The misuse of area in mapping census-type numbers', *Historical Methods Newsletter* 9, 213-16.

Wrigley, N. (1973) 'The use of percentages in geographical research', *Area,* 5, 183-6.

Wrigley, N. (1976) *An introduction to the use of logit models in geography,* Concepts and techniques in modern geography 10, Geo. Abstracts: Norwich.

Yule, G.U. (1910) 'On the interpretation of correlations between indices or ratios', *Journal of the Royal Statistical Society,* 73, 644-7.

Chapter 13

Exploratory data analysis

N.J. Cox and K. Jones

In *exploratory data analysis,* attempts are made to identify the major features of a data set of interest and to generate ideas for further investigation, whereas in *confirmatory data analysis,* attention is focused on model specification, parameter estimation, hypothesis testing and firm decisions about data. This distinction, made by the statistician J.W. Tukey, is essentially that between *descriptive* and *inferential* statistics. However, confusion can easily arise because Tukey has recently produced many special techniques for exploratory work, yet placed these new methods at some distance from classical statistics, whether descriptive or inferential. Tukey's innovations are now explained in a variety of texts (Tukey, 1977; Mosteller and Tukey, 1977; McNeil, 1977; Erickson and Nosanchuk, 1977) and they have attracted the interest of some geographers as methods which may be used in teaching and research (Cox and Anderson, 1978; Cox, 1978). Other geographers (e.g. Mather, 1976) march under the banner of exploratory data analysis yet do not employ Tukey's new procedures. In this review we adopt the wider sense of 'exploratory data analysis' and do not confine attention to Tukey's innovations. Four themes are apparent in exploratory data analysis: displays, residuals, transformations and resistance (Hoaglin, 1977). None of these is novel, either in statistics or in geographical data analysis, yet from a survey of the field it seems that each deserves greater emphasis in future geographical work.

Tukey's new exploratory methods have met a variety of reactions, ranging from wild enthusiasm (e.g. Wainer, 1977) to outright condemnation (Ehrenberg, 1979a, b). In particular, Ehrenberg criticized exploratory data analysis, especially as presented in the text by Tukey (1977), as poorly explained and motivated, and as fundamentally mistaken in its implication that data analysts need to work without prior knowledge, which is usually available and should be considered (cf. Ehrenberg, 1975). However, those who find Tukey's text unduly cryptic and idiosyncratic may readily be directed to other accounts, while proponents of exploratory methods do not suggest that prior knowledge should be ignored in exploratory work (Cox and Anderson, 1980). It is to be hoped that geographers will avoid unfounded extreme reactions to exploratory data analysis: an attempt to place recent work in a larger context, statistical and geographical, should help in this respect.

In this review we direct attention to those attitudes and procedures which appear most fruitful in exploratory work, and consider the relationship between exploratory (descriptive) and confirmatory (inferential) approaches.

Displays

'Plot both your data and the results of data analysis' is one of the basic attitudes of exploratory data analysis. Displays reveal the major features of data, help in the production of ideas for further investigation, and are useful in checking assumptions (Anscombe, 1973). However, while 'graphicacy' has long been an educational concern among geographers (e.g. Balchin, 1976), and many texts explain a limited standard set of graphical techniques (e.g. histograms, pie diagrams and scatter diagrams), graphical display remains neglected to some extent in quantitative geography. Since graphical inspection allows the identification of outliers, nonlinearities, discontinuities, skewness and other characteristics of the data which may make or mar the analysis, it is sensible to supplement data analyses with appropriate plots. The increasing availability of flexible computer graphics systems makes it easier to do this as a matter of routine, but it should be noted that, for data sets of moderate size, some new plots may be produced manually in a short time.

More telling, perhaps, than any general exhortation is a simple example given by Anscombe (1973, pp. 19-20), who devised four very different data sets which have the same univariate means and the same least squares regression results. It is clear from scatter diagrams whether bivariate regressions are appropriate, but relying on calculated summary measures in model evaluation would produce quite misleading interpretations in three out of four cases.

Here we draw attention to some novel plots for univariate data and (in the next section on Residuals) to appropriate displays for analysis of residuals (especially those from regression models). Lack of space precludes discussion of other kinds of plot, notably those designed specifically for the exploration of multivariate data (Gnanadesikan, 1977; Everitt, 1978).

Traditionally the histogram has pride of place for presenting univariate frequency distributions, and it will continue to be very useful. Three new kinds of displays deserve consideration, however, for this task. In a *rootogram,* not class frequencies but roots of class frequencies are plotted as ordinates, on the grounds that a square root transformation tends to stabilize variation in counts (Tukey, 1970, pp. 163-5; 1972, pp. 312-5; 1977, Ch. 17). In a *box plot,* minimum and maximum values are marked by point symbols and median and quartiles are marked by bars joined in a box. Thus range and interquartile range (or midspread, to use a Tukey term) are represented by distances between symbols. Further information can be added if desired and box plots for different sets of data juxtaposed for comparison (Tukey, 1972, pp. 301-3; 1977, Ch. 2; McNeil, 1977, Chs 1-2; Erickson and Nosanchuk, 1977, Ch. 4; McGill *et al.,* 1978: note variations in terminology). An example is given in Figure 13.1. Box plots are related to the dispersion diagrams once popular in geography, particularly for climatic data (e.g. Crowe, 1933; Gregory, 1978, pp. 147-50).

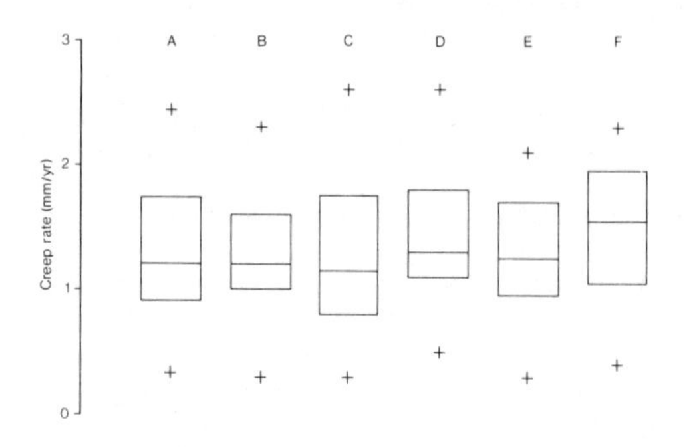

Figure 13.1 Multiple box plot of soil creep rates at Rookhope, upper Weardale, measured by A inclinometer pegs, B Anderson's tubes, C aluminium pillars, D Young's pits, E dowelling pillars, F Cassidy's tubes (Anderson and Cox, 1978)

In a *stem-and-leaf display,* values are represented by the combination of a stem (coarse) and a leaf (fine): the number of leaves on each stem corresponds to the class frequency of a histogram (Tukey, 1972, pp. 295-6; 1977, Ch. 1; McNeil, 1977, Ch. 1; Erickson and Nosanchuk, 1977, Ch. 2). As a simple case consider these annual rainfall figures (in mm) for Durham in the period 1952 to 1976 (Cox and Anderson, 1978, pp. 33-4): 575, 521, 778, 484, 628, 551, 615, 440, 782, 708, 574, 672, 515, 814, 756, 718, 742, 756, 562, 543, 563, 503, 589, 562, 683. The first five are plotted on the stem-and-leaf display below:

```
7 | 7
6 | 2
5 | 72
4 | 8
```

The numbers on the left of the vertical line are the stems: in this case 7, 6, 5, 4 (for 700, . . . , 400). The numbers on the right are the leaves, the leading digits after the stems. In this case the remaining digits have been dropped, although it would naturally be possible to round to the nearest digit. Adding the other figures to the display we obtain:

```
8 | 1
7 | 7805145
6 | 2178
5 | 72571646086
4 | 84
```

and the display can be tidied up by placing leaves on each stem in ascending order.

```
8 | 1
7 | 0145578
6 | 1278
5 | 01245666778
4 | 48
```

Hence numerical ordering produces a simple display which shows the form of the frequency distribution, costs less effort than a histogram yet contains more information, and helps when calculating measures based on ordered values, such as the median or midspread.

Residuals

Residuals are the remainders left after a model (any kind of summary description, from something simple like a measure of level to something more complex like a multiple regression) has been fitted to data. Usually we have a basic partition

$$\text{data} = \text{fit} + \text{residual}.$$

One common strategy is to assume that the model first thought of is so good that the residuals can be set aside safely as the amount unexplained, expressed indirectly as a standard error, a coefficient of determination or some other gross summary statistic.

A more realistic strategy, fundamental to exploratory data analysis and worth wider adoption in quantitative geography, is to doubt whether the model first thought of really is a good summary and to scrutinize the residuals carefully for any pattern which should be reflected in a revised model. The general approach is one of 'summarizing by fit and exposing by residuals' (Tukey and Wilk, 1966, p. 698).

Graphical display is the most useful weapon available for analysis of residuals. Some of the most valuable kinds of plots will be considered briefly in the specific context of regression analysis.

A general 'catch-all' plot shows the residuals e_i against the fitted values $\hat{Y}_i$ of the response or dependent variable (see, e.g., Chatterjee and Price, 1977, Ch. 2). Since a correctly specified regression model would account for all the systematic variation in the response, the corresponding residual plot would show no discernible pattern (e.g. Figure 13.2 (a)). Clear patterns of any kind indicate, however, that the model might need reformulation. A curved band of residuals (Figure 13.2 (b)) might be tackled by adding a square or a higher-order term or a cross-product term. A wedge shaped pattern (Figure 13.2 (c)) shows heteroscedastic variation suggesting the use of weighted least-squares or an appropriate transformation. A solitary point (Figure 13.2 (d)) indicates that the data set contains an outlier which needs further consideration.

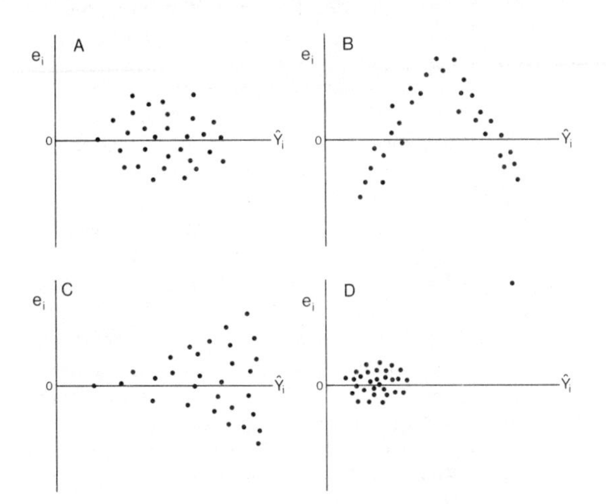

Figure 13.2 Residual plots – residuals v. predicted values of the dependent variable

Even if scatter diagrams were not used before regression, such residual plots would highlight the anomalies in Anscombe's (1973) data sets. While in one case there is no obvious pattern in the residuals, in the other instances a curvilinear pattern and definite outliers are very clear (Figure 13.3). The idiosyncrasies of these different data sets, hidden behind identical numerical summaries, are thus evident from graphical displays.

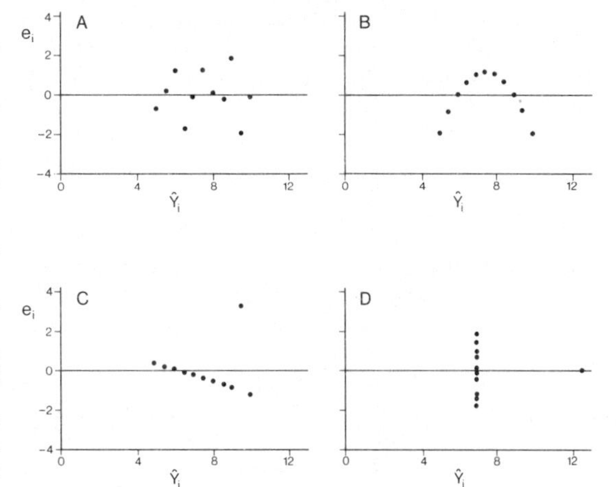

Figure 13.3 Residual plots for Anscombe's (1973) data sets

There are as yet only a few isolated examples of the use of residual plots with geographical data. Crewe and Payne (1971, 1976) used residual plots in their studies of voting behaviour in British elections. A first simple model regressed percentage Labour vote against percentage of manual workers. Fifteen constituencies (including the twelve Northern Ireland seats) were identified as true outliers on the basis of residual plots and substantive reasoning. The fifty largest positive and negative residuals remaining were used to derive a classification of constituencies and to suggest various explanatory variables that might be included in a revised model. After considerable trial and error, a final model was produced which accounted for nearly 90 per cent of the variation in the percentage Labour vote. Moreover, when the residuals from this model were examined, the apparent pattern was one of local effects which could not be reflected in a national model. Barnard (1978) attempted to develop a model of the distribution of the elderly in south-west Hampshire in terms of various dwelling and accessibility measures. A residual plot for the original regression model revealed a substantial outlier, found on investigation to be an Enumeration District in which nearly all the residents were military personnel renting Ministry of Defence accommodation. This outlier was subsequently deleted as anomalous.

Several other kinds of plots may be useful. Partial residual plots (Larsen and McCleary, 1972) essentially show the relationship between the response and a particular explanatory variable after the effects of other explanatory variables have been removed. They have been used by Jones (1980) in a study of geographical variations in mortality for tackling the difficult problem of identifying appropriate functional relationships between variables. It may also be worth plotting the frequency distribution of residuals, or residuals in series (as a graph or a map), or residuals against other variables (Tukey and Wilk, 1966, p. 699; Box *et al.*, 1978, pp. 182-7; Silk, 1979, pp. 245-7).

Resistance

Much of the theory behind statistical methods is concerned with optimal procedures, which are best according to specific criteria if particular assumptions about generating processes are satisfied. However, assumptions that relationships are additive and linear, that variables are identically and independently distributed and Gaussian, or that generating processes are stationary and isotropic - to name but a few prominent examples - are chosen at least in part for their mathematical convenience. They do indeed allow the derivation of many elegant and rigorous theorems. It is nevertheless rare in practice that such assumptions are exactly satisfied. Hence it is desirable that procedures be *robust,* and work well under a variety of conditions, not just under idealized specific conditions for which they have been proved optimal.

Robustness is especially important in exploratory work since it would be foolish to be confident that an unexamined set of data satisfies specific assumptions about behaviour. One kind of robustness particularly valuable in geography is *resistance* to data drawn from distributions which are longer-tailed than the Gaussian, and particularly to *outliers,* isolated values which are detached from the main body of data (cf. Wainer, 1976; Mosteller and Tukey, 1977, for introductions to resistance). This arises from the fact that while many of the procedures favoured by geographers (e.g. correlation, regression, analysis of variance, Student's t test) are most appropriate, if not optimal, for Gaussian variables, many distributions of geographical interest are longer-tailed and include outliers. These outliers are frequently genuine values, and not merely products of observational or experimental blunders. Dublin often appears as an outlier on plots of socio-economic data for the Irish Republic because it is genuinely different from the rest of the country! Such individual outliers can have a great distorting influence on any fitted model, and thus have both geographical and statistical importance.

It would be misleading to imply that resistant methods are the only means of dealing with outliers, which can be accommodated, incorporated, identified or rejected (Barnett, 1978), as seems appropriate. In general, graphical display is the most valuable way of checking for outliers, although this becomes more difficult as the number of data and the number of variables increase, and so resistant methods assume greater value (Cox and Hinkley, 1974, pp. 270-1; Wainer, 1976). If outliers can be identified, parallel 'with' and 'without' analyses can be conducted. Haggett *et al.* (1977, pp. 364-5), for example, repeated a regression of retail sales against personal income for Irish counties without values for Counties Cork and Dublin. It is also good practice to couple 'resistant' and 'unresistant' (or 'robust' and 'fragile') analyses (e.g. Wainer, 1976; Erickson and Nosanchuk, 1977).

The problem of robust estimation of the centre of a symmetric distribution has been the subject of extensive analytical and simulation studies by statisticians (e.g. Andrews *et al.*, 1972). It is now abundantly clear that the mean can perform very poorly with long-tailed distributions; that the median, among other estimators, is better in the presence of long tails or outliers; and that simple estimators exist which perform well under a wide range of conditions. Much of this has long been known to geographers, yet the discussions given in some geographical texts follow a comment about the resistance of the median with a dismissal on spurious grounds. Norcliffe (1977, p. 54) asserted generally that the mean is more efficient than the median, but this is not universally true (cf. Andrews *et al.*, 1972); Gregory (1978, pp. 25-6) unfavourably compared the median, possessing 'no real mathematical qualities' with the mean, 'based on sound mathematics', an invocation of mathematical respectability quite without foundation.

More positively, means and medians can be seen as limiting cases of the family of 'trimmed means' (see, e.g., Wainer, 1976, for further explanation). A p% trimmed mean is calculated by setting aside the p % largest values and the p % smallest values and taking the mean of the (100 - 2p) % of values which remain. p = 0 produces a mean, p = 50 a median and p = 25 a midmean. In principle p can be chosen according to the degree of resistance required and the character of the data, although there is little reason for using merely one value of p. One example of the use of trimmed means with air pollution data was given by Cleveland and Guarino (1976).

Moment-based measures of spread, asymmetry and tailedness (e.g. standard deviation and classical skewness and kurtosis) are generally unresistant, and there is much to be said for greater use of quantile-based measures (cf. Tukey, 1977; McNeil, 1977; Erickson and Nosanchuk, 1977). The interquartile range or 'midspread', for example, is more useful as a measure of spread than many geographers allow: common objections to it often boil down to prejudices that it is old-fashioned and not totally respectable.

Resistance is an important property not only for summary measures but also for other methods of data analysis. For example, two-way tables are often analysed using a model of the form

data = level + row effect + column effect + error.

The parameters are usually estimated via table mean, row means and column means. Tukey has devised a resistant method of iterative estimation for such tables known as median polish (Tukey, 1977, Chs 10 and 11; McNeil, 1977, Ch. 5; Erickson and Nosanchuk, 1977, Ch. 15). Anderson and Cox (1978) used median polish in a comparison of different instruments for measuring soil creep, and found a clear picture emerging from a fairly messy set of data, which was supported by an independent and more conventional analysis.

Perhaps the greatest need for resistant methods is in applications of correlation, regression and related multivariate analyses. Many resistant procedures have been devised by statisticians, and they deserve close attention from geographers. For example, Wainer (1976) outlined a method of estimating standard deviation, correlation and slope resistantly, and suggested that principal components and factor analyses be based on variance-covariance matrices derived in this way. There are simple methods for line fitting from bivariate medians of thirds of data sets (Erickson and Nosanchuk, 1977, Chs 11 and 12; McNeil, 1977, Ch. 3), while biweight estimation is an elegant resistant alternative to least squares (Mosteller and Tukey, 1977, Chs 10 and 14; McNeil, 1977, Ch. 7), used on air pollution data by Cleveland and Guarino (1976).

Transformations

Thus far we have used themes identified in exploratory data analysis by Hoaglin (1977) as headings for this review, and considered displays, residuals and resistance in turn. The remaining theme of transformations probably needs least emphasis for a geographical audience. Standard texts intended for geographers (e.g. Haggett *et al.*, 1977; Norcliffe, 1977; Taylor, 1977; Gregory, 1978) include sections on transformations, and transformations of variables of geographical interest are commonplace, at both elementary and research levels. Most of the basic transformations employed by statisticians (cf. Hoyle, 1973) are known to geographical data analysts: not only the common power, root and logarithmic transformations, but also transformations useful for categorical data such as the logit, probit and various others based on inverse trigonometric and hyperbolic functions.

Transformation of geographical data has usually been motivated by an inferential approach, and particularly by the idea that hypothesis testing requires data drawn from Gaussian (normal) distributions. One common aim of re-expressing variables has hence been normalization of frequency distributions. The degree of success of any transformation may be assessed by inspection of histograms or probability plots, calculation of skewness and kurtosis or performance of some distribution-free test such as chi-square or Kolmogorov-Smirnov. This approach has often been successful in its own terms. For example, many variables encountered by geographers are right-skewed (but not highly irregular) in distribution, and thus behave quite respectably if logarithms or square roots are taken. Investigators have then proceeded joyfully to inferential procedures; unfortunately they have frequently overlooked the fact that these may be inappropriate or irrelevant on other grounds (see below). Indeed it is common to find among geographers a notion that lack of normality is the basic statistical problem, and thus that paradise has been attained once a semblance of normality has been produced, whether by subterfuge or by honest means. This notion ignores the standard principle that other assumptions about data (especially mutual independence) are often crucial. The very term 'normal', still used by an overwhelming majority of geographers, continues to act as a misleading influence.

Some geographers have hoped that a common 'blanket' transformation will simultaneously normalize a variety of variables (perhaps a haphazard mixture of attributes about to be offered as ritual sacrifice in a principal components or factor analysis). No doubt data analysis would be simpler if life were easier, but there seem to be many empirical and statistical grounds for the contrary idea that each variable should be treated individually. Since right-skewness is a common kind of departure from normality, taking logarithms (for example) will usually improve a majority of variables, but neither logarithms nor any other simple transformation can serve as a panacea for non-normality (see, for example, results of Gardiner, 1973).

The view dominant in exploratory data analysis is that transformation of variables need be justified only by convenience. It need not be motivated by any specific inferential assumption, but merely by the aim of easier and more effective description. If pressed too far, however, this distinction of aims turns into a false antithesis. Transformations have been used frequently to achieve approximately linear relationships. It is well known, for example, that logarithmic transformation of one or both variables brings power function and exponential relationships into the family of linear relationships (cf. Tufte, 1974, pp. 108-31, for an especially lucid account). Here the aim of easier and more effective description is coupled with the hope of using the well-developed inferential machinery of the linear model.

The exploratory data analysis texts of Tukey (1977), Mosteller and Tukey (1977) and McNeil (1977) include not only many practical examples of a more liberal view of transformations, but also much valuable advice about specific issues such as the handling of zeros and the use of folded transformations. It is striking to find that Tukey regards additivity of effects and constancy of spread as more important in practice than normality (or even symmetry) of distribution, although any order of priority is confused by the happy circumstance that these three conditions frequently occur together when they do exist (Tukey and Wilk, 1966, p. 702).

Any reluctance by geographers to embark on transformation (e.g. Gould, 1970, pp. 442-3) is usually on one or both of two grounds: a feeling that transformed scales are unnatural, and an unease that transformation involves an unacceptable element of adhoc-ery, if indeed it does not verge upon statistical cheating. These objections are both exaggerated. The often cited case of pH, a logarithmic transformation of hydrogen ion concentration long accepted as a useful measure, serves as a reminder that 'naturalness' may reflect convention as much as reality. The thought that square roots of counts are less natural than raw counts is understandable, and accounts for reluctance to draw rootograms rather than histograms, but the feeling is reduced once preliminary rooting has been shown to produce a clearer picture in a few analyses. The appearance of adhoc-ery can also be avoided, and the choice of transformations made more systematic in a variety of ways. The standard power, root and logarithmic transformations can be seen to be members of a 'ladder of re-expressions' (Mosteller and Tukey, 1977, pp. 79-81): the analyst moves up and down the ladder as appropriate. The commutative property of such monotonic transformations

transform of quantile of raw data
= quantile of transform of raw data

can be used to reduce the work in choosing transformations from this ladder. (This property is an advantage of quantiles not possessed by means, and should be set against the fact that the additive property of means which allows them to be combined is not satisfied by quantiles: e.g. Ehrenberg, 1975, pp. 173-4.)

Smoothing

Smoothing is an approach used in exploratory data analysis of special interest to geographers, who are commonly faced with data in the form of spatial or time series. It is appropriate if data series may be regarded as a mixture of smooth and rough components, whereby

data = smooth + rough.

For example, data may be regarded as 'signal' mixed with 'noise', 'true values' mixed with 'measurement error', 'long-term trend' mixed with 'short-term fluctuation', or as 'regional trend' mixed with 'local deviation'.

Popular smoothing methods used in quantitative geography can be classified as either linear function fitting or local linear smoothing (Cox, 1979a). While such approaches are often useful and appropriate, their limitations (stressed here) justify interest in nonlinear smoothers introduced by Tukey as an alternative approach.

In linear function fitting we adopt a precisely specified model of the general form

data = linear function + stochastic error.

The linear function is usually a polynomial in the map coordinates (trend surface analysis). However, we thereby invoke particular assumptions about the structure of the data which may be unjustified (or unjustifiable) in exploratory work. Unless the linear function has some interpretation in terms of geographical processes, we might be making such interpretation more difficult. This approach leans heavily on the idea of linearity, whereas there are many grounds for expecting nonlinear behaviour of geographical responses. Usually least squares estimation is used: it is well known that this performs poorly in the presence of outliers or long-tailed data. The linear function is generally fitted globally to all the data: it may be more realistic not to assume that a single model is valid for every part of the data, but to move to local fitting.

In local linear smoothing, we compute a weighted average of the values in the neighbourhood of any particular value. One major advantage is then that weights can be chosen to suppress or magnify variations in particular frequency bands, at least if data are regularly spaced. Moving to local operations alleviates any doubts about the appropriateness of a global approach, and may reduce other difficulties, but we still lean heavily on linearity and (implicitly) on least squares.

Nonlinear smoothers based on running medians have been far less popular than linear smoothers, but they have been advocated in recent years by some statisticians (Beaton and Tukey, 1974; Tukey, 1977, Chs 7 and 16; McNeil, 1977, Ch. 6; Velleman, 1977) and geophysicists (Claerbout and Muir, 1973; Claerbout, 1976). When smoothing data series running medians tend to be more resistant than moving averages. A local median will be relatively uninfluenced by high or low spikes which usually cannot be regarded as part of the smooth, whereas a local mean will tend to mix such spikes in with the smooth.

Methods proposed by Tukey for smoothing one-dimensional series are based on taking running medians of successive trios of data values, often repeated until convergence and followed by Hanning, a linear smoother with weights ¼:½:¼. These methods can be generalized readily for cases of two-dimensional series, whether or not data are regularly spaced (Besag and McNeil, 1976; Cox, 1979a), but the one-dimensional methods have been far more widely used and, indeed, have been rather more successful.

This family of methods proposed by Tukey is designed for exploratory work. The attitude most fruitful in practice is to experiment with a variety of smoothers (setting aside any idea that one particular smoother might be 'best'), and to examine the resulting series of smooth and rough. This approach is easiest to implement when a computer library of smoothing routines is coupled with a cathode ray tube display (for immediate scanning) and a pen plotter or some other hard copy device.

In continuing work Cox (1979a, b) has employed non-linear smoothers on various data series of geographical interest: socio-economic data for the Irish Republic (Cliff and Ord, 1973); pollen abundances from a site in Papua and New Guinea (Walker and Wilson, 1978); soil and surface properties from a gilgai area in New South Wales (Webster, 1977) and hillslope angles from North Yorkshire (Cox, 1979b). In each case outliers and long tails are evident (and outliers were omitted rather cavalierly from their analyses by Cliff and Ord and by Walker and Wilson). The results proved of considerable interest: they cast doubt on previous interpretations in the first two cases; the methods provided a simple alternative to spectral analysis in identifying periodicities in the third; and some light was thrown on the difficult issue of scale variations in hillslope profile morphometry in the last. And this is what we should expect from exploratory methods: not only answers to some old questions, but also some interesting and provocative new questions to be considered in further work. Indeed nonlinear smoothers should be more widely used as exploratory methods in future, it being understood that the more popular linear methods will continue to be employed where shown to be appropriate.

Inference

Most of the literature on statistical methods in geography appears to be based, at least implicitly, on the view that descriptive statistics are rather obvious, if not trivial, while inferential statistics are both more challenging and more valuable intellectually. 'Confirmatory' approaches have generally been promoted at the expense of 'exploratory' approaches, in Tukey's terms. There are many reasons for this, some more respectable than others. Extended accounts of confirmatory methods in textbooks may be necessary if only because these more difficult ideas need to be explained at some length. It is more disturbing that both teaching and research in statistical geography have been strongly influenced by the recipes given in cook-books intended for other disciplines (e.g. biology, psychology, sociology) and the appropriateness of these recipes has generally received rather limited attention. A review of the basic ideas of statistical inference and of experience in quantitative geography leads us to the view that statistical inference has been oversold in geography, because it is often inappropriate or irrelevant for geographical problems (cf. Cox and Anderson, 1978, 1980). It follows that exploratory analyses deserve a greater proportion of teaching and research efforts in geographical data analysis.

It is necessary, therefore, to review the grounds for believing statistical inference to be oversold in geography. Since brevity may impart an air of dogmatism, let it be stressed that many complex issues are involved here which do remain controversial and unresolved.

In the first place, many test procedures assume normality (Gaussianity) of distributions; this may not be met by geographical data. On the other hand, recourse may be had to transformations, distribution-free procedures or outlier rejection in attempts to circumvent this difficulty. Secondly, many test procedures assume mutual independence of data or of stochastic disturbances, whereas it is natural to expect autocorrelation to be present in geographical data. Thirdly, not all geographical data sets may be usefully regarded as representative samples from a larger population. On the other hand, these two related problems may be attacked by invoking stochastic process theory or randomization procedures (but *not* distribution-free procedures). In short, since the assumptions behind simple inferential procedures are often not met in geographical problems, more complex procedures must be invoked, and these in turn may present difficulties (for further discussion cf. Gould, 1970; Cox and Hinkley, 1974; Box, 1976; Haggett *et al.*, 1977; Box *et al.*, 1978; Silk, 1979).

There are also grounds for doubting the relevance of the ideas of statistical inference in geographical data analysis. Although there are several schools of statistical inference, most geographers adhere to the Neyman-Pearson school which regards inference as essentially a matter of deciding between rival hypotheses. For instance, Johnston (1978, p. 14) cites as an example that 'our research hypothesis may be that south-facing slopes in Derbyshire have a more rapid growth rate for grass in April than do north-facing slopes, so that the null hypothesis is of no difference between the two types of slope'. The important question is whether it is really fruitful to regard data analysis in this way, reducing matters to a simple qualitative dichotomy (*either* some difference *or* no difference) with everything else set on one side. If one really were interested in the contrast between different aspects, the key question is estimating the magnitude of the difference, not establishing whether a difference exists. In any case a clear contrast between different situations is probably less likely in observational than in experimental studies. In general, testing hypotheses with the aim of producing firm decisions may be less valuable than attempts to summarize the quantitative evidence available and efforts to be open to the indications provided by the data (for further discussion cf. Edwards, 1972; Cox and Hinkley, 1974; Mosteller and Tukey, 1977; Cox and Anderson, 1978).

It must be admitted that there is something rather attractive about statistical inference. A moderate investment of intellectual effort yields a fair grasp of what is going on, yet the ideas are sufficiently abstruse to impart a feeling of sophistication, and (best of all) attention to the rules of the game provides simple definite answers: results are or are not significant at some conventional level, something easily recorded. Exploratory methods are much less satisfactory, for most are so simple that they appear suspiciously trivial, and (worst of all) practitioners are thrust into a messy and chaotic world where 'problems may often not have neat answers or a single correct solution' (Mosteller and Tukey, 1977, p. xi). Irony aside, it is to be hoped that quantitative geography in the 1980s will be less afflicted than in the past by a craving for the semblance of elegance, exactness and rigour exuded by inferential ideas, and that geographers will show more willingness to engage in uninhibited exploration of their data, guided but not dominated by the procedures devised by statisticians.

Acknowledgement

We are very grateful to Ian Evans for his comments on a draft of this article.

References

Anderson, E.W., and Cox, N.J. (1978) 'A comparison of different instruments for measuring soil creep', *Catena*, 5, 81-93.

Andrews, D.F., Bickel, P.J., Hampel, F.R., Huber, P.J., Rogers, W.H., and Tukey, J.W. (1972) *Robust estimates of location. Survey and advances*, Princeton University Press, NJ.

Anscombe, F.J. (1973) 'Graphs in statistical analysis', *American Statistician*, 27, 17-21.

Balchin, W.G.V. (1976) 'Graphicacy', *American Cartographer*, 3, 33-8.

Barnard, K.C. (1978) 'The residential geography of the elderly: a multiple-scale approach', unpublished PhD thesis, University of Southampton.

Barnett, V.D. (1978) 'The study of outliers: purpose and model', *Applied Statistics*, 27, 242-50.

Beaton, A.E., and Tukey, J.W. (1974) 'The fitting of power series, meaning polynomials, illustrated on band-spectroscopic data', *Technometrics*, 16, 147-85.

Besag, J.E., and McNeil, D.R. (1976) 'On the use of exploratory data analysis in human geography' [abstract], *Advances in Applied Probability*, 8, 652.

Box, G.E.P. (1976) 'Science and statistics', *Journal of the American Statistical Association*, 71, 791-9.

Box, G.E.P., Hunter, W.G., and Hunter, J.S. (1978) *Statistics for experimenters*, Wiley: New York.

Chatterjee, S., and Price, B. (1977) *Regression analysis by example*, Wiley: New York.

Claerbout, J.F. (1976) *Fundamentals of geophysical data processing*, McGraw-Hill: New York.

Claerbout, J.F., and Muir, F. (1973) 'Robust modeling with erratic data', *Geophysics*, 38, 826-44.

Cleveland, W.S., and Guarino, R. (1976) 'Some robust statistical procedures and their application to air pollution data', *Technometrics*, 18, 401-9.

Cliff, A.D., and Ord, J.K. (1973) *Spatial autocorrelation*, Pion: London.

Cox, N.J. (1978) 'Exploratory data analysis for geographers', *Journal of Geography in Higher Education*, 2 (2), 51-4.

Cox, N.J. (1979a) 'Nonlinear smoothing in one and two dimensions', Paper presented to Institute of British Geographers Annual Conference, Manchester.

Cox, N.J. (1979b) 'Models and methods in hillslope profile morphometry', unpublished PhD thesis, University of Durham.

Cox, N.J., and Anderson, E.W. (1978) 'Teaching geographical data analysis: problems and possible solutions', *Journal of Geography in Higher Education*, 2 (2), 29-37.

Cox N.J., and Anderson, E.W. (1980) 'In defence of exploratory data analysis', *Journal of Geography in Higher Education*, 4 (1), 85-9.

Cox, D.R., and Hinkley, D.V. (1974) *Theoretical statistics*, Chapman & Hall: London.

Crewe, I., and Payne, C. (1971) 'Analysing the census data' in D. Butler and M. Pinto-Duschinsky (eds), *The British general election of 1970*, 416-36, Macmillan: London.

Crewe, I., and Payne, C. (1976) 'Another game with nature: an ecological regression model of the British two-party vote ratio in 1970', *British Journal of Political Science*, 6, 43-81.

Crowe, P.R. (1933) 'The analysis of rainfall probability', *Scottish Geographical Magazine*, 49, 73-91.

Edwards, A.W.F. (1972) *Likelihood*, Cambridge University Press.

Ehrenberg, A.S.C. (1975) *Data reduction*, Wiley: London.

Ehrenberg, A.S.C. (1979a) [Review of Tukey, 1977], *Applied Statistics*, 28, 79-83.

Ehrenberg, A.S.C. (1979b) 'A note of dissent on data analysis', *Journal of Geography in Higher Education*, 3 (2), 113-16.

Erickson, B.H., and Nosanchuk, T.A. (1977) *Understanding data*, McGraw-Hill Ryerson: Toronto.

Everitt, B.S. (1978) *Graphical techniques for multivariate data*, Heinemann: London.

Gardiner, V. (1973) 'Univariate distributional characteristics of some morphometric variables', *Geografiska Annaler*, 54A, 147-53.

Gnanadesikan, R. (1977) *Methods for statistical*

data analysis of multivariate observations, Wiley: New York.

Gould, P.R. (1970) 'Is *Statistix inferens* the geographical name for a wild goose?', *Economic Geography*, 46, (supplement) 439-48.

Gregory, S. (1978) *Statistical methods and the geographer*, Longman: London.

Haggett, P., Cliff, A.D., and Frey, A. (1977) *Locational analysis in human geography*, Arnold: London.

Hoaglin, D.C. (1977) 'Mathematical software and exploratory data analysis' in J.R. Rice (ed.) *Mathematical software III*, 139-59, Academic Press: New York.

Hoyle, M.H. (1973) 'Transformations - an introduction and a bibliography', *International Statistical Review*, 41, 203-23.

Johnston, R.J. (1978) *Multivariate statistical analysis in geography*, Longman: London.

Jones, K. (1980) 'Geographical variation in mortality: an exploratory analysis', unpublished PhD thesis, University of Southampton.

Larsen, W.A., and McCleary, S.J. (1972) 'The use of partial residual plots in regression analysis', *Technometrics*, 14, 781-90.

McGill, R., Tukey, J.W., and Larsen, W.A. (1978) 'Variations of box plots', *American Statistician*, 32, 12-16.

McNeil, D.R. (1977) *Interactive data analysis*, Wiley: New York.

Mather, P.M. (1976) *Computational methods of multivariate analysis in physical geography*, Wiley: London.

Mosteller, F., and Tukey, J.W. (1977) *Data analysis and regression*, Addison-Wesley: Reading, Mass.

Norcliffe, G.B. (1977) *Inferential statistics for geographers*, Hutchinson: London.

Silk, J.A. (1979) *Statistical concepts in geography*, Allen & Unwin: London.

Taylor, P.J. (1977) *Quantitative methods in geography*, Houghton Mifflin: Boston, Mass.

Tufte, E.R. (1974) *Data analysis for politics and policy*, Prentice-Hall: Englewood Cliffs, NJ.

Tukey, J.W. (1970) 'Some further inputs' in D.F. Merriam (ed.), *Geostatistics: a colloquium*, 163-74, Plenum: New York.

Tukey, J.W. (1972) 'Some graphic and semigraphic displays' in T.A. Bancroft and S.A. Brown (eds), *Statistical papers in honor of George W. Snedecor*, 293-316, Iowa State University Press: Ames, Iowa.

Tukey, J.W. (1977) *Exploratory data analysis*, Addison-Wesley: Reading, Mass.

Tukey, J.W., and Wilk, M.B. (1966) 'Data analysis and statistics: an expository overview', *American Federation of Information Processing Societies Conference Proceedings*, 29, 695-709.

Velleman, P.F. (1977) 'Robust nonlinear data smoothers: definitions and recommendations', *Proceedings, National Academy of Sciences*, 74, 434-6.

Wainer, H. (1976) 'Robust statistics: a survey and some prescriptions', *Journal of Educational Statistics*, 1, 285-312.

Wainer, H. (1977) [Review of Tukey, 1977] *Psychometrika*, 42, 635-8.

Walker, D., and Wilson, S.R. (1978) 'A statistical alternative to the zoning of pollen diagrams', *Journal of Biogeography*, 5, 1-21.

Webster, R. (1977) 'Spectral analysis of gilgai soil', *Australian Journal of Soil Research*, 15, 191-204.

Chapter 14

Factor analysis

P.M. Mather

Introduction

Factor analysis would appear to resemble mathematics if, as Bertrand Russell wrote, 'mathematics is the only science where one never knows what one is talking about, nor whether what is said is true'. Despite this apparent disadvantage, factor analysis has proved to be a widely-used multivariate technique, particularly in urban geography, both in Britain and North America. As a generator of controversy, factor analysis has been unsurpassed, though one is a little less impressed by the number of new ideas or concepts which have been developed with its aid. Some critics have implied that users of factor analysis fall into the 'Lucky Jim' category (Kingsley Amis's eponymous hero defined his research as the 'casting of pseudo-light on non-problems').

While some of the criticism of factor analysis is undoubtedly deserved, I do not believe that the method itself is inherently deficient. It has often been misused and misunderstood, and its aims confused with those of other techniques. For a time the mere fact that a study employed factor analytic techniques was in itself considered sufficient to warrant publication. Naturally, a reaction against the profligacies of this period set in, and the conscious avoidance of the use of factor analysis was proposed as a course for the virtuous geographer (Williams, 1971). During the 1960s and 1970s a number of alternative factorial methods were proposed, adding confusion to an already muddled situation. Some of these new methods seem in retrospect to be largely exercises in bandwaggon construction. Cattell (1978, p. xi) is perhaps too caustic in his criticism of the mathematically-oriented factor analysts when he cites Macaulay's comment on Dr Johnson: 'How it chanced that a man who reasoned on his premises so ably, should assume his premises so foolishly, is one of the great mysteries of human nature.' Nevertheless, there does seem to be at least a grain of truth in this observation.

In the first part of this chapter a concise account of the factor model is given in order to clarify the aims and assumptions of the method. Next comes a brief review of applications (mainly in urban geography), and this is followed by a consideration of the philosophical basis of factor analysis. It is my belief that a statistical or mathematical technique cannot be used outside a particular framework of explanation, for it is this framework which provides definitions of questions concerning patterns, relationships and distributions. It also lays down the boundaries of a study by identifying the characteristics or variables that can legitimately be included. Thus, observations that are not objectively derived are excluded from a factor-analytic study. Other frameworks of thought permit such observations and, indeed, may accord them a high status. The interaction between the researcher's outlook and opinions and the techniques he uses is a close one which should not be disregarded.

The factor model

Most users of factor analysis begin with the intuitive feeling that there is some underlying reason why the variables that they have selected to describe their field of study exhibit some degree of correlation. Taken separately, however, these correlations will not necessarily reveal any underlying structure. Studied collectively, as a correlation matrix, an underlying structure may well be revealed, for the whole is often more informative than the sum of the individual parts, separately considered. This, of course, presumes that the variables have been carefully selected and that the factor model correctly represents the relationships between these variables.

Table 14.1 Symbols used in Chapter 14

p	number of variables
k	number of common factors
n	number of observations (cases, areas) for each variable
X	matrix of raw data (n rows, p columns)
F	matrix of factor scores (n rows, k columns)
Λ	matrix of factor loadings (p rows, k columns)
E	matrix of specific and error factor scores (n rows, p columns)
γ_{ij}	element on row i, column j of Λ
Σ	variance-covariance matrix of the x's

The common factor model

The aim of the common factor model is to define a number, k, of mathematical functions of the observed variables which, although fewer in number than the observed variables, are capable of reproducing the correlations among the variables. These mathematical functions or 'factors' are interpreted as mental constructs, or categories, which may be thought of as existing at a higher level of generalization than the basic variables. These functions need not be statistically independent in the sense that their intercorrelations are zero, but they should be conceptually distinct. The identification and naming of constructs is a function of the user's knowledge of his subject, and it is thus the criterion of meaningfulness which is decisive rather than any mathematical or statistical nicety.

In all generalizations, something of the uniqueness of the individual is lost. The factors, which are generalizations of the original variables, are constructs which represent or express that information which is common to some or all of the variables. Hence the common factor model can be written as

$$\mathbf{X} = \mathbf{F}\Lambda + \mathbf{E} \qquad (1)$$

where **X** is the matrix of observation on p variables for n cases, F is the n-by-k factor scores matrix, Λ the factor loadings matrix, and **E** is matrix of the unique parts of the variables. An individual measurement for case i on variable j is thus made up to two parts, one which is attributable to the effects of common factors, while the second part of the measurement is due to the effects of specific factors which are unique to the individual variables. Equation (1) can be rewritten in terms of a single observation to illustrate this apportioning of variance:

$$x_{ij} = \sum_{r=1}^{k} \lambda_{ir} f_{jr} + e_{ij} \qquad (2)$$

Here, the term $\Sigma\lambda_{ir} f_{jr}$ represents the cumulated effects of the r common factors while e_{ij} is the jth element of the unique factor, which could be expressed as

$$e_{ij} = \sum_{m=1}^{p} u_{im} g_{jm}$$ where the u's are the unique factor loadings and the g's are the corresponding scores.

If it is assumed that the common factors and the unique factors are uncorrelated, and that the unique factors are themselves uncorrelated, it follows that the variance-covariance matrix of the e's is diagonal. If k is the correct number of common factors, then Σ, the variance-covariance matrix of the x's, has rank k (i.e. its last p-k eigenvalues are zero) where k is less than p. The factor model is only defined when the x's are intercorrelated.

Indeterminacy of the factor model

One of the problems with factor analysis, and one which has been exposed to criticism in the literature, is that the statistical model given by equation (1) is not unique. The matrix Λ can be transformed to give another, different, matrix Λ^* without altering the mathematical equality (1):

$$\mathbf{X} = \mathbf{F}^* \Lambda^* + \mathbf{E} \qquad (3)$$

This transformation can be thought of as an orthogonal rotation of the factor axes. Equation (3) represents an infinite number of possible Λ^*'s and so there are any number of mathematically-equivalent matrices Λ^* which satisy (3). The criterion that is used to select one particular Λ^* from the infinite number of possible Λ^*'s is not mathematical but depends on the user's interpretation of the elements of the columns of Λ in terms of the p variables he has selected for study. Experience, knowledge and intuition play a part here. The indeterminacy of the factor model can be contrasted with the completely determined solution of the regression model, which is mathematically defined by the minimum sum of squares criterion (Σe_i^2 - minimum).

Orthogonal and oblique rotation

To overcome the problem of the lack of any mathematical or statistical criterion to define a unique solution, external criteria are used to fix the values of the elements of Λ. These criteria are based on the principle of simplicity, or Occam's Razor, and are designed to fix Λ in such a way that each column contains a number of high (significant) elements plus a number of near-zero (insignificant) elements, with few intermediate values. In addition, it is assumed that the columns of Λ are not necessarily orthogonal, though orthogonality could be thought of as a special case. In subject-matter terms, this means that the factors may become correlated. In the same way that the correlations among the original variables may lead to the speculation that these correlations were generated by a fewer number of common factors, so the correlations between the first-order factors leaves open the possibility that they in turn can be summarized by a set of second-order factors. This possibility will not be explored further but see Cattell (1978, Ch. 9).

Rotation methods include varimax, the most widely-used orthogonal rotation (i.e. the factors remain uncorrelated) and a host of 'oblique' rotations which allow the factors to become correlated. Early attempts at oblique rotation involved the plotting of scatter diagrams and manual trial and error movement of the factor axes. Automatic methods include oblimin, Promax, Harris-Kaiser, quartimin, biquartimin, covarimin, oblimax, and others. One reason for this proliferation is the fact that no single scheme can be effective for all kinds of data structure. Promax has been found to be a consistently good performer (Dielman *et al.*, 1972; Hakstian, 1971).

A third type of rotation, in addition to the orthogonal and oblique types mentioned above, is that involving the fitting of the initial factor matrix Λ to a specified target matrix in which the distribution of the factor loadings is hypothesized a priori. The fitting of the 'experimental' factor matrix Λ to the 'hypothesized' matrix Λ_H is achieved by the method of least-squares and is picturesquely termed the Procrustes method, after the legendary Greek character (Hurley and Cattell, 1962).

When dealing with oblique factors it is important that the distinction between the *coefficients* of the equations defining the factor-variable relationship and the *correlations* between the factors and variables should be maintained. The former are termed factor pattern coefficients and these are normally used in factor interpretation, while the latter are the factor structure coefficients and are used in the estimation of factor scores. Pattern and structure matrices are identical in the orthogonal case, but in the oblique case it is possible to obtain pattern coefficients which are larger than $\pm$ 1 while structure coefficients, being correlations, must lie between + 1 and − 1 inclusive. A further distinction between 'primary' and 'reference' factors should be borne in mind (see Harman, 1976, p. 274).

Factor scores

There are k common factors and p unique or specific factors in the common factor model. It is not possible to determine the inverse of a $(p + k) \times (p + k)$ matrix from p independent observations alone (unless $k = 0$, which is unlikely in this context) so factor scores cannot be directly calculated. Neither can they be estimated in the statistical sense. Factor scores, being measures of the influence of the factors on the individual observation units (such as EDs or census tracts), are of great interest to geographical users of the technique, and the fact that they cannot be directly calculated or statistically estimated is therefore of some importance. Several methods of approximating the values of the scores have been proposed; the two most widely-used methods are the regression method and Barlett's method. Details of these techniques are given by Harman (1976), Mather (1976) and Cattell (1978). It is sufficient to note here that factor scores calculated by computer packages such as SPSS are based on approximations, and that validity of these scores may vary from factor to factor. It has been shown (Guttman, 1955; see also Mather, 1976, p. 303) that the degree of congruence between the approximate factor scores and the true, but unknown, factor scores can be calculated, and users of factor scores should be careful to use the information that such calculations provide.

Relationship to principal components analysis

The common factor model involves the fitting of a statistical model to a set of observational data, much in the same fashion as the fitting of a regression model. If maximum-likelihood factor estimation procedures are used, a large-sample test of the goodness-of-fit of the model with k factors is available. The principal components technique is fundamentally different. Principal components analysis involves the re-expression of a set of measurements on p variables in terms of p orthogonal principal components chosen in such a way that the successive components each account for the maximum variance remaining after the extraction of the preceding components. Principal components analysis is a mathematical transformation and does not involve the fitting of a model. That is not to say that it does not have its uses as an information-compressing device; a good example is the reduction of the four channels of data obtained from Landsat imagery to two uncorrelated data sets, without any significant loss of information (Gonzales and Winz, 1977, p. 313). The model-fitting approach of factor analysis should therefore be carefully distinguished from the data transformation approach of principal components analysis; confusion over this point demonstrates a basic lack of appreciation of the nature of the two techniques. Not all geographical users of factor analysis appear to take this view (Davies, 1971, 1972; Mather, 1971, 1972), despite statements in the literature to the effect that 'component analysis is no good as a final scientific model' (Cattell, 1978, p. 17) or 'as is well known, common factors and components are incompatible both with respect to their mathematical properties and the purposes for which they are intended' (McDonald, 1977, p. 165).

Other issues in factor analysis

Given that most authors require a hundred pages to describe factor analysis, it is not surprising that this review is brief and cursory. Topics which are not covered here, including the choice of factoring technique, the related issues of communality estimation and the choice of the appropriate number of factors, the extraction of high-order factors, the measurement of the relationship between factorial solutions either for the same data set or for data sets referring to different time-periods, and the effects of scaling the variables, are described in Rummell (1971), Mulaik (1972), Harman (1976), Joreskog *et al.* (1976) and Cattell (1978).

Computer programs for factor analysis are provided by the well-known SPSS and BMD packages, while listings of Fortran subroutines for varieties of factor analysis methods are included in Mather (1976). A problem with the 'package' approach is that it encourages the profligate and thoughtless use of the computer, rather than the careful consideration of the nature of the problem and the characteristics of the method.

Factor analysis in geography

Clark, Davies and Johnston (1974) have provided a review of applications of factor analysis in geography up to 1974 and it would be pointless to duplicate their efforts here. Other valuable reviews are Rees (1971), Berry (1971) and Johnston (1976) who consider the subject of factorial ecology. This can be defined as the use of factor analysis in the study of urban areas and their differentiation in terms of social and economic characteristics. Typical of this approach in British geography is the work of Robson (1969), Giggs (1973), Davies and Lewis (1973), Goddard (1970), Johnston (1973), Rees (1972) and Davies and Musson (1978), while Taylor and Parkes (1975) have attempted to extend the method to include changes over time. Johnston (1976, p. 203) concludes that 'factorial ecology demonstrably offers a widely-accepted approach to the identification of underlying determinants of intra-urban residential patterns, as well as providing series of indices of value for further study of ecologies', which suggests the use of factor analysis as an exploratory device, useful in the search for the common features of a set of cities, with the implication that the discovery of such common features should lead to the development of hypotheses to explain their occurrence. Very few geographers have proceeded beyond the exploratory stage, although such developments were suggested by Mather and Openshaw (1974). The use of target rotations, which were described above, has been limited to textbook examples (Johnston, 1978a; Mather 1976; Taylor, 1977) with the exception of one paper by Perle (1978a). This study led to a disagreement between Perle and Johnston (Johnston, 1978b; Perle, 1978b; Johnston, 1979) not because target rotation itself was thought to be valueless, but because Johnston considered that the procedure was 'inappropriate as a means of monitoring progress towards policy goals for urban residential patterns'. (Johnston, 1979).

Before geographers move on to test hypotheses suggested by the 'determinants of intra-urban residential patterns' (Johnston, 1976) which are revealed by exploratory factor analysis, they would do well to consider the conclusions of Jones (1962), Armstrong (1967), Miesch (1969) and Mukherjee (1973), who demonstrate the inadequacy of the method when the correlation matrix has a known structure. Frequently the number of underlying dimensions, but not necessarily their nature, can be determined while in other cases the indicated number of factors is incorrect. Armstrong (1967) sets out a number of guidelines, while Mukherjee (1973) suggests the use of covariance structure analysis (see also Joreskog, 1970).

Philosophical basis of factor analysis

Hand-in-hand with the methods of statistical and mathematical analysis that were introduced into geography during the 1950s came the 'scientific method' with its 'rigorous dictates' which were necessary if geography, along with the other social sciences, was 'to be accepted and accorded an honored place in our society' (Burton, 1963). Gregory (1978), in a not untypical outburst of sarcasm, regards this attitude as a resurgence of 'the Victorian myth of the supremacy of the natural sciences' with its 'two ritualized expressions: the scientist as observer and the scientist as hero'. Factor analysis falls squarely within the scientific tradition with its emphasis on the development of verifiable (i.e. empirically-testable) propositions that can be considered to have existence independent of the observer. These propositions are derived from the study of observational data which have been collected within a framework that is not influenced by the observer's values or beliefs. They lead to explanations of spatial patterns and relationships in which characteristics of the individual (or the group) that are not objectively measurable are ignored. Geographical space is considered as an organizing principle (Eyles, 1978) with individual areas possessing attributes defined by housing quality characteristics, demographic structure, and descriptors of social status.

The scientific or positivist approach to explanation is no longer unchallenged; recent methodological developments include the realization that social patterns can be explained in terms other than those laid down by positivist-oriented geographers. Some of these alternative philosophies are rather tenuously defined, and are sometimes presented in an overblown and bombastic fashion; see, for example, Ley and Samuels (1978). One of these alternatives is the phenomenological viewpoint, which is summarized by Relph, who specifies that the range of permissible data should include 'man's immediate experience, including his actions, memories, fantasies and perceptions'. Phenomenology 'is *not* a method of analyzing or explaining some objective world through the development of prior hypotheses and theories' (Relph, 1970, p. 193). Although one would think that factor analysis would be out of place in such a context, Berry (1971) surprisingly denies that factorial ecology is inherently positivistic; he places it in the phenomenological camp. It would appear that Berry is confusing the trial and error methods of exploratory data analysis with what he terms the learning process: 'The essence of the phenomenological perspective . . . is the assumption that reflective knowledge can only be derived dialectically from the interplay of the world of our native experience and the structuring activities of our various perceptual and conceptual orientations' (Berry, 1971, p. 214).

Other approaches to the study of human geography have been proposed; one of the most widely publicized is radical geography (Harvey, 1973; Peet, 1977) with its Marxist base. Guelke (1971) has argued cogently against the scientific basis of explanation in favour of what he terms a humanistic view. It is important that the research worker should be aware of these different attitudes, for they inevitably determine the type of question to be asked, and the kind of technique that is relevant in the search for answers to these questions. The use of factor analysis presupposes the acceptance of the assumptions which underlie the scientific method, which has been described by Whitehead as 'to see what is general in what is particular and what is permanent in which is transitory' (quoted by Gould, 1979, p. 140). It would be dangerous to suggest, however, that there is a single, agreed set of rules and procedures that make up *the* scientific method.

Conclusion

Factor analysis, as described and applied by the geographers of the 1970s, is a valuable method of data analysis. Its mindless and automatic use has, however, led to charges that it is no more than a method of restating the obvious. Despite some technical advances, including the acceptance of oblique rotation methods (Hughes and Carey, 1972) and the comparative study of different factoring procedures (Giggs and Mather, 1975; Conway and Haynes, 1977; Davies, 1978) the geographical factor analyst has remained firmly committed to the purely exploratory use of the technique. In many instances, the aim has been to compress a multivariate data set into a set of mappable indices that show the spatial variability of a particular characteristic. If such indices are of general applicability (remembering Whitehead's definition of scientific thought) then the stage should be set for the derivation and testing of hypotheses to account for the patterns and relationships that recur in different studies. Confirmatory factor analysis and covariance structure analysis would seem to be useful in this regard.

Whether or not the hypotheses produced by such studies are of any value depends upon the individual's particular set of beliefs and principles. The 1970s have seen the fragmentation of geographical methodology and philosophy with the emergence of various sects, each with its own point of view. Like the break-up of western Christianity in the sixteenth century, this has left the orthodox, scientific establishment competing with alternative versions of the truth; these alternative versions are often preached with an evangelical zeal. One major benefit has been to bring to the attention of the individual researcher the importance of recognizing, and questioning, the basis of his view of the (geographical) world. It is to be hoped that, as a result of this increased awareness of the fundamental role of underlying philosophical assumptions, the use of factor analysis will in future take place within a more coherent and organized framework of thought. There is a danger, though, that too much time and energy will be devoted to the conversion of other geographers to one's own point of view, thus detracting from the advance of knowledge. As I have suggested in another context (Mather, 1979) there is more than one road to salvation (or explanation). It is not necessary to hold the idea that one 'paradigm' is dominant at one particular time, as Kuhn suggests. Blunden *et al.* (1978) summarize this opinion: '[we] do not accept that progress from one paradigm to a new one is an adequate representation of human geography today. Instead [we] . . . believe that a number of alternative paradigms coexist and that, because individual geographers differ over the fundamental value judgements, this diversity is likely to persist. Moreover, [we] see this as a good thing because adoption of multiple paradigms can offer an understanding of greater dimensionality . . .' (Blunden *et al.*, 1978, p. vii). Factorial ecology offers a means of studying one aspect of urban system. It cannot produce answers to all the questions that a student or urban systems may raise, because it is bounded by the domain of applicability of positivist principles. This is a point that should be appreciated both by users and critics of the method for, as Taaffe remarks ' we can no longer afford the exuberant confidence in current theories, models and techniques which dismisses values, societal utility, and the existence of alternative paths to Rome, including essentially verbal and essentially prescriptive paths' (Taaffe, 1974, p.12).

References

Armstrong, J.S. (1967) 'Derivation of theory by means of factor analysis or Tom Swift and his electric factor analysis machine', *The American*

Statistician, 21, 17-21.

Berry, B.J.L. (ed.) (1971) 'Introduction: the logic and limitations of comparative factorial ecology', *Economic Geography,* 47, No. 2 (supplement), 209-19.

Blunden, J., Haggett, P., Hamnet, C., and Sarre, P. (1978) *Fundamentals of human geography: a reader,* Harper and Row: London/Open University Press.

Burton, I. (1963) 'The quantitative revolution and theoretical geography', *Canadian Geographer,* 7, 151-62.

Cattell, R.B. (1978) *The scientific use of factor analysis in behavioral and life sciences,* Plenum Press: New York.

Clark, D., Davies, W.K.D., and Johnston, R.J. (1974) 'The application of factor analysis in human geography', *The Statistician,* 23, 259-81.

Conway, D., and Haynes, K.E. (1977) 'Advances in comparative factorial ecological analysis: parsimony, invariance, and homogeneity in factor analysis solutions', *Environment and Planning A,* 9, 1143-56.

Davies, W.K.D. (1971) 'Varimax and the destruction of generality: a methodological note', *Area,* 3, 112-18.

Davies, W.K.D. (1972) 'Varimax and generality: a reply to Mather', *Area,* 3, 254-9.

Davies, W.K.D. (1978) 'Alternative factorial solutions and urban social structure', *Canadian Geographer,* 22, 273-97.

Davies, W.K.D., and Lewis, G.J. (1973) 'The urban dimensions of Leicester', *Institute of British Geographers, Special Publication,* 5, 71-85.

Davies, W.K.D., and Musson, T.C. (1978) 'Spatial patterns of commuting in South Wales, 1951-1971: a factor analysis definition', *Regional Studies,* 12, 353-66.

Dielman, T.E., Cattell, R.B., and Wagner, A. (1972) 'Evidence on the simple structure and factor invariance achieved by five rotational methods of four kinds of data', *Multivariate Behavioral Research,* 7, 223.

Eyles, J. (1978) 'Social geography and the study of the capitalist city: a review', *Tijdschrift voor Economische en Sociate Geografie,* 69, 296-305.

Giggs, J.A. (1973) 'The distribution of schizophrenics in Nottingham', *Transactions of the Institute of British Geographers,* 59, 55-76.

Giggs, J.A., and Mather, P.M. (1975) 'Factorial ecology and factor invariance: an investigation', *Economic Geography,* 51, 366-82.

Goddard, J.B. (1970) 'Functional regions within the city centre: factor analysis of taxi flows in Central London', *Transactions. Institute of British Geographers,* 49, 161-82.

Gonzales, R.C., and Winz, P. (1977) *Digital image processing.* Addison-Wesley: Reading, Mass.

Gould, P.R. (1979) 'Geography 1957-1977: the Augean period', *Annals. Association of American Geographers,* 69, 139-51.

Gregory, D. (1978) *Ideology, science and human geography,* Hutchinson: London.

Guelke, L. (1971) 'Problems of scientific explanation in Geography', *Canadian Geographer,* 15, 38-53.

Guttman, L. (1955) 'The determinacy of factor score matrices with implications for five other basic problems of common factor theory', *British Journal of Statistical Psychology,* 8, 65-81.

Hakstian, A.R. (1971) 'A comparative evaluation of several prominent methods of oblique factor transformation', *Psychometrika,* 36, 175.

Harman, H.H. (1976) *Modern factor analysis,* University of Chicago Press, 3rd edn, revised.

Harvey, D. (1969) *Explanation in geography,* Arnold: London.

Harvey, D. (1973) *Social justice and the city,* Arnold: London.

Hughes, J.G., and Carey, G.W. (1972) 'Factorial ecology: oblique and orthogonal solutions', *Environment and Planning A,* 4, 147-62.

Hurley, J.R., and Cattell, R.B. (1962) 'The Procrustes program: producing direct rotation to test a hypothesised factor structure', *Behavioural Science,* 7, 258-62.

Johnston, R.J. (1973) 'Social area change in Melbourne, 1961-1966', *Australian Geographical Studies,* 11, 79-98.

Johnston, R.J. (1976) 'Residential area characteristics: research methods for identifying urban sub-areas - social area analysis and factorial ecology' in D.T. Herbert and R.J. Johnston (eds), *Social areas in cities: Volume 1, Spatial processes and form,* John Wiley: London.

Johnston, R.J. (1978a) *Multivariate statistical analysis in geography,* Longman: London.

Johnston, R.J. (1978b) 'On normative analyses and factorial ecologies - a response to Perle (1978)', *Environment and Planning A,* 10, 731-3.

Johnston, R.J. (1979) 'On Procrustean rotations and the monitoring of policies for residential patterns: a note', *Environment and Planning A,* 11, 463-72.

Jones, M.B. (1962) 'Practice as a process of simplification', *Psychological Review,* 69, 274-94.

Joreskog, K.G. (1970) 'A general method for analysis of covariance structures', *Biometrika,* 57, 239-57.

Joreskog, K.G., Klovan, J.E., and Reyment, R.H. (1976) *Geological factor analysis,* Elsevier: Amsterdam.

Ley, D., and Samuels, M.S. (1978) *Humanistic geography,* Croom-Helm: London.

McDonald, R.P. (1977) 'The indeterminacy of components and the definition of common factors', *British Journal of Mathematical and Statistical Psychology,* 30, 165-76.

Mather, P.M. (1971) 'Varimax and generality', *Area,* 3, 252-4.

Mather, P.M. (1972) 'Varimax and generality',

Area, 4, 27-30.
Mather, P.M. (1976) *Computational methods of multivariate analysis in physical geography*, John Wiley: London.
Mather, P.M. (1979) 'Theory and quantitative methods in geomorphology', *Progress in Physical Geography*, 3, 471-87.
Mather, P.M., and Openshaw, S. (1974) 'Multivariate methods and geographic data', *The Statistician*, 23, 283-308.
Miesch, A.T. (1969) 'Critical review of some multivariate procedures in the analysis of geochemical data', *Mathematical Geology*, 1, 171-84.
Mukherjee, B.N. (1973) 'Analysis of covariance structures and exploratory factor analysis', *British Journal of Mathematical and Statistical Psychology*, 26, 125-54.
Mulaik, S.A. (1972) *The foundations of factor analysis*, McGraw-Hill: New York.
Peet, J.R. (1977) 'The development of radical geography in the United States', *Progress in Human Geography*, 1, 240-63.
Perle, E.D. (1978a) 'Target rotation and normative policy analysis in urban ecology', *Environment and Planning A*, 10, 275-85.
Perle, E.D. (1978b) 'On normative analyses and factorial ecologies - a response to Johnston (1978)', *Environment and Planning A*, 10, 1207-8.
Rees, P.H. (1971) 'Factorial ecology: an extended definition, survey and critique', *Economic Geography*, 47, 220-33.
Rees, P.H. (1972) 'Problems of classifying sub-areas within cities' in B.J.L. Berry (ed.) *City classification handbook*, 265-330, John Wiley: New York.
Relph, E. (1970) 'An inquiry into the relations between phenomenology and geography', *Canadian Geographer*, 14, 193-201.
Robson, B.T. (1969) *Urban analysis*, Cambridge University Press.
Rummell, R.J. (1971) *Applied factor analysis*, Northwestern University Press: Evanston, Ill.
Taaffe, E.J. (1974) 'The spatial view in context', *Annals, Association of American Geographers*, 64, 1-16.
Taylor, P.J. (1977) *Quantitative methods in geography*, Houghton-Mifflin: Boston, Mass.
Taylor, P.J., and Parkes, D.N. (1975) 'A Kantian view of the city: a factorial ecology experiment in space and time', *Environment and Planning A*, 7, 671-88.
Williams, K. (1971) 'Do you sincerely want to be a factor analyst?', *Area*, 3, 228-9.

Chapter 15

Multidimensional scaling

A.C. Gatrell

Introduction

The proliferation of different concepts of space has placed new demands on the quantitative geographer and analytical cartographer for techniques to give tangible expression to such concepts. The realization that a relative, rather than absolute, view of space may offer a more appropriate context within which to map geographical objects and events has led to an interest in multidimensional scaling. This technique, regarded by statisticians as a weapon in their armoury of data analytical methods (Sibson, 1972), offers a graphical language for creating novel maps of relative spaces.

Multidimensional scaling (MDS) requires the definition of a set of objects that one wishes to map, together with the specification of a relation defined on that set. For instance, we might imagine a regional system of cities, related in terms of travel time on a rail network. Quite simply, MDS uses such dyadic information to create a map depicting a configuration of those cities in time space. Yet, as the applications reviewed below suggest, the method is applicable to diverse sets of objects and maps the structure of any relation, however defined. Most applications of geographical interest seek to compare the derived map with the more conventional representation of geographical space and this has prompted the investigation of methods for comparing point patterns (reviewed below).

MDS may be used, then, as a pragmatic device for map transformation (Forer, 1978b). This is to be contrasted with its use in behavioural geography, where an interpretation is solicited of the dimensions of the space in which the objects are thought to lie (Thrift, this volume).

The use of MDS by geographers in creating transformed spaces is not an established part of the British quantitative tradition. Its geographical use was pioneered in the United States by Waldo Tobler, and his work, and applications it has inspired, are not as widely known in Britain as they might be. Here, we review key features of the method before examining various applications, extensions and problem areas.

Locations from distances: the method of MDS

A relation R, defined on a set, S, of objects ($S \times S \rightarrow R$) provides a set of numbers representing the 'dissimilarity' or 'similarity' between pairs of objects. Given dissimilarities, δ_{ij}, MDS seeks to map the objects into a space of minimum dimensionality such that the distances, d_{ij}, in that space match as closely as possible the observed dissimilarities. It seeks a regression of the distances on the dissimilarities in which either the numerical values of the dissimilarities are employed or only the rank order of the values is used. The former is known as metric MDS, while the latter, based on monotonic regression, is called non-metric MDS.

The essence of non-metric MDS, which is more widely used, is captured graphically (Figure 15.1) where the distinction between δ_{ij}, d_{ij} and $\hat{d}_{ij}$ (the fitted distances on the monotonic regression line) is clear. As in standard regression, MDS seeks to minimize the sum of the squared deviations from the regression line. Through an iterative procedure the locations of points in the space are optimally adjusted to reduce badness-of-fit ('stress'). We emphasize here that each point on the graph corresponds not to one of the objects but to a (δ_{ij}, d_{ij}) pair and that residuals, $e_{ij} = (d_{ij} - \hat{d}_{ij})$ are vector rather than scalar quantities, showing which dissimilarities are represented well and which poorly by the regression.

The objects can, of course, be mapped into a space of m dimensions. Stress will be reduced as m increases, the plot of stress against dimensionality

Figure 15.1 Monotonic regression of configuration distances on dissimilarities

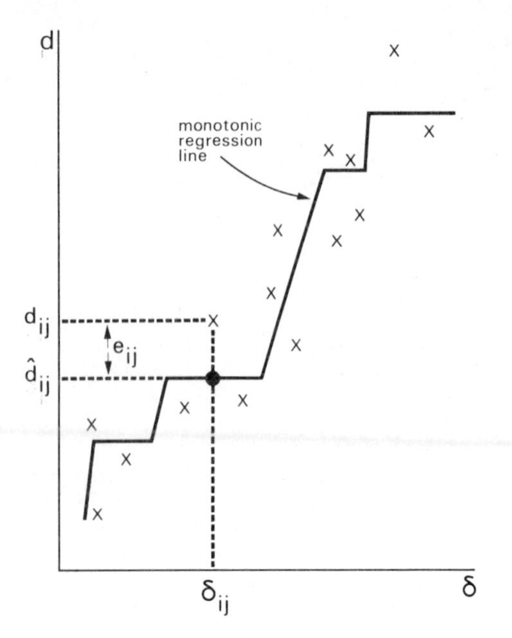

offering some guidance as to the selection of an appropriate number of dimensions. Other criteria include interpretability and ease of use and these frequently dictate the use of configurations of points in spaces where $m \leqslant 3$. Another criterion is the number of objects, since requirements of statistical stability dictate that m cannot be large when there are only a few objects.

MDS originated in psychometrics, in which metric procedures (Young and Householder, 1938) were augmented by non-metric methods (Shepard, 1962; Kruskal, 1964a, b). Algorithms have proliferated, as a perusal of the journal *Psychometrika* will testify. Clearly we cannot here discuss many technical details; the introduction by Kruskal and Wish (1978) provides a magnificently clear entry to the literature. The classic work in geography is by Golledge and Rushton (1972), though this is somewhat outdated now on applications. Tobler has exploited links between metric MDS and trilateration methods in surveying (Wolf, 1969) and, as indicated below, this provides a potentially rich analogy that has yet to be fully exploited in geographical applications (Tobler, 1976).

Finally, we note that dissimilarity matrices of geographical interest are frequently asymmetric: $\delta_{ij} \neq \delta_{ji}$. The distances in the configuration are symmetric, of course, so that substantial asymmetry in the input data will increase stress. Although one obvious solution is to symmetrize the dissimilarities, information concerning the asymmetries may be of interest. Later in the paper we discuss work on the problem of asymmetry.

'Maps from scraps': geographical applications of MDS

D.G. Kendall (1975) has coined the phrase 'maps from scraps' to denote the kinds of geographical problem discussed below. It seems appropriate, in view of the origins of MDS in psychology, to commence with an investigation of applications to environmental cognition.

Cognitive space

If the objects are urban landmarks or environmental cues and the relation is an estimate of distance by an individual then MDS yields a graphical model of that subject's cognitive map. The most extensive research along these lines has been conducted by Golledge and his students on the cognitive structure of Columbus, Ohio (Golledge, Briggs and Demko, 1969; Golledge, Rivizzigno and Spector, 1976; Golledge, 1978a, b). In a preliminary study, two groups of students, final year graduates and newcomers, made paired comparisons of the distances between eleven locations (Golledge, Briggs and Demko, 1969). While there was a tendency among both groups to underestimate distances away from the Central Business District (CBD) they overestimated distances towards the CBD, a function perhaps of perceived travel time. Subsequent work has adopted a careful experimental design, one study examining in detail the cognitive structure of the city as revealed by a control group of long-term residents (Golledge, Rivizzigno and Spector, 1976). The results (presented in the form of distorted graticules) suggest that distances along the major North-South axis are underestimated (Figure 15.2).

Figure 15.2 Cognitive distortion of geographical space in Columbus, Ohio

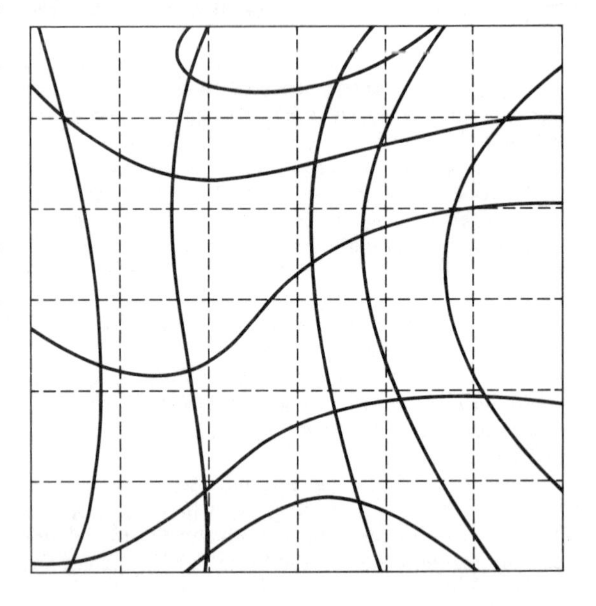

Cognitive spaces revealed by individual subjects display certain common features, though they necessarily reflect idiosyncratic travel behaviour and different residential locations.

Whether it is particularly fruitful to look at the individual MDS configurations is a moot point (Gould, 1976) and clearly more interesting questions involve studying the evolution of cognitive space and variations over different occupational or socio-economic groups. More recent work (Golledge, 1978a) has addressed the former topic, indicating that when newcomers are tested on three occasions over a six-month time period the correlations between the MDS and the locational coordinates increase; stress values too drop dramatically.

In view of the fact that there is some British work on urban environmental cognition (Pocock and Hudson, 1978) and locational search (Hudson, 1975) it seems surprising that British geographers have not conducted studies similar to those launched by Golledge. Careful consideration must be given to the selection of stimuli (objects). Golledge has given this recent attention (Golledge, 1978b), since his earlier sample of locations, while internally consistent, was highly clustered spatially (Golledge, 1978a, p. 85). Such clustered samples may make the task of distance ordering difficult for subjects (MacKay, 1976) and might make interpretation of the distorted grid problematic. Golledge's most recent work is based on locations distributed throughout the city of Columbus (Golledge, 1978b).

Marchand (1974) treated a much smaller number of urban landmarks in a study of pedestrians' cognition of a Paris suburb. A 'grab' sample of individuals located six landmarks on a hand-drawn map and the means of the distances measured from these maps were scaled. Underestimation of the distance between the two rail stations (Créteil and Champigny) indicated that the major transport axis plays a key role in causing cognitive distortion. In his paper, Marchand subsequently takes us a little way beyond the map transformation to speculate on the appropriate topology of the cognitive space and this would seem to provide a natural avenue for further research.

Other recent work in the USA, at both a local scale (MacKay, Olshavsky and Sentell, 1975) and national scale (Lloyd, 1976) has investigated relations between the locations of the objects (supermarkets and states, respectively) in cognitive space and spatial behaviour (shopping choice and migration). Lloyd's work in particular demonstrated the utility of the MDS configurations in predicting behaviour, and his work is elaborated below.

Time space

Maps are frequently drawn that depict time distance from one location to a set of other locations (Clark, 1977; Muller, 1978). MDS permits a time space map to be constructed that represents time distances between all pairs of locations, and recent work has applied the method in both inter-regional and intra-urban contexts. Marchand (1973) has analysed the changing structure of the Venezuelan road network, treating travel time relations defined on a set of 65 towns. He concluded, on the basis of stress values, that Venezuelan time space was seven-dimensional in 1936 and 1941, two-dimensional in 1950 and three-dimensional in 1961. He attributes the decline in dimensionality to 'homogenization of the network conditions' (Marchand, 1973, p. 519) but goes on to note that 'the full geographical consequences are not clear and need a more thorough theoretical study' (Marchand, 1973, p. 519). Marchand maps the 1950 solution, noting as others have done that long distances are better recovered by MDS than short ones (Kruskal and Wish, 1978, p. 46).

Forer (1973) has also expanded this research frontier, looking in detail at the evolving structure of the New Zealand airline network and in particular at how improvements in transport technology modify time space. Forer is interested less in dimensionality than in obtaining a cartographic representation in two dimensions and the stress values he obtains suggest that this is not unreasonable. Stress does not appear to vary systematically over the 24-year time period (1947-70), nor does stress vary substantially with variations in p, the Minkowski metric.

There is a dearth of work on intra-regional time spaces, though Kilchenmann (1972) mapped twenty-five towns in the Zürich region in a study that also looked at the centrality of towns in geographical and time space. At an intra-urban scale Ewing (1974) and Forer (1978a) have both used MDS to map travel time space, in Montreal and Christchurch respectively. Ewing goes beyond the simple configuration of points (centroids of traffic zones) produced by MDS to map vectors showing spatial mismatch between physical space and time space. He also indicates how a regular grid in physical space is distorted in time space as the presence or absence of urban motorways shrinks or stretches time space.

Forer (1978a), while providing little detail on the nature of his data, offers three representations of time space in Christchurch in 1916. These refer to three transport modes: walking, tramcar and car. We need some statistical comparisons of the three, but beyond this there is clearly some interesting work to be done on evolving urban time spaces and how these shape, and are shaped by, emerging urban form.

Cost space

Few attempts have been made to construct cost spaces using MDS, though Forer (1973) presents examples obtained from his study of New Zealand airlines (Figure 15.3). Both maps (for 1952 and 1970) indicate substantially lower stress values than the time spaces he obtained. This is not surprising, since we would expect the relationship between cost and physical distance to be very nearly monotonic, while high-speed links between geographically distant nodes lead to greater distortions of geographical space.

Figure 15.3 Cost space in New Zealand

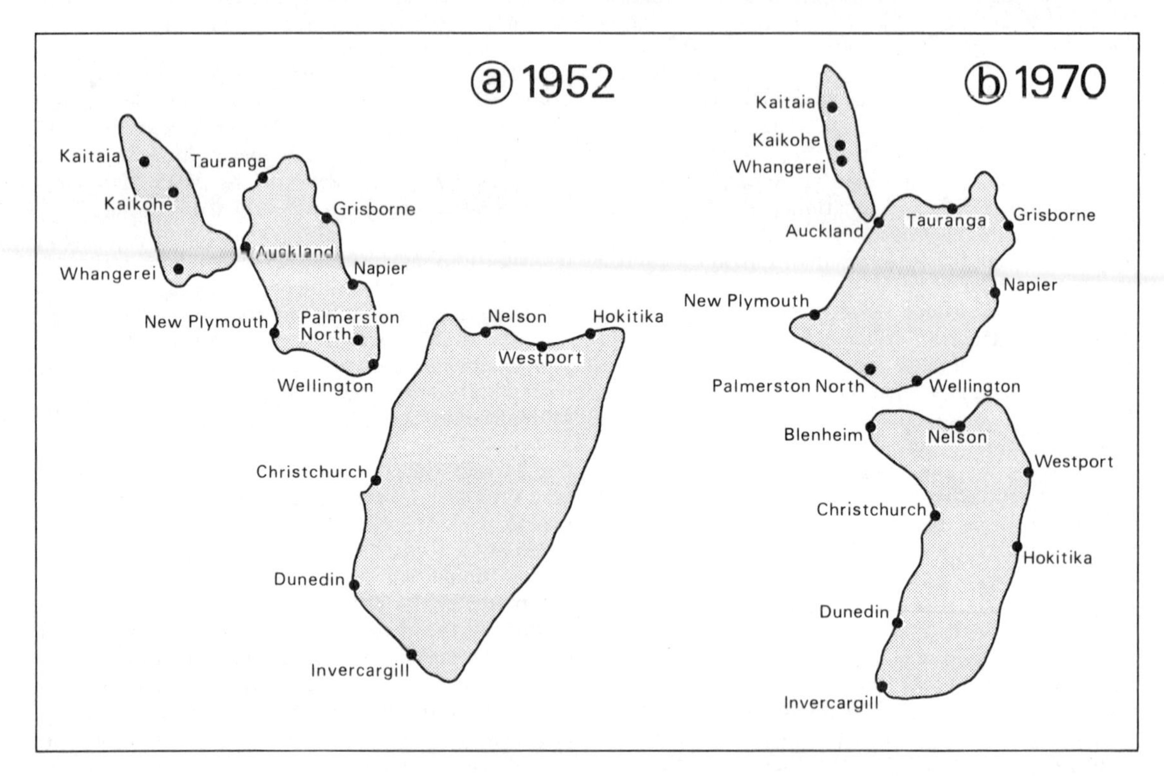

Pirie's work highlights the difficulties that emerge from ill-considered applications of nonmetric MDS (Pirie, 1977). His data comprise freight-cost matrices obtained for two years, the set of objects being fifty-one shipping points in South Africa. Regressing freight rate against distance permits an estimate of cost distance between all pairs of places. Yet because a curvilinear relationship between cost and distance is assumed, nonmetric MDS yields identical maps for both years and both tariff classes chosen for investigation. Pirie subsequently recognizes the problem and suggests that metric MDS would be more useful since it employs the interval measurements of cost rather than simply ordinal properties.

Applications in historical geography

Some of the most intriguing work using MDS comes from archaeology and historical geography, although the protagonists are not formally allied to either discipline. We cannot review archaeological applications here, but refer the interested reader to fascinating work by Tobler and Wineburg (1971), Kruskal (1971) and Kendall (1970). Of more direct relevance to the geographical subject-matter we are considering here are studies by Kendall. One problem he addresses (Kendall, 1971) is to construct a map of the eight 'Otmoor' parishes in Oxfordshire, simply from knowledge of the frequencies of marriages made between members of pairs of parishes (for the period 1600-1850). Kendall shows (Figure 15.4) that the marriage rates permit an accurate reconstruction of the configuration of parish centroids. The implication is that in circumstances where only similarity or dissimilarity data are available and we are ignorant of relative locations MDS provides as estimate of these. Some indication of absolute location is offered by orientating the MDS map to known locations (cf. Tobler and Wineburg, 1971).

At a micro-scale, Kendall has researched the problem of reconstructing the geographical structure of the manor of Whixley in medieval Yorkshire, where the only available information is provided by fragmentary manorial deeds (Kendall, 1975). The problem clearly requires much more than an algorithm that locates points from a single matrix of similarities, since the objects are parcels of land ('flatts') whose areal dimensions, where known, need to be incorporated. Further, some flatts have

Figure 15.4 The Otmoor parishes in geographical and 'marriage' space

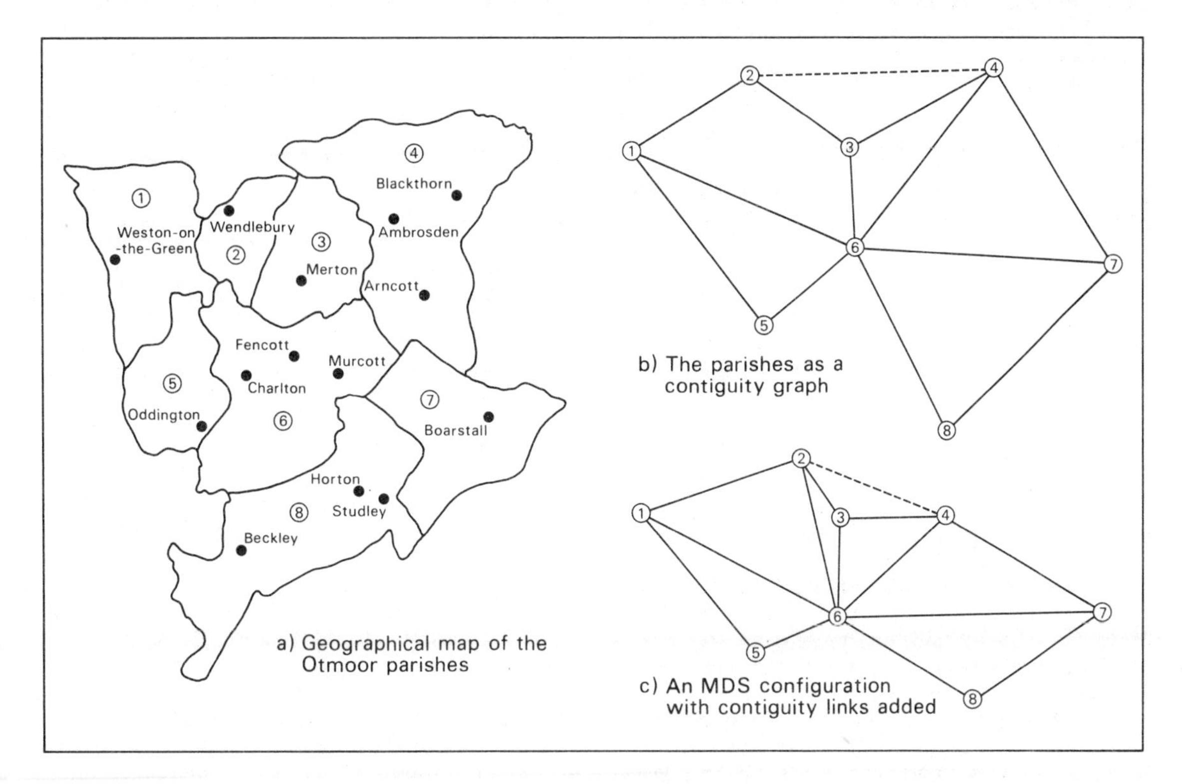

Detwyler (1970) on floral relationships among islands of the South Pacific. They derive a model of floral relations based on the number of species common to a pair of islands (C_{ij}), the total number of species per island (C_{ii}), island size and assumed migration of species. From the model, an expression for geobotanical distance is obtained:

$$\delta_{ij} = b^{-1}(\ln k + \ln(C_{ii} + C_{jj}) - \ln C_{ij}) \quad (1)$$

in which b and k are empirically determined constants. Multidimensional scaling of both the geobotanical and spherical geographical distances permits a graphical assessment of goodness-of-fit, while a linear regression of δ_{ij} on geographical distance indicates that 34 per cent of the variation in geobotanical distance is explained by great-circle distances. Again, particular pairs of islands reveal interesting features; for instance, North Island and South Island of New Zealand are much closer in geobotanical space than in geographical space. However, there are difficulties of course in representing areas by points so care must be exercised in interpreting such findings.

known locations, some pairs of flatts are known to be adjacent, while other pairs are not known to be contiguous. Needless to say, the problem calls for a special-purpose algorithm as well as meticulous historical investigation, and Kendall describes the substantial progress he has made in producing automated maps of the manorial topography.

Biogeographical spaces

Further imaginative work using MDS has emerged in biogeography, where Holloway and Jardine (1968) attempted to map geographical locations ('primary areas') in the Indo-Australian region in a transformed space. A measure of faunal dissimilarity is obtained for several classes, based on the number of species common to two primary areas. For instance, when species from the class *Aves* are considered, the two-dimensional configuration shows Australia and New Guinea to be relatively widely separated in zoogeographical space, more so than in untransformed geographical space. Holloway and Jardine suggest two plausible explanations: either that Australia once lay further south and has since moved northwards as a result of continental drift; or that there has been a long-standing ecological barrier to faunal diffusion.

In a similar vein is research by Tobler, Mielke and

Movement spaces

A relatively large literature exists on the investigation via MDS of structure in spatial interaction matrices. Such matrices of data (which can be assumed to define a similarity relation) differ from those considered above in that the relation is asymmetric. While detailed attention is given below to the problem of asymmetry, we review here some attempts to construct migration spaces before examining work on other flow data.

In an early paper, Lycan (1969) used metric MDS to represent the structure of migration between nine census regions in the USA. Defining the volume of migration between 1955 and 1960 as M_{ij} and population in 1960 as P_i, he postulates as a similarity measure:

$$m_{ij} = k \left\{ \frac{\min(M_{ij}, M_{ji})}{P_i P_j} \right\} \quad (2)$$

and uses the logarithm of m_{ij}^{-1} as a dissimilarity. Even with as few as nine regions Lycan finds it difficult to obtain an adequate two-dimensional representation.

Schwind (1971) has examined migration (considered as revealed spatial preferences) among five SEAs of Maine. He symmetrizes the data by taking

$$|m_{ij}| = (M_{ij} - M_{ji}) / (M_{ij} + M_{ji}) \quad (3)$$

and scales both these similarities and the two halves of the original migration frequency matrix (**M**). However, Schwind finds interpretation of the results somewhat problematic and clearly this type of work is more successful when set in a wider context. Thus Lloyd (1976) has obtained not only migration spaces but also cognitive and preference spaces and examines linkages between the three. All three spaces refer to the states of the USA and all are considered to be four-dimensional. The migration space is created by first transforming the actual flows to proportions ($P_{ij} = M_{ij} / \Sigma M_{ij}$) and then using as a dissimilarity measure

$$\delta_{ij} = \left| \left\{ \frac{P_{ij}}{P_{ij} + P_{ji}} \right\} - 0.5 \right| \quad (4)$$

This ensures symmetry and leads to a value of zero if neither state dominates and a maximum of 0.5 if there is no migration in one direction. Canonical correlation is used to relate the locational coordinates of the migration space to those of the preference and cognitive spaces of students from Pennsylvania and South Carolina. The results, indicating strong linkages between the three spaces, are most intriguing and the experiment could usefully be replicated in a British context.

More recently Slater (1978) and Slater and Winchester (1978) have employed metric MDS to explore the structure of migration flows within the USA and within the ninety *departments* of France. The migration matrices are initially adjusted to yield maximum entropy estimates of the movements that would have occurred if all regions had the same number of out-migrants and in-migrants. Size variations among the regions, that is, varying numbers of in-migrants and out-migrants are thus eliminated. The reciprocals of these standardized values are subsequently scaled, yielding a map that complements the dendrogram provided by clustering procedures. Both examples demonstrate that those regions with broad migration bases (receiving migrants from, and distributing migrants to, many regions) occupy the centre of migration space. Regions with a greater concentration of flows, however, are generally pushed away from the centre. Thus a dramatic visual representation of a 'core-periphery' regional system is realized. Beyond this, both papers emphasize an investigation of residuals (over- and under-estimation of functional distances) which as indicated below, provides a spatial goodness-of-fit assessment that may offer more illumination than a bald stress value.

In a exploratory study mapping flows of mail rather than people Gould (1975) has created a map of European 'postal space' after averaging the two halves of the matrix. MDS here offers a novel means of depicting graphically the structure of information flows (Figure 15.5). Countries that are peripheral in geographical space are peripheral in postal space, a feature that presumably reflects distance decay in the flows.

Goodchild and Kwan (1978) have looked at the spatial structure of information flows between seventeen Canadian cities. The data refer to a sample of the number of news events originating in city i and reported by a newspaper in city j. After estimating a model of the flows the model is inverted to solve for distance and the estimated distances scaled. Ottawa and Toronto lie at the centre of the configuration while cities which are less dominant as news sources occupy the periphery. In an interaction space, defined by business contracts in Sweden, Gatrell (1979) observed the same feature, with regional centres surrounding the core of the urban system (Stockholm, Göteborg and Malmö). Spatial autocorrelation was absent in the 'surface structure' of geographical space but detected in the 'deep structure' of the transformed map.

Other applications

Rushton (1972) has stated an interesting geographical problem:

> Given a spatial trend surface function expressing density of demand as a function of locational coordinates, determine the locations of supply points and their number such that in the proximal area surrounding each supply point approximately equal amounts of demand corresponding to a given 'threshold demand' will exist (p. 114).

Rushton derives an expression for the expected distance between supply points (central places) and uses metric MDS to scale the distances. This permits him to map the distribution of central places on a

Figure 15.5 European postal space

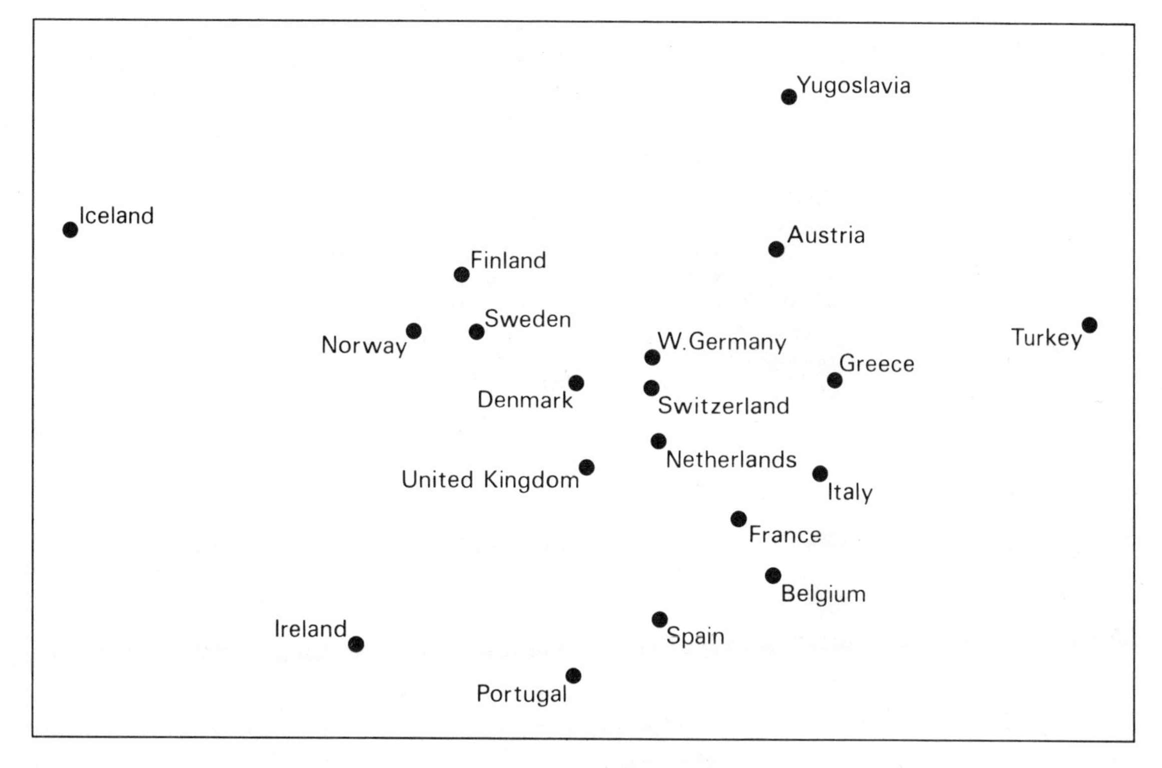

surface of non-uniform demand and a distorted hexagonal lattice may subsequently be sketched in (Figure 15.6). Curiously, Rushton's ingenious solution has not generated any empirical examination in a real-world context, but as Rushton notes a solution may not be easy to obtain if the demand surface is complex.

From a nonlinear mathematical programming model Odland (1976) derives an expression for the transport cost between pairs of zones in a city, based on production costs and disamenity costs (losses from real income). As cost data are unavailable, Odland takes employment density in i (E_i) as a surrogate for production intensity and residential density in j (R_j) as an index of crowding (disamenity) and fits the following equation to inter-zonal distances in four American cities:

$$d_{ij} = b_O - b_1 \ln E_i - b_2 \ln R_j + e_{ij} \quad (5)$$

Since the predicted distances are asymmetric, they are averaged and then scaled by metric MDS. There is a close correspondence between the scaled maps and the geographical locations of zone centroids, suggesting the validity of the original nonlinear programming model.

There have been exploratory uses of MDS in urban and regional planning (Hill and Tzamir, 1972; Massam, 1978). Given a set of n plans it is possible to calculate a 'concordance' index, c_{ij}, for all pairs of plans, where each plan is rated against multiple criteria. Massam (1978) scales the dissimilarity index

$$\delta_{ij} = |0.5 - c_{ij}| \quad (6)$$

to obtain a spatial representation of six urban transport plans for Singapore, in one and two dimensions. Three weighting schemes are applied to the criteria and each yields configurations that display the relative locations of plans with respect to each other.

Methodological issues

There has been little discussion by geographers of the relative merits of different scaling methods or algorithms for mapping purposes. Few authors discuss reasons for employing either metric or non-metric scaling and few emphasize the fact that different algorithms will yield different configurations. There is a need for systematic experimentation in order to assess how sensitive solutions are to the variety of methods available (cf. Spence, 1972). Further, we require a greater awareness of the sensitivity of the solution to the starting configuration, to the choice of dissimilarity relation and to the presence of error in input data. Below, we focus on two methodological issues that have attracted the attention of geographers and others before highlighting other appropriate research avenues.

Figure 15.6 Central places located on non-uniform demand surfaces

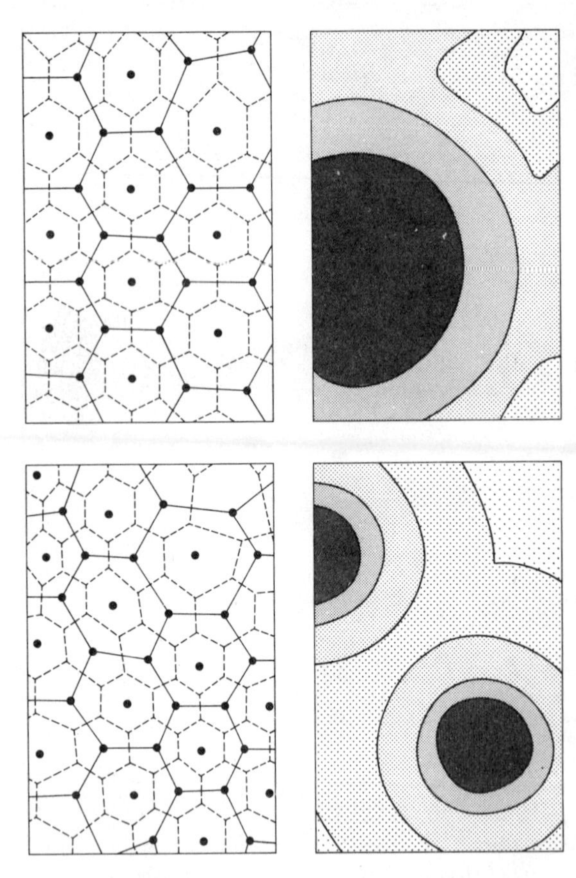

Map comparison

Reference has been made above to the frequent need to compare the point configuration generated by MDS with that realized in geographical space. Similarly, we may require some assessment of the difference between two configurations yielded by scaling and defined on the same set of points. In essence, both problems require estimation of the transformation (or 'spatial regression') that links one point pattern with the other. This transformation may involve dilation, translation, reflection and rotation.

Work by statisticians on the problem (known as 'procrustes analysis') generally assumes that dilation and translation have already been carried out and seeks a means of optimally rotating one configuration relative to the other (Schönemann, 1966; Gower, 1971; Sibson, 1978). The methods they have devised are for configurations in m dimensions and are confined to orthogonal rotation. Given two configurations, **A** and **B**, each represented as an (n x m) matrix, the problem is to find the transformation matrix **T** that minimizes the residual sum of squares (the 'Procrustes statistic'). We have:

$$\mathbf{B} + \mathbf{E} = \mathbf{AT} \quad (7)$$

where **E** is an error matrix and the criterion is to minimise tr $(\mathbf{E}'\mathbf{E})$ subject to $\mathbf{TT}' = \mathbf{I}$ (the orthogonality constraint). As both Schönemann (1966) and Gower (1971) show, the solution is provided by forming $\mathbf{S} = \mathbf{A}'\mathbf{B}$, diagonalizing the matrices

$$\begin{aligned} \mathbf{S}'\mathbf{S} &= \mathbf{V}\Lambda\mathbf{V}' \\ \mathbf{SS}' &= \mathbf{W}\Lambda\mathbf{W}' \end{aligned} \quad (8)$$

and obtaining $\mathbf{T} = \mathbf{WV}'$ (where Λ is a diagonal matrix of eigenvalues). The geographer would doubtless find much of interest in the matrix **E**, which for m = 2 permits a spatial goodness-of-fit assessment to complement the procrustes statistic. We return below to this theme.

In geography, Tobler's work pre-dates much of the statistical research and his treatment of the problem in terms both of complex numbers and matrix algebra (the two yield different transformations) is rather more general since the equations he derives include terms for translation and dilation, as well as rotation (Tobler, 1965). Tobler too expresses the problem as a linear mapping of one set (x, y) of coordinates onto another (u, v) and similarly adopts a least-squares approach that yields both an orthogonal rotation ('Euclidean regression' in Tobler's terminology) and an oblique transformation ('affine regression'). Regressing the (u, v) set on the (x, y) set provides a set of predicted coordinates $(\hat{u}, \hat{v})$ and permits calculation of errors (which are vectors). The predicted coordinates can be mapped as an orthogonal grid (Euclidean regression) or parallelograms (affine regression).

The methods generalize to m dimensions. In addition, extensions to multiple spatial regression are feasible. Could we not, for instance, model a time space as a function both of an earlier time space configuration and the true geographical coordinates? Or could we not model an individual's cognitive map (as obtained from MDS) as a function of several earlier estimates, providing a kind of temporal autoregression for point patterns? Surely there are intriguing possibilities for imaginative geographical research here.

Finally, as Tobler notes, we can obtain a correlation coefficient that measures the correspondence between the two point patterns. Given a set of point patterns (say, configurations from several individuals), could we not compute a matrix of such correlation coefficients and scale these as similarities? Individuals with similar cognitive maps should cluster in the same region of the new MDS map. In this way, we might make 'maps from maps'.

Tobler's observation that curvilinear spatial regression would be a useful extension (Tobler, 1965, p. 134) went unnoticed and it was left to Tobler himself to develop the idea (Tobler, 1978). If a square lattice is superimposed upon one configuration the method provides a means of computing the transformation that warps the lattice into a distorted grid that corresponds to the other map. Further spatial characteristics of the transformation are provided by mapping residuals ('displacement vectors') and ellipses of distortion that show graphically the amount of warping and stretching in any direction.

Ewing and Wolfe (1977) offer a heuristic procedure for obtaining such distorted grids based on a weighted sum of the displacement vectors. Their procedure has a less sound basis in analytical cartography than Tobler's work but none the less provides a relatively simple way of creating the transformed grid. In addition, it permits easy transference of other map features onto the transformed map. Ewing and Wolfe illustrate the method for a time space map of Toronto. Elsewhere in the subject only Golledge (*et al.*,1976; 1978a) has looked in any detail at deformed grids superimposed on MDS configurations.

Asymmetric relations

For those problems that yield $\delta_{ij} \neq \delta_{ji}$ the full square dissimilarity matrix may be scaled (though clearly this increases stress since $d_{ij} = d_{ji}$). Alternatively, the two dissimilarities for the ij pair may be averaged, a procedure which is acceptable if the dissimilarities are not substantially asymmetric. Kruskal and Wish (1978, p. 74-5) suggest removing row and column effects (as in ANOVA) to scale interactions, though this does not appear to have been adopted in the literature. Another of their suggestions is to split the matrix into n rows, perform a separate monotonic regression for each row and then combine the stress values into single stress; this procedure is known as multidimensional unfolding (MDU).

Gower (1977) and Constantine and Gower (1978) made use of MDU in seeking graphical displays of asymmetry. The method yields two sets of n points, one representing the rows and the other the columns, so that ij distances and ji distances are portrayed on the same map. Interest thus focuses on distances between objects in different sets rather than on objects within a set. If stress is not too high, the method highlights asymmetries (such that if $\delta_{ij} > \delta_{ji}$, $d_{ij} > d_{ji}$ on the map).

The second approach, clearly spelled out in Constantine and Gower (1978), is to write the asymmetric matrix as the sum of the symmetric and skew-symmetric matrix (**N**) and to analyse each separately. The symmetric component yields a map in the usual way, while a canonical decomposition of **N** offers another geometrical representation. If two dimensions account for a substantial part of the variation in **N** then we can approximate **N** as λ (**u v**′ − **v u**′) and plot the objects as points in (u, v) space. Because of the skew-symmetry the relationship between the points is interpreted in terms of areas of triangles, large areas indicating that the objects differ in their skew-symmetry with respect to other objects. Where the points are collinear the asymmetry is one-dimensional (one dimension accounts for

Figure 15.7 Asymmetry of state-to-state college attendance, 1968: 'winds of influence'

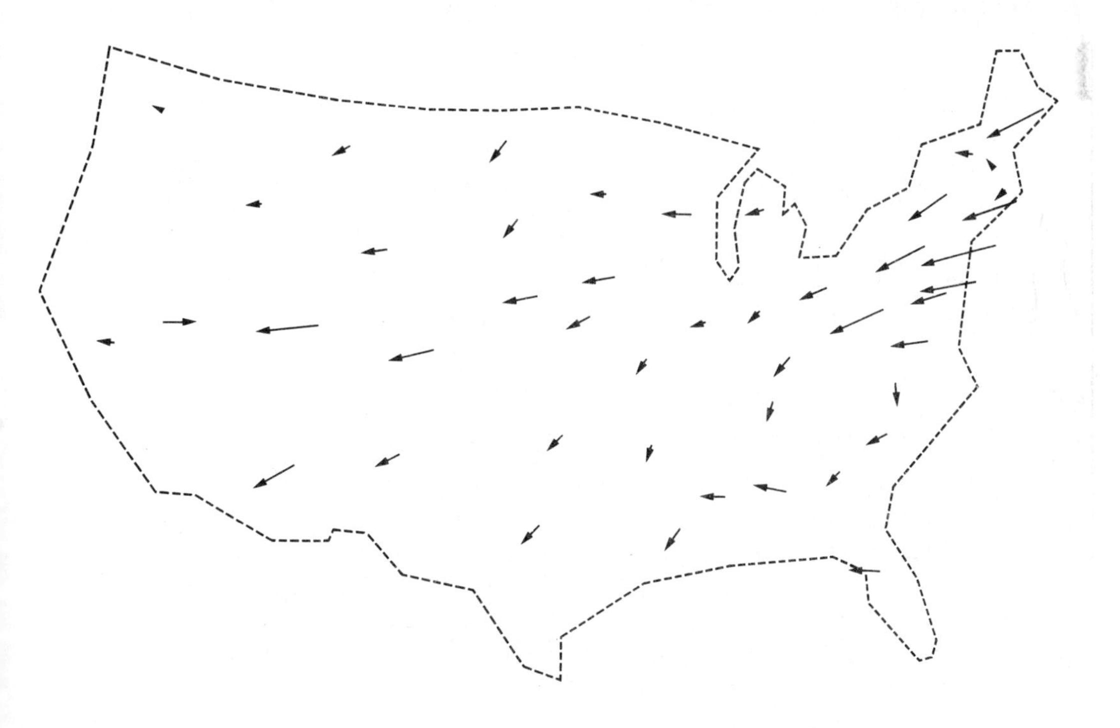

most of the variation in **N**) and this is useful since we can obtain (using MDS) a two-dimensional map of the symmetric matrix and simply plot asymmetry as a third dimension. Contours may be drawn, objects lying on the same contour bearing a symmetric relation while if i lies above j the implication is that $\delta_{ij} < \delta_{ji}$.

The possibility of representing both symmetric and skew-symmetric components on the same map is attractive, but is not always feasible. However, Tobler (1975) has described a novel method of analysing asymmetry that always yields a graphical representation of both components. From an asymmetric matrix **M** we may compute

$$c_{ij} = \frac{M_{ij} - M_{ji}}{M_{ij} + M_{ji}} \tag{9}$$

which may be interpreted as a 'current' aiding flow in one direction. Fixing i and computing c_{ij} for all j yields a cluster of vectors around i, the resultant defining a 'wind' at i. Repeating this for all i gives a vector field of 'winds of influence' (e.g. Figure 15.7) that provides a striking cartographical representation of asymmetry.

However, Tobler ingeniously carries the meteorological analogy further, asking if it is not possible to infer, from the winds, the pressure field (or 'forcing function') that generated them. Tobler shows, using vector differential calculus, that Poisson's equation can be applied to the vector field to yield two scalar fields, one the forcing function (which can be regarded as giving rise to the flows), the other a 'vector potential' or residual component. The forcing function that corresponds to Figure 15.7 is illustrated (Figure 15.8).

As Tobler notes, other definitions of c_{ij} are possible; furthermore, in obtaining the resultant we might weight each c_{ij} by the inverse of distance. We require some careful experimentation with asymmetric matrices in order to explore the sensitivity of Tobler's method to different options.

In the literature, Tobler's work has attracted few applications, partly because of the assumption of continuity embedded in the estimation of the forcing function (Goodchild and Kwan, 1978, p. 1315). However, in a micro-geographical application, Gatrell and Gould (1979) have mapped (using MDS) the players of team sports in two dimensions, depicted asymmetries in the movement of the ball between players as winds of influence, and obtained the corresponding forcing function.

Prospects

Clearly there is some exciting work to be done exploring other examples of the kinds of space discussed above and using MDS to map other novel spaces (cf. Gould, 1978). Secondly, there are possibilities of modelling the evolution of some spaces and using the coordinates of transformed spaces in evaluating geographical models. In addition, experimentation with a range of algorithms and artificial data sets (to investigate sensitivity to

Figure 15.8 Asymmetry of state-to-state college attendance, 1968: 'the forcing function'

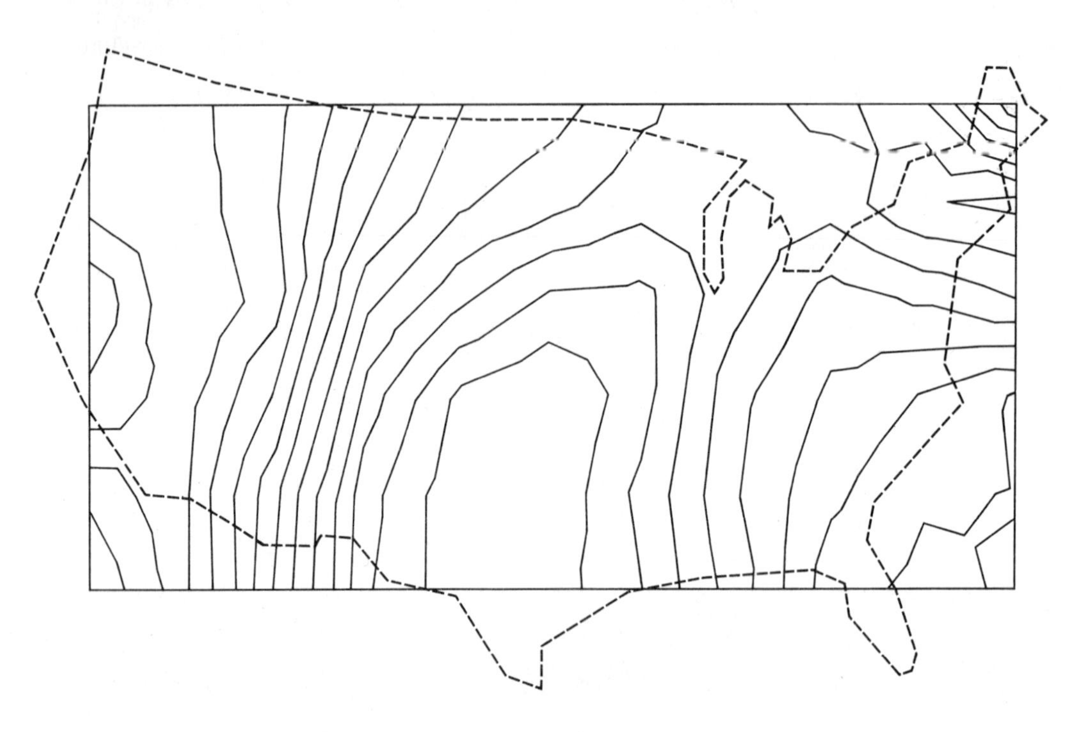

asymmetry, for instance) is required. Let us, however, consider one interesting geographical problem already alluded to above.

This concerns the notion of a residual in MDS, taking care to distinguish two kinds of residual that arise in applications. One arises from the regression of configuration distances on dissimilarities; the other appears in the map comparison problem. Both types of residual have a spatial representation as vectors.

In order to see how accurately a two-dimensional configuration reflects the original dissimilarity matrix we can, of course, compute a stress statistic. But, as with any aspatial goodness-of-fit statistic, this is frequently of less interest to the geographer than a mapping of residuals. As noted above, MDS yields a matrix of residuals, e_{ij}, which we can map as a cluster of vectors around i. Some objects are pulled closer together in the configuration while others are pushed further apart than they should be on the basis of the empirical relation (Kruskal and Wish, 1978, pp. 32-3). In trilateration, a procedure mathematically isomorphic with metric MDS (Tobler, 1976), the residuals may be used to map an ellipse around each i (Wolf, 1969; Wolf and Johnson, 1974). The ellipses provide a two-dimensional confidence interval for the estimated locations, their size reflecting the precision of location, while their shape and orientation reflect directional variations in precision. An attempt has been made to explore this idea in nonmetric scaling (Ramsay, 1978) and it would seem that the spatial information provided by such ellipses has much to offer the geographer interested in MDS.

It was noted above that the comparison of two MDS configurations, or an MDS map with geographical space, effected by means of spatial regression, also yields a residual. Here, however, there is a residual for each point, $e_i = (u_i - \hat{u}_i, v_i - \hat{v}_i)$ rather than for pairs of points. Now, while maps of such vectors are beginning to appear in the literature (Golledge, 1978b), there is little analytical work done as yet on the spatial pattern of such vectors. We need to tackle the subject of autocorrelation in vector maps, for this would surely illuminate the map comparison (e.g. in suggesting a curvilinear rather than a linear regression). Parenthetically, methods for the estimation of autocorrelation in vector maps would also aid in the statistical description of winds of influence.

As a device for creating graphical representations of the structure of a relation defined on a set of objects, MDS offers fruitful insights for the geographer. It can be regarded as a useful complement to more formal mathematical languages for the interpretation of structure, such as the polyhedral dynamics ('Q-analysis') of Atkin (Chapman, this volume).

References

Clark, J.W. (1977) 'Time-distance transformations of transportation networks', *Geographical Analysis*, 9, 195-205.

Constantine, A.G., and Gower, J.C. (1978) 'Graphical representation of asymmetric matrices', *Applied Statistics*, 27, 297-304.

Ewing, G.O. (1974) 'Multidimensional scaling and time-space maps', *Canadian Geographer*, 18, 161-7.

Ewing, G.O., and Wolfe, R. (1977) 'Surface feature interpolation on two-dimensional time-space maps', *Environment and Planning A* 9, 429-37.

Forer, P. (1973) 'Changes in the spatial structure of the New Zealand airline network', PhD dissertation, University of Bristol.

Forer, P. (1978a) 'Time-space and area in the city of the plains' in T. Carlstein, D. Parkes and N. Thrift (eds), *Timing Space and Spacing Time*, Vol. 1, Edward Arnold: London.

Forer, P. (1978b) 'A place for plastic space?', *Progress in Human Geography*, 3, 230-67.

Gatrell, A.C. (1979) 'Autocorrelation in spaces', *Environment and Planning A*, 11, 507-16.

Gatrell, A.C., and Gould, P.R. (1979) 'A micro-geography of team games: graphical explorations of structural relations', *Area*, 11, 275-78.

Golledge, R.G. (1978a) 'Learning about urban environments' in T. Carlstein, D. Parkes and N. Thrift (eds), *Timing Space and Spacing Time*, Vol. 1, Edward Arnold: London.

Golledge, R.G. (1978b) 'Representing, interpreting, and using cognized environments', *Papers of the Regional Science Association*, 41, 169-204.

Golledge, R.G., and Rushton, G. (1972) 'Multidimensional scaling: review and geographical applications', *Association of American Geographers Commission on College Geography, Technical Paper* No. 10.

Golledge, R.G., and Rushton, G. (eds) (1976) *Spatial Choice and Spatial Behaviour*, Ohio State University Press: Columbus, Ohio.

Golledge, R.G., Briggs, R., and Demko, D. (1969) 'The configuration of distances in intra-urban space', *Proceedings of the Association of American Geographers*, 1, 60-6.

Golledge, R.G., Rivizzigno, V.L., and Spector, A. (1976) 'Learning about a city: analysis by multidimensional scaling' in R.G. Golledge and G. Rushton, op. cit.

Goodchild, M., and Kwan, M. (1978) 'Models of hierarchically dominated spatial interaction', *Environment and Planning A*, 10, 1307-17.

Gould, P.R. (1975) 'Mathematics in geography: conceptual revolution or new tool?',

International Social Science Journal, 27, 303-27.
Gould, P.R. (1976) 'Cultivating the garden' in R.G. Golledge and G. Rushton, op. cit.
Gould, P.R. (1978) 'Concerning a geographic education' in D. Lanegran and R. Palm (eds), *Invitation to Geography*, McGraw-Hill: New York.
Gower, J.C. (1971) 'Statistical methods of comparing different multivariate analyses of the same data' in F.R. Hodson, D.G. Kendall and P. Tautu, op. cit.
Gower, J.C. (1977) 'The analysis of asymmetry and orthogonality' in J.R. Barra (ed.), *Recent Developments in Statistics*, North-Holland: Amsterdam.
Hill, M., and Tzamir, Y. (1972) 'Multidimensional evaluation of regional plans serving multiple objectives', *Papers of the Regional Science Association*, 29, 193-65.
Hodson, F.R., Kendall, D.G., and Tautu, P. (eds) (1971) *Mathematics in the Archaeological and Historical Sciences*, Edinburgh University Press.
Holloway, J.D., and Jardine, N. (1968) 'Two approaches to zoogeography: a study based on the distributions of butterflies, birds and bats in the Indo-Australian area', *Proceedings of the Linnean Society of London*, 179, 153-88.
Hudson, R. (1975) 'Patterns of spatial search', *Transactions of the Institute of British Geographers*, 65, 141-54.
Kendall, D.G. (1970) 'A mathematical approach to seriation', *Philosophical Transactions of the Royal Society of London A*, 269, 125-35.
Kendall, D.G. (1971) 'Maps from marriages: an application of non-metric multidimensional scaling to parish register data' in F.R. Hodson, D.G. Kendall and P. Tautu, op. cit.
Kendall, D.G. (1975) 'The recovery of structure from fragmentary information', *Philosophical Transactions of the Royal Society of London* A, 279, 547-82.
Kilchenmann, A. (1972) 'Quantitative Geographie als Mittel zur Lösung von planerischen Umweltproblemen', *Geoforum* 12, 53-71.
Kruskal, J.B. (1964a) 'Multidimensional scaling by optimizing goodness of fit to a non-metric hypothesis', *Psychometrika*, 29, 1-27.
Kruskal, J.B. (1964b) 'Non-metric multidimensional scaling: a numerical method', *Psychometrika*, 29, 115-29.
Kruskal, J.B. (1971) 'Multi-dimensional scaling in archaeology: time is not the only dimension' in F.R. Hodson, D.G. Kendall and P. Tautu, op. cit.
Kruskal, J.B., and Wish, M. (1978) *Multidimensional Scaling*, Sage University Papers, Series 07-011: London.
Lloyd, R.E. (1976) 'Cognition, preference, and behavior in space: an examination of the structural linkages', *Economic Geography*, 52, 241-53.
Lycan, R. (1969) 'Matrices of inter-regional migration', *Proceedings of the Association of American Geographers*, 1, 89-95.
MacKay, D.B. (1976) 'The effect of spatial stimuli on the estimation of cognitive maps', *Geographical Analysis*, 8, 439-52.
MacKay, D.B., Olshavsky, R.W., and Sentell, G. (1975) 'Cognitive maps and spatial behavior of consumers', *Geographical Analysis*, 7, 19-34.
Marchand, B. (1973) 'Deformation of a transportation surface', *Annals of the Association of American Geographers*, 63, 507-21.
Marchand, B. (1974) 'Pedestrian traffic planning and the perception of the urban environment: a French example', *Environment and Planning A*, 6, 491-507.
Massam, B.H. (1978) 'The search for the best alternative using multiple criteria: Singapore transit study', *Economic Geography*, 54, 245-53.
Muller, J-C. (1978) 'The mapping of travel time in Edmonton, Alberta', *Canadian Geographer*, 22, 195-210.
Odland, J. (1976) 'The spatial arrangement of urban activities: a simultaneous location model', *Environment and Planning A*, 8, 779-91.
Pirie, G.H. (1977) 'Charting transport cost surfaces by nonmetric multidimensional scaling', *South African Geographical Journal*, 59, 60-4.
Pocock, D.C.D., and Hudson, R. (1978) *Images of the Urban Environment*, Macmillan: London.
Ramsay, J.O. (1978) 'Confidence regions for multidimensional scaling analysis', *Psychometrika*, 43, 145-60.
Rushton, G. (1972) 'Map transformations of point patterns: central place patterns in areas of variable population density', *Papers of the Regional Science Association*, 28, 111-29.
Schönemann, P.H. (1966) 'A generalized solution of the orthogonal procrustes problem', *Psychometrika*, 31, 1-10.
Schwind, P.J. (1971) 'Spatial preferences of migrants for regions: the example of Maine', *Proceedings of the Association of American Geographers*, 3, 150-6.
Shepard, R.N. (1962) 'The analysis of proximities: multidimensional scaling with an unknown distance function', *Psychometrika*, 27, 125-40 and 219-46.
Sibson, R. (1972) 'Order invariant methods for data analysis', *Journal of the Royal Statistical Society Series B*, 34, 311-49.
Sibson, R. (1978) 'Studies in the robustness of multidimensional scaling: procrustes statistics', *Journal of the Royal Statistical Society Series B*, 40, 234-8.
Slater, P.B. (1978) 'Hierarchical clustering and robust multidimensional scaling: applications to interstate migration flows', unpublished manuscript, West Virgina University.
Slater, P.B., and Winchester, H.P.M. (1978) 'Clustering and scaling of transaction flow tables: a French interdepartmental migration example',

IEEE Transactions on Systems, Man and Cybernetics, 8, 635-40.
Spence, I. (1972) 'A Monte Carlo evaluation of three non-metric multidimensional scaling algorithms', *Psychometrika*, 37, 323-55.
Tobler, W.R. (1965) 'Computation of the correspondence of geographical patterns', *Papers of the Regional Science Association*, 15, 131-9.
Tobler, W.R. (1975) *Spatial interaction patterns*, International Institute for Applied Systems Analysis, Laxenburg, Austria Research Report RR-75-19.
Tobler, W.R. (1976) 'The geometry of mental maps' in R.G. Golledge and G. Rushton, op. cit.
Tobler, W.R. (1978) 'Comparison of plane forms', *Geographical Analysis*, 10, 154-62.
Tobler, W.R., and Wineburg, S. (1971) 'A Cappadocian speculation', *Nature*, 231, 39-41.
Tobler, W.R., Mielke, H., and Detwyler, T. (1970) 'Geobotanical distance between New Zealand and neighboring islands', *Bioscience*, 20, 537-42.
Wolf, P.R. (1969) 'Horizontal position adjustment', *Surveying and Mapping*, 29, 635-44.
Wolf, P.R., and Johnson, S.D. (1974) 'Trilateration with short range EDM equipment and comparison with triangulation', *Surveying and Mapping*, 34, 337-46.
Young, G., and Householder, A.S. (1938) 'Discussion of a set of points in terms of their mutual distances', *Psychometrika*, 3, 19-22.

Chapter 16

Point pattern analysis

R.W. Thomas

The term point pattern analysis embraces a variety of methods and techniques which, over the past twenty years, geographers have used to analyse properties of point distributions. These methods all employ probability theory to model some geometrical aspect of the spatial arrangement of observed point patterns. For instance, the simplest tests are based on theoretical models describing random or independent point patterns. The theory which underpins such methods has for the most part been developed outside geography and owes much to the work of mathematical statisticians such as Bartlett (1955, 1972) and quantitative plant ecologists such as Clark and Evans (1954), Greig-Smith (1964) and Pielou (1977). Indeed, the methods of point pattern analysis have been applied in such diverse disciplines as archaeology (Hodder and Hassall, 1971), astronomy (Neyman and Scott, 1958), cell biology (Barton, David and Fix, 1963) and geology (Switzer, 1976).

The methods of analysis reviewed in this chapter are all deduced from quite restricted geometric definitions of pattern. In order to classify these methods, Hudson and Fowler (1966) argue that a distinction should be made between methods measuring the dispersion of a point distribution and those methods which actually measure pattern. They argue that true pattern measures should refer to the internal geometrical properties of points, such as angles and distances between points, without regard to the region containing the points. In contrast, dispersion measures take the areal extent of the point distribution into account and, therefore, are dependent on the density of points within the study area. In fact the two most widely used methods of pattern analysis are both derived from measures of point dispersion. First, quadrat analysis measures dispersion in relation to the frequency distribution of the observed point pattern, that is the way in which the density of points varies over the study area. Second, nearest neighbour methods measure dispersion in relation to properties of the distribution of distances between each point and its nearest neighbouring point. These topics are treated in the first two sections of this review. The theory of true pattern methods is not so well developed and consists of techniques for modelling properties of areas, usually polygons, constructed around each point in the pattern. These methods are reviewed in the third section.

Quadrat analysis

The continuum of spatial pattern

Implicit in the derivation of both quadrat and nearest neighbour methods is the notion that all observed point patterns fall within a continuum of spatial pattern. The lower limit of this continuum (see Figure 16.1), is the totally clustered pattern where all points occupy the same location. At the other extreme is the totally uniform, or regular, pattern where each point is located on the vertex of a grid of equilateral triangles. Approximately halfway between these limits is the random pattern where the location of each point appears to be independent of all other points. All other point patterns are envisaged as mixtures of these basic types. Thus we may speak of patterns that are more regular than random or more clustered than random.

Quadrat methods analyse the frequency distribution of patterns over some pre-defined study area. A frequency array, $\{n_x\}$, is collected by laying a grid of n equal-sized cells over the study area and then counting the number of cells containing $x = 0, 1, 2, \ldots$ points (see Figure 16.1). This method of data collection is termed quadrat censusing. An alternative method of collecting the frequency array is quadrat sampling which involves randomly placing a quadrat n times within the study area, each time counting the number of points within the quadrat. The purpose of quadrat analysis is to compare this observed frequency array with a theoretical frequency array predicted by some postulated spatial point process which it is thought might generate the observed pattern. If a close correspondence exists between the observed and predicted frequencies, the postulated point process is taken as a possible explanation for the observed point pattern.

hypothesis of no significant difference between the observed and expected frequencies. The test is made with m-2 degrees of freedom where m is the number of classes of x used to carry out the test. An additional degree of freedom is lost for every model parameter which has to be estimated from the data. However, many statisticians, for example Mead (1974), are sceptical about the value of this test because it is low powered and consequently often fails to detect genuine differences between the observed and expected frequency distributions.

Figure 16.1 Types of point pattern: A perfectly clustered, B random (with census and frequency array), C perfectly regular

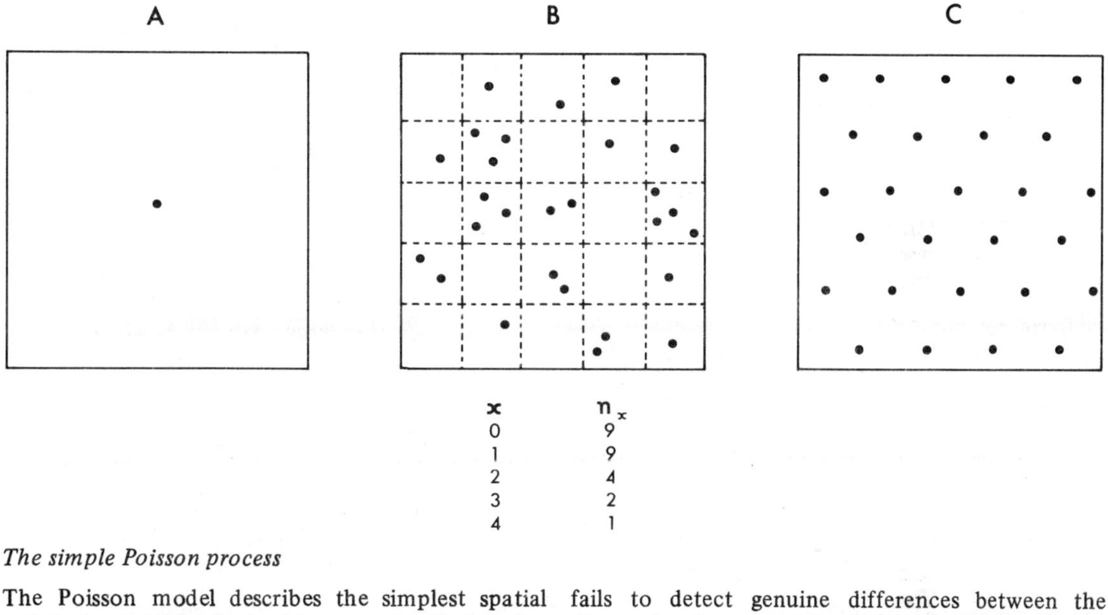

A B C

x	n_x
0	9
1	9
2	4
3	2
4	1

The simple Poisson process

The Poisson model describes the simplest spatial point process. It is obtained by assuming that during the evaluation of the pattern each cell has an equal and independent chance of receiving a point. This process generates a random pattern of points of the type shown in Figure 16.1b. The Poisson distribution is given by the formula

$$P_x = \frac{e^{-\lambda}\lambda^x}{x!}, \quad x = 0, 1, 2, \ldots, \qquad (1)$$

where P_x is the probability of finding x points in a specific quadrat of a pattern generated by a random process and where λ is the expected mean number of points per cell.

Fitting the model: The Poisson probabilities are obtained by estimating λ from the data as the observed average number of points per cell ($\lambda = r/n$, where r is the observed number of points) and then substituting in (1) for successive values of P_x. These probabilities are converted into predicted frequencies by multiplying each value by n. The X^2 goodness-of-fit test may be used to test the null

A useful property of the Poisson distribution is that its mean (λ) is always equal to its variance (σ^2). Therefore, the variance/mean ratio for such distributions is always equal to unity. This result allows observed frequency distributions to be tested for departure from the Poisson expectation using the difference between the observed variance/mean ratio ($S_x^2/\bar{x}$) and unity. The standard error of this difference is $[2/(n-1)]^{\frac{1}{2}}$ and its statistical significance may be tested against the t-distribution with a test statistic (Greig-Smith, 1964) given by

$$\text{calc.}t_{(n-1\ \text{d.f.})} = \frac{(S_x^2/\bar{x}) - 1.0}{\sqrt{2/(n-1)}} \qquad (2)$$

A result producing a variance/mean ratio significantly greater than unity is indicative of clustering in the observed pattern while a result significantly less than unity is indicative of a tendency towards clustering. An alternative significance test for this ratio has been given by Bartko, Greenhouse and Patlak (1968).

Applications: The earliest application of the simple Poisson model is Matui's (1932) analysis of house patterns on the Tonami plain, Japan. However, the model assumptions were rejected because the settlement pattern of nucleated villages caused clustering in the observed distributions of houses. Indeed, few geographical patterns have been found to follow the assumptions of the simple Poisson process. The most successful applications of the model have employed the variance/mean ratio to trace changes in the dispersion of a single pattern over time. Examples of this approach are provided by Sibley's (1972) study of changing shop location patterns in Northampton and Leicester, and Getis's (1964) study of changing land-use patterns in Lansing, Michigan, 1910–60.

Mixed Poisson process models

So far we have limited discussion to the way in which randomness or independence manifests itself in space. However, of greater intrinsic interest to the geographer are more complex probability distributions capable of modelling point processes which lead to either clustering or uniformity in the observed pattern. Many of these models can be deduced by mixing the Poisson assumptions with those of more complex point processes. The method of mixing may be either additive or multiplicative. An additive process is obtained by simply summing the effects of two or more sets of assumptions. Multiplicative processes are more complex and may be sub-divided into generalized and compound models. To obtain a generalized model one group of assumptions is conditioned by another such that the mixture assumes a fundamental spatial affinity between the individuals forming the point population. Compound models are obtained by joining independent sets of assumptions to describe some basic heterogeneity in the point population. The precise mathematical methods that are used to obtain such mixtures are described in Gurland (1957). Numerous probability distributions may be obtained by mixing, but what follows is a discussion of those models which have been applied to geographical problems.

Additive processes 1 – Dacey's county-seat model An ingenious additive model was derived by Dacey (1964, 1966) to describe the distribution of urban settlements in Iowa with populations greater than 2,500 people. The model is obtained by adding the Poisson assumptions to those of the Bernoulli distribution. Dacey used the following arguments to justify this mixture as an explanation for the frequency distribution of urban places over the counties of Iowa which are laid out in a square grid. First, those urban places which are county seats with populations greater than 2,500 will tend to be uniformly distributed one to a county (cell) in accordance with the Bernoulli (presence or absence) distribution given by

$$P_x = \theta^x (1-\theta)^{1-x} \qquad x = 0, 1 \ldots \qquad (3)$$

where θ takes on values between 0 and 1, and is the proportion of counties with county seats larger than 2,500 people. Second, it is assumed that all other urban settlements will be distributed according to the simple Poisson law with a density of γ places per county. Denoting the average number of all urban places per county by λ, then the addition of the Poisson and Bernoulli assumptions produces the following model to describe the probability that a specified county contains x towns

$$P_x = \frac{(1-\theta)\gamma^x e^{-\gamma}}{x!} + \frac{\theta x \gamma^{(x-1)} e^{-\gamma}}{x!} \quad x = 0, 1, \ldots (4)$$

The mean and variance of this distribution are given by

$$\lambda = \gamma + \theta, \quad \sigma^2 = \lambda - \theta^2, \qquad (5)$$

and if γ and θ are unknown their values may be give by the following moments estimates,

$$\hat{\theta} = (\bar{x} - S_x^2)^{\frac{1}{2}} \qquad \hat{\gamma} = \bar{x} - \hat{\theta} \qquad (6)$$

The model described by (4) predicts a distribution that is more regular than random, and the degree of regularity increases with the value of the parameter θ. The model gave good descriptions of the frequency distributions of large Iowa towns for all US census periods between 1840 and 1960.

Additive processes 2 – General double Poisson process model This model is obtained by adding two simple Poisson processes. It is assumed that some mechanism causes the density of points to be lower in one part of the study area than another, therefore two Poisson processes are operating at different densities which results in the appearance of clustering in the overall frequency distribution. The model for this double, heterogeneous point process is given by

$$P_x = a_1 \frac{e^{-\lambda_1}\lambda_1^x}{x!} + a_2 \frac{e^{-\lambda_2}\lambda_2^x}{x!} \quad x = 0, 1, \ldots (7)$$

where the products of the parameters $a_1 . \lambda_1$ and $a_2 . \lambda_2$ are roughly equivalent to the two point densities. Estimating these parameters is a lengthy procedure which is discussed in Schilling (1947) and Hinz and Gurland (1967). McConnell and Horn (1972) have used the general double Poisson to describe the frequency distribution of karst depressions on the Mitchell plain, Indiana. Their justification for the model is that karst depressions can be divided into two types: dolines, which are small features formed above the water table by solution along zones of the weakness in the rock and by ponding of surface run-off; and collapse sinks formed by cavern roof collapse which is due to the subsurface drainage system. They argue that both types of depression should be randomly distributed but at different densities and, indeed, the model described by (7) was found to predict the observed frequency distribution of all karst depressions quite closely.

Multiplicative processes: the negative binomial distribution This distribution is one of several models which have been used to describe contagious point processes. The generalized case of the negative binomial is obtained by a multiplicative mixing of the Poisson assumptions with a logarithmic growth law (see Anscombe, 1950). It is assumed that an initial set of points are distributed according to the logarithmic growth law. This process results in a random distribution of clusters where the number of points within each cluster follows the generalizing logarithmic growth law. The probability of finding a quadrat with *x* points is given by the expression

$$P_x = \binom{x+k-1}{k-1} p^k (1-p)^x \quad x = 0, 1, 2, \ldots \tag{8}$$

The mean and variance of the negative binomial are given by

$$\lambda = k(1-p)/p; \quad \sigma^2 = k(1-p)/p^2 \tag{9}$$

and the moments estimators of the parameters are

$$\hat{p} = \bar{x}/S_x^2 \; ; \; \hat{k} = \bar{x}\hat{p}/(1-\hat{p}). \tag{10}$$

A method for obtaining the more efficient maximum likelihood estimators of p and k is described in Bliss and Fisher (1953). The parameter k can take on values between 0 and infinity and is a measure of the degree of clustering generated by the contagious process. As k tends to infinity the clustering disappears and the negative binomial tends to the Poisson distribution, while as k tends to zero the clustering increases so that the negative binomial converges onto the logarithmic distribution.

Harvey (1966) has used the negative binomial to model Hägerstrand's (1953) maps showing the diffusion of various agricultural innovations among farmers in South Sweden. It was assumed that the initial adopters of the innovation would be randomly distributed while subsequent adopters would cluster around the initiators according to the logarithmic growth law. The good fit between the map and predicted frequencies suggested that these assumptions provided a reasonable description of the diffusion process. Similarly, Artle (1965), Rogers (1965, 1969) and Lee (1974) have all demonstrated that the negative binomial gives good descriptions for the clustering of various store locations within urban areas.

An alarming feature of the negative binomial distribution is that the model described by (8) may be deduced from a number of different starting assumptions. One alternative derivation of the model assumes each point is located independently but that some cells are more attractive than others. In this case the Poisson density parameter λ is not a constant but varies from place to place. The variation in λ from cell to cell is specified by a compounding distribution. If the compounding distribution is the gamma distribution then the appropriate description of the point process is the negative binomial. Moreover, while geographers have generally restricted their attention to the generalized and compound versions of the negative binomial, Boswell and Patil (1970) have shown that in mathematical statistics the distribution can be derived in at least a dozen other different ways. Among others, Getis (1967), Dacey (1968) and Cliff and Ord (1973) have suggested methods for distinguishing between generalized and compound negative binomial distributions. In particular, Cliff and Ord propose a method based on the analysis of how the k parameter varies when different sized cells are used to census the point pattern. Let s denote the number of the smallest quadrats that are combined to form larger cells. It can be proved that, if a generalized model is appropriate, the estimate of k obtained from the smallest cell frequencies should change to sk when the estimate is made from grouped frequencies. However, if a compound model is appropriate the estimate of k should remain unchanged after grouping. Applying this procedure to Matui's house pattern data, Cliff and Ord demonstrated that the estimates of k obtained from nine cell sizes tended to behave in accordance with the compound model.

Other mixed distributions

Many other distributions such as the Neyman A (Neyman, 1939), the Thomas double Poisson (Thomas, 1949) and the Polya-Aeppli (Getis, 1974) have been derived to model clustered point processes. However, most of these models have quite specialized ecological derivations which have limited their geographical applicability. For instance, the Neyman A, which was derived to model the distribution of insect larvae crawling away from recently hatched egg clusters, have been discussed quite frequently in the geographical literature (Olsson, 1967; Harvey, 1968a). But attempts to fit this model to patterns of innovation adopters (Harvey, 1966), vacation home locations (Aldskogious, 1969) and store locations (Rogers, 1965 and 1969) have not been successful.

A more complex approach to point clustering is provided by the bivariate probability distributions derived by mathematical statisticians such as Holgate (1964 and 1966) and Edwards and Gurland (1961). These distributions model two point patterns simultaneously and allow the researcher to study an object pattern in conjunction with a second variable thought to be responsible for areal heterogeneity. To date the only geographical application of bivariate models is Rogers and Martin's (1971) analysis of the distribution of foodstores in Ljubljana, Yugoslavia, in association with the underlying population distribution. They report some success in fitting a bivariate version of the negative binomial to this data but encountered technical difficulties with both parameter estimation and the application of goodness-of-fit tests.

The treatment of scale effects

All quadrat experiments are influenced by the selection of cell size, which often has a critical effect on the final interpretation of results. Although optimum definitions of cell size do exist (see Taylor, 1977), the final selection is always an arbitrary procedure. An ingenious scheme devised by Greig-Smith (1952) tackles this problem by subjecting the same pattern to analysis with different cell sizes. The test is based on the property of the Poisson distribution that its mean is equal to its variance and takes the form of a hierarchical analysis of variance. The idea is to test for randomness at a variety of scales within a square quadrat census where the number of cells on each axis is some power of 2. Scales are defined by successively grouping the smallest cells of size $s = 1$ into larger cells (blocks) of size $s = 2, 4, 8, 16, \ldots$ etc. The test is made on the variation between block frequencies of size s nested within block frequencies of size 2s and the variance ratios obtained for each scale are tested for departures from randomness using the F distribution.

Pielou (1977) lists a number of problems posed by this scheme: first, the divisions between scales are large and concomitantly the degrees of freedom fall rapidly with increasing block size; second, oblong blocks in the test consistently give mean squares less than square blocks; and most important, the successive mean squares are not independent and, therefore, F tests cannot be used to discriminate between different scales. However, a number of useful modifications to the scheme have been proposed. Mead (1974) has devised a test to decide whether the grouping of blocks increases the evidence for clustering between scales, while Zahl (1974) has suggested using overlapping blocks to estimate the sizes of point clusters. Goodall (1974) proposes a modification where variance estimates are obtained from random pairs of cells each spaced at several chosen distances apart. This modification allows a freedom to study different scales which is not present in the sequential grouping of the Greig-Smith scheme. Similarly, Liebetrau and Rothman (1977) have devised a method which overcomes the problem of scale interdependence. Their methods incorporate the co-variance structure of estimates obtained at different scales and also allows scales to be chosen and weighted in an arbitrary manner.

The entropy approach

The introduction of entropy measures for the study of observed point patterns is due to the Russian geographer Medvedkov (1967, 1970). He suggests that if the probabilities from the Poisson distribution are substituted in Shannon's measure of entropy, or uncertainty, given by

$$H = -\sum_{x=0}^{r} P_x \log P_x, \tag{11}$$

we obtain a quantity measuring the size of the random component in a point pattern. Further, using the fact that the entropy of probabilities derived from uniform patterns is zero, Medvedkov proposes techniques which allow the entropy of the observed frequency distribution to be separated into random and uniform components. These methods have been used by Semple and Golledge (1970) and Haynes and Enders (1975) to study changes in settlement patterns through time.

More recently, Thomas and Reeve (1976) have criticized the mathematics of Medvedkov's idea for failing to take account of the constraint imposed on the maximization of Shannon's entropy (11) by the observed point density, λ. They demonstrate that the probabilities predicted by the geometric distribution given by

$$P_x = \lambda^x/(1+\lambda)^{x+1} \qquad x = 0, 1, \ldots \tag{12}$$

maximize (11) subject to the constraints

$$\sum_{x=0}^{r} P_x = 1 \tag{13}$$

and

$$\sum_{x=0}^{r} P_x\, x = \lambda \tag{14}$$

This result has subsequently been confirmed by Webber (1976) and may be used to compute the redundancy of observed frequency distributions. Redundancy is calculated as one minus the ratio between the entropy of the observed probabilities and the maximum entropy obtained by substituting the geometric probabilities into (11). The index takes on values between zero and one. The larger its value the greater the evidence that the observed frequency distribution is restricted in its freedom to vary by some unassumed spatial mechanism. Redundancy may be used both to compare properties of different patterns and to evaluate the effects of cell size on a single pattern (see Thomas, 1979). Liebetrau and Karr (1977) have demonstrated that no single point-process is realized by the entropy-maximizing distribution and, therefore, the entropy approach is descriptive and cannot be used to test the operation of a specific spatial process.

Nearest neighbour methods

Most nearest neighbour methods compare properties of the distances between neighbouring points in some theoretical pattern with those found in an observed pattern. The techniques were devised as a response to problems in quantitative plant ecology and among the researchers who were instrumental in their development are Skellam (1953), Clark and Evans (1954, 1955), Morisita (1954), Moore (1954), Thompson (1956) and Pielou (1959, 1962). Dacey (1960a) was the first geographer to appreciate the potential of nearest neighbour methods and his early work built upon the results obtained by Clark and Evans (1954). Indeed, their results form a useful starting-point for a discussion of nearest neighbour methods in geography.

First order nearest neighbour distances

The Clark and Evans test Consider an infinite plane where points are located according to a simple Poisson process with point density λ. Clark and Evans proved that the mean and variance of the probability density function for nearest neighbour distances, r, in this pattern are given by

$$E(r) = \tfrac{1}{2}\lambda^{1/2}, \text{ and } \sigma^2(r) = (4-\pi)/4\lambda\pi \qquad (15)$$

The observed mean distance in a pattern of n points is given by

$$\bar{r}_O = \frac{1}{n}\sum_{i=1}^{n} r_i \qquad (16)$$

where r_i is the distance between the ith point and its nearest neighbour. Clark and Evans invoke the Central Limit Theorem to assert that the sampling distribution of $\bar{r}_O$ is normally distributed with a mean, E(r), and standard deviation $\sigma_{\bar{r}} = .26136/\sqrt{\lambda\pi}$, where λ is estimated as the observed number of points per unit area. Therefore, an appropriate test for the significance of the departure from randomness in a point pattern is the statistic

$$z = [\bar{r}_O - E(r)]/\sigma_{\bar{r}}, \qquad (17)$$

where z is the variate of the standard normal distribution. This method has been used by geographers to study the dispersion of settlements (King, 1962, and Birch, 1967) and the movement of intra-urban functions (Sherwood, 1970).

Boundary problems Empirical application of the Clark and Evans statistic poses a number of problems. The most serious of these are related to the study area boundary. Their statistic is derived for points located in an infinite plane whereas, in reality, distance estimates for the test are measured from points located in a finite study area. The most general effect of a boundary is to sever connections between nearest neighbours lying either side of the boundary and therefore raise the estimated mean nearest neighbour distance within the study region. This effect implies that for practical applications E(r) will be under-estimated by formula (15).

Pinder (1978) has proposed the following correction coefficient, C, for modifying the expected mean distance, E(r),

$$E(r) = C(a/n)^{1/2} \qquad (18)$$

where

$$C = 0.497 + 0.127\,(4/n)^{1/2} \qquad (19)$$

and *a* is the size of study area and n the number of points. The coefficient defined by (19) was obtained by applying linear regression to estimated mean nearest neighbour distances derived by Ebdon (1976) from simulated random patterns in a square for values of n ranging between 5 and 100. However Boots (1979), drawing upon theoretical work by Dacey (1975) on the Poisson's approximation to random measures in a square, suggests Pinder's correction factor is an overestimate. Dacey's theoretical approximation for C in a unit square is given by

$$C = 0.5 + .152n^{-1/2} + .37n^{-3/2}, \qquad (20)$$

which Boots suggests is preferable to the result (19) based on simulation. Correction coefficents have also been developed for the standard error of E(r) and for statistics measuring random point patterns in a circle (Hsu and Mason, 1974 and de Smith, 1977). Clearly these correction factors can only be applied to sample statistics obtained from regularly shaped study areas.

An alternative solution to the underestimation problem is Dacey's (1965) suggestion that the points should be mapped onto a torus (a doughnut shape with no boundaries) before distance measurements are taken. Ingram (1978) presents simulated results which suggest toroidal mapping is a good solution to the boundary problem, but admits that the procedures are difficult for irregularly shaped study areas.

Problems with significance tests It is well known (see De Vos, 1973) that the z-test proposed by Clark and Evans is often inaccurate because the sampling distribution of $\bar{r}_O$ approaches normality very slowly and, consequently, is only suitable for large values of n. An alternative strategy is to use nearest neighbour statistics where the sampling theory is more reliable. For example, the following statistic devised by Skellam (1953) is based on the squares of nearest neighbour distances and is given by

$$R = \pi\lambda \sum_{i}^{n} r_i^2/n. \tag{21}$$

The quantity 2R is distributed as χ^2 with 2n degrees of freedom and is appropriate for a wide range of values of n. However, Dacey (1975) suggests that for patterns distributed in a unit square this statistic is biased in favour of accepting the hypothesis of randomness. In this work Dacey also considers the performances of various other nearest neighbour statistics in a square and concludes that in most cases the tests are only reliable if n is large.

A further problem is posed by the Poisson assumption of independence (Hsu and Mason, 1974 and Cliff and Ord, 1975) which must be fulfilled for valid significance testing. It is apparent that point-to-point nearest neighbour distances are not independent of one another because once point j has been established as point i's nearest neighbour then j's nearest neighbour distance cannot exceed r_{ij}. For this reason distances should be measured from randomly sampled locations within the study area to the nearest data point. Such measurements are termed location to-point distances.

Indeterminacy This problem occurs because a one-to-one correspondence between pattern type and nearest neighbour index (the ratio $\bar{r}_o/E(r)$) does not exist (De Vos, 1973). Among others, Dawson (1975) and Vincent (1976) have illustrated the problem with maps depicting very different point patterns which nevertheless produce the same nearest neighbour index value. Such paradoxes are possible because the index only provides information about the distances between points and does not consider other features of pattern such as the angular arrangement of points.

Modifications and extensions of the nearest neighbour method

Ordered nearest neighbour distances One way of minimizing the effects of indeterminacy is to work with ordered nearest neighbour distances. The development of the technique is due to Skellam (1953), Moore (1954) and Thompson (1956) and involves measuring a set of distances $\{r_{ij}\}$ from point i to the j = 1, 2, 3 . . . nearest neighbouring points. If j = k measurements are taken from each point i and these distances are ordered such that (see Fig. 16.2)

$$r_{i1} \leqslant r_{i2} \leqslant r_{i3} \leqslant \;, \ldots, \leqslant r_{ik},$$

then the value r_{if} is referred to as the jth order distance. The observed mean distance for each order j is calculated from

$$\bar{r}_j = \sum_{i=1}^{n} r_{ij}/n \tag{22}$$

Thompson (1956) has shown that the expected mean distance for order j in a Poisson pattern of points is given by

$$E(r_j) = \frac{1}{\sqrt{\lambda}} \cdot \frac{(2j)!j}{(2^j{}_j!)^2} \tag{23}$$

Similarly, he obtains the standard deviation, $\sigma(r_j)$, for each order and proposes a significance test using the standard normal deviate. Alternatively, Skellam (1953) and Moore (1954) suggest working with the values r_{ij}^2 for each order j and using the nearest neighbour statistic defined by (21) as the test statistic. The value of $2R_j$ is distributed as χ^2 with 2jn degrees of freedom. Cowie (1968) and Roder (1974, 1975) have proposed an alternative method of significance testing which uses the frequency distribution for each of j neighbour distances. Their idea is to employ either the chi-square test or Kilmogorov-Smirnov one sample test to compare the goodness-of-fit between the observed frequencies derived from some probability density function.

The measurements of order distances is again influenced by the boundary problem. Dacey (1963) proposes that order distances should only be included in the test for the j neighbours which are closer to i than i is to the boundary. This convention implies that low order distances will include more sample measurements than high order distances.

Sectoral distances A variation on the order distance idea is the sectoral or regional method discussed in Clark and Evans (1954), Dacey (1962) and Dacey and Tung (1962). The method involves centering a circle divided into k equal-sized sectors over each data point and then measuring the distance from the centre point to the nearest neighbour in each sector. The k distances taken at each point are ordered from shortest, r_{i1}, to longest, r_{ik} (see Fig. 16.2), so that k regional or sectoral means may be calculated from

$$\bar{r}_{ik} = \sum_{i=1}^{n} r_{ik}/n \qquad k = 1, 2, \ldots \tag{24}$$

The expected mean sectoral distance in a Poisson pattern of points are given by

$$E(r_k) = \sqrt{k}\,/2\sqrt{\lambda} \tag{25}$$

Sectoral nearest neighbour methods have been used to study patterns of central places in Wisconsin (Dacey, 1962) and the distribution of grocery stores in Lansing, Michigan (Getis, 1964). Although no detailed study has been made of the relative power of the order and sectoral approaches, Dacey and Tung (1962) have commented on the types of pattern for which these tests seem appropriate. They suggest that the sectoral method seems to be more efficient at detecting both randomness and uniformity in observed point patterns, while the order method is more effective when the points tend to be clustered. More recently, Getis and Boots (1978) have concluded that research experience with sectoral methods suggests the technique is an unnecessary complication which poses difficult problems of application and interpretation.

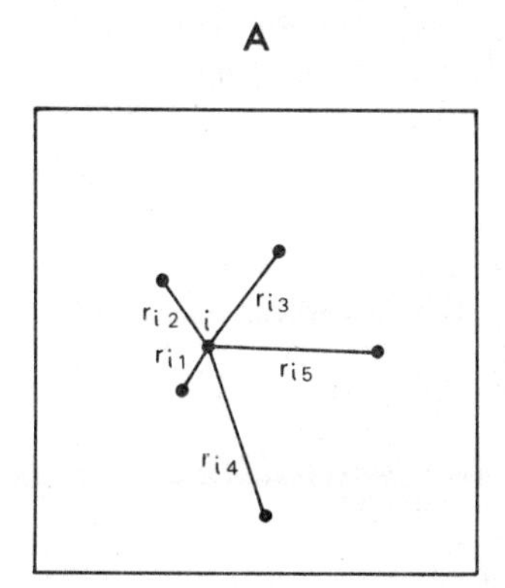

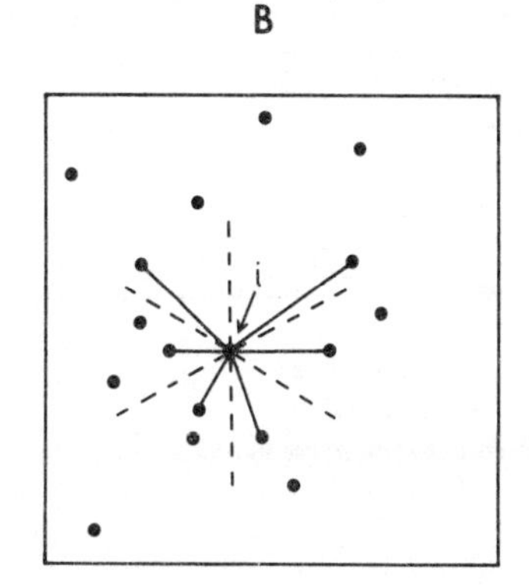

Figure 16.2 A: Order neighbour distances for point i
B: Sectoral distances for point i with k = 6 sectors

Related methods Geographers have experimented with a number of other variants of the nearest neighbour idea. The simplest of these methods is the analysis of reciprocal pairs, which are defined as pairs of points which are closer to each other than either is to a third point. Clark and Evans (1955) showed that the expected proportion of reciprocal pairs in a Poisson pattern is .6215 and Dacey (1969) has extended this idea to obtain expected proportions of jth order neighbours in a random pattern. However, applications of this method (Dacey, 1960b) have been restricted by the absence of significance tests for observed departures from the random expectation, and more detailed criticism of the method may be found in Porter (1960) and Pielou (1977). An alternative strategy has been to test directly for departures from uniformity (see Fig. 16.1) in observed patterns. The disturbed lattice models of Dacey and Tung (1962) and Jones (1971) test whether observed patterns are realizations of a uniform pattern where each point has received a random angular disturbance of some specified distance. Unfortunately, the significance tests do not provide direct evidence that the observed pattern has been disturbed. One recent statistical innovation, which geographers have not yet exploited, is the use of nearest neighbour statistics based on T-square distances (see Fig. 16.3) developed by Besag and Gleaves (1973) and Diggle, Besag and Gleaves (1976). These methods provide a powerful alternative to order neighbour statistics and are capable of dealing with heterogeneous environments where the point density varies from place to place.

Area methods

The measurement methods involved in quadrat and nearest neighbour analysis both entail a considerable loss of geometric detail in their description of observed point patterns. To overcome this information loss some geographers have suggested the use of mathematical techniques which consider properties of the area surrounding each point in the pattern.

The cell model

This model was derived by Meijering (1953) in the context of work on the structure of crystal aggregates in mineralogy, and in ecology the model is

Figure 16.3 The T-square distance v for point i (the line r is the nearest neighbour distance from i to j; the T-square distance v is the distance from j to the nearest neighbouring point k on the opposite side of the perpendicular r^T to the point i)

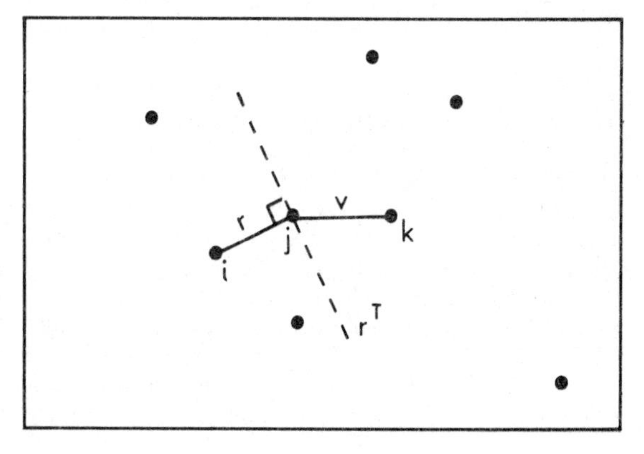

generally referred to as the S-mosaic (Matern, 1972; Pielou, 1977). To obtain the model it is first assumed that points are located in the plane according to a Poisson process. Around each point a convex polygon is constructed such that the area within each polygon is nearer to the enclosed point than any other point (see Fig. 16.4). Polygons constructed in this way are referred to as either Thiessen polygons, Dirichlet regions or Voronoi polygons. Among others, Evans (1945), Meijering (1953), Gilbert (1962), and Boots (1973) have derived parameter values measuring expected, average properties of Thiessen polygons centred on a Poisson pattern of points with density λ. In such a pattern the average number of sides (contact number) is 6, the expected perimeter length is $4/\sqrt{\lambda}$, the expected side length is $\frac{2}{3}\sqrt{\lambda}$ and the expected polygon area is $1/\lambda$. Unfortunately, higher order moments for these parameters cannot be derived analytically and for this reason significance tests for observed departures from randomness do not exist. However, simulated higher order moments for these parameters have been obtained by Crain (1972). A more damaging feature of the cell model has been unearthed by Boots (1977). Using simulation procedures he shows that the frequency distribution of contact numbers in a Poisson pattern is not markedly different from the distribution of contact numbers for points located according to a negative binomial point process, a result which suggests that the parameters of the cell model are not particularly sensitive to the underlying point process.

Boots (1973) has applied the cell model to the study of bus service centre hinterlands in Central Wales and South-Central England. In both cases the observed parameters of the hinterlands conformed quite closely to the expected values predicted by the cell model. This result suggests that service centres evolved as a function of a random growth process.

Simplicial graphs

Whenever Thiessen polygons are constructed for a point pattern, a second set of regions can automatically be obtained by connecting all pairs of points whose Thiessen polygons share a common edge. Such a set of regions are commonly referred to as simplicial graphs or Delaunay triangles (see Fig. 16.4). Working initially from results presented in Miles (1970), both Boots (1974) and Vincent, Haworth, Griffiths and Collins (1976) have proposed tests for randomness based on properties of simplicial graphs drawn around points generated by a Poisson process. In particular, Vincent *et al.* (1976) derive probability density functions describing l the link length between a typical pair of points and θ, the angle at a typical vertex of a neighbourhood triangle. For a Poisson pattern of points the probability density function for the link length is given by

$$f(l) = \frac{\pi\lambda l}{3}\left[l\sqrt{p}\ \exp\left(-\frac{\pi\lambda l^2}{4}\right) + \operatorname{erfc}\left(\tfrac{1}{2}\sqrt{\pi\lambda}\right)\right] \quad (26)$$

where λ is the point density and erfc denotes the complementary error function. Similarly, the probability density function for θ in a Poisson pattern is given by

$$f(\theta) = \frac{4}{3\pi}\sin\theta\left[\sin\theta + (\pi-\theta)\cos\theta\right] \quad (27)$$

By integrating (26) and (27) for chosen class intervals of l and θ respectively, predicted random frequencies are obtained which can be compared with observed frequencies using some goodness-of-fit test such as χ^2.

This approach has the advantage over quadrat and nearest neighbour methods that the parameters are density free and therefore independent of scale effects. However, once the hypothesis of randomness has been rejected, the test gives no indication about the process underlying the observed pattern.

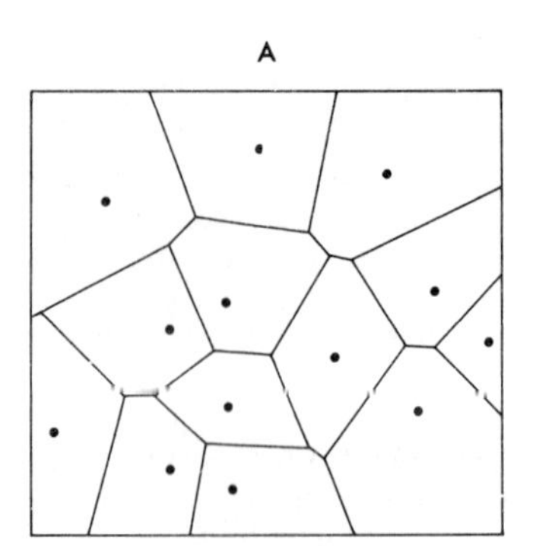

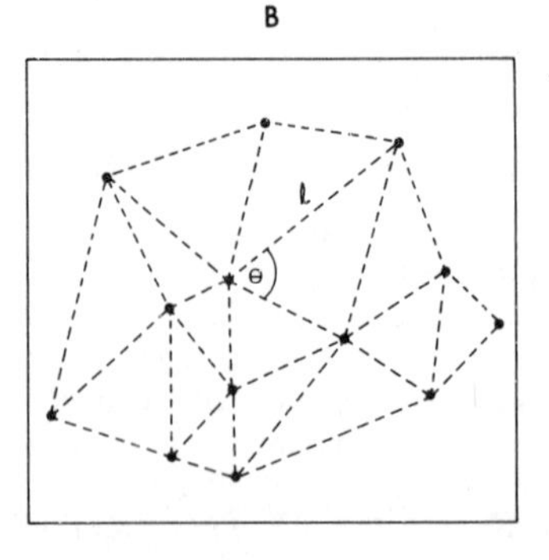

Figure 16.4 A: Thiessen polygons
B: A simplicial graph of the same pattern

Problems and prospects

Point pattern methods aroused the interest of geographers because they provided a means of testing theories of settlement spacing and information diffusion. In addition, there was the prospect of explaining point patterns in terms of stochastic generating processes. However, the initial enthusiasm for point pattern methods has been dampened by numerous technical difficulties encountered in their application. Quadrat analysis allows the testing of quite complex stochastic processes, yet the method is hamstrung by the difficulty of inferring process purely on the evidence of frequency data. In contrast, area models are derived from quite realistic definitions of pattern but, to date, are only capable of detecting randomness in point patterns. More generally, all point pattern methods are restricted by the fact that they reduce geographical problems almost entirely to the study of univariate distributions.

Although the future of point pattern analysis in quantitative geography is difficult to assess, many topics await further investigation. There is a clear need to develop area models for more complex point processes, and the work reported by Getis and Boots (1978) on geographical applications of the area-based growth models of Johnson and Mehl (1939) and Roach (1968) represents a small step in this direction. The analysis of scale effects is still in its infancy, and Hepple (1974) has suggested that useful insights might be gained from the two-dimensional analysis of the covariance structure of points. One hopeful sign is the increasing interest of statisticians in spatial patterns, and a useful review of recent statistical contributions is given by Ripley (1977). Indeed, if geographers follow the example of statisticians then future models will be constructed for specific geographical problems and, consequently, there will be less interest in the development of general pattern recognition techniques.

References

Alskogious, H. (1969) 'Modelling the evolution of settlement patterns: two studies of vacation house settlement', *Geografiska Regionstudier*, 6, Uppsala.

Anscombe, F.J. (1950) 'Sampling theory of the negative binomial and logarithmic series distributions', *Biometrika*, 37, 358–82.

Artle, R.K. (1965) *The structure of the Stockholm economy*, Cornell University Press: Ithaca, NY.

Bartko, J.J., Greenhouse, S.W., and Patlak, C.S. (1968) 'On expectations of some functions of Poisson variates', *Biometrics*, 24, 97–102.

Bartlett, M.S. (1955) *An introduction to stochastic processes*, Cambridge University Press.

Bartlett, M.S. (1972) 'Two-dimensional nearest neighbour systems and their ecological applications', *Statistical Ecology*, 1, 179–94.

Barton, D.E., David, F.N., and Fix, E. (1963) 'Random points in a circle and the analysis of chromosome patterns', *Biometrika*, 50, 23-9.

Besag, J.E., and Gleaves, J.T. (1973) 'On the detection of spatial pattern in plant communities', *Bulletin de l'Institut International de Statistique*, 45, 153–8.

Birch, B.P. (1967) 'The measurement of dispersed patterns of settlement', *Tijdschrift voor Economische en Sociale Geografie*, 58, 68–75.

Bliss, C.I., and Fisher, R.A. (1953) 'Fitting the negative binomial distribution to biological data', *Biometrics*, 9, 176–96.

Boots, B.N. (1973) 'Some models of the random subdivision of space', *Geografiska Annaler*, 55B, 34–48.

Boots, B.N. (1974) 'Delaunay triangles: an alternative approach to point pattern analysis', *Proceedings of the Association of American Geographers*, 6, 26–9.

Boots, B.N. (1977) 'Contact number properties in the study of cellular networks', *Geographical Analysis*, 9, 379–87.

Boots, B.N. (1979) 'Underestimation in nearest neighbour analysis', *Area*, 11, 208–10.

Boots, B.N., and Getis, A. (1977) 'Probability model approach to map pattern analysis', *Progress in Human Geography*, 1, 264–86.

Boswell, M.T., and Patil, G.P. (1970) 'Chance mechanisms generating the negative binomial distribution', in G.P. Patil (ed.), *Random counts in scientific work*, Vol. 1, 3–22, Pennsylvania State University Press.

Clark, P.J., and Evans, F.C. (1954) 'Distances to nearest neighbour as a measure of spatial relationships in population', *Ecology*, 35, 445–53.

Clark, P.J., and Evans, F.C. (1955) 'On some aspects of spatial pattern in biological populations', *Science*, 121, 397–8.

Cliff, A.D., and Ord, J.K. (1973) *Spatial Autocorrelation*, Pion: London.

Cliff, A.D., and Ord, J.K. (1975) 'Model building and the analysis of spatial pattern in human geography', *Journal of the Royal Statistical Society B*, 37, 297–348.

Cowie, S.R. (1968) 'The cumulative frequency nearest neighbour method for the identification of spatial patterns', University of Bristol, Department of Geography, Seminar Series A, No. 10.

Crain, I.K. (1972) 'Monte Carlo simulation of random Voronoi polygons: preliminary results',

Search, 3, 220–1.
Dacey, M.F. (1960a) 'A note on the derivation of nearest neighbour distances', *Journal of Regional Science*, 2, 81–7.
Dacey, M.F. (1960b) 'The spacing of river-towns', *Annals of the Association of American Geographers*, 50, 59–61.
Dacey, M.F. (1962) 'Analysis of central place and point pattern by a nearest neighbour method', *Lund Studies in Geography*, 24, 55–75.
Dacey, M.F. (1963) 'Order neighbor statistics for a class of random patterns in multidimensional space', *Annals of the Association of American Geographers*, 53, 505–15.
Dacey, M.F. (1964) 'Modified Poisson probability law for point pattern more regular than random', *Annals of the Association of American Geographers*, 54, 559–65.
Dacey, M.F. (1965) 'Numerical measures of random sets', Technical Report No. 5, Geographical Information Systems Project, Northwestern University, Evanston, Ill.
Dacey, M.F. (1966) 'A county-seat model for the areal pattern of an urban system', *Geographical Review*, 56, 527–42.
Dacey, M.F. (1968) 'An empirical study of the areal distribution of houses in Puerto Rico', *Transactions, Institute of British Geographers*, 45, 51–69.
Dacey, M.F. (1969) 'Proportion of reflexive n^{th} order neighbours in a spatial distribution', *Geographical Analysis*, 1, 385–8.
Dacey, M.F. (1975) 'Evaluation of the Poisson approximation to measures of the random pattern in the square', *Geographical Analysis*, 7, 351–67.
Dacey, M.F., and Tung, T. (1962) 'The identification of randomness in point patterns', *Journal of Regional Science*, 4, 83–96.
Dawson, A. (1975) 'Are geographers indulging in a landscape lottery?', *Area*, 7, 42–5.
de Smith, M.J. (1977) 'Distance distributions and trip behaviour in defined regions', *Geographical Analysis*, 9, 332–45.
De Vos, S. (1973) 'The use of nearest neighbour methods', *Tidjschrift voor Sociale en Economische Geografie*, 64, 307–19.
Diggle, P.G., Besag, J., and Gleaves, J.T. (1976) 'Statistical analysis of spatial patterns by means of distance methods', *Biometrics*, 32, 659–67.
Ebdon, D. (1976) 'On the underestimation inherent in the commonly used formulae', *Area*, 8, 165–9.
Edwards, C.B., and Gurland, J. (1961) 'A class of distributions applicable to accidents', *Journal of the American Statistical Association*, 45, 350–72.
Evans, U.R. (1945) 'The laws of expanding circles and spheres in relation to the lateral growth of surface films and the grain size of metals', *Transactions of the Faraday Society*, 41, 365–74.
Getis, A. (1964) 'Temporal land-use pattern analyses with the use of nearest neighbor and quadrat methods', *Annals of the Association of American Geographers*, 54, 391–8.
Getis, A. (1967) 'Occupancy theory and map pattern analysis', University of Bristol, Department of Geography, Seminar Paper Series A, 1.
Getis, A. (1974) 'Representation of spatial point processes by Polya methods' in M.H. Yeates (ed.), *Proceedings of the 1972 meeting of the IGU Commission on Quantitative Geography*, McGill-Queens University Press: Montreal.
Getis, A., and Boots, B.N. (1978) *Models of spatial processes*, Cambridge University Press.
Gilbert, E.N. (1962) 'Random sub-divisions of space into crystals', *Annals of Mathematical Statistics*, 33, 958–72.
Goodall, D.W. (1974) 'A new method for the analysis of spatial pattern by the random pairing of quadrats', *Vegetatio*, 29, 135–46.
Greig-Smith, P. (1952) 'The use of random and contiguous quadrats in the study of the structure of plant communities', *Annals of Botany (NS)*, 16, 293–312.
Greig-Smith, P. (1964) *Quantitative plant ecology*, Butterworth: London.
Gurland, J. (1957) 'Some interrelations among compound and generalized distributions', *Biometrika*, 44, 265–8.
Hägerstrand, T. (1953) *Innovationsforloppet ur korologisk synpunkt*, Gleerup: Lund.
Haggett, P., Cliff, A.D., and Frey, A. (1977) *Locational Analysis in Human Geography*, Arnold: London.
Harvey, D.W. (1966) 'Geographical processes and the analysis of point patterns', *Transactions, Institute of British Geographers*, 40, 81–95.
Harvey, D.W. (1968a) 'Some methodological problems in the use of the Neyman A and negative binomial probability distributions', *Transactions, Institute of British Geographers*, 44, 85–95.
Harvey, D.W. (1968b) 'Pattern, process and the scale problem in geographical research', *Transactions, Institute of British Geographers*, 45, 71–8.
Haynes, K.E., and Enders, W.T. (1975) 'Distance, direction, and entropy in the evolution of a settlement pattern', *Economic Geography*, 51, 357–65.
Hepple, L.W. (1974) 'The impact of stochastic process theory upon spatial analysis in human geography', *Progress in Geography*, 6, 89–142.
Hinz, P., and Gurland, J. (1967) 'Simplified techniques for estimating parameters of some generalized Poisson distributions', *Biometrika*, 54, 555–66.
Hodder, I.R., and Hassall, M. (1971) 'The non-random spacing of Romano-British walled towns', *Man*, 6, 391–407.

Holgate, P. (1964) 'Estimation for the bivariate Poisson distribution', *Biometrika*, 51, 241–5.
Holgate, P. (1966) 'Bivariate generalizations of Neyman's Type A distribution', *Biometrika*, 53, 241–4.
Holgate, P. (1972) 'The use of distance methods for the analysis of spatial distributions of points', in P.A.W. Lewis (ed.), *Stochastic point processes*, Wiley: New York.
Hsu, S., and Mason, J.D. (1974) 'The nearest neighbour statistics for testing randomness of point distributions in a bounded two-dimensional space', in M.H. Yeates (ed.), *Proceedings of the 1972 meeting of the IGU Commission on Quantitative Geography*, McGill-Queen's University Press: Montreal.
Hudson, J.C., and Fowler, P.M. (1966) 'The concept of pattern in geography', Discussion Paper No. 1, Department of Geography, University of Iowa.
Ingram, D.R. (1978) 'An evaluation of the procedures utilised in nearest neighbour analysis', *Geografiska Annaler*, 60B, 65–70.
Johnson, W.A., and Mehl, R.F. (1939) 'Reaction kinetics in processes of nucleation and growth', *Transactions of the American Institute of Mining, Metallurgical and Petroleum Engineers*, 135, 410–58.
Jones, A. (1971) 'An order neighbour approach to random disturbances on regular point lattices', *Geographical Analysis*, 4, 361–78.
King, L.J. (1962) 'A quantitative expression of the pattern of urban settlements in selected areas of the United States', *Tijdschrift voor Economische en Sociale Geografie*, 53, 1–7.
Lee, Y. (1974) 'An analysis of spatial mobility of urban activities in downtown Denver', *Annals of Regional Science*, 8, 95–108.
Liebetrau, A.M., and Karr, A.F. (1977) 'The role of Maxwell-Boltzmann statistics and Bose-Einstein statistics in point pattern analysis', *Geographical Analysis*, 9, 418–22.
Liebetrau, A.M., and Rothman, E.D. (1977) 'A classification of spatial distributions based on several cell sizes', *Geographical Analysis*, 9, 14–28.
McConnell, H., and Horn, J.M. (1972) 'Probabilities of surface karst' in R.J. Chorley (ed.), *Spatial Analysis in Geomorphology*, 111–34, Methuen: London.
Matern, B. (1972) 'Analysis of spatial patterns and ecological relations: the analysis of ecological maps as mosaics', *NATO Advanced Study Institute on Statistical Ecology*, Department of Statistics, Pennsylvania State University.
Matui, I. (1932) 'Statistical study of the distribution of scattered villages in two regions of the Tonami plain, Tayama prefecture', *Japanese Journal of Geography and Geology*, 9, 251–66.
Mead, R. (1974) 'A test for spatial pattern at several scales using data from a grid of contiguous quadrats', *Biometrics*, 30, 295–307.
Medvedkov, Y.V. (1967) 'The concept of entropy in settlement pattern analysis', *Papers, Regional Science Association*, 18, 165–8.
Medvedkov, Y.V. (1970) 'Entropy: an assessment of potentialities in geography', *Economic Geography*, 46, 306–16.
Meijering, J.L. (1953) 'Interface area, edge length, and number of vertices in crystal aggregates with random nucleation', *Philips Research Reports*, 8, 270–90.
Miles, R.E. (1970) 'On homogeneous planar Poisson point processes', *Mathematical Biosciences*, 6, 85–127.
Moore, P.J. (1954) 'Spacing in plant populations', *Ecology*, 35, 222–7.
Morisita, M. (1954) 'Estimation of population density by spacing methods', *Memoirs of the Faculty of Science*, Kyushu University, E.1, 187–97.
Neyman, J. (1939) 'On a new class of contagious distributions applicable in entomology and bacteriology', *Annals of Mathematical Statistics*, 10, 35–57.
Neyman, J., and Scott, E.L. (1958) 'Statistical approach to problems of cosmology', *Journal of the Royal Statistical Society*, Series B, 20, 1–43.
Olsson, G. (1967) 'Central place systems, spatial interaction and stochastic processes', *Papers, Regional Science Association*, 18, 13–46.
Pielou, E.C. (1959) 'The use of point to point populations', *Journal of Ecology*, 47, 607–13.
Pielou, E.C. (1962) 'The use of plant to neighbour distances for the detection of competition', *Journal of Ecology*, 50, 357–67.
Pielou, E.C. (1977) *Mathematical Ecology*, Wiley: New York.
Pinder, D.A. (1978) 'Correcting underestimation in nearest neighbour analysis', *Area*, 10, 379–85.
Pinder, D.A., and Witherick, M.E. (1973) 'Nearest neighbour analysis of linear point patterns', *Tijdschrift voor Economische en Sociale Geografie*, 64, 160–3.
Porter, P.W. (1960) 'Earnest and the Orephagians – a fable for the instruction of young geographers', *Annals of the Association of American Geographers*, 50, 297–9.
Ripley, B.D. (1977) 'Modelling spatial patterns', *Journal of the Royal Statistical Society*, Series B, 89, 172–212.
Roach, S.A. (1968) *The theory of random clumping*, Methuen: London.
Roder, W. (1974) 'Application of a procedure for statistical assessment of points on a line', *Professional Geographer*, 26, 283–90.
Roder, W. (1975) 'A procedure for assessing patterns without reference to area or density', *Professional Geographer*, 27, 432–40.
Rogers, A. (1965) 'A stochastic analysis of the spatial clustering of retail establishments',

Journal of the American Statistical Association, 60, 1094–103.

Rogers, A. (1969) 'Quadrat analysis of urban dispersion: 2. Case studies of urban retail systems', *Environment and Planning*, 1, 155–71.

Rogers, A. (1974) *Statistical analysis of spatial dispersion*, Pion: London.

Rogers, A., and Martin, J. (1971) 'Quadrat analysis of urban dispersion: 3. Bivariate models', *Environment and Planning*, 3, 433–50.

Schilling, W. (1947) 'A frequency distribution represented as the sum of two Poisson distributions', *Journal of the American Statistical Association*, 42, 407-24.

Semple, R.K., and Golledge, R.G. (1970) 'An analysis of entropy changes in a settlement pattern over time', *Economic Geography*, 46, 157–60.

Sherwood, K.B. (1970) 'Some applications of the nearest neighbour technique to the study of the movement of intra-urban functions', *Tijdschrift voor Economische en Social Geografie*, 61, 41–8.

Sibley, D. (1972) 'Strategy and tactics in the selection of shop locations', *Area*, 4, 151–6.

Sibley, D. (1976) 'On pattern and dispersion', *Area*, 8, 163–5.

Skellam, J.G. (1953) 'Studies in statistical ecology: 1. Spatial pattern', *Biometrika*, 39, 346–62.

Switzer, P. (1976) 'Applications for random process models to the description of spatial distributions of quantitative variables', in D.F. Merriam (ed.), *Random processes in geology*, Springer-Verlag: New York.

Taylor, P.J. (1977) *Quantitative methods in geography*, Houghton Mifflin: Boston.

Thomas, M. (1949) 'A generalization of Poisson's binomial limit for use in ecology', *Biometrika*, 36, 18–25.

Thomas, R.W. (1979) *An introduction to quadrat analysis*, Geo. Abstracts: Norwich.

Thomas, R.W., and Reeve, D.E. (1976) 'The role of Bose-Einstein statistics in point pattern analysis', *Geographical Analysis*, 8, 113–36.

Thompson, H.R. (1956) 'Distribution of distance to the nth neighbour in a population of randomly distributed individuals', *Ecology*, 37, 391–4.

Vincent, P.J. (1976) 'The general case: how not to measure point patterns', *Area*, 8, 161–3.

Vincent, P.J., Haworth, J.M., Griffiths, J.C., and Collins, R. (1976) 'The detection of randomness in plant patterns', *Journal of Biogeography*, 3, 373–80.

Webber, M.J. (1976) 'Elementary entropy maximising probability distributions: analysis and interpretation', *Economic Geography*, 52, 218–28.

Zahl, S. (1974) 'Applications of the S-method to the analysis of spatial pattern', *Biometrics*, 30, 513–24.

Part 4

Mathematical models

Mathematical models were first drawn to the attention of most British geographers in a synoptic fashion in the writings of Chorley and Haggett in the mid and late 1960s, particularly in volumes such as *Models in Geography* (1967). Clearly, however, much derived from North American origins, especially the work of Strahler, Krumbein, Dacey, Berry, Tobler, Garrison, Marble, Morrill, Curry and Olsson. In the 1970s interest in mathematical models has diffused widely and has had enormous impacts in some parts of the discipline. Perhaps the most coherently developed set of mathematical models has been used for urban and transport planning, but the same methodology is also used to a significant extent in population forecasting, hydrology, geomorphology and other areas.

The chapters presented in this section can only highlight some of the significant developments of the last two decades and suggest lines of development for the future. Moreover, overlaps of treatment require that the chapters on statistical methods and applications in Parts 3 and 5 of this volume be read jointly with the chapters on mathematical models presented below. However, the chapters presented here do highlight two major themes: on the one hand, the continued consolidation of research in certain areas and on certain topics, especially in urban planning and on the other hand, the creation of important new initiatives directing attention into the relatively uncharted waters (for the geographer) of bifurcation theory, stochastic process theory, Q-analysis, micro-simulation, and control theory. Taken together, these chapters present a formidable challenge to be taken up over the next decade.

Perhaps in no area related to geography, except that of urban planning, has the development of the methodology of mathematical modelling reached such a high degree of sophistication. Batty, in the opening chapter of this section, presents a formidable documentation of this methodology as it has evolved from North American origins and subsequently developed in Britain. In noting that this work has mainly been concerned with technical developments, rather than the formulation or testing of theory, he perhaps echoes a general criticism voiced also in the following chapters. Again, in a later chapter in this section, Senior draws heavily on urban planning applications in discussing static optimization techniques, and draws attention to the relation of linear and nonlinear programming to the entropy-maximizing method so extensively used in urban planning. There is no doubt that the contributions of the last twenty years' research in urban planning have now given us the best integrated understanding of mathematical modelling in geography in each of the areas of model design, calibration, forms of spatial representation, and linkage to design of objective functions. Hence it is not in this area where it might be expected that the research developments of future years will produce the widest-ranging implications for the subject: rather, it is outside of the range of established urban planning models where the broadest challenges remain.

The chapters in this section are not short of such challenges. At one extreme, Wilson urges geographers to respond to the stimulus of mathematical topology and bifurcation theory which he acknowledges as presenting some of the most difficult problems in modern science. At the other extreme, Chapman urges us to abandon models based on functional relations (which must include bifurcation theory) and instead to derive models from sets and relations between sets; or as he expresses it, 'to stop playing with functions and to start using relations'. Midway between these views Haggett and Cliff, Culling, Clarke *et al.*, and Chorley and Bennett urge the more extensive use of well-established existing theories such as network theory, the theories of stochastic processes and probability, and the theories of optimal control and dynamic programming.

Geography has always been a subject which has been subjected to various stimuli washing through it from the social and physical sciences. Within the area of mathematical models, these influences have been particularly marked. We might see the impetus of urban planning models as deriving from the systems analysis procedures developed in the aerospace and defence industry programmes; and the procedure of optimization as deriving from management science and operations research. In the future, research workers concerned with developing mathematical models in geography will be confronted with a range of new stimuli deriving from concepts formed outside the discipline. The developments of mathematical topology, discussed in Wilson's chapter, and drawn together by Thom, Zeeman, Hirsch and Smale, owe much to the stimulus of Lotka and Volterra working in biology at the turn of the century. Similarly, Culling's chapter draws on the lineage of probability theory deriving from Gauss, Wiener, Kolmogorov, Levy, Cramer, Doob, and, more recently, Kalman and Mandelbrot. Again, Chorley and Bennett's chapter draws on the developments of Clerk Maxwell, Pontryagin and Kalman. Finally, Chapman draws from the recent work of Atkin which itself derives from the network theories of mathematics which are also extensively employed by Haggett and Cliff. Recent stimuli of mathematics, probability theory, control engineering and management science are washing through the discipline of geography and viying for attention, and it is likely that each of these stimuli will generate important developments in the 1980s.

One possible prospect deserves isolation for particular attention: considerable linkages between methodologies are possible and likely. Geography is particularly well-suited to bringing together such otherwise disparate themes. Bifurcation theory throws particular emphasis, as Wilson notes, on the role of transitions between states: indeed it attempts an integrated explanation of phase, discontinuity and transition phenomena as a function of other exogenous phenomena or variables. These exogenous variables accord closely with the control variables and parameters of optimal control theory, and their modelling both as past processes, or as control instruments, can be very easily approached using the filtering technique deriving from Kalman. Hence there is a strong linkage between, on the one hand, the theories of mathematical topology, and, on the other hand, the more pragmatic algorithms of control, stochastic process theory and the statistical time series and spatial time series procedures discussed in earlier chapters. As a prospect for future research the development of such linkages must hold a great deal of promise, but are all the more specially significant because they hold a great deal of potential for overcoming the more fundamental weaknesses of the otherwise well-developed models of urban planning: namely the integration of model design with model application. Control theory, bifurcation theories of discontinuity, and the stochastic process theories of uncertainty hold the keys to a methodology of instrumentality, action, intervention and optimal planning. They extend the algorithms of optimal allocation from static to dynamic frameworks and link, as one integrated methodology, the stages of design, representation, calibration, simulation, forecasting, policy appraisal, and policy decision in model use.

References

Chorley, R.J., and Haggett, P. (1967) *Models in Geography*, Methuen: London.

Chapter 17

Urban models

M. Batty

Introduction

Urban models were first built in North America over twenty years ago as part of the growing realization that the burgeoning power of the digital computer might provide a means for getting to grips with complex systems manifesting equally complex problems. Thus urban models were policy-orientated from the beginning and the need to provide workable computer simulations has meant that such models have always been characterized by rudimentary theory. These models remain vehicles for developing policy, rather than testbeds for formulating scientific theory although a research tradition of sorts is being gradually established. The field has narrowed in focus as it has matured: from large-scale models to smaller scale and from an undiscriminating reliance on a battery of techniques to the development of those judged to be most promising.

This review will attempt to sketch the experience from a technical vantage point and in terms of British contributions. But to set the scene, the first generation of urban models pioneered in North America until the mid 1960s will be presented initially. The research tradition which subsequently evolved particularly in Britain, and which was based on the evolution of model-building principles and new model forms, will then be outlined, emphasizing estimation techniques, model structures, dynamics and disaggregation. Finally these contributions will be evaluated through applications in planning and some speculations on future directions will be made.

First generation urban models

A diverse array of styles, techniques and applications characterized the first generation models which were regarded by planners as providing artificial laboratories for experiments with urban structure (Dyckman, 1963). Most of these models were concerned with the location of land uses or economic activities in cities and regions, and there was a bias towards describing such systems in static, cross-sectional rather than dynamic terms. Furthermore, there was an emphasis on building models to simulate the existing structure of cities as a prelude to prediction, rather than on models of a more normative or optimizing type relating to the design of new cities. These models were based on elementary, often speculative, theories of location and transport economics, and on 'common-sense' interpretations of urban causation. Consequently, the emphasis was on techniques of simulation, rather than theoretical representations and this bias has dominated the field ever since.

These early models were implemented using three broad styles or techniques of simulation based on gravitational principles, linear analysis and mathematical programming. Gravitational models incorporated hypotheses about spatial interaction and location as a function of spatial impedance and attractiveness. Typically, interaction between any pair of places i and j notated by T_{ij} can be modelled as

$$T_{ij} \propto O_i D_j f(c_{ij}), \; i = 1, 2, \ldots, I; \; j = 1, 2, \ldots, J, \tag{1}$$

where O_i is the activity at or attraction of the origin of interaction i; D_j is the activity at or attraction of the destination j; and $f(c_{ij})$ is some function of spatial impedance between i and j, measured by c_{ij}. The location of activity, say population P_j at place j, can be derived by summing interaction in (1) over i

$$P_j \propto \sum_i T_{ij} \propto D_j \sum_i O_i f(c_{ij}). \tag{2}$$

The summation on the right-hand side (RHS) of

(2) is the accessibility or potential of j and such potential models were used by Hansen (1959), Lowry (1964) and Lathrop and Hamburg (1965) to simulate various land use patterns in their more comprehensive models.

The second technique was based on the linear model. This model admits a more inductive type of urban analysis to be developed and is easier to estimate than the nonlinear gravitational model. For example, activity or land use in zone i, Y_i, is modelled as a linear combination of exogenous variables X_{in}.

$$Y_i = \sum_n \beta_n X_{in}, \quad n = 0, 1, 2, \ldots, N, \tag{3}$$

where β_n is the weight of the independent variable X_{in} in the model and $X_{io} = 1, P_i$. Standard estimation procedures are available to fit the model, thus accounting for its frequent adoption in this field, and the Greensborough model due to Chapin and Weiss (1968) is typical.

The last major technique developed involved a shift in focus from positive modelling to normative, from predictive to prescriptive models based on optimization theory. Mathematical programming methods, in particular linear programming, were invoked and such models involved optimizing some objective function subject to constraints. For example, the location of activity l on site i, Y_{il}, $l = 1, 2, \ldots, L$, can be determined by optimization of an objective Z

$$\text{OPT } Z = \sum_i \sum_l g(c_{il}) Y_{il}, \tag{4}$$

subject to constraints on the total activities to be allocated, constraints on the interdependence between activities and constraints on the capacities or densities of location. Such constraints may be equalities or inequalities, and the cost function $g(c_{il})$ may be linear or nonlinear. The most straightforward model of this type is Schlager's (1967) land use plan design model based on cost minimization, in contrast to Herbert and Stevens's (1960) model in which optimization was used to model rational economic decision-making in the housing market.

Comprehensive modelling frameworks

The most interesting and influential first generation models, however, were based on mixing and replicating these different techniques in more comprehensive frameworks. Different economic activity sectors and land use patterns are distinct enough to enable urban models to be constructed in modular fashion, and thus various submodels can be coupled together to form more general structures. Moreover, such coupling leads to interaction between submodels of different sectors and enables essential feedbacks to be represented. Thus besides allocating activities to zones, these general models often include relationships required to derive the total levels of activities from one another. Two main types of general model emerged and these are worth illustrating.

Comprehensive linear models using well-established econometric techniques have been quite widely applied by replicating single equation linear models as in (3) for several different activities. Relationships for deriving and allocating activities are not separated out but an essential distinction is made between the endogenous variables Y_{il} and the exogenous X_{in}. A good example is the EMPIRIC model (Hill, 1965) which is stated as

$$Y_{im} = \sum_{\substack{l \\ l \neq m}} \alpha_l Y_{il} + \sum_n \beta_n X_{in}, \quad m = 1, 2, \ldots, L, \tag{5}$$

where α_l represent the coefficients describing the relationship between activities. The attraction of such models relates to the enormous effort devoted to understanding such simultaneous linear systems in econometrics, and considerable progress has been made in estimating and identifying appropriate mathematical model structures. Moreover, this framework can be highly disaggregated and, in principle, there is no limit to the number of activities which can be handled. Over thirty documented applications of this type of model have been made in North America with apparently considerable success (Pack, 1978).

The coupling of gravitational submodels together has also proved to be a useful method for developing more general urban models and this has led to structures with much more tightly ordered submodels than those of the econometric kind. Relationships between the submodels are usually based on macro-economic and population relations, resulting in models which are smaller but more transparent than those implied in (5). The best-known model and that which has fired the imagination of researchers in Britain was originally built for the Pittsburgh region by Lowry (1964) although a similar model was constructed for the Buffalo region by Lathrop and Hamburg (1965). Lowry's model is based on allocating two types of activity: population and services. The allocations are made using potential models as in (2), and these submodels are linked using the well-known relationships between population and services as contained in the economic base mechanism.

Formally, the model can be stated in two sets of two (four) equations, one involving derivation, the other allocation for each of the two sectors. Then

$$P = aE, \tag{6}$$

$$P_j \propto P \sum_i O_i f^1 (c_{ij}), \tag{7}$$

$$S = bP, \text{ and} \tag{8}$$

$$S_i \propto S \sum_j D_j f^2(c_{ij}), \qquad (9)$$

where P_j is population in j, S_i is service employment in i, $P\,(=\Sigma_j P_j)$ is total population, $S\,(=\Sigma_i S_i)$ is total service employment, and E is total employment; a is the activity rate and b the population-serving ratio; and f^1 and f^2 are the functional forms relating to the effect of spatial impedance in the population and service sectors respectively.

There are many ways of solving the simultaneous equations given in (6) to (9) but due to the non-linearity of the system, some iterative procedure is usually required. The procedure used by Lowry (1964) was direct. In equation (6), E can be computed, given the total basic employment in the system, B, and values for a and b, using the economic base equation $E = B(1 - ab)^{-1}$. The solution of equations (6) to (9) then follows immediately. However, Lowry's procedure was complicated by two features: first, the attractiveness variables O_i and D_j were themselves set as employment E_i and population P_j respectively, and second, constraints on the densities of population and services were met by arbitrary scaling procedures invoked after equations (7) and (9) had been solved. These features meant that a stable equilibrium to the system could only be found by iterating equations (6) to (9), starting with O_i as observed employment, and then substituting for D_j and O_i successively using predicted P_j and E_i $(= B_i + S_i$, where B_i is observed basic employment in i). This procedure is illustrated in Figure 17.1 and more detailed expositions of the original model can be found in Wilson (1974), Batty (1976) and Mohan (1979).

Developments in Britain

By the mid 1960s the optimistic mood responsible for the first generation of urban models in North America gave way to reaction (Lee, 1973) but at this time modelling caught the interest of planners in Britain. The perceived complexity of urban problems which gave rise to a new form of strategic or 'Structure' planning in Britain and the promulgation of an explicit systems approach to planning provided an ideal environment in which urban models could be built (McLoughlin, 1969). The well-developed planning system and the small scale of resources available for modelling led to an emphasis on positive models, simple in form but complex in structure which could be easily used for predictive purposes in the wider context of normative plan-making (Batty, 1978a). Thus came the concentration on the most successful type of first generation model – the Lowry model and gravitational variants – and this practical interest was bolstered by advances in spatial interaction modelling of a more theoretical nature pioneered by Wilson (1967, 1970).

The first attempts at urban modelling involved shopping models based on Lakshmanan and Hansen's

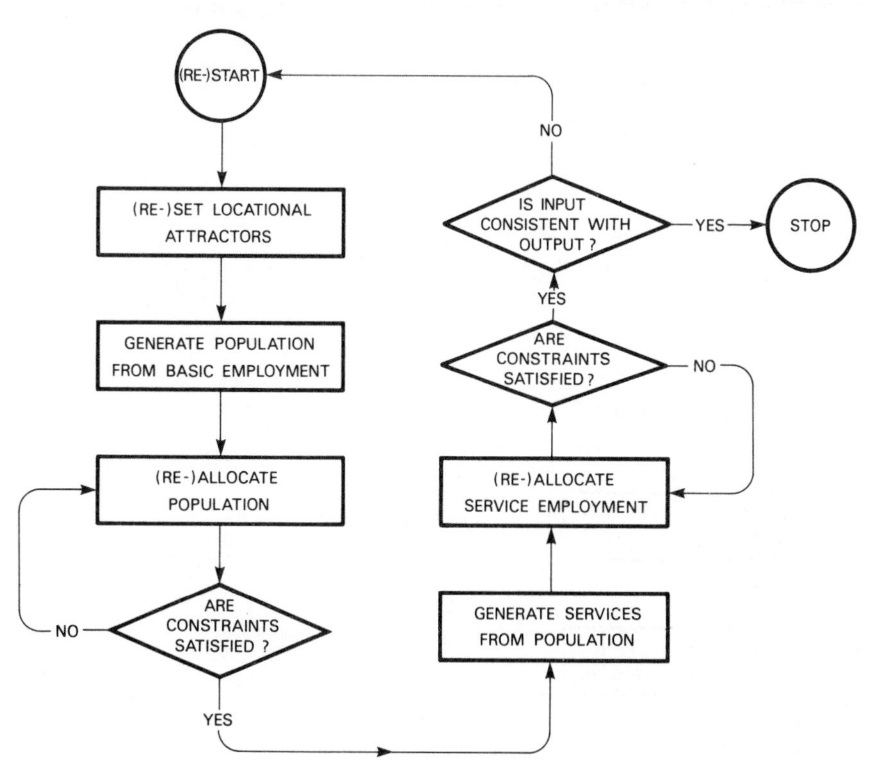

Figure 17.1 The structure and solution procedure of Lowry's model

(1965) Baltimore model, but hard on their heels came the first applications of Lowry-type models: in Central Lancashire (Batty, 1968), in Bedfordshire (Cripps and Foot, 1968) and in Reading (Echenique, 1968). The major departure from Lowry's (1964) original model involved the explicit treatment of interaction in the manner suggested by Garin (1966), Harris (1966) and Goldner (1968). Moreover, this enabled constraints to be handled more easily and wedded these models to the theoretical work being pursued by Wilson. The general structure of these Garin-Lowry models follows the original model in (6) to (9), but interaction is explicitly modelled and location and derivation can thus be regarded as a direct function of interaction. First define the probability of working in i and living in j as p_{ij} and the probability of living in j and being serviced in k as q_{jk}. Then

$$p_{ij} = \frac{D_j f^1(c_{ij})}{\sum_j D_j f^1(c_{ij})}, \quad \sum_j p_{ij} = 1, \text{ and} \tag{10}$$

$$q_{jk} = \frac{O_k f^2(c_{jk})}{\sum_k O_k f^2(c_{jk})}, \quad \sum_k q_{jk} = 1, k = 1, 2, \ldots, K. \tag{11}$$

The model can now be written as two sets of two equations, one dealing with interaction, the other dealing with derivation and location for each of the two sectors.

$$T_{ij} = E_i p_{ij}, \tag{12}$$

$$P_j = a \sum_i T_{ij}, \tag{13}$$

$$S_{jk} = P_j q_{jk}, \text{ and} \tag{14}$$

$$S_k = b \sum_j S_{jk}. \tag{15}$$

This system of equations in (12) to (15) is usually solved iteratively, starting with E_i in equation (12) as basic employment B_i, and then substituting S_k from (15) for E_i in (12), $i = k$, on subsequent iterations. Using this procedure, the increments of population and service employment generated form converging geometric series which in the limit sum to the totals P and S respectively. Other solution methods will be presented later.

This framework enables locational constraints and consistency to be more rigorously established than in the original model and this consists of finding attractors D_j and O_k which reflect the norms of

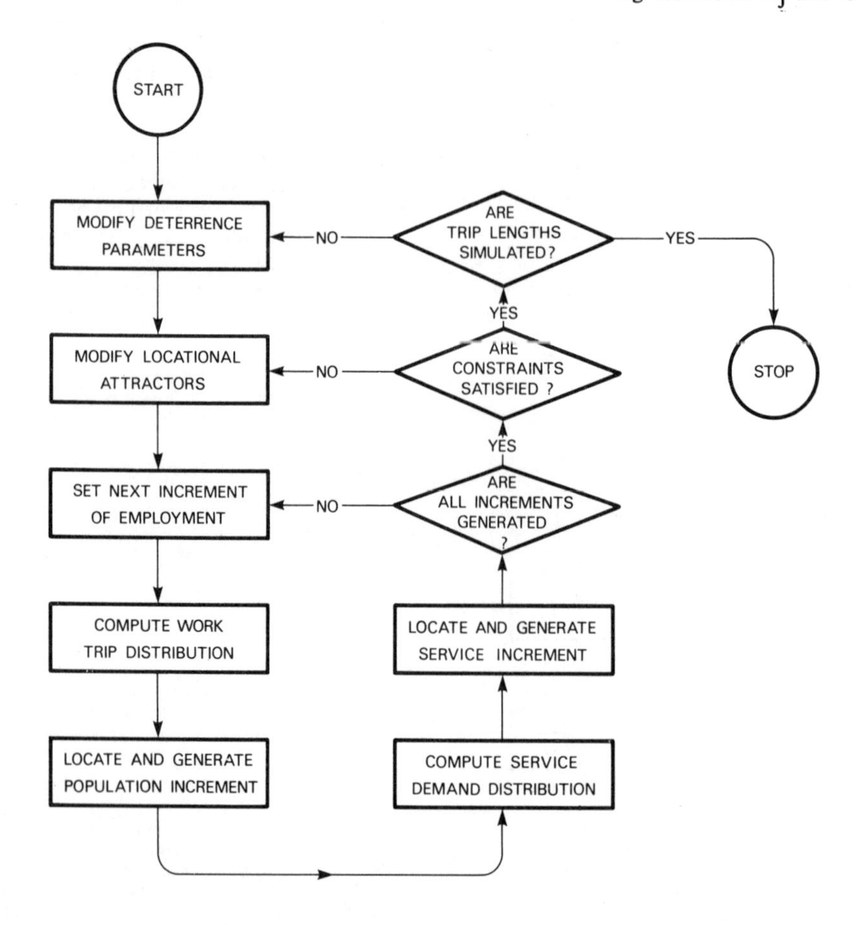

Figure 17.2 The structure and solution procedure of a typical activity allocation model

consistency and the constraints required. The general structure of the model is illustrated in Figure 17.2 and details can be found in Wilson (1974) and Batty (1976). The importance of this framework for urban modelling research in Britain cannot be overestimated for most of the contributions to be detailed below were motivated by problems emerging from this framework. In particular, questions of model design and estimation – fitting the model to data, and questions of model structure – solution and interpretation, directly relate to this model. These will now be dealt with and then more substantive issues involving new model forms evolving from this framework, and planning applications will be subsequently presented.

Model design and calibration

A major development, which has sustained interest in the gravitational approach, has involved devising frameworks for generating such models according to statistical principles. These methods known collectively as entropy-maximizing, have led to a consistency in both model formulation and estimation, thus enabling considerable improvements to be made on first generation model design. The methods, for long used in statistical physics (Levine and Tribus, 1979), first gained the attention of several researchers in the 1960s but it was Wilson (1967, 1970) who popularized the approach and who has done most to demonstrate its value. In essence, gravitational models can be derived by maximizing an uncertainty, information or entropy function subject to known constraints on the amount and distribution of interaction. Various forms of function can be optimized but that due to Williams (1976) is the most appropriate. The maximization program involves:

$$\max - \sum_i \sum_j T_{ij} \left\{ \log \frac{T_{ij}}{O_i D_j} - 1 \right\}, \qquad (16)$$

subject to constraints on origins, destinations and travel cost C stated respectively as

$$\sum_j T_{ij} = O_i, \qquad (17)$$

$$\sum_i T_{ij} = D_j, \text{ and} \qquad (18)$$

$$\sum_i \sum_j T_{ij} c_{ij} = C. \qquad (19)$$

Note that c_{ij} is now the travel cost from origin i to destination j.

Using Lagrange's method for maximizing a constrained function, the general interaction model is generated as

$$T_{ij} = O_i D_j \exp \left\{ -\lambda_i - \lambda_j - \lambda c_{ij} \right\}, \qquad (20)$$

where λ_i, λ_j and λ are the multipliers or parameters required to ensure that the model satisfies the constraints in (17), (18) and (19) respectively; $\exp\{-\lambda_i\}$ and $\exp\{-\lambda_j\}$ are referred to as balancing factors and are interdependent as can be seen if the model in (20) is substituted into (17) or (18) and the requisite simplifications made. The model can also be easily converted into conditional or total probability form if required. This framework has many other diverse features. By relaxing, modifying or extending the constraints, a whole family of models can be derived (Wilson, 1971). For example, the spatial impedance function in (20), $f(c_{ij})$, is $\exp\{-\lambda c_{ij}\}$ but any other can be derived by making a suitable transformation of (19). There are also relationships to utility-maximizing and maximum likelihood (Hyman, 1969; Wilson, 1970) but the real value of the framework in practical terms relates to the calibration of model parameters. Two methods are possible: the more usual will be described here, the other based on direct optimization of (16) will be detailed in the following section.

The conventional method of calibration involves separating the estimation of the balancing factors from the deterrence parameter λ. The properties of these balancing factors have been extensively studied (Evans, 1970; MacGill, 1977), and various biproportional (iterative) techniques have been devised for their estimation. These iterations are usually nested inside a higher-level iterative process for estimating λ, which in turn can be determined in several ways; by curve fitting (interpolation) (Hyman, 1969), by unconstrained optimization (search procedures) (Batty, 1971) or by nonlinear equation-solving (Gauss-Newton, Newton-Raphson methods) (Batty, 1976). When calibrating a series of gravity models embedded into a more general structure of the Lowry type, these methods still hold. For example, a biproportional procedure is used to satisfy any locational constraints on the model given in (12) to (15), and Newton-Raphson methods etc. used to calibrate the parameters λ^1 and λ^2 associated with the functions $f^1(c_{ij})$ and $f^2(c_{jk})$ in (10) and (11). Figure 17.2 illustrates the iterative procedures in which the general model is embedded.

Finally, in the quest to explore invariance in model forms, there has been considerable research into the effects of spatial aggregation on model calibration (Openshaw, 1977). This problem has been tackled by devising invariant forms of model (Beardwood and Kirby, 1975), by making models dimensionally consistent (Broadbent, 1970) or by devising optimal zoning systems (Masser, Batey and Brown, 1975). Furthermore models consistent with detail at different scales have been devised (Broadbent, 1971) and many of these developments have recently been synthesized in the book by Masser and Brown (1978) to which the reader is referred.

Analysis of model structure

The construction of more general models in modular fashion has also been the subject of research into principles for sound model design. Such research directions follow the econometric tradition of examining the equilibrium properties of models, and of deriving the equilibrium relations or reduced forms associated with given equilibrium conditions. For example the equilibrium conditions of the Lowry model in (12) to (15) can be restated more cogently: substituting (12) into (13), and (14) into (15) leads to

$$P_j = a\sum_i E_i p_{ij}, \quad \text{and} \tag{21}$$

$$S_k = b\sum_j P_j q_{jk}. \tag{22}$$

Equations (21) and (22) are subject to the accounting equation – the economic base condition

$$E_k = B_k + S_k, \tag{23}$$

from which the reduced form can be derived by substituting (22) into (23) and (21) into the result; that is, in equilibrium

$$E_k = B_k + ba\sum_i\sum_j E_i p_{ij} q_{jk}. \tag{24}$$

Equation (24) demonstrates the simultaneous nature of the model as a structure based on two interlocking spatial interaction sub-models.

As alluded to previously, there are several ways in which these types of model might be solved. Iteration on equations (21) and (22) starting with E_i in (21) and making successive substitutions from S_k in (22) for E_i, $i = k$, in (21) generates the incremental solution to the model. Iteration on equation (24) however starting with E_i as B_i or as zero leads to the same solution but with an accumulation of generated activity. This latter procedure gives much greater scope for efficient embedding of calibration procedures into the model; rather than embedding the model into these procedures which is current practice; these ideas have been explored by Baxter and Williams (1975) and Batty (1978b, 1979a). Other work relates to the examination of the convergence of interaction patterns generated by the model (Batty, 1979b; Schinnar, 1978), and to ways of using the reduced form and series expansion of the model solution to enable integration and feedback to occur between interaction and location, transport and land use (Berechman, 1976; Williams, 1979).

A rather different but equally fruitful line of inquiry has involved a more literal interpretation of the entropy-maximizing framework for deriving interaction models. It is quite possible to calibrate such models by maximizing the entropy directly, that is by solving the nonlinear program given in (16) to (19) using the optimality condition (20) in the objective function (16). Work along these lines indicates that a more efficient method of calibration involves minimizing the dual of (16) to (19). This can be derived as

$$\min \Big[\sum_i \lambda_i O_i + \sum_j \lambda_j D_j + \lambda C + \sum_i\sum_j O_i D_j \exp\left\{-\lambda_i - \lambda_j - \lambda c_{ij}\right\}\Big], \tag{25}$$

and as such, (25) can be solved for λ_i, λ_j and λ as a problem in unconstrained optimization (Wilson and Senior, 1974; Williams, 1976; Champernowne, Williams and Coelho, 1976). Emerging from such interpretations are two related themes: first that more general models can be treated and derived as joint entropy-maximizing problems, and second that the duals of such problems are of interest in interpretation, and for calibration (Coelho, 1977).

The Lowry model in (12) to (15) can be solved by joint or group entropy-maximization. Because the maximization of entropy relating to T_{ij} and that to S_{jk} is simultaneous, it is necessary to drop the specific notation for service location and regard the set of I zones as containing both basic employment and services, wherever appropriate. The joint function to be maximized is thus

$$\max \left\{ -\sum_i\sum_j T_{ji}\left\{\log \frac{T_{ij}}{D_j} - 1\right\} - \sum_i\sum_j S_{ji}\left\{\log \frac{S_{ji}}{O_i} - 1\right\} \right\}, \tag{26}$$

subject to the constraints on travel cost

$$\sum_i\sum_j T_{ij} c_{ij} = C^1 \quad \text{and} \quad \sum_i\sum_j S_{ji} c_{ji} = C^2, \tag{27}$$

and the appropriate form of the equilibrium conditions derived from (10) to (15) and (23)

$$\sum_j T_{ij} - \sum_j S_{ji} = B_i, \text{ and} \tag{28}$$

$$ba\sum_i T_{ij} - \sum_i S_{ji} = 0. \tag{29}$$

The two interaction models are derived as

$$T_{ij} = D_j \exp\left\{-\lambda_i - ba\lambda_j - \lambda^1 c_{ij}\right\}, \tag{30}$$

$$S_{ji} = O_i \exp\left\{+\lambda_i + \lambda_j - \lambda^2 c_{ij}\right\}, \tag{31}$$

where λ^1, λ^2, λ_i and λ_j are the parameters ensuring that (27), (28) and (29) are satisfied. These parameters can be estimated by solving the dual minimization problem which is unconstrained

$$\min \left\{ \sum_i \lambda_i B_i + \lambda^1 C^1 + \lambda^2 C^2 + \sum_i \sum_j D_j \exp\left\{ -\lambda_i - b a \lambda_j - \lambda^1 c_{ij} \right\} + \sum_i \sum_j O_i \exp\left\{ +\lambda_i + \lambda_j - \lambda^2 c_{ij} \right\} \right\} \quad (32)$$

Locational activities and interaction probabilities can be determined directly by substituting (30) and (31) into (12) to (15) but note that the solution given by this direct optimization is not equivalent to that implied by the conventional solution and calibration of the Lowry model according to Figure 17.2.

Besides viewing calibration as a formal optimization problem, this perspective enables such general models to be interpreted in normative terms. The primal or dual objective function in (26) or (32) can be augmented by other objectives involving profit or cost of land use development, thus linking the model to those of the linear programming type (Schlager, 1967; Herbert and Stevens, 1960) and to the TOPAZ models (Sharpe, Brotchie, Ahern and Dickey, 1974). Moreover, it is possible to design augmented entropy-maximizing frameworks which collapse to Herbert-Stevens, TOPAZ, Schlager or transport-cost minimization models as various constraints are relaxed. Finally, there are other ways of interpreting entropy functions as utility functions involving consumer surplus and access factors, and these extensions have led to considerable insights into the relationship between actual existing and designed spatial systems. Moreover, such functions can be used as spatial indicators in the development of these models in the plan-design comparative-static context (Coelho, 1977; Coelho and Williams, 1977; Leonardi, 1978). There are also extensions towards disaggregate transport demand modelling involving random utility. Thus in many senses the optimization paradigm for generating, estimating and using models to predict and prescribe is the most insightful and innovative of all the contributions reviewed here.

Disaggregation, dynamics and integration

Although the most distinct achievements in British modelling have been made in technical analysis of the kind presented so far, there have also been some interesting and noteworthy extensions building on the gravitational models outlined above. Disaggregation of such structures has attempted to capture greater detail of the spatial system and its activities, and, in particular, such attempts have been aided by the rigour of entropy-maximizing which allows quite complex but consistent disaggregation to take place. Some research has attempted to link disaggregated residential location models to more mainstream urban economics by introducing incomes and rents (Wilson, 1970) and several attempts have been made to develop such models empirically (Cripps and Cater, 1972; Baxter and Williams, 1973; Senior, 1978). But in general, such work has been restricted by data problems and by limits to this style of theorizing.

Of more central concern has been the question of extending the static model to a dynamic context. Forrester's (1969) model based on non-spatial 'systems dynamics' has been a source of inspiration to this quest and thus some Forrester-Lowry-like models have resulted (Sayer, 1975; Burdekin, 1977). But these models are pragmatically structured. Some extensions to the Lowry framework have been made using more theoretically considered notions, for example Wilson's (1970) mover pools, and dynamic multipliers (Batty, 1976). However, in general all these attempts are quite straightforward and sacrifice elegance and dynamic theory for operationality. Theoretical dynamics has been examined in terms of the internal dynamics of static models as mentioned above (Batty, 1979a, b) but of more substantive relevance are the dynamic structures for urban models based on catastrophe theory (Harris and Wilson, 1978) and space-time series reported elsewhere in this book.

Where disaggregation and dynamics have played the greatest part is in the design of integrated land-use/transport models. Indeed, this is the area where other modelling principles such as modularization and structural analysis also play an important role. The emphasis in such integrated models is on interdependence or feedback between activities which naturally involves dynamics. In particular, balancing of demand and supply is an important feature of such models in the land use and transport sectors leading to competition expressed through the price mechanism. Moreover the capacity and density of land and transport systems act as constraints on location, and thus such constraints are also handled in the temporal dimension. Extensions of these ideas to the theoretical realm have also been explored (Berechman, 1976; Wilson, 1974). In particular, the models designed by Mackett (1979), Echenique (1977), and Said and Hutchinson (1978) are operational, and these models form a useful means of linking Lowry-type frameworks to the conventional aggregate transport modelling process. In Figure 17.3, a typical integrated model based on Echenique's (1977) ideas gives some feeling as to how such frameworks are being developed.

Applications in land use planning

Much of the original interest in urban modelling in Britain was inspired by problems of land use planning and thus the field has not become the prerogative of any one discipline or academic subject area. Yet although motivated by policy problems, much of the research described in this review has been carried out in more academic confines, shielded from the very problems which led to the research in the first place. Indeed, in the late 1970s the research dimension to the field has been somewhat polarized within the areas of human and theoretical geography. Nevertheless, there exists a wealth of applications to real systems and planning problems which form part of the heritage of this field, and a crude count of well-documented applications of Lowry-type models, for example, reveals that since 1968 over thirty applications have been made in Britain (Batty, 1972; Batty, 1974). Probably as many again, built as part of various academic projects or in Structure Planning tasks, are inaccessible due to poor or non-existent documentation (Barras and Broadbent, 1979). And many others in South America, continental Europe and elsewhere have been directly influenced by British work (Mohan, 1979; Baxter, Echenique and Owers, 1975). In particular, British applications have been small-scale due to limits on human and monetary resources, and thus there has been a greater chance to emphasize questions of model structure and design than in the first generation of models. As has been stressed throughout this review, the British experience and contribution has been to urban modelling rather than urban models *per se*, to the use of models in planning rather than to the design of model-based land use plans.

The performance of these models in terms of

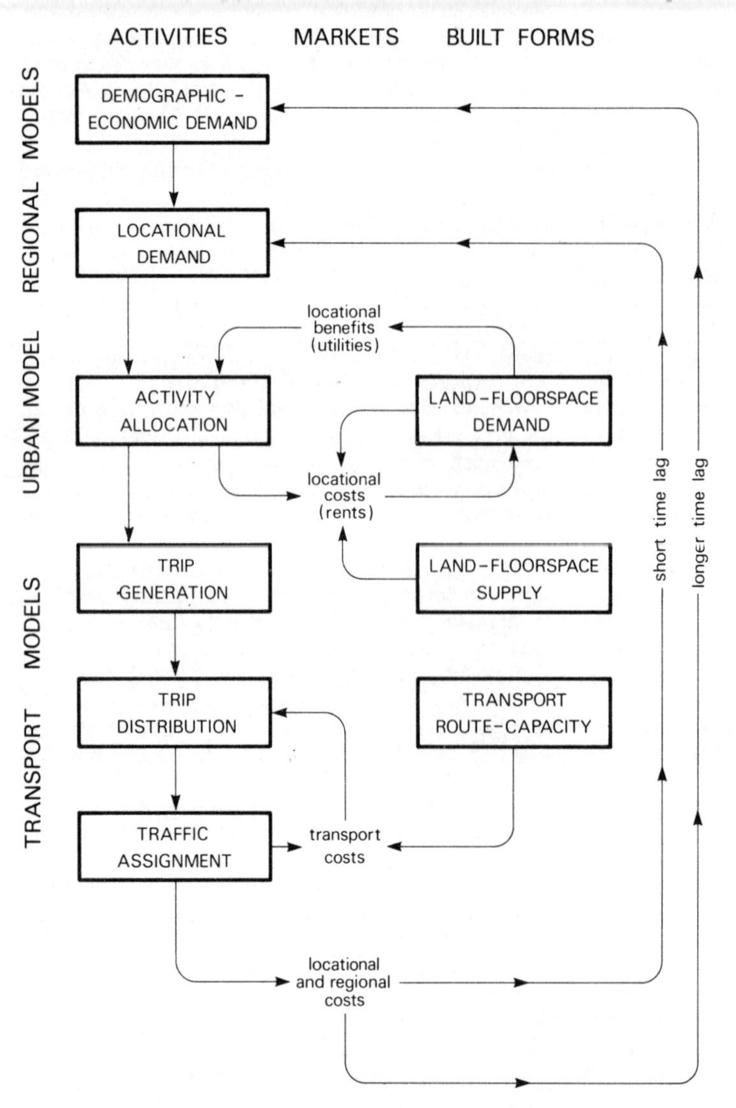

Figure 17.3 The structure of Echenique's integrated land-use transport model

their goodness-of-fit to existing data has been rather good but this is also ambiguous and there has been considerable debate as to the validity of the various performance statistics used (Openshaw, 1978). In fact, arguments about performance have been one of the reasons for the intensive research into calibration described earlier. Of the applications developed for real systems, it is also noteworthy that very few of these have eventually been used in positive plan-making. The causes of this dearth of 'planning' applications related largely to the difficulty of model development, the agencies involved in using models and their approach towards technical analysis, and the changing environment of planning. Models which have been used, however, have emphasized conditional prediction, particularly impact analysis of large-scale facilities – new towns, airports and transport systems (Batty, 1976; Cripps and Foot, 1970). Some work on using models to evaluate alternative plans has been done (Batty, 1978a; Breheny, 1978) but in general these types of model have been regarded as positive rather than normative tools. In short, the planning process itself in its technical phase is regarded by planners as a loosely-structured optimization procedure, and thus models which themselves are optimizing have sat uneasily in this process. Indeed the new paradigm of viewing models, whether positive or normative, as some form of optimization process, has raised awareness about these dilemmas to a new level. And in the future, research into this area involving the use of models in planning or for planning is likely to be extremely fruitful.

These issues in planning are still changing quite rapidly. For example, the limits to modelling posed by computational resources are disappearing, and the questions which many of these models are designed to address no longer seem so important. The philosophy of modelling as an aid to science or to design is being increasingly explored (Sayer, 1979) and thus the environment within which modelling takes place is likely to be quite different in the immediate future. There is no longer the certainty that computer models are of the utmost importance to planning, and thus the next decade, at least in terms of applications, will be fairly unpredictable.

Conclusions

The central contribution made by urban modellers in Britain during the last decade has been technical in every sense. The model design process has been thoroughly researched and the technical structure of existing models analysed in depth. Of similar importance has been the emphasis on empirical applications for many of the technical issues have been tackled from this standpoint. Limits on resources have also played their part for the large-scale disasters which were a feature of the first generation North American experience have been largely avoided. But least impressive has been the development of models based on sound urban theory. Only lip-service has been paid to the work of the urban economists and most models have avoided questions of social theory. In planning too, the experience has been mixed. More applications to planning problems could have been made and thus the experience gained has been fragmentary.

Future technical work is however clearer. The reinterpretation of model structures through the paradigm of optimization theory and the development of nonlinear econometrics pose powerful research frontiers. Moreover, these frameworks enable a host of hitherto unrelated issues – calibration, identification, design, dynamics etc. – to be considered simultaneously. The integration of model structures too, using notions concerning demand–supply balance and temporal feedback, seem useful and promising themes. But most important is the emerging synthesis of model frameworks, through optimization and integration, thus enabling the whole range of models developed so far to be seen as special cases of more general concepts. These are exciting and relevant issues and promise to inspire the field to produce new insights and achievements during the next decade.

References

Barras, R., and Broadbent, T.A. (1979) 'The analysis in English structure plans', *Urban Studies*, 16, 1–18.

Batty, M. (1968) 'A land use allocation model to distribute population and employment in a bounded subregion', *Computer Program Review Conference: Part III: Urban Planning*, Planning and Transport Research and Computation Ltd: London.

Batty, M. (1971) 'Exploratory calibration of a retail location model using search by golden section', *Environment and Planning*, 3, 411–32.

Batty, M. (1972) 'Recent developments in land use modelling: a review of British research', *Urban Studies*, 9, 151–77.

Batty, M. (1974) 'Computer models in structure planning', *Town and Country Planning*, 42, 453–57.

Batty, M. (1976) *Urban Modelling: Algorithms, Calibrations, Predictions*, Cambridge University Press.

Batty, M. (1978a) 'Urban models in the planning process' in D.T. Herbert and R.J. Johnston (eds), *Geography and the Urban Environment: Vol. 1,*

Progress in Research and Applications, 63–134, John Wiley: London.

Batty, M. (1978b) 'Operational urban models incorporating statics in a dynamic framework' in A. Karlqvist, L. Lundqvist, F. Snickars and J.W. Weibull (eds), *Spatial Interaction Theory and Planning Models*, 227–52, North-Holland: Amsterdam.

Batty, M. (1979a) 'Efficient calibration of an urban model with dynamic solution properties', *London Papers in Regional Science*, 9, 26–63.

Batty, M. (1979b) 'Invariant-distributional regularitics and the Markov property in urban models: an extension of Schinnar's result', *Environment and Planning A*, 11, 487–97.

Baxter, R., and Williams, I. (1973) 'The third stage in disaggregating the residential sub-model', *Land Use and Built Form Studies*, Working Paper 66, University of Cambridge School of Architecture.

Baxter, R., and Williams, I. (1975) 'An automatically calibrated urban model', *Environment and Planning A*, 7, 3–20.

Baxter, R., Echenique, M., and Owers, J.W. (eds) (1975) *Urban Development Models*, Construction Press: Lancaster.

Beardwood, J.E., and Kirby, H.R. (1975) 'Zone definition and the gravity model: the separability, excludability and compressibility properties', *Transportation Research*, 9, 363–9.

Berechman, J. (1976) 'Interfacing the urban land-use activity system and the transportation system', *Journal of Regional Science*, 16, 183–4.

Breheny, M.J. (1978) 'The measurement of spatial opportunity in strategic planning', *Regional Studies*, 12, 463–79.

Broadbent, T.A. (1970) 'Notes on the design of operational models', *Environment and Planning*, 2, 469–76.

Broadbent, T.A. (1971) 'A hierarchical interaction-allocation model for a two-level spatial system', *Regional Studies*, 5, 23–7.

Burdekin, R. (1977) 'The simulation and control of urban development', unpublished PhD thesis, University of Sheffield, Department of Control Engineering.

Champernowne, A.F., Williams, H.C.W.L., and Coelho, J.J.S.D. (1976) 'Some comments on urban travel demand analysis, model calibration and the economic evaluation of transport plans', *Journal of Transport Economics and Policy*, 10, 267–85.

Chapin, F.S., and Weiss, S.F. (1968) 'A probabilistic model for residential growth', *Transportation Research*, 2, 375–90.

Coelho, J.J.S.D. (1977) 'The use of mathematical optimisation methods in model based land use planning: an application to the new town of Santo Andre', unpublished PhD thesis, University of Leeds, School of Geography.

Coelho, J.J.S.D., and Williams, H.C.W.L. (1977) 'On the design of land use plans through locational surplus maximisation', Working Paper 203, University of Leeds, School of Geography.

Cripps, E.L., and Cater, E.S. (1972) 'The empirical development of a disaggregated residential location model: some preliminary results', *London Papers in Regional Science*, 3, 114-45.

Cripps, E.L., and Foot, D.H.S. (1968) 'Evaluating alternative strategies', *Official Architecture and Planning*, 31, 928–38.

Cripps, E.L., and Foot, D.H.S. (1970) 'The urbanisation effects of a third London Airport', *Environment and Planning*, 2, 153–92.

Dyckman, J.W. (1963) 'The scientific world of the city planners', *American Behavioral Scientist*, 6, 46–50.

Echenique, M. (1968) 'Urban systems: towards an exploratorive model', *Land Use and Built Form Studies*, Working Paper 7, University of Cambridge, School of Architecture.

Echenique, M. (1977) 'An integrated land use and transport model', *Transactions of the Martin Centre*, 2, 195–230.

Evans, A.W. (1970) 'Some properties of trip distribution models', *Transportation Research*, 4, 19–36.

Forrester, J.W. (1969) *Urban Dynamics*, MIT Press: Cambridge, Mass.

Garin, R.A. (1966) 'A matrix formulation of the Lowry model for intra-metropolitan activity location', *Journal of the American Institute of Planners*, 32, 361–4.

Goldner, W. (1968) *Projective Land Use Model*, BATSC Technical Report 219, Bay Area Transportation Study Commission: Berkeley, California.

Hansen, W.G. (1959) 'How accessibility shapes land use', *Journal of the American Institute of Planners*, 25, 73–6.

Harris, B. (1966) *Note on Aspects of Equilibrium in Urban Growth Models*, University of Pennsylvania, Department of City and Regional Planning: Philadelphia.

Harris, B., and Wilson, A.G. (1978) 'Equilibrium values and dynamics of attractiveness in production-constrained spatial-interaction models', *Environment and Planning A*, 10, 371–88.

Herbert, J.D., and Stevens, B.H. (1960) 'A model for the distribution of residential activity in urban areas', *Journal of Regional Science*, 2, 21–36.

Hill, D.M. (1965) 'A growth allocation model for the Boston region', *Journal of the American Institute of Planners*, 31, 111–20.

Hyman, G.M. (1969) 'The calibration of trip distribution models', *Environment and Planning*, 1, 105–12.

Lakshmanan, T.R., and Hansen, W.G. (1965) 'A

retail market potential model', *Journal of the American Institute of Planners*, 31, 134–43.
Lathrop, G.T., and Hamburg, J.R. (1965) 'An opportunity accessibility model for allocating regional growth', *Journal of the American Institute of Planners*, 31, 95–108.
Lee, D.B. (1973) 'Requiem for large-scale models', *Journal of the American Institute of Planners*, 39, 163–78.
Leonardi, G. (1978) 'Optimum facility location by accessibility maximizing', *Environment and Planning A*, 10, 1287–1305.
Levine, R.D., and Tribus, M. (eds) (1979) *The Maximum Entropy Formalism*, MIT Press: Cambridge, Mass.
Lowry, I.S. (1964) *A Model of Metropolis*, RM-4035-RC, The RAND Corporation: Santa Monica, California.
MacGill, S.M. (1977) 'Theoretical properties of biproportional matrix adjustments', *Environment and Planning A*, 9, 687–701.
Mackett, R.L. (1979) 'Modelling the impact of transport planning policy upon land use', Working Paper 115, University of Leeds, Institute for Transport Studies.
McLoughlin, J.B. (1969) *Urban and Regional Planning: A Systems Approach*, Faber & Faber: London.
Masser, I., and Brown, P.J. (eds) (1978) *Spatial Representation and Spatial Interaction*, Martinus Nijhoff: Leiden.
Masser, I., Batey, P.W.J., and Brown, P.J. (1975) 'The design of zoning systems for interaction models', *London Papers in Regional Science*, 5, 168–87.
Mohan, R. (1979) *Urban Economic and Planning Models*, Johns Hopkins University Press: Baltimore, Md.
Openshaw, S. (1977) 'Optimal zoning systems for spatial interaction models', *Environment and Planning A*, 9, 169–84.
Openshaw, S. (1978) *Using Models in Planning: A Practical Guide*, RPA Books, Corbridge.
Pack, J.R. (1978) *Urban Models: Diffusion and Policy Application*, Monograph Series No. 7, Regional Science Research Institute: Philadelphia.
Said, G.M., and Hutchinson, B.G. (1978) 'An urban systems model for the Toronto region', *Proceedings of the 2nd International Symposium on Large Engineering Systems*, 75–80, University of Waterloo: Ontario.
Sayer, R.A. (1975) 'Dynamic spatial models of urban and regional systems', unpublished DPhil thesis, University of Sussex, School of Social Sciences.
Sayer, R.A. (1979) 'Understanding models versus understanding cities', *Environment and Planning A*, 11, 853-62.
Schinnar, A.P. (1978) 'Invariant-distributional regularities of non-basic spatial activity allocations: the Garin-Lowry model revisited', *Environment and Planning A*, 10, 327–36.
Schlager, K.J. (1967) 'Land-use planning design models', *Proceedings of the American Society of Civil Engineers. Journal of the Highway Division*, 93–HW2, 135–42.
Senior, M.L. (1978) 'Approaches to residential location modelling: empirical and theoretical developments of disaggregated models', unpublished PhD thesis, University of Leeds School of Geography.
Sharpe, R., Brotchie, J.F., Ahern, P.A., and Dickey, J.W. (1974) 'Evaluation of alternative growth patterns in urban systems', *Computers and Operations Research*, 1, 345–62.
Williams, H.C.W.L. (1976) 'Travel demand analysis, duality relations and user benefit analysis', *Journal of Regional Science*, 16, 147–66.
Williams, I.N. (1979) 'An approach to solving spatial-allocation models with constraints', *Environment and Planning A*, 11, 3–22.
Wilson, A.G. (1967) 'A statistical theory of spatial distribution models', *Transportation Research*, 1, 253–69.
Wilson, A.G. (1970) *Entropy in Urban and Regional Modelling*, Pion: London.
Wilson, A.G. (1971) 'A family of spatial interaction models and associated developments', *Environment and Planning*, 3, 1–32.
Wilson, A.G. (1974) *Urban and Regional Models in Geography and Planning*, John Wiley: London.
Wilson, A.G., and Senior, M.L. (1974) 'Some relationships between entropy maximising models, mathematical programming models and their duals', *Journal of Regional Science*, 14, 207–15.

Chapter 18

Catastrophe theory and bifurcation

A.G. Wilson

The foundations of the methods

Background

Bifurcation theory is concerned with the way in which the nature of the solutions of differential equations change at certain critical values of the parameters of those equations. It has a long history, going back to the time of Poincaré, for example, but has only been incorporated into a broad framework relatively recently. Catastrophe theory is a newer development – of the last ten years or so – though with earlier pieces of mathematics being important in its development. The theory takes its name from the possibility of systems which it describes exhibiting jump behaviour.

In this brief introductory section, the main ideas of both catastrophe theory and bifurcation theory will be outlined. There is now a voluminous literature on these topics, but the reader is referred only to a number of key works which themselves include much more extensive bibliographies. For an account of bifurcation theory in an applied context, see Hirsch and Smale (1974). In the case of catastrophe theory, it is important to look at the seminal book by Thom (1975), but the more recent books by Zeeman (1977a) and Poston and Stewart (1978) offer a better introduction. Amson (1975) also provides an excellent introduction in an 'urban studies' context.

Gradient and general systems

Catastrophe theory is concerned with a special class of systems called gradient systems. They can be characterized as follows. Let such a system be described by a set of state variables (or endogenous variables), **x**, and a set of parameters (or exogenous variables, sometimes also known as control variables), **u**. Then, for a gradient system, there exists a potential function, F(x, u) say, which, when maximized (or, if appropriate, with the corresponding changes, minimized), determines the equilibrium state of the system. That is, **x** is given by

$$\max_{\mathbf{x}} F(\mathbf{x}, \mathbf{u}) \tag{1}$$

If the system is disturbed from equilibrium, its dynamics are described by the differential equations

$$\dot{\mathbf{x}} = -\frac{\partial F}{\partial \mathbf{x}} \tag{2}$$

and this type of system takes its name from the gradient vector on the right hand side of the simultaneous equation (2).

More typically, however, the dynamical behaviour of a system is given by a more general set of equations such as

$$\dot{\mathbf{x}} = G(\mathbf{x}, \mathbf{u}) \tag{3}$$

where the right hand side cannot be written in gradient form.

Catastrophe theory

Let us concentrate on equilibrium states. The maxima of F(**x**, **u**) will be given by the solutions of

$$\frac{\partial F}{\partial \mathbf{x}} = 0 \tag{4}$$

These form a set of simultaneous equations in **x** – with as many equations as there are **x**-variables – and so can be solved for **x** for each possible value of the parameters, **u**. These solutions can be represented geometrically by a surface in (**x**, **u**)-space representing the possible equilibrium states of the system. The two simplest cases with which we shall be mostly concerned involve a single state variable, x, and one or two parameters, given as either u (dropping the subscript, as with the x variable, when there is only one) or as u_1 and u_2. Thus the possible equilibrium states in these cases are represented by either a curve in the two-dimensional (x, u) space or a surface in the three-dimensional (x, u_1, u_2) space. Examples are shown in Figure 18.1 and 18.2.

Figure 18.1 Singularity in the potential function F with one control variable u

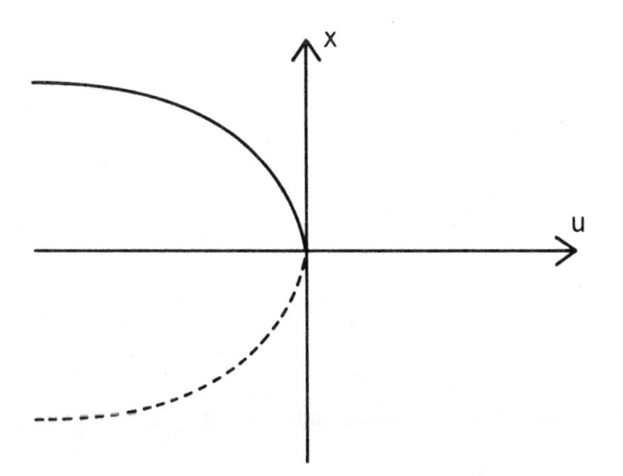

Catastrophe theory is concerned with the singularities of the function F. The maxima, minima and points of inflexion of F are called stationary points. In the example we have considered, the maxima represent stable equilibrium points, and the minima and points of inflexion, unstable points. The singularities are *degenerate* stationary points which occur when two or more stationary points coalesce. This gives another clue to the nature of the theory: the interesting features arise out of multiple solutions of (4) which in turn, of course, arise out of the non-linearities in the potential function, F. In the elementary diagrams of the form of Figures 18.1 and 18.2, the singular points exhibit themselves where the tangent to the curve or surface is perpendicular to the u axis (in Figure 18.1) or the (u_1,u_2)-plane (in Figure 18.2). This projection of the folds on to this axis or plane gives the set of critical parameter values.

Thom's main results on catastrophe theory are concerned with stability in two senses. First, they specify the nature of the singularities for a wide class of systems, and it is at these points that the system can be unstable in that it can exhibit 'jumps' in its type of behaviour. Second, he is concerned with the structural stability of the system as a whole and the form which this can take. For our present purposes, it will suffice to note that Thom's work tells us what the worst possible forms of singularity are for systems described by up to two state variables and four parameters (though, in practice, the number of state variables can be very much greater, but not the number of parameters). For a single state variable and one parameter, the worst kind of singularity is a fold – as shown in Figure 18.1; for one state variable and two parameters, the worst possible singularities are folds and cusp points – as shown in Figure 18.2. In the latter case, for $u_1 < 0$, a section of the equilibrium surface parallel to the

Figure 18.2 Singularities in the potential function F with two control variables

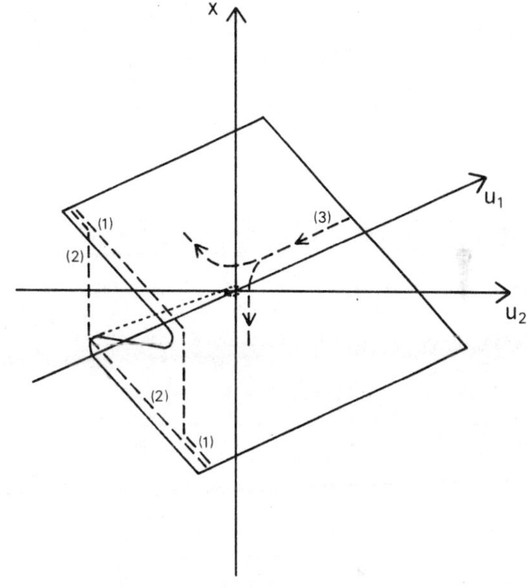

(x, u_2)-plane is folded, and the cusp point occurs where the fold in the surface ends. The possible shapes of the surface for the remaining elementary catastrophes are given in the texts cited earlier.

The behaviour of a system is characterized by a trajectory on the equilibrium surface, which is a surface of *possible* states. We can then immediately see by reference to the three sample trajectories on Figure 18.2 that unusual kinds of behaviour are possible. Suppose that the parameters are changing in such a way as to generate trajectory (1): the value of x changes smoothly until the fold in the surface is reached, and then it changes discontinuously because it must 'jump' to the lower surface. Trajectory (2) shows the reverse change, but the jump is not forced until a later point and so (1) and (2) together demonstrate a hysteresis effect. (This effect depends on a further assumption known as the 'perfect delay' convention.) Finally, trajectory (3) approaches the cusp point from above and a small change in either direction can take the system smoothly onto the upper or lower surface as shown.

The representations of systems in Figures 18.1 and 18.2 are canonical forms: they are generated from potential functions which are a standard form of polynomial. What Thom has shown is that, given some not-very-restrictive conditions, all gradient systems with one or two parameters have equilibrium systems which can be smoothly transformed into the shapes of Figures 18.1 and 18.2 respectively. This, potentially, is a powerful result for applied purposes.

Bifurcation theory

Differential equations such as (3) can have many types of solution. These include (i) stable equilibrium points, (ii) unstable equilibrium points, (iii) saddle points – a special type of unstable equilibrium, (iv) closed orbit periodic solutions, (v) limit cycles – which are convergent oscillations, (vi) divergent oscillations, or (vii) chaotic behaviour. Bifurcation theory is like catastrophe theory in one important respect: it is concerned with critical values of system parameters at which some unusual system behaviour can occur – in this case, the transition from one type of solution to another. There is one particular case when the results may appear similar to those of catastrophe theory: in the special cases involving the appearance or disappearance of stable equilibrium points. Indeed, it could be said that catastrophe theory is a special case of bifurcation theory in the sense of the latter used here, but it is useful to distinguish it because of the special results available.

The basis of applied work

Zeeman (1977a) usefully identifies six steps for building dynamic models in the sort of framework described here. They can be summarized as follows. *First*, construct the surface of possible equilibrium states and examine its singularities. This has the effect of making what has usually been called comparative statics more interesting than is often thought, because the behaviour of state variables can be discontinuous for small and smooth changes in parameters. *Second*, construct the 'fast equations'. These are of the form (4) and represent the return to equilibrium of the state variables, $\mathbf{x}$, for fixed values of the parameters. *Third*, specify the 'slow equation': the differential equations for the parameters, $\mathbf{u}$. The solution of these determines the particular trajectory or the equilibrium surface. The *fourth* step involves building in any feedback between the fast and the slow variables, and this has obvious connections to systems analytical approaches to model building. *Fifth*, recognize the existence of noise and possibly make the models stochastic. And, *sixth*, seek to build in diffusion effects, possibly by taking two or three space coordinates, together with time, as the 'parameters' of the model.

We will see in relation to examples below how far it has been possible for geographers to progress along this list (which is in developing order of difficulty). We also note one useful distinction which has been introduced implicitly here: that the state variables are the fast variables, and the parameters the slow. The way in which we categorize variables in this respect, therefore, is determined at least partly by their relative rates of change in time.

Applications in geography: general considerations

Do appropriate phenomena exist?

We can begin by asking whether the unusual types of behaviour discussed above are in fact observed for geographical systems. Brunet (1970), in work which must have largely preceded any knowledge of catastrophe theory, documented a range of examples of discontinuities in geographical phenomena – both physical and human. Amson (1975) identified a number of possibilities which include clustering in a previously dispersed society, the rapid increase in population of a city, the depopulation of inner city areas, the reversals of ethnic zoning patterns (the 'tipping' phenomenon discussed, for example, by Goering, 1978, and Woods, 1977). Atkinson (1976) and Wagstaff (1976) both explored the possibilities at this general level and the former even added a couple of examples of possible divergence phenomena (divergence from zero population growth) to the lists of possible jumps and hysteresis effects (the last of which, in Atkinson's paper, included a hysteresis-representation of spiralling, as for retail systems in Agergard, Olsen and Allpass, 1970). So we can conclude that there are certainly possibilities, especially if we recognize that discontinuous change, in applied work, need not be interpreted to mean instantaneous, but simply very rapid relative to 'before and after' rates of change.

Methods of application

There has been much criticism of catastrophe theory, notably by Zahler and Sussman (1977) and although they in turn can be criticized (Zeeman, 1977b), much of the response is in terms of applications in the natural sciences and there are some justified doubts about the nature of the applied work in the social sciences (Stewart, 1979). Wagstaff has also been criticized from a Marxist perspective (Day and Tivers, 1979), though they in turn may note the advocacy of the positive use of catastrophe theory in dialectical thought (Zwick, 1978). Many of the criticisms turn on the representation of systems in canonical form – as in Figures 18.1 and 18.2 – while the real systems may differ from such forms by very complicated transformations, and in failures to recognize the 'local' nature of Thom's theorem. Many of the applications of catastrophe theory in geography reviewed briefly below are open to this kind of criticism. What we can advocate is that catastrophe theory (and bifurcation theory which does not suffer from the same kinds of criticism) can be interestingly used in geography in at least two ways: firstly, to remind us that unusual forms of dynamic behaviour exist; and secondly, to attempt to construct directly the mechanisms of change within models which it may then be possible to interpret in terms of catastrophe theory.

Scale

It is nearly always important in defining geographical systems for modelling purposes to take care in the treatment of scale. Dynamic modelling is no exception. Indeed, in the case of applications of catastrophe theory, most examples in geography are at either micro or macro scales, and mostly the latter, because it is at coarse scales that it is more likely to be possible to characterize the system in terms of a small number of variables. The interesting geographical scale is often at the meso level, where spatial structure is distinguished in some detail, and where the problems of multiplicity of variables has to be faced. For these reasons, therefore, the accounts of geographical examples which follow are classified according to scale, beginning with micro and macro, in turn, and ending with meso.

Geographical examples

The micro scale

Many of the social science applications of catastrophe theory have been to individual behaviour (see Zeeman, 1977a, for examples). In the geographical and related literature, however, there are relatively few. Here we mention only two. One, due to the author Wilson (1976a), purports to represent individual modal choice, originally based on the fold catastrophe, but in a later minor extension, (Wilson, 1979), employing the cusp. In Figure 18.2, let x be a variable representing choice of mode so that if x is positive, this represents mode 1, if negative, mode 2. u_1 can be taken as a habit factor: when it is negative, habit effects occur and increasingly so as the modulus of u_1 increases; u_2 is taken as a measure of cost differences between the two modes (c_2-c_1, say). Then, when no habit effect occurs, any switch of mode will occur as soon as one mode becomes cheaper than another; but when u_1 is negative, the change will only occur after some interval in favour of the cheaper mode. This is equivalent to trajectory (1) on the figure. But if there is a reversal, and perfect delay applies, this will follow trajectory (2) and the habit factor has created a hysteresis effect. Such an effect has also been proposed from different theoretical foundations by Goodwin (1977) and by Williams and Ortuzar (1979). Empirical evidence for its existence has been discovered by Blase (1979). In this case, the example was developed to help the author to learn something about catastrophe theory rather than as a model with serious pretensions, but this does then have the useful effect of uncovering a phenomenon, in this case hysteresis, which can be built into other models.

The second example is due to Dendrinos (1978a). He shows that the well-known speed flow curve for traffic, which has a fold, can be considered to arise from assumptions about the utility maximizing behaviour of drivers in balancing the effects of speed and congestion.

The macro scale

Most of the examples at the macro scale are concerned fairly directly with modelling urban development, measured by appropriate state variables, in terms of a number of parameters.

The earliest work in this field seems to be that of Amson (1972a, 1972b, 1973a, 1973b, 1977) which is also reviewed by Kilmister (1976). He uses a continuous representation of the urban system and is thus able to reduce its characteristics to densities, rents and the like. His work is important in that he does construct the mechanisms of change directly. He offers alternative models of the equilibrium surfaces of his variables and is then able to interpret the results in terms of catastrophe theory. One particularly important feature is that one of his models is a non-canonical cusp surface where it is possible to give the transformation explicitly to transform it into standard form.

There are a number of authors who use standard catastrophe equilibrium surfaces without specifying in detail the way in which they are derived. Casti and Swain (1975) have as a state variable the level of a settlement in a central place hierarchy and their two parameters, generating a cusp surface, area population and disposable income per capita. Mees (1975) has population as a state variable and four parameters: the difficulty of transport, the average productivity, the difference in productivity between town and country and a crowding factor, and this involves him in using the butterfly catastrophe. Isard (1977) also has population as a state variable and his control variables are increases in productivity and marginal welfare per head. He attempts to interpret the potential function (which he takes in canonical form) as a welfare function in the tradition of economics. The most complicated of the applications of the standard catastrophe surfaces is that due to Dendrinos (1977b) and he has also worked with more explicit models of disequilibrium (1977a, 1978b) following work by authors like Richardson (1974, 1975), Fujita (1976) and Huang, Mueller and Vertinsky (1976). In his 'surface' model, he has two state variables (the quality of the housing stock and the utility of the residents) for a study of slum formation and four control parameters (*per capita* income, the rate of return on investment, the social rate of discount and the population to capital stock density), and this involves the use of one of the umbilic surfaces – the 'mushroom'.

Wagstaff (1978) applies the cusp catastrophe to a development problem in historical geography, and his study is perhaps most notable for his use of the value of the potential function as the variable he wants to predict in his model. The state variable is used to represent location and the parameters are threat of attack and quality of agricultural land respectively.

It is clear that most of the examples discussed briefly here involve theoretical work of a speculative nature and there is rarely, if ever, adequate data to test the theories. However, considerable ingenuity and variety of technique has been displayed and some of these explorations may prove fruitful. At present, of course, they cannot all be correct: if all the parameters which have been identified and used did play truly independent roles in urban or regional development, then they total many more than four and would thus take us outside the realms of elementary catastrophe theory.

The meso scale

The heart of geographical theory is concerned with location, flows and networks and these features need to be represented in models in some detail. This means that, inevitably, a large number of variables is involved and, more importantly, the interesting behaviour arises out of the interdependence of these variables. It is this, together with its non-linear nature, which generates bifurcation behaviour.

In this section, we concentrate mainly on one example as an illustration and then discuss its wider implications and some alternative ways of approaching it. There are two main possibilities for the spatial representation of a geographical system: to treat space continuously (as in Amson, 1973b), or to use a discrete zoning system. Here we will use the latter which avoids restrictive assumptions about exogenously given spatial quantities. The example is based on the Huff (1964) and Lakshmanan and Hansen (1965) model and builds on early work by Harris (1965). This argument is presented in detail by Harris and Wilson (1978) and some numerical work and further theoretical developments by Wilson and Clarke (1979).

The standard model can be taken as:

$$S_{ij} = \frac{e_i P_i W_j^{\alpha} e^{-\beta c_{ij}}}{\sum_k W_k^{\alpha} e^{-\beta c_{ik}}} \tag{5}$$

where S_{ij} is the flow of retail sales from i to j, e_i is the *per capita* income in zone i, P_i is the population of zone i, W_j is the size of shopping facilities in zone j and is taken as a measure of attractiveness, c_{ij} is the cost of travel from i to j and α and β are parameters. The total revenue attracted to j is D_j given by

$$D_j = \sum_i S_{ij} \tag{6}$$

The standard uses of this model are well known – see for example, Wilson (1974, Chapter 4). Here we are interested in adding a hypothesis which will enable us to analyse the dynamics of the development of the structural variables $\{W_j\}$ – the evolution of the pattern of shopping centres. Suppose for a particular zone j that W_j expands if D_j exceeds some cost of supply, kW_j (where k is a unit cost), and decreases otherwise. Then we have behaviour of the form (Wilson, 1976b):

$$\dot{W}_j = \epsilon(D_j - kW_j) \tag{7}$$

for a suitable constant, ϵ, with equilibrium conditions

$$D_j = kW_j \tag{8}$$

If we substitute for S_{ij} from (6) into (7) and for the resulting D_j into (8), we get a set of non-linear simultaneous equations to solve for the equilibrium values of W_j. These are

$$\sum_i \frac{e_i P_i W_j^{\alpha} e^{-\beta c_{ij}}}{\sum_k W_k^{\alpha} e^{-\beta c_{ik}}} = kW_j \tag{9}$$

The left hand side represents revenue and the right hand side cost. These can be plotted as separate functions of W_j. Their intersection then represents the equilibrium value of W_j and we can use (7) to investigate stability. (An investigation of the equilibrium properties of corresponding terms in an intervening opportunities model has been carried out by Chudzynska and Slodkowski, 1979.) This process is exhibited in Figure 18.3 for the case $\alpha > 1$.

Figure 18.3 Revenue and cost functions for two different values of

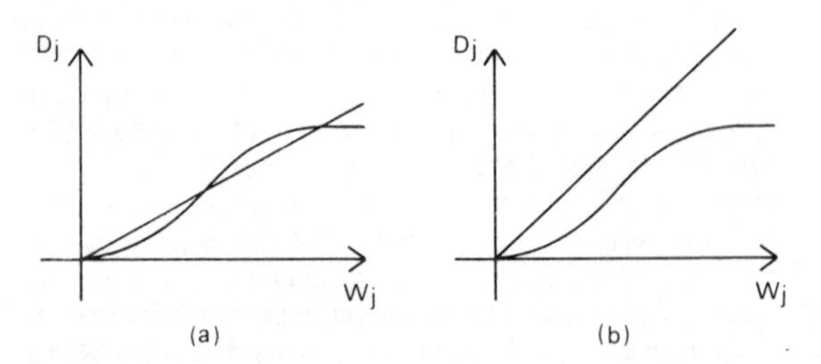

The ringed points are the stable equilibrium points. There are two cases divided by a critical value of k: for low values, the line always intersects the logistic revenue curve; for high values, it does not. The critical value, which is peculiar to this zone and hence is labelled k_j^{crit}, is exhibited when we plot the

Figure 18.4 Stable equilibria of W_j divided by the critical point k_j^{crit}

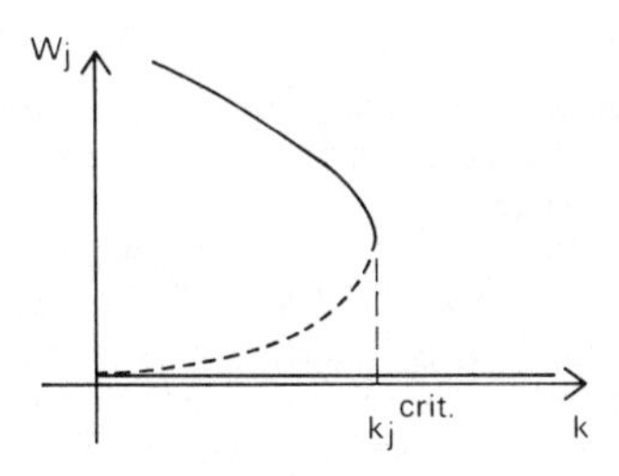

possible stable optima against k in Figure 18.4. This is reminiscent of the fold catastrophe in Figure 18.1.

Since zero is always a possible state, this means that there is not necessarily any development in a zone, but the reverse is not true. (And also note that the 'zero' states have been added to the fold diagram in Figure 18.4.) If k is higher than k_j^{crit}, then no development is possible. We might call these the DP (development possible) and NDP (no development possible) states respectively. Thus, what we get from this analysis is some insight into the conditions which must be satisfied in a zone as to whether development is possible or not and on the size of the development if it takes place. One can imagine an evolving urban system and that, as certain thresholds are passed, say either costs being lowered until the cost line intersects the revenue curve or population increasing so that the curve moves up to intersect the line as potential revenue increases, development occurs on the lines implied by this model.

The problem turns out to be a very complicated one for a number of reasons. First, k is not the only parameter which might change: α, β, $\{e_i\}$ and $\{P_i\}$, for example, all count as 'parameters', and change in any one of them implies that the curves in Figure 18.3 shift relative to each other. All these parameters have critical values for the particular zone, analogous to k_j^{crit}. $\alpha=1$ is also a critical point, and in this case, not a Thom-like cusp, but something more complicated (Amson, 1979). What is worse, however, is that the above analysis was presented on the assumption that all the other W_j's, that is W_k , $k \neq j$, are fixed. This will not be the case: they will all be varying simultaneously and affecting the situation in other zones through competition. It is argued that, for this reason, the basic problem of understanding the dynamics of the evolution of urban structure is a very difficult one, and something like simulation methods are necessary for further progress. In a sense, this can be taken as an encouraging development for theoretical geography, as it is now faced with problems which are just as hard, theoretically, as those of many other sciences.

Two concluding observations can usefully be made. First, this kind of argument can be applied much more widely: for example, to residential location modelling, though the problem is more difficult; and to a comprehensive model which results from stitching the various submodels together and in which retailing and residential 'fields' interact. This can lead to a 'domino effect' of the kind also suggested by Isard and Liossatos, (1978 – and see 1979 for a more detailed treatment.) When these investigations can be completed, then it may be possible to have a fully dynamic and evolutionary central place theory (Wilson, 1977a, 1979).

Secondly, it is appropriate to draw attention to alternative approaches to this particular problem, where other authors have used simulation methods and sometimes different theoretical frameworks, though there remains a family resemblance. White (1974) was an early exponent of this kind of approach to central place theory, and he examined the intersection of a logistic revenue curve and a cost curve though without investigating stability in the manner presented above. He has also carried out a number of interesting simulations (White, 1977, 1978). Another approach altogether is based on the work of Prigogine and his school in Brussels (see, for example, Nicholis and Prigogine, 1977). This involves setting up the appropriate differential equations, adding driving terms and solving them by simulation methods taking particular note to look for bifurcation points. This includes the building in of stochastic variables and is thus nearer to another of the points in Zeeman's list. These methods give particular attention to the logistic growth model (which involves another W_j factor on the right hand side of equation (7)) and this has been interestingly deployed in another context by Glaister (1976). For a detailed account, see Allen, Deneubourg, Sanglier, Boon and de Palma (1978), Allen and Sanglier (1979) and for an application in a different field – to transport mode choice, see Deneubourg, de Palma and Kahn (1979).

Future directions and prospects

Relationships with other disciplines

It is quite revealing to consider the treatment of similar problems in other disciplines both to seek analogies in technique and system similarity and to assess the status of the position reached in geography. The results are largely encouraging in both respects. The discipline with the most similar problems seems to be ecology. Many geographical problems can be seen as arising from interacting and competing populations, and this is also exactly the situation in ecology. For a review and examples, see Maynard Smith (1974), May (1971, 1976, 1978) and Oster and Guckenheimer (1976). Ecological equations exhibit a wide range of types of solution and a corresponding array of bifurcation behaviour. Geographical systems must be at least as complicated in general and so we can expect to make further progress in this direction. The retailing problem, for example, represents a set of producers of shopping centres competing for a fixed supply of customers who are distributed spatially in a non-random and non-regular way; this is analogous to a version of a system of several species competing for a fixed food supply (cf. Rescigno and Richardson, 1967). Both problems can be made more complicated by the fact that the 'food' may be of different types.

There is also an obvious affinity with the study of equilibrium problems in economics (see, for example, Balasko, 1978; and also Hotelling, 1929, provides an early example of an interest in locational stability), and also in physics and chemistry. We have already mentioned Prigogine's work in another context. Specifically, some of his 'order-from-fluctuations-far-from-equilibrium' structures are spatial as well as temporal and would seem to have clear geographical utility for formally similar systems, though the similarities have not yet been spelled out. In chemistry, the constituents of a mixture also appear like species competing for a fixed food supply, where the 'foods' are energy and materials.

It is interesting to take a specific example and consider the analogies across several disciplines which all involve the use of the logistic mechanism spelled out above for retailing structures. There are competition models in ecology working on this basis, an example of which is spelled out by Poston and Stewart (1978) for social and solitary bees. It arises in physical chemistry, where a transition across a critical point may be an explosion (Boddington, Gray and Wake, 1977). In archeology, it is at the basis of investigations of sudden transitions from dispersed to nuclear communities (Renfrew and Poston, 1978) which were once thought to be the result of attacks, but which can now be seen as arising from smooth changes in economic parameters. Economic examples arise in the study of fishery policies (Clark, 1976) and in the study of business cycles (Varian, 1979). There are also alternative geographical examples first concerned with centre size which can be used to help explain the transitions between 'corner-shop' and 'supermarket' retailing economies (Poston and Wilson, 1977) and secondly with mechanisms of urban development (Papageorgiou, 1979). A more complicated argument generating discontinuities from changing 'ease of travel' is that of Smith (1977).

The main conclusion to be drawn from this subsection is that geographical theorists are in the vanguard of an attack on interesting problems which are common to a number of disciplines – those which probably involve the modelling of what have elsewhere been called Weaver-III problems of organized complexity (Weaver, 1958; Wilson, 1977b).

Concepts to be developed

We saw in an earlier section that **x**-variables were distinguished from **u**-variables by their average time rates of change. It is reasonable to suppose that there may also be, say, **y**-variables which are 'slower' than the **u**-variables, and indeed a whole hierarchy of structure in this sense. These notions are also closely related to those of lags and the specifications of the delay conventions which are operating – above we have tended to assume perfect delay in the examples mentioned, and more complicated schemes are possible. Thus, the first area of broad concern can be summarized as related to the specification of the detailed dynamical structure of a model.

The second relates to Zeeman's six steps. Perhaps the two where least progress has been made in geographical modelling are the third and the sixth: we very rarely specify **u**-differential equations in any detail, nor have we pursued the construction of models which incorporate diffusion processes.

Finally, one of Thom's ambitions in the development of applied catastrophe theory was the construction of a theory of morphogenesis – the creation of new structures. Some simple examples have been explored in the geographical field (Wilson, 1978) but the underlying theoretical problems are very difficult. This is another case where explorations in other disciplines may offer new insights for geographers.

The next steps

The first important general conclusion to be drawn is that enough is now known about bifurcation phenomena to make it worth while to examine many of the equations which turn up in geographical theory in this respect. For example, the accounting or kinetic equations used by authors like Cordey-Hayes (1972) and Tomlin (1979) could be investigated in this way if their transition coefficients were made functions of other system variables. Then, the interdependencies and non-linearities which generate bifurcation behaviour would be made explicit. Similar comments apply to spatial population models in which the migration flows are made explicit (cf. Rees and Wilson, 1977) and to many of the equations developed in spatial time series analysis, with ultimate extensions to control theory (cf. Bennett and Chorley, 1978, Bennett,

1979).

In general, we can note that the next steps forward will be difficult for a variety of related reasons. First, the techniques involved are difficult, and just as geographers thought they were coping with the results of the last round of quantification, there is a demand to study dynamics from a topological viewpoint! Second, it is difficult to make progress unless some practical work is carried out. This will usually involve simulation methods and various ways of incorporating approximations to theoretical models. But in this way substantial new theoretical problems emerge as well as new insights (as in Wilson and Clarke, 1979). In the first instance, and perhaps for a considerable time, data will be woefully inadequate for these purposes. Not only is time series data needed, but data which is on a sufficiently fine temporal scale for jumps and other such phenomena to be identified. All this is particularly difficult in relation to potential applications in planning. The emphasis on criticality of parameters, the knowledge of which could then be used either to maintain systems in some state or to encourage them to develop, has an obvious value. But while insights as to potential mechanisms and changes of this type may be valuable, what would be even more so is a knowledge of the actual critical values of parameters, and this may be much harder to come by.

It is perhaps appropriate in a volume which is reviewing progress in the development of quantitative methods to conclude that there is at least one field in which, while building on the results of past work, there exist some of the most difficult problems in modern science.

References

Agergard, C., Olsen, P.A., and Allpass, J. (1970) 'The interaction between retailing and the urban centre structure: a theory of spiral movement', *Environment and Planning*, 2, 55–71.

Allen, P.M., and Sanglier, M. (1979) 'A dynamic model of growth in a central place system', *Geographical Analysis*, 11, 256–72.

Allen, P.M., Deneubourg, J.L., Sanglier, M., Boon, F., and de Palma, A. (1978) *The dynamics of urban evolution, Vol. 1: Interurban evolution, Vol. 2: Intraurban evolution*, Final Report to the US Department of Transportation, Washington DC.

Amson, J.C., (1972a) 'The dependence of population distribution on location costs', *Environment and Planning*, 4, 163–81.

Amson, J.C., (1972b) 'Equilibrium models of cities: 1. An axiomatic theory', *Environment and Planning*, 4, 429–44.

Amson, J.C. (1973a) 'Equilibrium models of cities: 2. Single species cities', *Environment and Planning*, 5, 295–338.

Amson, J.C. (1973b) 'Equilibrium and catastrophic modes of urban growth' in E.L. Cripps (ed.), *Space-time concepts in urban and regional models*, 108–28, Pion: London.

Amson, J.C. (1975) 'Catastrophe theory: a contribution to the study of urban problems?', *Environment and Planning, B*, 2, 177–221.

Amson, J.C. (1977) 'A note on civic state equations', *Environment and Planning A*, 9, 105–10.

Amson, J.C. (1979) Private communication.

Atkinson, G. (1976) 'Catastrophe theory in geography – a new look at some old problems', mimeo, Department of Geography, University of Cambridge.

Balasko, Y. (1978) 'The behaviour of economic equilibria: a catastrophe theory approach', *Behavioural Science*, 23, 375–82.

Bennett, R.J. (1979) *Spatial time series*, Pion: London.

Bennett, R.J., and Chorley, R.J. (1978) *Environmental Systems*, Methuen: London.

Blase, J.H. (1979) 'Hysteresis and catastrophe theory: empirical identification in transport modelling', *Environment and Planning, A*, 11, 675–88.

Boddington, T., Gray, P., and Wake, G.C. (1977) 'Criteria for thermal explosions with and without reactant consumptions', *Proceedings of the Royal Society of London, A*, 357, 403–22.

Brunet, R. (1970) *Les phénomènes de discontinuité en géographie*, Centre National de la Recherche Scientifique: Paris.

Casti, J., and Swain, H. (1975) 'Catastrophe theory and urban processes', RM-75-14, International Institute for Applied Systems Analysis: Laxenburg, Austria.

Chudzynska, I., and Slodkowski, Z. (1979) 'Comments on the urban spatial-interaction model based on the intervening-opportunities principle', *Environment and Planning, A*, 11, 527–39.

Clark, C.W. (1976) *Mathematical bioeconomics*, John Wiley: New York.

Cordey-Hayes, M. (1972) 'Dynamic framework for spatial models', *Socio-Economic Planning Sciences*, 6, 365–85.

Day, M., and Tivers, J. (1979) 'Catastrophe theory and geography: a Marxist critique', *Area*, 11, 54–8.

Dendrinos, D.S. (1977a) 'Short-run disequilibria in urban spatial structures', *Regional Science Perspectives*, 7 (2), 27–41.

Dendrinos, D.S. (1977b) 'Slums in capitalist urban

settings: some insights from catastrophe theory', *Geographica Polonica*, 42, 63-75.

Dendrinos, D.S. (1978a) 'Operating speeds and volume to capacity ratios: the observed relationship and the fold catastrophe', *Transportation Research*, 12, 191–4.

Dendrinos, D.S. (1978b) 'Urban dynamics and urban cycles', *Environment and Planning, A*, 10, 43–9.

Deneubourg, J.L., de Palma, A., and Kahn, D. (1979) 'Dynamic models of competition between transport modes', *Environment and Planning, A*, 11, 665–73.

Fujita, M. (1976) 'Spatial patterns of urban growth: optimum and market', *Journal of Urban Economics*, 3, 239–41.

Glaister, S. (1976) 'Transport pricing policies and efficient urban growth', *Journal of Public Economics*, 5, 103–17.

Goering, J.M. (1978) 'Neighbourhood tipping and racial transition: a review of social science evidence', *Journal of the American Institute of Planners*, 44, 68–78.

Goodwin, P.B. (1977) 'Habit and hysteresis in modal choice', *Urban Studies*, 14, 95–8.

Harris, B. (1965) 'A model of locational equilibrium for the retail trade', mimeo, Institute for Urban Studies, University of Pennsylvania, Philadelphia.

Harris, B., and Wilson, A.G. (1978) 'Equilibrium values and dynamics of attractiveness terms in production-constrained spatial-interaction models', *Environment and Planning, A*, 10, 371–88.

Hirsch, M.W., and Smale, S. (1974) *Differential equations, dynamical systems and linear algebra*, Academic Press: New York.

Hotelling, H. (1929) 'Stability in competition', *Economic Journal*, 39, 41–57.

Huang, C., Mueller, D., and Vertinsky, I. (1976) 'Urban systems dynamics: a comparative static analysis of system sizes with some policy implications', *Behavioural Science*, 21, 263–73.

Huff, D.L. (1964) 'Defining and estimating a trading area', *Journal of Marketing*, 28, 34–8.

Isard, W. (1977) 'Strategic elements of a theory of major structure change', *Papers, Regional Science Association*, 38, 1-14.

Isard, W., and Liossatos, P. (1978) 'A simplistic multiple growth model', *Papers, Regional Science Association*, 41, 7-13.

Isard, W., and Liossatos, P. (1979) *Spatial dynamics and optimal space-time development*, North-Holland: Amsterdam.

Kilmister, C.W. (1976) 'Population in cities', *Mathematical Gazette*, 60, 11–24.

Lakshmanan, T.R., and Hansen, W.G. (1965) 'A retail market potential model', *Journal of the American Institute of Planners*, 31, 134–43.

May, R.M. (1971) 'Stability in multi-species community models', *Mathematical Biosciences*, 12, 59–79.

May, R.M. (1976) 'Simple mathematical models with very complicated dynamics', *Nature*, 261, 459–67.

May, R.M. (1978) 'The evolution of ecological systems', *Scientific American*, 239, No. 3, 161–75.

Maynard Smith, J. (1974) *Models in ecology*, Cambridge University Press.

Mees, A.I. (1975) 'The revival of cities in medieval Europe – an application of catastrophe theory', *Regional Science and Urban Economics*, 5, 403–26.

Nicholis, G., and Prigogine, I. (1977) *Self-organisation in non-equilibrium systems*, John Wiley: New York.

Oster, G., and Guckenheimer, J. (1976) 'Bifurcation phenomena in population models' in J.E. Marsden and M. McCracken (eds), *The Hopf bifurcation and its applications*, Springer-Verlag: New York.

Papageorgiou, G. (1979) Private communication.

Poston, T., and Stewart, I.N. (1978) *Catastrophe theory and its applications*, Pitman: London.

Poston, T., and Wilson, A.G. (1977) 'Facility size vs. distance travelled: urban services and the fold catastrophe', *Environment and Planning, A*, 9, 681–6.

Rees, P.H., and Wilson, A.G. (1977) *Spatial population analysis*, Edward Arnold: London; Academic Press: New York.

Renfrew, A.C., and Poston, T. (1978), 'Discontinuous change in settlement pattern – a processual analysis' in A.C. Renfrew and K.L. Cooke (eds), *Transformations: a mathematical approach to culture change*, Academic Press: London.

Rescigno, A., and Richardson, I. (1967) 'Struggle for life I: two species', *Bulletin of Mathematical Biophysics*, 29, 377–88.

Richardson, H.W. (1974) 'Two disequilibrium models of regional growth' in E.L. Cripps (ed.), *Space-time concepts in urban and regional models*, 46–55, Pion: London.

Richardson, H.W. (1975) 'Discontinuous densities, urban spatial structure and growth: a new approach', *Land Economics*, 51, 305–15.

Smith, T.R. (1977) 'Continuous and discontinuous response to smoothly decreasing effective distance: an analysis with special reference to "overbanking" in the 1920s', *Environment and Planning, A*, 9, 461–75.

Stewart, I.N. (1979) Letter to the Editor, *Nature*, 270, 382.

Thom, R. (1975) *Structural stability and morphogenesis*, W.A. Benjamin, Reading, Mass.

Tomlin, S.G. (1979) 'A kinetic theory of urban dynamics', *Environment and Planning, A*, 11, 97–106.

Varian, H.L. (1979) 'Catastrophe theory and the

business cycle', *Economic Inquiry*, 17, 14–28.

Wagstaff, J.M. (1976) 'Some thoughts about geography and catastrophe theory', *Area*, 8, 316–20.

Wagstaff, J.M. (1978) 'A possible interpretation of settlement pattern evolution in terms of catastrophe theory', *Transactions, Institute of British Geographers*, NS 3, 165–78.

Weaver, W. (1958) 'A quarter century in the natural sciences', Annual Report, 7–122, Rockefeller Foundation, New York.

White, R.W. (1974) 'Sketches of a dynamic central place theory', *Economic Geography*, 50, 219–27.

White, R.W. (1977) 'Dynamic central place theory: results of a simulation approach', *Geographical Analysis*, 9, 226–43.

White, R.W. (1978) 'The simulation of central place dynamics: two sector systems and the rank-size distribution', *Geographical Analysis*, 10, 201–8.

Williams, H.C.W.L., and Ortuzar, J.D. (1979) 'Behavioural travel theories, model specification and the response error problem', Working Paper 116, Institute for Transport Studies, University of Leeds.

Wilson, A.G. (1974) *Urban and regional models in geography and planning*, John Wiley: Chichester.

Wilson, A.G. (1976a) 'Catastrophe theory and urban modelling: an application to modal choice', *Environment and Planning, A*, 8, 351–6.

Wilson, A.G. (1976b) 'Retailers' profits and consumers' welfare in a spatial interaction shopping model' in I. Masser (ed.), *Theory and practice in regional science*, 42–59, Pion: London.

Wilson, A.G. (1977a) 'Spatial interaction and settlement structure: towards an explicit central place theory' in A. Karlqvist, L. Lundqvist, F. Snickars and J. Weibull (eds), *Spatial interaction theory and planning models*, 137–56, North-Holland: Amsterdam.

Wilson, A.G. (1977b) 'Recent developments in urban and regional modelling: towards an articulation of systems' theoretical foundations', *Giornale di Lavoro*, AIRO, Parma.

Wilson, A.G. (1978) 'Towards models of the evolution and genesis of urban structure' in R.J. Bennett, R.L. Martin and N.J. Thrift (eds), *Towards the dynamic analysis of spatial systems*, Pion: London.

Wilson, A.G. (1979) 'Aspects of catastrophe theory and bifurcation theory in regional science', Working Paper 249, School of Geography, University of Leeds.

Wilson, A.G., and Clarke, M. (1979) 'Some illustrations of catastrophe theory applied to urban retailing structures' in M. Breheny (ed.), *London Papers in Regional Science*, Pion: London (Working Paper 228, School of Geography, University of Leeds).

Woods, R.I. (1977) 'Population turnover, tipping points and Markov chains', *Transactions, Institute of British Geographers*, 473-89.

Zahler, R.S., and Sussman, H.J. (1977) 'Claims and accomplishments of applied catastrophe theory', *Nature*, 269, 759–63.

Zeeman, E.C. (1977a) *Catastrophe theory*, Addison-Wesley, Reading, Mass.

Zeeman, E.C. (1977b) Letter to the Editor, *Nature*, 270, 381.

Zwick, M. (1978) 'Dialectics and catastrophe' in R.F. Geyer and T. van der Zouven (eds), *Sociocybernetics, Vol. 1*, 129–54, Martinus Nujhoff: Leiden.

Chapter 19

Stochastic processes

E. Culling

Introduction

Since 1940 probability theory has passed through a golden age. Although there were precursors the take-off point is marked by the presentation of Kolmogorov's axiomatic treatment in 1933. This memoir established the theory of probability as one aspect of measure theory and apart from certain ideas peculiar to probability such as independence the two theories are equivalent. As a branch of pure mathematics probability theory has no intrinsic connection with the real world and terms such as 'occurrence', 'event', 'coin', etc. are redundant. However, these terms were well established before measure theory was thought of and they provide an intuitive framework, so that it would be foolish, even if possible, to dispense with their use.

The Kolmogorov axioms characterize a field of probability which, briefly, can be described as setting up for every activity (observation, realization, etc.) in which there is a random element, an associated probability space or triple, (Ω, F, P) where Ω is an abstract space, F is a field of sub-sets of Ω (set of events), and P(E) is a measure (the probability of event E) (Kolmogorov, 1933; Cramer, 1936). Despite problems prompted by quantum mechanics Kolmogorov's model is still the seminal document for probability theory. The preferences of philosophers, psychologists and some statisticians notwithstanding, it is 'considered by workers in probability and statistics to be the correct one' (Bharucha-Reid, 1956, p.77).

One result of the establishment of probability theory as an accepted branch of mathematics has been the gradual disentanglement from statistics. Yet in geography the majority of students approach probability as a chapter in a statistics textbook, seemingly based on an empirical study of parlour games. The geographical tradition potentiated by this teaching bias tends to foster a descriptive, inferential use of stochastic processes.

The general idea of a stochastic (random) process is of any process (ordered set) in time, subject to probabilistic laws. More precisely a stochastic process is a family of random variables $\{x_t, t \in T\}$. The x_t are numerical values of the random variable x at time t within the range T. The symbols t and T are traditional. There is no necessity that they should represent time, in particular they can be sequences or intervals in Euclidean space. Thus the idea of a stochastic process depends upon that of a random variable and from the axiomatic standpoint, a random variable is defined as a measurable function and the theory of probability is entirely concerned with the measurable properties of such functions on various abstract spaces. Because of the paramount importance of the applications the spaces are in most cases appropriately termed sample spaces and their measurable point sets taken as abstractions of events (Doob, 1953, p.5).

We have no space to give a summary of the various classes of stochastic process. Short reviews in note form are often found as appendices (e.g. Anderson and Moore, 1979, p.316). A short readable review with an electrical engineering bias in Lin (1969). There are many excellent textbooks on stochastic processes available to the geographer and of these we mention Parzen (1962), Bartholomew (1967), Bartlett (1966) and Cox and Miller (1965). For probability theory in general the best text is still Feller (1957), and for Markov processes in particular Bharucha-Reid (1960). For the more advanced student there is the seminal work of Doob (1953) on stochastic processes and the encyclopaedic coverage of Loeve (1963 and 1977). Markov processes are dealt with by Chung (1967) and Dynkin (1965) and stationary processes by Yaglom (1963) and Cramer and Leadbetter (1967).

Stochastic processes in geography

It comes as somewhat of a surprise that probabilistic ideas have been used in geography for over twenty-five years. T. Hägerstrand and the Lund school were treating certain spatial problems by Monte Carlo methods within a broad diffusion framework in the early 1950s. Diffusion in geography has a wider connotation than is customary and is synonymous with spreading. Developments since those early days have been summarized in two recent articles, Gudgin and Thornes (1974) and Collins, Drewett and Ferguson (1974); the former is strong on the physical aspects, the latter concentrates on Markov methods in human geography. These articles carry full references and when these are followed up it is discovered that a large proportion are also reviews, a fact not unnoticed before (Gould, 1969, p.19). Rather than add to the burden the reader is referred to the two articles and to the earlier reviews of Gould (1969, pp.15–20), Collins (1973, pp.124–6), Brown and Moore (1969), Hepple (1974) and Cliff and Ord (1975). Before discussing future prospects we make a few comments prompted by Collins, Drewett and Ferguson (1974) and mention Brunsden and Thornes (1977).

Under the heading of attractiveness to geographers the authors state that probability models are appropriate 'because concepts and measurements tend to be inexact; relationships are often complicated and poorly understood and human behaviour is unpredictable'. They go on to say that it is often difficult to derive a deterministic model to take into account all the relevant variables and to treat each component individually as a random variable is also difficult so that it is advantageous to regard the process stochastically. This is an admirably clear statement of the motivation of most geographers, alas. In listing the advantages they reveal the other side of the coin.

Firstly, the impression is given that probabilistic methods are second-best. Earlier we dealt with probability theory as a mathematical theory in its own right and not a fuzzy substandard version of real variable theory. Furthermore, progress in geography as in science as a whole is never more than increasing indeterminacy (Neyman, 1960). Secondly, the motivation is to a black box approach. In practical applications such an approach centering on description and prediction is often appropriate but as scientists our primary aim is to understand. We live in a 'black-box' society and the temptation is to avoid the effort or to disguise from ourselves that the task is unfulfilled.

Since the review of Gudgin and Thornes (1974), a book has appeared entitled *Geomorphology and Time* by Brunsden and Thornes (1977), in which a chapter on stochastic models and form evolution is recommended to the reader as an introduction to stochastic methods in geomorphology. The authors distinguish five basic types of model as used in geomorphology: Monte Carlo techniques and digital stochastic simulation, stochastic processes, confined to discrete parameter and state space, continuous processes, and entropy maximization. This latter category requires a comment. Entropy models in geography depend upon analogies. It is almost a rule of methodology that arguments from analogy should be such that the degree of abstraction and generality be the same on both sides. To apply a well defined concept from the exact physical sciences to a broad area of geography at a lower level of abstraction is to court disappointment.

Two well-known applications in fluvial morphology, those of Leopold and Langbein (1962) and of Yang (1971), have been the subject of a recent re-evaluation (Davy and Davies, 1979). The conclusion reached is that the application of the specific entropy concepts to stream systems is illegitimate because the character of the systems lies outside the proper domain of the principles and assumptions governing entropy behaviour. A third well-known analogy is that between elevation and temperature (Scheidegger, 1964, 1970). I have formerly thought this analogy to be of value but would now find a closer analogy both in terms of abstraction and definition between the thermal vibration of molecules in a liquid or amorphous solid and 'vibrational' random activity of soil particles. Even so the parallel requires considerable theoretical justification that in practice amounts to an independent reformulation of the concept.

A more profound misgiving over the use of entropy ideas in irreversible processes is that the latter are founded in statistical mechanical ideas that ultimately reside in reversible behaviour. This gives rise to a pair of celebrated paradoxes, those of Zermelo and of Lotschmidt, that bid fair to wreck Boltzmann's efforts at the turn of the century to found a statistical mechanical theory of entropy. The generally accepted resolution is that of Smoluchowski who characterizes irreversible processes as reversible processes with enormously long recurrence period (Chandrasekhar, 1943, p.54). In the passage from a postulated elemental random behaviour to the apparent determinism of the differential equations governing the ensemble, some assumption has to be made to ensure irreversibility, even if, as is usually the case, it is made unconsciously.

Prospects

It is common practice among geographers as among practitioners of other disciplines to bewail the lack of communication between the various branches of the subject and to deplore the slow acceptance of new ideas and methods. This phenomenon is too widespread and of too remote an ancestry to be merely a matter of fashion, inertia or academic myopia, though all these have their contribution to make. Rather we must seek its roots in the idea of 'readiness'. The concept is Aristotelian. Taking the child as his model, by whom such natural actions as walking and talking will be undertaken when the child is physically and psychologically ready and not before, he applied the idea to the body politic. We can apply it to geography and no doubt we all have our own favourite examples. Bearing this in mind we turn to discuss prospects for the future. Rather than conduct a superficial survey across the board we focus upon two topics, the one close upon us, the other still some way below the horizon. In both, however, we come across some interesting examples of unreadiness.

Kalman filtering

The Kalman filter is a powerful recursive technique for estimating the state of a dynamic system and has been in existence for nearly two decades. It is in the mainstream of linear least-squares estimation and as such has an illustrious lineage extending back through Norbert Wiener to Gauss. Originating in control engineering, it soon received extensive and well publicized applications in aerospace. During the last decade its use has spread to environmental engineering, econometrics, management science and a wide spectrum of geophysics but in none of these has its application been as intensive as in hydrology and hydraulic research, wherein it is beginning to impinge upon the domain of geographers. In contrast its uptake by statisticians has been very slow and its development in general presents many curious episodes.

For its whole-hearted adoption by the hydrologists there appears to be a good reason. The invention of a new instrument or a new technique leads not only to a spate of fresh discoveries but also provokes new theoretical problems. The adoption of continuous and accurate recording presents in a highly visual form the detailed fluctuation in the intensity of natural phenomena. By the very nature of their material electronic engineers have been in the dynamic indeterminacy stage (Neyman, 1960) from the outset. In the last ten to fifteen years hydrologists have enjoyed an abundance of continuously recorded data and in some respects hydrolic research has been instrument-driven. Geomorphology and possibly other aspects of geography are now entering the stage of dynamic indeterminacy and it is as well to get the theoretical aspects clear lest the instruments take over. Furthermore the recording of natural phenomena is becoming more precise with consequent noise-to-signal problems. For these reasons alone linear least-squares estimation and the Kalman filter in particular will become of increasing importance in geography. To date I am aware of only one application by a geographer: the use of a derivation of Kalman filtering in an adaptive model of river channel change (Bennett, 1976).

Estimating procedures have been found amongst the Babylonians (Neugebauer, 1952) and Galileo inaugurated the 'modern' period of error minimization but it is to the young Gauss that priority is given for the solution of problems in planetary orbit measurement by linear least-squares methods. The Gauss-Legendre-Adrain priority squabble is covered by Sorenson (1970), together with a discussion, with extracts from the 'Theoria Motus', on Gauss's motivation (Gauss, 1963). The best mathematical summary is still Whittaker and Robinson (1926, Ch.IX). For a 'statistical' historical account see Seal (1967).

Briefly the linear least-squares procedure is to determine the most probable value of a parameter $\mathbf{x}$, ($\hat{\mathbf{x}}_n$), from a set of observations, $\mathbf{z}_k$, where the relationship is linear,

$$\mathbf{z}_k = \mathbf{H}_k\mathbf{x} + \mathbf{v}_k ,$$

and where the $\mathbf{v}_k$ are measurement errors. The most probable value is defined as that which minimizes the sum of the squares of the residuals,

$$\mathbf{r}_k = \mathbf{z}_k - \mathbf{H}_k\hat{\mathbf{x}}_n ,$$

that is,

$$L_n = \tfrac{1}{2} \sum_{k=o}^{n} \mathbf{r}_k{}' \mathbf{W}_k \mathbf{r}_k$$

is minimized. Gauss argued that the residuals would have the density function

$$f(r_k) = (2\Pi)^{-n/2} \; |W|^{1/2} \exp \left[-\tfrac{1}{2} r' W_k r_k \right] .$$

During the last war Wiener produced his famous 'yellow peril' report on linear mean-square estimation (Wiener, 1949). Independently Kolmogorov (1941a, b) reached the same result in the discrete case using Hilbert space methods. The estimation problem was generalized by Wiener to the case of continuous measurement and the most probable performance index replaced by a stochastic version of the least-squares criterion. Essentially the problem is that of disinterring a signal $\mathbf{s}_n$, assumed stationary, from noisy observations $\mathbf{z}_n$, where knowledge of the auto- and cross-correlations is known. The estimate of $\mathbf{s}_n$ after n observations, $\hat{\mathbf{s}}_{n/n}$, is assumed to be a linear combination of the $\mathbf{z}_i$,

$$\hat{\mathbf{s}}_{n/n} = \sum_{i=o}^{n} \mathbf{H}_{n,i} \, \mathbf{z}_i ,$$

where the $\mathbf{H}_{n,i}$, the 'filter gain' are to be chosen to minimize the mean-square error, i.e. to minimize,

$$M_n = E\left[(s_n - \hat{s}_{n/n})'(s_n - \hat{s}_{n/n})\right] .$$

As is well known the necessary and sufficient conditions for this optimal filter are that the errors in the estimate are orthogonal to the observations,

$$E\left[(s_n - \hat{s}_{n/n})\, z_n{}'\right] = 0 \qquad i = 0, 1, \ldots . n.$$

and this implies a Wiener-Hopf relationship,

$$E\left[s_n z_i{}'\right] = \sum_{j=o}^{n} H_{n,j}\, E\left[z_j z_i{}'\right] ,$$

$i = 0, 1, \ldots n.$

These equations are difficult to solve. In the discrete case, as above, the vector matrix equation is intractable for large n. While in the continuous case the solution of the resulting integral equation requires spectral factorization techniques (Masani, 1966). Towards the end of the 1950s the demand of the aerospace industry and the military, the availability of transistorized computers and despite many attempts at amelioration the computational difficulties of the Wiener filter, conspired to bring about the next step forward; a recursive solution to the least-squares estimation problem in a state space representation. This phase culminated in the Kalman filter (Kalman, 1960).

One again we witness a priority squabble. The parallel with 150 years before is furthered in that one of the claimants, Swerling (1958), was working on problems of tracking satellite orbits. Another possible claimant, Follin, proposed the idea of a recursive filter (Carlton and Follin, 1956), but this idea had already been used in a least-squares context by Plackett (1950). Plackett's paper has a statistical motivation and deals with the problem of adjusting an estimate upon the advent of fresh data. The worked example underlines the 'static experimental' approach (Neyman, 1960), and the paper is written as if Wiener and Kolmogorov had never existed. After two decades of neglect Plackett's paper gets a ritual mention from every British author though as far as I am aware has never been mentioned in an American paper on the topic. Despite several other claimants it is now generally agreed that Kalman is rightly credited as realizing the potential and presenting the ideas in a unitary format within the general framework of least-squares estimation. For reasons already given the topic was 'in the air' but the whole episode provides a fascinating commentary on scientific tunnel vision.

An historical survey of the Wiener-Kolmogorov predictor theory and subsequent developments is given in Kailath (1974). This paper is used as an introduction to a collection of all the relevant papers (Kailath, 1977), apart from Wiener's contribution (Wiener, 1949) and the Kalman and Bucy paper (1961) on the continuous parameter case, which is reprinted in another collection (Ephemerides and Thomas, 1973). The relationship between Plackett's recursive algorithm and Kalman filtering is covered by Young (1974).

Although the two approaches are mathematically equivalent the Kalman filter is much more convenient for geographers, the majority of whom are unlikely to have the mathematical equipment necessary for the Wiener filter. Firstly it is recursive and all that this implies in computational economy. Secondly the change from the Fourier domain to a state space representation implies the solution of a difference or differential equation rather than an integral equation. Furthermore, in the discrete case the equations can be translated directly into computer language, in fact the updating can be computed as a drill routine without any knowledge of the derivation. This is an example of 'algorithmic efficiency' and is unfortunate because the Kalman filter focuses attention on the underlying system rather than an input–output black-box formulation.

One further point before we give a brief outline of the Kalman filter. The Wiener filtering problem is not a statistical problem, it belongs to the theory of probability: 'Wiener's approach, as ours, requires that the probabilistic structure of the random processes be known exactly. Therefore confidence limits, statistical decision rules, etc., do not enter the picture' (Kalman, 1963, p.271). This is only to revert to Gauss's original intentions for linear least-squares estimation in that it was for the refinement of the estimate of the parameters of an orbit equation already known and not a technique of discovery.

We summarize the Kalman filtering problem and its solution for the discrete parameter case (Gauss-Markov sequence). The basic signal model comprises the state equation,

$$x_{k+1} = F_k x_k + w_k$$

and the observed equation,

$$z_k = H_k x_k + v_k ,$$

where w_k and v_k represent, respectively, the system noise and the observation noise and are assumed to be independent Gaussian processes of zero mean and,

$$E\left[v_k v_j{}'\right] = V_k \delta_{kj}$$

$$E\left[w_k w_j{}'\right] = W_k \delta_{kj}$$

$$E\left[v_k w_j{}'\right] = 0 \text{ for all k and j.}$$

The initial state x_o has mean value $x_{o/-1} = \bar{x}_o$ and the covariance matrix $P_{o/-1} = P_o$, independent of the system and observational noise sequences. Both $\bar{x}_o$ and P_o are given or estimated off-line.

The estimation problem is to determine

$$\hat{x}_{k/k-1} = E\left[x_k \mid z^{(k-1)}\right],$$

where $z^{(k-1)}$ denotes the sequence of observations $z_o, z_1, \ldots, z_{k-1}$. The estimate is measured by the error covariance matrix,

$$P_{k/k-1} = E\left\{(x_k - x_{k/k-1})(x_k - x_{k/k-1})' \mid z^{(k-1)}\right\}$$

of which the trace,

$$E\left\{\| x_k - x_{k/k-1} \|^2 \mid z^{(k-1)}\right\},$$

is the conditional error variance and is to be minimized by the conditional mean estimate.

The filter equations are then,

$$x_{k+1/k} = \left[F_k - K_k H_k'\right] \hat{x}_{k/k-1} + K_k z_k .$$

The gain K_k,

$$K_k = F_k P_{k/k-1} H_k \left[H_k' P_{k/k-1} H_k + V_k\right]^{-1}$$

is dependent upon the error covariance matrix given recursively by the Riccati type equation,

$$P_{k+1/k} = F_k \left\{P_{k/k-1} - P_{k/k-1} H_k \left[H_k' P_{k/k-1} H_k + V_k\right]^{-1} H_k' P_{k/k-1}\right\} F_k' + W_k.$$

The diagram (Fig. 19.1) has been drawn to show the Kalman filter as a copy of the original system driven by the estimation error (Kalman, 1963, p.302). This latter characteristic is brought out more clearly when we recast the calculations in algorithmic form (Fig. 19.2).

It can be shown that for the Gauss-Markov case the Kalman filter is the optimal filter. Deviation from the Gaussian case may provoke a non-linear optimum but if constrained to a linear filter then in all cases the Kalman filter is the optimal filter.

Paradoxically the Kalman filter has found its greatest application in non-linear systems and this applies especially to geophysical applications. In such cases an approximate solution is assumed and deviations are described by linear systems. The approximate linear model is then taken as the basis for the Kalman filter. Although Kalman has shown that the linear application is stable for quite wide conditions certain non-linear systems are liable to divergence. This is where the filter provides a small estimate of the error covariance matrix P_k which in turn causes the gain matrix K_k to be small with the result that fresh observations are given little weight and the filter runs on the system alone. This will be the case if W_k, the covariance of the system noise, is underestimated or ignored altogether with the result that errors cumulate within the system and estimate accuracy deteriorates. To reduce approximation errors the so-called extended Kalman filter is used in practice. Here the non-linear system is linearized by employing the best estimates of the state vector as the reference values used at each stage in the linearization. An example in a river quality study is given by Young (1974).

An important application of state space models in geography concerns the non-replicability of experiments on natural phenomena. The predictor can be used to estimate what would have otherwise happened and what the response of the chosen model is to intervention compared with observation (Bennett, 1976). In this connection we mention a development by Harrison and Stevens (1976) in

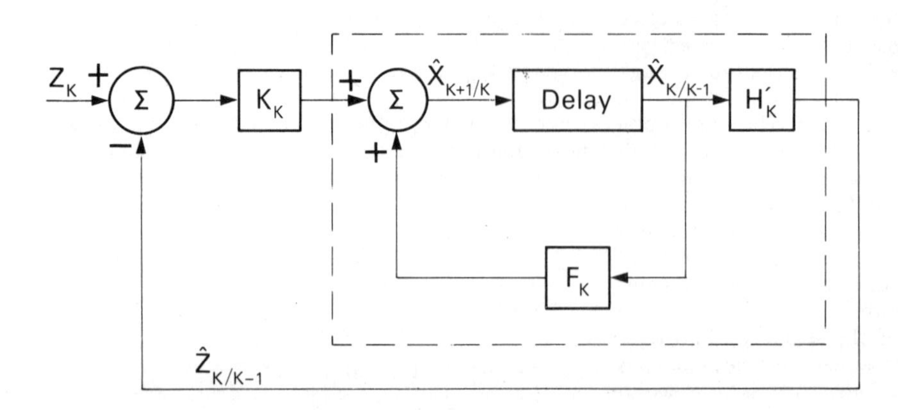

Figure 19.1 Kalman filter as a copy of the original system driven by the estimation error

which a set of response forms of pre-set probability are selected by the model in a Bayesian manner from the observational trace.

Figure 19.2 **Kalman filter algorithm (discrete Gauss- Markov case)**

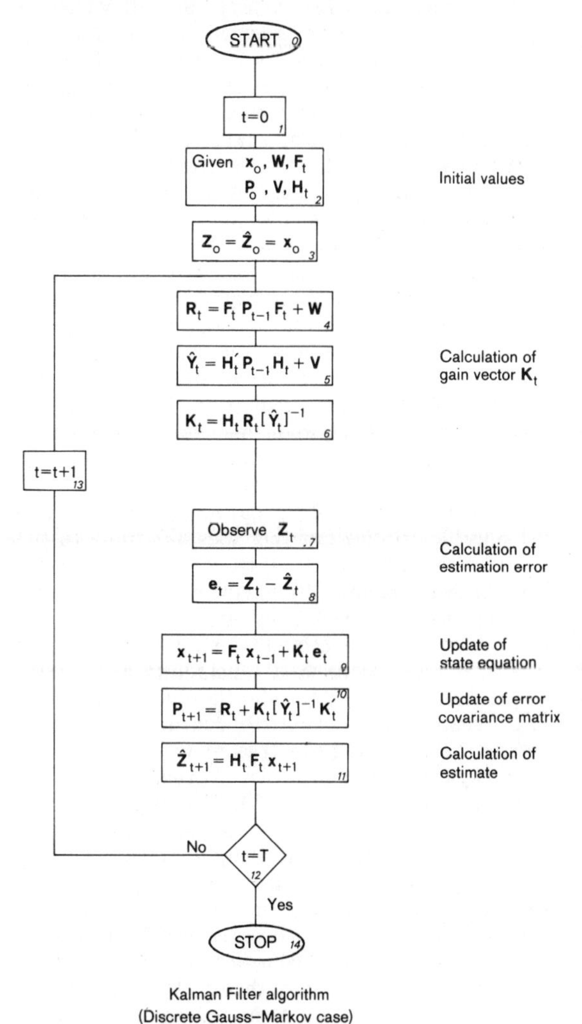

Kalman Filter algorithm
(Discrete Gauss–Markov case)

The literature on the Kalman filter is vast. The review by Kailath (1974) lists 390 items, but until now the geographer coming to the subject for the first time is poorly served. Textbooks dealing with the Kalman filter are all written for electrical engineers and require a knowledge of the jargon (Astrom, 1970; Jazwinski, 1970; and Anderson and Moore, 1979). Introductory texts on time series analysis are of no help; they range from the desultory (Chatfield, 1975, pp.228–30) to the disgraceful (Kendall, 1973, pp.126–8). However, a new book (Bennett, 1979) includes a section on Kalman filtering and it is pleasing to record that a geographer is to rectify the failings of statistical textbook writers. The easiest introduction for the student who merely wishes to use the algorithm is Harrison and Stevens (1974), but this is rather inaccessible. Their later paper (1976) is in a condensed format harder for the beginner to follow. A fairly readable introduction is the earlier part of the paper by Young (1974). In the latter part of the paper references are given to hydrological applications, particularly those of the control division of the Engineering Department at Cambridge. The proceedings of a recent American Geophysical Union (AGU) conference covering the full range of hydrologic and hydraulic research (Chiu, 1979) is not yet to hand but a preview of one of the contributions on flood management of the Arno has appeared (Panattoni and Wallis, 1979). A readable comparison of Box-Jenkins and Kalman techniques is found in Caines (1972), and if the past is anything to go by we are due for a resurgence in the 'comparison industry'. Finally, the serious student must master the three seminal papers, by Kalman alone (1960, 1963) and with R. Bucy (1961), of which the more expository (1963) is the one to start with and then perhaps go on to tackle Wiener's mathematics.

Fractals

Benoit Mandelbrot's book on fractals is the most exciting book on mathematics for many a long year and is almost surely destined to become a classic. Fractal is a neologism coined by Mandelbrot denoting an extremely irregular curve characterized by a fractional dimension (Mandelbrot, 1977, p.294). More precisely a fractal is a set in which

$$D_H > D_T$$

where D_H is the Hausdorff dimension and D_T the 'normal' topological dimension. Mandelbrot's book is written in a unique amalgam of technical and popular styles and sets out to explain how such 'an esoteric concept in pure mathematics as Hausdorff dimension can be applied to the empirical sciences'.

Although the name is new, fractal curves have been known for over a century, in fact ever since Weirstrass constructed an everywhere non-differentiable continuous function. The inadequacy of intuitive ideas of dimension became apparent when Cantor destroyed the common-sense idea that a plane is richer in points than a line and more obviously when Peano constructed a curve that completely fills a plane. The fractal dimension of such a curve is 2. Von Koch's well-known snowflake curve is a fractal of dimension 1.2618. This value can be arrived at as follows. At each stage of the construction each of the sides of unit length is replaced by four sides of length $1/3$. The Hausdorff dimension is then given by

$$D_H = {}^{\log 4}/_{\log 3} \simeq 1.2618 \ .$$

More generally if the transformation is into N sides of length r^{-1}, then

$$D_H = {}^{\log N}/_{\log r} .$$

For Cantor's celebrated ternary set we have

$$D_H = {}^{\log 2}/_{\log 3} \simeq 0.6309 .$$

The Koch and Cantor examples and all other fractals we mention and many more are illustrated in Mandelbrot's book. See also popular accounts by Gardner (1976, 1978) and by Mandelbrot (1978). A very enjoyable account of non-rectifiable curves is Vilenkin (1965).

Snowflake curves and the like were regarded as mathematical curiosities attracting the epithet 'pathological' and many great names refused to countenance them. Nevertheless it is now known that approximate examples exist in nature. We have to say approximate because natural examples tend to stop some way short of the limit because of physical limitations. Even so a prime example reaches the inconceivably fine texture of 10^{21} collisions a second.

A geographical example is provided by the question of the length of the coastline of Britain. One could start with an atlas and a pair of dividers and so arrive at a rough idea. For the next stage one could take a larger-scale map to derive a more 'accurate' figure. The process can then be repeated with larger-scale charts. Then recourse can be had to surveying the ground and reiterating at an ever larger scale; following every bay and headland and then every nook and cranny. At each stage the curve will get longer and the form more complicated, eventually verging on the pathologic. But the process has to stop somewhere, for at the pebble or mineral or molecular level the idea of a coastline is nonsensical.

This exercise was performed by L.F. Richardson who derived the empirical expression,

$$L = M G^{1-D} ,$$

where L is the length of the coastline, M a positive constant, G the length of the unit polygonal side (span of dividers etc.) and D a constant at least equal to unity, behaving as a 'similarity dimension'. Where it is possible to make the comparison the Richardson similarity dimension and the Hausdorff dimension provide the same values. Richardson found values ranging from 1.02 for the South African coast to 1.25 for the west coast of Britain (Richardson, 1960; Mandelbrot, 1967; 1977, Ch.II).

Richardson is perhaps best remembered for his parody of Swift's *ad infinitum* ode on the hierarchical structure of turbulence. Hierarchical or cascade structure is a special case of a characteristic property of fractal sets, that of statistical self-similarity. If a portion of the fractal is enlarged and the process repeated, at each stage the picture though different in detail is similar statistically. The distribution of stars is another example. It would not be surprising if the settlement pattern in the classic area south west of Chicago does not reveal several orders of self-similarity. Snowflake curves where the construction is exactly repeated at each stage are identically self-similar but it is characteristic of natural fractals that they are irregular and in fact constitute random processes. River meanders are another geographical example but a problem here is that in a reach of sufficient length to reveal the property the system is non-stationary. The ideal representative of self-similar meanders is the Minkowski sausage (Mandelbrot, 1977, p.33).

There are many more examples of fractals in nature: the diffusion of atmospheric turbulence, the percolation of a dye, the coagulation of whey, the alveolar surface of the lungs, the neurologic network of the brain, idealized forms of drainage basin, the structure of trees, the distribution of craters on the moon or of holes in gorgonzola cheese, the intervals between Nile floods, the distances between galaxies and, we are pleased to add, soil creep.

A great many of these natural examples are as yet only descriptive but this does not apply to turbulence (Mandelbrot, 1976; 1977, p.145) and to the best known fractal, Brownian motion. Brownian motion was discovered by the botanist Robert Brown. After many attempts, some verging on the occult, it received its first satisfactory explanation at the hands of Einstein and Smoluchowski in the early years of the twentieth century. To a large extent the physical properties of Brownian motion were well understood when Norbert Wiener approached the subject in 1923. The original motivation was to give a rigorous statement based upon Lebesgue measure and the Daniell integral. Wiener's paper on 'Differential Space' (1923) was a remarkable achievement inaugurating the rigorous study of continuous parameter processes based on measure theoretic ideas. Undertaken without the aid of the apparatus of the modern theory of probability it makes difficult reading even for top-rank mathematicians (Cramer, 1976, p.514). At the time no one was aware of the surprising mathematical properties and rich generalizations that a rigorous description would engender. Some idea of the scope of these generalizations can be gained from the paragraphs of Mandelbrot (1977, pp.279–84).

Standard normalized Brownian motion is a stochastic process $B(t)$: $0 < t < \infty$ on the triple (Ω, F, P), with the properties,

(i) $B(0, \omega) = 0$ for each ω.
(ii) $B(\cdot, \omega)$ is continuous for each ω.
(iii) The increments,
$B(t_1), B(t_2) - B(t_1), \ldots\ldots, B(t_n) - B(t_{n-1})$.
for $0 < t_1' < t_2 < \ldots\ldots < t_n$.

are independent and Gaussianly distributed with zero mean and unit variance.

The random function B(t) was shown by Wiener to be continuous but non-differential everywhere. There are no easy introductions to the rigorous theory of Brownian motion. Perhaps the best is the chapter by Martin (1966). See also Kac (1966) in a commemorative volume on Norbert Wiener. For the physical side there is the admirable review article by Chandrasekhar (1943).

The above definition refers to classical scalar Brownian motion sometimes called the Wiener process or more properly the Wiener-Levy-Bachelier process. For the tragic history of unreadiness that lies behind this see Mandelbrot (1977, pp.251–4, 265–8). To emphasize the fractal properties we re-write the definition in the form,

$$P\left\{\left[B(t+\Delta t) - B(t)\right] |\Delta t|^{-H} < x\right\} = F(x)$$

where $F(x)$ is the normalized Gaussian distribution function. For standard Brownian motion $H = \frac{1}{2}$. The generalization to a proper fractional Brownian process is given by $H \neq \frac{1}{2}, (0 < H < 1)$. For $H > \frac{1}{2}$ the process is termed persistent and there exists a positive correlation between the past increment and the future increment, reaching unity when $H = 1$. For $H < \frac{1}{2}$ the correlation is negative giving anti-persistent Brownian motion, reaching -0.5 when $H = 0$. The Hausdorff dimension of Brownian motion is given by $D_H = 2 - H$, with probability one (Orey, 1970) and the dimension of the level set is $D_H = 1 - H$ with probability one (Marcus, 1976). The fractional part of D_H gives a rough measure of the complexity of the trace of the fractal set. Thus a normal curve for which $D_H = D_T = 1.0$ gives the minimum and a Peano curve for which $D_H = 2.0$ gives the maximum complexity. We are now ready to examine the idea of a Brownian landscape.

The planar Brownian process or Brownian sheet is defined as a Gaussian random field $W(t_1, t_2)$, defined on the positive quadrant of a Euclidean plane, with zero mean and covariance,

$$E\left[W(s_1, s_2)\, W(t_1, t_2)\right] = \min(s_1, t_1) \times \min(s_2, t_2) \ .$$

A helpful analogy is that of Pyke (1972), who likens a Brownian sheet to a roughly shaken bed sheet pinned down along two adjacent sides. Actually it is even more irregular than this analogy suggests. Illustrations are to be found in Mandelbrot (1975; 1977, pp.201–36) and also in Adler (1978, Fig. 1, p.23). From a distance they appear to represent an Alpine-cum-lunar landscape. Upon closer inspection they are found to be too irregular on two counts. The Hausdorff dimension of the standard Brownian sheet is 2.5. If we take the zero level set (coastline), $W(t_1, t_2) = 0$, we have $D_H = 1.5$. This is too high; it out-Aegeans the Aegean (Adler, 1978, Fig. 2, p. 24). We need a dimension of about 1.25 to accord with Richardson's empirical data. By decreasing H to less than 0.5 we can construct the anti-persistent Brownian sheet and its zero level set which is much less irregular (more correlated) and more like the real thing.

The other source of regularity is geomorphological. The drainage net imposes a fair amount of systematic regularity on the landscape which cannot be reproduced by a Brownian sheet. We need some way of incorporating the drainage network or of separating the two aspects and then superimposing. One possible method is to employ a form of Fourier-Brownian analysis by which means we can conjecture the fitting of a series of Brownian functions to a profile if not to the landform. How one would calculate the coefficients is a problem for the future.

Mandelbrot extends the idea to the global scale (1975), and produces his own alternative to Pangea (1977, p.216; 1978, p.808). Towards the other end of the scale, the micro-relief of the non-vegetated slope is pathological and soil (voidage) structure is an irregular Sierpinski sponge ($D_H = 2.7628$) (Mandelbrot, 1977, p.166; 1978, p.810). Alas, when this idea was suggested over a decade ago an unready referee exclaimed that it was far too complicated for a geomorphologist to understand!

This review of the prospects for stochastic processes in geography has been highly selective yet has covered a lot of ground, from simple algorithms and Monte Carlo simulation to Hausdorff dimensions and Wiener-Hopf equations. We have mentioned a lot of names. Of these Gauss, Wiener, Kolmogorov, Levy, Cramer and Doob run through stochastic processes like a golden thread. Perhaps we can look forward to a little gold dust falling upon geography.

References

Adler, R.J. (1978) 'Some erratic patterns generated by the planar Wiener process', *Supplement Advances in Applied Probability*, 10, 22–7.

Anderson, B.D.O., and Moore, J.B. (1979) *Optimal Filtering*, Prentice-Hall: Englewood Cliffs, NJ.

Astrom, K.J. (1970) *Introduction to Stochastic Control Theory*, Academic Press: London.

Bartholomew, D.J. (1967) *Stochastic Models for Social Processes*, Wiley: London.

Bartlett, M.S. (1966) *An Introduction to Stochastic Processes*, 2nd edn, Cambridge University Press.

Bennett, R.J. (1976) 'Adaptive adjustment of channel geometry', *Earth Surface Processes*, 1, 131–50.

Bennett, R.J. (1979) *Spatial Time Series*, Academic Press: London.

Bharucha-Reid, A.T. (1956) 'Notes to supplementary bibliography: A.N. Kolmogorov, *Foundations of Theory of Probability*', Chelsea: New York.

Bharucha-Reid, A.T. (1960) *Elements of the Theory of Markov Processes and their Applications*, McGraw-Hill: London.

Brown, L.A., and Moore, E.G. (1969) 'Diffusion research in geography: a perspective', *Progress in Geography*, 1, 121–57.

Brundsen, D., and Thornes, J.B. (1977) *Geomorphology and Time*, Methuen: London.

Caines, P.E. (1972) 'Relationship between Box-Jenkins-Astrom control and Kalman linear regulator', *Proceedings IEEE*, 119, 615–20.

Carlton, A.G., and Follin, J.W. (1956) 'Recent developments in fixed and adaptive filtering', *AGARDograph*, 21, 295–300, reprinted in T. Kailath (1977).

Chandrasekhar, S. (1943) 'Stochastic problems in physics and astronomy', *Review of Modern Physics*, 15, 1–89.

Chatfield, C. (1975) *The Analysis of Time Series: Theory and Practice*, Chapman & Hall: London.

Chiu, Chao-Lin (1979) *Applications of Kalman Filter to Hydrology, Hydraulics and Water Resources*, University of Pittsburg, Pa.

Chung, K.L. (1967) *Markov Chains with Stationary Transition Probabilities*, Springer: Heidelberg.

Cliff, A.D., and Ord, J.K. (1975) 'Model building and the analysis of spatial patterns in human geography', *Journal of the Royal Statistical Society*, (B), 37, 297–348.

Collins, L. (1973) 'Industrial size distribution and stochastic processes', *Progress in Geography*, 5, 119–66.

Collins, L., Drewett, R., and Ferguson, R. (1974) 'Markov models in geography', *Statistician*, 23, 179–210.

Cox, D.R., and Miller, H.D. (1965) *The Theory of Stochastic Processes*, Chapman & Hall: London.

Cramer, H. (1936) *Random Variables and Probability Distributions*, Cambridge University Press.

Cramer, H. (1976) 'Half a century with probability theory: some personal recollections', *Annals of Probability*, 4, 509–46.

Cramer, H., and Leadbetter, M.R. (1967) *Stationary and Related Stochastic Processes*, Wiley: New York.

Davy, B.W., and Davies, T.R.H. (1979) 'Entropy concepts in fluvial geomorphology', *Water Resources Research*, 15, 103–6.

Doob, J.L. (1953) *Stochastic Processes*, Wiley: New York.

Dynkin, E.B. (1965) *Markov Processes*, 2 vols, Springer: Heidelberg.

Ephemerides, A., and Thomas, J.B. (1973) *Random Processes. Part I. Multiplicity Theory and Canonical Decompositions*, Dowden, Hutchinson & Ross, Stroudsburg, Pa.

Feller, W. (1957) *An Introduction to Probability Theory and its Applications*, 2 vols, Wiley: New York.

Gardner, M. (1976) 'Mathematical games', *Scientific American* (Dec.), 124–33.

Gardner, M. (1978) 'Mathematical games', *Scientific American* (Apr.), 16–31.

Gauss, K.F. (1963) *Theory of Motion of the Heavenly Bodies*, (trans. C.H. Davis), Dover: New York.

Gould, P.R. (1969) 'Methodological development since the 50's', *Progress in Geography*, 1, 1–49.

Gudgin, G., and Thornes, J.B. (1974) 'Probability in geographical research: applications and problems', *Statistician*, 23, 157–77.

Harrison, P.J., and Stevens, C.F. (1974) 'Bayesian forecasting: the Dynamic Linear Model', Warwick Statistical Research Report No. 11, University of Warwick, Coventry.

Harrison, P.J., and Stevens, C.F. (1976) 'Bayesian forecasting', *Journal of the Royal Statistical Society* (B), 38, 205–47.

Hepple, L.W. (1974) 'The impact of stochastic process theory upon spatial analysis in human geography', *Progress in Geography*, 6, 89–142.

Jazwinski, A.H. (1970) *Stochastic Processes and Filtering Theory*, Academic Press: New York.

Kac, M. (1966) 'Wiener and integration in function spaces', in 'Norbert Wiener 1894–1964', *Bulletin of the American Mathematical Society*, 72 (1), Part II, 52–68.

Kailath, T. (1974) 'A view of three decades of linear filtering theory', *IEEE Trans. Inf. Theory*, IT-20, 146–81.

Kailath, T. (1977) *Linear Least Squares Estimation*, Dowden, Hutchinson & Ross, Stroudsburg, Pa.

Kalman, R.E. (1960) 'A new approach to linear filtering and prediction problems', *Journal of Basic Engineering*, 82D, 35–45, collected in T. Kailath (1977).

Kalman, R.E. (1963) 'New methods in Wiener filtering theory', in *Proceedings 1st Symposium of Engineering Applications of Random Function Theory and Probability*, ed. J.L. Bogandorff and F. Kozin, Wiley: New York.

Kalman, R.E., and Bucy, R.S. (1961) 'New results in linear filtering and prediction theory', *Journal of Basic Engineering*, 83D, 95–108, collected in A. Ephemerides and J.B. Thomas (1973).

Kendall, M.G. (1973) *Time Series*, Griffin: London.

Kolmogorov, A.N. (1933) *Grundbegriffe der Wahrscheinlichkeitsrechnung*, Springer: Berlin (trans. N. Morrison); *Foundations of Theory of Probability*, Chelsea: New York.

Kolmogorov, A.N. (1941a) 'Interpolation and extrapolation of stationary random sequences' (trans. W. Doyle and I. Selin), RAND Corporation (1962), collected in T. Kailath

(1977).

Kolmogorov, A.N. (1941b) 'Stationary sequences in Helbert space', *Bulletin of Mathematics of the University of Moscow* (trans. J.F. Barrett), collected in T. Kailath (1977).

Leopold, L.B., and Langbein, W.R. (1962) 'The concept of entropy in landscape evolution', US Geological Survey Professional Paper 500A.

Lin, Y.K. (1969) 'Random processes', *Applied Mechanics Revue*, 22, 825–31.

Loeve, M. (1963) *Probability Theory*, 3rd edn, Van Nostrand, NY.

Loeve, M. (1977) *Probability Theory I*, 4th edn, Pringer, NY.

Mandelbrot, B.B. (1967) 'How long is the coastline of Britain? Statistical self-similarity and fractal dimension', *Science*, 155, 636–8.

Mandelbrot, B.B. (1975) 'Stochastic models for earth relief, the shape and the fractal dimension of the coastlines and the number-area rule for islands', *Proceedings of the National Academy of Science*, 72, 3825–8.

Mandelbrot, B.B. (1976) 'Intermittent turbulence and fractal dimension; kurtosis and the spectal exponent 5/3 + B', in R. Temam (ed.), *Turbulence and Navier-Stokes Equations*, Lecture notes in mathematics 565, 121–45, Springer: New York.

Mandelbrot, B.B. (1977) *Fractals: Form, Chance and Dimension*, Freeman: San Francisco.

Mandelbrot, B.B. (1978) 'Getting snowflakes into shape', *New Scientist*, 78, No. 1108 (22 June), 808–10.

Marcus, M.B. (1976) 'Capacity of level sets of certain stochastic processes', *Zeitschrift für Wahrscheinlichkeitstheorie und verwandte Gebiete*, 34, 279–84.

Martin, W.T. (1966) 'The Brownian motion process and differential space' in B. Rankin, (ed.), *Differential Space, Quantum Systems and Prediction*, MIT Press: Cambridge, Mass.

Masani, P. (1966) 'Wiener's contribution to generalised harmonic analysis, Prediction Theory and Filter Theory' in 'Norbert Wiener 1894–1964', *Bulletin of the American Mathematical Society*, 72, Part II, 73–125.

Neugebauer, O. (1952) *The Exact Sciences in Antiquity*, Princeton University Press, NJ.

Neyman, J. (1960) 'Indeterminacy in science and new demands on statisticians', *Journal of the American Statistical Association*, 55, 625–39.

Orey, S. (1970) 'Gaussian sample functions and the Hausdorff dimension of level crossings', *Zeitschrift für Wahrscheinlichkeitstheorie und verwandte Gebiete*, 15, 249–56.

Panattoni, L., and Wallis, J.R. (1979) 'The Arno River flood study, 1971–76', *Transactions of the American Geophysical Union*, 60 (1), 1–15.

Parzen, E. (1962) *Stochastic Processes*, Holden-Day: San Francisco.

Plackett, R.L. (1950) 'Some theorems in least squares', *Biometrika*, 37, 19–57.

Pyke, R. (1972) 'Partial sums of matrix arrays and Brownian sheets' in E.F. Harding and D.G. Kendall (eds), *Stochastic Geometry and Analysis*, Wiley: New York.

Richardson, L.F. (1960) 'The problem of contiguity', and appendix to *Statistics of deadly quarrels*, ed. Q. Wright and C.C. Lienau, Boxwood Press, Pittsburgh. Also in *General Systems Yearbook*, 6, 139–187, Society for General Systems (1961).

Scheidegger, A.E. (1964) 'Some implications of statistical mechanics in geomorphology', *Bulletin International Association Scientific Hydrology*, 9, 12–16.

Scheidegger, A.E. (1970) *Theoretical Geomorphology*, Springer: New York.

Seal, H.L. (1967) 'The historical development of the Gauss linear model', Studies in the History of Probability and Statistics, XV, *Biometrika*, 54, 1–24.

Sorenson, H.W. (1970) 'Least squares estimation from Gauss to Kalman', *IEEE Spectrum*, 7, 63–8.

Swerling, P. (1958) 'A proposed stagewise differential correction procedure for satellite tracking and prediction', RAND Corporation, also published in *Journal of Astronautical Science*, 6, 46–52, and collected in T. Kailath (1977).

Vilenkin, N.Y. (1965) *Stories about Sets. Trans Scripta Technica*, Academic Press: London.

Whittaker, E.T., and Robinson, G. (1926) *The Calculus of Observations*, Blackie: London.

Wiener, N. (1923) 'Differential space', *Journal of Mathematics and Physics*, 58, 131–74, included in N. Wiener (1976).

Wiener, N. (1949) *The Extrapolation, Interpolation and Smoothing of Stationary Time Series*, Wiley: New York, reissued as *Time Series* (1964), MIT Press, Cambridge, Mass.

Wiener, N. (1976) *Collected Works*, ed. P. Masani, Vol. I, MIT Press, Cambridge, Mass.

Yaglom, A.M. (1963) *An Introduction to the Theory of Stationary Random Functions* (trans. R.A. Silverman), Prentice-Hall: Englewood Cliffs, NJ.

Yang, C.T. (1971) 'Potential energy and stream morphology', *Water Resources Research*, 7, 311–22.

Young, P.C. (1974) 'Recursive approaches to time series analysis', *Bulletin of the Institute of Mathematical Analysis*, 10, 209–24.

Chapter 20

Optimization: static programming methods

M.L. Senior

The scope of British research

Uses and developments of static optimization methods by British geographers over the past two decades have been distinctly patchy. Content-wise, Linear Programming (LP) methods have commanded a disproportionate amount of attention (Haggett *et al.*, 1977), and but for significant contributions from geographers now employed in academic planning departments (e.g. Openshaw, 1978; Robertson, 1976) and from 'non-geographers' recruited to the discipline (e.g. Wilson, 1974; Batty, 1976; Williams, 1979) research achievements might have been especially bleak.

To foster a broader appreciation of research potential in this field, this British work is reported under headings identifying major trends in the use of optimization methods in studies of spatial and resource problems drawn from various disciplines.

Programming of commodity flows and the space economy

Whereas a variety of LP applications have come from American geography and regional science, British geographers have concentrated on inter-regional commodity flows. Freight compared with passenger transport is a much under-researched area, so such work has been generally welcomed. Yet close scrutiny of Chisholm's and O'Sullivan's (1973) benchmark study reveals crucial inadequacies.

First, regional transport cost advantages in the space economy, measured directly by the dual variables, are interpreted and exploited casually. Only recently has more serious attention been directed to this issue (Pitfield and Benabi, 1978). Second, their contribution to the 'Transportation-Problem-versus-gravity-model' debate (Gordon, 1979) is of dubious technical quality. They seem ignorant of the questionable properties of the unconstrained gravity models with which they experiment. They then prefer the production-constrained gravity model to the production-attraction-constrained one, because the latter requires more computation and assumes a closed flow system. Yet their Transportation Problem (TP) with equality demand and supply conditions, namely:

Minimize total interaction costs:

$$C = \sum_i \sum_j T_{ij} c_{ij} \qquad (1)$$

where T_{ij} and c_{ij} are interaction flows and costs respectively, subject to:

$$\sum_j T_{ij} = O_i \text{ (supply quantities)} \qquad (2)$$

$$\sum_i T_{ij} = D_j \text{ (demand requirements)} \qquad (3)$$

$$T_{ij} \geqslant 0, \qquad (4)$$

requires the same closed system. Moreover, the production-attraction-constrained model:

$$T_{ij} = A_i B_j O_i D_j \exp(-\beta c_{ij}) \qquad (5)$$

$$A_i = \left\{ \sum_j B_j D_j \exp(-\beta c_{ij}) \right\}^{-1} \qquad (6)$$

$$B_j = \left\{ \sum_i A_i O_i \exp(-\beta c_{ij}) \right\}^{-1} \qquad (7)$$

is the only gravity form capable of satisfying constraints (2) and (3) and of replicating a TP solution when the travel impedance parameter, β, is large (Evans, 1973a). Indeed, this theoretical link suggests that this version of the gravity model is more flexible than the TP and equally capable of supplying dual variable information (Wilson and Senior, 1974). Furthermore, the empirical problems of model comparison in terms of what constitute satisfactory calibration and goodness-of-fit criteria, are major issues which have just begun to receive the attention they deserve (Pitfield, 1978, 1979).

Fusion of these strongly empirical commodity flow models into the more operational regional science models of the space economy (Paelinck and Nijkamp, 1976; Gordon, 1974) remains an intriguing, if difficult, research challenge.

Entropy maximizing as nonlinear programming

When first presented (Wilson, 1967, 1970) the most notable features of the entropy maximizing (EM) methodology were the generation of internally consistent gravity models and the interpretation of entropy as uncertainty. That the methodology involved a Nonlinear Program (NLP), with concave objective function:

Maximize entropy:

$$S = -\sum_i \sum_j T_{ij} \ln T_{ij} \quad (8)$$

subject to linear constraints (2), (3) and:

$$\sum_i \sum_j T_{ij} c_{ij} = C, \quad (9)$$

failed to arouse significant interest until the works of Evans (1973a, b) who proved theoretically the link between gravity models and Transportation Problems; of Nijkamp and Paelinck (1974) who equated EM and Geometric Programming models; and of Wilson and Senior (1974) who developed and explored the limiting and duality properties of conceptually equivalent LP and EM models (Senior and Wilson, 1974) by treating an LP objective function, like (1), as an additional constraint, like (9), on entropy maximization. This implies that the LP objective need not attain its optimal value in the EM problem, thus giving rise to interpretations of sub-optimal behaviour.

However, unlike many applied mathematical programs in the social sciences, EM programs are essentially non-behavioural. Paradoxically, in transport planning, EM travel demand models often provide inputs to economic evaluation procedures assuming utility maximizing trip-makers (Williams, 1977). To bridge this behavioural consistency gap Neuberger (1971) and Champernowne *et al.* (1976) derive almost identical gravity demand forms using an NLP with an objective function interpretable as the maximization of economic surplus, B:

Maximize:

$$B = -\frac{1}{\beta} \sum_i \sum_j T_{ij} \ln T_{ij} - \sum_i \sum_j T_{ij} c_{ij} \quad (10)$$

subject to constaints (2), (3) and (4). Although this problem involves only a structural modification of the EM program, its interpretation is fundamentally different as the trip-makers are endowed explicitly with economic rationality, β measures their dispersion of preferences, and the 'entropy' term becomes a utility function with probabilistic connotations at the micro-level.

This line of research has spawned promising developments. Connections have been made to probabilistic choice theories to structure travel demand models and generate consistent evaluation measures (Williams, 1977). Moreover, the behavioural approach facilitates the interpretation of dual variables as spatial benefit indicators (Williams and Senior, 1978), which in turn can form inputs to optimal land use design models (Coelho and Williams, 1978). The full implications of such work has yet to be developed in other contexts; housing and commodity flow systems provide two suitable candidates.

Optimal spatial representation

A particularly British offshoot from the application of entropy-derived spatial models has been a concern for 'appropriate' zonal and study area representation. Several systematic design procedures have been evolved since Broadbent (1970) initiated the subject.

Multi-criteria aggregation involves grouping the least aggregate spatial units at which data are available into operational zoning systems according to several partially incompatible criteria. A sequential and heuristic optimization procedure has been developed by Masser, Batey and Brown (1975). First the study area boundary and number of zones are established using Broadbent's crude 'rules' as a guideline. Second, the data units are aggregated to the required number of zones to optimize on the equality of spatial description variables, such as population, using electoral districting algorithms. Batty (1974) has developed an alternative spatial entropy criterion for this purpose. Third, the previous aggregation is assessed in terms of an interaction intensity criterion, and modifications effected with reference to an independent aggregation minimizing intrazonal interaction. Finally further modifications can be undertaken in the light of optimal zoning systems for secondary variables, for example as indicated by cluster analyses of population characteristics.

Multi-level specification involves reducing the dimensionality of the original data matrix by partitioning it into spatial subsystems. The details of intra-subsystem interaction are preserved while inter-subsystem flows are treated in summary form. Two optimization aspects are involved: deciding on the number of partitions to attain maximum data set reduction and using maximum intra-subsystem interaction as a partitioning criterion. Masser and

Brown (1977) review optimization methods for this process.

Radical departures from the above schemes are developed by Openshaw (1977) whose main contention is that zoning systems should be designed to optimize model performance. Perhaps the ultimate solution to this problem is to abandon spatially aggregate models! In the meantime it is surprising that Masser and Brown (1977) advocate research on consolidating existing methods, when a wide variety of multi-objective optimization techniques could be brought to bear on this spatial representation problem.

Optimal facility location

Optimizing facility locations has attracted operational research and management scientists concerned with warehouse and branch plant location, and subsequently American geographers and regional scientists specializing in public sector problems (Revelle *et al.*, 1970).

For location problems in continuous space Weber's simple transport cost minimizing model has been a starting-point, whereas location on a network has no obvious origins in classical location theory. Eilon *et al.* (1971) and Lea (1973) review the solution algorithms available and comment on the relative merits of continuous and network representations. In the distribution management field (Eilon *et al.*, 1971) the longer-term plant location decisions are complemented by shorter-term transport operation and management issues, namely travelling salesmen, vehicle scheduling and fleet size problems. Despite the overtly spatial nature of this private sector optimization, geographers have long since relinquished concern for the normative locational aspects of the firm's internal organization.

Forerunners of geographers' interest in public facility location were the studies of Yeates (1963) and Gould and Leinbach (1966) on school districting and hospital location respectively. Interest snowballed as the technical foundations for solving spatially structured combinatorial problems (Scott, 1971) and the diversity of public sector problems (Massam, 1972, 1975) were communicated to a wider geographic audience. Yet Hodgart's (1978) review of this field reveals the minuscule contribution of British geographers. Robertson (1974, 1976, 1978) has pursued rather specialist research on adapting the Tornqvist algorithm for facility location in remote rural areas, where road distance is far greater than linear distance, and has exploited its potential for optimizing various recreational strategies. Her persistence with this particular algorithm is rather curious however; it is not very efficient and alternative network heuristics seem more relevant to her rural studies. Yet her work does raise some issues which should be incorporated in a research agenda.

A recurring theme is the conflict between accessibility and catchment size (facility viability) criteria. This raises pertinent questions concerning the adaptability of public services, in terms of range and quality, to spatial variations in demand and the possibility of hierarchical provision. It also suggests that multi-objective programming techniques should be used to explore criteria tradeoffs. Furthermore, socioeconomic disaggregation of consumers and greater attention to demand elasticities (Wagner and Falkson, 1975) would bring the equity implications of supply strategies into sharper focus. It is important too, as Hodgart (1978) stresses, not to confuse demand elasticity effects with consumer locational choice behaviour. At least three situations should be distinguished: first, inelastic demand plus allocation of all consumers to the nearest facility, subject to capacity constraints if applicable; second, inelastic demand combined with consumer choice of facility, with the implication that some consumers patronize facilities beyond their nearest facility, either because they perceive some net advantage in doing so or because they behave suboptimally; and third, elastic demand situations where travel costs to facilities plus any entrance charges affect the total level of demand in the system. The first situation is the one assumed in most studies; the third is still much under-researched. Research on the second situation has been initiated recently using gravity models (Hodgson, 1978) and notions of endogenous attraction, where consumer choice is influenced by levels of patronage (Goodchild, 1978). It appears that in this area a promising synthesis between traditional location-allocation models and approaches to facility location using accessibility or surplus maximizing concepts (Coelho and Wilson, 1976; Leonardi, 1978) is in the making.

Optimal plan design: unidimensional approaches

Whilst geographers have specialized in the optimal spatial design of public facilities, others have looked to optimization methods as a more general means of automating cumbersome design and evaluation stages characteristic of many types of planning process. Such hopes stem from the thoughts of Harris (1965) and the pioneering work of Schlager (1965), who implemented an LP model to minimize land use development costs subject to demand constraints and design standards.

Schlager has been criticized for his use of crude empirical constraints on the coexistence of land use types and his neglect of spatial interactions. Subsequent LP models have been more comprehensive. Ben-Shahar *et al.* (1969) sketched the issues involved in constructing a multi-component social welfare function, but then concentrated on only monetary items thus collapsing their multidimensional problem to a unidimensional one. However, costs of replacing existing infrastructure and of trip-making were incorporated. In practice though, minimizing trip costs in a linear objective is usually unsatisfactory (Local Government Operational Research Unit, 1970).

To avoid simplistic representations of interactions and to incorporate scale economies and sundry externalities, NLP formulations are needed. Of these Quadratic Programming problems are the most tractable, and Andersson (1978) and Hopkins (1977, 1979) discuss their appropriateness for handling spatial interactions and interdependencies. The most common models are the Koopmans-Beckmann (1957) quadratic assignment problem, and certain versions of the TOPAZ model family (Brotchie *et al.*, 1971) of the general form (Williams, 1979):

$$\underset{X}{\text{Maximize:}}\ B_1(X) + B_2(X)\ . \tag{11}$$

$B_1(X)$ is a linear net benefit term for establishing and operating activities, X, in zones, and $B_2(X)$ is a quadratic (or higher order) net benefit term for activity interactions. Geometric Programming too has a growing reputation as one of the more tractable NLP methods (Beightler and Phillips, 1976), and engineering design applications suggest its potential relevance to spatial planning (Openshaw, 1978). Its ability to handle polynomial functions have permitted the representation of capital and labour scale economies, agglomeration economies and Cobb-Douglas production functions in regional economic planning models (Nijkamp, 1972; Kadas, 1975). The method has also been applied to environmental and resource planning issues (Nijkamp and Paelinck, 1973; McNamara, 1976).

Other NLP problems may be less tractable, such as Wilson's (1976) consumer surplus objective for optimizing shopping centre size subject to a gravity model of consumer patronage. However, if these problems are convertible to convex programs a range of solution methods are applicable. For Wilson's model this conversion embeds the optimization program generating the consumer model in the shopping centre design program (Coelho and Wilson, 1976; Coelho *et al.*, 1978). Furthermore, the dual of this converted problem may be even more manageable (Coelho and Williams, 1978).

Despite implementation difficulties there is growing interest in more comprehensive optimization schemes. Lundqvist (1975) has pioneered work on coupling quadratic building stock and optimal network design models, producing a nonlinear mixed integer program, which is segmented and transformed to an all-integer form for solution (cf. Los, 1978). Lundqvist (1978a, b) also raises the as yet little explored, but fundamental, issues concerning the short-run adaptivity of users to an optimally-designed infrastructure which is flexible only in the long run. He proposes freedom of action as a more useful criterion than social welfare for planning purposes, and raises the spectre of multi-objective optimization when he defines freedom of action operationally as a tradeoff between access and space utilization.

Optimal plan design: multi-objective approaches

The types of spatial and resource planning which interest geographers typically involve multiple, conflicting and incommensurate objectives. A single plan can rarely optimize all objectives simultaneously, while unidimensional optimization demands the selection of a single objective or the conversion of multiple objectives to a common metric. Dissatisfaction with the traditional objective of maximizing some aggregate measure of economic efficiency, and a growing concern for environmental standards has prompted a marked interest in multi-objective optimization in water resource planning (dating back to Maass *et al.*, 1962) and in environmental approaches to regional science (Nijkamp, 1977).

Multi-objective optimization recognizes the existence of multiple optima, and some multi-objective methods involve prior identification of these non-inferior solutions from the feasible set (Cohon and Marks, 1975). In this way decision-makers, who ultimately must choose a compromise optimum by specifying preferred tradeoff levels between objectives, need not agonize over inferior solutions whose performance can be bettered on at least one objective without impairing the achievement on any other objective. Furthermore, the display of non-inferior solutions has a potentially considerable information value for decision-makers, as Miller and Byers (1973) demonstrate for watershed resource planning. However, as non-inferior solutions may be numerous, there are many multi-objective methods which seek to shortcut their complete identification by proceeding more directly to a compromise solution. Cohon and Marks (1975) classify these according to whether decision-maker preferences are specified *a priori*, as with goal programming (Wereczberger, 1976), or are progressively articulated by interactive methods (Monarchi *et al.*, 1973). Methods vary considerably as to their computational efficiency, clarification of objective tradeoffs and demands on decision-maker rationality (Krzysztofowicz *et al.*, 1977). Additionally, there are multi-criteria decision-making techniques not necessarily involving mathematical programming (MacCrimmon, 1973; Keeney and Raiffa, 1976; Nijkamp and van Delft, 1977), but which can be thought of as heuristic multi-objective optimization.

The recognition of many geographical issues, both spatial and ecological, as multi-criteria decision-making problems, and the employment of a considerable armoury of methods for treating them as such directly, represents a significant advance on traditional unidimensional optimization and a most promising research area for the 1980s. Two British examples serve to illustrate this research potential.

Pearman (1977) has developed a novel procedure for network design and evaluation in the presence of multiple criteria. Because of the computational expense of optimizing a network design for just a single objective, he generates a range of good but suboptimal designs by heuristic methods. The performance of each network for multiple criteria are calculated and input to a weighted maximin procedure to select the preferred design.

Equally intriguing is Openshaw's and Whitehead's (1975) Decision Optimizing Technique, a zero–one integer LP model, which defines an optimal strategy by selecting one policy from a set of mutually exclusive policies in each decision area, subject to various constraints such as the mutual compatibility of options from different decision areas. The method is very much geared to practical decision-making, although its zero–one formulation may prove somewhat restrictive in practice. Openshaw and Whitehead contemplate its use in multiobjective-multipreference situations, where the politicians must not only decide on preferred tradeoffs between objectives, but simultaneously take into account varying preferences for each objective among the interest groups they represent.

Problems and prospects

Optimization methods have not had a good press in geography. As a means of explaining geographic behaviour their normative implications have been dismissed as superrationality, a misconception caused by their frequent use for operationalizing neoclassical economic theory. 'It is not normative modelling which is at fault but the kind of norms built into such models' (Harvey, 1973, p. 96). It would seem profitable to interpret much geographic behaviour as optimizing behaviour, but involving multiple objectives and subject to many and diverse constraints.

There has also been insufficient appreciation by geographers of the design potential of optimization methods for those subsystems of the geographic environment subject to centralized control. This control is operated at a variety of spatial scales by anything from a local firm or public authority to national institutions or multinational corporations.

As for the prospects, one can discern a natural progression from the application of basic LP forms, through the use of heuristic, integer and nonlinear convex methods, to the future employment of stochastic and non-convex programming techniques, although developments in the latter field will depend on advances in mathematical programming theory (Boyce, 1979). It would be surprising too if dynamic optimization did not figure more prominently, but such considerations lead to optimal control issues which are reviewed elsewhere.

As for British geography, remoteness from a number of research frontiers is a handicap. Its future health may be best served by more British geographers' fingers in an expanding optimization pie.

References

Andersson, A.E. (1978) Introduction to the special issue on quadratic models of spatial allocation, *Regional Science and Urban Economics*, 8, 1–4.

Batty, M. (1974) 'Spatial entropy', *Geographical Analysis*, 6, 1–31.

Batty, M. (1976) *Urban modelling: algorithms, calibrations, predictions*, Cambridge University Press.

Beightler, C.S., and Phillips, D.T. (1976) *Applied geometric programming*, John Wiley: New York.

Ben-Shahar, H., Mazor, A., and Pines, D. (1969) 'Town planning and welfare maximization: a methodological approach', *Regional Studies*, 3, 105–13.

Boyce, D. (1979) Introduction to the special issue on transportation network design, *Transportation Research*, 13B, 1–3.

Broadbent, T.A. (1970) 'Notes on the design of operational models', *Environment and Planning*, 2, 469–76.

Brotchie, J.F., Toakley, A.R., and Sharpe, R. (1971) 'A model for national development', *Management Science*, 18, B14–B18.

Champernowne, A.F., Williams, H.C.W.L., and Coelho, J.D. (1976) 'Some comments on urban travel demand analysis, model calibration, and the economic evaluation of transport plans', *Journal of Transport Economics and Policy*, 10, 267–85.

Chisholm, M., and O'Sullivan, P. (1973) *Freight flows and spatial aspects of the British economy*, Cambridge University Press.

Coelho, J.D., and Williams, H.C.W.L. (1978) 'On the design of land use plans through locational surplus maximization', *Papers of the Regional Science Association*, 40, 47–59.

Coelho, J.D., and Wilson, A.G. (1976) 'The optimum location and size of shopping centres', *Regional Studies*, 10, 413–21.

Coelho, J.D., Williams, H.C.W.L., and Wilson, A.G. (1978) 'Entropy maximizing submodels within overall mathematical programming frameworks: a correction', *Geographical Analysis*, 10, 195–201.

Cohon, J.L. and Marks, D.H. (1975) 'A review and evaluation of multiobjective programming techniques', *Water Resources Research*, 11, 208–20.

Eilon, S., Watson-Gandy, C.D.T., and Christofides, N. (1971) *Distribution management: mathematical modelling and practical analysis*, Griffin: London.

Evans, S.P. (1973a) 'A relationship between the gravity model for trip distribution and the transportation problem in linear programming', *Transportation Research*, 7, 39–61.

Evans, S.P. (1973b) 'Some applications of mathematical optimisation theory in transport planning', PhD thesis, University of London.

Goodchild, M.F. (1978) 'Spatial choice in location-allocation problems: the role of endogenous attraction', *Geographical Analysis*, 10, 65–72.

Gordon, I.R. (1974) 'A gravity flows approach to an interregional input-output model for the UK' in E.L. Cripps (ed.) *Space-time concepts in urban and regional models*, Pion: London.

Gordon, I.R. (1979) 'Freight distribution model predictions compared: a comment', *Environment and Planning A*, 11, 219–21.

Gould, P., and Leinbach, T.R. (1966) 'An approach to the geographic assignment of hospital services', *Tijdschrift voor Economische en Sociale Geographie*, 57, 203–6.

Haggett, P. Cliff, A.D., and Frey, A. (1977) *Locational analysis in human geography*, 2nd edn, Arnold: London.

Harris, B. (1965) 'New tools for planners', *Journal of the American Institute of Planners*, 31, 90–5.

Harvey, D. (1973) *Social justice and the city*, Arnold: London.

Hodgart, R.L. (1978) 'Optimizing access to public services: a review of problems, models and methods of locating central facilities', *Progress in Human Geography*, 2, 17–48.

Hodgson, M.J. (1978) 'Toward more realistic allocation in location-allocation models: an interaction approach', *Environment and Planning A*, 10, 1273–85.

Hopkins, L.D. (1977) 'Land-use plan design – quadratic assignment and central facility models', *Environment and Planning A*, 9, 625–42.

Hopkins, L.D. (1979) 'Quadratic versus linear models for land-use plan design', *Environment and Planning A*, 11, 291–8.

Kadas, A.S. (1975) 'An application of geometric programming to regional economics', *Papers of the Regional Science Association*, 34, 95–106.

Keeney, R.L., and Raiffa, H. (1976) *Decisions with multiple objectives: preferences and value tradeoffs*, John Wiley: New York.

Koopmans, T.C., and Beckmann, M.J. (1957) 'Assignment problems and the location of economic activities', *Econometrica*, 25, 53–76.

Krzysztofowicz, R., Castano, E., and Fike, R.L. (1977) Comment on J.L. Cohon and D.H. Marks, op. cit., *Water Resources Research*, 13, 690–2.

Lea, A.C. (1973) 'Location-allocation systems: an annotated bibliography', University of Toronto, Department of Geography, Discussion Paper 13.

Leonardi, G. (1978) 'Optimum facility location by accessibility maximizing', *Environment and Planning A*, 10, 1287–1305.

Local Government Operational Research Unit (1970) 'Systems design project. Land use plans by linear program – towards a solution', Reading.

Los, M. (1978) 'Simultaneous optimization of land use and transportation: a synthesis of the quadratic assignment problem and the optimal network problem', *Regional Science and Urban Economics*, 8, 21–42.

Lundqvist, L. (1975) 'Transportation analysis and activity location in land-use planning – with applications to the Stockholm Region' in A. Karlqvist, L. Lundqvist and F. Snickars (eds), *Dynamic allocation of urban space*, Saxon House: Farnborough.

Lundqvist, L. (1978a) 'Planning for freedom of action', in A. Karlqvist, L. Lundqvist, F. Snickars and J.W. Weibull (eds), *Spatial interaction theory and planning models*, North-Holland: Amsterdam.

Lundqvist, L. (1978b) 'Urban planning of locational structures with due regard to user behaviour', *Environment and Planning A*, 10, 1413–29.

Maass, A., Hufschmidt, M.M., Dorfman, R., Thomas, H.A., Marglin, S.A., and Fair, G.M. (1962) *Design of water-resource systems: new techniques for relating economic objectives, engineering analysis, and governmental planning*, Macmillan: London.

MacCrimmon, K.R. (1973) 'An overview of multiple objective decision making' in J.L. Cochrane and M. Zeleny (eds), *Multiple criteria decision making*, University of South Carolina Press: Columbia.

McNamara, J.R. (1976) 'An optimization model for regional water quality management', *Water Resources Research*, 12, 125–34.

Massam, B.H. (1972) 'The spatial structure of administrative systems', Resource Paper 12, Association of American Geographers, Commission on College Geography, Washington, DC.

Massam, B.H. (1975) *Location and space in social administration*, Arnold: London.

Masser, I., and Brown, P.J.B. (1977) 'Spatial representation and spatial interaction', *Papers of the Regional Science Association*, 38, 71–92.

Masser, I., Batey, P.W.J., and Brown, P.J.B. (1975) 'The design of zoning systems for interaction models' in E.L. Cripps (ed.), *Regional science – new concepts and old problems*, Pion: London.

Miller, W.L., and Byers, D.M. (1973) 'Development and display of multiobjective project impacts', *Water Resources Research*, 9, 11–20.

Monarchi, D.E., Kisiel, C.C., and Duckstein, L. (1973) 'Interactive multiobjective programming in water resources: a case study', *Water Resources Research*, 9, 837–50.

Neuberger, H. (1971) 'User benefit in the evaluation of transport and land use plans', *Journal of Transport Economics and Policy*, 5, 52–75.

Nijkamp, P. (1972) *Planning of industrial complexes by means of geometric programming*, Rotterdam University Press.

Nijkamp, P. (1977) *Theory and application of environmental economics*, North-Holland: Amsterdam.

Nijkamp, P., and van Delft, A. (1977) *Multicriteria analysis and regional decision-making*, Martinus Nijhoff: Leiden.

Nijkamp, P., and Paelinck, J.H.P. (1973) 'An interregional model of environmental choice', *Papers of the Regional Science Association*, 31, 51–71.

Nijkamp, P. and Paelinck, J.H.P. (1974) 'A dual interpretation and generalization of entropy maximising models in regional science', *Papers of the Regional Science Association*, 33, 13–31.

Openshaw, S. (1977) 'A geographical solution to scale and aggregation problems in region-building, partitioning and spatial modelling', *Transactions, Institute of British Geographers*, NS 2, 459–72.

Openshaw, S. (1978) *Using models in planning: a practical guide*, Retail and Planning Associates: Corbridge.

Openshaw, S., and Whitehead, P.T. (1975) 'A decision optimizing technique for planners', *Planning Outlook*, 16, 19–32.

Paelinck, J.H.P., and Nijkamp, P. (1976) *Operational theory and method in regional economics*, Saxon House: Farnborough.

Pearman, A.D. (1977) 'Problems of optimising investment in road networks' in P. Bonsall, Q. Dalvi and P.J. Hills (eds), *Urban transportation planning: current themes and future prospects*, Abacus Press: Tunbridge Wells.

Pitfield, D.E. (1978) 'Freight distribution model predictions compared: a test of hypotheses', *Environment and Planning A*, 10, 813–36.

Pitfield, D.E. (1979) 'Freight distribution model predictions compared: some further evidence', *Environment and Planning A*, 11, 223–6.

Pitfield, D.E., and Benabi, B. (1978) 'Transportation problem duals for freight flows in Britain: a guide to industrial location?', mimeo, Department of Transport Technology, University of Technology, Loughborough.

Revelle, C., Marks, D.H., and Liebman, J.C. (1970) 'An analysis of private and public sector location models', *Management Science*, 16, 692–707.

Robertson, I.M.L. (1974) 'Road networks and the location of facilities', *Environment and Planning A*, 6, 199–206.

Robertson, I.M.L. (1976) 'Accessibility to services in the Argyll district of Strathclyde: a locational model', *Regional Studies*, 10, 89–95.

Robertson, I.M.L. (1978) 'Planning the location of recreation centres in an urban area: a case study of Glasgow', *Regional Studies*, 12, 419–27.

Schlager, K.J. (1965) 'A land use plan design model', *Journal of the American Institute of Planners*, 31, 103–11.

Scott, A.J. (1971) *Combinatorial programming, spatial analysis and planning*, Methuen: London.

Senior, M.L., and Wilson, A.G. (1974) 'Explorations and syntheses of linear programming and spatial interaction models of residential location', *Geographical Analysis*, 6, 209–38.

Wagner, J.L., and Falkson, L.M. (1975) 'The optimal nodal location of public facilities with price sensitive demand', *Geographical Analysis*, 7, 69–83.

Werczberger, E. (1976) 'A goal-programming model for industrial location involving environmental considerations', *Environment and Planning A*, 8, 173–88.

Williams, H.C.W.L. (1977) 'On the formation of travel demand models and economic evaluation measures of user benefit', *Environment and Planning A*, 9, 285–344.

Williams, H.C.W.L. (1979) 'Mathematical programming approaches to activity location – some theoretical considerations', mimeo, University of Leeds, School of Geography.

Williams, H.C.W.L., and Senior, M.L. (1978) 'Accessibility, spatial interaction and the spatial benefit analysis of land use-transportation plans' in A. Karlqvist, L. Lundqvist, F. Snickars and J.W. Weibull (eds), *Spatial interaction theory and planning models*, North-Holland: Amsterdam.

Wilson, A.G. (1967) 'A statistical theory of spatial distribution models', *Transportation Research*, 1, 253–69.

Wilson, A.G. (1970) *Entropy in urban and regional modelling*, Pion: London.

Wilson, A.G. (1974) *Urban and regional models in geography and planning*, John Wiley: London.

Wilson, A.G. (1976) 'Retailers' profits and consumers' welfare in a spatial interaction shopping model' in I. Masser (ed.), *Theory and practice in regional science*, Pion: London.

Wilson, A.G., and Senior, M.L. (1974) 'Some relationships between entropy maximizing models, mathematical programming models and their duals', *Journal of Regional Science*, 14, 207–15.

Yeates, M. (1963) 'Hinterland delimitation: a distance minimising approach', *Professional Geographer*, 15, 7–10.

Chapter 21

Optimization: control models

R.J. Chorley and R.J. Bennett

Aspects of systems and control

Control theory possesses four main characteristics. First, it is a *dynamic* optimization technique, capable of allowing for system changes, modified forecasts and interperiod tradeoffs. Second, it permits optimal allocation over long time horizons, consistent in length with the accuracy of the underlying model. Third, it emphasizes system goals and poses the question: to achieve a given target, what instrument settings must be used? Thus it shifts emphasis from mere model construction to model use. Fourth, it allows the incorporation of uncertainty of both measurements and system specification into optimization solutions. Because of these characteristics, increasing attention has been directed by geographers to control theory over the last five years, and it can be expected that this will be a significant research area within the subject over the next ten years.

The concept of control in geography is intrinsically bound up with that of systems modelling (Chorley and Kennedy, 1971; Bennett and Chorley, 1978; Bennett, 1979). In view of the recent literature summarizing systems principles and control techniques, it is only necessary to mention briefly the following four aspects of systems modelling.

First, although all geographical systems are *open*, in the sense that they exchange mass, energy and information with their surroundings, modelled systems are assumed to be closed, in the sense that only the action of a specified group of variables is considered to be dominant.

Second, within such limitations, systems may be specified in two major ways:

(a) *Morphological systems*, consisting of a network of structural relationships, defined by correlation-type techniques, between assumed constituent parts. Ecological systems have been traditionally so defined.

(b) *Input/output (cascading) systems*, in which inputs of mass, energy and information (which may be previous system states) are subjected to mathematical operators to produce outputs. The technology of systems analysis has developed to treat this class of systems. Some modern economic systems and hydrological ones have been so specified.

Third, geographers encounter both ***hard systems***, which are susceptible to rigorous specification, quantification and mathematical prediction in terms of their response, and *soft systems* which involve individual and group human behaviour and which can be handled at present only by experience, rule of thumb and intuition. In fact, most of the systems with which geographers are concerned are composed of interfaced hard and soft systems which means that their specification is as yet vague and their control principles largely unknown.

Fourth, systems of geographical interest can be further categorized in terms of their treatment of time and space:

(a) *Temporal systems* assume a minimum of spatial variation and concentrate on the relations of streams of inputs and outputs through time. Most formalized systems analysis is of this type and links with time series analysis are close.

(b) *Purely spatial systems* assume the instantaneous spatial diffusion of process. Purely spatial processes are abstract constructs which describe either very special systems or unique measurement conditions. They are useful in certain contexts but they permit functional interpretation only by the use of *a priori* hypotheses – for example, maximum entropy assumptions and constraints on interaction between locations (Wilson, 1974; Bennett, 1975) – and do not allow the definition of the optimal control systems.

(c) *Space-time systems.* In these spatial specificity is

introduced – i.e. when the variables are considered to be distributed over a spatial domain or region of interest. Such systems, termed distributed parameter systems by the engineer, are structured according to three main elements (Butkovskiy, 1969, p.28): (1) The connections between the spatial elements over time; (2) the dimensionality of the space considered; (3) the form of measurement of the space-time system which is available.

Most developments of space-time models in the environmental, engineering and physical sciences have been concerned with processes defined as continuously diffentiable variables, whereas most socio-economic and 'geographical' systems have been developed with N-regional, or N-sample irregular data arrays. The time stream of inputs and outputs is made up of either a surface of stimuli and responses (continuous systems) or an array of inputs and outputs (discrete systems), with dimensions equal to the number of discrete spatial locations. Impulse, step, ramp, periodic and other inputs may characterize this surface or array, and the output may be linear or nonlinear, stationary or nonstationary, giving sets of spatial as well as temporal trends and periodicities. These may be analysed by a range of trend-surface, two-dimensional fourier and spectral techniques, and other methods.

Aims of control

Ashby's law of requisite variety states that for an intelligence to gain control over a system it must be able to take at least as many distinct actions as the system can exhibit. In other words, it must be able to take a sufficient number of counter measures so as to force down the variety of possible outcomes to within a restricted range commensurate with the objectives of the controller. This law teaches us that, in order to control a system, one must either increase the possible variety of actions available to one, or operate in some manner so as to restrict the variety of responses which the system can produce (Chorley and Kennedy, 1971, p. 299).

The overall decision-making process to achieve this is fundamentally determined by the components and information upon which the decisions must be based – i.e. for what, for whom, and by whom is control intended? The overall goals and objectives of system response govern the general trajectory or time path of system evolution. These goals may be of two forms: exogenously set, or self-regulatory. Where the system goals are determined outside the system of interest the controller attempts to keep the system at the exogenously determined reference input set point by continuous feedback of deviations, or by more complex feedback-feedforward systems. In environmental systems this corresponds to subsystem regulation within nested control loops. Within the socio-economic system exogenous goals are set under idealist design only in totalitarian regimes. These contrast with self-regulating systems in which goals are determined by consensus as a sum of component interaction involving political, social, economic and environmental feedbacks. Such are the social goals in democracies and anarchies which maintain an internal homeostasis or constancy of operation in the presence of external fluctuations. It will be difficult to forecast the consensus goal derived, since this would depend upon knowing each of the individual democratic forces – i.e. the socio-political power of each component. The resulting effect is easily recognized, however, as a dynamic equilibrium, often in the face of many lagged dynamic responses, according to Le Chatelier's principle.

Thus control systems can be interpreted in three main ways:

(1) As wholly or partly autonomous feedback mechanisms that control the environment. These systems are designed to operate on the input so as to yield the required output. This is most clearly exemplified by the early control systems which were designed to monitor a changing reference quantity and to keep the output equal to a fixed reference input in the presence of external disturbances (disturbance inputs) (as in a thermostat) (Grodins, 1970, pp.2–6). This commonly implies the use of tools (control efficators) and energy sources to control the environment by changing the future course of events towards a goal or selected future target state (reference input). This consists of perception, conception, selection, programming and execution.

(2) As an extension of the life process of man's conscious control over the environment by means of man-machine systems. Such systems combine in a complex way the capabilities of man and machine for optimal decision-making. In this the machine carries out routine tasks that can be described and optimized mathematically and repeated automatically, the man intervening only in infrequent strategic and policy decisions as part of a 'mixed scanning' process. The human hand is an example of such a multilevel system. At a very low level there is automatic adjustment of the hand to the shape of an object to be grasped by feedback from the skin to the actuators, and the adjustment of the prehension force measuring weight (Tomovic, 1969). At a higher (algorithmic) level coded coordination movements are employed, such as the grasping actions to make a fist or sign. At the highest level of decision, the man is required only to determine the algorithm to be used – fist, sign, open hand, etc. – and the rest of the commands proceed automatically (Tomovic, 1969).

(3) As the ability to change the system itself so that it yields the desired output from the existing input (Grodins, 1970). Such changes are effected either by a planned reorganization of the system or by manipulating the disturbance inputs so that the system is forced through an irrecoverable threshold and subjected to internal reorganization. This represents a restructuring of the decision levels, of the algorithms to be employed in achieving given system response, and of the nature of the goals towards which the system is being drawn. Hence, within this control system man–machine and autonomous control systems are contained as lower-level cases.

The common characteristics of most control problems are shown in Figure 21.1 where the system which it is desired to control is made up of a transfer function process with two system inputs: an uncontrolled input sequence $\{X_t\}$ which represents stimuli to system action and an additional controlled input sequence $\{U_t\}$. The controlled input sequence is the one which normally receives the fullest attention. This represents a variable (or set of variables in multivariate systems) that can be manipulated by the controller and decision maker to influence the system under consideration and induce the output to approach more closely to 'desired' output settings. Variables amenable to such manipulation are termed policy instruments. In the overall control system any given output is compared with a reference input or target (r_t) which represents the desired response of the system. The difference between the real and desired output generates an 'error' sequence which, when weighted by the chosen performance index or objective function, is used by the controller to manipulate those variables which are amenable to control (or policy instruments $\{U_t\}$). The performance index is a 'figure of merit' by which deviations from the desired response are scaled (Bennett and Chorley, 1978, pp.94–6).

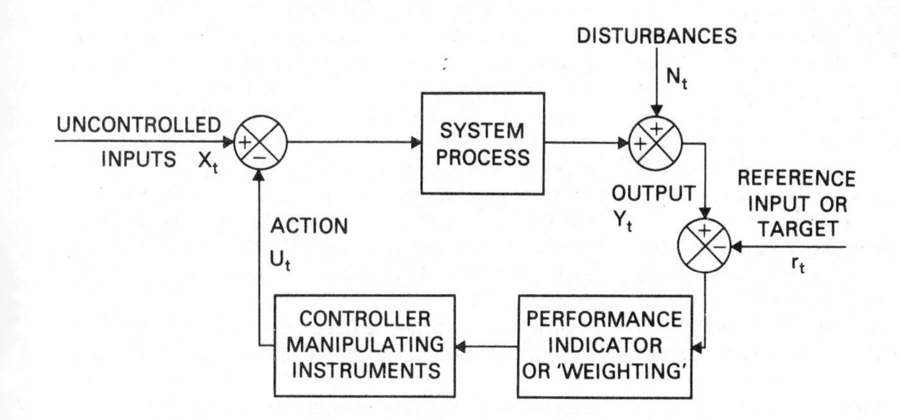

Figure 21.1 The basic components of a control system (from Bennett and Chorley, 1978, p. 96)

Applications of control

As we can see, systems of interest to the geographer can be broadly divided into the morphological and the input/output. These two groups differ greatly in their ambition in circumscribing parts of the real world, in their complexity of structure and in their susceptibility to prediction and control.

Morphological systems are complex webs of variables connected by empirical non-linear mathematical relationships, the operations of which are closely specified by sets of parameters (exponents and constants). Nonlinear systems are extremely ambiguous in their behaviour (Patten, 1975, p.217). The nonlinearity of physico-ecological systems derives from the time variance of the system state, the special variance of inputs and transfers, and from the thresholds and discontinuities which produce sharp disjunctions in the behaviour of the subsystems (Amorocho, 1973, pp.209–10; Patten, 1975, pp.215–6). However, the most obvious source on nonlinearity is the nonstationarity of the storages, in that they operate relatively slowly and their rate of intake depends not only on the rate of supply but also on the state of the storage (Dooge, 1968, p.67). In hydrological systems, for example, a major source of nonlinearity is the behaviour of soil moisture storage and the decay of infiltration capacity with moisture content. In ecosystems, too, nonlinearity results from the fact that rates of mass or energy transfer (fluxes, defined by non-dynamic state variables) between subsystems (storage compartments – e.g. biomass – defined by dynamic state variables) are dependent upon the simultaneous effects of both the 'donor control' (e.g. resources of food limited) and of 'recipient control' (e.g. predator limited) (Kowal, 1971, p.140; Patten, 1975, p.218).

In geography and related disciplines, morphological systems have been applied to particularly extensive and complex relationships, notably ecosystems, and symbiotic models (Bennett and Chorley, 1978, pp.336–55, 382–6, 527–33). The difficulties of understanding the operation of these systems, let alone controlling them, have been set out in detail elsewhere. Suffice it to be said here that their ambitious complexity, empirical specification, non-linear linkages and problems in parameter estimation ensure that their behaviour is largely indeterminate (the unhelpful phrase 'counter-intuitive' has even been employed) when equilibrating and explosive when non-equilibrating. Although recent research is leading to more realistic specification of this type of system, it is true to say that sufficient problems remain to ensure that they are not yet susceptible to the type of control techniques which have been previously outlined.

Studies of interest to the geographer which are susceptible to the more orthodox methods of control are of the much more limited input/output kind. The modelling of economic systems (Bennett and Chorley, 1978, pp. 416–63) has long provided an example of such methods, but a distinctive geographical flavour has been imparted by the increasing introduction of linear or areal elements into time-series modelling. Important examples of this latter work include modelling hydrological flood flow (Bennett and Chorley, 1978, pp.319–31, 342-52 and 356-68), river and air pollution (Nijkamp, 1975; Nijkamp and Verhage, 1976; Bennett and Chorley, 1978, pp.332–4, 368–78 and 496–9; Young, 1978; Whitehead, Young and Michell, 1978; Whitehead, Young and Hornberger, 1979; Whitehead and Young, 1979), sediment movement (Bennett and Chorley, 1978, pp.334–5 and 353–9), plant growth and irrigation management (Lai, 1979), and disease spread (Bennett and Chorley, 1978, pp.378–82; Cliff *et al.*, 1975). However, it is in the field of regional space-time modelling that the most interesting current applications of geographical control system models are occurring (Mackinnon, 1975; Evtushenko and Mackinnon, 1976; Bennett and Tan, 1979b; see summary by Tan, 1979a, pp. 1.3–1.13).

Most of these applications derive in part from the stimulus of work in macro-economics concerned with Keynesian stabilization of national economies (see e.g. Tustin, 1953; Theil, 1966; Pindyck, 1973; Vishwakarma, 1974; Chow, 1975; Aoki, 1976; and Turnovsky, 1977). These applications draw on the mathematical results of Pontryagin *et al.* (Pontryagin, 1962) and Kalman (1960) developed for engineering systems. Seminal contributions in economic geography have been those of Mackinnon (1975), Evtushenko and MacKinnon (1976), Mehra (1975), Lee (1978) and Tan (1979b) concerned with controlling migration in national settlement systems. In addition Williams and Wilson (1976), Coelho and Wilson (1976) and Phiri (1979) have considered the application of control theory to the allocation of retail and other stock in urban systems. Recently, the problem of financial resource allocation has also been a major source of attention, and Bennett and Tan (1979a, b) and Tan 1979a, b) have considered the problem of allocating central government grants between local authorities.

Prospects

It is clear that control theory, although potentially a very important field of dynamic optimization technique, has as yet been developed to only a limited extent in geography. To a large extent the development of further applications of optimal control theory will depend on resolving five outstanding issues (Bennett, 1977, Bennett and Tan, 1979a).

First, the quadratic cost criterion, although having the advantage of simplicity, is an arbitrary weighting; especially in treating both positive and negative deviations from targets with equal weight. For example, increases or decreases from a target level of employment, or from a target for satisfaction of social welfare needs, are not equally tolerable. Moreover, we would not expect positive and negative deviations to be equally tolerable over all regions; some regions (the more developed) might be able to tolerate positive deviations from employment levels (reducing over-heating) but might be less able to tolerate negative deviations (which generate excessive wage-push inflation).

A second issue concerns the fact that system models are usually not known with certainty but are derived from parameter estimates using time series or other data. The important consequence is that a separation principle usually does not hold between parameter estimation and control (Åstrom and Wittenmark, 1971). Attempts at separate treatment of the two problems will not yield a global minimum of the objective function (see Bennett and Chorley, 1978). In response to this problem we may adopt the Åstrom and Wittenmark 'self-tuning regulator' (a special form of Kalman filter) or the learning algorithm given by Chow (1975, Chapter 11). Further developments of this method in geography are still required.

Related to the separation problem is a third issue of identification in a 'closed loop'. The problem is that, by keeping a system at a set point target, exogenous inputs are manipulated as policy instruments in order that the output is unvarying from the target. If this control is perfect, there will be no variation in system variables at all, and by using only time series of data on system inputs and outputs we will not be able to determine the character of the regional system model. The difficulty has been highlighted as of particular significance in road and air traffic modelling by Bennett (1978).

A fourth issue is how to make allowances for changes in coefficients and structure of the system model. Most economic and environmental systems are dominated by evolution and shifts in coefficients in their describing models. There has been growing awareness of this problem (Chow, 1975), and it is relatively simple to extend the structure of the control solutions given above to systems with parameter dynamics. What is lacking, as yet, is a satisfactory method of identifying the parameter models and learning what occurs, when these are not known *a priori*. Again, the self-tuning regulator of Åstrom and Wittenmark (1971) or Chow (1975) learning method must be modified to apply to geographical control systems.

The final outstanding issue of the control approach arises from the property that geographical systems are usually open systems with many cross-flows between regions. If we require to specify goals independently for each region, new ways of thinking with regard to economic and environmental regulators will be required. For example, in almost all decentralized states macro-economic regulation is maintained as a central government function, but we know from mathematical controllability conditions that we cannot achieve n optimal spatial targets without n independent control instruments. We are drawn to accept, therefore, that some measure of disaggregation of macro-economic regulators is required if we are to maintain control of decentralized economic systems.

The prospects for the application of control theory to geographical problems clearly depend on the solution to a number of technical issues, but research to date clearly demonstrates that these are being overcome. We may expect that, as this work becomes better known, control theory methods will be much more widely applied. In particular their application to inter-regional pollution control, catchment and reservoir management, effluent disposal, inter-area resource allocation and urban planning have been, and will be, significant areas for important applications.

References

Amorocho, J. (1973) 'Nonlinear hydrologic analysis', *Advances in Hydroscience*, 9, 203–51.

Aoki, M. (1976) *Optimal control and system theory in dynamic economic analysis*, North Holland: Amsterdam.

Åstrom, K.J., and Wittenmark, B. (1971) 'Problems of identification and control', *Journal of Mathematical Analysis and its Applications*, 34, 90–113.

Bennett, R.J. (1975) 'Dynamic systems modelling of the North West Region', *Environment and Planning, A*, 7, 525-38, 539-66 and 617-36.

Bennett, R.J. (1977) 'Optimal control models of regional economies', *IFAC Workshop on urban, regional and national planning: environmental impacts, Kyoto*, Pergamon: Oxford.

Bennett, R.J. (1978) 'Forecasting in urban and regional planning closed loops: the examples of road and air traffic forecasts', *Environment and Planning*, A, 145–62.

Bennett, R.J. (1979) *Spatial time series: Analysis, forecasting and control*, Pion: London.

Bennett, R.J., and Chorley, R.J. (1978) *Environmental systems: Philosophy, analysis and control*, Methuen: London.

Bennett, R.J., and Tan, K.C. (1979a) 'Stochastic control of regional economies' in C.A.P. Bartels and R.H. Ketellapper (eds), *Exploratory and explanatory statistical analysis of spatial data*, Martinus Nijhoff: Leiden.

Bennett, R.J., and Tan, K.C. (1979b) 'Allocation of the UK Rate Support Grant using the methods of optimal control', *Environment and Planning*, A, 11, 1011–27.

Butkovskiy, A.G. (1969) *Distributed Control Systems*, Elsevier: New York.

Chorley, R.J., and Kennedy, B.A. (1971) *Physical Geography: A Systems Approach*, Prentice-Hall: London.

Chow, G.C. (1975) *Analysis and control of dynamic economic systems*. Wiley: New York.

Cliff, A.D., Haggett, P., Ord, J.K., Bassett, K.A., and Davies, R.B. (1975) *Elements of Spatial Structure*, Cambridge University Press.

Coelho, J.D., and Wilson, A.G. (1976) 'The optimum location and size of shopping centres', *Regional Studies*, 10, 413–21.

Dooge, J.C.I. (1968) 'The hydrologic cycle as a closed system', *Bulletin of the International Association of Scientific Hydrology*, 13, (1), 58–68.

Evtushenko, Y., and Mackinnon, R.D. (1976) 'Nonlinear programming approach to optimal settlement system planning', *Environment and Planning*, A, 8, 637–54.

Grodins, F.S. (1970) 'Similarities and differences between control systems in engineering and biology' in E.O. Attinger (ed.), *Global System Dynamics*, pp. 2–6, S. Karger: Basel.

Kalman, R.E. (1960) 'A new approach to linear filtering and prediction problems', *Transactions ASME, Journal of Basic Engineering*, D82, 35–45.

Kowal, N.E. (1971) 'A rationale for modelling dynamic ecological systems', in B.C. Patten (ed.), *Systems Analysis and Simulation in Ecology*, Vol. 1, pp.123–94, Academic Press: New York and London.

Lai, P.W. (1979) 'A transfer function approach to short term stochastic models for soil tension in a sandy soil', unpublished PhD thesis, University of London.

Lee, R. (1978) 'Application of optimal control theory to urban and regional planning', Paper presented at Association of American Geographers Conference, New Orleans (Department of Geography, University of Iowa).

Mackinnon, R.D. (1975) 'Controlling interregional migration processes of a Markovian type', *Environment and Planning*, A, 7, 781-92.

Mehra, R.K. (1975) *An optimal control approach to National Settlement Planning*, IIASA Report Rm-75-58, Laxenburg, Austria.

Nijkamp, P. (1975) 'Spatial interdependencies and environmental effects' in A. Karlqvist, L. Lundqvist and F. Snickars (eds), *Dynamic Allocation of Urban Space*, Saxon House: Farnborough.

Nijkamp, P., and Verhage, C. (1976) 'Cost benefit analysis and optimal control theory for environmental decisions: a case study of the Dollard estuary' in M. Chatterji and P. van Rompuy (eds), *Environment, Regional Science and Interregional Modelling*, Springer: Berlin.

Patten, B.C. (ed.) (1975) *Systems Analysis and Simulation in Ecology*, Vol. 3, Academic Press: New York and London.

Phiri, P.A. (1979) 'Equilibrium points and control problems in dynamic urban modelling', unpublished PhD thesis, University of Leeds.

Pindyck, R. (1973) *Optimal planning for economic stabilisation*, North-Holland: Amsterdam.

Pontryagin, L.S. (1962) *The mathematical theory of optimal processes*, Wiley: New York.

Tan, K.C. (1979a) 'Optimal control of linear econometric systems with linear equality constraints on the control variables', *International Economic Review*, 20, 253-8.

Tan, K.C. (1979b) 'Constrained control of spatial systems: an application to allocation of the Rate Support Grant', unpublished PhD thesis, University of London.

Theil, H. (1966) *Applied Economic Forecasting*, North-Holland: Amsterdam.

Tomovic, R. (1969) 'On man-machine control', *Automatica*, 5, 401-4.

Turnovsky, S.J. (1977) 'Optimal control of linear systems with stochastic coefficients and additive disturbances' in J.D. Pitchford and S.J. Turnovsky (eds), *Applications of control theory to economic analysis*, North Holland: Amsterdam.

Tustin, A. (1953) *The mechanism of economic systems*, Harvard University Press: Cambridge, Mass.

Vishwakarma, K.P. (1974) *Macro-economic regulation*, Rotterdam University Press.

Whitehead, P.G., and Young, P. (1979) 'Water quality in river systems: Monte-Carlo Analysis', *Water Resources Research*, 15, 451-9.

Whitehead, P.G., Young, P., and Hornberger, G. (1979) 'A systems model of stream flow and water quality in the Bedford-Ouse River', *Water Resources Research*, 15.

Whitehead, P.G., Young, P., and Michell, P.A. (1978) 'Some hydrological and water quality modelling studies in the ACT region', *Hydrology Symposium, Canberra, 5-7 September*, 1-7.

Williams, H.C.W.L., and Wilson, A.G. (1976) 'Dynamic models for urban and region analysis' in T. Carlstein, P.N. Parkes and N. Thrift (eds), *Timing space and spacing time*, Arnold: London.

Wilson, A.G. (1974) *Urban and Regional Models in Geography and Planning*, Wiley: London.

Young, P. (1978) 'General theory of modelling for badly defined systems' in G.C. Vansteenkiste (ed.), *Modelling, Identification and Control in Environmental Systems*, 103-35, North Holland: Amsterdam.

Chapter 22

Graph theory

A.D. Cliff and P. Haggett

Introduction

There are few areas of British quantitative geography where it is harder to be chauvinistic than in graph theory. This branch of mathematics owes its origins to an eighteenth-century Swiss mathematician, and many of its most fundamental developments in this century have been made by French topologists. The recent entry of graph-theoretical concepts into geography is unequivocally American. In physical geography, the entry point was Columbia University, New York City, where in the 1950s a group under A.N. Strahler examined the relationship between stream ordering, watershed morphology and basin hydrology. A simultaneous independent development took place on the West Coast, where a group under W.L. Garrison at the University of Washington, Seattle, explored the applications of graph theory to regional transport networks.

Much of the British work reviewed in this chapter (where *British* is loosely defined as those working in institutions in the United Kingdom and/or publishing in British journals) is therefore largely a lagged response to impulses starting elsewhere. There are, however, important exceptions, in that the work of A.G. Wilson and R.H. Atkin (discussed elsewhere in this volume) has important potential applications in a graph-theoretical context.

The approach followed in this chapter is to set up a taxonomy of graphs in a tabular form. The structure of this table and its individual cells are used to lead the discussion in the remainder of the paper. Throughout, the emphasis is on giving representative examples of work; for more comprehensive accounts with substantial bibliographies, see the early text by Haggett and Chorley (1969) and the more recent reviews by Tinkler (1977, 1979) and by Cliff, Haggett and Ord (1979).

A taxonomy of graphs

A graph is a representation of a network that can be used as a means of studying its structural properties. In the simplest graph, this consists of two sets of elements, the vertices and the edges, and the only information we have about any edge or vertex is its existence (1) or its absence (0). By contrast, a complex graph will have information about the direction and values (say, capacity, flow, frequency) of any edge and a separate set of information about any vertex.

Table 22.1 A taxonomy of graphs

EDGE / VERTEX		PRESENCE OR ABSENCE	INFORMATION LEVEL ON EDGES	
			CATEGORICAL	CONTINUOUS
PRESENCE OR ABSENCE		CELL I	CELL II (a)	CELL III (a)
INFORMATION LEVEL ON VERTICES	CATEGORICAL	CELL II (b)	CELL II (c)	
	CONTINUOUS	CELL III (b)		CELL III (c)

Table 22.1 presents the relationship between the level of information available about edges and vertices in order to build up a taxonomy of graphs. Data may be available at three different levels:

(1) dichotomous (presence or absence);
(2) categorical (dichotomous, polychotomous, or ordered data on the edges or vertices present in the graph);
(3) continuous (interval or ratio data on the edges or vertices present in the graph).

Table 22.1 is organized in such a way that as we move from Cell I to higher-order cells, the level of information about elements in the graph increases; that is, we move from an abstract graph *sensu strictu* towards a real-world network, as illustrated by journey-to-work trips from origin areas to destination areas within a city. The sequence is useful in so far as it links those areas of quantitative spatial analysis which are conventionally labelled as *graph theory* to those which are discussed in the existing literature under such headings as integer programming or maximum-entropy modelling. In the following sections, we will adopt the sequence in Table 22.1 as the framework for discussion. In each case we give first some examples of the contributions by British geographers, and then turn to identify some promising research areas.

Elementary graphs

Geographical contributions

Much of the geographical work on elementary graphs (Cell I in Table 22.1) has focused upon the physical interpretation of the eigenvalues and eigenvectors of transportation networks whose topological structure is presented in the form of a connection or shortest-path matrix (Gould, 1967). A convenient algorithm for obtaining the eigenvalues and eigenvectors of matrices, and one which is highly suggestive of the physical interpretation of these quantities, has been described by Kendall (1957, pp.19–26) and extended by Cliff and Ord (1977). Given any matrix with real eigenvalues, this algorithm will, in general, extract these eigenvalues in order of descending absolute value. If $\mathbf{A}$ is the matrix in question (which we shall initially assume to be symmetric), then corresponding to each eigenvalue λ_j is an eigenvector $\mathbf{v}_j$ of unit length satisfying

$$\mathbf{A}\mathbf{v}_j = \lambda_j\mathbf{v}_j \ .$$

The algorithm to extract these eigenvalues and eigenvectors has the following form:

Step 1 Choose $\mathbf{v}^{(0)}$ to be an initial approximation to an eigenvector. Here, it is adequate to choose $\mathbf{v}^{(0)}$ to be $(1, 1, \ldots, 1)^T$. Scaling the largest component to unity is convenient during the iterative cycle, and quicker than scaling the sum of squares to unity, provided that we check that the largest element is not too close to zero before scaling. If the largest element is close to zero, this implies that the initial solution is close to an eigenvalue already extracted. In such situations, the algorithm proceeds by changing the signs of individual elements of $\mathbf{v}^{(0)}$ until the deadlock is broken. Since the vectors

$$\mathbf{v}_j^{(0)} = (\overbrace{-1, -1, \ldots, -1,}^{j-1} \ \overbrace{1, 1, \ldots, 1}^{p-j+1})^T$$

form a basis for p-dimensional space, this procedure will always find an appropriate starting-point unless the remaining eigenvalues are very close to, or equal to, zero.

Step 2 For $i = 1, 2, 3 \ldots$, let $\mathbf{v}^{(i)} = \mathbf{A}\mathbf{v}^{(i-1)}$, and continue until the solution has converged.

Once the first eigenvector $\mathbf{v}_1$ has been found and normalized, we form the new matrix $\mathbf{A}^{(1)} = \mathbf{A} - \lambda_1\mathbf{v}_1\mathbf{v}_1^T$, where λ_1 is the eigenvalue corresponding to $\mathbf{v}_1$. We can then repeat the above steps to obtain the second eigenvector. If the original choice of $\mathbf{v}^{(0)}$ leads to the problems described above, then this original choice can be modified as before. We continue in this way until all of the eigenvectors are known. Further details for non-symmetric matrices are given in Cliff, Haggett and Ord (1979).

The value of this algorithm in analysing graphs was considered by Tinkler (1972). The algorithm makes clear the fact that the components of the first eigenvector are directly proportional to the sum of the entries in the corresponding row or column of the matrix under consideration, and that the eigenvalue is then directly proportional to the largest row or column sum. This implies that when a network is being analysed by means of its adjacency matrix, the more highly connected the network is, the larger will be its eigenvalues; and the higher the valency of each vertex, the larger will be the components of the corresponding eigenvector. However, if the shortest-path matrix is analysed, the good overall connectedness of the graph will be reflected in a low first eigenvalue, and well-connected individual vertices will have low eigenvector components. Badly-connected networks and vertices will exhibit the reverse characteristics – namely, low eigenvalues and eigenvector components for the adjacency matrix, and high eigenvalues and eigenvector components for the shortest-path matrix.

The fact that the largest eigenvalue is a measure of the overall connectedness of a transportation network can be established by defining bounds on it (see Cliff and Ord, 1977). If λ_1 is the largest eigenvalue of the adjacency matrix $\mathbf{A} = (a_{ij})$, then

$$\frac{1}{n}\sum_{i,j} a_{ij} \leq \lambda_1 \leq \max_i \sum_j a_{ij} \ .$$

The lower bound is simply the average number of edges per vertex, and the upper bound is the maximum vertex-valency.

Similarly if $S = (s_{ij})$ is the shortest-path matrix, then

$$\frac{1}{n}\sum_{i,j} s_{ij} \leq \lambda_1 \leq \max_i \sum_j s_{ij} \ .$$

Here the lower bound is the average number of edges travelled per journey, and the upper bound is the longest journey (that is, the longest shortest-path) in the network.

The physical interpretation of the second and subsequent eigenvalues and eigenvectors has been the subject of some controversy (see, for example, Hay, 1975, and Tinkler, 1975). However, there is some evidence to suggest that the outvalencies for each vertex may be used to separate out well-connected subgraphs within the main graph.

In order to illustrate these points, we have computed the eigenvalues and eigenvectors of the adjacency and shortest-path matrices of several artificial networks (Figure 22.1), and of the graphs of the airline networks connecting the seven largest cities in each of four countries (Figure 22.2). In

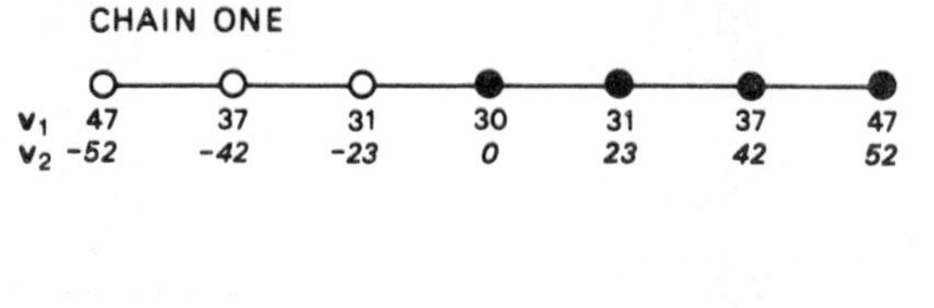

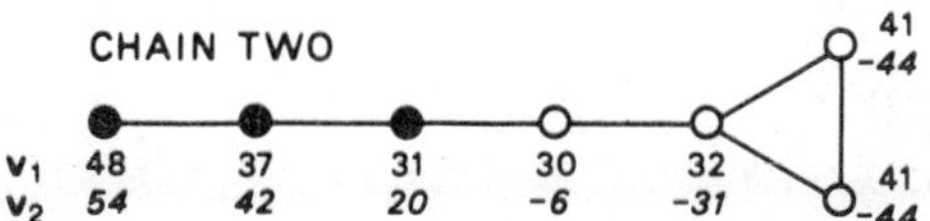

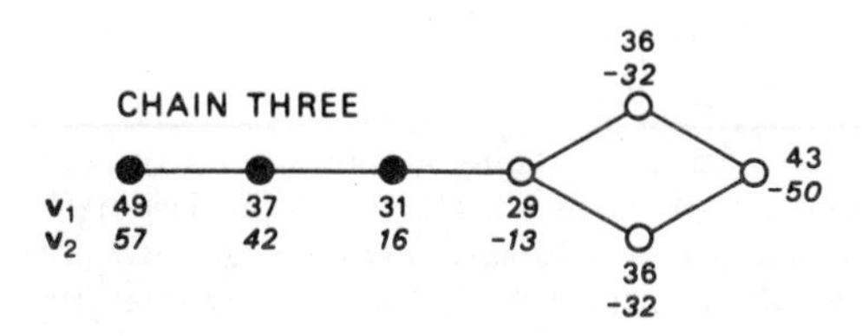

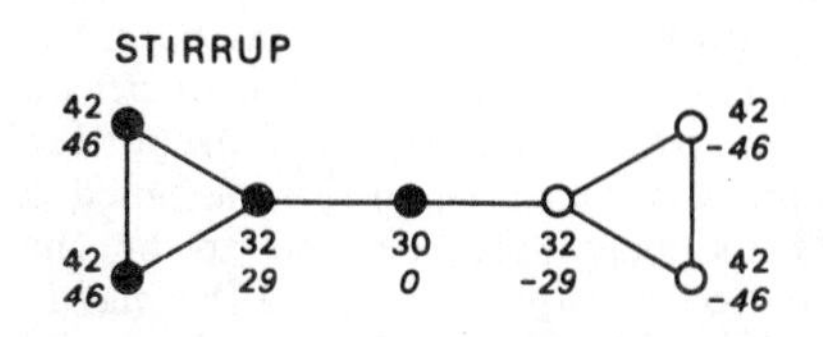

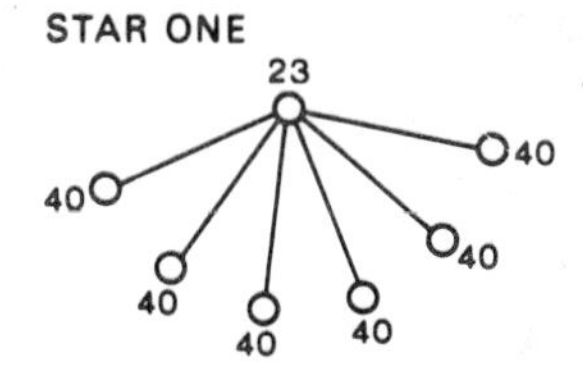

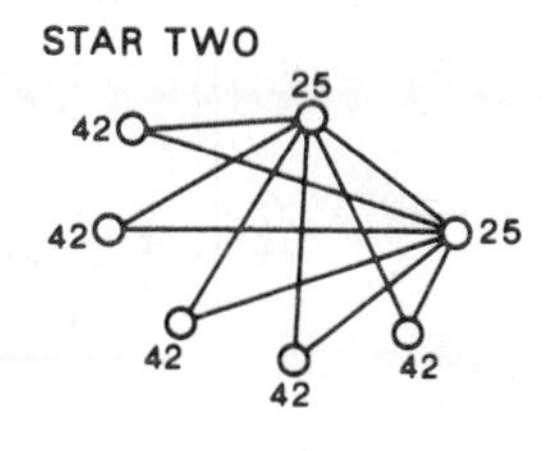

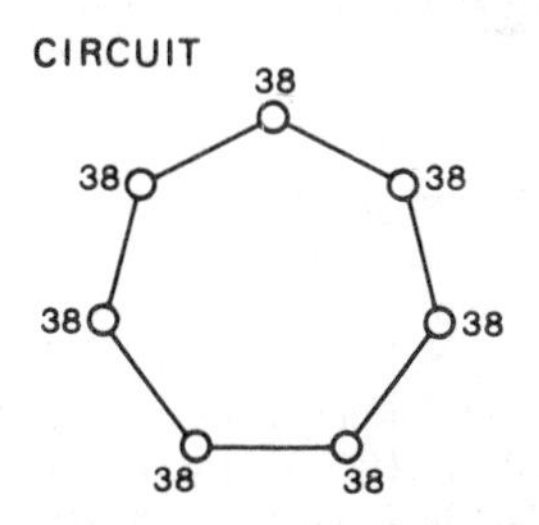

Figure 22.1 First (Roman numbers) and second (italicized numbers) eigenvector components for some hypothetical networks (Cliff, Haggett and Ord, 1979, p. 314)

these diagrams we have given next to each vertex the components of the eigenvector corresponding to the largest and, in most cases, the second-largest eigenvalue of **S** in absolute value. In general, the larger the valency of the vertex, the lower is the corresponding component of the first eigenvector. In the networks of Figure 22.1, the components of the second eigenvector (in italics) seem to delineate distinct regions, with the 'between-regions' links being relatively weak compared to the 'within-region' links. In Figure 22.2, the regions which arise are indicated by shading the corresponding groups of vertices; the sizes of the vertices indicate the ranking of the cities according to population size. Multiple eigenvalues tend to occur whenever two vertices are 'interchangeable' – that is, when they are joined to the same vertices. For such networks the second eigenvector is meaningless. Although empirical networks are less prone to give rise to multiple eigenvalues, even here eigenvalues of similar magnitude may arise, thereby making regional interpretations rather difficult.

Areas for further research

A recent collection of essays on *Applications of Graph Theory* by Wilson and Beineke (1979), allows the contribution of geographers to be compared directly with those in ten other applied fields ranging from chemistry to operations research. What is striking about the geographical work is its pragmatic and empirical nature, with a strong emphasis on description and cross-regional comparisons.

Exploitation of the known mathematical structure of graphs is common in a number of applied fields. Analysis of electrical networks commonly uses the Kirchhoff and Maxwell rules which give topological formulae for the determinants of the various matrices. In chemistry, where vertices and edges represent, respectively, the atoms and the covalent chemical bonds between them, complex reactions can be studied in terms of Desargues-Levi graphs. In linguistics, the known properties of cyclic trees are used in the decomposition of sentence structure. In these and similar cases (Wilson and

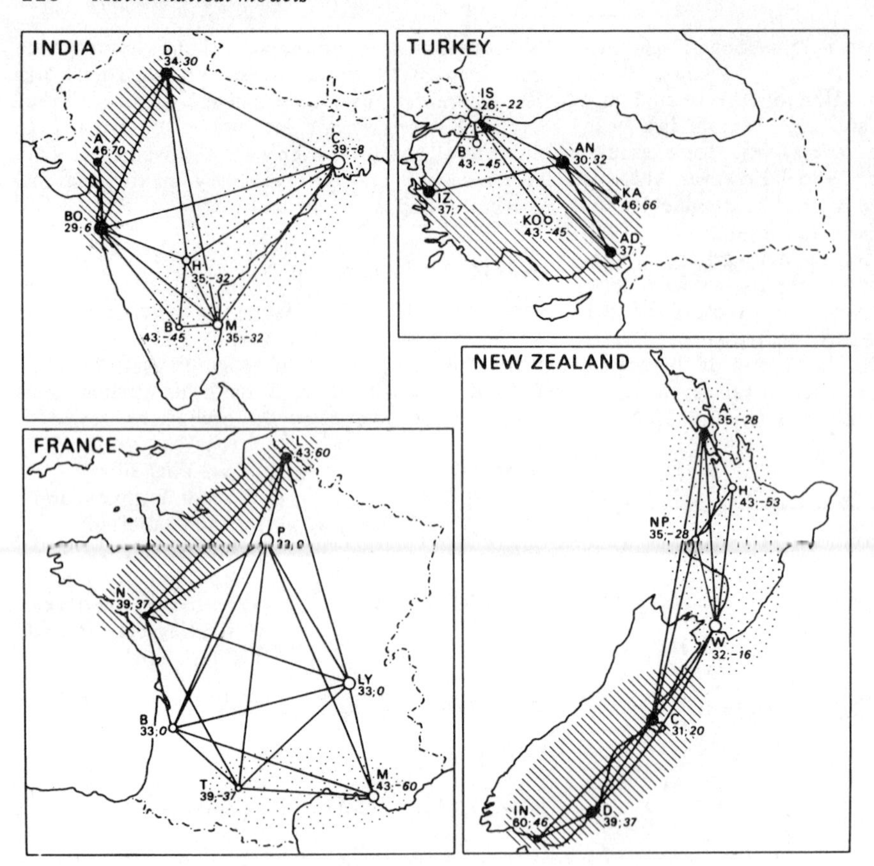

Figure 22.2 First (Roman numbers) and second (italicized numbers) eigenvector components for the airline networks between the seven largest cities of India, Turkey, France and New Zealand. Network 'regions' are shaded; nodes are proportional to population size (Cliff, Haggett and Ord, 1979, p. 316)

Beineke, 1979), different areas of graph theory which have been rather fully explored from a mathematical viewpoint, provide starting-points for new research insights in applied fields.

Directed, chromatic and ordered graphs

Geographical applications

The second tier of cells in Table 22.1 includes graphs where either the edges (IIa), or the vertices (IIb), or both (IIc) are labelled with information measured in a categorical manner. The categories may be dichotomous (such as the *direction of flow* along an edge), or polychotomous (for example, vertices might be *coloured* according to some categorization) or *ordered* where the several categories are in a fixed relationship to each other. Graphs labelled in these three ways are termed directed graphs (or digraphs), chromatic graphs, and ordered graphs, respectively. Each of the three types may occur in any of the three cells.

Cell IIa: Streams as hierarchical and random graphs. Apart from any anastomosing section (that is, a stream section which forms a closed loop), stream channels show all the topological properties of a rooted tree. Such branching graphs have attracted the attention of earth scientists for the last half-century, as part of their search for answers to questions about stream evolution and the relationship between the morphology of a stream network and its hydrology (for example, the flood characteristics). One significant research area has been the development of random topology models of stream systems based on their combinational properties. Under this approach, each channel is assigned a number called its magnitude, defined to be the total number of sources upstream. Thus the magnitude of each edge is the sum of the magnitudes of the edges leading into it. The magnitudes of stream channels have been used by Strahler (1964) to define the *order* of a stream system. Channels with magnitude 1 are called first-order channels. Where these merge, a second-order channel is formed; where two second-order channels merge, a third-order channel is formed, and so on. The highest order stream segment fixes the order of the system. The amount of branching in the system is summarized in the *bifurcation ratio*. If N_i is the number of streams of order i, the ith bifurcation ratio, R_i, is defined as $R_i = N_i/N_{i+1}$. The average bifurcation ratio, $\bar{R}$, is the average over all R_i for the stream system under consideration.

Stream networks with the same magnitude may

Figure 22.3 The forty-two topologically distinct channel networks with six sources and eleven links (Cliff, Haggett and Ord, 1979, p. 298)

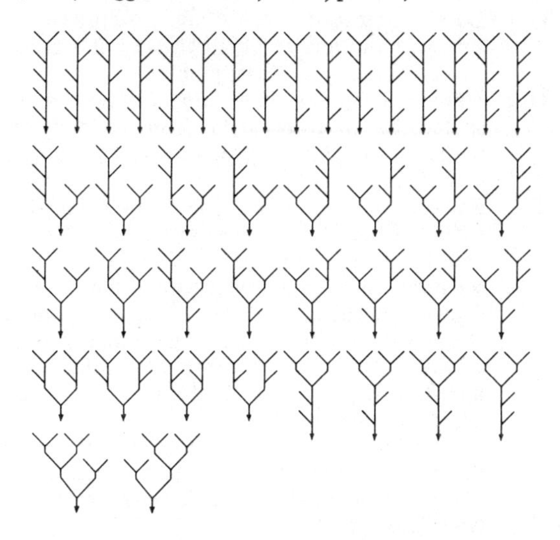

have different topological structures, as is illustrated in Figure 22.3. This shows the forty-two topologically distinct channel networks (TDCNs) with six sources and eleven links. If there are k sources, then the number N(k) of TDCNs is given by the Catalan number

$$N(k) = \frac{1}{2k-1}\binom{2k-1}{k}.$$

For networks with one to nine sources, these numbers are 1, 1, 2, 5, 14, 42, 132, 429 and 1430, respectively.

Given the large number of possible TDCNs, geographers have been interested in two related questions: (i) which is the 'most probable' pattern?, and (ii) do naturally-occurring stream networks approximate to this most probable pattern?

Consideration of Figure 22.3 shows that it is possible to divide the forty-two TDCNs there into two groups in terms of their Strahler-order – namely, sixteen second-order networks (in the first row), and twenty-six third-order networks (in rows 2-5). In Table 22.2, we classify the TDCNs with $k \leqslant 9$ according to their Strahler-order. Thus for $k = 9$, a third-order network is the most likely type, with a probability of $\frac{1288}{1430}$ ($\triangleq 0.90$), whereas the second-order and fourth-order systems have probabilities of about 0.09 and 0.01, respectively.

With a large number of sources, the number of TDCNs becomes very large indeed. For example, a stream system with fifty sources has $\frac{1}{99}\binom{99}{50} \triangleq 5.10 \times 10^{26}$ possible patterns. At one extreme, the fifty first-order streams might be flowing into a single elongated second-order channel to give the highly improbable average bifurcation ratio, $\bar{R} = 50$. At the other extreme, we can envisage a highly-branched structure with $\bar{R} \triangleq 2.22$, arising from a sixth-order system with a sequence of 50, 25, 12, 6, 3, 1 stream segments of ascending Strahler-order. Werrity (1972) has shown that the most probable stream number set with fifty sources has the sequence 50, 12, 3, 1, giving a fourth-order system with $\bar{R} = 3.83$; the probability of such a stream system arising is approximately 0.097. Fourth-order networks of all kinds make up the overwhelmingly likely set (about 60 per cent) of all possible stream systems with fifty sources.

Empirical studies from a number of different climatic and geological environments show a clustering of bifurcation ratios in the range 3.5 to 4.0. One of the most interesting results of such work on random networks is that the most probable network of a given order is that in which the bifurcation ratio R_1 is close to 4.0. Werrity's detailed fieldwork in south-west England has confirmed the general agreement between the observed topological characteristics of stream systems and those predicted by the random TDCN model. Divergences from the random model, where they occur, can be attributed to specific geological controls or erosional history. Thus we can conclude that most stream systems have a branching structure which satisfies the most probable TDCN with $R_1 \triangleq 4.0$. By regarding stream systems as random graphs in this way, a yardstick has been obtained which allows a clearer identifica-

Table 22.2. Topologically distinct channel networks

Number of sources k	*Number of TDCNs with Strahler order* *1*	*2*	*3*	*4*	*Total number of TDCNs*
1	1				1
2		1			1
3		2			2
4		4	1		5
5		8	6		14
6		16	26		42
7		32	100		132
8		64	364	1	429
9		128	1288	14	1430

tion of departures from randomness generated by specific environmental controls.

Cell IIb: Spatial diffusion in chromatic graphs. Insights into the structure of diffusion processes can often be gained by reducing the map to a graph, in which the areas become the nodes or vertices and the links between them the edges. Nodes are then appropriately coloured depending upon whether or not adoption has occurred at a given node. Such an approach was taken by Haggett (1976) in an analysis of measles epidemics in Cornwall. Haggett constructed seven different graphs of the twenty-seven local authority (General Register Office) areas in Cornwall (Figure 22.4); each graph was designed to correspond as closely as possible with a hypothetical diffusion process, namely:

G–1 Local contagion with the assumption of spread only between contiguous GRO areas. Thirty-four edges. Planar graph.

G–2 Wave contagion with the assumption of spread by shortest paths from an endemic reservoir area (Plymouth). Twenty-eight edges. Planar graph.

G–3 Regional contagion with the assumption that spread is locally contagious and not on a country-wide basis, but rather within two separate regional subsystems (east and west Cornwall). Thirty-two edges. Planar subgraphs.

G–4 Urban-rural contagion with the assumption of spread within sets of urban and rural communities treated as separate subgraphs. 181 edges. Nonplanar subgraphs.

G–5 Population size with the assumption of spread down the hierarchy of population size from largest to smallest centre. Twenty-six edges. Nonplanar graph.

G–6 Population density with the assumption of spread through the density hierarchy. Twenty-six edges. Nonplanar graph.

G–7 Journey-to-work contagion with the assumptions that (a) these flows provide a surrogate for spatial interaction between areas, and (b) that spread follows high interaction links. Fifty-eight edges. Nonplanar graph.

To discriminate between the seven graphs, all the 222 weekly maps of measles outbreaks in Cornwall, 1966–70, studied by Haggett were translated into outbreak/no outbreak terms. Vertices on each of the 1554 (222 x 7) graphs were colour coded black (B = outbreak) or white (W = no outbreak) and the Moran BW statistic under non-free sampling (see Chapter 10 on autocorrelation in space and time in this volume) was evaluated to measure the degree of contagion present in the graph. The greater the degree of correspondence between any graph and the transmission path followed by the diffusion wave, the larger (negative) should become the standard score for BW. In practice, there may be practical problems of common links as discussed in Haggett (1976, p. 145).

As a result of testing the spread patterns on the seven different graphs, the following preliminary observations could be made.

(1) The spatial diffusion process was readily separated into inter-epidemic and epidemic periods, with different spread processes dominant in each.

(2) During the long periods between main epidemics there was a lower level of contagion on the graphs based upon spatial structure (G–1 through G–4 and G–7), so that any diffusion currents recorded were extremely weak and poorly structured. Infections persisted in the larger population clusters (G–5, G–6) and moved slowly through the low-

Figure 22.4 Level of autocorrelation among Cornish GRO areas reporting measles cases, weeks 1–40 of the series studied by Haggett (1976). This period covers one major epidemic. Definitions of graphs G_1 to G_7 are given in text. Vertical pecked lines indicate epidemic peak in terms of number of reported cases. Horizontal peaked line indicates $\alpha = 0.05$ significance level (one tail) in a test of positive spatial autocorrelation (Cliff and Haggett, 1979, p. 23)

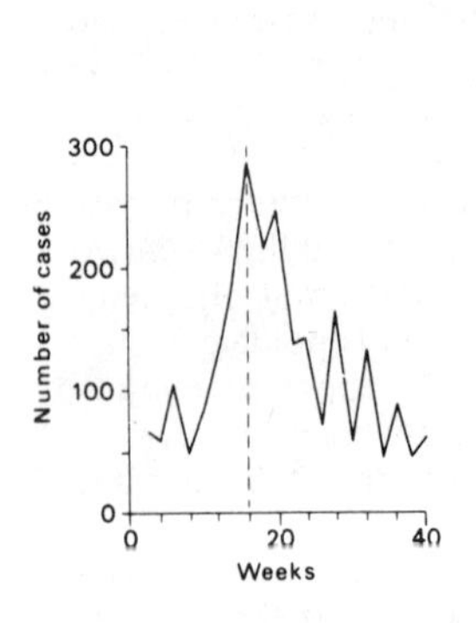

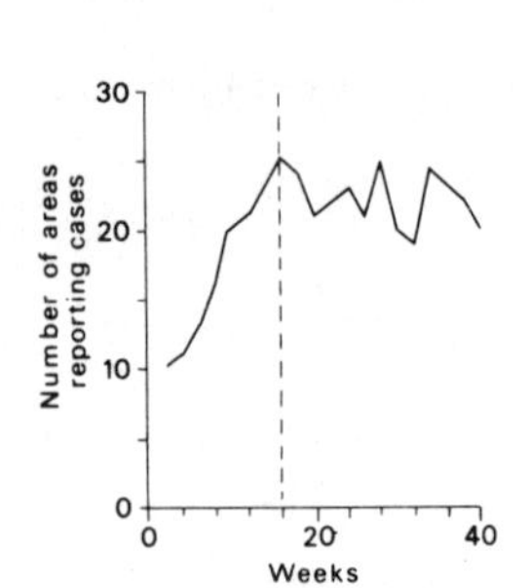

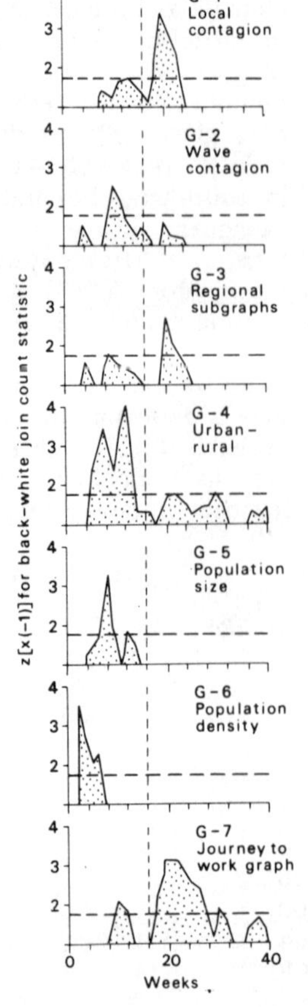

density rural areas in a sporadic manner.

(3) During the shorter epidemic periods, the general level of contagion was about one-and-a-half times higher, so that spatial processes were more distinctive and easier to monitor (see Figure 22.4).

(3a) The advance phase of the epidemic, which started about week 8 in Figure 22.4, was marked by a rapid increase both in intensity and spread. Population size became a less important determinant; wave and town–country effects increased sharply (G–2, G–4).

(3b) At the peak of the epidemic wave, local contagion and regional effects became important, setting up strong areal contrasts between compact clusters of infected/non-infected areas (G–1, G–3).

(3c) During the subsequent retreat phase, the falloff in intensity was not associated with a corresponding reduction in geographical area, so that the epidemic appeared to decay spatially *in situ*. Contagion in G–1, G–2 and G–3 fell steadily during the retreat phase: one exception was the spatial interaction model, G–7, which showed somewhat higher values for a greater length of time after the peak; this implies that longer-range contacts between population centres may come later in the history of the epidemic wave.

Although this suggested history of the diffusion process of an epidemic must remain speculative for the reasons given in Cliff and Haggett (1979), this analysis of a spread process using chromatic graphs identifies for a childhood disease the hierarchical and contagious elements generally associated with innovation diffusion; it further indicates that there is a strong contrast between the early build-up (population size dominates) and fade-out phases (spatial structure dominates) of an epidemic.

Cell IIc: Market cycles as ordered graphs One major area of geographical research concerned with the ordered structure of human settlements is central-place theory. Settlements may be conceived as an irregular planar lattice in which functions are provided on a permanent or periodic basis. For example, rural centres may hold markets on different days of the week. It will be clear (see Figure 22.5) that such periodic functions may form a 'market cycle' in time and a 'market ring' in space. In the simple example shown, the six smaller settlements form a regular once-a-week cycle, with a seventh vacant day for rest or restocking: arrows indicate the ring in space formed by the presumed movement of an individual trader. Empirical regional studies show a wide variety of cycles which are unrelated, or only loosely related, to the conventional seven-day week. Two-day and four-day systems have been described in East and West Africa, and ten-day cycles have been noted in China.

Tinkler (1973) has approached the periodic market as a map-colouring problem, in which no market town in spatially adjacent (contiguous) counties can have the same colour (market day). Obviously, direct competition would be self-defeating for the markets in such a situation (akin to running summer fêtes in adjacent parishes on the same Saturday in June!). An 'efficient colouring' of the set of counties is one in which no two contiguous counties have the same colour. What is the minimum number of colours (days) needed to satisfy

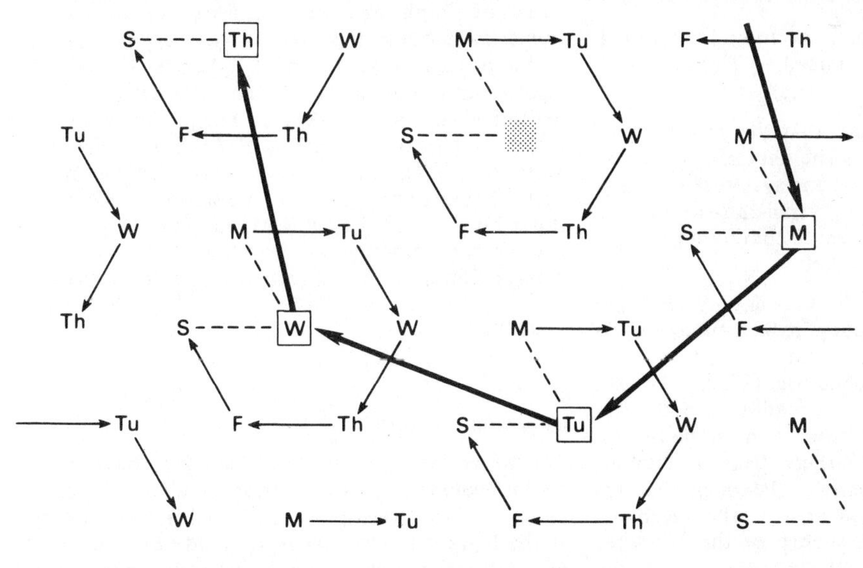

Figure 22.5 Hypothetical market cycles in time and space based upon a six-day week (Cliff, Haggett and Ord, 1979, p. 322)

this condition for all counties?

This problem was solved by the Appel-Haken theorem which proved in 1976 the long-held conjecture that every map can be coloured with four colours in such a way that neighbouring counties are assigned different colours (Appel and Haken, 1977). It is evident, therefore, that four days will ensure that there are no conflicts between adjacent markets. Figure 22.6 shows some possible four-day periodic systems. Note that we can conceive of both a closed and an open system, but that greater

Areas for further research

Each of the three fields illustrated above is already the subject of continuing empirical research. The combinational properties of both stream networks and of central-place systems are actively pursued; likewise diffusion processes operating on chromatic graphs are being studied further by Cliff *et al.* (1981) in an attempt to isolate the trackways along which epidemics spread.

In statistical analysis, the major advances made in

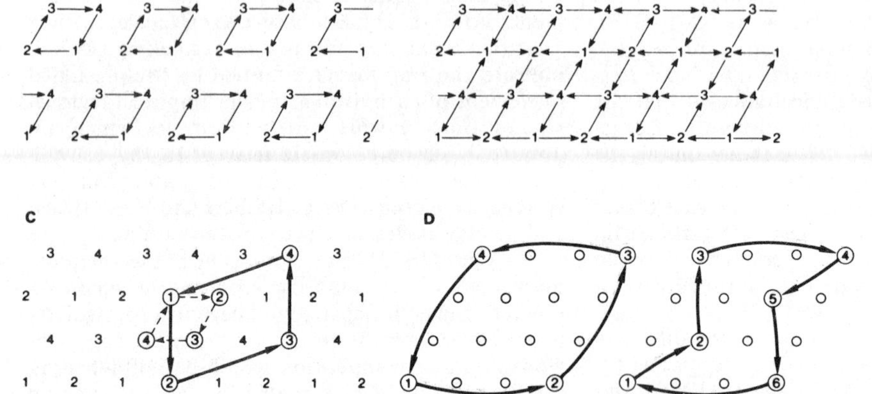

Figure 22.6 Four-day periodic market systems based upon A closed systems, B open systems. C shows the integration of lower and higher order circuits and D the way in which circuits are modified to absorb additional centres (Cliff, Haggett and Ord, 1979, p. 323)

advantages (in terms of exchange of goods) accrue under the open system. Higher and lower order periodic systems may be linked together, as shown in Figure 22.6C.

The four-colour system is of special interest as a theoretical lower bound. Indeed, as Tinkler (1973) speculates:

> Is it too much to suggest that the widely found four day market week is the natural response to the topology of actual market systems? In reality this will probably arise from trial-and-error adjustments of the market system to eliminate same-day conflicts for adjacent markets.

One further advantage of the four-day cycle is that it is relatively easy to modify it to a six-day cycle by including new markets within the centres of the triangles in the lattice structure (Figure 22.6D). With this modification, it fits readily into the six-working-day calendar common in most countries. As a counterbalance to these theoretical ring studies, Skinner (1964/65) summarizes the main features of a ten-day cycle for rural areas in the Szechwan province of China. The resilience of these market structures is attested by the breakdown of the more centralized state trading centres superimposed on this area in the early 1950s, and the total re-establishment of the traditional market cycle a decade later.

the study of categorical data in the 1970s are now being introduced into geography (Wrigley, 1979), and may be expected to bring side benefits to the study of graph structures measured in categorical terms. In mathematics, one of the more interesting methods to emerge in the 1970s was 'Q-analysis' ('polyhedral dynamics' is an alternative term) developed by R.H. Atkin at Essex University. A major empirical application of Atkin's methods is currently under way at Cambridge University as part of a general study of international television flows by a team led by P.R. Gould, and the publication of their findings in the early 1980s may well open an important new direction for applied research.

Higher-order graphs

Geographical contributions

The outer tier of cells in Table 22.1 spans those graphs where the information loaded on edges (IIIa), vertices (IIIb), or both (IIIc) is measured at the highest levels; that is, on continuous interval or ratio scales. While each of the three cells is well represented in British quantitative geography over the review period, they are not analysed separately here since they overlap directly with other important research problems and are thus covered in other chapters of this book. Topics in Cell IIIa, such as Forer's (1974) use of cost and time information on edges to produce space-time maps of airline graphs of New Zealand and the Pacific involved multi-dimensional scaling. It is thus properly discussed by A.C. Gatrell in Chapter 15. Likewise Cell IIIb topics, such as Cliff and Ord's use of connection matrices for spatial autocorrelation and correlogram analysis, are referenced in Chapter 10. Finally Cell IIIc problems, where both edges and vertices are fully measured, are fundamental inputs to gravity models, maximum entropy models, and linear programming (transportation problem) models; these are described by M.J. Batty in Chapter 17 on urban modelling.

Areas for further research

The implication of the foregoing paragraph is that while graphs have formed an *intrinsic* part of the studies described in these other chapters, their structure was not the prime purpose of the analysis. There are grounds for arguing that an *extrinsic* emphasis on the underlying nature of the graph itself should be given more weight in future studies, namely: (i) that the spatial property of the graph affects our estimates of the results obtained in behavioural studies; and (ii) that a graph, rather than a continuous map surface, provides a more faithful representation of geographical reality in spatial allocation studies.

We have already seen attention being paid to the first argument, (i) in the running trans-Atlantic debate in the pages of *Regional Studies* on the interpretation of the distance-decay coefficients in spatial interaction models. Curry (1972), in his original paper, argued that spatial autocorrelation among the mass terms in such models (which he described as a *map pattern effect*) was confounded with a behavioural component describing how interaction is modified by increasing separation between places. Curry suggested that this rendered invalid attempts to interpret the parameter of the distance term in gravity models as a measure of 'friction of distance'. In a series of articles involving Cliff, Martin, Ord, Johnston, Curry, Griffith and Sheppard, published in *Regional Studies* (1974–6), general agreement was eventually reached that if the interaction model is mis-specified (that is, the data really conform to one kind of gravity model, but mistakenly a different, wrong, model is fitted), any estimators of the friction of distance parameter will be biased and inconsistent; for correctly specified models, however, ordinary least squares provide best linear unbiased estimates. Given the poverty of geographical theory on interaction processes, mis-specification is likely to be a common problem. Parallel with the map pattern debate is the demonstration by Yule and Kendall (1957) followed up by Openshaw and Taylor (1979) that the choice of regional structure, or weighted graph in our terminology, affects the results of correlation and regression studies.

In human geography, the distorting effects of transport costs in tearing and twisting the simple picture of isotropic accessibility given on a conventional map are well known. Given the increasing ability of computers to handle large matrices – and, by implication, the complex graphs they describe – it becomes increasingly advantageous to set many geographical problems into a network rather than a continuous-space format. We must note here a major gap in British research, in that so little of the literature of operations research has found its way into our teaching or research programmes. While S.L. Hakimi's important findings on location within a graph (the so-called p-median problem) were published in the early 1960s, they remain largely unapplied to relevant geographical problems. This neglect has arisen partly because much closer links exist between geography and statistics than with normative fields, but it reflects more generally the fact that, for a long time, computer storage and speeds restricted study to desk-top-sized problems. Both the augmenting of computing capacity and more flexible software for the decomposition of graphs into interconnecting spatial units should allow more work to be undertaken on this important set of graph location problems.

Conclusion

Because of the chameleon-like tendency of graphs (as a spatial form) to dissolve into matrices (non-spatial) they are sometimes overlooked as a distinctive phenomenon for quantitative geographical research. In this chapter, we have argued for a taxonomy of graphs which runs from their familiar and overt elementary form to their embedded and covert role in spatial interaction models. It would be hard to argue that British geographers (even when broadly defined) have been making the running in the formal analysis of graphs; over the last decade and a half indeed it has tended to be a backwater topic in comparison with more classical statistical areas and time-series modelling. Nevertheless, we believe its potential importance is high; this is beginning to be recognized as geographers see it as increasingly useful as a shorthand for describing a world of increasing spatial complexity. To grasp fully its potential will mean building new bridges towards colleagues working in fields such as operations research and electrical engineering. Significant advances may also come from a future capacity to represent complex geographical networks in microchip formats. Perhaps the graphs and maps of the late 1980s may be silicon-based!

References

Appel, K., and Haken, W. (1977) 'Every map is four-colorable, I, II', *Illinois Journal of Mathematics*, 429-527.

Cliff, A.D., and Haggett, P. (1979) 'Geographical aspects of epidemic diffusion in closed communities' in N. Wrigley (ed.) *Statistical Applications in the Spatial Sciences*, 5-44, Pion: London.

Cliff, A.D., and Ord, J.K. (1977) 'Latent roots and vectors of an arbitrary real matrix', *Environment and Planning*, A, 9, 703-14.

Cliff, A.D., Haggett, P., and Ord, J.K. (1979) 'Graph theory and geography' in R.J. Wilson and L.W. Beineke, op. cit., 293-326.

Cliff, A.D., Haggett, P., Ord, J.K. and Versey, G.R. (1981) *Spatial Diffusion: an Historical Geography of Epidemics in an Island Community*, Cambridge University Press.

Curry, L. (1972) 'A spatial analysis of gravity flows', *Regional Studies*, 6, 131-47.

Forer, P. (1974) 'Changes in the Spatial Structure of the New Zealand Airline Network', unpublished doctoral thesis, University of Bristol.

Gould, P.R. (1967) 'On the geographical interpretation of eigenvalues', *Transactions and Papers, Institute of British Geographers*, 42, 63-92.

Haggett, P. (1976) 'Hybridizing alternative models of an epidemic diffusion process', *Economic Geography*, 52, 136-46.

Haggett, P., and Chorley, R.J. (1969) *Network Analysis in Geography*, Edward Arnold: London.

Hay, A. (1975) 'On the choice of methods in the factor analysis of connectivity matrices: a comment', *Transactions and Papers, Institute of British Geographers*, 66, 163-7.

Kendall, M.G. (1957) *Multivariate Analysis*, Griffin: London.

Openshaw, S., and Taylor, P.J. (1979) 'A million or so correlation coefficients: three experiments on the modifiable areas unit problem' in N. Wrigley (ed.), *Statistical Applications in the Spatial Sciences*, 127-44, Pion: London.

Skinner, S.W. (1964/65) 'Marketing and social structure in rural China', *Journal of Asian Studies*, 24, 3-399.

Strahler, A.N. (1964) 'Quantitative geomorphology of drainage basins and channel networks' in V.T. Chow (ed.), *Handbook of Applied Hydrology*, 4.40-4.74, McGraw-Hill: New York.

Tinkler, K.J. (1972) 'The physical interpretation of eigenvalues of dichotomous matrices', *Transactions and Papers, Institute of British Geographers*, 55, 17-46.

Tinkler, K.J. (1973) 'The topology of rural periodic market systems', *Geografiska Annaler*, B, 55, 121-33.

Tinkler, K.J. (1975) 'On the choice of methods in the factor analysis of connectivity matrices: a reply', *Transactions and Papers, Institute of British Geographers*, 66, 168-71.

Tinkler, K.J. (1977) *An introduction to graph theoretical methods in geography*, Concepts and techniques in modern geography, 14, Geo Abstracts: Norwich.

Tinkler, K.J. (1979) 'Graph theory', *Progress in Human Geography*, 3, 85-116.

Werrity, A. (1972) 'The topology of stream networks' in R.J. Chorley (ed.), *Spatial Analysis in Geomorphology*, 167-96, Methuen: London.

Wilson, R.J., and Beineke, L.W. (eds) (1979) *Applications of Graph Theory*, Academic Press: London.

Wrigley, N. (1979) 'Developments in the statistical analysis of categorical data', *Progress in Human Geography*, 3, 315-55.

Yule, G.U., and Kendall, M.G. (1957) *An Introduction to the Theory of Statistics* (13th edn), Griffin: London.

Chapter 23

Q-analysis

G.P. Chapman

Introduction

This paper is about R.H. Atkin's (1974, 1977, 1981) methodology of Q-Analysis and its prospective contribution to geography, although the irony is that the prospect may build upon the lists of capes, bays and principal towns that bored our grandparents stiff in their 'geography' lessons. To turn anything like that into something interesting, let alone stimulating and ultimately quite revealing, takes a considerable force, but that is what we are being offered. But it is important to realize that the methodology is new, is still developing, and as yet within geography it is without a major widely acknowledged achievement. Neither will such an achievement necessarily come quickly, because we are not dealing with *a* technique with *a* programme: there will be no instant gratification from the latest factor rotation. Instead we are dealing with a methodology that addresses a range of issues from the nature of data to the nature of the space within which we live, from the absurdities of existing classificatory techniques, to the holist/reductionist debate in science. There are algorithms associated with the methodology, to operationalize some of the concepts, but the application of the algorithm alone does not constitute an application of the methodology, no more than the calculation of a correlation coefficient guarantees the validity of the data used.

Recent technical innovation in geography has rather left the real world some way behind. Patterns are analysed to see the extent to which they depart from a random expectation: where randomness is defined in some perfect space of equal *a priori* probabilities. Autocorrelation procedures acknowledge that there should be some weighting of the possible adjacencies, but leave the matter rather arbitrary as to what they are. A conspiracy of silence has banished the inconvenient idiosyncrasies of the real world from the considerations of the modern technocrats and indeed from the universe of discourse of the new radicals, busy reconstructing each others' images of how the world ought to be. But as O-level students of geography know, the lie of the land is a major structural backcloth to the human geography created on it, and as any player of Diplomacy knows, the configuration of Europe has a significant bearing on military strategy. Can we say anything sensible about how these structures affect traffic that flows over them? More significantly, can we be less hidebound by physical sensibility and say anything about the structures of the human world that are the backcloth to other forms of traffic? The institutions that our societies have created constitute different structures that permit different activities. Journalists commonly report that the structure of the trade union movement in the UK generates traffic which is very unlike that generated by the structure of the trade union movement in Germany. In the UK persons with guaranteed salaries have access to a different kind of housing market compared with those who may earn as much but in unguaranteed wages. In this age of inflation a man may find on paper that the asset value of his house increases faster than the debts he incurs because of inflation, but he finds that money in different forms is not equivalent money, that money on one part of a structure does not necessarily connect with a lack of money on other parts. One of Atkin's major points is that unless we develop a method for the description of these structures and then relate the traffic to that description we will remain within the realms of the soft and woolly, with only an r^2 of variance explained to cuddle in the darkness of the black box. His approach to these problems is based on set theory and algebraic topology, mathematics not necessarily well known or used within geography, but some understanding of which is essential to the ideas that follow, and hence some basic concepts are considered now (see also Johnson, 1977a).

The definition of sets

A *set* is a collection of well-defined objects: commonly a set is indicated either by listing the objects, which are then known as the *elements* of the set, or by indicating the property of objects which defines set membership – this is known as the set builder notation.

$A = \{\text{drawing pin, cabbage, Shakespeare}\}$

is the first method;

$B = \{x|x \text{ is a cinema movie}\}$

is the second.

For the sets to be well defined, it is necessary in the first case that the objects listed as elements are well defined, and in the second case that the possible candidates for membership of the set are well defined, and also that for every possible candidate it is possible to say whether or not they belong to the set. Without hard *a priori* definition of the elements, the set

$C = \{\text{the centre of the town, the thrilling programme}\}$

is not well defined. Without clear definition of possible candidates and the criterion of set membership

$D = \{z|z \text{ is a town}\}$

is not well defined either. Broadly speaking this runs parallel to comments I have made on the priority of entitation over quantification in Chapman (1977). If we do not define what we are dealing with clearly, then we end in apparent paradoxes. For example 'The Golden Mountain does not exist' seems paradoxical, in that we have to mention something first in order to say it does not exist, so how could we mention it in the first place? Russell resolved the paradox by noting that 'There is no entity c such that 'x is golden and mountainous' is true when x is c, and not otherwise.' In other words x cannot be uniquely and unambiguously equated with an entity c: and if it cannot, then we have not said what it was that did not exist. This allows many social scientists to escape from threatening situations, by claiming that they did not mean 'this' or 'that' exactly, having given no clear definition in the first place.

In algebraic terms, the elements of sets are usually noted by lower-case letters, sets by capital letters, and sets of sets by script letters, thus we have b, B, and β. The null set, i.e. the set containing no elements, is denoted by $\emptyset$. A subset of B is defined as any set all of whose elements are also members of B: and the power set of B is defined as the set of all possible subsets of B, including the null set. Thus if

$B = \{1, 2, 3\}$

the power set P(B) is

$P(B) = \{\{1, 2, 3\}, \{1, 2,\}, \{1, 3\}, \{2, 3\}, \{1\}, \{2\}, \{3\}, \emptyset\}$

The *Cartesian Product* of two sets is the set containing all the ordered pairs of all their elements: the word ordered being used to indicate that the order of the elements is significant.

Suppose we have a set

$E = \{\text{big, small}\}$

then the product with set A above

$F = A \times E = \{$(drawing pin, big), (drawing pin, small), (cabbage, big), (cabbage, small), (Shakespeare, big), (Shakespeare, small)$\}$

Note that $A \times E \neq E \times A$.

Relations between sets

A relation R between any two sets may be defined as a subset of their Cartesian Product. For example $R \subset A \times E = \{$(drawing pin, small), (cabbage, big), (Shakespeare, small)$\}$, which may be conveniently illustrated by an incidence matrix, where we place a 1 to indicate that one of the ordered pairs is in the subset R, and an O otherwise

	drawing pin	cabbage	Shakespeare
big	0	1	0
small	1	0	1

A relation R between two sets A or E can be used to define subsets of either A or E. We may define as example

$M = \{x|x \text{ is R related to small}\} = \{\text{drawing pin, Shakespeare}\} \subseteq A$

and

$N = \{x|\text{ drawing pin is R related to } x\} = \{\text{small}\} \subseteq E$

Now it is clear in the latter case that there is only one element in the set so defined, but I wish to underline that $\{$drawing pin$\}$ and *drawing pin* are not the same concepts. Consider a terrible geography exam in which all candidates except Mr Monadnock fail. The set of all successful candidates (=$\{$Mr Monadnock$\}$) and Mr Monadnock are clearly not expressing the same concepts. Confusion between elements and sets is responsible for the following celebrated paradox, whose consequences for the way in which we organize data is profound.

First we make a statement: 'A certain village contains a man who is the village barber. The barber shaves all those men, and only those men, who do not shave themselves.' Now we ask a question: 'Does the barber shave himself?' (Atkin, 1974).

The answer to the question results in a paradox: 'If he shaves himself, then he is not shaved by the barber: hence he does not shave himself' or alternatively 'If he does not shave himself, then he is shaved by the barber, which means he shaves himself'.

The paradox arises because the word barber can be defined in two ways: either to refer to an element, or to a set – in actual fact a subset defined by a relation. And in the paradoxical example above we are using both meanings simultaneously.

First we will look at the barber defined simply as an element. Suppose there are a set of men in the village $M = \{m_1, m_2, m_3, m_4, m_5, m_6, m_7, m_8, m_9\}$, and since we are assuming that *the* barber is *an* (a single) element, let us arbitrarily say he is m_5. I wish to define a relation between the set M and this element according to whether or not the elements of M are shaved by m_5; to do so I simply define a set $F = m_5$ purely in order that I may define the relation $R \subseteq M \times F$

		F
		m_5
	m_1	1
	m_2	0
	m_3	1
	m_4	1
M	m_5	?
	m_6	0
	m_7	1
	m_8	1
	m_9	0

In this example I have put a ? in the space which will say whether or not the barber (m_5) shaves himself. It is a matter of observation to find out if he does or does not.

Now let us define the barber as a set, in fact a subset of M. Suppose we have the relation $R \subseteq M \times M$, where mRn is m is shaved by n, as represented by the following incidence matrix.

		Shaves							
		m_1	m_2	m_3	m_4	m_5	m_6	m_7	m_8
Shaved	m_1					1			
	m_2		1						
	m_3			1					
	m_4	1							
	m_5	1							
	m_6					1			
	m_7					1			
	m_8					1			

The barber is then the set $B = \{x \mid (x, m) \in R, m \subseteq M, x \neq m\} = \{m_1, m_5\}$

It could be that there is only one man in the village who shaves other people – but *still* according to this definition we would be dealing with 'barber' as a set.

Now we can see that there are two meanings of barber in the original statement, and two meanings of 'barber' in the answers to the question.

'A certain village contains a man (i.e. an element, a single entity) who is the village barber (therefore the barber is an entity). The barber (now we have a definition based on a relation) shaves all those men and only those men who do not shave themselves (note that this does not automatically mean any longer that there is a single man as barber).'

'Does the barber (ill-defined – which meaning of barber?) shave himself?'

First answer: 'If he (a man) shaves himself, then he is not shaved by the barber (a subset of men defined by the shaving relation) so he (a man) does not shave himself.'

Second answer: 'If he (a man) does not shave himself, then he is shaved by the barber (a subset) which means he shaves himself.'

Atkin (1974) concludes that we must never allow the notions of set membership and relations-between-sets to become confused, and perhaps the most important conclusion is that sets are hierarchically arranged, and that we must never confuse an element of a set with a member of the power set.

Hierarchies of sets

Consider a geographical land-use classification scheme:

{pastoral, arable, horticultural, industrial, residential, commercial, other}

Clearly 'other' is not on the same hierarchical level, since it is an undefined member of the power set, which also remains undefined because 'other' is undefined. Thus an immediate and *extremely* demanding result of the methodology is that such wave-of-the-hand magic is inadmissible.

There are a number of different types of relation that may be defined, some with particular significance to the relation in a set, i.e. to itself, such as $R \subseteq M \times M$. A relation that is symmetric, reflexive and transitive is an *equivalence relation* – something which cannot be explored here for lack of space – but it leads to a *partition* of a set into *disjoint subsets*. If the sets B_i, $i=1, \ldots, n$, are all members of the power set P(A) of A, and if $U_i B_i = A$, and if for all i and j either $B_i \cap B_j = \emptyset$ or $B_i = B_j$, then the set $\{B_i\}$ is a partition of A. For example, if

$$A = \{1,2,3,4,5,6,7,8,9,10\}$$

and $B_1 = \{1, 3\}$, $B_2 = \{7, 8, 10\}$, $B_3 = \{2, 5, 6\}$, $B_4 = \{4, 9\}$ then

$\beta = \{B_i\} = \{B_1, B_2, B_3, B_4\}$ is a partition of A.

Clearly this is a very restricted and special kind of relation, but it is just this restricted and special form which has dominated classification for hundreds of years – it is reductionist and breaks up a whole set into disjoint pieces. The wide and uncritical concern with partitions probably stems from the early scientific successes of evolutionary biology, but their extrapolation into other fields has rarely been justified. Fortunately for us humans such partitioning is rarely observed in the relations on which we depend. For example, a set of people a,b,c,d,e,f, g,h may be observed to have a relation $R \subseteq M \times P(M)$, which produces the following subsets of M: $\{a\}$, $\{c, d, f\}$, $\{b, g, h\}$, $\{a, b, c\}$, $\{d, e, f, g, h\}$ which are clearly not disjoint, since $B_i \cap B_j \neq \emptyset$ for all i and j. Perhaps these subsets have names: Shop A, Factory B, School C, Family 1, and Family 2, and the relation R is defined as 'spends part of each working day at'. The 'overlapping' (Figure 23.1) produces a connectivity between people which provides

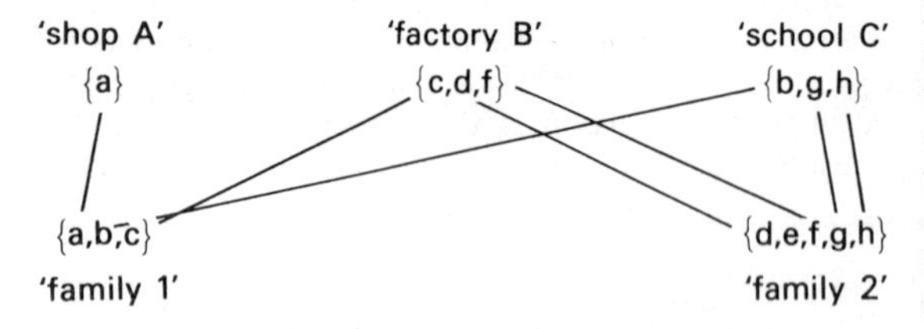

Figure 23.1 Intersecting member of the power set

the channels of communication in society, and provides the glue that holds society together. If we relax the disjoint condition and simply say: if for all i, $B_i \epsilon P(A)$, and $U_i B_i = A$, then $\beta = \{B_i\}$ is a cover set of A. To many, the relaxation of the partition rule may appear an open invitation to unstructured freedom – but in practice this proves to be the opposite of the case.

Suppose we have a set of countries, USA, USSR, Chile, Mexico, Brazil, Germany, China, France, UK, Japan which we wish to group regionally. The obvious relation to draw is indicated in Figure 23.2, and it appears that the cover set approach allows us to say all that we ever wished to about Germany being in both Eastern and Western Europe, and the USSR being both European and Asian. But in fact we have broken the rules for defining elements at the same levels. It is patently not true that the whole of the USSR is both European and Asian, and patently not true that the whole of Germany is both East and West European. To keep the elements at the same hierarchical level we are forced to name the two Germanies FRG and GDR, and to name the two parts of the USSR, European USSR and Asiatic USSR. This should not be surprising since the geographers' use of continental names was always supposed to be a partition of the earth's surface. We *can* introduce Germany as a member of the power set, and also the USSR, to produce the relation shown in Figure 23.3. There is nothing wrong in this, but we cannot then define a yet higher level to include such words as 'Europe' since there will still be problems over the assignation of the USSR at the $P^2(A)$ level, even though Germany can be defined satisfactorily. Note that the usefulness of the hierarchy so defined is another matter entirely.

This brief example illustrates how the whole hierarchy rests on its foundations, the lowest level set, whose elements must be well defined. Most of the time we ignore these strictures, and end with a confused soft science. The land-use classifications, using words such as 'commercial' or 'residential', often have dubious distinctions drawn. Many uses might in fact be both commercial and residential – a hotel, for example. Q-analysis responds by saying, fine, that's just how the world is, but if you are dealing with a cover set, then define the elements of the set at the next lowest level and then define those members of the power set which produce the intersections.

A relation is not a function

Since a relation may be any set of ordered pairs in the Cartesian Product, it is obvious that we are dealing with what may simply be called many–many 'assignments'. With a function there is the severe restriction that all members of A are assigned

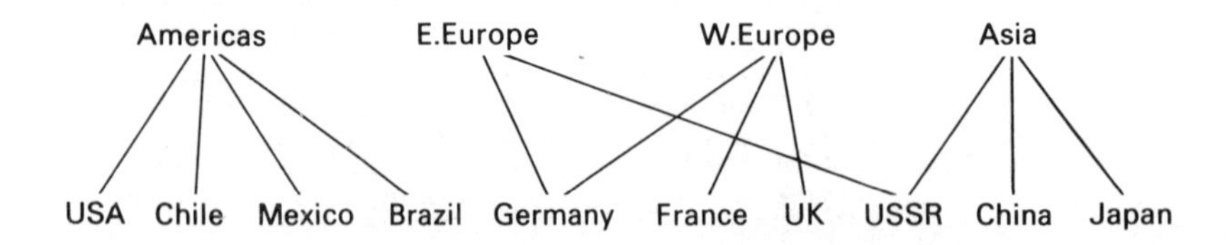

Figure 23.2 An incorrect hierarchy

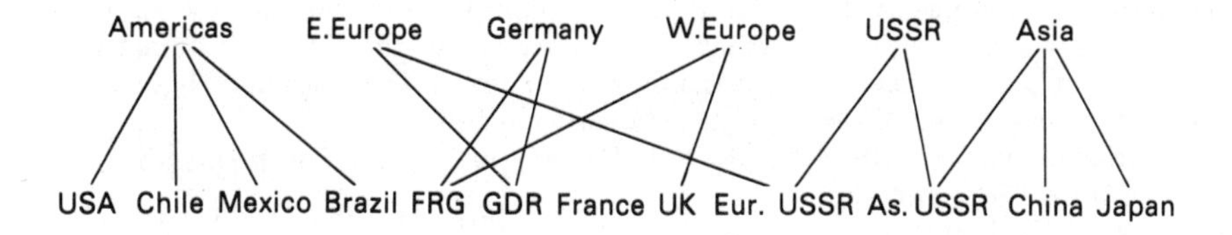

Figure 23.3 A correct hierarchy

uniquely to a single member of B. The mathematics we are brought up on tends to be dominated by functions, which happen to be the special and restrictive case. For example

$$Y = a + bX$$

is a function, since for any set of numbers X there exists a set of numbers Y such that each x is assigned to one and only one y.

We force data to fit such models, even though they may be wildly inappropriate. Consider a simple plot of paired values of data, to which a linear function is to be fitted. Suppose the fit is not a good one. We are then forced into one of two positions: a) admit the model is inappropriate, or b) claim that the data is subject to error but the model is still appropriate. In the first case we are then duty bound to try and find something better: in the second to get hold of some proper data. The more sophisticated and complex the procedure, the more one tends to hope that a 'real' answer will emerge from the – for example – factor analysis of 20 variables on 40 observations. Errors and confusions are assumed to cancel out, to leave us with a picture of something that does exist the other side of the glass through which we see so darkly. Much of the problem stems from the fact that the variables are deemed superior to the observations, i.e. that a set of values for variable X is taken from the entities, and then separately a set of values for variable Y. The holistic nature of entities is lost as all are stripped in turn of first one variable (e.g. money) and then another (e.g. age).

The answer given here is, where appropriate, to stop playing with functions and to start using relations. But to do what? At least with existing techniques there is some process, a set of motions through which the data passes, so that raw material is digested and then reissued.

The simplicial complex

Before considering Atkin's Q-analysis of the structure of a relation, it is important to remember that an investigator might erroneously assume that any given matrix represents a relevant relation between two sets, and that such a relation does represent the structure of the backcloth, and not simply the traffic. But if his assumptions are invalid, then so too is his analysis. 'Answers' will come out of the computer, but with little real meaning. (Usually it quickly becomes apparent that one is dealing with nonsense.)

Consider the relation $\lambda \subseteq M \times X$, where X is the set {Golf Club, Liberal Party, Labour Party, Conservative Party, City Council, Chamber of Commerce, Tennis Club} in a small town, and M is a set of men $\{m_1 \ldots\ldots\ldots\ldots m_7\}$ in the town, and the relation $m\lambda x$ stands for 'is a member of' as illustrated in the incidence matrix at the foot of the page. We say that each man constitutes a simplex. One of several ways of representing the simplex is by a convex polyhedron whose vertices are recorded in

	Golf Club	*Liberal Party*	*Labour Party*	*Conservative Party*	*City Council*	*Chamber of Commerce*	*Tennis Club*
m_1	1	1					1
m_2			1		1		
m_3				1		1	1
m_4	1						1
m_5			1		1		1
m_6	1			1		1	
m_7	1			1	1		

the incidence matrix. The simplicial complex is the set of polyhedra, complete with any shared faces or edges, which represent the total relation. Thus we can see in Figure 23.4 that the first row of the matrix, representing m_1, is a triangle whose vertices are Golf Club, Liberal Party, Tennis Club, and that m_4 is an edge on this triangle. Algebraically we write $KM(X;\lambda)$ to indicate the simplicial complex listing of the components formed by simplices which are connected at the different dimensional levels, known as 'q' levels. The relation of Figure 23.4 produces a listing as follows:

$$q = 2\ Q_2 = 5\ \{m_1\}\cdot\{m_3\}\ \{m_6\}\ \{m_7\}\ \{m_5\}$$
$$= 1\ Q_1 = 3\ \{m_1, m_4\}\ \{m_3, m_6, m_7\}\ \{m_5, m_2\}$$
$$= 0\ Q_0 = 1\ \{m_1, m_4, m_3, m_6, m_7, m_5, m_2\}$$

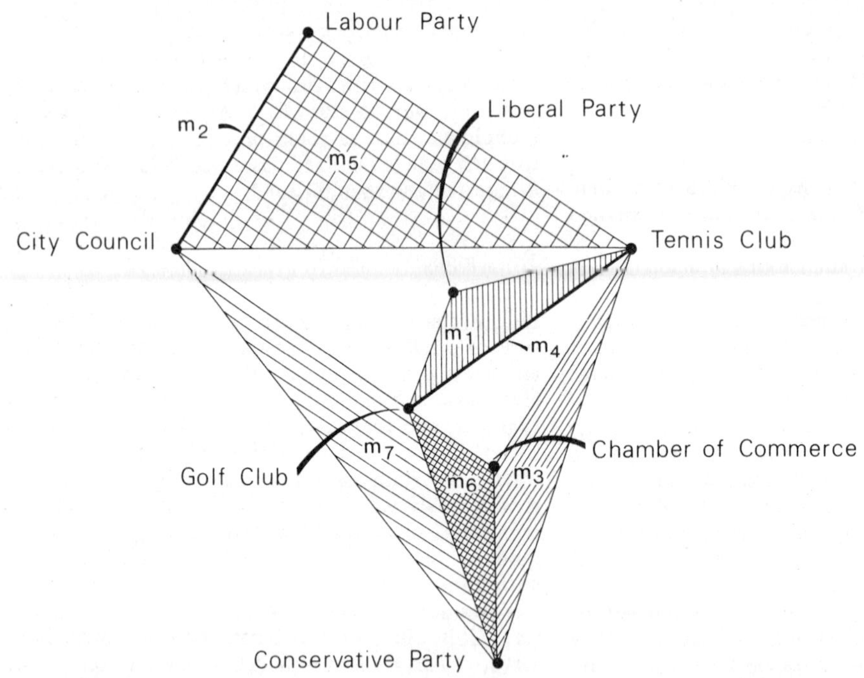

Figure 23.4 A simplicial complex

of the polyhedra of M with vertices in X defined by λ.

We cannot draw without perspective more than a two-dimensional shape on this paper, and even with perspective to help can we draw no more than a three-dimensional representation, but many relations will contain simplices of much higher dimension. These can be handled algebraically, but not graphically, since the physical world is bounded by three dimensions, unlike the world of relational structures, which is truly multidimensional.

In the view of Q-analysis, the person is represented as a whole, by the whole simplex, and is far from being the mere sum of the vertices of the simplex. The person is then able to relate to adjacent persons through their shared faces. Thus m_3 relates to m_6 through the shared face (in this case an edge) of the two vertices the Conservative Party and the Chamber of Commerce – which we refer to as a one-dimensional connection, since the edge is one-dimensional. We note that m_6 in turn shares a face with m_7, but that within this component beyond m_7 there is no further one-dimensional connection, only the zero-dimensional connection (a single shared vertex) with m_5 and m_2.

The Q-analysis of the whole relation produces a

At $q = 2$ we are considering polyhedra which are triangles, and there are five of these, none of which shares a 2-dimensional face with any of these others. At $q = 1$ we find that there are three components, in each of which the simplices are connected by one-dimensional edges. Note that all the simplices are now in a component, since there is no simplex with a dimension less than 1. At $q = 0$ there is a single component. The structure vector $Q = \left\{\begin{smallmatrix} 2 & & 0 \\ 5 & 3 & 1 \end{smallmatrix}\right\}$ summarizes the connectivity pattern.

The significance of this concept of structure is that it does actually show the configuration of connections implicit in the relation, while treating each simplex as a whole. It also shows the changing dimensional way in which the simplices are connected with each other. There is a landscape of connections – and in fact it is possible to produce contour sketches of the structures as in Figure 23.5. It is on this structure that traffic may flow, and some kinds of traffic require higher dimensionality than others. How often does one go to a party where there are many new persons, to find that conversation with another guest is restricted because 'we only have one thing in common'? The best conversations live on the highest dimensional shared faces – where the vertices are not separately adding to the conversation, but contribute as a whole.

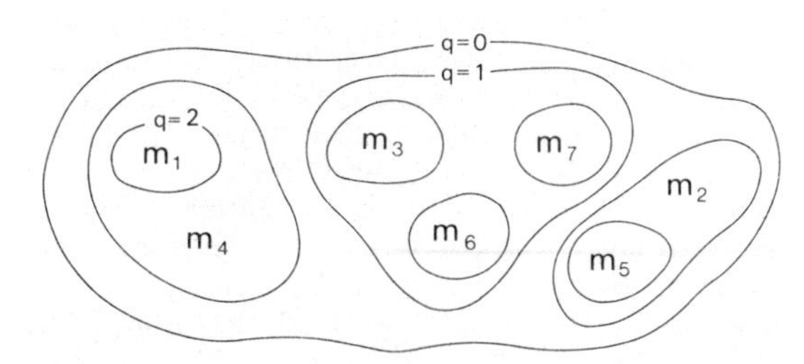

Figure 23.5 A contour 'landscape' representation of Figure 23.4

A similar analysis can be performed on the conjugate of the relation, $KX(M;\lambda^{-1})$, where λ^{-1} denotes transpose, that is, treating the columns as simplices instead of the rows. The resulting graphical representations and the components are shown in Figure 23.6. The two components at q=1 are particularly interesting, suggesting that a single dimension discourse can move over each of the two, but not between them. At a zero-dimensional level everyone is connected.

This representation does mirror the way in which the world works. Any given committee that wishes some business to be sounded out with another committee will inevitably choose that group of persons that is a face to both committees, and failing the existence of any such persons, those persons who have the greatest number of vertices in common with individuals on the other committee (such as membership of clubs and personal interests).

It is obvious that the connectivity of the pieces on a chess board is in essence what the game is all about – pieces in combination having power which singly they do not possess, and other pieces being pivotal in holding an attack or defence together, and in the idea of a domino-like collapse. It is also an obvious statement about the world of international trade – that ultimately we are all connected. Those students who wish their colleges not to invest in firms which have subsidiaries in South Africa neglect the fact that the college may hold shares in firms in America which trade with firms in France which trade with firms in South Africa.

The connectivity analysis produces a grouping of the simplices for the different dimensional levels. It is an unusual kind of grouping, in that a polyhedron at 'one end' of a component may share no vertices with a polyhedron at the 'other end' with which it is connected. But consider any two societies, within each of which there may be considerable variation among the people – between the upper and lower classes, and between the upper and lower castes, for example. Yet these two societies are held together by connectivity within each, and are distinct from each other by a break in the connectivity. In that sense it is a realistic grouping in many circumstances. This is, however, far from the more significant aspect of the way in which traffic (defined below) may flow on the structure. We all speak of the interdependence of complex systems, such as ecosystems, in which disease in one species may have a chain reaction in other species, which ultimately depend on the first for food. The actual nature of the initial disaster may change as we look at each species: for the first, disease, for a second, starvation, for another

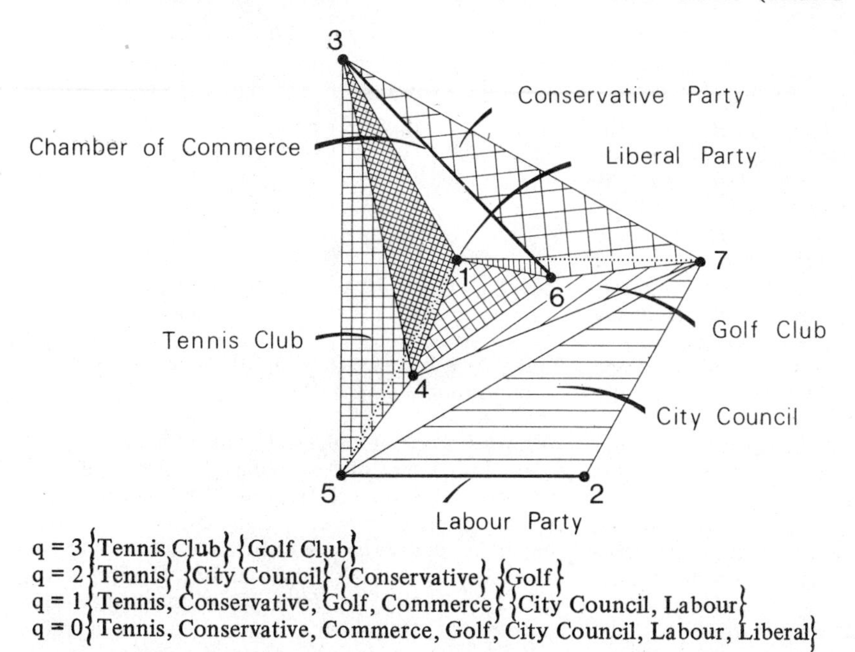

Figure 23.6 The conjugate simplicial complex of Figure 23.4

a specific nutritional deficiency which causes another disease, and so on. But the disaster is in fact one disaster within a certain connectivity structure.

If we have any problems in grasping this, it is because following the way in which we have usually grouped entities, according to the similarities of their vertices, we have naturally looked for traffic which is related to each such group, and is therefore dependent on specific vertices. Therefore it might seem that entities which have different vertices should have different traffic. But inflation and recession, language and fashion, flow across connected structures, often changing in their specific nature as they move. In fact it is well known that if a sound changes in a given dialect in any given language, and causes ambiguity because of its similarity with another sound, that latter will change too, in a chain of displacement, so that a dialect is distinguished ultimately not simply by one or two changes, but a whole set of changed sounds. This is necessary because a language is a highly connected structure. In a good novel the characters are all well connected with each other, and the effects of what happens to one will usually spread far and wide in effects on the others. Atkin's analysis of 'A Midsummer Night's Dream' makes this kind of point quite elegantly (Atkin, 1981).

Weighted relations

The dimensions of the polyhedra are defined by the number of vertices in any simplex. These in turn are defined by any given relation between two sets. When this relation is a binary one, the geometry of the backcloth is immediately defined. But some relations are weighted, i.e. they are composed of numbers other than ones and zeros in the matrix. It is then necessary to convert such a matrix into a binary form for any connectives to be studied. The non-zero entries in a matrix are sliced and made one or zero according to whether they exceed or are less than a given threshold value.

If we consider a set of successive threshold values from 1 to the maximum number in the matrix, say 'n', then we have a set of nested structures $K_n \subset K_{n-1} \subset \ldots\ldots K_3 \subset K_2 \subset K_1$. Altering the slicing parameter therefore selects a different structure with different dimensions and connectivities. This is a major step, and theoretically should not be undertaken without strong justification for the threshold values, which may even be different for the different rows and/or columns of the matrix. With road traffic flows the threshold values may be those beyond which the carrying capacity is exceeded and congestion occurs: for a business it may be a debt/asset/earnings ratio beyond which creditors will foreclose. In practice it is not always easy to establish these values, but the relation may nevertheless be sliced at differing values, simply to see the kinds of structures that are implied. In Table 23.1, the inter-area world oil flows for 1976 are tabulated. With a slicing parameter = 1, the structure is as shown in Figure 23.7. Figure 23.7a is a statement of the connectives of the exporters, and indicates how their interests in importing regions are sufficiently well connected at all levels of q that change in the level of exports by any one of them must ultimately connect with all the others. The conjugate describes a similarly highly connected structure for the importers: although here there are some slight disconnections between the USA and

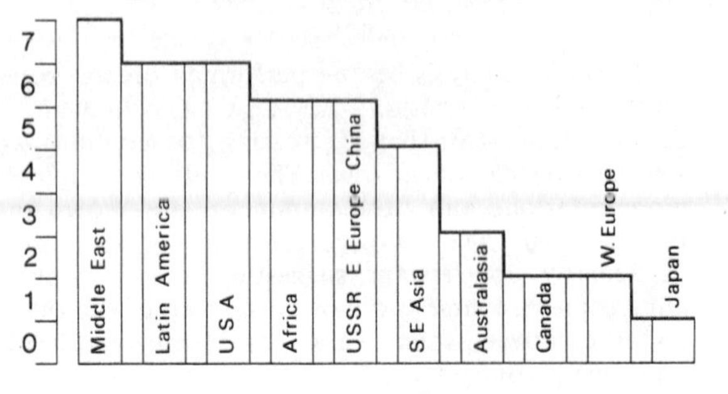

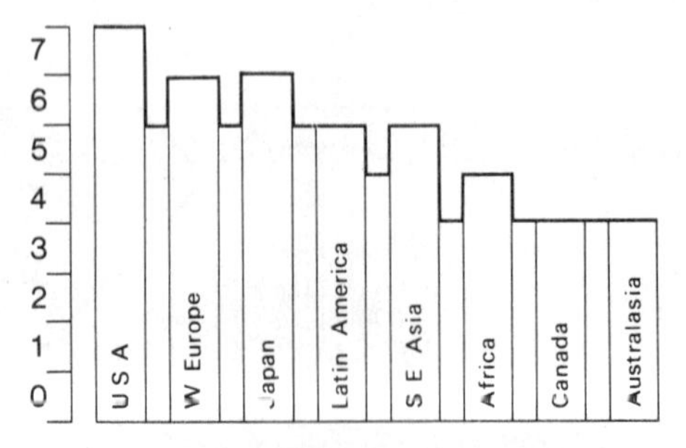

Figure 23.7 Connectivities of (a) oil exporters, (b) oil importers, slicing at 1

Western Europe at high levels, indicating some small room for divergent oil policies. Slicing at 100 reinforces this picture (Figure 23.8): it represents the 'main frame' connectives: on the exporters' side the major change is a significant reduction in the dimension of the East Block and on the importing side a reduction in Japan's dimensionality and eccentricity. (Eccentricity is defined as a ratio $(\hat{q} - \check{q})/(\check{q} + 1)$ where $\hat{q}$ (top-q) is the dimension of a simplex, and $\check{q}$ (bottom-q) is the highest dimension at which the simplex is connected with any other simplex.)

I have a suspicion that whenever we are unable to specify non-arbitrary slicing parameters we are dealing with traffic (defined properly in the next section) on a backcloth, rather than the backcloth

Table 23.1 World inter-area oil movement 1976 10^5 tonnes

From \ To	USA	Canada	Latin America	Western Europe	Africa	Japan	Australasia	South East Asia
USA		6	39	45	6	2	17	2
Canada	300			3				
Latin America	1050	162	70	175	5		5	2
Western Europe	78				49			
Middle East	946	195	857	4665	242	808	1962	123
Africa	981	25	191	1232		19	35	
South East Asia	280	7					515	38
Japan						1		
Australia	5					1	21	
USSR and China and Eastern Europe	10		82	696	10	51	69	

Source: BP Statistical Review of World Oil Industry, 1976

itself. This is certainly true in this example – the actual structure on which the oil traffic flows is quite clearly the immensely complex structure of oilfield ownership, multinational capital and expertise, political affiliation etc. By looking at the traffic in the way that we have we are hinting at the kinds of structure that may be supporting it, but certainly we have made no definitive description of it. It is a merit of Q-analysis that the researcher is forced to admit this.

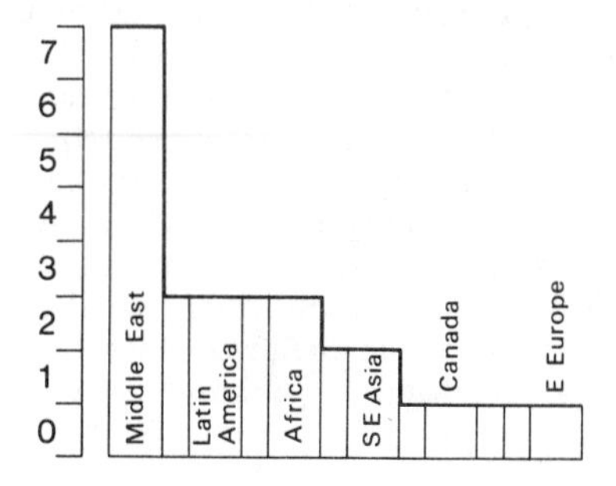

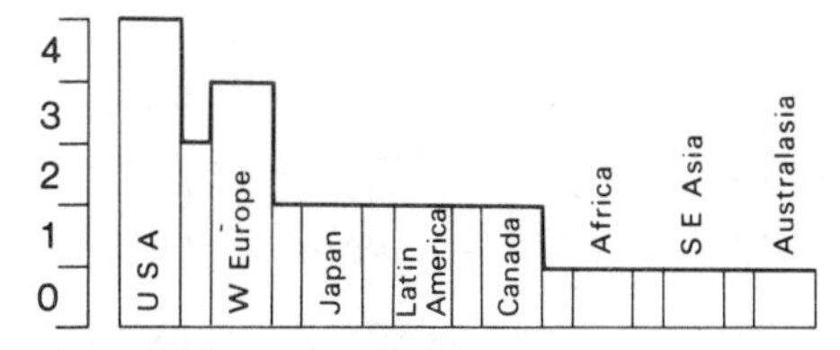

Figure 23.8 Connectivities of (a) oil exporters, (b) oil importers, slicing at 100

Traffic on a backcloth

Let us denote a simplex by σ, and qualify this by q to indicate the dimension of the simplex, and by an index i to denote each simplex in a simplicial complex, σ_q^i. For example, suppose we have five TV programmes $V = \{v_1, v_2, v_3, v_4, v_5\}$ and five programme descriptors $D = \{d_1, d_2, d_3, d_4, d_5\}$ which describe the subject matter of programmes (for example – crime, economics, human relationships, art, environment), then v_2 might be the simplex

$$\sigma_1^2 = \langle \text{crime, environment} \rangle \quad \text{(illegal fishing?)}$$

Traffic can be represented by a pattern of numbers, viz, a mapping of the simplices into a set, which is usually a set of numbers. So in this context we might have a mapping between the simplices and numbers indicating the duration of programmes in minutes, e.g.

$$\pi : \{\sigma_q^i, q=0 \ldots\ldots \quad , i=1 \ldots\ldots 5\}$$
$$\rightarrow \{20, 50, 30, 40\}$$

For any given t-dimensional simplex in a complex a change in the values of a given traffic mapping is defined as a t-force. For example the scheduling officer of the channel might decide to repeat the programmes, but abridge some of them. In this case the changes in the value of the mapping constitute a t-force. Suppose the first transmission is:

$$30 \langle \text{crime} \rangle + 20 \langle \text{art, human relations} \rangle$$
$$+ 50 \langle \text{crime, environment} \rangle +$$
$$30 \langle \text{crime, economics} \rangle + 40 \langle \text{art, human relations, environment} \rangle$$

and the second is:

$$30 \langle \text{crime} \rangle + 20 \langle \text{art, human relations} \rangle$$
$$+ 50 \langle \text{crime, environment} \rangle +$$
$$10 \langle \text{crime, economics} \rangle + 40 \langle \text{art, human relationships, environment} \rangle$$

then the t-forces are

$$-10 \langle \text{crime} \rangle \qquad -20 \langle \text{crime, economics} \rangle$$

In some circumstances t-forces may be (but are not necessarily) transmitted through the structure. Change in the value on one simplex may initiate changes elsewhere in the complex. The connectivities of the simplices then become important; relatively highly connected simplices being much more likely, *ceteris paribus*, to facilitate transmission than those less highly connected. Unfortunately there is insufficient space to pursue this line of argument in proper detail here, but consider the way in which the build-up of traffic at one point in a city's road network often transmits to other points in the network. Although there was a phase of building more highly connected road systems in this country, recently there has been a dawning realization of the value of disconnecting things, and of having limited access roads, which inhibit the transmission of t-forces (Johnson, 1977b).

Atkin's ideas, as the reader may be surprised to note, are concerned with clear and well-organized description. But these two qualities are exactly what we need in the social sciences. In physics there is no explanation for gravity. It is what 'causes' things to fall to earth, but one's knowledge of it is only in the movement of things that fall. In the end there is nothing but the observations, coherently and consistently made as long as the observation language is coherent and consistent. What we are being offered is a language to define structure, and traffic, and to relate the two together, and to differentiate between the different kinds of change. Far from inhibiting imagination and speculation it encourages it, since the border between what we know and what we want still to know is that more clearly defined.

Atkins takes the view that 'kinematics', that is a consistent system of description, has to precede the 'kinetics', i.e. lawful explanatory statements. It was when Galileo took to measurement rather than the prevalent concern with the 'intentions' of falling bodies, that progress in understanding was made.

An illustrative comparison

I wish to contrast the thinking implied by Q-analysis with the 'traditional' approach to multivariate problems as exemplified by Factor Analysis. My starting-point is interest in the growth of large Indian cities (over 100,000 in 1971) between 1961 and 1971. The data matrix is for 142 observations on 17 variables, the latter being log size, specialization, % scheduled castes, % scheduled tribes, % agriculture, % livestock, % mining, % domestic industry, % organized industry, % construction, sex ratio, % literacy, % growth 1961–71, % trade, % transport, % services and % non-workers.

The first stage of factor analysis is to pull the variables off the 142 entities, and to treat the variables as the prime units of concern, each of them having an independent existence. Thus the concept of each city being a holistic polyhedron is lost immediately. The second stage is to produce the correlation matrix between all variables. The pattern of connection between the variables implied by the correlation matrix is illustrated in Figure 23.9, where all values ≥ |.24| (significant at the 1 per cent level) have been plotted. One can surmise that organized industry, growth and specialization may lead to the identification of one factor, and that sex ratio, trade, livestock, literacy and non-workers to another, defined orthogonally and independently. But to split these variables into separate and additative factors is clearly absurd when ultimately they are all connected. What of the chains that say that domestic industry is inversely connected with

Table 23.2 Factor loadings

		F_1	F_2	F_3	F_4	F_5	F_6	F_7
1	log size	.48	–.01	–.04	.02	–.08	.00	.06
2	specialization	.75	–.09	.05	.22	–.26	–.20	–.02
3	% scheduled castes	.06	.15	–.09	.13	–.37	.02	.08
4	% scheduled tribes	–.03	–.07	.50	.15	–.02	–.04	.02
5	% agriculture	–.70	–.02	.15	.10	.01	–.05	–.04
6	% livestock	–.15	.26	–.05	.23	.76	.05	.08
7	% mining	.06	–.04	.05	.08	–.12	.00	.95
8	% domestic industry	–.45	–.10	–.17	–.11	–.07	–.16	.00
9	% organized industry	.43	–.72	.05	.31	.00	–.32	–.20
10	% construction	–.16	.06	.82	.03	.01	.06	–.09
11	sex ratio	–.30	–.25	–.29	–.22	.52	.05	–.17
12	% literacy	.37	.01	–.05	–.14	.43	–.05	–.08
13	% growth 1961–71	.16	–.10	.33	.45	.01	–.22	.06
14	% trade	–.01	.01	–.16	–.90	.12	–.06	–.06
15	% transport	.09	.04	–.13	–.03	–.01	.96	.00
16	% services	.22	.94	–.17	.04	–.05	–.06	–.15
17	% non-workers	–.13	.20	–.56	–.08	.04	.22	–.18

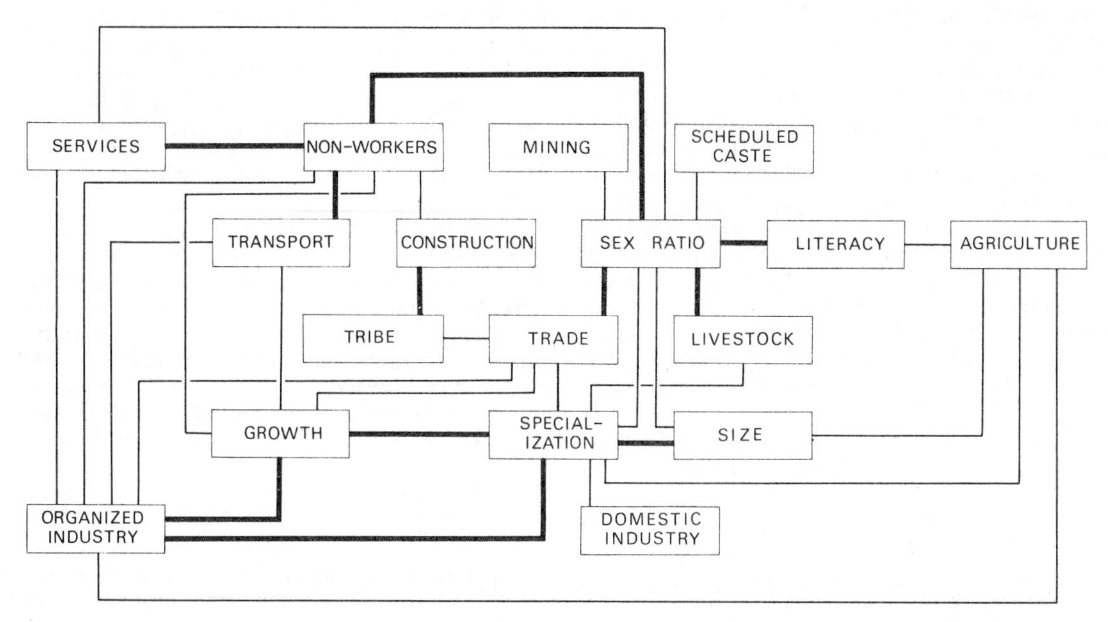

Figure 23.9 Correlations significant at 1% level between 17 variables over 142 cases – Indian cities in 1971. Heavy lines show positive correlations, lighter lines show negative correlations

specialization, which is inversely connected with livestock, which is correlated with sex ratio . . . ? And any one who knows India knows that 'construction' and 'tribal people' are associated with growth and 'specialization', though here the connection appears only via an inverse relation with trade and a subsequent inverse relation between that and growth.

At first glance the rotated varimax solution (Table 23.2) appears to be reasonable. The factors can be named as follows:

F_1: Large, highly specialized, organized industry, little agriculture, little domestic industry (the new industrial factor).

F_2: Little organized industry, lots of services (the traditional service factor).

F_3: High construction, low unemployment, lots of tribals, high growth (the growth factor).

F_4: Little trade, organized industry, high growth (?).

F_5: High livestock, high literacy, good sex ratio (the traditional Hindu factor – Brahmin South?).

F_6: Transport, no organized industry (transport factor).

F_7: Unambiguously mining.

But there are some puzzles: why is Factor 4 different from Factor 1, and why does growth appear in Factors 3 and 4? If we look at the loading of the variables on the factors, we find that organized industry, sex ratio, literacy and growth, also possibly livestock, are stretched across several factors. What the Indian Census joined together, Factor Analysis has rent asunder.

The next stage of factor analysis is to classify the observations in terms of their factor scores. To do this the factor scores are calculated from the factor score coefficients and the z values of the observations on each variable in the manner of a multiple regression equation.

Space does not permit a full presentation of the results here, but one example will be enough to raise the eyebrows. Jamshedpur, a city well known to all O-level geographers for its established steel and engineering industries and progressive industrial nature happens to have a very low score on Factor 1. So it is not the new industrial city we thought it was. The reason, of course, is that it has a high negative loading on Factor 2, that is, 'it is noticeably a town *which lacks not having organized industry*, and simply lacks services'. The answer must therefore be to classify cities according to a composite score on several factors: so we must reunite those variables torn asunder. Parsimony, wherefore art thou?

Another example is the association in the original correlation matrix between trade and stagnation (inverse correlation with growth). Simple plotting of cities with high growth rates and with high levels of specialization soon showed that there is a group of cities in the Punjab which are fast growing and which have high employment in trade, the opposite of the stated general trend. To people who know of the dynamic rural economies there, this comes as no surprise. But clearly the greater number of cities with traditional traders in areas of stagnating rural economies results in the kind of correlation calculated.

Inexorably, we are forced back to look at each city as a polyhdedron, unique, and whole. The methodology would require us to analyse a well-defined relation between well-defined sets. One of these is the set of 142 cities defined by the Indian census, and the other is a set of attributes. Here we have a problem, since the 17 variables are not so well-defined. If we accept them as they are, then to produce a binary relation they have to be sliced. To do that we need to know threshold values, and at this moment I have not thought of any that are non-arbitrary. Also, to include the effects of the absence of attributes, I have doubled the list of variables, by including an inverse of each. Then for each city this row of 34 attributes has been sliced at ⩾1 standard deviation of each attribute.

A full representation of the results is available in Chapman (1979). The following is a précis of some of the interesting features of the results.

The structure vector of the connectivities of the cities is:

$$Q = \{\overset{9}{2}\ 3\ 6\ 9\ 14\ 17\ 23\ 21\ 6\ \overset{0}{1}\}$$

indicating that although they all form one component at q=0, the number of separate components rapidly increases. This reflects the variety and individuality of these cities, and in effect is an instruction not to try to crunch them into a few partitioned classes. Coming down from the highest dimension (q=9) the first components of connected cities appear at q=5, when Dhanbad, Singanallur and Bokaro form a component, and Calcutta, Bombay and Thana form another component. The former is a mining component, and the latter a modern industrial component. At q=4 the two coalesce, and expand to include Calcutta, Bombay, Thana, Surat, Dhanbad, Chandigarh, Durgapur, Dehra Dun, Malegaon, Rourkela, Singanallur, Bokaro and Bhubaneswar.

To those who know India this appears a strange list at first sight. Chandigarh does *not* have much industry, and therefore appears out of place. But inspection of the data shows that it does have a very high growth rate, a very high degree of specialization, and virtually no domestic industry. On the basis of these characteristics it is very much part of the modern Indian urban scene, and connected with other modern urban centres whose base may be industry rather than government service.

To illustrate this we present below a diagram of the conjugate analysis, showing the way in which the variables connect. Compare Figure 23.10 with the diagram of the correlations in Figure 23.9, and with the Table 23.2 showing factor loadings of the variables.

This analysis is far from complete. It is quite obvious that the variables are of two kinds: most are about structure – but growth is clearly traffic on the structure. Further analysis has to take this into account. Ahmad's (1966) Factor Analysis of Indian cities included latitude and longitude as two measures of location. So far in this analysis there is no such

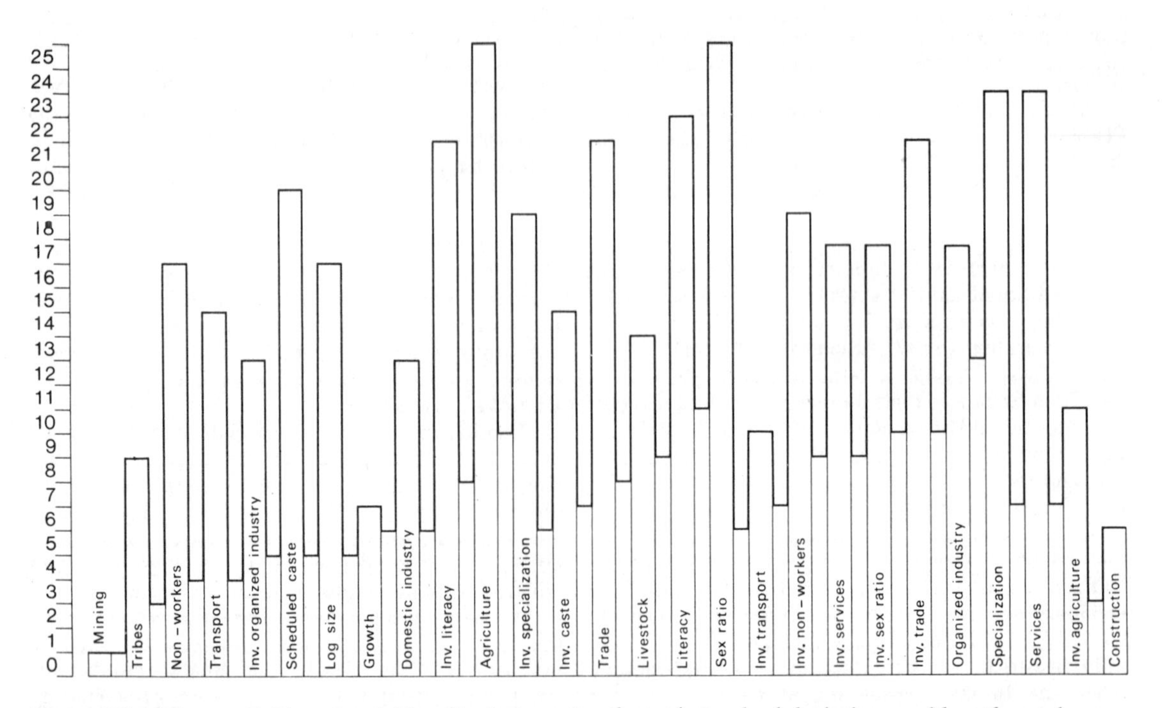

Figure 23.10 Connectivities of variables sliced at greater than +1 standard deviation, and less than −1 standard deviation

data: as a result cities which are far apart in physical space may appear highly connected. This is not a problem which would worry a Factor Analyst, since the topology he produces is supposedly universally applicable. However, where we have long chains of connection linking places far apart on the basis of their attributes only, Q-analysis forces us to ask whether this is the description we require. It is a comparatively simple matter to build into this analysis distances from defined regional centres, presence or absence of particular linguistic groups, dietary differences, and the like, to improve the description, if this is what we want.

Out of this comes one policy statement that bears emphasis. Governments have recently tried to discriminate between areas within their borders in order to dispense regional aid. The wrangling over definitions is interminable, because there is always one area which is excluded from aid which considers that it is similar to one that is given aid; and because you cannot partition a connected structure.

In India Factor Analysis and Factor Typologies have been suggested as a solution to the problems of definition: but it is apparent that the violence of the approach results in absurdities, such as the misclassification of Jamshedpur and the Punjab cities reported above. A single linear additive equation cannot cope with the subtleties of regional individuality. A description which preserves individual entities and then details the chains of connection between them is far preferable.

To my mind the outstanding problem of geographical methodology is the relationship between uniqueness and generality – a problem which is dusted off once in a while as a curious paradox, and then allowed to rest again, unsolved. In Atkin's methodology there are the tools at hand to make some sensible statements about it and about many other matters. Let me conclude by quoting some words from myself (Chapman, 1977, p.409) written some time in 1975 before I had heard of Q-analysis.

> since geographers always insist on theorising about some billiard ball planet . . . it seems to me that we need a period in which we try and develop a consensus on what in empirical reality we are trying to study, and how we go about it. We need a theory of empirical enquiry which embraces entitation and observation, and one that will help us in comparative studies far more than any that currently exist. Since it also seems likely that such a theory will call our attention to complex interdependent systems, . . . [it has to have] methods of understanding and explanation that are not partial and reductionist.

I think I now know at which door to knock.

References

Ahmed, Q. (1966) 'Indian cities: characteristics and correlates', *Research Papers*, no. 102, Department of Geography, University of Chicago.

Atkin, R.H. (1974) *Mathematical Structure in Human Affairs*, Heinemann: London.

Atkin, R.H. (1977) *Combinatorial Connectivities in Social Systems*, Birkhäuser Verlag: Basle.

Atkin, R.H. (1981) *Multi-dimensional Man*, Penguin: Harmondsworth.

Chapman, G.P. (1977) *Human and Environmental Systems: A Geographer's Appraisal*, Academic Press: London and New York.

Chapman, G.P. (1979) 'The impact of city function, city location and government expenditure on urban growth in India, 1961–1971', a Paper presented to the 3rd Indo-British Seminar of Geographers, Mysore, mimeo, Cambridge.

Johnson, J.H. (1977a) 'An introduction to Q-analysis and Related Topics', mimeo, Regional Research Project, Department of Mathematics, University of Essex, Colchester.

Johnson, J.H. (1977b) 'A Study of Road Transport', *Research Report XI*, Regional Research Project, Department of Mathematics, University of Essex, Colchester.

Chapter 24

Micro-simulation

M. Clarke, P. Keys and H.C.W.L. Williams

Introduction

One of the more significant developments in analytic planning methods over the last decade has been the refinement and application of models specified and estimated at the level of the individual decision-making unit (individual, household, etc). This is well illustrated in the case of travel demand analysis in which there has been a significant shift, particularly in the United States, away from the conventional aggregate approach to policy analysis, involving models specified in terms of trip groups, to one embracing probabilistic consumer choice concepts in which theoretical statements are married to a representation of behaviour at the micro-level. This particular transition was prompted not only by the inefficiencies, expense and insensitivity of the modelling frameworks developed through the 1960s, but also by the new planning contexts and policy measures which have emerged in the 1970s.

While the changes in modelling style were initiated to a large extent by technical considerations – notably the exploitation of the full variability between units in a given sample size – it was apparent to some of the early exponents of micro-analysis that when the framework was endowed with an appropriate theoretical base the models were more likely to accord with a behavioural explanation of statistical patterns, which some would argue is a necessary precondition for the development of stable forecasting relations.

While the behavioural paradigm was already fashionable in human geography in the late 1960s, much of the early work was of an empirical nature and this style tended to dominate work in the field. It was notably Hägerstrand (1970) and Thomas (1969) who urged the reconciliation of aggregate 'macro-level' descriptions of socio-economic phenomena and their spatial manifestations with disaggregate micro-level concepts. A discussion of the development of these ideas is given by Leigh and North (1978) and Thrift (1981; Chapter 32, this volume).

The formal aspects of the aggregation process have, of course, long been of interest to economists (see, for example, Green, 1964) but it was not until the advent of high-speed electronic computers that the representational and computational aspects involved in aggregation have been studied in depth.

Orcutt *et al.* (1961, 1976) recognized that for many socio-economic phenomena and policy-related issues of interest, several interrelated attributes of the relevant behavioural units had to be specified if a significant amount of variability between these units was to be realized. That is, a greater amount of *information* was required to discriminate realistically between the actions and behaviour of individuals in particular problem contexts. The focus of these developments was, as Guthrie (1972) has discussed, on the high degree of heterogeneity which exists in a given population.

Orcutt and his colleagues (1961) noted the possibility of the efficient manipulation of this information by storing and computationally processing samples of the relevant population multiply classified by the various attributes of interest. This approach was then used in conjunction with Monte Carlo techniques for the solution of models. Although simulation models based on Monte Carlo methods have been used very widely over the past thirty years in disciplines associated with Operational Research, to the authors' knowledge Orcutt's work, dating back to the mid 1950s, was the first attempt to build specific operational models of socio-economic systems, and in particular one of household dynamics, based on micro-simulation.

In this paper we review the basic principles of micro-analytic simulation, together with some recent applications of the approach. Many of the important issues in model construction relate to the transformation of information and to the aggregation process itself. These aspects are considered below as a basis for a discussion of several examples presented later in the paper. It is not our aim here to provide a comprehensive survey of model applications but to illustrate the variety and potential of the micro-simulation approach. The applications reveal several important representational and analytical issues which are then treated more formally. In a final section we offer some views on the prospects and limitations of the method.

It is worth remarking that, in a number of the applications, the advantages and some of the principles of the micro-simulation approach involving list processing appear to have been recognized independently. One could speculate that this is partly due to the specialist nature of the applications dispersed over several disciplines, and the fact that much of the literature is of the form of 'in-house' documentation (e.g. publications of the Urban Institute in Washington, United States Government reports) which has perhaps failed to reach a wide audience. Moreover, for much of the last two decades during which micro-simulation methods have been developed, there appeared to be no great incentive to jettison conventional, and easily accessible, planning techniques.

It will become clear that the process of aggregation over micro-relations in both static and dynamic formats is a central theme of the paper. It is then also of significance to note that the development of travel choice of models specified at the level of the individual is also traceable to the late 1950s. Because until fairly recently little emphasis has been given in studies of travel and housing demand to the detailed process of aggregation over micro-relations, it is perhaps inevitable that these lines of model development attributable to Orcutt *et al.* (1961) and Warner (1962) have not been generically related. The recent and increasing use of simulation in travel demand forecasting indicates a unified model building strategy.

Representation, models, and the aggregation problem

Whether models are meant to provide a causal representation of behaviour or are simply intended as a device for the formal description of variability in a data set, the modelling process, in essence, consists of a series of *transformations* on *information*. We express this process simply as

$$\{\text{information out}\} \leftarrow T\{\text{information in}\}$$

in which the transformation operator $T\{\ldots\}$ may be in the form of simple data manipulation procedures; may express an analytic dependence between variables; or may assume the form of an algorithm or heuristic process.

All information received and transformed in the modelling process is specified with reference to a set of states relating to how a system can be described and observed. The input information sought may either be suggested by the theoretical framework within which the model has been developed, or, and more usually, the form of the model is conditioned by the information available.

Of the many particular questions to which models are addressed a great number fall into three, often related, categories which we label: *dynamics, response* and *control*. We shall formally describe the first two categories by means of the equations for system output $\mathbf{y}(t)$, and its response $\delta\mathbf{y}(t)$ to a stimulus $\delta\mathbf{s}(t)$, as follows

$$\mathbf{y}(t) = \mathbf{f}(\mathbf{x}(t), \mathbf{u}(t), t,: \boldsymbol{\theta}(t)) \qquad (1)$$

$$\delta\mathbf{y}(t) = \mathbf{g}(\mathbf{x}(t), \mathbf{u}(t), t,: \psi(t))\, \delta\mathbf{s}(t). \qquad (2)$$

In general the outputs will consist of several components and have been given vector designations. $\mathbf{x}(t)$ and $\mathbf{u}(t)$ are time dependent vectors of independent and controllable input variables respectively, while $\boldsymbol{\theta}(t)$ and $\psi(t)$ are vectors of parameters to be estimated. In general, the stimulus $\delta\mathbf{s}(t)$ is expressed as a function of changes in $\mathbf{x}$ and $\mathbf{u}$. The functional dependencies, $\mathbf{f}$ and $\mathbf{g}$ may in turn be derived empirically or express an analytic relationship between variables consistent with theoretical statements embracing tested or untested hypotheses. The traditional method of solving the response problem (2) and obtaining the elasticities with respect to policy or other variable changes is essentially through differentiating a typically cross-sectionally estimated functional relationship, $\mathbf{f}$.

The third category into which many modelling applications fall, that of control, may be expressed in terms of an extremal problem in which it is sought to determine the vector, $\mathbf{u}$, which optimizes a given objective function, subject to the system dynamics (1) and other constraints. It will become clear that this latter category is the least well developed in micro-simulation applications.

In order to introduce the micro-simulation approach and various aspects of the aggregation issue, we shall restrict our discussion to a study of households and the individuals comprising them. A micro-description of the household involves the provision of multiply-classified demographic, economic, social and activity/travel characteristics $(\mathbf{c}_1, \mathbf{c}_2, \mathbf{c}_3, \mathbf{c}_4)$ for each of its members. At any time we may identify with a household H_j the following list

$$H_j \left\{\ldots, I^i_j\ (c^i_1, \ldots, c^i_4), \ldots i = 1, \right.$$

$$\left. n_j : \mathbf{c}^j_1, \ldots \mathbf{c}^j_4 \right\} \qquad (3)$$

in which I^i_j refers to the ith member of a household containing n_j members. We emphasize that the vectors **c** will, in general, refer to lists of attributes and in certain cases it may be appropriate to define aggregate summary quantities $\mathbf{c}^j$ pertaining to the household H_j. Traditional surveys (e.g. Census, General Household Survey, Family Expenditure Survey, Transportation Surveys, etc.) collect overlapping subsets of individual and household attributes which for the purpose of confidentiality or presentation are often published in a form in which quantities are aggregated both over individuals *within* a household and those *between* households.

We may regard the household and/or individual as existing in a multiply-classified state defined by the levels of the various sets of attributes which characterize them. Thus, if $c_1 \ldots c_m$ is a full list of relevant attributes and $\ell_1 \ldots \ell_m$ the corresponding levels associated with each, then the state of an individual i may be represented by a vector α_i the components of which refer to the occupied levels of the various attributes. The state of the system as a whole may now be represented by the vectors $\{\alpha_1, \alpha_2, \ldots \alpha_i, \ldots \alpha_Q\}$ which correspond to the individual states of the Q members of a population or sample S.

It is the province of theory to understand how individuals come to be associated with the particular combinations of attributes which characterize a state. Behaviouralists argue that this association and individual transitions between states, which constitute events, are the outcome of *decision contexts* and attempt to formulate model relationships in terms of the choices, constraints and preferences of all relevant decision-making units. A natural starting-point for this analysis is a consideration of a decision context D

$$[D^{\mathbf{a}} \{\mathbf{A}(\mathbf{Z}) : \mathbf{R}\} p]_i \qquad (4)$$

in which **a** represents the relevant attributes of a decision-maker i confronted by a choice between a set of alternatives $\mathbf{A}^i$ each member of which A_μ is endowed by a set of attributes $\mathbf{Z}^i_\mu$ deemed to be of importance in the decision making process p_i. **R** represents a set of constraints which dictate the options which are feasible. Note that **a** will in general be a subset of **c**. The choice set **A** may be composed of discrete options as in location decisions, or continuous as in the mortgage incurred in house purchase. Further, we may think of this choice context as endowed with very general characteristics. For example, each element of D may be dependent on the actions of one or more other decision units (through **A**, **Z**, **R** or *p*), whether this refers to the actions of fellow purchasers in a housing market, or to the constraints imposed by public sector housing management (Thrift and Williams, 1981).

The issue of aggregation appears in many aspects of model development from the definition of alternatives, the measurement of attributes, the grouping of observations, to the summation over micro-relations. In a behavioural approach the aggregation process thus provides the essential link between the output of a model and the decision models pertaining to individuals within a population Q.

In an empirical study the modeller as an observer of a system to be represented, will note in an adopted frame of reference (e.g. the way in which D is described) dispersion or variability between the actions of individuals who are endowed with apparently identical attributes, and may seek to explain this dispersion by the existence of unobserved attributes/constraints, dispersion of preferences, heterogeneity in choice sets etc. (Williams and Ortuzar, 1979). The probability that an individual with *observable* characteristics $\mathbf{a}^i$ and $\mathbf{R}^i$ will be associated with an alternative A_k will in general be expressed in terms of a parametric relation

$$P_k(\mathbf{a}^i, \mathbf{R}^i) = h(\mathbf{Z}^i ; \phi) \, . \qquad (5)$$

This relation may either be stated simply to accord with the empirical regularities in a micro-data set or be consistent with a specific model of the decision process by which the alternative A_k was selected from **A**. Such a model may for example be underpinned by extremal principles such as utility maximization or by satisficing behaviour (Thrift and Williams, 1981). We shall comment on the form and application of analytic micro-models below.

The total number of individuals or decision units N_k in Q associated with the alternative k may now be obtained by general summation (addition over discrete variables, integration over continuous variables) as follows

$$N_k = \underset{\mathbf{a},\mathbf{R}}{S} \rho(\mathbf{a}, \mathbf{R}) P_k(\mathbf{a}, \mathbf{R}) \qquad (6)$$

in which $\rho(\mathbf{a}, \mathbf{R})$ represents the probability density of the set of characteristics over the population Q.

In practice it is usually necessary to perform this aggregation process S by numerical methods. One such approach is Monte Carlo simulation in which a representative sample of the population is formed according to the probability density $\rho(\ldots)$ and each unit assigned to an alternative $A_k \epsilon \mathbf{A}$ according to a comparison between the value of a random number t $(0 \leqslant t \leqslant 1)$ and a cumulative probability index formed from P_k. If the sample is sufficiently large a statistically reliable estimate of N_k may be obtained.

The above example is a simple illustration of aggregation over a probabilistic relation estimated at the micro-level. The alternative courses of action **A** may be defined in a dynamical context corresponding to a wide range of processes, and the Monte Carlo method used to aggregate over individual actions whether or not they are provided with a behavioural basis.

Before describing some recent applications of

micro-analytic simulation, we emphasize a number of distinctions in terminology. *Micro-analysis* is an approach to problem solving and model building based on a smallest unit representation, while the *Monte Carlo technique* is used for the experimental realization or simulation of a process expressed by means of a set of equations or operations. It is also used, as described above, as a numerical method for the integration of functions, and as such forms an important approach to the aggregation process. In turn *List Processing* is simply a device for the computational storage and manipulation of information.

An overview of model applications

The applications of micro-simulation fall broadly into the two categories of dynamics and response described above. We include static considerations as a special case of the former. To our knowledge there have been no applications of the approach in the social sciences which involve aspects of control.

The use of micro-simulation has in each case been partly justified on the basis that it was necessary to include a large amount of information about individual decision units. This aspect will form a prominent part of the discussion in the next section. Here we discuss a number of recent examples which deal successively with demographic, economic and spatial issues.

It is appropriate to start with a brief discussion of the work of Guy Orcutt and his colleagues, because of its pioneering nature. In his original book (Orcutt *et al.*, 1961) a framework was provided for the micro-simulation of the United States demographic/economic system. In subsequent work particular emphasis was paid to the dynamics of households and the derivation of income. This effort culminated in a flexible set of computer programs known as DYNASIM (Dynamic Simulation of Income, Orcutt *et al.*, 1976), designed as a policy testing instrument applied at the national level. Although the degree of detail is high, requiring a large number of sub-models, the essential elements of the approach involve the economic and demographic characteristics of individual members of a population sampled at a given time. The empirical complexity of the model is largely a consequence of the need to establish the dependence of a wide variety of processes (e.g. household formation/dissolution, birth, death, marriage, entry into the labour market, etc.) on a multitude of individual and household attributes. For example, labour force participation is derived as an empirical function of an individual's race, sex, age, marital status, disability status, the previous year's labour force participation, the household's last year's transfer income, number of children under six, last year's unemployment rate. Estimation of this and other functions was achieved by resorting to a wide variety of data sources and special-purpose studies.

This work, undertaken at the Urban Institute in Washington, has spawned a number of interesting applications both in academic and government institutions. A typical use of DYNASIM and its predecessor TRIM (Transfer of Income Model) was the identification of particular classes of the population eligible for certain services and welfare provision. An overview of the use of static and dynamic micro-simulation models for public welfare and demographic analyses and their joint interaction is given by Harris (1978).

Many of these applications have been performed at the national level, though a few regional applications do exist, including a model of household energy consumption developed by Caldwell *et al.* (1979) for New York State, and a model of household production and consumption for the Yorkshire and Humberside region developed by Clarke, Keys and Williams (1979, 1980). In the latter application the authors were concerned with the projection up to 1982 of changes in the economic circumstances of households under alternative regional macro-economic assumptions. In this example, explicit models of matching are developed in multi-sector labour and housing systems.

The model developed by Kain and his colleagues at the United States National Bureau of Economic Research is one of the most detailed representations of a metropolitan housing system. The explicit spatial representation of household residential and workplace location combined with the socio-economic identifiers produced very large data storage and retrieval problems in the earlier version of the model (as in the Detroit prototype, see Ingram *et al.*, 1972). List processing was seen as a means of alleviating these problems and was incorporated into the most recent version of the model (Kain *et al.*, 1976). While the model may address many aspects of the private housing system, a prominent use has been in the analysis of housing allowances.

There are a number of alternative, and less ambitious, activity location/travel demand models that employ micro-simulation concepts. Of considerable interest are the applications of Wilson and Pownall (1976) and Kreibich (1979) who explicitly use Monte Carlo methods for examining micro-level interdependencies. In the former the interdependency between workplace, residential and shopping location was examined, under conditions of change (relocation of households in the public sector) while the latter investigated the relationship between location behaviour, car availability and modal choice using a heuristic process in which time budget constraints played a prominent role. An alternative and related treatment of constraints in activity/trip analysis is provided by Lenntorp (1976) in the time-geographic tradition of Hägerstrand (1970).

Recently, Monte Carlo simulation has been used

for the purpose of aggregating micro-analytic travel demand relations (see, for example, Watanatada and Ben-Akiva, 1978), and for the purpose of examining travel decision processes. The car-pooling model developed by Bonsall (1979) is a good example of this type of work. The decision to share a ride (car pool) was found to be dependent on a large number of detailed characteristics of the driver, passenger and trip. Simulation proved to be a convenient tool for the aggregation over the decision processes of a highly heterogeneous population group.

The solution of models by micro-analytic simulation: some formal considerations

The applications discussed above combine a number of methodological characteristics which can be subject to individual scrutiny. We now examine several of these issues in the context of model development and solution. Although sometimes of a rather technical nature they are of prime importance in the implementation of models developed within the micro-simulation framework.

List processing and the storage of information

A number of authors, including Orcutt *et al.* (1961), Wilson and Pownall (1976) and Kain *et al.* (1976), have commented on the advantages of list processing for the efficient storage and retention of information. Some comments on this important aspect are therefore necessary. A central feature here is the computational representation of the state of a system. With reference to the categorical definitions used above it is easily seen that each individual or unit may exist in any one of $\prod_{i=1}^{m} \ell_i$ possible states. The system as a whole may then be described in terms of the traditional occupancy matrix, **N**, the elements of which, **N**(α), record the number of units in each alternative state, α. Typically, for large m, $\prod_i \ell_i$ will be an exceedingly large number and, moreover, the matrix **N** will be very sparse – a large proportion of states (cells) will not be occupied by *any* individual.

An alternative means of recording the state of the whole system and storing equivalent information is to list the individual states of each component member $\alpha_1, \ldots \alpha_i, \ldots \alpha_Q$. For a population of size Q, this will involve recording QM quantities. Which of these methods is the more efficient in its capacity for storage of information appears to depend therefore on the relative sizes of $\prod_i \ell_i$ and QM. For units which are described by a large number of categories the former may well be greater, and often considerably greater, than the latter. The efficiency of the list representation is further enhanced when it is noted that for many purposes in which simulation is adopted for the numerical solution of models, a satisfactory computation of the *required outputs* may be achieved with a sample, q, which will be a small fraction of Q. In summary, typically

$$\prod_i \ell_i \gg qM. \tag{7}$$

One method of overcoming this dimensionality problem with the occupancy matrix is to form summary matrices from **N**, in which summation is performed over certain categories, and this is often done in government and other survey publications (e.g. Census, Family Expenditure Survey, General Household Survey, etc.) in which it is required to portray large amounts of information with due regard for confidentiality. It is commonly remarked that considerable information is lost through this process of aggregation.

It appears that both the *storage efficiency* and *information loss* arguments have been used for the strong justification of the list processing approach, which retains full information in a compact form. There are, however, a number of crucial points to be made here relating both to the nature of a model and the information involved. If it is desired to record and process information on the attributes of each member of a real population of individuals and examine the full individual information as output (e.g. for tax or health service purposes) the above arguments hold. For simulation models in which the aim is to *represent* and *replicate* real world processes we are not interested in any *particular* member of the real world but in information pertaining to samples, representative in terms of the joint distribution of characteristics over that population. The properties of this joint distribution and the nature of model operations rather than the characteristics of any *particular* sample itself are then the decisive determinants of both the storage efficiency and information retention issues. It is clear that the structure of dependence between the attributes is at the heart of the matter. If, for example, in an extreme case, the attributes $c_1, \ldots c_m$ were independent in the sense that the joint density function, $\rho(c_i, \ldots, c_m)$, could be factored

$$\rho(c_1, \ldots, c_m) = \prod_{i=1}^{m} \rho_i(c_i) \tag{8}$$

then the storage of the probability matrix, ρ, may be reduced to that of the right hand side of equation (8). The $\sum_{i=1}^{m} \ell_i$ storage requirement will typically be considerably less than that required to store the attributes of a list of individuals sampled from ρ. This argument will hold in suitable modified form when partial dependencies occur.

Equally, it is illusory to regard information to be lost in the formation of summary matrices or summary lists, if the attributes 'separated' in the aggregation process are independent. In any given application involving many attributes the arguments of efficiency and information must therefore be related to the structure of the model itself. Often considerable interdependencies *will* exist between the attributes and list processing may well be a strong contender on the above grounds. It is possible to employ mixed representations in which some information is stored in an occupancy matrix form while other partial probability matrices are recreated from list processing after model operations have been employed. This method is useful in interfacing micro- and macro-models and has been adopted by Clarke, Keys and Williams (1979).

The generation of an initial population

The issue of attribute dependency is also an important ingredient in the provision of a sample population, which is adopted in all the above examples. In dynamical models this represents a starting-point for the forward iterative solution of dynamical equations.

The information required for the construction of the multiply-classified sample $q(\mathbf{c})$ may be sought directly in a survey, or indeed sampled from another survey, as in the applications of Orcutt *et al.* (1976), Kain *et al.* (1976), and Caldwell *et al.* (1979). In the absence of a suitable survey an alternative strategy is to attempt to generate $q(\mathbf{c})$ by sampling from a *synthesized* joint probability matrix, $\rho(\mathbf{c})$, or corresponding contingency table. The theoretical basis for generating entries to a full contingency table consistent with available conditional and marginal probability distributions is long established and discussed in detail in the synthetic sampling procedure outlined by McFadden *et al.* (1977). In general the more complex the structure of relationships between the variables the more available information in the form of marginal probabilities will be required to reproduce this structure in a synthesized joint distribution $\rho(\mathbf{c})$.

A strategy which has been used in sample generation is to express the joint probability as a product of conditional distributions:

$$\rho(\mathbf{c}) = \rho(c_1)\,\rho(c_2|c_1)\ldots\;\rho(c_m/c_{m-1}\ldots c_1) \tag{9}$$

and adopt approximations which impose a simplified structure on the conditional dependencies. That is, assumptions are made about important interrelationships between the characteristics. Examples of this approach have been discussed by Wilson and Pownall (1976), MacGill (1978) and by Bonsall (1979).

In the applications examined the sample size q adopted was found to vary considerably from about 4,000 in the study of Orcutt *et al.* (1976) to 78,000 in the housing market study of Kain *et al.* (1976). It is difficult to give specific guidelines for the required size. This clearly depends on a number of factors which include: the nature of the model output; the number of attributes and the levels selected; and, importantly, on the structure of dependency between the attributes.

Models of characteristic (attribute) dependence and matching

Models of dependence between individual characteristics may appear both in the formation and manipulation of the samples. This process of association usually involves a parametric function specified in terms of a set of attributes which include policy variables, as in the expression (5). In the application of Wilson and Pownall (1976), for example, spatial interaction models were employed to express an association of workplace, residence and shopping locations in terms of the generalized cost of interzonal travel.

As we noted in that section, models of association at the micro-level may or may not involve a behavioural rationale, but where a representation of the decision process *is* sought the resultant model may be of very general form and include a complex set of constraints and preference relations. Individual choice models, particularly those belonging to the logit family (which is comprised of particular functional forms which may be adopted in the expression (5)) underpinned by random utility theory (see, for example, Williams and Ortuzar, 1979) have now been widely used in travel demand and residential location models, and indeed have been employed in the housing study of Kain *et al.* (1976) and car pooling model of Bonsall (1979).

Many of the micro-simulation examples involve the matching of items of 'demand' and 'supply', each of which are endowed with lists of characteristics $\mathbf{c}_d$ and $\mathbf{c}_s$, respectively. The labour and housing systems provide obvious examples in terms of the respective matches Individual $\leftrightarrow$ Job and Household $\leftrightarrow$ House . Two more examples include the matching of individuals (passenger and driver) in car pooling schemes, and the merging of sexes in the marriage union, central to the process of family/household formation (see Orcutt *et al.*, 1976; Clarke, Keys and Williams, 1979).

Certain rules of association which represent real world processes of choice or allocation are adopted by the modeller in the implementation of this matching process. From a computational viewpoint it is required to match items stored on two lists according to rules which entail a measure of association $B(\alpha_d, \alpha_s)$ between demand and supply micro-states. The elements of B may represent prior probabilities in information theoretic approaches, and utilities or costs of association in allocation models based on choice processes. Equally, each element of B could be expressed as a series of constraints in multicriterion satisficing approaches. The algorithm used to implement the process may thus embed probabilistic micro-models of the type discussed above and be suitably tailored to reflect 'scarcity' on the demand or supply side. In the computational operation it is necessary to reflect competition between micro-units without introducing bias due to the position of each item in its respective list. In practice this bias may be rather expensive to completely eradicate – the successive double random sampling from the separate lists and probabilistic (Monte Carlo) matching according to an index based on $B(\alpha_d, \alpha_s)$ can be very time-consuming. It is, however, often possible to employ less expensive procedures at the risk of introducing some inaccuracy. One such approach involves the solution of an aggregate allocation process at a lower level of resolution and sampling from its outcome. Such allocation procedures may be devised for interfacing individual circumstances with a macro-environment, and are of considerable theoretical interest (see, for example, the papers by Snickars and Weibull, 1977; Clarke, Keys and Williams, 1979; Anas, 1979; Los, 1979).

The matching process thus frequently embodies key elements of a model which determine, and ideally explain, the association of an individual or item with a particular state.

Dynamics and the solution of integrated models

While it is sometimes possible to provide a formal statement of the solution to a system of equations expressing socio-economic system dynamics, whether these are of a stochastic or deterministic form, it is usually necessary in applications to resort to some means of numerical integration of a set of differential (or difference) equations to achieve the required model outputs. Monte Carlo simulation is one approach to the solution of such equations and has found application in household models embracing demographic and economic processes. These include DYNASIM (Orcutt *et al.*, 1976), the National Bureau of Economic Research (NBER) housing model (Kain *et al.*, 1976) and the model of household dynamics developed by Clarke, Keys and Williams (1979).

The household and its component individuals will in general partake in many dynamical processes, and indeed some induce others. This correlation between individual state transitions or events provides a dynamical interaction between individual (and household) characteristics. The complexity of dynamical behaviour derives precisely from this micro-level interdependence reflecting interaction between the various processes.

We can write a straightforward accounting identity for the change in state occupancy $\Delta N_\alpha(t, t+\Delta t)$ in the discrete time interval $[t, t+\Delta t]$ in terms of the matrix $\mathbf{F}(t, t+\Delta t)$ of flows or transitions between different states as follows:

$$\Delta N_\alpha(t, t+\Delta t) = \sum_{\alpha'} F(\alpha', t : \alpha, t+\Delta t) - F(\alpha, t : \alpha', t+\Delta t) \ . \qquad (10)$$

By a judicious choice of characteristics and associated levels (which may, for example, incorporate 'duration of stay' effects in a particular state) the use of a Markovian assumption can often be justified in transition models

$$\Delta N_\alpha(t, t+\Delta t) = \sum_{\alpha'} r(\alpha', \alpha : t)\, N_{\alpha'}(t) - r(\alpha, \alpha' : t)\, N_\alpha(t)\, \Delta t, \qquad (11)$$

and allows the onus of model building to be transferred to the matrix of rates $\mathbf{r}(t)$. The number of characteristics which are 'active' in any model operation will, of course, be dependent on the particular process involved.

The dynamics of a household may thus be formally reduced to a set of coupled differential equations in which the several processes and their respective time dependent rate matrices typically refer to overlapping sets of characteristics as discussed by Clarke, Keys and Williams (1979). This interdependency and the sparseness of the rate matrices commend the use of list processing in conjunction with the Monte Carlo method. In the sequential consideration of individual items in a list it is computationally expedient to arrange for an examination of the participation of each member in as many processes (giving birth, death, job/house change, etc.) as possible consistent with the causal restrictions imposed in the model, to avoid wasteful and repetitious examination of list information.

The most challenging aspect of dynamic modelling involves an attainment and understanding of the dependence of the rate matrices

$$\mathbf{r} = \mathbf{r}(\mathbf{Z}(t), t) \qquad (12)$$

on time and (time dependent) policy instruments $\mathbf{Z}(t)$. Time series information may be available, and again, the explicit derivation of models from decision contexts may be entertained. Although much progress has been made in deriving certain of these functional dependencies (see, for example, Orcutt *et al.*, 1976) the establishment of a comprehensive model of demographic-economic interactions at the micro-level remains a considerable challenge.

Prospects and problems

We have described some aspects of the development and application of simulation models based on a smallest unit representation. It is clear from the above discussion that the method is interwoven by several themes, and the framework we have outlined is one in which data can be organized, transformation equations established and a model solution method devised.

In order to discuss the merits and limitations of the approach in any modelling context we must compare its advantages with those of its competitors. In addition we may ask whether subject areas which have not traditionally been associated with the approach would benefit from the use of micro-simulation. In the applications described, which involve the dynamics or statics of demographic and economic/activity systems, alternative approaches have involved some form of aggregation. As emphasized previously the aggregation issue embraces many features. Some, for example the aggregation over functional relationships, concern the bias of parameter estimates, others relate rather more closely to model design, and the type of information required. One can always point to the worst excesses of the aggregate approach, and this form of criticism has adorned the literature in recent years (cf. transportation models). However, we would point out that the arguments in favour of a highly disaggregate micro-approach are not as obvious as some of the proponents might suggest. For some purposes, as in the case of short-term population forecasts, it might be simply unnecessary to develop a micro-approach – trend forecasts developed within an appropriate accounting framework may be sufficient. This is equally true of the production sector where the firm, as a decision unit, bears many formal similarities with the household (Thrift and Williams, 1981). While the household sector is demonstrably complicated in its range of possible micro-behaviour it has considerably less variety both in its characteristics and its potential response and adaptive behaviour. Although some micro-models do exist at the firm level drawing on the initial impetus provided by Clarkson and Simon (1960), Shubik (1960) and Cyert and March (1963), and a mass of empirical work has been undertaken, there has been considerably less progress than in the household sector. Again for conventional forecasting issues econometric and input-output models may suffice, and indeed it was primarily the complexity of the economic behaviour of individual and interacting firms which inspired Leontief's aggregate approach.

In practice the problems of data acquisition and measurement can often frustrate the development of micro-models of the type discussed. It is not uncommon to have some aggregate information available in time-series form and some micro-information available at one cross-section. By generating aggregate forecasts from both macro- and micro-models one could hope to combine the distinct and complementary features of both. Aggregate and disaggregate approaches should not necessarily be regarded as competitors but formal strategies for achieving complementarity await full development. This goal may be brought nearer by the wider availability of longitudinal data sets.

In any approach to problem-solving or policy analysis computational issues and questions of efficiency should be clearly distinguished from representational and theoretical aspects. Many factors will determine the formulation of a model, including the available inputs, the required outputs and the resources at hand. An assessment of a model must not be divorced from the context within which it is created or applied. The excessive promotion of a particular methodological perspective has in the past tended, ultimately, to be to its detriment. With regard to micro-simulation as an *approach* to problem-solving we would, however, conclude on a note of guarded optimism. The smallest unit representation discussed here appears to be an extremely flexible one in the study of emerging and important themes, notably the interrelationship between an individual and its environment, whether this be of a social, economic, institutional or spatial nature. Moreover, it does appear, as Wilson and Pownall (1976) have suggested, highly suitable for an examination of extensive interdependencies between attributes which arise from a variety of processes and constraints. Our own work at Leeds is concerned with a number of applications concerning these interdependencies. Of particular interest is the mutual interaction both between individuals within a household, and between individuals and an environment created by the labour and housing systems, local and central government.

Many of our remarks have been implicitly concerned with traditional forecasting approaches. What might prove to be of particular value is the use of micro-simulation in a context of relatively new approaches to problem and policy analysis. The method is wholly complementary to gaming-simulation approaches in which the mutual interaction of decision-makers is subject to experimental scrutiny, and may readily be embedded within control frameworks. The approach is equally suitable for application in a range of different planning frameworks as diverse as traditional cost-benefit analysis and planning for freedom of action in which the range of feasible alternatives available to individuals with widely different personal and environmental circumstances is the central focus of interest.

The method we have described can be viewed as a 'first principles' approach to the representation of systems and as such provides the basis for the design of models and appropriate data sets, and the study of interdependencies and systemic effects originating at the micro-level.

References

Anas, A. (1979) 'The impact of transit investment on housing values. A simulation experiment',

Environment and Planning, A, 19, 239–55.

Bonsall, P. (1979) 'Microsimulation of mode choice: a model of organised car sharing', PTRC proceedings, Summer Annual Meeting, University of Warwick, July 1979.

Caldwell, S., *et al.* (1979) 'Forecasting regional energy demand with linked macro/micro models', Working Papers in Planning, No. 1, Department of City and Regional Planning, Cornell University, Ithaca, NY.

Clarke, M., Keys, P., and Williams, H.C.W.L. (1979) 'Household dynamics and socio-economic forecasting: a micro-simulation approach', Paper presented at the European Congress of the Regional Science Association, Working Paper 257, School of Geography, University of Leeds.

Clarke, M., Keys, P., and Williams, H.C.W.L. (1980) 'Models of matching and mobility in the labour and housing system', Working Paper, University of Leeds.

Clarkson, G.P.E., and Simon, H.A.S. (1960) 'Simulation of individual and group behaviour', *American Economic Review*, 50, 920–32.

Cyert, R.M., and March, J.G. (1963) *A Behavioural Theory of the Firm*, Prentice-Hall, Englewood Cliffs, NJ.

Green, H.A.J. (1964) *Aggregation in Economic Analysis*, Princetown, University Press, NJ.

Guthrie, H.W. (1972) 'Microanalytic simulation of household behaviour', *Annals of Economic and Social Measurement*, 1/2, 141–69.

Hägerstrand, T. (1970) 'What about people in regional science?', *Papers and Proceedings of the Regional Science Association*, 24, 7-21.

Harris, R. (1978) *Microanalytic Simulation Models for Analysis of Public Welfare Policies*, The Urban Institute, Washington, DC.

Ingram, G.K., Kain, J.F., and Ginn, F.R. (1972) *The Detroit Prototype of the NBER Urban Simulation Model*, National Bureau of Economic Research, New York.

Kain, J.F., Apgar, W.C., and Ginn, J.R. (1976) *Simulation of the Market Effects of Housing Allowances, Vol. 1: Description of the NBER Urban Simulation Model*, Research Report, Department of City and Regional Planning, Harvard University, Cambridge, Mass.

Kreibich, V. (1979) 'Modelling car availability, modal split and trip distribution by Monte Carlo simulation: a short way to integrated models', *Transportation*, 8, 153–66.

Leigh, R., and North, D.J. (1978) 'The potential of the micro behavioural approach to regional analysis' in P.W.J. Batey (ed.), *London Papers in Regional Science*, 8, Pion: London.

Lenntorp, B. (1976) 'Paths in space-time environments', Land Studies in Geography, Series B, Human Geography, No. 44.

Los, M. (1979) 'Discrete choice modelling and disequilibrium in land use and transportation planning', Publication 137, Centre de Recherche sur les Transports, University of Montreal.

McFadden, D., Cosslett, S., Duguay, G., and Jung, W. (1977) 'Demographic data for policy analysis', The Urban Travel Demand Forecasting Project, Phase 1, Final Report Series, Vol. 8. Institute of Transportation Studies, University of California, Berkeley.

Macgill, S.M. (1978) 'Multiple classifications in socio-economic systems and urban modelling', Working Paper 212, School of Geography, University of Leeds.

Orcutt, G., Greenberger, M., Rivlin, A., and Korbel, J. (1961) *Microanalysis of Socio-economic Systems: A Simulation Study*, Harper & Row, New York.

Orcutt, G., Caldwell, S., and Wertheimer, R. (1976) *Policy Exploration through Micro-analytic Simulation*, The Urban Institute, Washington, DC.

Shubik, M. (1960) 'Simulation of the industry and the firm', *American Economic Review*, 50, 908–19.

Snickars, F., and Weibull, J. (1977) 'A minimum information principle – theory and practice', *Regional Science and Urban Economy*, 7, 137–68.

Thomas, M.D. (1969) 'Regional economic growth: some conceptual aspects', *Land Economics*, 45, 43–51.

Thrift, N. (1981) 'Behavioural geography: a paradigm in search of a paradigm' in R.J. Bennett and N. Wrigley (eds), *Quantitative Geography in Britain: Retrospect and Prospect*, Routledge & Kegan Paul, London.

Thrift, N. and Williams, H.C.W.L. (1981) 'On the development of behavioural location models', *Economic Geography* (in press).

Warner, S.L. (1962) *Stochastic Choice of Mode in Urban Travel: A Study in Binary Choice*, Northwestern University Press, Evanston, Ill.

Watanatada, T., and Ben-Akiva, M. (1978) 'Spatial aggregation of disaggregate choice models: an areawide urban travel demand sketch planning model', Presented at the Transportation Research Board Annual Meeting, January 1978.

Williams, H.C.W.L., and Ortuzar, J.D. (1979) 'Behavioural travel, theories, model specification and the response error problem', Paper presented at the PTRC Annual Summer Meeting, University of Warwick.

Williams, H.C.W.L., and Wilson, A.G. (1979) 'Some comments on the theoretical and analytic structure of urban and regional models', Working Paper, School of Geography, University of Leeds.

Wilson, A.G., and Pownall, C.E. (1976) 'A new representation of the urban system for modelling and for the study of micro-level interdependence', *Area*, 8, 246–54.

Part 5

Applications: physical and human geography

In this section the statistical methods and mathematical models discussed in Parts 3 and 4 are considered within the context of the various subdisciplines within geography, and the broader themes which were introduced in Chapter 1 are elaborated. The opening chapters discuss the development of quantitative research in four major branches of physical geography. Two themes stand out. First, the changing nature of inductive statistical modelling in most areas of physical geography. Second, the growth of a deductive, deterministic, mathematical modelling tradition.

In a recent review of statistical methods in physical geography, Unwin (1977) contrasts the approach to statistical analysis adopted by physical geographers in the 1960s with that which they increasingly began to adopt in the mid to late 1970s. In the 1960s most British research of this type involved the use of classical univariate and multivariate statistical methods to investigate the morphology of environmental systems which were assumed to be at equilibrium. That is to say, physical geography was concerned to a large extent with the description and analysis of form rather than the modelling of process; the development of environmental systems over time was largely ignored, and classical statistical models which assume independence of observations over time and space were used. In the 1970s, however, physical geographers in Britain became more concerned with environmental systems whose morphologies were time dependent. To analyse these systems they increasingly turned to time series models and to a general consideration of the nature of stochastic process theory. In this way their work became linked to the space-time series statistical modelling (see chapters 7 to 10) which had become an important area of research in British quantitative human geography.

This pattern of development in statistical modelling is a common feature of the reviews of quantitative applications in climatology, hydrology and geomorphology, provided by Unwin, Anderson, Richards, Thornes and Ferguson. Models for time, distance, and spatial series are shown now to play an important role in each of these fields, and they share a common concern with topics such as autocorrelation functions, series length and quality, differencing, transfer function modelling and adaptive parameter modelling. Only in the field of biogeography does this trend in statistical modelling appear to be absent. Here, Matthews shows that classification and ordination of plant communities using a variety of multivariate techniques remains the dominant concern, though there are some recent examples of the adoption of stochastic process models to study vegetation change through time, and exploratory uses of autocorrelation measures to study spatial dependence.

Alongside these changes in the nature of statistical modelling in physical geography since the early 1960s has come a growth in the amount of deductive mathematical modelling. Unwin is able to outline a hierarchy of deductive mathematical models of climate for systems of increasing complexity, and Thornes and Ferguson adopt a similar approach in their consideration of deductive models of geomorphic systems. Despite the increase in British research in this area, however, applications of statistical analysis far outnumber attempts at deductive mathematical modelling, and Thornes and Ferguson criticize British quantitative geomorphologists for relying too heavily on statistical description and analysis of phenomena for which the basic mechanics are already established. They believe that there has been too much naive inductive or probabilistic modelling of environmental systems exhibiting order which could be modelled deterministically, and they argue that more mathematical training of physical geographers is required. In terms of the mathematical modelling challenges in the next decade, Unwin, Thornes and Ferguson agree with Wilson (see Chapter 18) that particular emphasis should be placed on system dynamics, transitions between states, and discontinuities, and as Wilson demonstrates, this is likely to involve the consideration of catastrophe and bifurcation theory.

The four chapters on quantitative applications in physical geography are then followed by eight chapters which review quantitative orientations in the major branches of human geography. Once again, many of the themes introduced in the earlier sections of the book stand out.

First, most branches of human geography are shown to have a tradition of inductive statistical modelling which goes back to the early 1960s. Although much of this work has been rather pedestrian there are areas in which a creative link has been forged between pure and applied quantitative research. For example, Martin and Spence show how the concern of British quantitative geographers with the statistical modelling of time and space series has resulted in applied research in economic geography concerned with regional economic fluctuations, structural models of regional and urban labour markets, and the geography of wage inflation, and how this has involved the use of time and space-time models of the type reviewed in Chapters 8 and 9 by Hepple and Bennett, together with time-varying factor analyses. Similarly, Thrift, Johnston and Wrigley show how discrete choice modelling, which is based on the computationally tractable statistical models for categorical data reviewed in Chapter 11 by Wrigley has become, or is likely to become, an important research methodology in the many fields of human geography which are concerned with spatial behaviour and choice. This also stimulates interest in further links with the micro-simulation approaches being developed at Leeds by Clarke, Keys and Williams (see Chapter 24). In addition Taylor and Gudgin show how study of the geography of elections has been advanced by a statistical orientation and has contributed to a better understanding of the modifiable areal unit problem discussed in Chapter 5.

Second, as in physical geography, applications of statistical analysis are seen to outnumber attempts at mathematical modelling. In this respect, urban models which are reviewed in Chapter 17, and which form part of a more widely defined urban geography, represent the most coherently developed set of mathematical models. However, as Hay demonstrates, there are also important uses of network analysis and the properties of matrices in transport geography, and Rees shows how British population geographers have developed spatial demographic accounting methods to a high level of sophistication. Once again, many of the mathematical modelling challenges of the next decade are seen to be within the area of dynamic modelling, structural change, discontinuities, and transitions between states, and in this context Johnston and Wrigley draw attention to the importance of Wilson's pedagogic illustrations of the use of catastrophe theory to model the evolution of urban retail structures.

Third, most of the authors of these chapters consider it necessary to give explicit consideration to the recent criticisms in human geography of the positivist-quantitative paradigm. It is clear that the authors of these chapters accept the validity of many of these criticisms, certainly as far as they apply to the deficiencies of much of the early quantitative research in geography. Nevertheless, they do not accept that quantitative orientations are necessarily positivist, nor that quantitative research has no future in the various branches of human geography with which they are concerned. Instead, the impression given by many of these chapters is that quantitative research in human geography has recently come through the fire of criticism and reappraisal, and, having emerged, is now beginning to develop many of the characteristics its critics saw as lacking in much of the quantitative work of the 1960s and early 1970s.

As we have previously stated, the result of this reappraisal is likely to be a more 'applicable' quantitative geography. Hints of this are to be found in many of these chapters, particularly those by Johnston and Bennett on political and policy-orientated geography. Both authors feel some disillusionment with much of the earlier work in these areas, and they call for a new orientation in which geographers will address questions of social distribution. Bennett sees this as leading to a greater concern with social goals, policy instruments, policy assessment and policy recommendation, each of which require approaches which are, in part, necessarily quantitative in formulation. In this respect his chapter echoes a theme which is pervasive in many of the chapters in this section: namely a new emphasis on 'applied' rather than 'pure' quantitative analysis. It is likely, therefore, that as quantitative geography proceeds through the 1980s, the techniques and modes of analysis used in given subject areas will arise more naturally from the specific characteristics of the problem in hand, rather than as a result of a need to evidence the potential of a particular method of analysis.

Reference

Unwin, D.J. (1977) 'Statistical methods in physical geography', *Progress in Physical Geography*, 1, 185–221.

Chapter 25

Climatology

D.J. Unwin

Introduction

Climatology has always sat uneasily within British geography departments yet to date it has not been convincingly taken up by any other discipline. In consequence it has been depressed, a sort of Cinderella to the ugly sisters of geomorphology and biogeography. The reasons for this depression are both internal and external. Internally, worthwhile climatic research is likely to be relatively expensive, requiring technical skills that few, if any, departments possess and the 'hard science' orientation of most courses (Atkinson, 1978a, b; Musk and Tout, 1979) means that it has attracted few students, making the acquisition of expensive hardware and the necessary staff difficult to justify. According to the 1978 *Third Directory of British Climatologists* (Smithson and Gregory, 1978) there are only 138 people in Britain who so describe themselves, about half of whom are based in geography departments. The research output has been correspondingly small, at least in so far as it is reported in the geographical literature. Over the past decade less than three per cent of articles published in the *Transactions* have been climatological, and, other than the occasional conference report and a two-page note on econoclimatology by Perry (1971), the first ten years of *Area* have seen not a single climatological contribution! Moreover, if the most recent list of thesis completions in North America (Raup, 1979) is a reliable guide to research activity, this depression is by no means confined to Britain. Externally, traditional climatology, especially 'geographical' climatology, has failed to establish itself as serious science. Clearly, what is hard science to our students looks very simple indeed to the meteorological community. The most recent statement by the Natural Environment Research Council on *Research in Applied and World Climatology* (NERC, 1976) almost goes out of its way to avoid mentioning climatology as most of its geographical practitioners would recognize it, and, contrary to the opinion expressed in the first half of the same report, support in the form of research studentships and funds is laughably low (Thornes, 1978). There are, of course, exceptions to these generalizations and geographical contributions have been recognized as valuable but these have invariably been the work of exceptional men working in fields somewhat removed from day-to-day meteorology. Examples include work using satellite imagery, in synoptic, agricultural, urban and econoclimatology and climatic change. Recent years have also seen a succession of well-received climatology texts from British geographer-climatologists (Barry and Perry, 1973; Stringer, 1972a, 1972b; Lockwood, 1974 and 1979; Crowe, 1971; Chandler and Gregory, 1976; Lamb, 1972 and 1977).

Whatever the exceptions, the consequences for the scope and nature of work in climatology have been profound. The lack of any real concentration of manpower and facilities has meant that research has been fragmentary and limited in its objectives whilst the general absence of funding and technical expertise has given rise to a tendency for work to be restricted to the analysis of data collected by other agencies. Given these facts, and the self-evident consideration that the very notion of climate is a statistical one, it is hardly surprising that virtually all published work has been statistical in nature, and, indeed, as I have argued elsewhere (Unwin, 1977), was so long before any so-called 'quantitative revolution' in British geography.

Weather, climate and climate models

Over a decade ago, the influential volume *Models in Geography* (Chorley and Haggett, 1967) contained an essay by R.G. Barry on 'Models in meteorology and climatology' (Barry, 1967). The 'models' cited – the general circulation, long waves, jet streams and so on – were without exception meteorological. True, they were, and are, used by climatologists in the explanatory description of weather but none are intrinsically climatic. Reading Barry's essay, one senses that he was hard pressed to find anything to stand alongside the shining new models elsewhere in the Cambridge store but, writing in 1979, things are much easier. This is because the last decade has seen the birth of a truly model-based climatology that uses true climatic models (Terjung, 1976; Lockwood, 1979). Before describing some of these, it is wise to clarify exactly what is meant by the terms 'weather', 'climate' and 'climate models'. *Weather* is the instantaneous state of the atmosphere as defined by the values of any number of suitable elements. Although the human being may perceive it and react to it in different ways it is manifestly a real phenomenon. By contrast, *climate* is an invention of man that has been variously defined, but most definitions are along the lines of Lorenz's succinct 'the statistics of weather' (Lorenz, 1975), with 'statistics' understood to include not only the element time-averages but also their frequency distributions, autocorrelations and covariances in both time and space. A *climate model* is therefore any model which describes these statistics or whose output yields them. Viewed in this way, two distinct approaches to climate modelling may be recognized. The first uses inductive statistical analyses of varying complexity to synthesize a climate model from observed weather data, and a great deal of traditional climatology can in retrospect be categorized as modelling with this approach. The second, more recent, approach develops deductive physical mathematical models incorporating the processes thought to be important and whose output is the required climate. Of course, the two approaches are complementary in that empirical results are both incorporated into the deductive models and are used in validation studies.

Within each approach, a variety of models have been produced, characterized by increasing complexity and fidelity to the real atmosphere as the *dimensionality* of the problem is increased and giving an hierarchy of climate models. Whatever the approach, the simplest models are for single weather elements at single places. More complex one-dimensional models incorporate a dimension of variability in time or space; at a higher level still two-dimensional models examine variability in, say, altitude and latitude considered simultaneously, and the most complex of all incorporate the time and space dependence of numerous weather elements.

Inductive statistical models

Classical statistics Numerous statistical methods have been used to model the weather at a single place (see, for example the standard texts by Brooks and Carruthers, 1953; Conrad and Pollak, 1956; Crowe, 1971, pp.507–31; Barry and Perry, 1973, pp. 214-90). Indeed, useful climatological information can be obtained by simply acquiring and tabulating summary weather data, as, for example, that reported in *The Climate of London* (Chandler, 1965). Disaggregation of the data for each weather element according to month, season or weather type often reveals quite distinctive frequency distributions to which one or other of the standard forms may be fitted (Thom, 1966; Suzuki, 1967; Essenwanger, 1976) and where more than one element is considered correlation and contingency analysis have been used (Murray, 1968; Hay, 1968). Although the usual product-moment and chi-square statistics are often used, a number of indices based on the ratios of elements on and off the principal diagonal of a contingency table have been proposed (Cehak, 1970) and used (Lee, 1977).

Time series Most meteorological data constitute a time series and it is the expected sequence that forms one of the most important characteristics of a climate. In consequence, time series models have often been used in the hope of discriminating amongst possible driving processes, in statistical forecasting (Craddock, 1965) or in the study of long- and short-period climatic change (Curry, 1962). When ordered by time, most series show distinct daily or seasonal periodicities, so that harmonic analysis has often been employed (Horn and Bryson, 1960; Sabbagh and Bryson, 1962; Maddon, 1977). Where the evidence of periodicity is less striking, spectral methods have been used to test for the existence of cycles of varying length and importance such as a two-year quasi-biennial oscillation and an eleven-year 'sunspot cycle'. With the notable exception of a cycle at about two years, spectral analysis shows that these postulated medium-term cycles are illusory (see for example Alter, 1933; Polowchak and Panofsky, 1968; Bain, 1976; Brinkworth, 1977; Wagner, 1971; Johnsen *et al.*, 1970; Angell and Korshover, 1968; Tabony, 1979). The very long records of climatic change from tree-ring widths (Fritts, 1965) and deep ocean cores (Hays, Imbrie and Shackleton, 1976) have been subject to a particularly rigorous scrutiny as a very direct test of the Milankovitch hypothesis of climatic change.

A second, related, climatological interest in time series has been the derivation of Markov chain models of sequences of designated events such as warm and cold (Caskey, 1964) and wet and dry days (Fitzpatrick and Krishnan, 1967; Green, 1970). Generally, meteorological persistence is such that first-order models are adequate for daily data but shorter time periods or regions of great variability seem to require higher-order models (Lowry and

Guthrie, 1968).

Despite the existence of many climatological studies of time series, the actual range of models employed has been rather limited, and, by modern standards, old fashioned. This is especially so in relation to the ready availability of long series of equally spaced data and is in marked contrast to the extensive and sophisticated use of time series models in hydrology and even human geography. Yet the possibilities are endless and will almost certainly prove scientifically worthwhile. We await, for example, autoregressive models of short-period climatic change, transfer function models relating urban heat islands to external forcing conditions and so on.

Spatial series It might be expected that geographer-climatologists would have much to contribute to the analysis of the spatial fields of meteorological elements, yet, as with time series, the full potential seems not to have been realized. For example, a commonly-used technique is to draw iso-correlates of a single reference station with all other stations in order to derive optimum observational networks (Huff and Shipp, 1969), to examine generating processes (Felgate and Reed, 1975; Elsom, 1978) or in objective interpolation (Bertoni and Lund, 1963). Spatial analysts will be surprised to read that despite its obvious relevance to these kinds of problem, there have been very few studies of the overall spatial autocorrelation structure of these fields or in their spectra. Exceptions are the work of Hutchinson (1974) in rain-gauge network design and Rayner (1967) and Wallace (1971) using spectral decomposition.

In contrast, the regression analysis of spatial data using some measure of location in the independent variables has been used extensively in the study of rainfall distribution (Gregory, 1965; El Thom, 1969, Jackson, 1969). Formal polynomial trend surface models have also been used to tackle similar problems (Unwin, 1969; Smithson, 1969; Mandeville and Rodda, 1970; Anderson, 1970). All these studies attempt to model the spatial distribution of means of weather elements and so can fairly be regarded as climatic, but mention should also be made of the widespread meteorological use of surface-fitting techniques to model instantaneous fields (Eddy, 1967). These range from high-order polynomial fits (Dixon, 1969) to a variety of mathematical interpolation procedures (Fritsch, 1971; Shaw and Lynn, 1972; Lynn, 1975).

Space-time series At the highest level of dimensionality, there have been several attempts to model the space-time dependence of weather elements. To date, most workers have used what in the meteorological literature is called eigenvector analysis or empirical orthogonal function analysis (Craddock, 1973) but which readers will recognize as a variation on the principal components theme in which the time sequences of individual elements are correlated, station by station, and the eigensystem of the resulting covariance matrix extracted. Each vector summarizes a characteristic pattern in the space/time field and in virtually all applications the more important patterns have been found to have a simple direct substantive interpretation or to be useful in classification (Christensen and Bryson, 1966). The method was originally developed by meteorologists (Lorenz, 1956), but subsequent applications have been to monthly temperature anomaly data (Grimmer, 1963), pressure (Kutzbach, 1967), rainfall (Perry, 1970), urban temperatures (Preston-Whyte, 1970; Clarke and Petersen, 1973) and tree ring data (Lamanche and Fritts, 1971). Recently, theoretical studies have examined the objective determination of how many vectors to extract (Bell, 1978) and in the analysis of relationships between two fields (Prohaska, 1976).

To date, however, there has been surprisingly little interest shown by climatologists in Britain and elsewhere in the various families of models (STAR (Space-time Autoregressive model), STIMA (Space-time integrated moving average model), etc.) developed by spatial analysts. In view of their obvious utility in climatic analysis and the existence of numerous long series of data with which to identify and calibrate such models, this is perhaps one of the most fruitful avenues for progress in statistical climate modelling.

Classificatory models A final inductive approach in climate modelling is one which has a long history and which in every sense is the most ambitious. This is climatic classification. Prior to the 1960s virtually all climatology texts dealt at length with classification and a great many competing schemes were proposed. Characteristically, these involved the use of two or three readily available weather elements and a divisive strategy based upon arbitrary, but sensible, thresholds. It is very evident that most authors of these schemes had a very clear idea of what regions they wanted to emerge and that an almost universal error was to confuse the resulting models with the real observed facts. An obvious alternative is to apply agglomerative methods from numerical taxonomy to data collected from as many stations as possible and to see what kind of classification emerges. The pioneer work in this was that by Steiner (1965) who extracted the principal components from a 16 x 16 correlation matrix of data from a sample of 67 US stations. It was found that almost 90 per cent of the variance was explained by only four components, and that these were interpretable on scales such as 'wetness', 'turbidity' and 'continentality'. Combinations of the component scores defined climatic regions. Following this work McBoyle has performed similar analyses for European (McBoyle, 1972) and Australian (McBoyle, 1971) data. The European Study is of interest because it uses an identical set of variables to that used by Steiner. However, for Europe, only three components were needed to summarize 90 per cent of the variance and, needless to say, their interpretation differs from those in the USA. In all cases, these automatic procedures give regionalizations similar to those obtained by the traditional methods and it is difficult to see what can be obtained from further work of this kind, unless, as in the most recent work, the new schemes are put to some further use as, for example, in ecology (White and Lindley, 1976; Paterson, Goodchild and Boyd, 1978) or the study of climatic change (Gregory, 1975).

Deductive physical-mathematical models

Introductory remarks

From the preceding sections it can be seen that geographers have played a limited part in the inductive modelling of climate but that, to date, the level of analysis has not been very sophisticated. In addition, there are a number of urgent questions that cannot be readily answered by the analysis of past data. These are questions of *climatic forecasting* which, following Lorenz (1975), can be of two kinds. Predictions of the 'first kind' concern the natural evolutionary change of the present climatic system whereas predictions of the 'second kind' examine hypothetical, but important, questions on the likely effects of such natural and artificial factors as changes in the solar constant, atmospheric composition or the character of the earth's surface. Given the evidence from time series analysis of little regular variability except on a geological timescale, it is evident that purely statistical models are unlikely to provide useful predictions of the first kind. In view of their almost universal assumption of long-term stationarity, they are even less likely to produce accurate predictions of the second kind. Instead, attention has been directed towards climate models characterized by a substantial content of deductive physical and mathematical reasoning but often tempered with the addition of well-established semi-empirical relationships and empirical coefficients. The growth of deductive climate modelling in the last decade has been very rapid, but geographic contributions to this growth have been very limited. The most influential papers were published ten years ago, yet by 1974 the authors of a major review were able to cite over two hundred references (Schneider and Dickinson, 1974; what follows is based to a large extent on that work) and in the last five years this momentum has been sustained. Today, virtually no issue of *Journal of the Atmospheric Sciences, Journal of Applied Meteorology* or *Monthly Weather Review* is without at least one paper on this theme.

The problem and the hierarchy of models

The climate system is one of enormous complexity involving factors such as the solar radiation input, atmospheric composition, the earth's surface character and numerous coupling mechanisms that all involve processes of very different types and time scales. A complete model should incorporate not only the relatively well understood atmospheric factors, but also little understood long-term processes associated with changes in the oceans, ice sheets and biosphere. There is therefore a need to simplify the physics involved and a characteristic of virtually all models is their resort to parametric representation, or 'parameterization'. In this, small-scale processes that cannot be directly modelled are related to average conditions that are computed explicitly. A second characteristic of deductive climate models is that it is possible to rank them in a hierarchy characterized by its increasing dimensionality on a spectrum from grossly oversimplified global models that parameterize virtually all the internal processes to very complex three-dimensional general circulation models (GCMs) of the earth/atmosphere system that parameterize as little as possible. Schneider and Dickinson (1974) recognize a six-level hierarchy of climate models, but, for the purposes of review, a simpler, four-level, classification is more appropriate. This begins with zero-dimensional energy balance models and progresses through one-dimensional latitudinal energy balance models to two-dimensional 'intermediate' models and, at the end of the scale, fully-fledged, three-dimensional GCMs. In considering these models, a temptation is to automatically assume that the more complex models are the best for all problems. This is not necessarily the case. Each type of model has its range of application and utility and it remains to be convincingly demonstrated that in long-term climatic work the complex GCMs are better than quite elementary energy balance approaches.

Zero-dimensional energy balance models The simplest climate models are based upon the numerical solution of the surface energy balance equation:

$$R_N = Q_I(1 - a) + L_a - L_e = G + H + LE$$

R_N = net radiation
Q_I = incident global radiation
L_a = counter radiation
L_e = terrestrial radiation
G = flux into substrate
H = sensible heat flux
LE = latent heat flux
a = surface albedo

As it stands, this is an instantaneously-valid statement of energy conservation, but integration over longer time-periods forms the basis of energy balance climatology (Hare, 1976; Hare and Thomas,

1974). In an important paper published in 1969, Myrup was able to show that if one is prepared to make a number of simplifying assumptions every term in the balance can be calculated as a function of the known time of day and the unknown surface radiation temperature T_S. The model was made dynamic by systematically varying Q_I – the forcing function – through a diurnal cycle and solved for T_S. By changing the values of parameters such as the albedo, transmissivity, wet fraction and so on, it is possible to simulate the likely effects of land use changes and atmospheric pollution on the resulting climate. Despite criticism of its assumptions (Miller, Johnston and Lowry, 1972), Myrup's model seems to give moderately realistic results. The original form used an analogue computer to find T_S, but Outcalt (1971, 1972a) has programmed it for a digital machine and has subsequently used it to investigate the general temperature effects of urban land use (Outcalt, 1972b) and arctic soil temperatures (Outcalt, 1972c; Brazel and Outcalt, 1973a, b). In this final form the Myrup/Outcalt model makes a crude attempt to model the soil heat flux and the vertical temperature profile and so is not strictly zero-dimensional, but, despite this, and the developments reported by Jenner (1975) and by Dozier and Outcalt (1979) it remains relatively crude. Similar energy balance models have been reported by a number of American geographers, notably Terjung (1970), Terjung and Louie (1974), Morgan *et al.* (1977) and in relation to human comfort conditions by Morgan and Baskett (1974) and Tuller (1975).

It is apparent that this type of model is best suited to modelling short-period, local, climates under specific, settled, conditions when it is possible to ignore advective effects, so that rather fewer attempts have been made to extend the energy balance approach to study climate over wide areas or long-time periods. Notable exceptions are the so-called *climatonomy* models of Lettau (1969; Lettau and Baradas, 1973; see also Greenland, 1973; Hare and Hay, 1971) for drainage basins, and a global model due to Fraedrick (1978). As these authors show, zero-dimensional models have the advantages of simplicity, elegance, ease of computation, over their appropriate range of application, reasonable fidelity to the real world and are excellent pedagogic devices but, as a means of studying global climatic variation, they are obviously unsatisfactory.

One-dimensional models The real atmosphere varies with latitude, longitude, altitude and time. At the next level in the hierarchy of climate models are one-dimensional models which calculate either the vertical or the latitudinal variation as part of their output. In local studies, or when examining the horizontally averaged global climate, it is often reasonable to neglect horizontal advection and to concentrate attention on the vertical profile. One-dimensional vertical models have a long history and have been much used in both the study of local urban effects (Atwater, 1972; Bergstrom and Viskanta, 1973) and global changes due to changes in atmospheric composition (Manabe and Wetherald, 1967; Rasool and Schneider, 1971). As might be expected, geographic interest in such models has been limited. In contrast, one-dimensional latitudinally-varying models have given rise to a great deal of literature and comment, the best-known being those proposed independently by Sellers (1969) and Budyko (1969). Both solve a surface energy balance equation at all latitudes but include a term to represent the horizontal flux of energy, parameterized either in relation to the difference between the zonal and planetary mean temperatures or to the temperature difference between successive latitudinal belts. Both incorporate empirical constants in this parameterization and both strongly couple the albedo and temperature through 'ice-albedo feedback'. The results obtained by these authors differ in detail, but both strongly suggest a global climate that is very sensitive to changes in the external, forcing, solar radiation. As they stand, both are also equilibrium models yielding a single solution for each set of boundary conditions but Schneider and Gal-Chen (1973) report a time-dependent version that gives similar results and there have been numerous other reported modifications (Budyko, 1972; North, 1975; Frederiksen, 1976). To the meteorologist, the parameterization of atmospheric dynamics in these energy balance approaches must seem an unjustifiable oversimplification, but the fact remains that the Budyko/Sellers model is able to reproduce the known latitudinal climatic variation and has highlighted the possible critical role of ice-albedo feedback in climatic change.

Two-dimensional 'intermediate' models The next obvious step in the hierarchy of models is occupied by those which explicitly calculate two-dimensional variation. At the local scale, numerous models of this type have been formulated for urban conditions (Bornstein, 1975; Vukovich, Dunn and Crissman, 1976) and at the global scale two approaches have been used. The first adapts energy balance models by adding a term to incorporate longitudinal energy transfer either by examining 'land' and 'sea' masses at each latitude independently (Sellers, 1973), or by specifying longitudinal variations in the parameters (Saltzman, 1967), or by parameterizing the energy transfer in advance (Adem, 1970). The second approach is more meteorological in flavour. Energy balance models are able to reproduce the diabatic atmospheric heating largely because this heating is closely related to surface conditions. Unfortunately, the redistribution of this energy depends on large-scale adiabatic meteorological processes that show much less surface dependence and it is therefore necessary to model this motion explicitly, or to find a suitable parameterization. A number of zonally-averaged two-dimensional dynamic models in which meridional motions are parameterized in various ways (these are reviewed by Schneider and Dickinson, 1974, pp.478–86; see also Ohring and Adler, 1978) have been proposed.

Three-dimensional general circulation models (GCMs) A final approach to the problem of climate modelling incorporates all three spatial dimensions and is one in which the three-dimensional motions are all explicitly computed. Primarily intended for long-range weather forecasting, such models have been produced for the atmosphere itself, that is, with a prescribed distribution of cryosphere and ocean temperatures, or they can interact with and change these critical climatic factors. To synthesize a climate from such a model, all that need be done is to run it with a single set of initial and boundary conditions and to produce a statistical summary of the results. Without exception, GCMs have been developed by large teams of physicists and mathematicians working at institutions such as the UK Meteorological Office, the National Centre for Atmospheric Research, University of California at Los Angeles, the RAND Corporation and the Princeton Geophysical Fluid Dynamics Laboratory (see Manabe, 1975; Chang, 1977; Gilchrist, 1979). Typically, the spatial resolution is a few degrees of latitude and longitude and, say, 1 km in the vertical and the time step is a few minutes. Ideally, all processes occurring at scales larger than this are modelled explicitly, those at a smaller scale are parameterized. There can be no doubt that the more complex of these models are able to accurately simulate today's climate. For example, Manabe (1975, Figure 2) presents maps comparing the actual global pattern of Koeppen climatic types with those synthesized from the model output, and, with the exception of limited areas of China and the southern USA, the correspondence is remarkably good. Whether this justifies confidence in simulations using boundary conditions for other climates is, of course, another matter, but numerous experiments of this type have been carried out and the models themselves have become more and more refined. Of particular importance is the discovery that because of complex, non-linear interactions in the modelled system, results are often what Mason (1979) calls 'counter-intuitive', that is, they would not have been foreseen by recourse to elementary theory or even common sense.

At first sight, this 'integration of a GCM' approach to climate modelling seems to be the most likely to succeed, but there are several reservations to be made. First, it is by no means certain that explicit modelling of all motion is necessary, yet the amount of computation involved is vast, involving simulation: real time ratios that are no better than 1:100. As Ohring and Adler (1978, p.187) point out 'Most of this time is spent moving parcels of air around the globe with new synoptic patterns obtained every few minutes or so . . . not very efficient . . . for conducting experiments related to climatic change'. A related problem is their relatively crude treatment of longer-term climate processes such as air/sea interaction, ice-albedo feedback and so on. Third, in these rather complex experiments it is often difficult to isolate cause–effect relationships, and, finally, there is a severe problem of establishing the statistical significance of the results, summarized in Chervin (1978). Chervin points out that, despite their seemingly deterministic character, GCMs invariably contain 'noise' from sources such as uncertainty in the initial conditions, the use of finite differences and the empirical parameterizations, all of which can be propagated and enlarged by the model. Any individual result must therefore be treated as but one realization from a distribution of possible outcomes and if 'real' effects are to be inferred it is necessary to evaluate their statistical significance. Although this argument was developed for one particular type of systems model, it is likely to be of more general importance.

Some applications

Whatever their type and complexity, climate models have been used extensively in the investigation of a number of important problems and have highlighted a number of fundamentals concerning the nature of the earth's climate system. A first area of application has been to try to answer 'what if' questions involving Lorenz's predictions of the second kind. Given a model that accurately simulated the present climate, an obvious next step is to study its response to changes both natural and man-induced in the boundary conditions. Examples of this type of study are those by Manabe (1971) examining the influence of changes in atmospheric CO_2, Rasool and Schneider (1971) on dust, Manabe and Terpstra (1974) on orography, Charney (1975) on desertification due to vegetation removal, Kutzbach *et al.* (1977) on sea surface temperatures (SSTs), Washington and Chervin (1979) on thermal pollution, and a number of studies concerned with solar radiation (Budyko, 1969; Sellers, 1969, 1973).

A second use of climate models has been to post-dict past climates using boundary conditions thought representative of previous periods of earth history. Given the multidisciplinary background needed to set up such simulations this is a field in which physical geographers have much to contribute and, indeed, much of the pioneer work has been based upon INSTAAR at Boulder in collaboration with the nearby NCAR modellers (Williams, Barry, and Washington, 1974; Gates, 1976). The results, reviewed in Barry (1975) and Williams (1978) are consistent in their predictions of the glacial climate at the time of the last glacial maximum and are in accord with those which might be made as a result of examining other, more conventional, independent evidence.

A third use of climate models has been in the investigation of the climate system, particularly its so-called *ambiguity*. This is the possibility, first raised by Lorenz (1968), that the climate system does not have a unique, stable set of infinitely long-term statistics. Rather, it behaves as an almost intransitive system, that is, a system in which one set of statistical properties may persist for a long period but then suddenly switch to another state. The system can therefore yield several climates, and, importantly, these climates may refer to the same external conditions. If Lorenz is correct, it follows that climatic change should not be viewed as a long-term, evolutionary change in the statistics of weather but as a rapid switch from one virtually stable state into another. Moreover, if the ambiguity property can be proved, such a switching process need not necessarily involve any change in the external conditions. One major property of the Sellers/Budyko energy balance models is that they exhibit the branching behaviour implied by Lorenz's ideas, notably through the positive feedback. An enormous literature has already built up examining this property (Chylek and Coakley, 1975; Su and Hsieh, 1976; Ghil, 1976; Frederiksen, 1976; Drazin and Griffel, 1977; Lian and Cess, 1977; Crafoord and Källén, 1978). Ghil, for example, finds that a Sellers model gives three stable solutions with global mean temperatures of 288°K (present), 267°K (ice-age earth) and 175°K (white earth, completely ice covered). The transition between states is amenable to study using catastrophe theory (Fraedrick, 1978), but we still do not know whether or not the atmospheric circulation will allow all such states to occur.

Concluding remarks

During the 1960s it was fashionable to talk about revolutions in geographic technique and content but, other than to put a final nail in the coffin of excessive concern for global classification, these changes, revolutionary or not, have had only limited impact on climatology, and the subject has continued its relative decline within the discipline. Regrettably, this demise of climatology in its traditional geographical home has gone on alongside a rediscovery of its importance both by society at large and by the meteorological community. Partly this rediscovery results from very real fears about the economic future in the face of a more variable future climate but it is also a result of general movements towards greater environmental awareness. Geographers were prominent in the initial stages of both developments yet the meteorological response, especially the physical-mathematical modelling of the climate system, has largely passed us by, and, as this review will have made clear, we have played very little part in this work. What work has been undertaken has been in the very traditional statistical analysis of weather data, as users of simple local zero-dimensional energy balance models, or as eclectic data-gatherers for input to GCMs designed and maintained elsewhere. In the UK, largely, one suspects, as a result of perceived, if not real, pressure from the meteorological community through the research councils, there has been a rush by climatologists to make themselves appear respectable scientists. As was noted in the introduction, this has led to their becoming very limited analysers of data in applied fields that, if anything, can be seen to be even less respectable science. In this rush we have tended to teach our students and research into anything but climatology as usually defined. From the teaching point of view, Lockwood's recent, model-based, text on *Causes of Climate* (Lockwood, 1979) will do something to redress the balance but there remains a lot of lost ground to be regained. Although it is unlikely that UK geographers will have much to contribute to deductive modelling, it remains true that what could have been and still is a great strength, a reasonable expertise in statistical modelling, has hardly been exploited (Gregory, 1976). As was noted at several points in this review, the opportunities for worthwhile and realistic statistical modelling of climate are many, and it is to be hoped that they will be taken up.

References

Adem, J. (1970) 'Incorporation of advection of heat by mean winds and by ocean currents in a thermodynamic model for long-range weather prediction', *Monthly Weather Review*, 98, 776–86.

Alter, D. (1933) 'Correlation periodogram investigation of English rainfall', *Monthly Weather Review*, 61, 345–52.

Anderson, P. (1970) 'The uses and limitations of trend-surface analysis in studies of urban air pollution', *Atmospheric Environment*, 4, 129–47.

Angell, J.K., and Korshover, J. (1968) 'Additional evidence for quasi-biennial variations in tropospheric parameters', *Monthly Weather Review*, 96, 778–84.

Atkinson, B.W. (1978a) 'What should we be teaching about the atmosphere?', *Journal of Geography in Higher Education*, 2, 38–44.

Atkinson, B.W. (1978b) 'The atmosphere: recent observational and conceptual advances', *Geography*, 63, 283–300.

Atwater, M.A. (1972) 'Thermal effects of urbanisation and industrialisation in the boundary layer: a numerical study', *Boundary Layer Meteorology*, 3, 229–45.

Bain, W.C. (1976) 'Power spectrum of temperature in central England', *Quarterly Journal of the Royal Meteorological Society*, 102, 464–6.

Barry, R.G. (1967) 'Models in meteorology and climatology' in R.J. Chorley and P. Haggett, op. cit., 97–144.

Barry, R.G. (1975) 'Climate models in palaeoclimatic reconstruction', *Palaeogeography, Palaeoclimatology and Palaeoecology*, 17, 123–7.

Barry, R.G., and Perry, A.H. (1973) *Synoptic Climatology: Methods and Applications*, Methuen: London.

Bell, C.E. (1978) 'The number of significant proper functions of two-dimensional fields', *Journal of Applied Meteorology*, 17, 717–22.

Bergstrom Jr, R.J., and Viskanta, R. (1973) 'Modelling of the effects of gaseous and particulate pollutions in the urban atmosphere. Part I: Thermal structure', *Journal of Applied Meteorology*, 12, 901–12.

Bertoni, E.A., and Lund, I.A. (1963) 'Space correlations of the height of constant pressure surfaces', *Journal of Applied Meteorology*, 2, 539–45.

Bornstein, R.D. (1975) 'The two-dimensional URBMET urban boundary layer model', *Journal of Applied Meteorology*, 14, 1459–77.

Brazel, A.J., and Outcalt, S.I. (1973a) 'The observation and simulation of diurnal evaporation contrast in an Alaskan Alpine pass', *Journal of Applied Meteorology*, 12, 1134–43.

Brazel, A.J., and Outcalt, S.I. (1973b) 'The observation and simulation of diurnal surface thermal contrast in an Alaskan Alpine pass', *Archives of Meteorology, Geophysics and Bioklimatology*, B 21, 157–74.

Brinkworth, B.J. (1977) 'Autocorrelation and stochastic modelling of isolation sequences', *Solar Energy*, 19, 343–7.

Brooks, C.E.P., and Carruthers, N. (1953) *Handbook of Statistical Methods in Meteorology*, HMSO: London.

Budyko, M.I. (1969) 'The effect of solar radiation variation on the climate of the earth', *Tellus*, 21, 611–19.

Budyko, M.I. (1972) 'The future climate', *Transactions, American Geophysical Union*, 53, 868–74.

Caskey, J.E. (1964) 'Markov chain model of cold spells at London', *Meteorological Magazine*, 93, 136–8.

Cehak, K. (1970) 'Some examples of climatological processing for technical purposes', *Building Climatology*, Technical Note 109, 235–9, World Meteorological Office: Geneva.

Chandler, T.J. (1965) *The Climate of London*, Hutchinson: London.

Chandler, T.J., and Gregory, S. (1976) *The Climate of the British Isles*, Longman: London.

Chang, J. (ed.) (1977) *General Circulation Models of the Atmosphere*, Methods in Computational Physics 17, Academic Press: New York.

Charney, J. (1975) 'Dynamics of deserts and droughts in the Sahel', *The Physical Basis of Climate and Climate Modelling*, 171–6, World Meteorological Office: Geneva.

Chervin, R.M. (1978) 'The limitations of modelling: the question of statistical significance' in J. Gribbin (ed.), *Climatic Change*, 191–201, Cambridge University Press.

Chorley, R.J., and Haggett, P. (eds) (1967) *Models in Geography*, Methuen: London.

Christensen, W.I., and Bryson, R.A. (1966) 'An investigation of the potential of component analysis for weather classification', *Monthly Weather Review*, 94, 697–709.

Chylek, P., and Coakley, J.A. (1975) 'Analytical analysis of a Budyko type climate model', *Journal of the Atmospheric Sciences*, 32, 675–79.

Clarke, J.F., and Petersen, J.T. (1973) 'An empirical model using eigenvectors to calculate the temporal and spatial variations of the St Louis heat island', *Journal of Applied Meteorology*, 12, 195–210.

Conrad, V., and Pollak, L.W. (1956) *Methods in Climatology*, Harvard University Press: Cambridge: Mass.

Craddock, J.M. (1965) 'The analysis of meteorological time series for use in forecasting', *The Statistician*, 15, 167–90.

Craddock, J.M. (1973) 'Problems and prospects for eigenvector analysis in meteorology', *The Statistician*, 12, 133–45.

Crafoord, C., and Källén, E. (1978) 'A note on the conditions for existence of more than one steady-state solution in Budyko-Sellers type models', *Journal of the Atmospheric Sciences*, 35, 1123–5.

Crowe, P.R. (1971) *Concepts in Climatology*, Longman: London.

Curry, L. (1962) 'Climatic change as a random series', *Annals, Association of American Geographers*, 52, 21–31.

Dixon, R. (1969) 'Orthogonal polynomials as a basis for objective analysis', *UK Meteorological Office*, Scientific Papers 30, HMSO: London.

Dozier, J., and Outcalt, S.I. (1979) 'An approach toward energy balance simulation over rugged terrain', *Geographical Analysis*, 11, 65–85.

Drazin, P.G., and Griffel, D.H. (1977) 'On the branching structure of diffusive climatological models', *Journal of the Atmospheric Sciences*, 34, 1696–1706.

Eddy, A. (1967) 'The statistical objective analysis of scalar data fields', *Journal of Applied Meteorology*, 6, 597–609.

Elsom, D.M. (1978) 'Spatial correlation analysis of

urban air pollution data in an urban area', *Atmospheric Environment*, 12, 1103-7.

El Thom, M.A. (1969) 'A statistical analysis of the rainfall over the Sudan', *Geographical Journal*, 135, 378-87.

Essenwanger, O.M. (1976) *Applied Statistics in Atmospheric Science: Part A: Frequencies and Curve Fitting*, Elsevier: Amsterdam.

Felgate, D.G., and Reed, D.G. (1975) 'Correlation analysis of cellular structure of storms observed by raingauges', *Journal of Hydrology*, 24, 191-200.

Fitzpatrick, E.A., and Krishnan, A. (1967) 'A first-order Markov model for assessing rainfall discontinuity in central Australia', *Archives of Meteorology Geophysics and Bioklimatology*, B 15, 242-59.

Fraedrick, K. (1978) 'Structural and stochastic analysis of a zero-dimensional climate system', *Quarterly Journal of the Royal Meteorological Society*, 104, 461-74.

Frederiksen, J.S. (1976) 'Non-linear albedo-temperature coupling in climate models', *Journal of the Atmospheric Sciences*, 33, 2267-72.

Fritsch, J.M. (1971) 'Objective analysis of a two-dimensional data field by the cubic spline technique', *Monthly Weather Review*, 99, 379-86.

Fritts, H.C. (1965) 'Tree ring evidence for climatic change in western North America', *Monthly Weather Review*, 93, 421-43.

Gates, W.L. (1976) 'The numerical simulation of ice-age climate with a global general circulation model', *Journal of the Atmospheric Sciences*, 33, 1844-73.

Ghil, M. (1976) 'Climate stability for a Sellers-type model', *Journal of the Atmospheric Sciences*, 33, 3-20.

Gilchrist, A. (1979) 'Concerning general circulation models', *Meteorological Magazine*, 108, 35-51.

Green, J.R. (1970) 'A generalized probability model for sequences of wet and dry days', *Monthly Weather Review*, 98, 238-241.

Greenland, D.E. (1973) 'Application of climatonomy to an Alpine valley', *New Zealand Journal of Science*, 16, 811-23.

Gregory, S. (1965) *Rainfall over Sierra Leone*, Research Papers 2, University of Liverpool, Department of Geography.

Gregory, S. (1975) 'On the delimitation of regional patterns of recent climatic fluctuations', *Weather* 30, 276-87.

Gregory, S. (1976) 'On geographical myths and statistical fables', *Transactions of the Institute of British Geographers*, NS 1, 385-400.

Grimmer, M. (1963) 'The space filtering of monthly surface temperature anomaly data in terms of pattern, using empirical orthogonal functions', *Quarterly Journal of the Royal Meteorological Society*, 89, 395-408.

Hare, F.K. (1976) 'Energy-based climatology and its frontier with ecology' in R.J. Chorley (ed.) *Directions in Geography*, 171-92, Methuen: London.

Hare, F.K., and Hay, J.E. (1971) 'Anomalies in the large scale annual water balance over northern North America', *Canadian Geographer*, 15, 79-94.

Hare, F.K., and Thomas, M.K. (1974) *Climate of Canada*, Wiley: Toronto.

Hay, F.M. (1968) 'Relationships between autumn rainfall and winter temperature', *Meteorological Magazine*, 91, 278-82.

Hays, J.D., Imbrie, J., and Shackleton, N.J. (1976) 'Variations in the earth's orbit: pacemaker of the Ice Ages', *Science*, 194, 1121-32.

Horn, H.L., and Bryson, R.A. (1960) 'Harmonic analysis of the annual march of precipitation over the United States', *Annals, Association of American Geographers*, 50, 157-71.

Huff, E.A., and Shipp, W.L. (1969) 'Spatial correlations of storm, monthly and seasonal precipitation', *Journal of Applied Meteorology*, 8, 542-50.

Hutchinson, P. (1974) 'Progress in the use of the method of optimum interpolation for the design of the Zambian raingauge network', *Geoforum*, 20, 49-62.

Jackson, I.J. (1969) 'Pressure types and precipitation over north-east England', *Research Series* 5, University of Newcastle upon Tyne, Department of Geography.

Jenner, J.B. (1975) 'Simulation of Land-use Effects on the Urban Temperature Field', unpublished PhD Thesis, University of Maryland, USA.

Johnsen, S.J., Dansgaard, W., Clausen, H.B., and Langway, C.C. (1970) 'Climatic oscillations, 1200-2000 AD.', *Nature*, 227, 482-3.

Kutzbach, J.E. (1967) 'Empirical eigenvectors of sea level pressure, surface temperature and precipitation complexes over North America', *Journal of Applied Meteorology*, 6, 791-802.

Kutzbach, J.E., Chervin, R.M., and Houghton, D.D. (1977) 'Response of the NCAR General Circulation Model to prescribed changes in ocean surface temperatures. Part I: Mid latitude changes', *Journal of the Atmospheric Sciences*, 34, 1200-13.

Lamanche, V.C., and Fritts, H.C. (1971) 'Anomaly patterns of climate over the western United States, 1700-1930, derived from principal components analysis of tree-ring data', *Monthly Weather Review*, 99, 138-42.

Lamb, H.H. (1972) *Climate: Present Past and Future. Vol. 1: Fundamentals and Climate Now*, Methuen: London.

Lamb, H.H. (1977) *Climate: Present Past and Future. Vol. 2: Climatic History and the Future*, Methuen: London.

Lee, D.O. (1977) 'Urban influence on wind

direction over London', *Weather*, 32, 162–70.

Lettau, H. (1969) 'Evapotranspiration climatonomy. 1: A new approach to numerical prediction of monthly evapotranspiration, runoff and soil moisture storage', *Monthly Weather Review*, 97, 691–9.

Lettau, H., and Baradas, M.W. (1973) 'Evapotranspiration climatonomy II: Refinement of parameterisation, exemplified by application to the Mabacon River watershed', *Monthly Weather Review*, 101, 636–49.

Lian, M.S., and Cess, R.D. (1977) 'Energy balance climate models: a re-appraisal of ice-albedo feedback', *Journal of the Atmospheric Sciences*, 34, 1058–62.

Lockwood, J.G. (1974) *World Climatology: an environmental approach*, Arnold: London.

Lockwood, J.G. (1979) *Causes of Climate*, Arnold: London.

Lorenz, E.N. (1956) 'Empirical orthogonal functions and statistical weather prediction', *Science Report* 1, Dept of Meteorology, MIT, Cambridge, Mass.

Lorenz, E.N. (1968) 'Climatic determinism', *Meteorological Monographs*, 8, 1–3.

Lorenz, E. (1975) 'Climatic predictability', *The Physical Basis of Climate and Climate Modelling*, 132–6, World Meteorological Office: Geneva.

Lowry, W.P., and Guthrie, D. (1968) 'Markov chains of order greater than one', *Monthly Weather Review*, 798–801.

Lynn, P.P. (1975) 'Rainfall interpolation using multiquadric surfaces', *Computer Applications*, 2, 321–34.

McBoyle, G.R. (1971) 'Climatic classification of Australia by computer', *Australian Geographical Studies*, 9, 1–14.

McBoyle, G.R. (1972) 'Factor analytic approach to a climatic classification of Europe', *Climatological Bulletin*, McGill-McQueen's University Press: Montreal.12.

Maddon, R.A. (1977) 'Estimates of the autocorrelations and spectra of seasonal mean temperatures over North America', *Monthly Weather Review*, 105, 9–18.

Manabe, S. (1971) 'Estimates of future change of climate due to increase of carbon dioxide concentration in the air' in W.H. Matthews, W.W. Kellogg and G.D. Robinson (eds), *Man's Impact on Climate*, 249–64, MIT Press: Cambridge, Mass.

Manabe, S. (1975) 'The use of comprehensive general circulation modelling for studies of climate and climate variation', *The Physical Basis of Climate and Climate Modelling*, 148–62, World Meteorological Office: Geneva.

Manabe, S., and Terpstra, T.B. (1974) 'The effects of mountains on the general circulation of the atmosphere as identified by numerical experiments', *Journal of the Atmospheric Sciences*, 31, 3–42.

Manabe, S., and Wetherald, R.T. (1967) 'Thermal equilibrium of the atmosphere with a given distribution of relative humidity', *Journal of the Atmospheric Sciences*, 24, 241–59.

Mandeville, A.N., and Rodda, J.C. (1970) 'A contribution to the objective assessment of areal rainfall amounts', *Journal of Hydrology* (NZ), 9, 281–91.

Mason, B.J. (1978) 'Recent advances in the numerical prediction of weather and climate', *Proceedings of the Royal Society*, London, A 363, 297–333.

Mason, B.J. (1979) 'Computing climatic change', *New Scientist*, 82, 196–8.

Miller, E.L., Johnston, E., and Lowry, P. (1972) 'The case of the muddled metromodel', *Preprints, Conference on Urban Environment and Second Conference on Biometeorology*, American Meteorological Society: Philadelphia.

Morgan, D.L., and Baskett, R.L. (1974) 'Comfort of man in the city: an energy balance model of man-environment coupling', *International Journal of Biometeorology*, 18, 184–94.

Morgan, D., Myrup, L., Rogers, D., and Baskett, R. (1977) 'Microclimate within an urban area', *Annals, Association of American Geographers*, 67, 55–65.

Murray, R. (1968) 'Some predictive relationships concerning seasonal rainfall over England and Wales and seasonal temperature in Central England', *Meteorological Magazine*, 97, 303–10.

Musk, L.F., and Tout, D.G. (1979) 'Climatological teaching in British universities and polytechnics', *Geography*, 64, 21–5.

Myrup, L.O. (1969) 'A numerical model of the urban heat island', *Journal of Applied Meteorology*, 8, 908–10.

Natural Environment Research Council (1976) *Research in Applied and World Climatology*, Publication Series B 17, NERC: London.

North, G.R. (1975) 'Theory of energy-balance climate models', *Journal of the Atmospheric Sciences*, 32, 2033–43.

Ohring, G., and Adler, S. (1978) 'Some experiments with a zonally averaged climate model', *Journal of the Atmospheric Sciences*, 35, 186–205.

Outcalt, S.I. (1971) 'A numerical surface climate simulator', *Geographical Analysis*, 3, 379–92.

Outcalt, S.I. (1972a) 'The development and application of a simple digital surface climate simulator', *Journal of Applied Meteorology*, 11, 629–36.

Outcalt, S.I. (1972b) 'A reconnaissance experiment in mapping and modelling the effect of land use in urban thermal regimes', *Journal of Applied Meteorology*, 11, 1369–73.

Outcalt, S.I. (1972c) 'The simulation of subsurface effects on the diurnal surface thermal regime in cold regions', *Arctic*, 25, 306–8.

Paterson, J.G., Goodchild, M.A., and Boyd, W.J.R. (1978) 'Classifying environments for sampling purposes using a principal components analysis of climatic data', *Agricultural Meteorology*, 19, 349–62.

Perry, A.H. (1970) 'Filtering climatic anomaly fields using principal components analysis', *Transactions, Institute of British Geographers*, 50, 55–72.

Perry, A.H. (1971) 'Econoclimate - a new direction for climatology', *Area*, 3, 178–9.

Polowchak, V.M., and Panofsky, H.A. (1968) 'The spectrum of daily temperatures as a climatic indicator', *Monthly Weather Review*, 96, 596–600.

Preston-Whyte, R.A. (1970) 'A spatial model of an urban heat island', *Journal of Applied Meteorology*, 9, 571–3.

Prohaska, J.T. (1976) 'A technique for analysing the linear relationships between two meteorological fields', *Monthly Weather Review*, 104, 1345–53.

Rasool, S.I., and Schneider, S.H. (1971) 'Atmospheric carbon dioxide and aerosols: effects of large increases on global climate', *Science*, 173, 138–41.

Raup, H.G. (1979) 'Current trends in geographic research', *Professional Geographer*, 31, 99–100.

Rayner, J.N. (1967) 'A statistical model for the explanatory description of large-scale time and spatial climate', *Canadian Geographer*, 11, 67–86.

Sabbagh, M.E., and Bryson, R.A. (1962) 'Aspects of the precipitation climatology of Canada investigated by the method of harmonic analysis', *Annals, Association of American Geographers*, 52, 426–40.

Saltzman, B. (1967) 'On the theory of the mean temperature of the earth's surface', *Tellus*, 19, 219–29.

Schneider, S.H., and Dickinson, R.E. (1974) 'Climate modelling', *Reviews of Geophysics and Space Physics*, 12, 447–93.

Schneider, S.H., and Gal-Chen, T. (1973) 'Numerical experiments in climate modelling', *Journal of Geophysical Research*, 78, 6182–94.

Sellers, W.D. (1969) 'A global climatic model based on the energy balance of the earth atmosphere system', *Journal of Applied Meteorology*, 8, 392–400.

Sellers, W.D. (1973) 'A new global climatic model', *Journal of the Atmospheric Sciences*, 12, 241–54.

Sellers, W.D. (1976) 'A two-dimensional global climate model', *Monthly Weather Review*, 104, 233–48.

Shaw, E.M., and Lynn, P.P. (1972) 'Areal rainfall evaluation using two surface fitting techniques', *Bulletin, International Association of Hydrological Sciences*, 17, 419–33.

Smithson, P.A. (1969) 'Regional variations in the synoptic origin of rainfall across Scotland', *Scottish Geographical Magazine*, 85, 182–95.

Smithson, P., and Gregory, S. (eds) (1978) *The Third Directory of British Climatologists*, Association of British Climatologists: Sheffield.

Steiner, D. (1965) 'A multivariate statistical approach to climatic regionalisation and classification', *Tijdschrift van het Kononklijk Nederlandsch Aardrijkskundig Genootschap*, 82, 329–47.

Stringer, E.T. (1972a) *Foundations of Climatology*, Freeman: San Francisco.

Stringer, E.T. (1972b) *Techniques of Climatology*, Freeman: San Francisco.

Su, C.H., and Hsieh, D.Y. (1976) 'Stability of the Budyko climate model', *Journal of the Atmospheric Sciences*, 33, 2273–5.

Suzuki, E. (1967) 'A statistical and climatological study on the rainfall in Japan', *Papers in Meteorology and Geophysics* (Tokyo), 18, 103–82.

Tabony, R.C. (1979) 'A spectral and filter analysis of long-period rainfall records in England and Wales', *Meteorological Magazine*, 108, 97–118.

Terjung, W.H. (1970) 'Urban energy balance climatology', *Geographical Review*, 60, 31–53.

Terjung, W.H. (1976) 'Climatology for geographers', *Annals, Association of American Geographers*, 66, 199–222.

Terjung, W.H., and Louie, S.S.-F. (1974) 'A climatic model of urban energy budgets', *Geographical Analysis*, 6, 341–67.

Thom, H.C.S. (1966) 'The distribution of annual tropical cyclone frequency', *Journal of Geophysical Research*, 65, 213–22.

Thornes, J.E. (1978) 'Applied climatology: atmospheric management in Britain', *Progress in Physical Geography*, 2(3), 481–93.

Tuller, S.E. (1975) 'The energy budget of man: variations with aspect in a downtown urban environment', *International Journal of Biometeorology*, 19, 2–13.

Unwin, D.J. (1969) 'The areal extension of rainfall records - an alternative model', *Journal of Hydrology*, 7, 404–14.

Unwin, D.J. (1977) 'Statistical methods in physical geography', *Progress in Physical Geography*, 1 (2), 185–221.

Vukovich, F.M., Dunn III, J.W., and Crissman, B.W. (1976) 'A theoretical study of the St Louis heat island: the wind and temperature distribution', *Journal of Applied Meteorology*, 15, 417–40.

Wagner, A.J. (1971) 'Long-period variations in seasonal sea-level pressure over the northern hemisphere', *Monthly Weather Review*, 99, 49–66.

Wallace, J.M. (1971) 'Spectral studies of tropospheric wave disturbances in the tropical western Pacific', *Reviews of Geophysics and Space*

Physics, 9, 557–612.
Washington, W.M., and Chervin, R.M. (1979) 'Regional climatic effect of large-scale thermal pollution: simulation studies with the NCAR General Circulation Model', *Journal of Applied Meteorology*, 18, 3–16.
White, R.J., and Lindley, D.K. (1976) 'The reduction of climatological data for ecological purposes: a preliminary analysis', *Journal of Environmental Management*, 4, 161–82.
Williams, J. (1978) 'The use of numerical models in studying climatic change' in J. Gribbin (ed.), *Climatic Change*, 178–96, Cambridge University Press.
Williams, J., Barry, R.G., and Washington, W.M. (1974) 'Simulation of the atmospheric circulation using the NCAR global circulation model with ice age boundary condition', *Journal of Applied Meteorology*, 13, 305–17.
World Meteorological Office (1975) *The Physical Basis of Climate and Climate Modelling*, Global Atmospheric Research Program, Publication Series 16, Geneva.

Chapter 26

Hydrology

M.G. Anderson and K.S. Richards

Introduction

Hydrology has a tradition of quantification reaching back to Perrault's measurement of rainfall and run-off in the Seine basin in the early seventeenth century. This long period of quantitative research has always maintained a healthy balance between the deductive and inductive, theoretical and empirical, and deterministic and stochastic approaches. Emphasis in research has, in recent years, ranged from pre-1945 determinism based on Fourier analysis (Yevjevich, 1974) through the use of regression methods predicting flow characteristics from basin parameters in the 1960s, to the stochastic modelling of the early 1970s, and the recently developing partially distributed models of catchment behaviour based upon deterministic process specification. Much of the impetus for quantification in hydrology is provided by the practical importance of an understanding of the hydrological cycle and its components, which in itself creates a dilemma between the needs for physical understanding and for forecasting of system behaviour, which demand different quantitative approaches. A further dilemma in application of quantitative methods in geographical hydrology, discernible in recent literature but yet to be fully recognized, arises because of increased field measurement and computing capability. It is desirable to explain process mechanics for spatial units having geomorphological or geographical meaning, but deterministic process modelling often deals with the experimental plot scale which, because of boundary conditions, cannot be easily aggregated to either the hillslope or catchment scales. These two dilemmas are highlighted in this review, which draws attention to recent developments and possible future prospects in the principal areas of interest for quantitative, geographical hydrology.

Probability models

The magnitude-frequency properties of extreme hydrological events influence both geomorphological process and engineering design, and their assessment requires the fitting of a probability density function to a distribution of extreme event magnitudes. NERC (1975) define the functions commonly used in flood frequency analysis, which include log-normal, Gumbel types I and III (log-transformed), Pearson type III and log-Pearson type III. In the absence of theoretical justification for a particular model, selection is influenced particularly by statistical problems encountered during calibration, such as those related to the length and homogeneity of the data record, the data series employed, and the independence of events in the series.

Simulation experiments (Benson, 1960a) demonstrate that a forty-year record is required to estimate mean annual flood within ± 10 per cent of the true value with 95 per cent confidence. Longer records still are needed to estimate higher magnitude events with comparable accuracy, and the higher moments of extreme event distributions (e.g., skewness) needed for the calibration of some models. Calibration is particularly difficult when a rare high magnitude outlier event occurs in a short record; the example in Figure 26.1 shows the significant effect on estimated flood magnitudes of excluding one outlier. However, a long record is of limited value if it is heterogeneous. Changing climate and land use affect rainstorms and floods continuously, so that any period of data may include a time trend. Rodda (1970) showed that ten-year daily rainfall in Oxford was *c*.50 mm for 1901–25, but *c*.65 mm for 1941–65, adding to existing evidence of increasing frequency of heavy rain in urban areas. Slight trends in flood magnitudes are difficult to interpret, how-

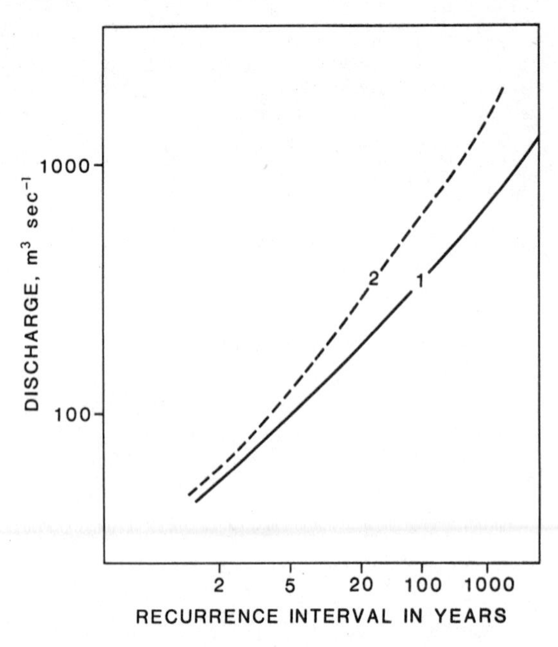

Figure 26.1 Log-Pearson Type III flood frequency curve for Western Run, Maryland (gauging station record, 33 years). Curve 1 shows frequencies excluding the 1972 Tropical Storm Agnes flood event of 1076 cumecs. Curve 2 includes the 1972 event (after Costa, 1978)

ever, because both land use and climatic changes can occur simultaneously (Howe *et. al.*, 1967). Alternative data series do give a satisfactory sample size while avoiding heterogeneity; for example, the annual exceedance or maximum series may be abandoned in favour of the partial duration series, which also has a smaller sampling variance if at least 1.65 floods per year are included (Cunnane, 1973). However, there is some evidence of a non-Poisson, clustered distribution of events in such a series if the sequence of events is viewed as a point process (Cunnane, 1979).

No single probability distribution is universally accepted, and Reich (1970) suggests that different parts of a single data set may indicate preference for alternative models. This recalls the evidence for dog-leg effects in magnitude-frequency plots (Potter, 1958), which may reflect outlier events in a short record, or members of two distinct event distributions. Harvey (1971) has recognized two hydrograph types in some British catchments; a sharp peak associated with summer convectional storms, and a flatter, attenuated hydrograph of lower intensity, winter, storms. Thus a mixed extreme event distribution may arise, and the physical justification for Singh and Sinclair's (1972) approach is clearer. This involves fitting two, combined, normal distributions with different means and variances, rather than a skewed distribution requiring higher moment estimates. Similar approaches are used to model broken or mixed sediment size distributions in geology (Mundry, 1972). NERC (1975) demonstrate that the preference in the USA for the log-Pearson type III distribution (US Water Resources Council, 1967) reflects the use of mean absolute deviation as a criterion of goodness of fit, and argue that a general extreme value distribution is preferable because it is less sensitive to variations in the criteria employed. In one sense, however, the question of goodness of fit is less important than the limitations presented by data quality. Extreme rainfalls are often of limited spatial extent, and point gauges inevitably fail to capture every event in a given area. A similar, but potentially remediable, problem is presented by the greater availability of mean daily rather than instantaneous peak streamflow values (Dury, 1959).

Despite these limitations, magnitude-frequency analysis is widely exploited in studies of channel process and engineering structures. The frequency of floodplain inundation is particularly critical for sediment transport and floodplain occupance, and distinctions between in-channel benches (Woodyer, 1968), active bank tops, floodplains and terraces have been based on the distributions of probabilities of their exceedance. Homogeneity of these distributions is improbable (Williams, 1978), and Figure 26.2 shows the wide range of return periods of bank-full floods inundating the active floodplain. Harvey (1969) has demonstrated systematic downstream variation in the frequency of bank-full flow, with an increase or decrease depending on catchment hydrological regime and lithology. Furthermore, the significance of an event is now recognized to reflect more than simply its return period. Anderson and Calver (1977) and Wolman and Gerson (1978) have focused attention on landscape recovery rates, and argue that the sequence of events is as important

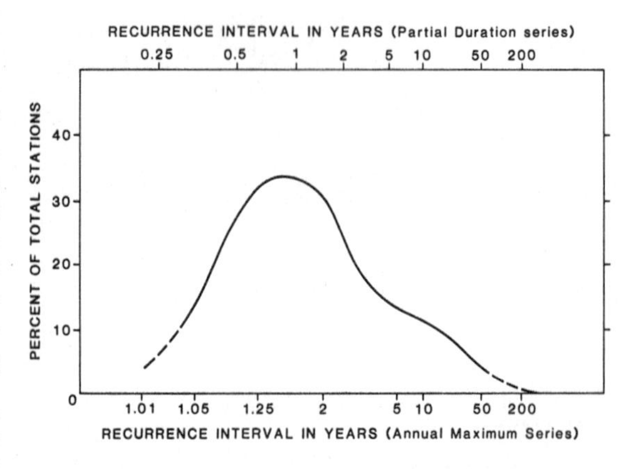

Figure 26.2 Frequency distribution of recurrence intervals for bank-full flow (36 stations in USA) (after Williams, 1978)

as the occurrence of specific events. In this context, the inter-arrival times between events of different magnitude demands more attention. For example, Anderson (1976) outlines the Erlang distribution

$$A_o(t) = e^{-\lambda kt} \sum_{n=0}^{k-1} \frac{(k\lambda t)^n}{(n!)} \quad (1)$$

where $A_o(t)$ is the probability that an inter-arrival time is $>t$, λ is the mean arrival rate, and $k^{-1/2}$ is the coefficient of variation. Application of this model to inter-arrival time distributions in the Snowy Mountains, Australia, shows that significant changes in flood peak inter-arrival times occur downstream as floods are routed through the channel network.

Regression models

Estimation of the magnitude of a T-year return period flood at ungauged sites is commonly based on regression models. In England and Wales, Cole (1966) has defined six regions which are hydrologically homogeneous (Dalrymple, 1957) and within which simple regressions of mean annual flood ($Q_{2.33}$) on basin area (A) can be used to predict this event in ungauged basins. Anderson (1972) has suggested that the exponent in a power function relationship is itself area-dependent, with an envelope curve for flood intensity (q = Q/A) being

$$q = 20000\,A.^{(0.9A^{-0.5}-1)} \quad (2)$$

However, this is an empirical convenience reflecting constraints on the range of sampled basins, and a more rigorous multivariate approach is preferable. Benson (1960b) exemplified this in an early application, in New England, of exploratory analysis, mapping residuals to identify spatial autocorrelation and necessary additional independent variables. Regression models for flood prediction in Britain include those by Nash (1959), Nash and Shaw (1966) and Rodda (1967); the last of these is of some interest physically because one of the independent variables predicting $Q_{2.33}$ is mean annual maximum daily rainfall, $P_{2.33}$, in a regression taking the form

$$Q_{2.33} = 12A^{0.77}\,P_{2.33}^{2.92}\,D^{0.81} \quad (3)$$

the third predictor being drainage density. The Flood Studies Report (NERC, 1975) provides a complete empirical analysis of British data, with national and regional equations for $Q_{2.33}$ based on several independent variables thoroughly investigated for redundancy by principal components analysis and parameter stability by ridge regression. Hydrograph indices other than peak discharge have been correlated with catchment variables to permit estimation of synthetic unit hydrographs for ungauged catchments. For example, Painter (1971) and Lowing and Newson (1973) have developed statistical models to predict time-to-peak, and Nash (1960) related the moments of the instantaneous unit hydrograph to catchment variables.

This empirical approach encounters numerous statistical problems. Regression coefficients only reflect the sample data, and sampling bias often arises because of the difficulty of ensuring, for example, that large basins with steep slopes are measured. Bias may also arise because nesting of catchments results in autocorrelation. The independent variables are invariably intercorrelated, and multicollinearity limits physical interpretation of the coefficients (Nash, 1959). Statistically optimal prediction models are identified by using principal components analysis to define orthogonal independent variables (Wong, 1963), but have been criticized for their lack of physical rationale and their dimensional imbalance (Court, 1972). Multiple regression relationships demand the assumption of linearity in the 'statistical theory' sense that transformation can linearize them (Clarke, 1973). Theory alone can satisfactorily indicate the existence of nonlinearity, which otherwise may simply reflect the effects of omitted variables. A novel approach

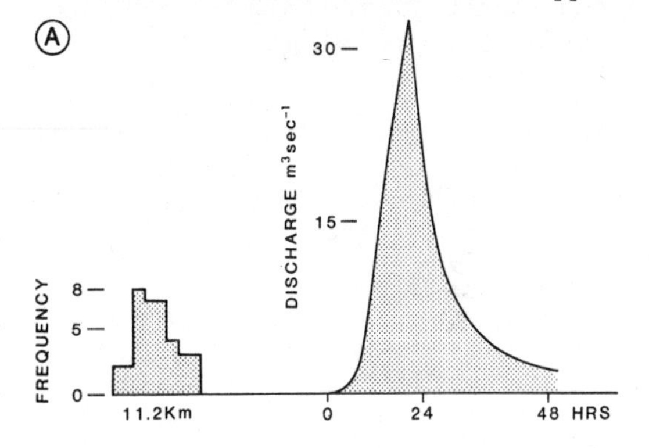

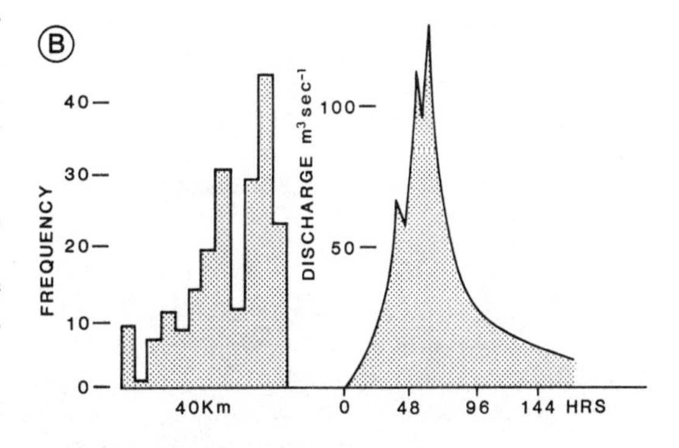

Figure 26.3 Relationships between frequency distributions of exterior link distances and hydrograph forms (after Rogers, 1972)

incorporating nonstandard functional relationships is the use of graphical coaxial correlation by Bogardi (1974) in which goodness of fit is measured by minimum root mean square deviation, improving the linear multiple regression R^2 considerably.

The major limitation of regression approaches is the lumping of parameters. Particularly in the context of partial area runoff generation, distributed models are required, to be explicitly based on network structure and spatial patterns of runoff production in relation to variable soil infiltration, storm direction, and rainfall intensity (Beven and Kirkby, 1979). Surkan (1969) and Rogers (1972) have both related hydrograph form to network shape, and Figure 26.3 illustrates the relation between the frequency distribution of exterior link distances from the basin mouth and the shape of a hydrograph generated by the network concerned. Although this correlation is possibly simplistic (Smart, 1978), Kirkby (1976) has also emphasized the combined influence of network topology and geometry on hydrograph form. Since flow paths to the outlet strongly influence hydrographs, and other spatially distributed variables can be 'tied' to network structure, these approaches seem likely to be of major significance. They may be paralleled by developments in flood routing methods, which also provide a means of predicting hydrograph form (peak discharge and shape) at various points through a drainage basin. These have a firm theoretical foundation in kinematic wave models (Price, 1973), and require solution of a continuity equation for which input is the only known quantity. To obtain output from a reach, some empirical data on storage-discharge characteristics are necessary, which again may be obtained from a distributed model relating channel size to location in the network. Alternatively, Gregory (1976) suggests a method whereby channel volume is obtained by integrating the channel-capacity-channel-length relationship between required limits.

Time series models

Input, system state and output variables of most hydrological systems form time series in which deterministic and stochastic components occur to varying degrees. In using standard regression methods,

$$\hat{Y}_t = \hat{\alpha} x_t + u_t, \tag{4}$$

to model input–output relationships, the residual series u_t is often autocorrelated and the dynamic system behaviour averaged in the parameter $\hat{\alpha}$. These problems are obviated by the family of transfer-function models of Box and Jenkins (1970), which are concerned with systems developing through time according to some probabilistic law which can be identified from the pattern of autocorrelation observed in the series (Carlson *et al.*, 1970). The success achieved by this approach depends on the objective of analysis, an evident distinction being between modelling for simulation and prediction, or for physical explanation.

In hydrological design problems, knowledge of the statistical structure of the series is required to identify the generating process for simulation or forecasting. A given series Y_t can be subdivided into

$$Y_t = T_t + S_t + E_t \tag{5}$$

where T_t is the trend, modelled by low order polynomial regression; S_t is a seasonal component fitted by harmonics; and E_t is a residual series modelled by an autoregressive moving average stochastic process (Quimpo, 1968). Kottegoda (1972) analysed five daily streamflow means of some British rivers by this method, which suffers from the disadvantages of clumsiness because of the number of parameters necessary, and of the critical importance of successful subdivision into components. Removal of the deterministic T_t and S_t components can result in awkward patterns of autocorrelation in E_t (Kavaas and Delleur, 1975), so that seasonal differencing may be preferred. Harmonic analysis averages amplitude and phase over a period, and demodulation is a useful guide to variability in seasonal components (Anderson, 1975; Quimpo and Cheng, 1974). Table 26.1 shows the variability in harmonic amplitude and phase fitted

Table 26.1 Amplitude (A) and phase (ϕ) estimates from 6 simulated series (N = 960). After Quimpo and Cheng (1974).

Series	*A*	ϕ
1	0.3716	–0.544
2	0.3511	–0.750
3	0.3495	–0.730
4	0.3577	–0.647
5	0.3925	–0.726
6	0.3116	–0.866
Mean	0.3557	–0.710
'True' value	0.3535	–0.785

to series of n=960 values generated by a known harmonic plus Markov process. Clearly, long series are needed to obtain reliable estimates. Estimators for the parameters of stochastic processes in the E_t series are also unreliable, resulting in downward bias (Wallis and O'Connell, 1972).

Garcia-Martinez (1972) has shown that, perversely, the estimator with least bias has the largest variance, and that two popular bias corrections fail to eliminate bias. Thus, simulation of hydrological series using equation (5) may underestimate the short lag memory of the sequence, with potentially serious consequences for practical problems such as reservoir operating policies (Lloyd, 1963). Alternative modelling strategies exist, but their value remains unclear. For example, Mandelbrot and Wallis (1968, 1969) have developed fast fractional Gaussian noise as the basis for self-similar models with small but non-negligible interdependency over long time lags. Potter (1979), however, argues that annual precipitation exhibits no long memory except in non-homogeneous series rendered non-stationary by processes such as micro-climatic change resulting from urbanization round the gauge. However, the shot noise model used to simulate streamflow of the River Nene (Weiss, 1977) has considerable potential because of the physical realism of the generated series, with saw-tooth characteristics including steep rises and exponential decay.

Time series modelling to aid explanation of physical behaviour of hydrological systems compounds the four-dimensional nature of streamflow generation with the statistical problems of trend removal and estimation. Gudgin and Thornes (1974) observe that including a spatial component in attempts to develop space-time models of hill-slope hydrology is limited by considerable boundary problems, and Anderson and Burt (1978) have emphasized the importance of the additional dimension of the thickness of the zone of soil profile saturation. Hydrological versions of space-time models for mapped data on a plane (Bennett, 1975) are a remote possibility. However, Lai (1977) has successfully applied transfer function models to dynamic relationships of soil moisture tension, in the form

$$Y_t = \alpha X_{t-b} + \beta Y_{t-1} + u_t \qquad (6)$$

where Y_t is soil moisture tension at 50 cm depth at time t, X_t is tension at 10 cm, b is the response delay and u_t the residual series. Thus Y_t depends on X_{t-b} and Y_{t-1}. Such an approach can be extended to multiple input models and frequency domain analysis, for example in studies of water quality variables. Whilst Edwards and Thornes (1973) identify leads and lags between discharge and water quality variables, Dowling (1974) however, argues that phase estimates in the frequency domain cannot be automatically interpreted as simple leads or lags in the time domain. Numerous different distributed-lag generating processes can produce the same lag in physical time. This is a common problem in physical interpretation; time series models apply a non-unique synthesis method to unknown systems, and equivalent fits by different models according to statistical criteria complicate physical understanding. Progress thus requires prior development of a theoretical basis, which also emphasizes the links between discrete stochastic models and continuous natural processes. The best illustration is Quimpo's (1973) theoretical derivation of autocorrelation functions for streamflow series based on cascading a random input through a series of linear reservoirs.

Models for spatial data analysis

Spatially distributed hydrological data are invariably measurements of continuous phenomena (rainfall or evapotranspiration) at discrete locations. Records from adjacent sites are spatially autocorrelated, which complicates assessment of confidence limits of mean areal rainfall estimates (Hutchinson, 1973) and severely limits the effective information content of an individual gauge record. Matalas and Langbein (1962) show that minimal cross-correlations between gauges (averaging $r_{jk} = 0.1$) result in an infinite number of gauges yielding the information of only ten with uncorrelated records. Practical problems also arise when regionalized variables are sampled at discrete sites; for example, estimation of total or average areal values, and optimum location of sampling points.

Numerous approaches to the areal averaging problem exist, and even traditional methods are automated. Thiessen polygon raingauge coefficients (weights) have been defined by Monte Carlo techniques (Diskin, 1969), and this is also possible for the isohyetal methods. However, more sophisticated quantitative approaches have been applied, often using rainfall-height relationships and trend surface models. Rodda (1962), for example, estimated mean annual catchment rainfall for the Ystwyth basin using a rainfall-height linear regression and a topographic map, and Mandeville and Rodda (1970) suggested the method of integrating the three-dimensional polynomial equation of a trend surface. Unwin (1969) standardized rainfall values to mean sea level values using a rainfall-height relationship, then estimated regional trends using trend surfaces. However, since the data used to derive a rainfall-height relation initially incorporated spatial information, it is preferable to adopt Edwards's (1972) approach employing a 'trivariate hypersurface' in which rainfall is predicted directly from spatial coordinates U and V, and elevation H. Such a multivariate spatial model is often statistically efficient in comparison with conventional trend surfaces; the trivariate quadratic surface, for example, has comparable R^2 values to a cubic trend surface, but fewer parameters. Statistical inefficiency also limits the usefulness of two-dimensional Fourier series analysis for generalizing the pattern of regionalized hydrological variables for integration or interpolation, as parameters are often unstable and excessively numerous (Edwards, 1972). A more subjective, graphical approach is attempted by Collinge and Jamieson (1968), who demonstrate that discrepancies between raingauge and radar estimates of rainfall vary regionally and with wind speed and direction, in a systematic manner that allows satisfactory estimates to be made from four radar stations and detailed wind data. Given the limited accuracy of raingauges, this points to a means of rationalizing rainfall monitoring by regional calibration and limited point measurement, and is a radically different approach to the regression methods which demand a dense gauge network.

More recently alternative methods have been developed based on numerical analysis and geostatistics. The former include a generalization to two dimensions of quadratic and cubic spline interpolation (Fritsch, 1971) and the fitting of multiquadric surfaces such as circular hyperboloids to local point sets (Hardy, 1971). Shaw and Lynn (1972) compare these methods, and obtain satisfactory areal rainfall estimates using multiquadric surfaces, which are applicable to irregularly spaced data, and which require a matrix of gauge coordinates which can be converted into a set of weights. This is a practical advantage; weights for trend surfaces (Chidley and Keys, 1970) are not solely dependent on gauge locations, but also vary with the order of fitted surface, which cannot be assumed to be constant (Edwards, 1972). Numerical methods fit observed data points exactly, which assumes no measurement error. However, the residuals obtained when using regression models are not solely measurement errors, so this cannot be considered a significant criticism. Methods based on geostatistics (Matheron, 1970) are explicitly concerned with regionalized variables whose values are known at a few discrete points but which require interpolation (for example, at grid intersections for computer contour plotting; Huijbregts, 1975). The interpolation method, Kriging, requires information on the spatial correlation field around each point, as measured by the semi-variogram, an inverse correlogram (Agterberg, 1970). The variogram, γ_s, is

$$\gamma_s = \tfrac{1}{2}E(X_{k+s} - X_k)^2 \quad (7)$$

for a random variable X measured at point k and at points lagged spatially by a distance s. Variance between points increases with lag, while correlation decreases, and a typical rainfall variogram is that in Figure 26.4, showing a 'nugget' effect – finite variance at the origin reflecting small scale variability. Problems exist in defining variograms or spatial correlograms; data are rarely available at equal spacing, which demands a specific sampling exercise or complex estimation procedures for irregular data spacing (Agterberg, 1970). Also, spatial auto-

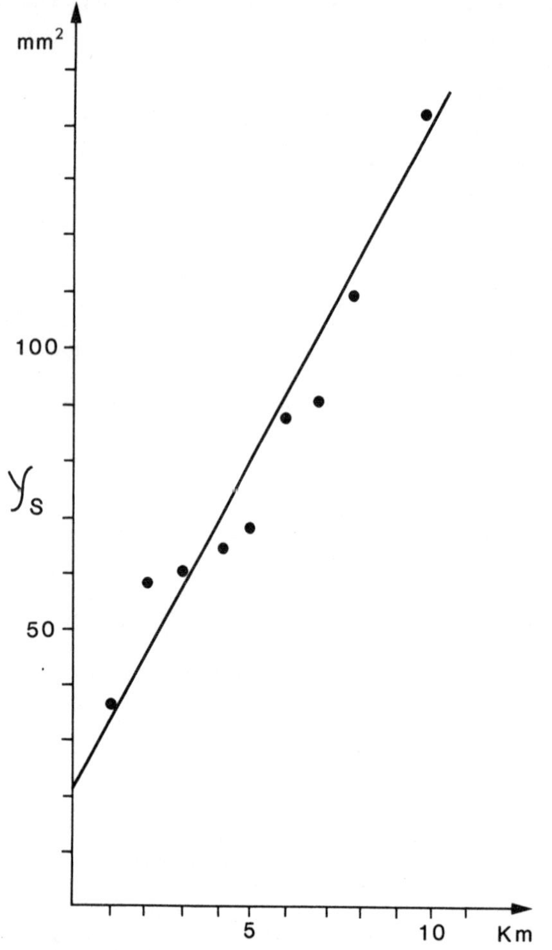

Figure 26.4 A variogram for a spatial field of rainfall amounts (from Delfiner and Delhomme, 1975

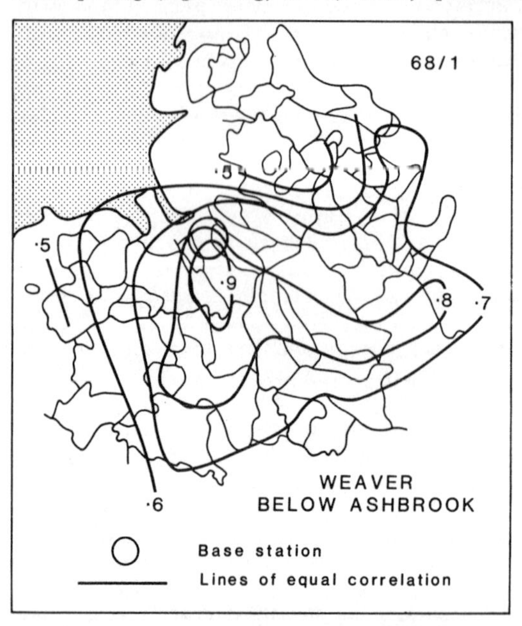

Figure 26.5 Isocorrelates of discharge data at gauging stations surrounding that on the River Weaver, below Ashbrook (after NERC, 1975)

correlation varies with relief (Hutchinson, 1970), and anisotropy of the correlation field is normal, as Figure 26.5 shows. However, because Kriging provides a smooth surface with improved root mean square deviation compared to trend surface models, and an estimate of interpolation error, it has considerable potential. Delhomme and Delfiner (1973) demonstrate its use in optimizing location of new raingauges (Figure 26.6), and similar applications in design of gauge networks are provided by Hutchinson (1974) and McCullagh (1975).

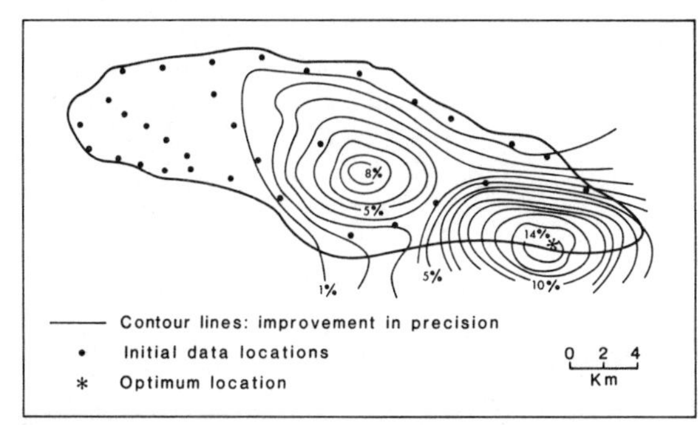

Figure 26.6 Optimum location of an additional raingauge assessed by applying Kriging to measure improvement in precision (from Huijbregts, 1975)

Error analysis

Geographical hydrologists rarely include error analysis in reporting results, and spatial and temporal trends are thus often not convincingly established. All measurements, as with statistical estimates, vary in bias and precision, and are subject to observer, instrument and sampling errors (Davidson, 1978). Operator variance is commonly considered in relation to groups in which opposing deviations cancel and convergence of group mean to the true mean occurs (Chorley, 1958). However, since measurements are made by individuals, the significance of this is debatable. Instrument error is of increasing interest as automated, continuous monitoring of hydrological variables develops. Rainfall can be measured by radar, streamflow by ultrasonic or electromagnetic methods, solutes by conductivity meter, suspended load by turbidity meters, and bedload by acoustic methods. These indirect approaches provide standards against which errors of measurement and discrete sampling by traditional methods (rain gauges, current meters, bedload traps) can be assessed.

Sampling errors resulting from discrete measurement of continuously varying phenomena are particularly important. Walling (1975) emphasizes the need to match sampling frequency, length of observation period, and scale of study. Edwards and Thornes (1973) obtain useful results from twenty years of weekly sampled water quality data, but for shorter observation periods more frequent sampling is desirable, particularly as the scale of investigation changes from seasonal to storm period variations. Information theory may conveniently summarize the relative information content of different sampling schemes, such as the continuous record and monthly averages of specific conductance in Figure 26.7. Walling (1977) compares spot sample measurements of suspended sediment, used to define rating curves, with continuous turbidity meter measurement calibrated by suspensions of known concentration. Four separate ratings (winter and summer, rising and falling stages) give errors in estimating total annual sediment transport of 21–44 per cent when combined with hourly flow data and –80 per cent to +400 per cent using a monthly time base. Rules governing sampling frequency are not clearly established. Reynolds (1975) discusses a method based on the t-distribution, but temporal and spatial autocorrelation reduce the effective sample size. Sanders and Adrian (1978) have suggested a relationship between sampling frequency per year and the confidence interval of the mean of the residual random component of streamflow after removal of deterministic seasonal and autoregressive persistence elements (Figure 26.8). However, redundancy of data collected automatically during uniform conditions remains a problem unless economic sampling is permitted by equipment sensing rates of change of the phenomenon measured (Claridge, 1973). Spatial location of measurement sites adds to the temporal sampling problem. Anderson and Burt (1978), for example, optimize design and siting of throughflow troughs after controlled laboratory study using Darcy's law to predict slope discharge from soil horizons, locating troughs in the field so that measured flow accords with the Darcy prediction.

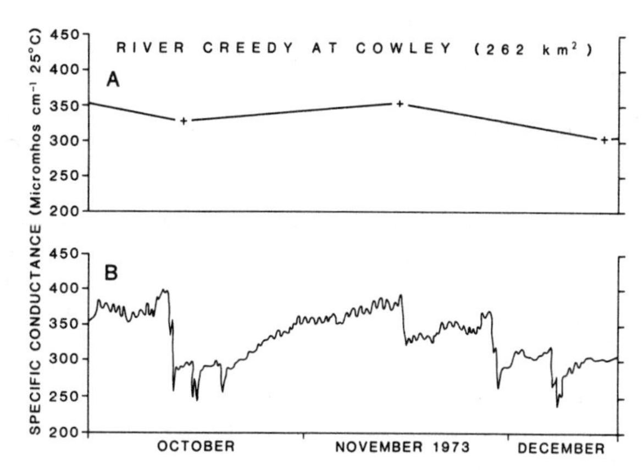

Figure 26.7 Temporal variation in specific conductance, River Creedy. A Interpolated from weekly spot measurements averaged over one month, B Continuous record (from Walling, 1975)

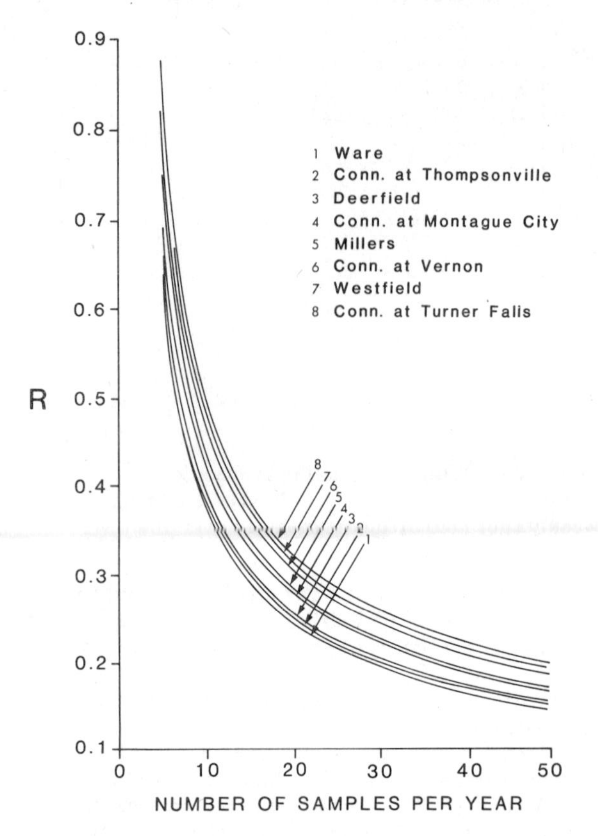

Figure 26.8 The magnitude (R) of the half-width of the confidence interval of the mean of the random component of the annual mean log flow versus the number of samples per year for each sampling station in the Massachusetts portion of the Connecticut River basin (after Sanders and Adrian, 1978)

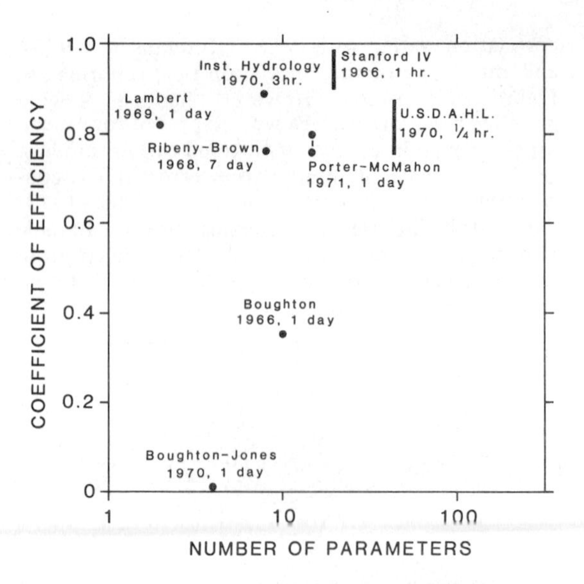

Figure 26.9 Assessment of efficiency of a number of hydrological models (from Kirkby, 1975, data from Aitken, 1973, and Lambert 1969)

Error analysis also becomes significant in assessment of the parsimony of multivariate hydrological models, which should balance goodness of fit with simplicity. Kirkby (1975) uses an efficiency index, E, defined as

$$E = 1 - \frac{\text{Residual sum of squares}}{\text{Total sum of squares}} \tag{8}$$

which is plotted against the number of parameters in the model (Figure 26.9). An efficient (parsimonious) model maximizes E for the number of parameters. This is partially achieved in significance testing by the progressive loss of degrees of freedom with increasing model complexity. Stepwise regression, or trend surface fitting using Chayes's (1970) method for testing successively higher order surfaces, provide examples. However, Baird *et al.* (1972) note in the context of trend surface analysis that such an approach may encourage the researcher to stop increasing model complexity at a low order, when a high order model fits the data very closely; only prior theoretical reasoning can obviate this problem and point to a specific level of complexity.

Conclusion

The dilemmas of scale and detail raised in the introduction and illustrated in the review will not, in all probability, be resolved by adoption of stochastic process models. Rather, the indications are that partially distributed mathematical models will eventually resolve the problem. Beven's (1977) numerical model of hillslope hydrographs, and partial area process models such as those suggested by Beven and Kirkby (1979) are examples. The fact that increasing application of quantitative methods in all aspects of hydrology has rarely been paralleled by error analysis of either field data or model parameter estimates has increased the difficulty of assessing the relative success of different quantitative approaches. Where errors have recently been examined, such as in sediment load estimation and field throughflow trough design, they have been shown to be intolerably large and wholly at odds with the use of models with reasonable sensitivity. It is clear that research in quantitative hydrology must address itself more to the study of model sensitivity at a range of scales, and to field error analysis, since failure to do so may well prejudice the selection of the appropriate quantitative method for a particular problem.

References

Agterberg, F.P. (1970) 'Autocorrelation functions in geology' in D.F. Merriam (ed.), *Geostatistics*, 113-41, Plenum: New York.

Aitken, A.P. (1973) 'Assessing systematic errors in rainfall-runoff models', *Journal of Hydrology*, 20, 121-36.

Anderson, G.N. (1972) 'Effect of catchment area on flood magnitude', *Journal of Hydrology*, 16, 225-40.

Anderson, M.G. (1975) 'Demodulation of streamflow series', *Journal of Hydrology*, 26, 115-21.

Anderson, M.G. (1976) 'Peak flow inter-arrival times', *Journal of Hydrology*, 30, 105-12.

Anderson, M.G., and Burt, T.P. (1978) 'Toward more detailed field monitoring of variable source areas', *Water Resources Research*, 14, 1123-31.

Anderson, M.G., and Calver, A. (1977) 'On the persistence of landscape features formed by a large flood', *Transactions, Institute of British Geographers*, NS 2, 243-54.

Baird, A.K., Baird, K.W., and Morton, D.M. (1971) 'On deciding whether trend surfaces of progressively higher order surfaces are meaningful: discussion', *Geological Society of America, Bulletin,* 82, 1219-33.

Bennett, R.J. (1975) 'The representation and identification of spatio-temporal systems: an example of population diffusion in North-West England', *Transactions, Institute of British Geographers*, 66, 73-94.

Benson, M.A. (1960a) 'Characteristics of frequency curves based on a theoretical 1000-year record', in T. Dalrymple (ed.), *Flood Frequency Analyses*, United States Geological Survey, Water Supply Paper, 1543-A.

Benson, M.A. (1960b) 'Areal flood-frequency analysis in a humid region', *Bulletin of the International Association of Hydrological Sciences*, 19, 5-15.

Beven, K.J. (1977) 'Hillslope hydrographs by the finite element method', *Earth Surface Processes*, 2, 13-28.

Beven, K.J., and Kirkby, M.J. (1979) 'A physically based, variable contributing area model of basin hydrology', *Hydrological Sciences Bulletin*, 24, 43-69.

Bogardi, J. (1974) *Sediment transport in alluvial streams*, Akademiai Kiado: Budapest.

Box, G.E.P., and Jenkins, G.M. (1970) *Time series analysis - forecasting and control*, Holden-Day: San Francisco.

Carlson, R.F., MacCormick, A.J.A., and Watts, D.G. (1970) 'Application of linear random models to four annual streamflow series', *Water Resources Research*, 6, 1070-8.

Chayes, F. (1970) 'On deciding whether trend surfaces of progressively higher order are meaningful', *Geological Society of America, Bulletin*, 81, 1273-8.

Chidley, T.R.E., and Keys, K.M. (1970) 'A rapid method of computing areal rainfall', *Journal of Hydrology*, 12, 15-24.

Chorley, R.J. (1958) 'Group operator variance in morphometric work with maps', *American Journal of Science*, 256, 208-18.

Claridge, G.G.C. (1973) 'Studies in element balances in a small catchment at Taita, New Zealand' in *Results of research on representative and experimental basins*, Vol. 1, 523-40, IASH - UNESCO, Wellington symposium.

Clarke, R.T. (1973) 'A review of some mathematical models used in hydrology, with observations on their calibration and use', *Journal of Hydrology*, 19, 1-20.

Cole, G. (1966) 'An application of the regional analysis of flood flows', in *River Flood Hydrology*, Institute of Civil Engineers, 39-57.

Collinge, V.K., and Jamieson, D.G. (1968) 'The spatial distribution of storm rainfall', *Journal of Hydrology*, 6, 45-57.

Costa, J.E. (1978) 'Holocene stratigraphy in flood frequency analysis', *Water Resources Research*, 14, 626-32.

Court, A. (1972) 'Heterodox hydrology: discussion', *Geographical Analysis*, 4, 194-203.

Cunnane, C. (1973) 'A particular comparison of annual maxima and partial duration series methods of flood frequency prediction', *Journal of Hydrology*, 18, 257-71.

Cunnane, C. (1979) 'A note on the Poisson assumption in partial duration series models', *Water Resources Research*, 15, 489-94.

Dalrymple, T. (1957) 'Flood frequency relations for gauged and ungauged streams', *International Association of Hydrological Sciences* (IAHS), Toronto symposium, Vol. 3, 268-79.

Davidson, D.A. (1978) *Science for physical geographers*, Edward Arnold: London.

Delfiner, P., and Delhomme, J.P. (1975) 'Optimum interpolation by Kriging' in J.C. Davis and M.J. McCullagh (eds), *Display and Analysis of Spatial Data*, 38-53, Wiley: London.

Delhomme, J.P. and Delfiner, P. (1973) 'Application du krigeage a l'optimisation d'une campagne pluviometrique en zone aride', *Design of Water Resources Projects with Inadequate Data*, 193-210, UNESCO - WMO - IAHS Symposium, Madrid.

Diskin, M.H. (1969) 'Thiessen coefficients by a Monte Carlo procedure', *Journal of Hydrology*, 8, 323-35.

Dowling, J.M. (1974) 'A note on the use of spectral analysis to detect leads and lags in annual cycles of water quality', *Water Resources Research*, 10, 343-4.

Dury, G.H. (1959) 'Analysis of regional flood frequency on the Nene and Great Ouse', *Geographical Journal*, 125, 223-9.

Edwards, A.M.C., and Thornes, J.B. (1973) 'Annual cycle in river water quality: a time series approach', *Water Resources Research*, 9, 1286–95.

Edwards, K.A. (1972) 'Estimating areal rainfall by fitting surfaces to irregularly spaced data' in *Distribution of precipitation in mountainous areas*, WMO, Geilo symposium, Vol. 2, 565–87.

Fritsch, J.M. (1971) 'Objective analysis of a two-dimensional data field by the cubic spline technique', *Monthly Weather Review*, 99, 379–86.

Garcia-Martinez, L.E. (1972) 'Parameter estimation for first-order autoregressive models', *American Society of Civil Engineers, Proceedings, Journal of Hydraulics Division*, 98, 1343–9.

Gregory, K.J. (1976) 'Stream network volume: An index of channel morphometry', *Geological Society of America, Bulletin*, 88, 1075-80.

Gudgin, G., and Thornes, J.B. (1974) 'Probability in geographical research – applications and problems', *Statistician*, 23, 157–77.

Hardy, R.L. (1971) 'Multiquadric equations of topography and other irregular surfaces', *Journal of Geophysical Research*, 76, 1905–15.

Harvey, A.M. (1969) 'Channel capacity and the adjustment of streams to hydrologic regime', *Journal of Hydrology*, 8, 82–98.

Harvey, A.M. (1971) 'Seasonal flood behaviour in a clay catchment', *Journal of Hydrology*, 12, 129–44.

Howe, G.M., Slaymaker, H.O., and Harding, D.M. (1967) 'Some aspects of the flood hydrology of the upper catchments of the Severn and Wye', *Institution of British Geographers, Transactions*, 41, 33–58.

Huijbregts, C. (1975) 'Regionalised variables and quantitative analysis of spatial data' in J.C. Davis and M.J. McCullagh (eds), *Display and Analysis of Spatial Data*, 96 114, Wilcy: London.

Hutchinson, P. (1970) 'A contribution to the problem of spacing raingauges in rugged terrain', *Journal of Hydrology*, 12, 1–14.

Hutchinson, P. (1973) 'The accuracy of estimates of areal main rainfall', in *Results of research on representative and experimental basins*, Vol. 1, IASH – UNESCO, Wellington symposium, 203–18.

Hutchinson, P. (1974) 'Progress in the use of the method of optimum interpolation for the design of the Zambian raingauge network', *Geoforum*, 20, 49–62.

Kavaas, M.L., and Delleur, J.W. (1975) 'Removal of periodicities by differencing and monthly mean subtraction', *Journal of Hydrology*, 26, 335–53.

Kirkby, M.J. (1975) 'Hydrograph modelling strategies' in R. Peel, M. Chisholm and P. Haggett (eds), *Processes in Physical and Human Geography*, 69–90, Heinemann: London.

Kirkby, M.J. (1976) 'Tests of the random network model and its application to basin hydrology', *Earth Surface Processes*, 1, 197–212.

Kottegoda, N.T. (1972) 'Stochastic five-daily stream flow model', *American Society of Civil Engineers, Proceedings, Journal of Hydraulics Division*, 98, 1469–85.

Lai, P.W. (1977) 'Stochastic-dynamic models for some environmental systems: transfer function approach', London School of Economics, Discussion Paper 61.

Lambert, A.O. (1969) 'A comprehensive rainfall-runoff model for an upland catchment', *Journal of the Institute of Water Engineers*, 23, 231–8.

Lloyd, E.H. (1963) 'A probability theory of reservoirs with serially correlated inputs', *Journal of Hydrology*, 1, 49–128.

Lowing, M.J., and Newson, M.D. (1973) 'Flood event data collation', *Water and Water Engineering*, 77, 91–5.

Mandelbrot, B.B., and Wallis, J.R. (1968) 'Noah, Joseph and operational hydrology', *Water Resources Research*, 4, 909–18.

Mandelbrot, B.B., and Wallis, J.R. (1969) 'Computer experiments with Fractional Gaussian Noises, Part 1, 2 and 3', *Water Resources Research*, 5, 228–41, 242–59, 260–7.

Mandeville, A.N., and Rodda, J.C. (1970) 'A contribution to the objective assessment of rainfall amounts', *Journal of Hydrology* (New Zealand), 9, 281–91.

Matalas, N.C., and Langbein, W.B. (1962) 'The relative information of the mean', *Journal of Geophysical Research*, 67, 3441–8.

Matheron, G. (1970) 'Random functions and their application in geology', in D.F. Merrian (ed.) *Geostatistics*, 79–87, Plenum: New York.

McCullagh, M.J. (1975) 'Estimation by Kriging of the reliability of the proposed Trent telemetry network', *Computer Applications*, 2, 357–74.

Mundry, E. (1972) 'On the resolution of mixed frequency distributions into normal components', *Mathematical Geology*, 4, 55–60.

Nash, J.E. (1959) 'Discussion of "A study of bankfull discharge", by Nixon, M.', *Proceedings, Institute of Civil Engineers*, 14, 395–425.

Nash, J.E. (1960) 'A unit hydrograph study with particular reference to British catchments', *Proceedings, Institute of Civil Engineers*, 17, 249–82.

Nash, J.E., and Shaw, B.L. (1966) 'Flood frequency as a function of catchment characteristics', in *River Flood Hydrology*, Institute of Civil Engineers, 115–35.

Natural Environment Research Council (1975) *Flood Studies Report*, 5 volumes, Natural Environment Research Council, London.

Painter, R.B. (1971) 'The hydrograph time-to-peak method of flood prediction', *Water and Water Engineering*, 75, 235–7.

Potter, K.W. (1979) 'Annual precipitation in the

Northeast United States: Long memory, short memory or no memory?', *Water Resources Research*, 15, 340–6.
Potter, W.D. (1958) 'Upper and lower frequency curves for peak rates of runoff', *Transactions of the American Geophysical Union*, 39, 100–5.
Price, R.K. (1973) 'Flood routing methods for British rivers', *Proceedings, Institute of Civil Engineers*, 55, 913–30.
Quimpo, R.G. (1968) 'Autocorrelation and spectral analyses in hydrology', *American Society of Civil Engineers, Proceedings, Journal of Hydraulics Division*, 94, 363–73.
Quimpo, R.G. (1973) 'Link between stochastic and parametric hydrology', *American Society of Civil Engineers, Proceedings, Journal of Hydraulics Division*, 99, 461–70.
Quimpo, R.G., and Cheng, M.-S. (1974) 'On the variability of seasonal parameters in hydrologic time series', *Journal of Hydrology*, 23, 279-87.
Reich, B.M. (1970) 'Flood series compared to rainfall extremes', *Water Resources Research*, 6, 1655–67.
Reynolds, S.G. (1975) 'A note on the relationship between the size of area and soil moisture variability', *Journal of Hydrology*, 24, 71–6.
Rodda, J.C. (1962) 'An objective method for the assessment of areal rainfall amounts', *Weather*, 17, 54–9.
Rodda, J.C. (1967) 'The significance of characteristics of basin rainfall and morphometry in a study of floods in the United Kingdom', *International Association of Hydrological Sciences*, Publication No. 85, 834–45.
Rodda, J.C. (1970) 'Rainfall excess in the United Kingdom', *Transactions, Institute of British Geographers*, 49, 49–59.
Rogers, W.F. (1972) 'New concept in hydrograph analysis', *Water Resources Research*, 8, 973–81.
Sanders, T.G., and Adrian, D.D. (1978) 'Sampling frequency for river quality monitoring', *Water Resources Research*, 14, 569–76.
Shaw, E.M., and Lynn, P.P. (1972) 'Areal rainfall evaluation using two surface fitting techniques', *Bulletin of the International Association for Scientific Hydrology*, 17, 419–33.
Singh, K.P., and Sinclair, R.A. (1972) 'Two-distribution method for flood-frequency analysis', *American Society of Civil Engineers, Proceedings, Journal of Hydraulics Division*, 98, 29–44.
Smart, J.S. (1978) 'The analysis of drainage network composition', *Earth Surface Processes*, 3, 129–70.
Surkan, A.J. (1969) 'Simulation of storm velocity effects on flow from distribution channel networks', *Water Resources Research*, 10, 1149–60.
United States Water Resources Council (1967) *A Uniform technique for determining flood flow frequencies*, Bulletin 15, Water Resources Council, Washington, DC.
Unwin, D.J. (1969) 'The areal extension of rainfall records: an alternative model', *Journal of Hydrology*, 7, 404–14.
Walling, D.E. (1975) 'Solute variations in small catchment streams', *Transactions, Institute of British Geographers*, 64, 141–7.
Walling, D.E. (1977) 'Limitations of the rating curve technique for estimating suspended sediment loads, with particular reference to British rivers', *International Association of Hydrological Sciences*, Publication No. 122, 34–48.
Wallis, J.R., and O'Connell, P.E. (1972) 'Small sample estimation of ρ_{1i}' *Water Resources Research*, 8, 707–12.
Weiss, G. (1977) 'Shot noise models for the generation of synthetic streamflow data', *Water Resources Research*, 13, 101–8.
Williams, G.P. (1978) 'Bank-full discharge of rivers', *Water Resources Research*, 14, 1141–8.
Wolman, M.G., and Gerson, R. (1978) 'Relative scales of time and effectiveness of climate in watershed geomorphology', *Earth Surface Processes*, 3, 189–208.
Wong, S.T. (1963) 'A multivariate statistical model for predicting mean annual flood in New England', *Annals, Association of American Geographers*, 53, 298–311.
Woodyer, K.D. (1968) 'Bankfull frequency of rivers', *Journal of Hydrology*, 6, 114–42.
Yevjevich, V.M. (1974) 'Determinism and stochasticity in hydrology', *Journal of Hydrology*, 22, 225–38.

Chapter 27

Geomorphology

J.E. Thornes and R.I. Ferguson

Introduction

Geomorphology is concerned with the explanation of topographic forms at the earth's surface. This involves the mechanics of process-form relationships, explanation of spatial variations in terms of the agents controlling the processes, and investigation of the sequential development of particular landforms or landscapes.

Until the 1960s British geomorphologists (then as now predominantly geographers) were engaged mainly in the third of these activities and felt little need for quantitative techniques or methodology. In the United States, on the other hand, the more obvious practical relevance of geomorphology and a closer alliance with engineering hydraulics and geology stimulated a more mechanical and theoretical approach to geomorphology – first in the late nineteenth century exploration of the West when the Huttonian fundamentals of the subject were extended by Powell, Gilbert, and Davis, and then again from the 1930s on when the importance of soil and water conservation led to basic research on fluvial processes by engineers (Horton, Lane), hydrologists (Langbein, Leopold), and geologists (Krumbein, Strahler). The innovative contribution of these scientists was the use of a variety of deductive and inductive quantitative techniques, and both these and the general approach were taken up in British geography departments in the late 1950s by Chorley and other geomorphologists quick to see the potential of the new approach.

The main application of quantitative methods in geomorphology has been and still is the description and analysis of field data on particular sites, events, or areas, in a mainly inductive framework. The problems of sampling and sampling theory, curve or surface fitting, and hypothesis testing, dominated the first major review of quantitative geomorphology (Chorley, 1966) and rapidly became incorporated in undergraduate teaching. Chorley (1972) could thus describe 1971 as the year which marked 'the coming-of-age of the application of modern statistical techniques to geomorphology'. Particular statistical methods continue to be adopted and, less often, discarded but the major methodological trend in the 1970s has been the embedding of individual studies and statistical analyses in a deductive model-building context. A flourishing school of geomorphological research, with British workers in the forefront, is now primarily concerned with quantitative modelling using a variety of techniques.

In this review we discuss both inductive and deductive methods, but particularly the latter for three reasons: the techniques are more varied and distinctive, they are less commonplace than the inductive application of statistics which is well covered in standard texts (e.g. Agterberg's (1974) *Geomathematics*), and they appear to offer the most exciting prospects for future advances.

Our review follows Wilson (1979) in adopting Weaver's (1958) typology of systems and models of systems. Weaver distinguished three kinds of system:

simple systems which involve not more than three or four variables and can be handled by relatively simple conventional techniques;

systems of complex disorder which involve large numbers of components and therefore of state variables but with only weak linkages between them, and are handled by the probabilistic methods of statistical mechanics; and

systems of complex order which again involve many components but with strong, organized interaction. Such systems are not intrinsically different from the first type, but their complexity – the number of possible interactions increases as the square of the number of components – rules out many simple techniques of study.

Some geomorphological problems fall quite clearly into one only of these categories, but many topics can be tackled in different ways according to the number of variables and linkages recognized. Often this in turn depends on the time or space scale of interest (Schumm and Lichty, 1965). To some extent complexity of approach must be constrained by the availability of appropriate quantitative techniques, but there is still ample scope in geomorphology for the development or importation of new techniques and there is certainly willingness to recognize the complexity of earth surface systems.

Models of simple systems

Relationships among a few variables may be investigated either deductively or inductively. Most studies of simple geomorphic systems are inductive, using techniques of statistical analysis many of which are familiar throughout the natural and social sciences. Deductive models in geomorphology are less widespread but more satisfactory. They are derived from physical or chemical principles or assumptions, often very simple ones, by mathematical reasoning which in contrast may not be simple. This is especially so in models of transient behaviour rather than system equilibrium. We consider here first equilibrium models, then transient models, then inductive statistical analysis.

Equilibrium models in geomorphology are relationships between process and response derived by some kind of static or steady-state analysis. A good example is the analysis of the competence of a river, i.e. the biggest size of bed particle it can set into motion, by equating the various forces and resistances acting on a solitary particle on a plane bed. The equation can be solved for any one variable when the others are known. By focusing not just on critical particle size but also on lateral slope, British sedimentologists working near the ill-defined boundary of geomorphology have put this simple approach to good use in modelling the three-dimensional form and sediment size distribution of meander point bars (Allen, 1970; Bridge, 1977).

This example involves only simple algebra. In other cases the derivation of an equilibrium state requires the solution of a differential equation, as in the calculation of zones of basal freezing or melting beneath an ice sheet of known physical properties given assumptions about atmospheric and geothermal heat inputs. This particular equilibrium model was again developed outside geography but has recently been applied by Sugden (1977) to the Quaternary ice sheet covering North America, using ice thicknesses and creep velocities obtained from another equilibrium model, of plastic flow. The computed thermal zones compare well with geographical patterns of erosion and deposition (Sugden, 1978).

Many other applications of physical or chemical principles to geomorphic situations exist, but the methodology of mathematical deduction is common to all. Similarly, although each application is open to individual criticism as regards the assumptions made, certain general technical problems can be recognized. One is that simple systems do not usually remain simple when juxtaposed with other similar systems. For example, the behaviour of a solitary pebble or sand grain on a plane stream bed is unlikely to match closely the behaviour of a bed full of mutually interfering particles; in this case the whole is definitely more than the sum of the parts, and modelling should enter a more complex domain. Secondly, in the real world there are temporal and spatial variations in boundary conditions and controlling variables: thus Sugden (1977) notes the possibility that the Laurentide icesheet did not stand at its maximum extent for long enough to attain a thermal equilibrium. Dynamic behaviour requires either a difficult and mathematically more advanced style of simple modelling, examined next, or a shift to a more complex approach altogether.

Models of transient behaviour in simple systems almost always take the form of partial differential equations, i.e. relationships involving rates of change with respect to two or more variables. Such equations are generally more difficult mathematically than those describing equilibrium situations, but the difficulty varies greatly according to the precise form of the model and thus on the assumptions made. An example is the variation in concentration of a water quality determinant as streamflow varies downstream and over time. For conservative substances such as dissolved solids the partial differential equation reduces to a simple dilution or mixing model which explains much of the variation in solute concentrations during floods (Gregory and Walling, 1973). If sources and sinks exist and dispersion must be allowed for, as with dissolved oxygen, the equation is more complicated and gives a qualitatively different solution, the sag curve well known to biologists.

Much modelling activity in geomorphology and related fields has been towards the joint solution under unsteady conditions of the differential equations for mass conservation or continuity and mass transport or flow. The simplest approach, dropping the energy and momentum terms to consider only kinematic rather than dynamic behaviour, was applied to glacier fluctuations by Nye (1960) and has been used extensively in hydrology to model overland flow and the propagation of floods down rivers. In a more specifically geomorphological context Kirkby (1971) combined downslope sediment continuity with a generalized equation for sediment transport in terms of slope and contributing distance to obtain a deductive model of hillslope evolution.

The emphasis in geomorphological applications of simple dynamic or kinematic models has been on analytic solutions for ideal cases, and to a lesser extent discrete computer simulation (e.g. Armstrong, 1976, Kirkby, 1978a). An intermediate approach with great potential is the numerical solution of differential equations. Numerical methods can cope with a much wider range of equations and boundary conditions, and are widely used in related fields such as engineering hydrology (see for example the review by Liggett and Cunge, 1975), but their potential is only just being realized by geomorphologists. It seems likely that applications of finite difference methods (e.g. Butcher and Thornes, 1978) and finite element analysis (e.g. Beven, 1977) will become widespread by the mid 1980s.

The major obstacle to advance in this direction is that building a deductive model requires precise thought about relationships and assumptions, and solving it requires some mathematical expertise. Formal theory is weakly developed in geomorphology (Thornes, 1978) and many practitioners lack advanced mathematical training. The rate of adoption of this style of modelling will thus depend heavily on initiatives in individual self-tuition and postgraduate training. The mathematical demands are reduced somewhat by the use of numerical methods, now widely available in computer packages such as the NAG library, but these methods introduce fresh problems concerning the choice of time and space resolutions.

A more general problem in dynamic modelling is the fixing of parameter values. Ideally parameters should have a physical meaning and be easy to measure in the real world, but this is not always the case and other methods of estimation may have to be used. Optimization by minimizing an objective function such as the mean squared prediction error – a common procedure in hydrology – may give erratic results when the objective function is broad, especially if several parameters are involved. At the other extreme, some models are extremely sensitive to changes in individual parameters. A good example is the second-generation US soil erosion model of Foster and Meyer (1972), which produces feasible results that nevertheless have little meaning because the sensitivity of the model is heavily dependent on unknown parameter values.

Inductive models Simple geomorphological systems have also been identified or decomposed using inductive, inferential techniques. Indeed, applications of statistical analysis far outnumber attempts at deductive modelling, very probably because of the lesser requirements for formal physical theory or mathematical technique. But this is not to say that statistics has become popular just as an easy option: a disproportionate number of advances in statistical technique in British geography appear to have originated in geomorphology as indeed did the very use of statistics back in the 1950s. Moreover, several of the elementary statistical techniques which entered British geomorphology from America at that time have now fallen by the wayside because of theoretical objections or for lack of convincing results.

A case in point is the geological technique of trend surface analysis, initially hailed as a valuable addition to the geographical repertoire and applied in the 1960s to a variety of geomorphological patterns including those of soil texture (Chorley *et al.*, 1966) and corrie elevations (e.g. Robinson *et al.*, 1971) as well as topography itself (e.g. Thornes and Jones, 1969). The decline in its popularity can probably be attributed to the arbitrariness in most geomorphological contexts of imposing a particular functional form and the consequent lack of geomorphological, as opposed to statistical, significance of either the order or the coefficients of the fitted surface. One of the few exceptions to this criticism is the application to raised beaches (e.g. Andrews, 1970; Gray, 1978) in which a quadratic surface makes sense in terms of differential glacio-isostatic rebound.

A technique which has proved even less popular is the analysis of variance, one of the few applications being Kennedy and Melton's (1972) comparison of valley-side slope angles in different situations. It is perhaps fair to say that experimental design has been neglected by British (and other) geomorphologists, but in many cases the subject does not lend itself to this technique of analysis: those nominal variables that are relevant, such as rock type, often affect geomorphic processes and responses through ratio-scale properties such as infiltration capacity whose influence can be studied by other types of analysis. For the same reason the analysis of contingency tables is rare in geomorphology. Qualitative variables have however been used in discriminant analysis (e.g. Ferguson, 1973) and covariance analysis (Trenhaile, 1974), in the first case by the dummy variable technique which throws these lesser known techniques back into the more familiar ground of regression analysis.

Correlation and regression analysis has undoubtedly been the best stayer amongst the statistical techniques first imported into British geomorphology. Indeed, perhaps the best known of all quantitative applications in geomorphology is the regression exercise termed 'hydraulic geometry', which is the fitting of log-log relationships between stream discharge and dependent variables such as channel width, depth, and velocity as they vary either over time or downstream (Leopold and Maddock, 1953). British workers have used this model to infer channel changes (Park, 1977a) and attempted to relate its slope coefficients to environmental variables (Park, 1977b), and on the other hand have questioned the linearity of the relationships

(Richards, 1973; Knighton, 1979) and the neglect of spatial dependency (Thornes, 1977a). Hydraulic geometry thus continues to provide a useful frame of reference for investigating and discussing the behaviour of stream channels – and, as we shall see later, not only within the frame of reference of simple models.

Other applications of regression analysis have been of more specific value in providing either parameter estimates for the physically-based models discussed previously, or black-box forecasting ability. A typical example of the first kind is the regression of strain against stress to provide mechanical parameters for models of rock, soil, or ice failure. The predictive use of regression is illustrated by Walling and Webb's (1975) study of spatial variations in water quality in Devonshire. A close relationship between total dissolved solids and electrical conductivity allowed a large number of reconnaissance measurements of the latter to be used in compiling a regional map of water chemistry.

More sophisticated forecasting applications of correlation and regression analysis have involved the identification of leads and lags in temporal and spatial series (Edwards and Thornes, 1973; Richards, 1976a) but this is fraught with problems where complex feedback mechanisms are at work, and success depends on careful matching of observation frequency with the physical scale of the processes operating.

Finally we may note the original, exploratory use of regression analysis. There have been many applications of stepwise regression in British geomorphology as a simple means of inferring the relative and overall importance of different predictors but remarkably few examples of causal model building or testing whether by linkage analysis as pioneered in America by Melton (1958) or by Blalock-style partial correlation analysis (e.g. Ferguson, 1973).

Although regression analysis has been extensively used in British geomorphology for well over a decade it still cannot be said that it is always used most effectively or correctly. Geomorphologists are well aware that few relationships are linear in the real world, but the appropriate nonlinear functional forms are not always apparent even if they are sought for. The assumption of homoscedasticity is equally doubtful but seldom tested. Fortunately the widespread use of logarithmic transformation to normalize the marginal distributions of variables, under the false impression that this is necessary for statistically valid regression analysis, has beneficial side-effects on typical nonlinear heteroscedastic relationships. The danger of spurious correlation is frequently unheeded or misunderstood, despite the existence of a comprehensive review for hydrologists (Benson, 1965), and many of the relationships reported or discussed in the literature have a common factor on both sides, some other definitional link, or a common external control. In hydraulic geometry, for example, discharge is defined as the product of width, depth, and velocity and thus cannot fail to have a non-zero log-log correlation with at least one of them. This is but one of several reasons why correlation analysis (and the equivalent testing of $Ho : b = 0$ in regression analysis) should be, and slowly is, giving way to regression analysis with the emphasis on estimation or testing of coefficients in deductive relationships.

Two last causes for concern in current statistical practice in British geomorphology are the use of estimated statistics (such as regression coefficients) in further statistical analyses without thought for their heteroscedastic nature, and the use of time-mean variables (e.g. streamflow or water quality indices) in spatial correlation studies without regard to the often very skewed shape of the distributions they purport to summarize.

Models of complex disorder

By 'disorder' Weaver (1958) referred mainly to molecular systems whose elements interact only weakly and are extremely large in number. The techniques for modelling such systems come from statistical mechanics and probability theory. These ideas were introduced into geomorphology more or less simultaneously in Britain and America, though subsequent work has taken place mainly in the US. The rationale for treating landform development in a probabilistic way was variously held to be the general analogy with the behaviour of ideal closed and open systems (Chorley, 1962), the intrinsic randomness of particular geomorphic processes (Culling, 1963), or the aggregate randomness of complex landscape systems in which a large number of individually deterministic relationships interact (Leopold and Langbein, 1963).

Various strategies may be adopted in trying to model systems of complex disorder. Both intrinsic and aggregate randomness may be tackled explicitly by deducing or postulating some kind of probabilistic model. The behaviour of the system may then be deduced analytically or determined by Monte Carlo simulation. Such models may involve purely chance behaviour, as in a random walk, or chance fluctuation about or input into a deterministic kernel. Another way of incorporating any relevant deterministic theory is to use stochastic constraints to make up a sufficient set of relationships to allow a unique solution of an otherwise indeterminate system. Finally many geomorphological systems of complex disorder comprise time series or one- or two-dimensional spatial fields whose mix of regular and random variation can be studied inductively by techniques of statistical series analysis. We now consider these different strategies in turn, concentrating on applications by British geomorphologists. Early American work is discussed at greater length in Thornes and Brunsden (1977).

Probabilistic models are well illustrated by Culling's (1963) pioneering analysis of hillslope development by soil creep. By assuming that each soil particle moves randomly subject to the availability at different depths of void spaces, and to overall downslope drift, Culling deduced a diffusion-equation model for the expected slope configuration under variable boundary conditions.

A second application is to drainage networks, whose aggregate randomness was first postulated by the American earth scientist Shreve (1966). In this and subsequent papers Shreve and Smart demonstrated that Horton's well-known 'laws' of drainage composition can be explained by assuming that all topologically distinct ways of combining N headwater streams into one main river are equally likely to occur. British geomorphologists have shown Shreve's descriptive indices to be theoretically more informative than Horton's (Jarvis and Werritty, 1975), but on the whole have not found them particularly good discriminators of geological peculiarities (Werritty, 1972a) or hydrological response (Kirkby, 1976a). As Kirkby suggests, the combination of topological with metric properties offers greater promise here but problems remain in the definition of stream networks and their metric properties from maps (Werritty, 1972b). The probabilistic approach has also failed so far to lead to an explanation of the growth and development of stream networks, which has recently been treated instead by the deterministic analogy of allometric growth (see Faulkner, 1976, and Cox, 1977, for conflicting British views on this).

A wide range of advanced mathematical techniques have been used in deducing probabilistic models of geomorphic systems: calculus in Culling's diffusion analogy for soil creep, combinatorial algebra and random walk theory in the analysis of drainage networks, and series expansion and the moment generating function in Ferguson's (1977a) derivation of a nonlinear relationship between meander sinuosity and direction variance assuming a normal distribution of direction. As with the deductive modelling of simple deterministic systems, the mathematical demands on the geomorphologist can be reduced by the use of numerical techniques, in this case Monte Carlo simulation, instead of deductive analysis. Applications by British geomorphologists include King and McCullagh's (1971) simulation of the development of a coastal spit subject to randomly-varying winds and waves, and the use of a stochastic runoff model to drive a deterministic process model of intermittent headwater erosion (Calver, 1978; Thornes, 1977b, derives a similar, two-stage model analytically rather than by simulation).

Apart from the mathematical demands, the major drawback to deductive probabilistic modelling in geomorphology is the problem of testing. By its very nature a probabilistic model can produce a variety of responses in the same conditions, so testing must necessarily be statistical. The apparent realism of a single simulation may be indicative but should be supplemented by comparison of observed and predicted distributions, or at least means, of dependent variables. This makes heavy demands on empirical data, but one cannot escape the fact that with but a single realization we cannot tell whether the model is in error or the observation is an extreme case.

Stochastic constraints The use of some assumed probabilistic tendency to render determinate an otherwise underspecified system has been restricted to fluvial geomorphology, but has been quite popular there. River behaviour and channel morphology combine regular elements rooted in physical laws with quasi-random variability at various time and space scales, from local fluctuations in turbulent flow to rather arbitrary variations in discharge over time and surface materials over space. Some workers maintain that deterministic modelling is possible, at least in principle (Hey, 1978), but so far the use of stochastic constraints instead of or alongside physically-based relationships has proved more popular.

The constraint first suggested (Chorley, 1962) was entropy maximization, as subsequently used in human geography (Wilson, 1970), but geomorphologists have generally preferred to think in terms of minimizing variance. Two early applications by American geomorphologists were Langbein's (1964) justification of the empirical hydraulic geometry relationships (discussed above) as minimizing the variance of channel geometry, and Langbein and Leopold's (1966) random model of meander geometry.

Some British geomorphologists have criticized both the philosophy and the details of these early models (Richards, 1973; Ferguson, 1976). Others have sought alternative models of the same kind, for example Knighton's (1977) derivation of a hydraulic geometry from the perhaps equally dubious constraint that the magnitude of adjustment of the exponents decreases with time. In an analysis of modal states of channel pattern (meandering or braided) Kirkby (1977) chose to apply the criterion of maximum efficiency of sediment transport, though his analysis proceeds through classical rather than quantum methods. Similar work in America by Chang (1979) has produced particularly interesting but as yet untested results.

The general deductive approach to systems of complex disorder has been criticized on the grounds that the models build on our own ignorance. In some cases physical relationships do appear to have been wilfully ignored. But the underlying strategy of this form of modelling is soundly scientific:

progressively constraining the simplest possible configurations of the system until an acceptable pattern emerges, proceeding from the unknown to the known. At the largest and smallest scales (drainage patterns and soil creep, for example) a high degree of randomness does appear to exist. At the intermediate level of river channels stronger physical regularities are present and purely probabilistic models are less appropriate than stochastic constraints as just discussed, or stochastic disturbances as for example invoked by Ferguson (1976) to explain the irregularity of meander geometry.

Inductive stochastic models Several of the geomorphological problems which have been tackled by deductive stochastic modelling have also been considered inductively, particularly by the use of series analysis. Three kinds of model have attracted particular interest among British geomorphologists: discrete Box-Jenkins models of univariate time or, particularly, distance series, and subsequently their continuous counterparts and general multivariate transfer function models.

The application of discrete series analysis to measurements equally spaced down river channels (Ferguson, 1975; Richards, 1976b) or hillslope profiles (Thornes, 1972) had its roots in earlier work by American and Australian geomorphologists and hydrologists, but benefited also from the systematization of autoregressive-moving average model identification by Box and Jenkins (1970). The primary aim of such modelling is to achieve a parsimonious description of the autodependence in the series under study, but this may also be of use in quantifying periodicities such as meander wave length (Ferguson, 1975). Moreover, it is hoped that the nature of the best-fit model, and the values of its parameters, may lead to insight into the operation of the system itself. One problem here is the question of scale of resolution (Thornes, 1973) when an essentially continuous series is measured at discrete intervals. Ferguson (1976, 1977b) has shown that discrete analysis can be used in conjunction with a continuous model (stochastic differential equation) to bypass some of the conceptual problems, but the sampling theory of continuous-model parameter estimation is unknown. An even more fundamental conceptual difficulty relates to the assumption of one-way causation: water and sediment flow only downhill but disturbances to hillslopes or river channels can work back uphill. At a more technical level noise may be introduced by measurement errors, trend removal or other transformation may be necessary to reduce nonstationarity in the series, and parameters too may be nonstationary. So far only the second of these problems has received much attention, although Bennett (1976) has experimented with adaptive-parameter modelling of river profiles. The extension from univariate series analysis to multivariate transfer function modelling (Lai, 1979) is also in its infancy but should prove useful.

Models of complex order

Many geomorphological problems involve not only the strong relationships implicit in simple physical models but also the recognition that distinct subsystems are integrated into a wider environment. The nesting, cascading, or other interaction of subsystems means that feedback is an important orderly feature of complex systems. For example, the flow of water in a rigid channel of specified form can be modelled as a simple physical system and so can the entrainment of loose particles at the channel boundary by a specific flow, but fluvial geomorphologists must also seek to understand the two-way interaction of these processes in the formation and maintenance of a stable channel. At a qualitative level this is well known – see for example the recent book by West (1978) – but quantitative modelling is another matter.

Philosophy and techniques Some deductive geomorphological models already described here as 'simple' ones incorporate feedback (e.g. the hillslope evolution model of Kirkby, 1971), and negative feedback is implicit in the assumption of dynamic equilibrium underlying much statistical analysis and stochastic modelling (Langbein and Leopold, 1964). The distinctive features of the complex-order approach which go beyond these lower levels of modelling are a concern for the overall stability or instability of a system and a desire to model its transient behaviour as well as its equilibrium state. The emphasis is on change and susceptibility to change in realistically complex systems, whose overall structure must be considered in a sufficiently general way for structural and behavioural discontinuities to be accommodated in the analysis.

The greater flexibility of this approach, particularly in the study of long-term landform development, is stressed by Brunsden and Thornes (1979). As argued by Chorley and Kennedy (1971) and elaborated by Bennett and Chorley (1978), dynamical systems analysis can include both the dynamic aspects of some simple physical models and the equilibrium elements of statistical and stochastic models as special cases of a more general methodological framework.

Stability and thresholds The major focus so far of attempts to investigate quantitatively the complex order of geomorphological systems has been the recognition of stable and unstable areas within process domains. Empirical thresholds have been known for some time, particularly in relation to river channel patterns. Their conceptual importance in geomorphology has been publicized by Schumm (1973, 1977) and had previously been discussed more closely by Brunet (1968) whose work deserves to be better known among English-speaking geographers. Quantitative stability analysis became known to geomorphologists through work by applied mathematicians on the initiation of meanders (Callander, 1969) and of hillslope hollows and drainage channels (Smith and Bretherton, 1972).

The empirical, conceptual, and analytical approaches to system stability evidently involve different techniques, from graphical discriminant analysis to catastrophe theory and mathematical analysis of perturbations.

The first has already been considered under the heading of inductive modelling of simple systems. Catastrophe theory has been known to British geomorphologists for a decade but its qualitative application is of very doubtful value (see Alexander, 1979). A tentative quantitative application has been sketched by Thornes (1979) to describe the hysteresis of discharge-sediment relations and depletion phenomena in ephemeral channels. The onset of entrainment on the rising limb of the hydrograph of discharge per unit width, and the onset of deposition on the falling limb, induce channel instabilities which can be modelled as a cusp catastrophe.

Perturbation analysis proceeds by setting up dynamic equations that describe the equilibrium state of the system, introducing a small spatial perturbation and discovering by analysis or numerical simulation whether the perturbation dies away or is amplified – i.e. whether the system is stable or unstable. In the examples cited above Callander (1969) deduced that a straight river capable of eroding its bed and banks is unstable and will tend to meander or braid, and Smith and Bretherton analysed the conditions in which a hollow in a hillslope subject to erosion will grow into a gully or channel. This second example is obviously relevant to the explanation of drainage density, and Kirkby (1978b) in the first British application of quantitative stability analysis in geomorphology has extended Smith and Bretherton's analysis into a model for drainage density. Essentially, a hollow grows if the additional water it collects can move more than the additional sediment carried in from upslope by creep and wash. These processes in turn can be modelled by combining sediment transport and mass balance equations (Kirkby, 1971) with a simple model of slope hydrology in relation to climatic characteristics (Kirkby, 1976b). Critical slope lengths and angles for drainage initiation can thus be calculated for different climates.

A noteworthy feature of both catastrophe theory and perturbation analysis models is that they combine two timescales of change, the 'fast dynamics' of almost instantaneous equilibration between flowing water and channel or hillslope sediment and the 'slow dynamics' of changes in the controlling variables (tracks across the manifold in a geometric view of catastrophe models) and evolution of the system. The simpler modelling strategies discussed in earlier sections of our review are generally unable to combine these short-term and long-term viewpoints. Since unstable landforms tend to be short-lived they are statistically less important but genetically highly significant and the quantification of stability analysis thus opens up important avenues of research for the 1980s.

Concluding remarks

Our classification of models and review of published applications suggests that quantitative geomorphologists in Britain have relied rather too heavily on statistical description and analysis of phenomena for which the basic mechanics, at least, are already established. In other words there has been too much naïve inductive or probabilistic modelling of systems exhibiting order which could be modelled deterministically. We count amongst this activity some of our own work. Only one British geographer-geomorphologist (M.J.Kirkby) has achieved a quantitative understanding of the relationships between form and process to match that attained by British non-geographers in closely cognate areas of earth science such as fluvial sedimentology (for example J.R.L. Allen, 1977) and glacial geology (for example G.S. Boulton, 1974, 1975) to name but two.

This situation indicates a gap between our conceptual insight into landscape systems – and here the British record is good – and our ability, or willingness, to set up and analyse quantitative models. The unwillingness is a matter of the geographer's excessive regard for real-world complications and reluctance to generalize, most clearly articulated by Wooldridge (1958) and Pitty (1971); the inability is a reflection on the continuing lack of attention to mathematical training of physical geographers. In geomorphology at least there is less need for advanced statistical skills or computer programming ability than for basic mathematical tools, notably the calculus of several variables and principles of numerical analysis. There is however a brighter side to the picture: geomorphologists have by and large been able to focus critically on problems rather than techniques for solving them. In this respect we do not wholly agree with Alexander's (1979) criticism of 'the geographer's insatiable appetite for new methods of analysis'. The search for novelty will not restrict the advance of quantitative geomorphology; failure to come to grips with already well-established techniques in appropriate modelling frameworks may do.

References

Agterberg, F.P. (1974) *Geomathematics*, Elsevier: Amsterdam.

Alexander, D. (1979) 'Catastrophic misconceptions', *Area*, 11, 228-30.

Allen, J.R.L. (1970) 'A quantitative model of grain size and sedimentary structures in lateral deposits', *Geological Journal*, 7, 129-46.

Allen, J.R.L. (1977) 'Changeable rivers: some aspects of their mechanics and sedimentation' in K.J. Gregory, op. cit., 15-45.

Andrews, J.T. (1970) 'A geomorphological study of post-glacial uplift with particular reference to arctic Canada', *Institute of British Geographers*, Special Publication 2.

Armstrong, A. (1976) 'A three-dimensional simulation of slope forms', *Zeitschrift für Geomorphologie*, Supplement, 25, 20-8.

Bennett, R.J. (1976) 'Adaptive adjustment of channel geometry', *Earth Surface Processes*, 1, 121-50.

Bennett, R.J., and Chorley, R.J. (1978) *Environmental Systems: Philosophy, Analysis and Control*, Methuen: London.

Benson, M.A. (1965) 'Spurious correlations in hydraulics and hydrology', *Journal of the Hydraulics Division, American Society of Civil Engineers*, 91, HY4, 35-42.

Beven, K. (1977) 'Hillslope hydrographs by the finite element method', *Earth Surface Processes*, 2, 13-28.

Boulton, G.S. (1974) 'Processes and patterns of glacial erosion' in D.R. Coates (ed.) *Glacial Geomorphology*, 41-87, State University of New York: Binghamton.

Boulton, G.S. (1975) 'Processes and patterns of subglacial sedimentation: a theoretical approach' in A.E. Wright and F. Moseley (eds), *Ice Ages: ancient and modern*, 7-42, Seel House: Liverpool.

Box, G.E.P., and Jenkins, G.M. (1970) *Time Series Analysis – Forecasting and Control*, Holden-Day: San Francisco.

Bridge, J.S. (1977) 'Flow, bed topography, grain size and sedimentary structure in open channel bends: a three dimensional model', *Earth Surface Processes*, 2, 401-16.

Brunet, R. (1968) 'Les phénomènes de discontinuité en géographie', *Mem. Doc. Centre de Docum. Cartog. du CNRS*, NS 7.

Brunsden, D. and Thornes, J.B. (1979) 'Landscape sensitivity and change', *Transactions of the Institute of British Geographers*, NS 4, 463-84.

Butcher, G.C., and Thornes, J.B. (1978) 'Flow depth monitoring in an ephemeral channel and its relationship to channel changes', London School of Economics, Department of Geography, Discussion Paper No. 71.

Callander, R.A. (1969) 'Instability and river channels', *Journal of Fluid Mechanics*, 36, 465-80.

Calver, A. (1978) 'Modelling drainage headwater development', *Earth Surface Processes*, 3, 233-42.

Chang, H.H. (1979) 'Minimum stream power and river channel patterns', *Journal of Hydrology*, 41, 303-28.

Chorley, R.J. (1962) 'Geomorphology and General Systems Theory', *US Geological Survey, Professional Paper*, 500-B.

Chorley, R.J. (1966) 'The application of statistical methods to geomorphology' in G.H. Dury (ed.), *Essays in Geomorphology*, 275-387, Heinemann: London.

Chorley, R.J. (1972) *Spatial Analysis in Geomorphology*, Methuen: London.

Chorley, R.J., and Kennedy, B.A. (1971) *Physical Geography: A Systems Approach*, Prentice-Hall: London.

Chorley, R.J., Stoddart, D.R., Haggett, P., and Slaymaker, H.O. (1966) 'Regional and local components in the areal distribution of surface sand facies in the Breckland, Eastern England', *Journal of Sedimentary Petrology*, 36, 209-20.

Cox, N. (1977) 'Allometric change of landforms: discussion and reply', *Geological Society of America Bulletin*, 88, 1199-202.

Culling, W.E.H. (1963) 'Soil creep and the development of hillside slopes', *Journal of Geology*, 71, 127-61.

Edwards, A.M.C., and Thornes, J.B. (1973) 'Annual cycle in river water quality: a time series approach', *Water Resources Research*, 9, 1286-95.

Faulkner, H. (1976) 'An allometric growth model for competitive gullies', *Zeitschrift für Geomorphologie*, Supplement, 20.

Ferguson, R.I. (1973) 'Channel pattern and sediment type', *Area*, 5, 38-41.

Ferguson, R.I. (1975) 'Meander irregularity and wavelength estimation', *Journal of Hydrology*, 26, 315-33.

Ferguson, R.I. (1976) 'Disturbed periodic model for river meanders', *Earth Surface Processes*, 1, 337-47.

Ferguson, R.I. (1977a) 'Meander sinuosity and direction variance', *Geological Society of America Bulletin*, 88, 212-14.

Ferguson, R.I. (1977b) 'Meander migration: equilibrium and change' in K.J. Gregory, op. cit., 235-48.

Foster, G.R., and Meyer, L.D. (1972) 'A closed-form soil erosion equation for upland areas' in H.W. Shen (ed.), *Sedimentation: Symposium to Honour H.A. Einstein*, 99-102, H.W. Shen: Fort Collins, Colorado.

Gray, J.M. (1978) 'Low-level short platforms in the south-west Scottish Highlands: altitude, age and

correlation', *Transactions of the Institute of British Geographers,* NS 3, 151-64.
Gregory, K.J. (ed.) (1977) *River Channel Changes,* John Wiley: Chichester.
Gregory, K.J., and Walling, D.E. (1973) *Drainage Basin Form and Process,* Arnold: London.
Hey, R.D. (1978) 'Determinate hydraulic geometry of river channels', *Journal of the Hydraulics Division, American Society of Civil Engineers,* 104, 869-85.
Jarvis, R.S., and Werritty, A. (1975) 'Some comments on testing random topology stream network models', *Water Resources Research,* 11, 309-18.
Kennedy, B.A., and Melton, M.A. (1972) 'Valley asymmetry and slope forms of a permafrost area in the Northwest territories, Canada', *Transactions of the Institute of British Geographers, Special Publication* 4, 107-21.
King, C.A.M., and McCullagh, M.J. (1971) 'A simulation model of a complex recurved spit', *Journal of Geology,* 79, 22-37.
Kirkby, M.J. (1971) 'Hillslope process–response models based on the continuity equation', in *Transactions of the Institute of British Geographers,* Special Publication 3, 15–30.
Kirkby, M.J. (1976a) 'Tests of the random network model and its application to basin hydrology', *Earth Surface Processes,* 1, 197–212.
Kirkby, M.J. (1976b) 'Hydrological slope models: the influence of climate' in E. Derbyshire (ed.), *Geomorphology and climate,* 247–67, Wiley: Chichester.
Kirkby, M.J. (1977) 'Maximum sediment efficiency as a criterion for alluvial channels' in K.J. Gregory, op. cit., 429–42.
Kirkby, M.J. (1978a) 'Soil development models as a component of slope models', *Earth Surface Processes,* 2, 203–30.
Kirkby, M.J. (1978b) 'The stream head as a significant geomorphic threshold', University of Leeds, Working Paper 216.
Knighton, A.D. (1977) 'Alternative derivation of the minimum-variance hypothesis', *Geological Society of America Bulletin,* 88, 364–6.
Knighton, A.D. (1979) 'Comments on log-quadratic relationships in hydraulic geometry', *Earth Surface Processes,* 4, 205–9.
Lai, P.W. (1979) *Transfer function modelling,* Concepts and Techniques in Modern Geography No. 22, Geo Abstracts: Norwich.
Langbein, W.B. (1964) 'Geometry of river channels', *Journal of the Hydraulics Division, American Society of Civil Engineers,* 90, 301–13.
Langbein, W.B., and Leopold, L.B. (1964) 'Quasi-equilibrium states in channel morphology', *American Journal of Science,* 262, 782–94.
Langbein, W.B., and Leopold, L.B. (1966) 'River meanders – theory of minimum variance', *US Geological Survey Professional Paper* 422–H.
Leopold, L.B., and Langbein, W.B. (1963) 'Association and indeterminacy in geomorphology' in C.C. Albritton (ed.), *The fabric of geology,* 184-92.
Leopold, L.B., and Maddock, J.T. (1953) 'The hydraulic geometry of stream channels and some physiographic implications', *US Geological Survey Professional Paper* 252.
Liggett, J.A., and Cunge, J.A. (1975) 'Numerical methods of solution of the unsteady flow equations' in K. Mahmood and V. Yevjevich (eds), *Unsteady Flow in Open Channels,* Vol. 1, 89-182, Water Resources Publications: Fort Collins, Colorado.
Melton, M.A. (1958) 'Correlation structure of morphometric properties of drainage systems and their controlling agents', *Journal of Geology,* 66, 442–60.
Nye, J.F. (1960) 'The response of glaciers and ice sheets to seasonal and climatic changes', *Proceedings of the Royal Society Lond.* A256, 559–84.
Park, C.C. (1977a) 'Man induced changes in channel capacity' in K.J. Gregory, op. cit., 121–44.
Park, C.C. (1977b) 'World-wide variations in hydraulic geometry exponents of stream channels: an analysis and some observations', *Journal of Hydrology,* 33, 133–46.
Pitty, A.F. (1971) *Introduction to geomorphology,* Methuen: London.
Richards, K.S. (1973) 'Hydraulic geometry and channel roughness – a nonlinear system', *American Journal of Science,* 273–96.
Richards, K.S. (1976a) 'Complex width-discharge relations in natural river sections', *Geological Society of America. Bulletin,* 87, 199–206.
Richards, K.S. (1976b) 'The morphology of riffle-pool sequences', *Earth Surface Processes,* 1, 77–88.
Robinson, G., Petersen, J.A., and Anderson, P.M. (1971) 'Trend surface analysis of corrie altitudes in Scotland', *Scottish Geographical Magazine,* 87, 142–6.
Schumm, S.A. (1973) 'Geomorphic thresholds and complex response of drainage systems' in M. Morisawa (ed.), *Fluvial Geomorphology,* Proceedings 4th Ann. Geomorphology Symposium, 299–309, Binghamton, New York.
Schumm, S.A. (1977) *The Fluvial System,* Wiley: New York.
Schumm, S.A., and Lichty, R.W. (1965) 'Time, space, and causality in geomorphology', *American Journal of Science,* 263, 110-19.
Shreve, R.L. (1966) 'Statistical law of stream numbers', *Journal of Geology,* 74, 17-37.
Smith, T.R., and Bretherton, F.P. (1972) 'Stability and the conservation of mass in drainage basin evolution', *Water Resources Research,* 8, 506-24.
Sugden, D.E. (1977) 'Reconstruction of the morphology, dynamics and thermal character-

istics of the Laurentide ice sheet at its maximum', *Arctic and Alpine Research*, 9, 21–47.

Sugden, D.E. (1978) 'Glacial erosion by the Laurentide ice sheet', *Journal of Glaciology*, 20, 367–91.

Thornes, J.B. (1972) 'Debris slopes as series', *Arctic and Alpine Research*, 4, 337–42.

Thornes, J.B. (1973) 'Markov chains and slope series: the scale problem', *Geographical Analysis*, 5, 322–8.

Thornes, J.B. (1977a) 'Hydraulic geometry and channel change' in K.J. Gregory, op. cit., 91–100.

Thornes, J.B. (1977b) 'Channel changes in ephemeral streams: observations, problems and models' in K.J. Gregory, op. cit., 317–32.

Thornes, J.B. (1978) 'The character and problems of theory in contemporary geomorphology' in C. Embleton, D. Brunsden and D.K.C. Jones (eds), *Geomorphology: Present Problems and Future Prospects*, 14–24, Oxford University Press.

Thornes, J.B. (1979) 'Sediment transport in ephemeral channels and its wider implications', London School of Economics, School of Geography, Discussion Paper 72.

Thornes, J.B., and Brunsden, D. (1977) *Geomorphology and Time*, Methuen: London.

Thornes, J.B., and Jones, D.K.C. (1969) 'Regional and local components in the physiography of the Sussex Weald', *Area*, 1, 13–21.

Trenhaile, A.S. (1974) 'The geometry of shore platforms in England and Wales', *Transactions of the Institute of British Geographers*, 62, 129–42.

Walling, D.E., and Webb, B.W. (1975) 'Spatial variations of river water quality: a survey of the River Exe', *Transactions of the Institute of British Geographers*, 65, 155–71.

Weaver, W. (1958) 'A quarter century in the natural sciences', Annual Report, Rockefeller Foundation, New York, 7–122.

Werritty, A. (1972a) 'The topology of stream networks' in R.J. Chorley (ed.), *Spatial Analysis in Geomorphology*, 167–96, Methuen: London.

Werritty, A. (1972b) 'Accuracy of stream link lengths derived from maps', *Water Resources Research*, 8, 1255–64.

West, E. (1978) *The Equilibrium of Natural Streams*, Geo Abstracts: Norwich.

Wilson, A.G. (1970) *Entropy in Urban and Regional Modelling*, Pion: London.

Wilson, A.G. (1979) 'From comparative statics to dynamics in urban systems theory', *Geoforum*, 10, 283-96.

Wooldridge, S.W. (1958) 'The trend of geomorphology', *Transactions of the Institute of British Geographers*, 25, 29-35.

Chapter 28

Vegetation studies

J.A. Matthews

Introduction

Vegetation is composed of an assortment of plant species, growing together in assemblages or communities, interacting with each other and with their environment, changing or maintaining particular combinations through time, and forming spatial patterns in the landscape. The value of a quantitative approach to understanding such a complex multivariate system has been widely acknowledged by British biogeographers (e.g. Pears, 1968; Harrison, 1971; Shimwell, 1971; Matthews, 1975; Goldsmith and Harrison, 1976; Randall, 1978), although many do not employ quantitative techniques routinely and almost all new technical developments have been initiated by others, especially by plant ecologists.

The last fifteen years have seen trials of a wide range of quantitative techniques in the context of vegetation. As applications have progressed beyond the exploratory stage the potential for non-trivial substantive contributions has increased, as has the influence of techniques of vegetation analysis on other subjects. However, a number of techniques formerly believed to be ideal for the analysis of vegetation are now known to be inappropriate, or at best suboptimal, and many technical problems remain. The quantitative approach to vegetation study has, therefore, yet to reach full bloom.

In this review, most emphasis is given to the technical problems outstanding from the user's viewpoint. Comprehensive coverage is attempted only in relation to descriptive and statistical analysis at the level of the plant community; this reflects the interests of the present author, rather than the inherent importance of other levels of integration (e.g. whole ecosystems, species populations and individual plants) and other methodologies (e.g. ecosystem simulation, theoretical mathematical models and laboratory or field experimentation).

Essential preliminaries

Data collection: description and sampling

Vegetation data normally consist of a species x site matrix, a decision based firmly on the genetic uniqueness of the species, which finds expression in its morphology and ecology. This state of affairs is likely to continue, although there is some interest in the use of physiognomic data (Webb *et al.*, 1967, 1970, 1976; Werger, 1978) and higher taxonomic categories (van der Maarel, 1972; Dale and Clifford, 1976), which may find wider application in extensive (large areas) surveys and in complex kinds of vegetation. At each site, the quadrat is most widely used, but transects and 'plotless' methods (involving point-to-plant or plant-to-plant distances) are alternatives.

Although it has been long recognized that use of different attributes – such as presence, density, cover, frequency and biomass – can give rise to different results during subsequent analysis (Greig-Smith, 1964; Orlóci, 1975), only recently have attempts been made to examine systematically the nature and magnitude of such effects. A number of tests were made by Smartt *et al.* (1974; 1976) on thirteen different data types derived from the same area of vegetation. Great differences were found between the various data types and their behaviour during analysis, particularly between presence–absence ('qualitative') data and the interval-scale ('quantitative') measures. Some support was found for the suggestion (Lambert and Dale, 1964; Austin and Greig-Smith, 1968; Norris and Barkham, 1970) that quantitative data yield little additional useful information to presence–absence data, particularly in the detection of ecological groupings that reflect major divisions in the habitat. In view of the large number of zero entries in most vegetational data matrices, presence–absence data would be expected to contain significant information. Nevertheless, quantitative data types were found more useful at a low level of the classificatory hierarchy where finer vegetational detail was involved, a conclusion in agreement with the finding of Noy-Meir (1971) who also found quantitative data to be important at large quadrat sizes. Clearly there is scope for further investigations of the basic types of data, perhaps involving simulated data sets incorporating strictly controlled characteristics.

Four well-known probabilistic sampling designs – random, systematic, stratified random and stratified systematic unaligned – have been re-examined in relation to real and simulated vegetation maps (Smartt and Grainger, 1974). Particular attention was given to the effects of size and degree of fragmentation of sampled patches, thus providing some of the information necessary in order to choose an appropriate design in relation to the scale of vegetation pattern. The study was followed by field tests of a new flexible systematic technique involving the distribution of available sample sites in such a way that areas of greatest vegetational heterogeneity are sampled most intensively (Smartt, 1978). A proportion of the total number of sites is used to estimate the heterogeneity; the remaining sites are then interpolated in proportion to this. According to Smartt, superior classifications and vegetation maps result. It would appear, however, that there is not one optimal flexible design for all purposes and that this kind of design is most appropriate if communities that occur only locally are regarded as worthy of notice. It has, therefore, implications for the apparent necessity for judgment sampling in large-area vegetation surveys.

Indeed, all the recent studies on data collection that have been mentioned, together with some further observations on the effects of quadrat size (Noy-Meir *et al.*, 1970; Matthews, 1979a, b), indicate the need for an integrated approach to questions of data type, sampling design and quadrat size, with due regard for the nature of the vegetation sampled (particularly the commonness and rarity of species and the scale of vegetation pattern on the ground) and for the purpose of study.

Resemblance functions and transformations

Goodall (1973a) reviewed vegetational applications of some seventy measures of species or site resemblance, and made some recommendations for general use. He was forced to conclude, however, that although the purpose of study and type of data had limited the range of choice in some studies, the selection of measures appeared to be a matter of taste. Recent theoretical and empirical work has clarified the transformations that are implicit in the most commonly used resemblance functions and has revealed relationships between them. Thus the range of choice has been narrowed somewhat and, more importantly, the appropriateness of particular functions for particular purposes has been elucidated (Williams and Dale, 1965; Gower, 1966; Orlóci, 1967a, 1972; Williams, 1971; Noy-Meir, 1973a; Noy-Meir *et al.*, 1975; Noy-Meir and Whittaker, 1977; van der Maarel, 1979).

The two most common kinds of transformation involve centering (Dagnelie, 1960) and standardization (Austin and Greig-Smith, 1968). Centering is the subtraction of the mean for the species or site; standardizations are divisions by totals, norms, maxima or measures of dispersion for species or sites or both. For example, conventional **R**-mode principal component analysis involves centering by species mean without standardization (between-species covariance) or centering by species with standardization by species standard deviation (between-species correlation). Basically, different standardizations result in *a priori* weighting of variables. For example, standardization by species gives equal weight to all species, irrespective of their abundances, which means that emphasis is given to species that are relatively uncommon in the raw data. According to Noy-Meir *et al.* (1975) resemblance functions incorporating this kind of transformation are suitable for 'species-oriented' studies, such as those emphasizing the flora, species adaptation, conservation and evolution; standardization by site is suitable for 'site-oriented' studies, such as those with a strong interest in the habitat or indicators of environmental information; no standardization gives most weight to the most abundant species (highest variance in centered analyses) and is appropriate for resource surveys and studies in production ecology.

It is, therefore, no surprise to find that many 'unsatisfactory' results reported in the literature can be attributed to the use of inappropriate resemblance functions (Noy-Meir and Whittaker, 1977). Furthermore, the beginnings of a *rapprochement* between the methods of quantitative vegetation science, as practised in Britain and America, and the various continental schools of plant sociology, owes much to a better understanding of the resemblance functions preferred (cf. Noy-Meir and Whittaker, 1977; van der Maarel, 1979).

Classification and ordination of communities

The simplification or structuring (Lambert and Dale, 1964) of data with respect to the resemblances between species or sites provides at least the initial aim of most vegetation studies and may provide also the basis for inference, experiment or management. Two contrasting approaches to structuring are commonly recognized in the literature on vegetation – classification and ordination – and tend to be associated with particular philosophies on the nature of vegetational variability in the abstract

(Goodall, 1954b, 1963, Ponyatovskaya, 1961; Whittaker, 1963; McIntosh, 1967; D.J. Anderson, 1971).

Classification provides a simplification of an original n x s data matrix by reducing n entities to $n^{*}<n$ classes, whereas ordination reduces the dimensionality of the variable space from s to $s^{*}<s$ variables (Cormack, 1971). The distinction is very clearly represented in the general geometric model proposed by Goodall (1963), which has been represented as a three-dimensional diagram by Pears (1968). In the model, sites are normally considered to be entities (points in a species space) and species are the variables (axes defining the space). Degree of resemblance between sites is represented by distance in species space. Thus classification can be viewed as arranging sites into a small number of plant assemblage-types or community-types (Whittaker, 1963, 1973a) while ordination arranges them according to a small number of assemblage-gradients or community-gradients. In practice there are, of course, many more than three dimensions: there should be a large, representative sample of sites, and it is equally possible to classify the species into categories (species groups) or to ordinate the species (inverse analyses). Some methods, such as nodal analysis (Lambert and Williams, 1962) and inosculate analysis (Dale and Anderson, 1973) involve combined normal and inverse analyses.

Approaches via classification

The first formal numerical classificatory technique to be used widely in vegetation study was association analysis, originated by Goodall (1953), modified by Williams and Lambert (1959, 1960) and described in many textbooks (Kershaw, 1964; Greig-Smith, 1964; Shimwell, 1971; Goldsmith and Harrison, 1976). This technique produces groups by a hierarchical, divisive, monothetic sorting process. Certain of its weaknesses have been overcome in divisive information analysis (Lance and Williams, 1968; Field, 1969; Bottomley, 1971), a technique that has been applied to many different kinds of vegetation (Moore, 1971; Moore and Beckett, 1971; Holland, 1971; Austin, 1972; Matthews, 1979a).

Polythetic techniques (involving group definition by the simultaneous consideration of more than one, preferably all, species) are generally superior to monothetic techniques, although they may not be so readily interpreted. They were developed later and are, at present, mostly agglomerative on account of the computational difficulties in obtaining 'global' optimum solutions by division (Pielou, 1969; Goodall, 1973b). Polythetic agglomerative techniques that have been used with some success in vegetation studies include those based on single-linkage clustering (Jardine and Sibson, 1971; Jancey, 1974), group-average clustering (Williams *et al.*, 1966; Orlóci, 1975) and minimum-variance clustering (Ward, 1963; Orlóci, 1967b; Crawford *et al.*, 1969). A number of divisive polythetic techniques, incorporating 'short cuts' to avoid an unfeasible amount of computation, have been proposed (Lambert, 1972; Dale and Anderson, 1973; Noy-Meir, 1973b; Hill *et al.*, 1975), and these may prove preferable when fully developed and tested. However, the irrevocability of steps, a drawback of all hierarchical procedures, remains a feature of rigid divisive sorting in which an inappropriate early partition fails to be corrected later. An additional limitation of most existing techniques is their tendency to seek spherical clusters in species space, which may split natural clusters that are elongate.

The problem of suboptimality has led to the development of reallocation procedures which, in principle, can be used to check for misclassified individuals at any stage (Goodall, 1966a, 1968; Greig-Smith *et al.*, 1967; Crawford and Wishart, 1968; Moral, 1975). For example, Matthews (1979a) classified a large number of sites by minimum-variance agglomeration and carried out an iterative reallocation at four levels of the classification. An alternative to a hierarchical sorting process is reticulate classification (seeking groups at a given level of complexity directly) which recognizes the artificiality of imposing a rigid hierarchy on reality (Webb, 1954; Kershaw, 1964). Plexus techniques (McIntosh, 1973) and other ordinations have been used in an attempt to produce reticulate classifications, but formal numerical classifications based on this principle are in the early stages of development (Williams, 1971; Cormack, 1971). A related limitation of existing techniques is that nearly all seek non-overlapping (mutually exclusive) groups, whereas overlapping clusters may be more appropriate, particularly in view of the importance of 'transitional' communities in vegetation (Yarranton *et al.*, 1972; van der Maarel, 1979).

Given the present state of knowledge, definitive answers cannot be given to two important questions that must be posed by all intending users of numerical classifications. The first question is what technique(s) to use? Many users of particular techniques have found them satisfactory in so far as they have produced interpretable results. However, there have been too few comparisons of different techniques on the same data set or under other controlled conditions (e.g. Lambert and Williams, 1966; Williams *et al.*, 1966, Crawford *et al.*, 1969; Moore *et al.*, 1970; Pritchard and Anderson, 1971; Dale and Anderson, 1972; Lambert, 1972; Frenkel and Harrison, 1974) and the visual and descriptive tests that have been performed have failed to yield unequivocal results. The short-term answer appears to be twofold: use techniques with relatively well-known theoretical properties and proven usefulness in empirical studies (Sibson, 1971; Everitt, 1974); and do not rely too heavily on one technique.

Moreover, it should be possible to define group 'cores' that are consistently recognized in a range of classifications and, by exploiting differences between classifications, to shed light on different aspects of complex real-world vegetation.

The second question is how many groups to extract? Almost all sorting procedures involve an arbitrary stopping rule, which often seems satisfactory because interpretable groups are produced. Goodall (1966b, 1973b) has argued that the decision ought to be based on probabilistic considerations, but few attempts have been made to test the statistical significance of differences between groups. Of relevance here is the belief that, in most areas of vegetation, community-types are not 'natural groups' (Gilmour and Walters, 1963; Whittaker, 1963, 1973b). If this is so, then the community-types defined in any classification of vegetation can only be a convenience and one should not expect very marked differences between them. Again, in the short term, until reliable techniques for defining the *number* of groups are developed, it would be sensible to carry out analyses at more than one level of classification (unless there is a compelling reason for preferring a particular number of groups). Such an approach has an additional advantage in that different scales of vegetational pattern may be manifest at different levels of resolution (Matthews, 1979a).

Approaches involving ordination

The 1970s have seen an upsurge of interest in numerical ordination techniques as the limitations and difficulties of classification of plant communities have become more widely appreciated (Orlóci, 1973; Noy-Meir and Whittaker, 1977; van der Maarel, 1979). Ordination is an *a priori* alternative to classification and is particularly useful when: (i) knowledge is lacking as to whether or not distinct groups actually exist in a data set; or (ii) it is known that they do not exist. Ordination may also be combined profitably with classification when (iii) it is known that distinct groups are present but there is uncertainty as to how many groups there are; or (iv) intra- and/or inter-group differences are of interest.

Polar ordination and formal linear techniques Although ordination of plant communities has earlier origins, largely graphical and not wholly objective plexus diagrams (de Vries, 1953; McIntosh, 1973) and Wisconsin ordinations, particularly polar ordination (Curtis and McIntosh, 1951; Bray and Curtis, 1957; Cottam *et al.*, 1973), represent the first widely-used methods. The first formal technique of multivariate analysis to be applied for this purpose was principal component analysis (Goodall, 1954a). For some time, a kind of contest appeared in the literature between polar ordination and principal component analysis (Noy-Meir and Whittaker, 1977). Lack of generality in the former, the assumption of linearity in the latter, relative distortion and interpretability were major bones of contention (Orlóci, 1966; Austin and Orlóci, 1966; A.J.B. Anderson, 1971; Gauch and Whittaker, 1972; Beals, 1973). Despite some difficulties, particularly concerning application to and interpretation of nonlinear structures in species space (Noy-Meir and Austin, 1970; Austin and Noy-Meir, 1971; van Groenewoud, 1976; Nichols, 1977), careful use of different variants of principal component analysis continues to yield useful insights into the structure of vegetation (van Groenewoud, 1965; Yarranton, 1967a; Austin, 1968; Goldsmith, 1973a; Noy-Meir, 1973a; Bouxin, 1976; Feoli, 1977; Matthews, 1979b).

Related linear ordination techniques, such as factor analysis (Dagnelie, 1960, 1973; Fresco, 1969) and principal coordinate analysis (Flenley, 1969; Birks, 1976; Adam, 1978) have been less used in vegetation study. Recently, reciprocal averaging (correspondence analysis), essentially a non-centred eigenvector analysis standardized by site and species totals, has been advocated by Hill (1973a, 1974). This technique appears, in the light of tests on simulated data sets, to distort nonlinear gradients to a lesser extent than principal component analysis (Gauch *et al.*, 1977; Fasham, 1977). However, it is susceptible to small numbers of sites with unusual species combinations (Hatheway, 1971; Hill *et al.*, 1975; Bouxin, 1976).

Nonlinear techniques Two categories of nonlinear ordination, catenation (Noy-Meir, 1974a) or seriation (Dale, 1975) have been explored in relation to vegetational data: curvilinear ordinations make less stringent assumptions about the nature of the data under analysis and are generally applicable to a wide variety of data, including vegetational data; Gaussian ordinations are specifically designed to be applied to data with the particular attributes of vegetational data.

So far, nonmetric multidimensional scaling techniques have been the most used curvilinear ordinations (A.J.B. Anderson, 1971; Birks and Deacon, 1973; Birks and Walters, 1973; Fasham, 1977; Prentice, 1977; Matthews, 1978a, 1978b). These techniques aim to represent a set of points in a space of reduced number of dimensions, while preserving the rank-order of the true resemblances between the points (e.g. Kruskal, 1964; Guttman, 1968). From an arbitrary configuration, an iterative procedure minimizes a measure of badness-of-fit ('stress') between the 'perfect' multidimensional resemblances and the configuration in the space of reduced dimensionality. Comparative tests on simulated data sets suggest that the ordinations produced contain less distortion than is the case with either principal component analysis or reciprocal averaging (Fasham, 1977). Prentice (1977) has emphasized the appropriateness of these techniques for making ecological inferences, if used in combination with a suitable resemblance function, while Matthews (1978a) has shown how they may be used to objectivize the construction of a species plexus. Limitations include: the possibility of local minimum solutions; the necessity to specify the number of dimensions (analogous to the stopping rule problem in classification); and the likelihood that in some complex real-world vegetation contexts the monotonicity assumption may not be flexible enough.

Continuity analysis (parametric mapping) is characterized by even greater flexibility (Shepard and Carroll, 1966; Noy-Meir, 1974a, b). Whereas non-metric multidimensional scaling can be viewed as a technique for fitting rising or falling (monotonic) curves or surfaces in a multidimensional species space, continuity analysis is capable of fitting more complex curves or surfaces (e.g. 'spirals' or 'corkscrews') constrained only by the assumption of 'smoothness' or 'continuity'. A number of trials have been made on real and simulated vegetational data by Noy-Meir (1974a, b), who concluded that the technique usually succeeded in accurately recovering simulated gradients which were strongly curved in Euclidean species space. Difficulties (Noy-Meir and Whittaker, 1977) include: computational problems with large matrices (also a problem with nonmetric multidimensional scaling); failure to converge to a plausible solution in some cases (particularly when strong discontinuity and/or random variability are present); some distortion at the extremes of gradients; and the arbitrary choice of parameters in the model (which influence the complexity of the curves or surfaces fitted).

Gaussian ordination Gaussian ordinations are based on a mathematical model of the way in which species respond to environmental gradients. The Gaussian equation (Gauch and Chase, 1974) assumes a symmetric, unimodal bell-shaped response curve, a form found for many species by direct gradient analysis and by ecological experiment (Whittaker, 1967; van Groenewoud, 1976). Gaussian ordination attempts to recover multivariate vegetational gradients which reflect the behaviour of large numbers of species on underlying environmental gradients. Using an iterative procedure from a random initial configuration, Gauch *et al.* (1974) recovered the true gradient from simulated single-gradient data. They also performed successful tests on real vegetational data with a predominant gradient, starting from a 'reasonable initial guess'. Ihm and van Groenewoud (1975) have also claimed good recovery of simulated gradients for a related method.

These preliminary results are promising. Nevertheless, a note of caution has been introduced by Austin (1976a, b) who has questioned the generality of Gaussian species-response curves. He has also shown that neither Gaussian ordination, nor continuity analysis, nonmetric multidimensional scaling or reciprocal averaging can cope with bimodal or strongly asymmetrical response curves. The lack of a link between curvilinear ordination and vegetational theory on the one hand, and the oversimplified theory underlying Gaussian ordination on the other, suggest why it is not yet possible to point to a 'best' ordination technique.

Classification and ordination in series

Combined analyses, involving classification and ordination, enable the retention of some of those aspects of vegetational complexity that are lost in the assumptions of one or other approach taken in isolation. Moreover, the benefits appear to outweigh the possible disadvantages from compounding of errors. Ordination as a preliminary to or as a way of improving classifications has been mentioned above (e.g. Hill *et al.*, 1975; Moral, 1975; Matthews, 1979b). Greig-Smith *et al.* (1967), Crawford and Wishart (1968), Kershaw (1968), Flenley (1969), Birks (1976), Adam (1978) and Matthews (1979b, d) provide examples of various techniques used for the ordination of previously classified sites or group centroids. In these ways, relationships within and between types can be examined.

Multiple discriminant analysis, although a linear technique, appears to be useful for the ordination of community-types and the analysis of classifications generally. The technique seeks axes in species space which differentiate as clearly as possible between previously defined types (Norris and Barkham, 1970; Grigal and Goldstein, 1971; Matthews, 1979c). Furthermore it can be used to represent within-type variability, either graphically or by a probabilistic measure of group membership (Matthews, 1979c).

Environmental relationships

Descriptive quantitative studies of the environmental relationships of vegetation have been approached from two general directions: either via a preliminary vegetational structuring (cf. 'vegetational stratification' of Kellman, 1974, and 'indirect gradient analysis' of Whittaker, 1973a); or via a direct environmental analysis (cf. 'environmental stratification' of Kellman, and 'direct gradient analysis' of Whittaker). In practice there has been rather slow progress, due to the difficulties of obtaining representative measurements of meaningful environmental factors (Billings, 1952, 1974) and, of more concern here, to the limitations of the quantitative techniques available.

Explanation via vegetational structuring

Many of the papers cited in the sections on classification and ordination aimed to explain community-types or community-gradients in terms of environmental factors or factor complexes. Least progress has been made in the context of community-types (Greig-Smith, 1971a). Grunow (1967) showed that it may be possible to relate the major categories of a hierarchical classification of communities to environmental categories, but such a complete matching is rarely possible (see, for example, the results of Grigal and Arneman, 1970). In addition, little attention has been paid to either the quantitative evaluation of environmental differences between community-types, or to the establishment of whether or not a significant difference in environment exists between them (by use, for example, of analysis of variance).

Although more progress has been made with community-gradients, most studies rely on graphical representations of two- or three-dimensional vegetational spaces in which habitat types or environmental variables may be plotted. For example, Goldsmith (1973b) and others have plotted values of environmental variables in spaces defined by the first two principal components. If an environmental factor shows a pattern in the space, then some kind of relationship between the factor and the community-gradients is indicated. Formal quantitative techniques that have been used to establish more precisely the nature of such relationships may be grouped into four categories: (i) simple correlation of community-gradients with individual factors; (ii) multiple regression with individual environmental factors as independent variables; (iii) correlation with environmental complexes (the complexes defined independently of the community-gradients); and (iv) canonical correlation of sets of vegetational and environmental variables.

Simple correlation Attempts to correlate individual axes of an ordination with individual environmental variables have been largely unsuccessful, because of the non-linear and multi-factor nature of species response to environment. Tests on simulated community-gradients, representing species distributed with overlapping bell-shaped response curves along environmental gradients, have shown that any community-gradient (known to be related linearly to an environmental gradient) will be presented as a complex curve by a linear ordination technique (Noy-Meir and Austin, 1970; Austin and Noy-Meir, 1971; Whittaker and Gauch, 1973; Gauch *et al.*, 1977). In retrospect, therefore, it is clear that any correlation between an axis of a linear ordination and an underlying environmental gradient will be nonlinear, and possibly very convoluted indeed. Some improvement may be brought about by use of transformations and/or nonparametric resemblance functions, but some relationships are too complex (e.g. nonmonotonic relationships) to be treated in this way. It can be anticipated that nonlinear ordinations will be more useful in this respect as they are more likely to yield axes that are linearly correlated with environmental variables. Prentice's (1977) results using nonmetric multidimensional scaling appear to support this contention.

Multiple regression Multiple regression is a possible method for relating community-gradients to more than one environmental variable and hence factor-complexes (Yarranton, 1977a, 1970; Greig-Smith, 1971a; Austin, 1972; Noy-Meir, 1974b). The related technique of path analysis, in which multi-stage considerations allow a dependent variable in relation to one group of variables to be an independent variable in relation to other variables, also has attractions for the investigation of ecological relationships (D. Scott, 1973, 1974). A detailed application of stepwise multiple regression by Noy-Meir (1974b) is particularly noteworthy as he used community-gradients derived by the nonlinear technique of continuity analysis. Although some transformation of environmental variables was necessary, strong and approximately linear relationships were found between community-gradients and one or a few environmental variables and, in general, the environmental variables had much higher correlations with community-gradients than with individual species!

Correlation with independently-derived factor-complexes Similar problems of nonlinearity have beset those who have attempted to define environmental factor-complexes by principal component analysis (Austin, 1968; Barkham and Norris, 1970) or factor analysis (Dagnelie, 1960, 1965a, b), and then to compare the factor-complexes with community-gradients (see also Webster, 1977). Nonlinear ordinations, yet to be used in this way, are likely to produce more meaningful results.

Canonical correlation Canonical correlation is an elegant linear technique, which seeks to reveal the ways in which two sets of variables are related (Dagnelie, 1965a, b). In vegetational context it has been described as measuring the extent to which sites occupy the same relative position in species space as in environmental-factor space (Barkham and Norris, 1970). An analysis by Austin (1968) produced uninterpretable results; difficulties were encountered by Barkham and Norris (1970) who could only interpret one canonical correlation in their application; Huntley and Birks (1979) produced largely negative results, concluding that over 80 per cent of the variability in their vegetation set was related to other environmental variables than those measured. The technique, which appears to have many unsatisfied assumptions when applied to vegetation and environmental variables, requires further testing.

Explanation via direct environmental analysis

It has been suggested that the initial classification or ordination of communities, prior to the search for correlations between vegetation and environment, adds unnecessary complexity and is inefficient (Whittaker, 1973d; Kellman, 1974). The alternative approach, of explanation via a direct environmental analysis, can be regarded as seeking vegetational patterns in an environmental space; that is, within a space the axes of which are environmental factors or factor-complexes. The pioneering work in this field by Whittaker and co-workers (Whittaker, 1956, 1967, 1973c, d) has emphasized environmental gradients, although the use of environmental categories is also possible (Kellman, 1974, 1975). An overriding limitation of the studies so far attempted has been their subjectivity, both in the selection of environmental gradients thought to be of significance, a criticism that applies to the selection of single environmental factors or factor-complexes (such as the environmental scalars of Loucks, 1962), and in the selection of vegetational attributes. Greatest difficulties arise if; (i) effective environmental factors are not anticipated; or (ii) other vegetational attributes respond differently to environment. On the other hand, as Kellman (1974) has pointed out, the examination of changes in species populations, species richness and other simple vegetational attributes (often operationalized as elevation and complex moisture gradients) has provided some of the most profitable substantive results of vegetation study to date (see, for example, Beschel and Webber, 1962; Beals, 1969; Whittaker and Gauch, 1973; J.T. Scott, 1974).

The present author does not, therefore, take as pessimistic a view as Yarranton (1967b), who advocated the study of environmental relationships at the species level. His opinion that environmental relationships should not be pursued at the community level was based on the claim that to analyse species collectively is at variance with the principle (Gleason, 1926, 1939) of the individualistic behaviour of species in relation to environmental gradients. If this claim is true, then much of the justification for the quantitative analysis of vegetation as such is removed. Fortunately, the results of analysis of model gradients with *individualistic* bell-shaped species response curves (referred to above) refute the claim. Specifically, these analyses have shown that: (i) wherever the number of controlling environmental factors or factor-complexes is small in relation to the number of species, an appropriate classification or ordination of communities will reflect the underlying environmental gradient(s); and (ii) only if the occurrence of each species is related to a *different* environmental gradient will the collective analysis of species fail to elucidate environmental relationships.

Temporal and spatial coordinates

Vegetation change

Most studies of vegetation change through time have tended to be qualitative. One quantitative approach involves use of the time factor in much the same way as an environmental factor. Some of the possibilities for this kind of approach were outlined explicitly by Williams *et al.* (1969) within a small area of rain forest experimentally cleared of vegetation and examined twelve times over the following ten years (see also Austin, 1977). The greatest problem in such descriptive studies lies in obtaining data representative of longer-term change. The difficulty can be circumvented to some extent by inference of changes; for example, based on surfaces available for plant colonization for differing periods of time (Matthews, 1978b, c, d; Elven, 1978) or on *in situ* deposits (Walker, 1970). The former may be appropriate for quantitative studies of changes on time-scales of tens and hundreds of years, the latter on time-scales of hundreds and thousands of years. In addition to the multivariate descriptive approaches, recent years have seen increasing interest in two groups of quantitative techniques specifically concerned with modelling dynamic processes. These are: Markovian-type models based on transition probabilities; and simulation models involving birth, growth and death processes.

Markovian-type models A first-order Markov chain is a stochastic process in which transitions among various states occur with characteristic probabilities that depend only on the current states and not on the way those states developed. If a process can be regarded as Markovian, it is possible to predict changes over a relatively long period of time based only on a knowledge of present states and their transition probabilities. Stephens and Waggoner (1970; Waggoner and Stephens, 1970) were the first to apply this kind of analysis to successional change in vegetation. They worked with over 300 sites which were classified into community-types (states) according to their dominant tree species. Transition probabilities were estimated from vegetation surveys carried out in 1927, 1937, 1957 and 1967. Multiplication of the transition matrix for the first decade predicted remarkably well the observed transitions over the forty-year period. Furthermore, the steady state predicted from the transitions in the first decade were in good agreement with the steady state predicted from the fourth decade transitions. In these ways the stationarity assumption (that transition probabilities remain constant through time) was tested and found to be acceptable.

A related approach has been adopted by Horn for a tree-by-tree replacement process during secondary succession (Horn, 1975a, b, 1976). He derived transition probabilities by assuming that every sapling under a canopy tree has an equal chance of replacing that tree. In one analysis, with the probability matrix weighted to take account of the longevity of each species, the predicted steady-state forest was found to agree closely with the actual proportions of species in the oldest and least disturbed stands of forest. In some other successions, however, the numerical abundance of saplings may not be a reasonable estimate of transition probabilities, and more complex (and perhaps technically impossible) measurements of the probabilities may be needed (Horn, 1975a; see also, Austin, 1980).

Where the process of vegetation change does exhibit Markovian properties, further useful analyses are possible (Slatyer, 1977). These include: (i) determination of transient, closed or absorbing states, each of which have analogies in various concepts of vegetation succession (Connel and Slatyer, 1977); (ii) calculation of rates of change and persistence of states; (iii) partitioning of the transition matrix and investigation of the several components separately; and (iv) extension to encompass second- and third-order processes, spatial dependence, and an element of nonstationarity. These possibilities for elaboration of the basic approach have yet to be realized in vegetation studies.

An alternative to the use of conventional Markov chains is the use of techniques of feedback and control systems by means of differential equations that describe the trajectories from one community-type to another. The rate constants of a system of differential equations can be calculated from the same data set as transition probabilities and, if the rates remain constant through time with no dependence on previous conditions, then the response of a model based on differential equations should be identical to the output of a conventional Markov chain (Slatyer, 1977).

The main limitations of Markovian-type models are those of definition, validation and mechanism. First, meaningful states and reliable transition probabilities or transfer rates are required. Second, there are severe difficulties to be faced in testing the models over relatively long time-periods (when assumptions such as stationary or constant linear transfer rates may not hold). Third, these models lack dependence on functional mechanisms. Nevertheless, they are one of the most promising groups of techniques for quantitative vegetation analysis to emerge over the last decade.

Functional simulation models Simulation models of vegetation change have been based on the dynamics of establishment and death of dominant species together with their physiological responses throughout their life-cycles. These models utilize a large amount of empirical information and are strongly functional in character, although a number of relatively poorly understood compartments are often incorporated. The main advantages of simulation models are: the orientation towards process; the lack of a requirement of long historical records for parameter estimation; and the lack of formal mathematical constraints. However, they are often highly deterministic, lack simplicity and need a great amount of computer time. Although they are easier to test than Markov models in the short-term, they have similar problems of validation in the long term (Slatyer, 1977).

A good example of this kind of model is that developed in the Hubbard Brook Forest Ecosystem Study (Botkin *et al.*, 1972a, b; Botkin and Miller, 1974). A minimal number of assumptions were introduced and the simplest realistic mathematical expressions were used throughout. The model was developed step by step, beginning with optimal growth for single trees, the effect of less than optimal light and moisture levels, and the allocation of resources amongst competing trees. Annual growth increment was provided deterministically for each tree; new saplings were added and old trees eliminated stochastically. In all, the model requires less than twenty parameters to define a species and its environment, and it appears capable of reproducing successional changes and predicting vegetational response to altitudinal variation, climatic change, natural perturbations and management practices.

Further examples of simulation models are given in Patten (1971). This kind of model is being very actively developed, based on research that originated as part of the International Biological Programme. Although directed towards an understanding of whole ecosystems, these models will have important implications for the study of vegetation.

Spatial patters – a vacant niche

Even though there is a strong interest in the detection of spatial patterns of species within plant communites, using grids or transects of contiguous quadrats (Greig-Smith, 1964; Hill, 1973b) or nearest neighbour distances (Clark and Evans, 1954; 1979), less interest has been shown in the quantitative study of spatial patterns at the community level. The reason for this seems to lie in the ecologist's primary concern with environmental relationships. Indeed, Greig-Smith (1971b) has suggested that position on the ground is irrelevant to the main problem of relating vegetational composition to environmental factors. The suggestion may be valid from the viewpoint of an ecologist, but it ignores: (i) geographical aspects of vegetation of interest for their own sake; and (ii) the usefulness of spatial relationships as a basis for inference (particularly in the absence of direct information on environment and/or time).

In the context of vegetation mapping, there are many examples involving the spatial relationships of community-types (e.g. Grunow, 1967; Frenkel and Harrison, 1974; Matthews, 1979a) although a contiguity constraint has yet to be incorporated in a classification of vegetation. Quite wrongly, the tendency has been to assume that if mapping is an objective of study, then the classification of communities is the only possible approach to the structuring of vegetational data. There has been a very small number of published attempts to map community-gradients. Most of these have involved principal component scores (Goodall, 1954a; Barkham and Norris, 1970; Matthews, 1979b), although reciprocal averaging (Hatheway, 1971), continuity analysis (Noy-Meir, 1974b) and multiple discriminant analysis (Matthews, unpublished) have been used for this purpose.

Trend surface analysis is one of the few explicitly spatial numerical methods that have been applied to vegetational data. The basic procedure involves the least-squares fitting of linear and higher order polynomial surfaces to the distribution of a spatial variable. Gittins (1968) introduced the technique into the plant ecological literature and also provided a simple example of four-variable trend analysis using altitude as the additional variable. He stressed the usefulness of trend surface analysis for distinguishing between regional and local scales of spatial pattern (based on trends and residuals, respectively). The technique has also been used to detect very weak spatial patterns in species diversity (Matthews, 1978b) and to describe the closeness of fit of simple surfaces to community-gradients defined by principal component scores (Matthews, 1979b). In this way, it was shown that the most important vegetational gradients in species space (highest variability accounted for in principal component analysis) were the simplest, and hence largest-scale, spatial gradients on the ground (highest variation accounted for by trend surface analysis).

Hill (1973b) has suggested using spectral analysis for the investigation of the intensity of spatial pattern in one-dimensional data from transects. Sokal and Oden (1978a, b) have introduced measures of autocorrelation into biology and included an ecological example. However, the explicit treatment of location has been greatly neglected in *quantitative* vegetation study, which points clearly to a vacant niche that could be filled appropriately by biogeographers.

Conclusion

Vegetational data are relatively simple in comparison with those in many other fields of interest to geographers. Furthermore, there are well-defined substantive problems of overriding interest, which the student of vegetation is content to puzzle over. It is not surprising, therefore, that substantial advances are being made in the application of quantitative techniques in the context of vegetation. These advances form the background to the present review. In absolute terms, however, and in the light of the techniques available, vegetational data are complex; which accounts for many of the difficulties that have been encountered during the last fifteen years. The period can be characterized as coming to grips with the complexity of real-world vegetation after initial over-optimism. In particular, there has been an increasing realization that, for applications to be useful, quantification (in all its guises) must be combined sensibly with the existing body of vegetational concepts and theory. This I have made the main theme of the present review.

Limitations have been emphasized because it is by overcoming them that progress will be made, at least in the short term. There appear to be no insurmountable obstacles to the steady improvement of our understanding of techniques of data collection and transformation, the somewhat neglected 'preliminaries' in quantitative vegetation study. Perhaps the highest priority in these areas should be given to research into the meaning that can be attached to resemblance functions. Further progress can be predicted in the development of a number of kinds of optimal classification technique. This development is likely to be most rapid if classifications are tested against simulated data sets (a method employed so profitably in comparing ordination techniques). Although it is difficult, if not dangerous, to recommend longer-term priorities, less reliance on classification in general and on hierarchical sorting in particular appear to be called for. Considerable importance should be attached to the development of a set of optimal ordination techniques, to permit the realistic structuring of vegetational data. The results of recent work suggest that nonlinear techniques are likely to be most useful in this respect.

Prospects for the quantitative investigation of environmental and temporal relationships of vegetation are less promising. Major advances in explanation via vegetational structuring are likely once more realistic structuring techniques have been perfected. Partial explanations, by for example the imaginative use of direct environmental or temporal analysis, can proceed, but are hindered by dependence on the choice of variables. Various kinds of stochastic and functional modelling have considerable potential, particularly in the study of vegetation change over relatively long time-periods. These are no substitute, however, for well designed vegetation surveys and monitoring of long-term changes (which are necessary for model generation and testing). Although the subject has not been covered explicitly in this review, the writer concurs with Kellman (1974) who has advocated strongly the case for an increase in formal field experimentation in the investigation of environmental and temporal relationships.

There is a need for a clearer distinction between vegetation patterns in species space, in environmental space, in earth-surface space, and along the time dimension; not only in order to quantify patterns within these 'spaces' separately, but also in order that interrelationships between the several spaces can be investigated quantitatively. By neglecting earth-surface space and by failing to direct their expertise towards quantification of this aspect of vegetation study, geographers may be missing their vocation and failing to contribute towards forthcoming syntheses.

References

Adam, P. (1978) 'Geographical variation in British saltmarsh vegetation', *Journal of Ecology*, 66, 339–66.

Anderson, A.J.B. (1971) 'Ordination methods in ecology', *Journal of Ecology*, 59, 713–26.

Anderson, D.J. (1971) 'Classification and ordination in vegetation science: controversy over a non-existent problem?', *Journal of Ecology*, 53, 521–6.

Austin, M.P. (1968) 'An ordination of a chalk grassland community', *Journal of Ecology*, 56, 827–44.

Austin, M.P. (1972) 'Models and analysis of descriptive vegetation data' in J.N.R. Jeffers (ed.), *Mathematical models in ecology*, Blackwell Scientific Publications: Oxford.

Austin, M.P. (1976a) 'On non-linear species response models in ordination', *Vegetatio*, 33, 33–41.

Austin, M.P. (1976b) 'Performance of four ordination techniques assuming three different non-linear response models', *Vegetatio*, 33, 43–9.

Austin, M.P. (1977) 'Use of ordination and other multivariate descriptive methods to study succession', *Vegetatio*, 35, 165–75.

Austin, M.P. (1980) 'An exploratory analysis of grassland dynamics: an example of a lawn succession', *Vegetatio*, 43, 87–94.

Austin, M.P., and Greig-Smith, P. (1968) 'The application of quantitative methods to vegetation survey II. Some methodological problems of data from rain forest', *Journal of Ecology*, 56, 827–44.

Austin, M.P., and Noy-Meir, I. (1971) 'The problem of non-linearity in ordination: experiments with two-gradient models', *Journal of Ecology*, 59, 763–73.

Austin, M.P., and Orlóci, L. (1966) 'Geometric models in ecology II. An evaluation of some ordination techniques', *Journal of Ecology*, 54, 217–27.

Barkham, J.P., and Norris, J.M. (1970) 'Multivariate procedures in an investigation of vegetation and soil relations of two beech woodlands, Cotswold Hills, England', *Ecology*, 51, 630–9.

Beals, E.W. (1969) 'Vegetation change along altitudinal gradients', *Science*, 165, 981–5.

Beals, E.W. (1973) 'Ordination: mathematical elegance and ecological naïveté', *Journal of Ecology*, 61, 23–35.

Beschel, R., and Webber, P.J. (1962) 'Gradient analysis of swamp forest', *Nature*, 194, 207–9.

Billings, W.D. (1952) 'The environmental complex in relation to plant growth and distribution', *Quarterly Review of Biology*, 27, 251–65.

Billings, W.D. (1974) 'Environment: concept and reality' in B.R. Strain and W.D. Billings (eds), *Vegetation and environment*, Junk: The Hague.

Birks, H.J.B. (1976) 'The distribution of European pteridophytes: a numerical analysis', *New Phytologist*, 77, 257–87.

Birks, H.J.B., and Deacon, J. (1973) 'A numerical analysis of the past and present flora of the British Isles', *New Phytologist*, 72, 877–902.

Birks, H.J.B., and Walters, S.M. (1973) 'The flora and vegetation of Barno Jezero, Durmator, Montenegro', *Glasnik Republikog Zavodaza Zaštu Prirode i Prirodnjackog Muzeja u Titogradu*, 5, 5–23.

Botkin, D.B., and Miller, R.S. (1974) 'Complex ecosystems: models and predictions', *American Scientist*, 62, 448–53.

Botkin, D.B., Janak, J.F., and Wallis, J.R. (1972a) 'The rationale, limitations and assumptions of a north east forest growth simulator', *IBM Journal of Research and Development*, 16, 101–16.

Botkin, D.B., Janek, J.F., and Wallis, J.R. (1972b) 'Some ecological consequences of a computer model of forest growth', *Journal of Ecology*, 60, 849–72.

Bottomley, J. (1971) 'Some statistical problems arising from the use of the information statistic in numerical classification', *Journal of Ecology*, 59, 339–42.

Bouxin, G. (1976) 'Ordination and classification in the upland Rugege forest (Rwanda, Central Africa)', *Vegetatio*, 32, 95–115.

Bray, J.R., and Curtis, J.T. (1957) 'An ordination of the upland forest communities of southern Wisconsin', *Ecological Monographs*, 22, 217–34.

Clark, P.J., and Evans, F.C. (1954) 'Distance to nearest neighbour as a measure of spatial relationships in populations', *Ecology*, 35, 445–53.

Clark, P.J., and Evans, F.C. (1979) 'Generalization of a nearest neighbour measure of dispersion for use in k dimensions', *Ecology*, 60, 316-17.

Connel, J.H., and Slatyer, R.O. (1977) 'Mechanisms of succession in natural communities and their role in community stability and organization', *American Naturalist*, 111, 1119-44.

Cormack, R.M. (1971) 'A review of classification', *Journal of the Royal Statistical Society*, Series A, 134, 321-67.

Cottam, G., Goff, F.C., and Whittaker, R.H. (1973) 'Wisconsin comparative ordination' in R.H.Whittaker, op. cit.

Crawford, R.M.M., and Wishart, D. (1968) 'A rapid classification and ordination method and its application to vegetation mapping', *Journal of Ecology*, 56, 385-404.

Crawford, R.M.M., Wishart, D., and Campbell, R.M. (1969) 'A numerical analysis of high altitude scrub vegetation in relation to soil erosion in the eastern cordillera of Peru', *Journal of Ecology*, 58, 173-91.

Curtis, J.T., and McIntosh, R.P. (1951) 'An upland forest continuum in Wisconsin', *Ecology*, 32, 476-96.

Dagnelie, P. (1960) 'Contribution à l'étude des communautés végétales par l'analyse factorielle', *Bulletin du Service de la carte phyto-géographique*, Ser. B, 5, 7-71, 93-195.

Dagnelie, P. (1965a) 'L'étude des communautés végétales par l'analyse statistique des liasons entre les espèces et les variables écologiques: principes fondamentaux', *Biometrics*, 21, 345-61.

Dagnelie, P. (1965b) 'L'étude des communautés végétales par l'analyse statistique des liasons entre les espèces et les variables écologiques: un exemple', *Biometrics*, 21, 890-907.

Dagnelie, P. (1973) 'L'analyse factorielle' in R.H. Whittaker op. cit.

Dale, M.B. (1975) 'On objectives of methods of ordination', *Vegetatio*, 30, 15-32.

Dale, M.B., and Anderson, D.J. (1972) 'Qualitative and quantitative information analysis', *Journal of Ecology*, 60, 639-54.

Dale, M.B., and Anderson, D.J. (1973) 'Inosculate analysis of vegetation data', *Australian Journal of Botany*, 21, 253-76.

Dale, M.B., and Clifford, H.T. (1976) 'On the effectiveness of higher taxonomic ranks for vegetation analysis', *Australian Journal of Ecology*, 1, 37-62.

Elven, R. (1978) 'Association analysis of moraine vegetation at the glacier Hardangerjøkulen, Finse, South Norway', *Norwegian Journal of Botany*, 25, 171-91.

Everitt, B. (1974) *Cluster analysis*, Heineman: London.

Fasham, M.J.R. (1977) 'A comparison of non-metric multidimensional scaling, principal components analysis and reciprocal averaging for the ordination of coenoclines and coenplanes', *Ecology*, 58, 551-61.

Feoli, E. (1977) 'On the resolving power of principal component analysis in plant community ordination', *Vegetatio*, 33, 119-25.

Field, J.G. (1969) 'The use of the information statistic in the numerical classification of heterogeneous systems', *Journal of Ecology*, 57, 565-9.

Flenley, J.R. (1969) 'The vegetation of the Wabag region New Guinea Highlands: a numerical study', *Journal of Ecology*, 57, 465-90.

Frenkel, R.E., and Harrison, C.M. (1974) 'An assessment of the usefulness of phytosocio-logical and numerical classificatory methods for the community biogeographer', *Journal of Biogeography*, 1, 27-56.

Fresco, L.F.M. (1969) 'Factor analysis as a method of synecological research', *Acta Botanica Neerlandica*, 18, 477-82.

Gauch, H.G., and Chase, G.B. (1974) 'Fitting the Gaussian curve to ecological data', *Ecology*, 55, 1377-81.

Gauch, H.G., and Whittaker, R.H. (1972) 'Comparison of ordination techniques', *Ecology*, 53, 868-75.

Gauch, H.G., Chase, G.B., and Whittaker, R.H. (1974) 'Ordination of vegetation samples by Gaussian species distributions', *Ecology*, 55, 1382-90.

Gauch, H.G., Whittaker, R.H., and Wentworth, T.R. (1977) 'A comparative study of reciprocal averaging and other ordination techniques', *Journal of Ecology*, 65, 157-74.

Gilmore, J.S.L., and Walters, S.M. (1963) 'Philosophy and classification' in W.B. Turrill (ed.), *Vistas in botany*, 4, Pergamon Press: Oxford.

Gittins, R. (1968) 'Trend surface analysis of ecological data', *Journal of Ecology*, 56, 845-69.

Gleason, H.A. (1926) 'The individualistic concept of the plant association', *Bulletin of the Torrey Botanical Club*, 53, 7-26.

Gleason, H.A. (1939) 'The individualistic concept of the plant association', *American Midland Naturalist*, 21, 92-110.

Goldsmith, F.B. (1973a) 'The vegetation of exposed sea cliffs at South Stack, Anglesey I. The numerical approach', *Journal of Ecology*, 61, 787-818.

Goldsmith, F.B. (1973b) 'The vegetation of exposed sea cliffs at South Stack, Anglesey II. Experimental study', *Journal of Ecology*, 61, 819-29.

Goldsmith, F.B., and Harrison, C.M. (1976) 'Description and analysis of vegetation' in S.B. Chapman (ed.), *Methods in plant ecology*, Blackwell Scientific Publications: Oxford.

Goodall, D.W. (1953) 'Objective methods for the classification of vegetation I. The use of positive interspecific correlation', *Australian Journal of Botany*, 1, 39-63.
Goodall, D.W. (1954a) 'Objective methods for the classification of vegetation III. An essay on the use of factor analysis', *Australian Journal of Botany*, 2, 304-24.
Goodall, D.W. (1954b) 'Vegetational classification and vegetational continua', *Angewandte Pflanzensoziologie*, 1, 168-82.
Goodall, D.W. (1963) 'The continuum and the individualistic association', *Vegetatio*, 11, 297-316.
Goodall, D.W. (1966a) 'Deviant index – a new tool for numerical taxonomy', *Nature*, 210, 216.
Goodall, D.W. (1966b) 'Hypothesis testing in classification', *Nature*, 211, 229-30.
Goodall, D.W. (1968) 'Affinity between an individual and a cluster in numerical taxonomy', *Biométrie – praximétrie*, 9, 52-5.
Goodall, D.W. (1973a) 'Sample similarity and species correlation' in R.H. Whittaker, op. cit.
Goodall, D.W. (1973b) 'Numerical classification' in R.H. Whittaker, op. cit., 1973a.
Gower, J.C. (1966) 'Some distance properties of latent root and vector methods used in multivariate analysis', *Biometrika*, 53, 325-38.
Greig-Smith, P. (1964) *Quantitative plant ecology*, Butterworth: London.
Greig-Smith, P. (1971a) 'Analysis of vegetation data: the user viewpoint' in G.P. Patil *et al.*, op. cit.
Greig-Smith, P. (1971b) 'Comments on a paper by W.H. Hatheway' in G.P. Patil *et al.*, op. cit.
Greig-Smith, P., Austin, M.P., and Whitmore, T.C. (1967) 'The application of quantitative methods to vegetation survey I. Association analysis and principal component ordination of rain forest', *Journal of Ecology*, 55, 483-503.
Grigal, D.F., and Arneman, H.F. (1970) 'Quantitative relationships among vegetational and soil classifications', *Canadian Journal of Botany*, 48, 555-66.
Grigal, D.F., and Goldstein, R.A. (1971) 'An integrated ordination-classification analysis of an intensively sampled oak-hickory forest', *Journal of Ecology*, 59, 481-92.
Groenewoud, H. van (1965) 'Ordination and classification of Swiss and Canadian forests by various biometric and other methods', *Bericht über das Geobotanische Forschungsinstitut Rübel in Zürich*, 35, 28-102.
Groenewoud, H. van (1976) 'Theoretical considerations on the covariance of plant species along ecological gradients with regard to multivariate analysis', *Jouranl of Ecology*, 64, 837-47.
Grunow, J.O. (1967) 'Objective classification of plant communities: a synecological study in the sourish mixed bushveld of Transvaal', *Journal of Ecology*, 55, 691-710.
Guttman, L. (1968) 'A general nonmetric technique for finding the smallest coordinate space for a configuration of points', *Psychometrika*, 33, 469-506.
Harrison, C.M. (1971) 'Recent approaches to the description and analysis of vegetation', *Transactions of the Institute of British Geographers*, 52, 113-27.
Hatheway, W.H. (1971) 'Contingency-table analysis of rain forest vegetation' in G.P. Patil *et al.*, op. cit.
Hill, M.O. (1973a) 'Reciprocal averaging. An eigenvector method of ordination', *Journal of Ecology*, 61, 237-49.
Hill, M.O. (1973b) 'The intensity of spatial pattern in plant communities', *Journal of Ecology*, 61, 225-35.
Hill, M.O. (1974) 'Correspondence analysis: a neglected multivariate method', *Journal of the Royal Statistical Society*, Series C, 23, 340-54.
Hill, M.O., Bunce, R.G.H., and Shaw, M.W. (1975) 'Indicator species analysis, a divisive polythetic method of classification, and its application to a survey of native pinewoods in Scotland', *Journal of Ecology*, 63, 597-613.
Holland, P.G. (1971) 'Seasonal change in plant patterns of deciduous forest in southern Quebec, (Canada)', *Oikos*, 22, 137-48.
Horn, H.S. (1975a) 'Forest succession', *Scientific American*, 232, 90-8.
Horn, H.S. (1975b) 'Markovian properties of forest succession' in M.L. Cody and J.M. Diamond (eds), *Ecology and evolution of communities*, Harvard University Press: Cambridge, Mass.
Horn, H.S. (1976) 'Succession' in R.M. May (ed.), *Theoretical ecology*, Blackwell Scientific Publications: Oxford.
Huntley, A., and Birks, H.J.B. (1979) 'The past and present vegetation of the Morrone Birkwoods National Nature Reserve, Scotland II. Woodland vegetation and soils', *Journal of Ecology*, 67, 447-67.
Ihm, P., and Groenewoud, H. van (1975) 'A multivariate ordering of vegetational data based on Gaussian type gradient response curves', *Journal of Ecology*, 63, 767-77.
Jancey, R.C. (1974) 'Algorithm for detection of discontinuities in data sets', *Vegetatio*, 29, 131-3.
Jardine, N., and Sibson, R. (1971) *Mathematical taxonomy*, Wiley: New York.
Kellman, M.C. (1974) 'Modes of vegetation analysis: retrospect and prospect', *Professional Geographer*, 26, 38-43.
Kellman, M.C. (1975) *'Plant geography'*, Methuen: London.
Kershaw, K.A. (1964) *Quantitative and dynamic ecology*, Arnold: London.

Kershaw, K.A. (1968) 'Classification and ordination of Nigerian savannah vegetation', *Journal of Ecology*, 56, 467–82.

Kruskal, J.B. (1964) 'Multidimensional scaling by optimizing goodness of fit to a nonmetric hypothesis', *Psychometrika*, 29, 1–27.

Kruskal, J.B., and Carroll, J.D. (1969) 'Geometric models and badness-of-fit functions' in P.R. Krishnaiah (ed.), *Multivariate analysis*, 2, Academic Press: New York.

Lambert, J.M. (1972) 'Theoretical models for large-scale vegetation survey' in J.N.R. Jeffers (ed.), *Mathematical models in ecology*, Blackwell Scientific Publications: Oxford.

Lambert, J.M., and Dale, M.B. (1964) 'The use of statistics in phytosociology', *Advances in Ecological Research*, 2, 55–99.

Lambert, J.M., and Williams, W.T. (1962) 'Multivariate methods in plant ecology IV. Nodal analysis', *Journal of Ecology*, 50, 775–802.

Lambert, J.M., and Williams, W.T. (1966) 'Multivariate methods in plant ecology VI. Comparison of information analysis and association analysis', *Journal of Ecology*, 54, 635–64.

Lance, G.N., and Williams, W.T. (1968) 'Note on a new information statistic classificatory program', *Computer Journal*, 11, 195.

Loucks, O.L. (1962) 'Ordinating forest communities by means of environmental scalars and phytosociological indices', *Ecological Monographs*, 32, 137–66.

Maarel, E. van der (1972) 'Ordination of plant communities on the basis of their plant genus, family and order relationships' in R. Tuxen (ed.), *Gundfragen und Methoden in der Pflanzensoziologie*, Junk: The Hague.

Maarel, E. van der (1979) 'Multivariate methods in phytosociology, with reference to the Netherlands' in M.J.A. Werger (ed.), *The study of vegetation*, Junk: The Hague.

McIntosh, R.P. (1967) 'The continuum concept of vegetation', *Botanical Review*, 33, 99–187.

McIntosh, R.P. (1973) 'Matrix and plexus techniques' in R.H. Whittaker, op. cit.

Matthews, J.A. (1975) 'The gletschervorfeld: a biogeographical system and microcosm', University of Edinburgh Department of Geography, *Research Discussion Papers*, 2, 1–44.

Matthews, J.A. (1978a) 'An application of non-metric multidimensional scaling to the construction of an improved species plexus', *Journal of Ecology*, 66, 157–73.

Matthews, J.A. (1978b) 'Plant colonisation patterns on a gletschervorfeld, southern Norway: a meso-scale geographical approach to vegetation change and phytometric dating', *Boreas*, 7, 155–78.

Matthews, J.A. (1979a) 'The vegetation of the Storbreen gletschervorfeld, Jotunheimen, Norway I. Introduction and approaches involving classification', *Journal of Biogeography*, 6, 17–47.

Matthews, J.A. (1979b) 'The vegetation of the Storbreen gletschervorfeld, Jotunheimen, Norway II. Approaches involving ordination and general conclusions', *Journal of Biogeography*, 6, 133–67.

Matthews, J.A. (1979c) 'A study of the variability of some successional and climax assemblage-types using multiple discriminant analysis', *Journal of Ecology*, 67, 255–71.

Matthews, J.A. (1979d) 'Refutation of convergence in a vegetation succession', *Naturwissenschaften*, 66, 47–9.

Moore, J.J., Fitzsimons, P., Lambe, E., and White, J. (1970) 'A comparison and evaluation of some phytosociological techniques', *Vegetatio*, 23, 323–38.

Moore, P.D. (1971) 'Computer analysis of sand dune vegetation in Norfolk, England, and its implications for conservation', *Vegetatio*, 23, 323–38.

Moore, P.D., and Beckett, P.J. (1971) 'Vegetation and development of Llyn, a welsh mire', *Nature*, 231, 363–5.

Moral, R. del (1975) 'Vegetation clustering by means of ISODATA: revision by multiple discriminant analysis', *Vegetatio*, 29, 179–90.

Nichols, S. (1977) 'On the interpretation of principal components analysis in ecological contexts', *Vegetatio*, 34, 191–7.

Norris, J.M., and Barkham, J.P. (1970) 'A comparison of some Cotswold beechwoods using multiple-discriminant analysis', *Journal of Ecology*, 58, 603–19.

Noy-Meir, I. (1971) 'Multivariate analysis of desert vegetation II. Qualitative/quantitative partitioning of heterogeneity', *Israel Journal of Botany*, 20, 203–13.

Noy-Meir, I. (1973a) 'Data transformations in ecological ordination I. Some advantages of non-centering', *Journal of Ecology*, 61, 329–41.

Noy-Meir, I. (1973b) 'Divisive polythetic classification of vegetation data by optimized division on ordination components', *Journal of Ecology*, 61, 753–60.

Noy-Meir, I. (1974a) 'Catenation: quantitative methods for the definition of coenoclines', *Vegetatio*, 29, 89–99.

Noy-Meir, I (1974b) 'Multivariate analysis of the semiarid vegetation of south-eastern Australia II. Vegetation catenae and environmental gradients', *Australian Journal of Botany*, 22, 115–40.

Noy-Meir, I., and Austin, M.P. (1970) 'Principal component ordination and simulated vegetational data', *Ecology*, 51, 551–2.

Noy-Meir, I., and Whittaker, R.H. (1977) 'Continuous multivariate methods in community analysis: some problems and developments', *Vegetatio*, 33, 79–98.

Noy-Meir, I., Tadmor, N.H., and Orshan, G. (1970)

'Multivariate analysis of desert vegetation I. Association analysis at various quadrat sizes', *Israel Journal of Botany*, 19, 561-69.

Noy-Meir, I., Walker, D., and Williams, W.T. (1975) 'Data transformations in ecological ordination II. On the meaning of data standardization', *Journal of Ecology*, 63, 779-800.

Orlóci, L. (1966) 'Geometric models in ecology I. The theory and application of some ordination methods', *Journal of Ecology*, 54, 193-215.

Orlóci, L. (1967a) 'Data centering: a review and evaluation with reference to component analysis', *Systematic Zoology*, 16, 208-212.

Orlóci, L. (1967b) 'An agglomerative method for the classification of plant communities', *Journal of Ecology*, 55, 193-206.

Orlóci, L. (1972) 'On objective functions of phytosociological resemblance', *American Midland Naturalist*, 88, 28-55.

Orlóci, L. (1973) 'Ordination by resemblance matrices' in R.H. Whittaker, op. cit.

Orlóci, L. (1975) *Multivariate analysis in vegetation research*, Junk: The Hague.

Patil, G.P., Pielou, E.C., and Water, W.E. (eds), (1971) *Statistical ecology*, 3, Pennsylvania State University Press: University Park.

Patten, B.C. (ed.) (1971) *Systems analysis and simulation in ecology*, Academic Press: New York.

Pears, N.V. (1968) 'Some recent trends in classification and description of vegetation', *Geografiska Annaler*, 50(A), 162-72.

Pielou, E.C. (1969) *An introduction to mathematical ecology*, Wiley: New York.

Ponyatovskaya, V.M. (1961) 'On two trends in phytosociology', *Vegetatio*, 10, 373-85.

Prentice, I.C. (1977) 'Non-metric ordination methods in ecology', *Journal of Ecology*, 65, 85-94.

Pritchard, N.M., and Anderson, A.J.B. (1971) 'Observations on the use of cluster analysis in Botany with an ecological example', *Journal of Ecology*, 59, 727-47.

Randall, R.E. (1978) *Theories and techniques in vegetation analysis*, Oxford University Press.

Scott, D. (1973) 'Path analysis: a statistical model suited to ecological data', *Proceedings of the New Zealand Ecological Society*, 20, 79-95.

Scott, D. (1974) 'Description of relationships between plants and environment' in B.R. Strain and W.D. Billings (eds), *Vegetation and environment*, Junk: The Hague.

Scott, J.T. (1974) 'Correlation of vegetation with environment: a test of the continuum and community-type hypothesis' in B.R. Strain and W.D. Billings (eds), *Vegetation and environment*, Junk: The Hague.

Shepard, R.N., and Carroll, J. (1966) 'Parametric representation by non-linear data structures' in P.R. Krishnaiah (ed.), *Multivariate analysis*, Academic Press: New York.

Shimwell, D.W. (1971) *Description and classification of vegetation*, Sidgwick & Jackson: London.

Sibson, R. (1971) 'Discussion on Dr Cormack's paper', *Journal of the Royal Statistical Society*, Series A, 134, 355-6.

Slatyer, R.O. (ed.) (1977) *Dynamic changes in terrestrial ecosystems: patterns of change, techniques for study and applications to management*, MAB Technical Note 4, UNESCO: Paris.

Smartt, P.F.M. (1978) 'Sampling for vegetation survey: a flexible systematic model for sampling location', *Journal of Biogeography*, 5, 43-56.

Smartt, P.F.M., and Grainger, J.E.A. (1974) 'Sampling for vegetation survey: some aspects of unrestricted, restricted, and stratified techniques', *Journal of Biogeography*, 1, 193-206.

Smartt, P.F.M., Meacock, S.E., and Lambert, J.M. (1974) 'Investigations into the properties of quantitative vegetational data I. Pilot study', *Journal of Ecology*, 62, 735-59.

Smartt, P.F.M., Meacock, S.E., and Lambert, J.M. (1976) 'Investigations into the properties of quantitative vegetation data II. Further data type comparisons', *Journal of Ecology*, 64, 41-78.

Sokal, P.R., and Oden, N.L. (1978a) 'Spatial autocorrelation in biology I. Methodology', *Biological Journal of the Linnean Society of London*, 10, 199-228.

Sokal, P.R., and Oden, N.L. (1978b) 'Spatial autocorrelation in biology II. Some biological implications and four applications of evolutionary and ecological interest', *Biological Journal of the Linnean Society of London*, 10, 229-49.

Stephens, G.P., and Waggoner, P.E. (1970) 'The forests anticipated from 40 years of natural transitions in mixed hardwoods', *Bulletin of the Connecticut Agricultural Experiment Station*, New Haven, 707, 1-58.

Vries, D.M. de (1953) 'Objective combinations of species', *Acta Botanica Neerlandica*, 1, 497-9.

Waggoner, P.E., and Stephens, G.R. (1970) 'Transition probabilities for a forest', *Nature*, 225, 1160-1.

Walker, D. (1970) 'Direction and rate in some British post-glacial hydroseres' in D. Walker and R.G. West (eds), *Studies in the vegetation history of the British Isles*, Cambridge University Press.

Ward, J.H. (1963) 'Hierarchical grouping to optimize an objective function', *Journal of the American Statistical Association*, 58, 236-44.

Webb, D.A. (1954) 'Is the classification of plant communities either possible or desirable?', *Botanisk Tidsskrift*, 51, 362-70.

Webb, L.J., Tracey, J.G., and Williams, W.T. (1976) 'The value of structural features in tropical forest typology', *Australian Journal of Ecology*, 1, 3-28.

Webb, L.J., Tracey, J.G., Williams, W.T. and

Lance, G.N. (1967) 'Studies in the numerical analysis of complex rain-forest communities II. The problem of species sampling', *Journal of Ecology*, 55, 525–38.

Webb, L.J., Tracey, J.G., Williams, W.T., and Lance, G.N. (1970) 'Studies in the numerical analysis of complex rain-forest communities V. A comparison of the properties of floristic and physiognomic-structural data', *Journal of Ecology*, 58, 203–32.

Webster, R. (1977) *Quantitative and numerical methods in soil classification and survey*, Oxford University Press.

Werger, M.J.A. (1978) 'Vegetation structure in the southern Kalahari', *Journal of Ecology*, 66, 933–41.

Whittaker, R.H. (1956) 'Vegetation of the Great Smoky Mountains', *Ecological Monographs*, 26, 1–80.

Whittaker, R.H. (1963) 'Classification of natural communities', *Botanical Review*, 28, 1–239.

Whittaker, R.H. (1967) 'Gradient analysis', *Biological Reviews*, 42, 207–64.

Whittaker, R.H. (ed.)(1973a) 'Introduction', *Ordination and classification of communities*, Junk: The Hague.

Whittaker, R.H. (1973b) 'Approaches to classifying vegetation' in R.H. Whittaker, op. cit.

Whittaker, R.H. (1973c) 'Direct gradient analysis: techniques' in R.H. Whittaker, op. cit.

Whittaker, R.H. (1973d) 'Direct gradient analysis: results' in R.H. Whittaker, op. cit.

Whittaker, R.H., and Gauch, H.G. (1973) 'Evaluation of ordination techniques' in R.H. Whittaker, op. cit.

Williams, W.T. (1971) 'Principles of clustering', *Annual Review of Ecology and Systematics*, 2, 303–26.

Williams, W.T., and Dale, M.B. (1965) 'Fundamental problems in numerical taxonomy', *Advances in Botanical Research*, 2, 35–68.

Williams, W.T., and Lambert, J.M. (1959) 'Multivariate methods in plant ecology I. Association analysis in plant communities', *Journal of Ecology*, 47, 83–101.

Williams, W.T., and Lambert, J.M. (1960) 'Multivariate methods in plant ecology II. The use of an electronic digital computer for association analysis', *Journal of Ecology*, 48, 689–710.

Williams, W.T., Lambert, J.M., and Lance, G.N. (1966) 'Multivariate methods in plant ecology V. Similarity analysis and information analysis', *Journal of Ecology*, 54, 427–45.

Williams, W.T., Lance, G.N., Webb, L.J., Tracey, J.G., and Dale, M.B. (1969) 'Studies in the numerical analysis of complex rain forests communities III. The analysis of successional data', *Journal of Ecology*, 57, 515–35.

Yarranton, G.A. (1967a) 'Principal components analysis of data from saxicolous bryophyte vegetation at Steps Bridge, Devon I-III', *Canadian Journal of Botany*, 45, 93–115, 229–58.

Yarranton, G.A. (1967b) 'Organismal and individualistic concepts and the choice of methods of vegetation analysis', *Vegetatio*, 15, 113–6.

Yarranton, G.A. (1970) 'Towards a mathematical model of limestone pavement vegetation III. Estimation of the determinants of species frequency', *Canadian Journal of Botany*, 48, 1387–1404.

Yarranton, G.A., Beasleigh, W.J., Morrison, R.G., and Shafi, M.I. (1972) 'On the classification of phytosociological data into nonexclusive groups with a conjecture about determining the optimum number of groups in a classification', *Vegetatio*, 24, 1–12.

Chapter 29

Population geography

P. Rees

Quantitative methods in population geography

In Britain a relatively small band of geographers are actively engaged in research into the ways populations behave in space. The Population Geography Study Group has been one of the smaller, though active, study groups of the Institute of British Geographers. Members figure prominently in the survey of current population research in the United Kingdom in Jackson (1979) and some nine out of thirty-one organizations listed as conducting 'demographic/social policy' research are Departments of Geography (pp. 39–41 in Jackson, 1979).

Almost all population geographers use numbers in their work, and so employ quantitative methods to analyse those numbers. If we adopt a rather tighter definition of quantitative methods as methods of analysis going beyond the simple construction of rates of incidence of population phenomena to a concern with how those rates of incidence can be explained quantitatively and with the consequences of those rates we find a much smaller number of active workers. If attention were to be confined to those studies which have gained attention in the wider field of demographic studies, and to those studies originating in Britain, my guess would be that we would have a rather small collection of work to examine.

This review of the questions posed and work carried out on 'populations in space' will therefore, of necessity, have to step outside the narrow confines of quantitative population geography in Britain in the 1970s. The connections to a wider world must be made.

However, this survey is likely to satisfy neither the traditional demographer nor the person interested in quantitative methods *per se*. Relatively little will be said, for example, about fertility patterns and trends, a dominant concern in demographic studies, because with a few exceptions (Compton, 1978; Jones, 1975) geographers have left this field to others. Quantitative geographers will also be rather dissatisfied with the way attention is given to 'space' as a variable of import: space is treated in discrete chunks called regions in a rather practical fashion in most population studies. Distance between regions does enter into migration models but not perhaps in an entirely satisfactory way.

What kind of questions have population geographers addressed? How have they sought to answer these questions and with what results? Of general and local British interest have been the following list of questions and associated problems.

(1) How do regional and local populations grow (or shrink) over time?
(2) Given a proper understanding of the first question, how are regional and local populations likely to grow in the future? How do we forecast the likely population changes in subnational spatial units?
(3) How can we investigate answers to the first two questions for many populations of complex character simultaneously? We need to study the evolution of multidimensional populations.
(4) A number of questions need to be answered in the course of constructing models to handle the first three questions. How should our systems be closed? How should conflicting forecasts at different scales be resolved? What role should time (and time series analysis) play in the investigation? Can we test our forecasting models?
(5) Recognizing that the movement of people over space plays a crucial role in regional and local population change, how can we predict the scale and direction of movement? Does our extensive experience of the modelling of migration really help in its prediction?
(6) What influence does migration have on the life expectancies of people living in regions or local

areas? What problems does the tendency of populations to remember their past have on models that forget past history?

The sections of this review that follow look at the answers suggested to these by population geographers and others, and an attempt is made to expose those that remain unsolved.

How do regional and local populations grow?

Simple components of growth

Several researchers (Department of the Environment, 1971; Lawton, 1977, 1980; Champion, 1976; Eversley, 1971; Stillwell, 1979) have described in detail the pattern of population growth in British standard regions, subregions, or counties in the recent past, and others (Kennett, 1977, 1980; Department of the Environment, 1976; Gleave and Cordey-Hayes, 1977) have described patterns for a system of city regions. The methods are familiar to most geographers and so will not be detailed here.

The major empirical finding of the investigation of the components growth for British city regions (Department of the Environment, 1976; Kennett, 1977, 1980) and standard regions (Rees, 1979a) was that net migration was the most influential component in determining the spatial variation in population growth rates in Britain. The simple correlation between net migration and natural increase rate for British standard regions for single years between 1965 and 1966 and 1975 and 1976 averaged +0.20 whereas the average correlation of growth rate and net migration rate was +0.96 (Rees, 1979a, Table 6, p. 24). Such results provide 'moral' support for the geographer's view of the world against the overwhelming emphasis placed by demographers on the fertility component of natural increase, which at the national scale is much more important, of course.

A multiregional view of population change

It has long been recognized that the simple view of components of population growth covers a multitude of population movements, into and out of regions.

The Rogers (1966, 1968) model for example has been used by Compton (1969) in an analysis of the population dynamics of Hungary, and in modified form for a set of British regions (Rees, 1976, 1977a). Although the model has been superseded by more sophisticated life-table or accounts based versions it still constitutes a useful tool for learning about the dynamics of multiregional growth (see Rogers, 1976 for examples of its use in combination with other models).

One of the major issues that has exercised planners and civil servants concerned with population projection is whether a 'net migration' perspective was adequate. In a survey of contemporary local authority practice in 86 authorities Woodhead (1979) found that 48 per cent (Tables 3, 4) still employ a projection model incorporating net migration compared with 48 per cent using gross flows (and 4 per cent not taking migration into account at all). The official subnational projections (Campbell, 1976) still employ a methodology incorporating net migration although the Department of the Environment has recently let out to contract the design of a model for projection incorporating gross flows between areas (Girling, 1979). More will be said on this issue in the next section in connection with projection.

An accounting view of population change

The representation of population change in components of growth methodology involves sets of events counts (births, deaths) and move counts (migrations). An alternative representation is to define a variable the superscripts of which define transitions of a population between states over a period. A matrix of such variables is called an accounts matrix, the rows of which may represent initial states in a period, the columns of which may represent final states in a period. The initial states will include 'birth' and the final states 'death', both of which will be classified in multiregional population accounts by location.

The theory of population accounts originates with Stone (1965, 1971, 1972, 1975a) and has been developed in a spatial unit context by Rees and Wilson (1973, 1975, 1977), Wilson and Rees (1974) and by Illingworth (1976). Examples of accounts for British regions are given in Rees (1976, 1977a and 1979a).

Once an accounts matrix has been estimated (see Rees and Wilson, 1977; Rees, 1978a; Rees, 1980a; and Illingworth, 1976, for details of methods), the matrix model of population growth can then easily be derived (see Rees and Wilson, 1977, Chapter 6). What then is the advantage of first estimating an accounts matrix before beginning a projection exercise? The advantages are as follows:

(1) Construction of the accounts matrix forces the analyst to check and adjust his base period data and to iron out inconsistencies. Illingworth (1976), Jenkins and Rees (1977) and Rees (1980a) give details of methods and provide evidence of the importance of this step. A variety of balancing factor, biproportional and RAS methods (Macgill, 1977; Bacharach, 1970; Stone, 1975b; and Byron, 1978) can be applied to improve estimates of the accounts matrix of regional population change.

(2) The analyst is forced to close the system under study. Frequently, multiregional matrix models of population change are mis-specified if they lack a rest of the world region.

(3) The analyst can use a variety of different models for projecting the different components of growth embodied in the accounts. He or she is not necessarily tied to the linear matrix model (see Rees and Wilson, 1977, Chapter 7; Jenkins and Rees, 1977; and Rees, 1980c, for various alternatives).

The disadvantages from which accounting has been said to suffer (Baxter and Williams, 1978; Cordey-Hayes, 1975) are: (i) a great deal of work of estimation is involved in preparing the data for input to the model that estimates the accounts; (ii) the framework is purely involved with data assembly – no insights into the behaviour of the population are provided.

Both criticisms have validity but the conclusions which I would draw from them are rather different from those of the authors cited above. That a great deal of estimation work is necessary means that official agencies such as the Office of Population Censuses and Survey need to improve the organization and reporting of their data bases. Population accounts constitute a plan for so doing.

The second criticism can be regarded as a strength. No particular model of population behaviour nor hypothesis about trends is imposed. But they can be tested, historically, against the data assembled in accounts.

The linkage of natural increase and net migration

So far, we have viewed natural increase and net migration as separate components. Of course, they are linked in that migrant flows will have associated with them natural increases (some migrants will die, and some will give birth to children). If the net balance of migrant flows is into a region, natural increase will be higher than it would otherwise have been, and it will be lower if net migration is on balance out of a region. This linkage can clearly be demonstrated over several time-periods in a projection model. The effect can also be detected within one period using accounts since the births and deaths accruing to stayers, out-migrants and in-migrants can be separately distinguished. However, the effects are not necessarily those one might have at first supposed by looking at the net migration figures. A sophisticated set of projections is probably needed to unravel all of the likely effects.

Views of population change: prospect

What are the problems that remain for our views of population change, and how likely is it that descriptions of the population geography of Britain which will use the results of the 1981 census will adopt the multiregional or accounting framework rather than the simple components-of-growth?

There are still a number of theoretical problems needing resolution. These concern the ways in which judgment about the statistical reliability, the variances of each of the sets of terms that enter accounts-population estimates (some error variance), population census counts (less error variance), migrant counts (sampling error, misallocation error), births and deaths data (small errors) – can be incorporated in the methods of estimation. Current practice is outlined in Rees (1980a) but this has been rightly criticized by Keyfitz (1980). The solution probably lies in the adaptation of methods suggested some time ago by Stone, Champernowne and Meade (1942) and elaborated more recently by Stone (1975b) and Byron (1978). A best estimate of accounts matrix is made, taking into account the size of the error variances about the estimates of the elements of the accounts. The problem still remains, however, of making judgmental estimates of those error variances.

There are also implementation problems. Computer programmes for developing multiregional accounts need improvement before they can be widely used, particularly in more complex age-sex disaggregated forms. The data series also need upgrading, particularly those for migration. Census type retrospective migrant data are the correct data for use in accounts (see Rees, 1977b) but in Britain we get this data only with the census. It would be very simple, cheap and effective for the Office of Population Censuses and Survey to give some spatial detail in their General Household Survey reports (for example, OPCS, Social Survey Division, 1978). The purist might say that the sample sizes for any elements in a migration table would yield standard errors of the estimates of rates or proportions that would be too broad. But any numbers would be useful in this context since in between censuses all that the researcher has to go on are indirect partial and unsatisfactory sources such as the National Health Service Register moves (to be used in the Department of Environment project, Girling, 1979) or statistics on electoral role numbers or housing starts and completions.

However, it is interesting to note that a full registration system such as that operated in the Netherlands, Denmark, Sweden, Norway, Finland, Czechoslovakia, Hungary, and other countries does not actually yield the number of *transitions* (migrants) between two regions over a period but rather the number of *moves* (see Ledent, 1978a, b, 1980, for a discussion of the consequences of using one set of data or another). Transition data are the data required in multiregional population models of the Rogers type, or in models derived from population accounts. From the point of view of stock projection use of movements data makes little difference: moves 'surplus' to transitions (see Illingworth, 1976, and Rees, 1977b, for a discussion of the concept of 'surplus' moves) cancel out. However, any statistics based on the probability of making a transition between states which use movement data must be in error (see Rees, 1978b).

Whether the multiregional and accounting frameworks will be adopted for analysis of the 1981 census data really depends on the models adopted for forecasting the population of spatial units. To these attention is now turned.

How are regional and local populations likely to grow?

Preliminaries

It is planners (national, regional and local) who have been most concerned with forecasting population rather than geographers, so we must step outside population geography *per se* to understand the issues involved. The terms 'forecasting' and 'projection' are here used fairly interchangeably as endeavours to guess what will happen in the future, though they have often been given separate meanings (Pittenger, 1976; Brass, 1974) of 'best guess' and 'conditional prediction' respectively.

Multiregional models versus single region models with net migration

The principal arguments for the adoption of multiregional methods have been outlined by Rogers and Philipov (1979). The most important may be put as follows. Starting from the same base period data and making the same assumption that the migrant pattern remains constant, we observe migration flows in the multiregional case responding to the fluctuating size of origin region population stocks. In the net migration case there are still net migrants even if the populations of other regions have disappeared. This effect is even more serious if net flows are used rather than net rates.

The long-run behaviour of the two kinds of models is quite different. In the multiregional models the regional populations all adopt the same long-run growth rate and assume constant shares of the population. In the single region model regions continue to grow at the rates initially observed, and some regional populations may disappear.

Which model should we choose to project the regional populations? The answer must be the multiregional model which clearly exhibits more reasonable behaviour in the long run. Whether the multiregional model adopted should close the system using net immigration rates or vectors of immigrant flows depends on the view taken of immigration into a country from outside. This flow is subject in part to legislative control and there might be a case for representing immigration explicitly as a flow vector rather than a rate. The internal migration terms are not, in Britain, subject to such legislative control and there is no corresponding case for representing them as flows, although there are other contexts in which a flow representation might be used (see section on population change below).

Multiregional matrix models versus multiregional accounting models for forecasting

Three varieties of multiregional models for population projection can be distinguished. The first is the multiregional matrix version of the cohort survival model in which the required rates are calculated directly from data; the second is a similar model in which the rates are computed from corresponding accounts; and the third variety is the accounts-based model.

The mis-specifications that can occur in the first case have been discussed in Rees and Wilson (1977). The degrees of mis-specification of the input variables varies from study to study. Liaw's projections of the Canadian multi-provincial system (Liaw, 1978a, b, 1980) are very well specified, for example, although they were not developed from an accounting base. Other studies, however, neglect minor migrant flows or external migration, although the authors usually claim that the omissions are not important (Gleave and Cordey-Hayes, 1977, p. 29; Baxter and Williams, 1978, p. 43).

The second and third variations of the multiregional model exhibit relatively small differences in outcome (see the results in Rees, 1977a). The mis-specification errors sometimes associated with the first variety are avoided.

However, another and stronger argument can be put forward for first calculating an accounts matrix before projecting the population. This is that in estimating accounts the consistency and reliability of the input data are assessed, and often the initial accounts matrix is adjusted. Migration rates, for example, may be changed in the adjustment process, and hence the projection outcomes will change. In essence, if an accounts matrix is prepared for a base period and the accounts are adjusted so as to balance correctly to reliable estimates of the initial state totals (start of period, populations, birth, immigrants) and the final state totals (end of period, populations, deaths, emigrants), the corresponding population projection model will at least reproduce observed change in the base period. Without using constrained accounts there is no such guarantee. These arguments are set out in detail in Rees (1979b).

How can we handle many populations of complex character simultaneously?

General issues

So far our discussion has been centred on the classification of populations by their regional location. However, to project these populations properly we have to disaggregate by age and sex, and this is done for the preferred multiregional models by the authors involved (Rogers, 1968, 1975; Willekens and Rogers, 1978; Rees and Wilson,

1977; Plessis-Fraissard and Rees, 1976) in both theoretical and applied contexts. The details of the models involved are given in these references.

There are problems involved in such disaggregation, however. The number of variables involved and the size of the matrices employed are increased. If five-year age groups are used the number of variables is increased by say 18 X 2 or 36. If one-year age groups are used then the number of variables goes up by say 85 X 2 or 170. There may be 170 separate versions of the multiregional accounts matrix as a result, for example, of one very large and empty accounts matrix. This may be acceptable for a table containing $2 \times (N^2-1)$ elements when N is 3, making 170 X 16 or 2,720 variables in total, but when N is 108 (total number of non-metropolitan counties (39), metropolitan districts (36) and London boroughs (33) in England and Wales) explicit representation of all elements $(170 \times 2(108^2-1) = 3{,}965{,}420)$ will result in an unreliably estimated matrix with a majority of elements being zero and the average being about 15.

Several approaches to this 'dimensionality' problem have been tried. The first is the development of aggregated models and an investigation of their performance in comparison with disaggregated models. The second is the use of chained probability equations, often in a hierarchic fashion. Both approaches involve information loss but gain through reduction in the number of elements involved.

Experiments with aggregated and decomposed models

Rogers (1976) has experimented with a wide variety of aggregated and decomposed models for a system consisting of the populations of the nine census divisions of the United States with 1955–60 as the base period. Projections of the population were carried out for a full nine by nine multiregional model and then eight aggregated or decomposed or combined alternatives were compared with the full nine by nine model.

In the first model Rogers aggregates the 9 X 9 census division system to a 4 X 4 census region system with relatively good results for those four aggregated units. The problem, of course, is that this doesn't provide predictions of the nine census division populations. The second model, a simple components-of-growth model (with no age groups distinguished) underpredicts, as one might expect, but does relatively well in projecting the regional shares.

The third model involves aggregation of the 9 X 9 model into nine sets of 2 X 2 models. This model performs much better than the second, because, of course, age information is introduced, and it also outperforms the next three models that involve tearing the 9 X 9 system apart in various ways and compensating for this with net migration terms. The seventh model, in which a components-of-growth model is combined with a single-region, natural-increase-only model, appears to be a compromise that works relatively well although it may not produce regional agè breakdowns that properly reflect the influence of migration. The final model involves tearing of the 9 X 9 system into three sets of 3 X 3 systems compensated by aggregating the other regions into a 'rest of the country' category. Although this last model does better than either the biregional aggregation model and single region with components-of-growth models, the gain does have its price. Model (8) demands migration data tables in the same degree of detail as the full 9 X 9 model. Model (7) gets by with an aggregate multiregional migration table only, the model (3) gets by with total in- and out-migration disaggregated by age. These latter two data tables are much more likely to be available than full age–sex disaggregated multiregional data tables, especially when there are large numbers of regions (zones). Another advantage of models (3) and (2) over those involving decomposition – (5), (6) and (8) – is that they are general procedures, applicable to any multiregional system, irrespective of the pattern of interregional flows. Design of the best decomposition involves careful investigation of the particular flow system and design of a model and computer program appropriate to that system.

A migration accounting framework

However, when modelling population systems with large numbers of regions this careful investigation and specific model building is clearly worth while. Masser (1976) outlines a hierarchical accounting system that could well be adapted for large multiregional projection models.

The small units, the migration patterns of which are being investigated or for which projections are to be carried out, are grouped into blocks of regions. Within each block all interregional flows are accounted for explicitly. The flows from regions in one block to regions in another are only accounted for in terms of subtotals, but consistency is maintained.

Model number (8) in Rogers (1976) – Decomposition B with biregional aggregation – turns out to be a slightly greater aggregation than Masser's in that migration from a region to the rest of the system outside the region's block is substituted for migrations from a region to other regional blocks taken individually.

Considerable savings can be made by the aggregations carried out in Masser's accounting framework. These savings will be important, even if fully explicit data exist, in reducing the zero elements in the migration matrix, making it less sparse and therefore more reliable from a statistical point of view. The savings will also make the job of translating migration or population models into operational computer programs easier. Expressions can be derived using simple calculus that give the number of blocks which will yield the maximum savings for a given number of regions (see Rees, 1980c, for details).

Choice of blocks of regions

Choice of the number and composition of blocks would not solely be determined by the savings in elements, of course. Masser and Brown (1975) have discussed the adaptation of Ward's grouping procedure for flows systems. The distance measure adopted is the difference between observed and expected flows from region i to region j plus the difference between observed and expected flows from region j to region i. The flows are converted to proportions of the total migration in the system to prevent the grouping procedure from being dominated by large flows to a 'nodal' region, and a contiguity constraint is also applied.

Two criteria can be used in the grouping procedure. Either the proportion of the total interaction that takes place within the blocks can be maximized or it can be minimized. In the case discussed above maximum intrablock interaction is the desired objective as it is these flows that will be represented in most spatial detail (by origin and destination region). Frequently, however, other criteria, external to the migration analysis, intervene in the choice of blocks, or in the initial regionalization of the data base units. Population projections are normally prepared for administrative or planning purposes, and their boundaries may or may not be optimal from a migration analysis point of view.

Probability chain models

The issues of dimensionality and aggregation had to be faced by the Population Studies Section of the Greater London Council (Gilje and Campbell, 1973). A model was required that would handle population projections for the thirty-three boroughs, four external zones (the Outer Metropolitan Area, the Outer South East, the Rest of Great Britain and the Rest of the World), disaggregated by single years of age and sex.

The solution involved grouping the boroughs into blocks in a fashion similar to that of Masser together with a set of four external zones. The model uses multiregional 'rates' within Greater London but models the interaction to and from external zones in the form of exogenously specified flows, apart from flows to the Outer Metropolitan Area and Outer South East which are modelled in the same way as flows between boroughs in different blocks. The migrant flow between boroughs is modelled not by a migration rate multipled by a population risk (say, the initial population) but by two chains of more aggregated probabilities, one chain applying to intrablock flows, another to interblock flows.

By substituting a chain of more aggregate probabilities for the disaggregated transition probabilities in a simpler population model, the data requirements of the projection model can be reduced, although in the London boroughs' model this is not the case because interblock flows are only available as sums of the detailed interborough flows. Even if we have not reduced the data requirements there may still be a case for using more aggregate probabilities in a chain if these aggregate terms have a lower error variance than the elements of a larger, sparse fully disaggregated matrix.

The penalty that has to be paid is that although many conditional probabilities are included in the chain, at some point the hypothesis of independence between two characteristics of the population must be invoked. Otherwise there would be no modelling gain. What Gilje and Campbell do not do is to test the goodness of fit of the resulting model. No idea is given of the sacrifices in reproducing the observed data that have been made to gain an increase in reliability.

Migration estimation models

Willekens and his colleagues (Willekens, 1977; and Willekens, Por and Raquillet, 1979) have developed efficient methods for estimating the detailed migration data needed in age–sex disaggregated population models from more aggregate data. These estimation methods may help in solving the 'dimensionality' problem.

For example, in the UK the migrant figures are sample numbers only, and we can apply sampling theory. A typical problem would be to generate estimates of $m_{ij}(x)$, migrants from region i to region j by age x, from the marginals, $m_{ij}(.)$, all age interregion migrants, $m_{i.}(x)$, total outmigrants by age, and $m_{.j}(x)$, total immigrants by age. Then the standard errors of $m_{ij}(.)$, $m_{i.}(x)$, and $m_{.j}(x)$ will, relatively speaking, be much smaller than those of $m_{ij}(x)$. The standard errors of a predicted set of $m_{ij}(x)$'s should also be lower than the observed set. Extreme values in the observed $m_{ij}(x)$ set will be dampened in the predicted $m_{ij}(x)$ set – the data will have been smoothed. In the long term the values predicted by a model using aggregated data may be more 'reliable' than those using disaggregated data. These notions undoubtedly implicitly underpin the Gilje and Campbell model, and need careful further investigation.

Migration age profiles

One of the reasons that multiregional models may run into dimensionality problems is that it is important to include age disaggregation in the models. The local planner undoubtedly would like this age disaggregation to be as fine as single years where possible so that the $(2N)^2$ elements of a multi-regional population model must be multiplied by 85 or more.

Rogers, Raquillet and Castro (1978) have investigated the way in which the migration rate by age schedule may be modelled, thereby reducing the number of varying quantities in an associated population model. For R migration rates (where R may be 85+) are substituted n model parameters. They decompose the age specific migration rate, profile or function into four components: pre-labour force function, the labour force function, the retirement function and the base level function.

The pre-labour force component is modelled as a simple negative exponential function of age. The labour force component is modelled as a 'double exponential' as is the retirement function. The base level function is a constant. The full model has eleven parameters and a reduced form without the retirement component would have seven. The 'savings' of this model representation would not be great for five-year age group projection models but would effect significant 'savings' in the case of single year of age models. As in the previous section the model would tend to smooth the migration rate profiles and this might again be regarded as an advantage if the migration rates were based on samples. The gain in model element representation would, of course, be even greater if a national age profile model were combined with a gross inter-regional migration model (one of the estimation cases discussed by Willekens, 1977; and Willekens, Por and Raquillet, 1979).

Other issues in population projection

How should systems be closed?

One advantage of building a multiregional population model from a set of accounts which was identified earlier was that explicit consideration had to be given to proper closure of the system being studied. A variety of different options for model construction have been outlined by Jenkins and Rees (1977) and Rees (1979b). The option that best reflects the degree to which external migration is controlled must be adopted. For example, immigration into Britain is subject to many controls and is probably best represented as a flow, subject to the occasional surprise (as, for example, in 1972 when large numbers of Ugandan Asians were admitted). Although in the 1960s and earlier emigration from Britain was a fairly free process, restrictions have been introduced in the 1970s which have limited the numbers emigrating.

Top down or bottom up in forecasting?

It is usual for both central and local governments (and sometimes intermediate regional bodies) to carry out forecasts. Inevitably, these will differ because of differences in models and assumptions and the differences will be a matter of methodological and political debate. However, even if there were no such differences the central and local forecasts would still differ because of problems involved in aggregation. The populations projected for the whole system of interest for a more disaggregated system will always be greater than those projected by an aggregate model (Rogers and Philipov, 1979) unless the conditions for 'perfect' aggregation are met (Rogers, 1969). Given this discrepancy, should forecasts for lower-level regions be adjusted to those for higher-level regions, or should the higher-level region forecasts be simply an aggregation of the lower-level? I am assuming that one body is carrying out the forecasts in this case. If both higher- and lower-level bodies are carrying out forecasts, the figures of the more important, higher-level body usually win out.

The answer to the question probably depends on a judgment or investigation as to the reliability of lower-level figures as opposed to higher. Thus, the question is a cousin to those raised at the end of the previous section and merits further research.

What role should time series analysis play?

So far we have assumed that population projections employ the base period rates *ad infinitum*. Clearly, this is not reasonable and most forecasters study the trends in key rates over the past and try to foresee what will happen in the future. Key rate forecasting is carried out in a fairly *ad hoc* manner and several scenarios are mapped out. High, middle high, middle low and low scenarios are proposed and the projections conditional on those scenarios are worked out. The usual idea in formulating scenarios is to assume continued change in the short run along the same path as in the past followed by asymptotic shifting towards a long-term equilibrium value. The long-run values of the rates may be in the same direction as the observed short-run trend or they may be in the opposite direction.

Formal time or space-time series analysis has been carried out on population stock numbers themselves (see, for example, Pearl and Reed, 1920; Pittenger, 1976; or Bennett, 1975a, b). This works well as long as the underlying processes of change continue along smooth paths, but usually a components view of population change is taken (as outlined earlier) and the rates associated with the components are forecast separately.

Negative exponential equations have been used to model recent mortality trends by age with some success (see OPCS, 1978; Rees, 1977a, for examples) but the same success has not attended fertility rate forecasting: indeed the major errors in British regional forecasts in the 1960s have been the result of getting the rates for this component wrong (Rees, 1980b). The main problem has been a failure to predict when trends (upwards or downwards) will end and new ones occur. In the 1970s forecasters (for example OPCS, 1978) have been at pains to predict an end to the post-1964 fall in fertility rates, although the turning-point had to be postponed for several years. In 1978 this approach was vindicated (see OPCS, 1979, Table 9, for the latest statistics).

Application of econometric time series techniques to the problem of fertility rate forecasting has not been very fruitful as Passel's (1976) extensive analysis demonstrates. The reason is that, since the end of the demographic transition in developed countries, populations have varied in their desire to have children in a fairly complex way, responding to long-term shifts in social attitudes to the family and the role of women, to short-term shifts due to economic conditions and to propaganda on behalf of family limitation (Simons, 1980). The forecaster is thus faced with the daunting task of predicting all these influential variables in order to forecast the fertility rate schedules. The work of Easterlin (1980), however, suggests that we may be able to predict fertility cycles from the age structure of the existing population.

When we turn to the forecasting of migration rates, a distressing situation reveals itself in Britain. For interregional migration rates the forecaster must depend on the Census as a source since the official continuous social survey – the General Household Survey – fails to give any regional breakdown of migration results. There are just three interregional migration tables of one-year duration (1960–1, 1965–6, 1970–1) and two interregional migration tables of five-year duration (1961–66, 1966–71) upon which to base conclusions as to trend. Comparisons are hampered by substantial and repeated changes in the areal units of measurement over the period.

Can population projections be tested?

In formal time series analysis the time series model is always tested. This is rarely done with the population projection models discussed in the two previous sections. A cross-sectional test is often carried out of migration models involved (as in Masser, 1976; but not in Gilje and Campbell, 1973). A cross-sectional test of the multiregional cohort survival type model yields information only on the degree of rounding produced by the number of decimal places adopted in the rates, rather than a test. Stillwell (1979, Chapter 9) has carried out a number of migration model tests using 1961–66-based data to predict 1966–71 migration flows. One of the key elements in the projection that has to be got right is the overall change in the rate of migration in the system – no matter how sophisticated the distribution model it will not solve this problem. Good projections therefore depend on having a good series of migration data.

So, in this important aspect of applied population geography, reliance must be placed on 'after the fact' testing of projections – did the projections of five years ago turn out to be true? – rather than 'before-the-fact' testing with a good historical series. Several authors have suggested that it would be useful to planners if forecasters assigned *a priori* probabilities to rate projections and thence to population projections. One might, for example, adopt a mean long-term total fertility rate (TFR) of 2.1 and assume that the range 1.8 to 2.4 represented the range minus and plus one standard deviation about the mean. The probability that the long-term TFR would fail below 1.8 would therefore be 0.16, for example. These probabilities might be based on a survey of informed opinion (the Delphi technique) or on a careful survey of family intentions.

The uncertainties associated with population projections will, of course, tend to vary with the age groups being considered and the future period of projection, as the Central Policy Review Staff (1976) have shown for the national projections. However, their conclusions may not apply to subnational areas because of variability in migration.

Critical investigation of the issue of testing population projections is indicated for the 1980s using the new migration data of the 1981 Census.

Why do people move around?

So far the discussion of population change and movement has focused on measurement and model representation issues. This emphasis has been adopted because the issues of measurement and formal model representation have attracted much less attention than they probably deserve. Before we can understand a phenomenon we must properly measure it.

However, researchers have always been eager to know the reasons for measured patterns and perhaps the most popular activity among population geographers in the 1970s and earlier has been the construction and cross-sectional testing of migration models. This field has been reviewed many times over the past decade from a variety of different viewpoints by Shaw (1975), Willis (1955), Stillwell (1979, Chapter 6) and Gleave and Cordey-Hayes (1977). Detailed mathematical frameworks have been proposed by Alonso (1978), Wilson (1971, 1980) and investigated in detail by Ledent (1978c). Remarks here will therefore be brief, and suggestive rather than definitive.

Three frameworks appear to characterize the migration modelling field: (i) the Markovian framework; (ii) the spatial interaction framework; and (iii) the econometric framework.

These can, of course, be linked. Alonso (1978)

integrates the Markovian and spatial interaction approaches. Wilson (1980) shows that Alonso's mathematical framework is very close to the one developed earlier in the transport field (Wilson, 1971). In this framework econometric modelling of origin interaction propensities and destination attractiveness has been carried out.

A key problem in Markovian analysis is, of course, to establish a state space within which the assumptions of the analysis – that transitions between states in a population depend only on the current state and that the probabilities of transitions apply uniformly to the population in the current state – are reasonably correct. In this respect return and repeat migration disturb both assumptions, and new models are required to handle these phenomena in non-survey contexts.

In migration models built from experience in the transport field insufficient attention has been given to features of the migration process that differ from that of intra-urban journeys. Virtually all migration modelling efforts of this kind have been of an historical nature (with the exception of Stillwell, 1979, Chapter 9) and yet projective migration models are what planners demand. Spatial interaction models have the potential for connection to policy sensitive variables, but have yet to displace Markovian models in multiregional population analysis.

Detailed comparison of the long-run behaviour of the linear (Markovian) population model and nonlinear models incorporating spatial interaction model features has been carried out by Ledent (1978c). He carries out a detailed mathematical investigation of the existence and convergence properties of a long-run equilibrium of the nonlinear model. His conclusions are that several equilibrium states of a multiregional system may exist, that they may be dependent on the initial system state, and that one or more zero states may exist at equilibrium. These features, he argues, make the nonlinear model unsatisfactory compared with the linear model for stable state investigations.

Such a conclusion would, of course be unacceptable to those attempting to verify hypotheses about migration behaviour derived from economic theory. A continual search for better regression models characterizes much of the 'regional science' effort in migration study. Much of this effort might benefit from embedding the econometric equations describing the 'attractiveness' of destinations or 'repulsiveness' of origins in a consistent spatial interaction framework.

However, the importance of making robust connections between the population system and the economy is such that fruitful interaction of this style of modelling and that of the two other frameworks is bound to develop. A vigorous literature in the interface between demographic and economic studies is emerging, and is extensively and perceptively reviewed in Ledent (1978d), recent British examples with an applied planning orientation being Madden (1976) and Breheny and Roberts (1978).

How long do people live and where?

Before closing this review, something should be said about one of the major achievements of the past decade in population studies: we can now begin to add to the traditional answers to the question 'how long do people live?', some notion of where their lives are to be spent.

The long established life table model of demographic science has been expanded into a multiregional life table model by Rogers (1975) and developed as an operational tool by Ledent (1978a, 1978b, 1980) (probability definitions), by Willekens and Rogers (1978) (computer programs), by Rees and Wilson (1975, 1977) and by Rees (1978b) (connection with accounts) and Willekens (1980) (application to working life tables). These methods have been applied in a variety of countries by scholars working in collaboration with the International Institute for Applied Systems Analysis in Austria.

Application of the methods has revealed – and not for the first time – some of the difficulties of the underlying Markovian model. These difficulties include dependence of migration probabilities on the time period of measurement because of return and repeat migration (Rees, 1977b; Long and Hansen, 1977); the dependence of current migration probabilities on place of birth (Long and Hansen, 1975), possibly to be solved by expansion of the state space; and the problem of converting movement data into transition probabilities (Ledent, 1980), perhaps insoluble. These problems resemble equivalent ones faced in population projection methodology except that they occur in more acute form in multistate life tables, for the simple reason that the principal results are expressed in matrix form, for example, the matrix of life expectancies by place of residence and by place of birth, rather than in the vector form of population projections, populations by place of current residence.

Conclusion

What then, have been the main achievements of the past decade in population geography and what are the problems it still faces?

The theme that runs through work on population geography is a concern to account for, to build into models and to explain the interactions (migration) among regions within nations. A common belief was that these explicit connections had to be represented in the system models that were built. Success in this aspect of population studies probably meant that less attention was paid to the interactions between the population system and other systems involving human activity, especially the economy and the housing market. The neglect of intersystem interactions compared with interregion interactions could, however, be regarded simply as an 'apparent' effect, a product of the neglect of the reviewer. To repair this neglect or this defect is one of the challenges for population geography in the 1980s.

As the reader was warned in the introduction this review has strayed far beyond the traditional concerns of population geography, although much more could have been said on all the topics covered here. And yet the list of research areas cursorily covered in or omitted from this review – marriage patterns, simulation of populations, household patterns and formation processes, population policies . . . – is legion, and testimony surely to the strength of interest in population matters. Problems in model construction and testing, data gathering and empirical analysis abound, but the variety of contribution from sundry disciplines surely represents a good omen for the coming decade. Population geographers should feel privileged to be part of this ferment.

References

Alonso, W. (1978) 'A theory of movement' in N.M. Hansen (ed.), *Human settlement*, Ballinger: Cambridge, Mass.

Bacharach, M. (1970) *Biproportional matrices and input-output change*, Cambridge University Press.

Baxter, R., and Williams, I. (1978) 'Population forecasting and uncertainty at the national and regional scale', *Progress in Planning*, 9 (1), Pergamon Press: Oxford.

Bennett, R.J. (1975a) 'Dynamic systems modelling of the North West region: 1. Spatio-temporal representation and identification', *Environment and Planning A*, 7, 525–38.

Bennett, R.J. (1975b) 'Dynamic systems modelling of the North West Region: 2. Estimation of the spatio-temporal policy model', *Environment and Planning A*, 7, 539–66.

Brass, W. (1974) 'Perspectives in population prediction: illustrated by statistics of England and Wales', *Journal of the Royal Statistical Society A*, 137 (4), 532–70.

Breheny, R.J., and Roberts, A.J., (1978) 'An integrated forecasting system for structure planning', *Town Planning Review*, 49 (3), 306–18.

Byron, R. (1978) 'The estimation of large social account matrices', *Journal of the Royal Statistical Society A*, 141 (3), 359–67.

Campbell, R.A. (1976) 'Local population projections', *Population Trends*, 5, 9–12.

Central Policy Review Staff (1976) *Population and the social services*, HMSO: London.

Champion, A.G. (1976) 'Evolving patterns of population distribution in England and Wales', *Transactions, Institute of British Geographers*, NS 1, 401–20.

Compton, P.A. (1969) 'Internal migration and population change in Hungary between 1959 and 1965', *Transactions, Institute of British Geographers*, 47, 111–30.

Compton, P.A. (1978) 'Fertility differentials and their impact on population distribution and composition in Northern Ireland', *Environment and Planning A*, 10 (12), 1343–456.

Cordey-Hayes, M. (1975) 'Migration and the dynamics of multiregional population systems', *Environment and Planning A*, 7, 793–814.

Courgeau, D. (1973) 'Migrants et migrations', *Population*, 28 (1), 95–129.

Department of the Environment (1971) *Long term population distribution in Great Britain – a study*, HMSO: London.

Department of the Environment (1976) *British cities: urban population and employment trends 1951–71*, Research Report 10, Department of the Environment: London.

Easterlin, R.A. (1980) *Birth and Fortune*, Grant McIntyre: London.

Eversley, D.E.C. (1971) 'Population change and regional policies since the war', *Regional Studies*, 5, 211–28.

Gilje, E.K., and Campbell, R.A. (1973) 'A new model for projecting the population of the Greater London boroughs', Research Memorandum 408, Greater London Council: London.

Girling, P.D. (1979) 'Developing the migration component of the official subnational projections: specification of research provided with invitation to tender', unpublished document, Department of the Environment: London.

Gleave, D., and Cordey-Hayes, M. (1977) 'Migration dynamics and labour market turnover', *Progress in Planning*, 8 (1), 1–95, Pergamon Press: Oxford.

Hobcraft, J.N., and Rees, P.H. (eds) (1980) *Regional demographic development*, Croom Helm: London.

Illingworth, D.R. (1976) 'Testing some new concepts and estimation methods in population accounting and related fields', PhD dissertation, School of Geography, University of Leeds.

Jackson, D. (1979) 'Current population research in the United Kingdom; a classified inventory of current research projects and selected bibliographies on the consequences of contemporary population trends', CPS Working Paper No. 79–2, Centre for Population Studies, London School of Hygiene and Tropical Medicine.

Jenkins, J.C., and Rees, P.H. (1977) 'Computer programs and aggregate accounts-based and associated forecasting models of population', Working Paper 205, School of Geography, University of Leeds.

Jones, H. (1975) 'Spatial analysis of human fertility in Scotland', *Scottish Geographical Magazine*, 91, 102–13.

Kennett, S. (1977) 'Migration trends and their contribution to population trends in the urban system (1961–1971)', Working Report No. 50, Urban Change Project, Department of Geography, London School of Economics.

Kennett, S. (1980) 'Migration within and between the metropolitan areas of Britain, 1966–71' in J.N. Hobcraft and P.H. Rees, op. cit.

Keyfitz, N. (1980) 'Multidimensional demography and its data, a comment', *Environment and Planning A*, 12, 615–22.

Lawton, R. (1977) 'People and work' in J.W. House (ed.) *The UK space: resources, environment and the future*, 2nd edn, Weidenfeld & Nicolson: London.

Lawton, R. (1980) 'Regional population trends in England and Wales, 1750–1971' in J.N. Hobcraft and P.H. Rees, op. cit.

Ledent, J. (1978a) 'Some methodological and empirical considerations in the construction of increment-decrement life tables', Research Memorandum RM-78-25, International Institute for Applied Systems Analysis: Laxenburg, Austria.

Ledent, J. (1978b) 'Temporal and spatial aspects in the conception, estimation and use of migration rates', Paper presented at the IIASA Conference on Analysis of Multiregional Population Systems; Techniques and Applications, 19–22 September 1978.

Ledent, J. (1978c) 'Stable growth in the non-linear components of change model of interregional population growth and distribution', Research Memorandum, RM-78-28, International Institute of Applied Systems Analysis: Laxenburg, Austria.

Ledent, J. (1978d) 'Regional demoeconomic growth: a survey of theories and empirical models', Working Paper, WP-78-51, International Institute for Applied Systems Analysis: Laxenburg: Austria.

Ledent, J. (1980) 'Multistate life tables: movement versus transition perspectives', *Environment and Planning A*, 12, 533–62.

Liaw, K-L. (1978a) 'Dynamic properties of the 1966–71 Canadian spatial population system', *Environment and Planning A*, 10, 389–98.

Liaw, K-L. (1978b) 'Sensitivity analysis of discrete-time, age-disaggregated interregional population systems', *Journal of Regional Science*, 18, 263–81.

Liaw, K-L. (1980) 'Multistate dynamics: the convergence of an age-by-region population system with a time invariant structural matrix. A Canadian case study', *Environment and Planning A*, 12, 589–613.

Long, L.H., and Hansen, K.A. (1975) 'Trends in return migration to the South', *Demography*, 12 (4), 601–14.

Long, L.H., and Hansen, K.A. (1977) 'Interdivisional primary, return, and repeat migration', *Public Data Use*, 5 (2), 3–10.

Macgill, S.M. (1977) 'Activity-commodity spatial interaction models and related applications', PhD thesis, School of Geography, University of Leeds.

Madden, M. (1976) 'A multilevel forecasting framework', unpublished paper, Department of Civic Design, University of Liverpool.

Masser, I. (1976) 'The design of spatial systems for internal migration analysis', *Regional Studies*, 10, 39–52.

Masser, I., and Brown, P.J.B. (1975) 'Hierarchical aggregation procedures for interaction data', *Environment and Planning A*, 7, 509–23.

OPCS, Social Survey Division (1978) *The general household survey, 1975*, HMSO: London.

OPCS (1978) *Population projections, 1976–2016*, Series PP2, No. 8, HMSO: London.

OPCS (1979) *Population trends 17, Autumn 1979*, HMSO: London.

Passel, J. (1976) 'Population projections utilising age-specific and age-parity-specific birth rates predicted with time series models: projections with confidence limits', PhD thesis, Johns Hopkins University, Baltimore, Md. Thesis No. 76–22, 942, Xerox University Microfilm, Ann Arbor, Michigan, 48106.

Pearl, R., and Reed, L.J. (1920) 'On the rate of growth of the population of the United States since 1790 and its mathematical representation', *Proceedings of the National Academy of Sciences*, 6, 275-88.

Pittenger, D.B. (1976) *Projecting state and local populations*, Ballinger: Cambridge, Mass.

Plessis-Fraissard, M., and Rees, P.H. (1976) 'A computer programme for constructing age–sex disaggregated multiregional population accounts (DAME): a full description', Working Paper 169, School of Geography, University of Leeds.

Rees, P.H. (1976) 'Modelling the regional system: the population component', Working Paper 148, School of Geography, University of Leeds. Revised version published in J.I. Clarke and J. Pelletier (eds) (1978) *Régions géographiques et régions de aménagements*, Editions l'Hermès: Lyon.

Rees, P.H. (1977a) 'The future population of East Anglia and its constituent counties (Cambridge,

Norfolk and Suffolk)', a report prepared for the East Anglia Economic Planning Council by the University of Leeds Industrial Services Ltd, under Department of the Environment Contract No. DGR/461/23.

Rees, P.H. (1977b) 'The measurement of migration from census data and other sources', *Environment and Planning A*, 9, 247–72.

Rees, P.H. (1978a) 'Problems of multiregional population analysis: data collection and demographic accounting', Working Paper 229, School of Geography, University of Leeds. *Sistemi Urbani*, 3, 3-32.

Rees, P.H. (1978b) 'Increment-decrement life tables: some further comments from a demographic accounting point of view', *Environment and Planning A*, 10, 705–26.

Rees, P.H. (1979a) *Migration and settlement: 1. United Kingdom*, Research Report RR-79-3, International Institute for Applied Systems Analysis: Laxenburg, Austria.

Rees, P.H. (1979b) 'Regional population projection models and accounting methods', *Journal of the Royal Statistical Society*, Series A, 142 (2), 223–55.

Rees, P.H. (1980a) 'Multistate demographic accounts: measurement and estimation procedures', *Environment and Planning A*, 12, 499–531.

Rees, P.H. (1980b) 'Population forecasts for British regions: a comparison' in J.N. Hobcraft and P.H. Rees, op. cit.

Rees, P.H. (1980c) 'Population geography: a review of model building efforts', Working Paper 268, School of Geography, University of Leeds.

Rees, P.H., and Wilson, A.G. (1973) 'Accounts and models for spatial demographic analysis 1: aggregate population', *Environment and Planning*, 5, 61–90.

Rees, P.H., and Wilson, A.G. (1975) 'Accounts and models for spatial demographic analysis 3: rates and life tables', *Environment and Planning A*, 7, 199–231.

Rees, P.H., and Wilson, A.G. (1977) *Spatial population analysis*, Edward Arnold: London.

Rogers, A. (1966) 'Matrix methods of population analysis', *Journal of the American Institute of Planners*, 32, 40–4.

Rogers, A. (1968) *Matrix analysis of interregional population growth and distribution*, University of California Press: Berkeley and Los Angeles.

Rogers, A. (1969) 'On perfect aggregation in the matrix cohort-survival model of interregional population growth', *Journal of Regional Science*, 9, 417–24.

Rogers, A. (1975) *Introduction to multiregional mathematical demography*, John Wiley: New York.

Rogers, A. (1976) 'Shrinking large-scale population-projection models by aggregation and decomposition', *Environment and Planning A*, 8, 515–41.

Rogers, A., and Philipov, D. (1979) 'Multiregional methods for subnational population projections', Working Paper WP-79-40, International Institute for Applied Systems Analysis: Laxenburg, Austria.

Rogers, A., Raquillet, R., and Castro, L. (1978) 'Model migration schedules and their applications', *Environment and Planning A*, 10, 475–502.

Shaw, R.P. (1975) *Migration theory and fact: a review and bibliography of current literature*, Bibliography Series Number Five, Regional Science Research Institute: Philadelphia.

Simons, J. (1980) 'Developments in the interpretation of recent fertility trends in England and Wales' in J.N. Hobcraft and P.H. Rees, op. cit.

Stillwell, J.C.H. (1979) 'Migration in England and Wales: a study of inter-area patterns, accounts, models and projections', PhD dissertation, School of Geography, University of Leeds.

Stone, R. (1965) 'A model of the educational system', *Minerva*, 3, 172–86.

Stone, R. (1971) *Demographic accounting and model-building*, OECD: Paris.

Stone, R. (1972) 'A Markovian education model and other examples linking social behaviour to the economy', *Journal of the Royal Statistical Society A*, 135 (4), 511–43.

Stone, R. (1975a) *Towards a system of social and demographic statistics*, Department of Economic and Social Affairs, Statistical Office, Studies in Methods, Series F, No. 18. United Nations, New York.

Stone, R. (1975b) 'Direct and indirect constraints in the adjustment of observations', Paper, Department of Applied Economics, University of Cambridge.

Stone, R., Champernowne, D.G., and Meade, J.E. (1942) 'The precision of national income estimates', *Review of Economic Studies*, 9, 111–25.

Willekens, F. (1977) 'The recovery of detailed migration patterns from aggregate data: an entropy maximising approach', Research Memorandum RM-77-58, International Institute for Applied Systems Analysis: Laxenburg, Austria.

Willekens, F. (1980) 'Multistate analysis: tables of working life', *Environment and Planning A*, 12, 563–88.

Willekens, F., and Rogers, A. (1978) *Spatial population analysis: methods and computer programs*, Research Report RR-78-18, Institute for Applied Systems Analysis: Laxenburg, Austria.

Willekens, F., Por, A., and Raquillet, R. (1979) 'Entropy, multiproportional and quadratic techniques for inferring detailed migration patterns from aggregate data. Mathematical theories, algorithms, and computer programs', Working Paper WP-79-88, International Institute for Applied Systems Analysis: Laxenburg, Austria.

Willis, K.G. (1975) *Problems in migration analysis*, Saxon House: Farnborough.

Wilson, A.G. (1971) 'A family of spatial interaction models', *Environment and Planning*, 3, 1–32.

Wilson, A.G. (1980) 'Comments on Alonso's "theory of movement" ', *Environment and Planning A*, 12, 727–32.

Wilson, A.G., and Rees, P.H. (1974) 'Accounts and models for spatial demographic analysis 2: age-sex disaggregated populations', *Environment and Planning*, 6, 101–16.

Woodhead, K. (1979) 'Computer programs for population projections: a study of current local authority practice and requirements', Paper, Department of Planning, Cambridgeshire County Council, Cambridge.

Chapter 30

Economic geography

R.L. Martin and N.A. Spence

Introduction

This review of quantitative methods used in studies of economic geography comes at a time when the subject as reflected in its literature seems to be questioning the efficacy of such methods (Duncan, 1979; and Sayer, 1978). Re-evaluation is naturally highly desirable and there are numerous lessons to be learned from mistakes made in the past, but the philosophy of totally abandoning the approach seems to the present authors to be inappropriate. A discussion of the reasons for this view might usefully form a short introduction to the review proper and might with profit be returned to in the concluding remarks.

It is clear that in the 1960s and 1970s research in economic geography, typical of most other areas of the subject, adopted a positivist philosophy. The methodologies attempted to replicate the approaches of the natural sciences. Some research was purely deductive with formal mathematical treatments (Denike and Parr, 1970; Beavon and Mabin, 1975; and Parr and Denike, 1970) and, typically, sought to extend the classical location theory of central places as developed by Christaller and Losch by incorporating new aspects, such as the changing composition of the urban hierarchy or the phenomenon of the multiplicity of establishments within central places, by use of the same methodology. Some research was inductive in the sense of formal hypothesis-testing (e.g. Todd, 1974; 1977), and, for example, attempted to specify the relationships between spatial economic polarization and regional development by posing an analytical framework in postulate form. Most research, however, was simply quantitatively descriptive (e.g. Stillwell, 1969; Fothergill and Gudgin, 1979), often with the objective of identifying the various components of regional industrial change, usually as expressed in terms of employment, using methods such as shift-share analysis. In this approach, a 'regional share' is defined as the amount by which a given region would grow if it developed at the national rate. An 'industrial mix component', defines the contribution due to the region's specialization in national growth or decline industries. Finally, a 'differential shift component' identifies the contribution due to each industry in the region growing at a faster or a slower rate than its national growth rate. Once defined, the components are calculated for a specific spatio-temporal system and the results described and interpreted, though rarely explained.

Little research was normative in the sense of making prescriptions about what should be, but often, as in the case of each of the references in the previous paragraph, a consideration of public policy issues came to the fore. Reaction against the positivist approach is based on this interface between research, and especially policy-oriented research, and the society on which and for which it is developed. It is argued that a positivist approach requires a detachment from the subject- matter that is impossible when man is researching society. This is the theme of an important essay in economic geography by L.J. King called *Alternatives to a Positive Economic Geography* (King, 1976). King argues that positivists traditionally search for generalizations, attempting to make 'law-like' statements and claiming to use a methodology which makes no concession to personal involvement. Often such research avoids a policy contribution, arguing that pure research is for others, politicians and the like, to evaluate and make use of. Critics of this view claim that this 'value-free' or 'pure' research does not and cannot exist. According to King these critics usually belong to one of two schools. The first group, arguing a case which he claims to be a mere distraction from the fundamental debate on economic geography as a social science, believe in alternative philosophies of phenomenology and existentialism. This approach calls for an altogether different interpretation of experience in the acquisition of knowledge. The second group, much more fundamental to King, calls for a socialist involvement in research and is usually associated with the writings of Marx and Engels and their philosophical descendants. King argues a case for a middle course between the extremes of the positivists and the phenomenologists/Marxists. Undoubtedly this is unsatisfactory to the scholars taking such views but it may appeal to the wider group of individuals searching for a methodological/philosophical home. A discussion of this middle ground should form a relevant conclusion to this review.

Reviews of wide subject areas often tend to be merely extensive bibliographies which although useful in their own right tend to be unevaluative. For this reason only a small amount of the limited space available will be used in general remarks. The main review will consist of an evaluation of three major areas of inquiry in economic geography with potential for growth and development. The first area may be described as methods of analysing regional economic fluctuations. This may be subdivided into those, mainly time series, methods employed in locational analysis and those mathematical equilibrium equation models of formal location theorizing. Second, there is a growing body of research concerned with structural models of regional and urban labour markets. The third area involves the use of matrix methods of analysis in a variety of demographic and economic accounting frameworks, of much value in forecasting and impact assessment.

Before embarking on a discussion of these main fields some mention might be made of other methods currently used in economic geography which are not to be given a full treatment here due to the particular and biased interests of the authors. First by way of explanation note that there is of course a varied and profuse literature quantitatively analysing transportation and spatial interaction and, related to this, the modelling of the urban activity system. Although arguably this is the preserve of the economic geographer it has been excluded here as it should more appropriately be dealt with elsewhere in this volume. The variety of multivariate methods of analysis has in recent years also seen much use in this area of geography as in others. The methods range from the relatively simple ordination procedures of principal components and factor analyses, through much more complex three-dimensional factoring methods, to the relational and hypothesis-testing procedures of canonical and discriminant analyses. The methodology has indeed advanced so far as to make available a 'general field theory' accounting for the interrelationships between structure and movement in economic space (Berry, 1966). However, it is perhaps fair to say that some of the early descriptive applications of multivariate techniques to measure regional socio-economic health are still the most valuable (e.g. Berry, 1965; and Smith, 1968). Subsequent to this it is clear that the methodology has advanced beyond the capabilities of empirical interpretation (Ray, 1971; Cant, 1971; Willis, 1972; Clark, 1973a, b). Perhaps the main exceptions to this are the time-varying factor analyses which are now becoming available and which will be reviewed in the next section.

Regional economic fluctuations

Since the late 1960s, regional fluctuations in employment, income and the general level of economic activity have received increasing attention as a subject for empirical and theoretical analysis. This recent interest in the 'geography of business cycles' has been stimulated in part by economic events over the past few years, especially the emergence of high and rising unemployment in most capitalist economies, but also in part by the development within 'locational analysis' of statistical techniques for the comparison, modelling and forecasting of regional time-series data. The notion that the economic cycle has a differential impact across the space-economy can be traced back to the work of American economists in the 1940s and 1950s on the nature of regional responses to national business fluctuations, using an industrial-mix, export-base or interregional trade theory frame of reference (for a review of these studies see Conroy, 1975). Drawing upon this earlier work, the recent and current geographical literature on regional economic fluctuation has focused on three interrelated themes: (i) the relative *sensitivity* of different regions to cyclical changes in economic activity; (ii) the relative *timing* of cyclical response amongst such local areas; and (iii) the interregional *transmission* of economic impulses. Both time-domain and frequency-domain methods of analysis have been used to investigate these issues.

The spatio-temporal structure of cyclical sensitivity

If the sensitivity of a region or urban area to fluctuation is a function of its industrial composition to some significant extent (Thompson, 1967, pp. 133–72), then different regions will have different mixtures of long-term and short-term cycles, and it should be possible to differentiate regions on the basis of their temporal behaviour. Several statistical procedures have been used for this purpose. One method of identifying the (unknown) cyclical structure of a given regional time series is to compute its auto-spectrum. This yields a decomposition which gives the relative importance of different periodic components, such as seasonality, short-term cycles and long-term trend movements. The significance of these components in different regions can then be compared by calculating the relative (percentage) contribution of each component to the total variance of each regional auto-spectrum. A more direct comparison of regions, for example ranking them with regard to cyclical or seasonal sensitivity, can be obtained from the absolute values of the auto-spectra. This frequency decomposition of regional economic behaviour has been a common theme in the exploratory analysis of the regional incidence of unemployment, where, typically, regions or local areas are classified (indexed) according to their sensitivity to unemployment fluctuations of different wavelength (e.g. Cliff, Haggett *et al.*, 1975; Van Duijn, 1972; Bartels, 1977a, b).

A different approach is to formulate a relationship that decomposes a region's fluctuations into components associated with certain hypothesized influences. Let y_{it} denote the value of some index of economic activity in region i at time t. Then the typical hypothesis has been that

$$y_{it} = y_{it}^{S} + y_{it}^{A} + y_{it}^{R} \qquad (1)$$

where y_{it}^{S} is the slowly-moving long-term component of change specific to the region, y_{it}^{A} is the aggregate cyclical component resulting from the region's response to fluctuations in the national level of activity, and y_{it}^{R} is the regional cyclical component, again specific to each region and reflecting, among other things, the peculiarities of the region's industrial structure. This specification derives from Brechling's (1967) study of British regional unemployment, in which he postulated that the overall level of unemployment in a region is a composite of local structural factors, national cyclical forces, and a regional cyclical effect. Using a quadratic function of time to represent the first component, and the national rate of unemployment to represent the second, the regional effect in Brechling's model is identified simply by the pattern of serial autocorrelation in the residuals $(U_{it} - \hat{U}_{it})$ from the regression

$$U_{it} = b_{i0} + b_{i1}t + b_{i2}t^2 + b_{i3}U_{N,t+k_i} + e_{it} \qquad (2)$$

where U_{it} and U_{Nt} are the local and national umemployment rates, and k is the lead or lag in the relative timing of aggregate cyclical fluctuations in the region. If b_{i3} is interpreted as the coefficient of cyclical sensitivity, then regions may be differentiated according to whether this coefficient is greater than or less than unity. Variants of this approach have been used to identify the spatial impact of unemployment cycles at several scales and in diverse settings (e.g. Jeffrey and Webb, 1972; Cliff, Haggett *et al.*, 1975; Frost and Spence, 1978a; King and Clark, 1978; Pedersen, 1978).

However, as descriptions of the spatial-temporal structure of cyclical sensitivity, Brechling-type components-models have distinct limitations. First, the specification is overly simplistic; in reality the trend and cycle are intimately linked (Kalecki, 1971, Ch. 15; Pasinetti, 1974, Ch. 3). Second, the regional structure of cyclical response (the set of b_{i3} coefficients) is assumed to be constant over time; that is, cyclical response is assumed to be invariant both from one cycle to another and over the different phases of each cycle. Third, the regional effect, being measured by the regression residuals, is influenced by specification and estimation errors, and may not therefore provide a valid basis for seeking ordered groupings of regions that share similar irregularities in their response and timing behaviour. An alternative version of (1) is the S-mode factor-analytic model, in which the nature of regional fluctuations is to be revealed by analysis rather than postulated in advance as in (2). The hypothesis is that the level of economic activity in each of the n regions, $y_1, y_2, .. y_n$, can be explained by a smaller number of underlying common 'behaviour patterns', $z_1, z_2, ... z_n$ which account for the intercorrelations of the regional series, i.e.

$$y_{it} = a_{i1} z_{1t} + a_{i2} z_{2t} + ... a_{ip} z_{pt} + e_{it} \qquad (3)$$

or,

$$y_t = Az_t + e_t \qquad (4)$$

where Z_t is the (px1) vector of behaviour types (factors), **A** is the (nxp) matrix of 'factor loadings', and e_t the (nx1) vector of error terms. The common factors are themselves linear functions of the original series,

$$z_t = Dy_t \qquad (5)$$

where **D** is the (pxn) of 'factor scores'.

The p factors extracted from the (nxn) matrix **R** of interregional correlations, $r_{ij} = \text{corr}(y_{it}, y_{jt})$, i, j = 1, . . . n permit groups of regions with similar time profiles to be identified by treating each factor as a group on which it has the highest loading. The factors themselves do not indicate what the behaviour types are, but merely the subsets of regions (or cities) which share a particular pattern. The behaviour types are summarized by the time-series plots of factor scores.

The application of this procedure to regional and urban employment and unemployment data for the US (Casetti, Jeffrey, King and Odland, various studies, 1969–1974; Pigozzi, 1975), Denmark (Pedersen, 1978), Netherlands (Bartels, 1977a) and Britain (Frost and Spence, 1978a) has consistently yielded a small number of key components of cyclical variability which differentiate systematically between peripheral (depressed) and central (growth) regions (e.g. Britain, Netherlands, Denmark), or between groups of subregions or cities that form distinct regional production systems (e.g. the US).

Although, unlike the other two methods, this factor-analytic model does take into account the intercorrelations between regions, nevertheless it too disregards the possibility that the spatial structure of economic fluctuation has a more complex intertemporal nature because of lead-lag relationships between regions. A second theme, therefore, has been the application of cross-correlation and cross-spectral techniques to detect geographical differences in the relative timing of economic impulses.

Timing of response and lead-lag relationships

The spatial pattern of the timing of cyclical sensitivity will depend not only on local industrial mixes, but also on the linkages between regions (cities) with common industries and the linkages between different industries. Many of these interdependencies may be temporally distributed, operating with various leads and lags. The presumption is that the empirical analysis of timing relationships can be used to identify the linkages between regional-industrial complexes, and thus cast light on the spatial transmission of given impulses.

The simplest approach is to measure the behaviour of individual regions relative to some chosen base or reference series, such as the national series in the Brechling model (2). The latter has been utilized to determine interregional leads and lags by estimating regression equations for each region i against national levels for positive and negative values of k_i, and simply choosing that value which produces the best-fitting equation as measured by the coefficient of determination, R^2 (Haggett, 1971; Northern Region Strategy Team, 1975; Frost and Spence, 1978b). Alternatively, cross-spectral analysis (Granger and Hatanaka, 1964; Jenkins and Watts, 1968) can be used to provide at each frequency the coherence square, gain and phase statistics between the crossed (regional) and reference (national) series. Phase indicates the time displacement that best aligns the frequency components of the two series at a particular frequency f, and in the time domain can be interpreted as the average lead or lag in time between the two series isolated for that specific frequency, i.e. the average time difference between the two time series components that have a cyclical period approximately 1/f.

Estimates of the timing of regional fluctuations relative to a national input series, obtained by time-domain regression or spectral methods, have rarely revealed leads/lags of more than six months; more often they are only one or two months, or in many instances, effectively zero (simultaneous cyclical response) (Haggett, 1971; Hepple, 1975; Cho and McDougall, 1978; Frost and Spence, 1978b). However, there must be some slight doubt over the technique used to identify leads/lags here, particularly concerning the use of the national series as a bench-mark against which to measure the performance of individual sub-areas. To the extent that an area's response timing is a function of its industrial mix, the national series will be a complex mixture of timings reflecting the sum of the sub-regional structures and their consequent response patterns. So to compare a sub-regional series with the national series may not give the clearest indication of timing differences since at least some of that sub-region's pattern will inevitably be included in national events (see Johnston, 1979, on this point). The alternative to this is to compare sub-regions directly one with one another in the hope that differences between them will be more clearly revealed.

King, Casetti and Jeffrey (1969) have suggested the following vector difference equation to represent such interregional dependencies:

$$\mathbf{y}_t = \mathbf{B}_1\mathbf{y}_{t-1} + \mathbf{B}_2\mathbf{y}_{t-2} + \ldots + \mathbf{B}_L\mathbf{y}_{t-L} + \mathbf{C}\mathbf{x} \qquad (6)$$

where **y** is the (nx1) vector of regional or urban unemployment rates, **B** is the (nxn) matrix of linkage coefficients, and **x** is an (mx1) vector of current and lagged national forces which, through C (nxm), may influence all areas simultaneously but differentially. In this system, interest centres on the pattern and magnitude of the b and c coefficients. King *et al.* tested this model for a system of urban areas in Midwest USA. Residual city series, obtained by the regression of the urban series against national unemployment, were subjected to lagged cross-correlation analysis in an attempt to identify the structure of the linkage matrices, $\mathbf{B}_k$. Results indicated three sub-systems of cities centred on Pittsburgh-Youngstown, Detroit and Indianapolis, with the latter two sub-systems lagging three to five months behind the first. In general, however, interregional lagged cross-correlation studies of this sort have provided little evidence to suggest that pronounced lead-lag relationships between individual sub-regions are a consistent feature of the space-economy; rather coincident behaviour would seem to be the normal pattern (Bartels, 1977a; Frost and Spence, 1978b; see also Engerman, 1965).

There are, however, difficulties in detecting relationships by cross-correlating two series which are themselves autocorrelated (Box and Newbold, 1971; Pierce, 1977). For example, the estimated lagged cross-correlations can have high variance and the estimates at different lags can be highly correlated with one another (Bartlett, 1935). One may be misled by attributing some significance to apparent patterns in the cross-correlation function, which are really the result of the sampling properties of the estimates used (Box and Newbold, 1971). The approach that has been suggested to alleviate these interpretative problems of cross-correlation analysis is a two-stage one in which, first, the two time series being compared are 'pre-whitened' by fitting univariate ARIMA models (Box and Jenkins, 1970) to each, and then, second, cross-correlating the residual series from these fits (Box and Jenkins, 1970; Martin and Oeppen, 1975; Haugh, 1976; Haugh and Box, 1977). The statistical properties of this procedure are not yet fully known, however, and there is some evidence that it may well bias the results towards finding *no* cross-correlation relationships (Granger, 1977). In general, pair-wise cross-spectral analysis has yielded more easily interpretable results, particularly since it gives an indication of lead-lag relationships between different regions at each of their constituent cyclical components (Trott, 1969; Bassett and Haggett, 1971; Haggett, 1971; Cliff, Haggett *et al.*, 1975; Bartels, 1977a). But as the size of the regional or urban system studied increases, it becomes increasingly difficult to determine the structure of regional interdependencies from the (nxn) matrices of cross-spectra, **S**(f), because of the large number of pair-wise comparisons involved, and a more condensed presentation of the information contained in the cross-spectral matrices is required.

In the same way that factor analytic and principal components methods may reveal common underlying patterns of regional cyclical sensitivity as described above, recent advances in the spectral theory of vector time series and in business cycle research suggest that the output of an interregional cross-spectral analysis can be used as input to a principal components analysis to study common lead-lag patterns in the multi-region system. Consider the *dynamic* factor model (Brillinger, 1975; Geweke, 1976), or 'unobservable index' model (Sargent and Sims, 1977) for the n-region series, $\mathbf{y}_t$,

$$\mathbf{y}_t = \sum_{k=0} \mathbf{A}_k \mathbf{z}_{t-k} + \mathbf{e}_t = \mathbf{A}^* \mathbf{z}_t + \mathbf{e}_t, \qquad (7)$$

where $\mathbf{z}_t$ is a (px1) vector of common factors or indices, $\mathbf{A}_k$ is an (nxp) matrix, $\mathbf{e}_t$ is an (nx1) vector of disturbances, and the common-factor series is linked to the observable series by

$$\mathbf{z}_t = \sum_{k=0} \mathbf{D}_k \mathbf{y}_{t-k} \qquad (8)$$

The spectral density of $\mathbf{y}_t$ may then be written as

$$\mathbf{S}_y(f) = \widetilde{\mathbf{A}}\, \mathbf{S}_z(f)\, \widetilde{\mathbf{A}}' + \mathbf{S}_e(f), \quad |f| \leqslant \pi, \qquad (9)$$

where $\widetilde{\mathbf{A}}$ is the Fourier transform of $\mathbf{A}^*$. The frequency domain principal components analysis of the (stationary) series $\mathbf{y}_t$ is a standard principal components analysis carried out on the individual frequency components of $\mathbf{y}_t$ and their Hilbert transforms. Let $\lambda_j(f)$ and $v_j(f)$ j = 1, . . . p denote the first p eigenvalues and eigenvectors of $\mathbf{S}_y(f)$. Then the j-th principal component series, z_{jt}, has the power spectrum $\lambda_j(f)$, $|f| \leqslant \pi$, while the moduli and arguments of the elements of the associated eigenvectors, $|v_{ij}(f)|$ and arg $v_{ij}(f)$, i = 1, . . . n and j = 1, . . . p give the gains and phase shifts between the common-factor series as inputs and each of the regional series as outputs.

Although the theory of frequency domain principal components analysis is still in its early stages, Bartels (1977a) has already demonstrated its potential usefulness in the simultaneous analysis of sensitivity and timing relationships in interregional fluctuations. In addition to 'exploratory' analysis of this sort, it should be possible to use this approach for 'confirmatory' analysis by allowing constraints on the operator **A** in (7) or permitting some or all of the z-variates to be correlated in different ways. Developments by Joreskog (1969) for the confirmatory factor model ought to be readily adaptable to our problem, so long as restrictions on **A** take the form $a_{ij,k} = 0$ for all k corresponding to certain combinations of i and j. Besides making possible a wide variety of tests, sufficient restrictions on **A** allow a substantive interpretation of individual z's. If the regional economic theory behind the zero restrictions is valid, the restrictions provide a rigorous way of averting the dynamic analogue of the 'rotation problem' of conventional factor analysis.

Modelling the spatial transmission of economic impulses

It is evident that while these different statistical models may provide precise descriptions of the spatio-temporal structure of economic fluctuation, they are essentially taxonomic, or 'measurement without theory', and as such throw little light on the transmission mechanisms and causal processes underlying observed behaviour. Theoretical rationale has traditionally been sought in the form of Keynesian-type interregional multiplier-accelerator models (see Airov, 1963; Van Duijn, 1972), in which regional output is determined by dynamic consumption and investment relations and by autonomous expenditures. Such systems typically reduce to an n-region vector difference equation, the solution of which gives the interregional time paths of adjustment (exponential growth, explosive cycles, damped cycles, steady contraction) following a change in private or public consumption. In an integrated space-economy the regional system will be indecomposable (Gandolfo, 1971, Ch. 8) so that all regions will exhibit the same qualitative behaviour, determined by the eigenvalues of the system, but with different amplitudes and timings depending on the pattern of consumption and capital coefficients. Similar overall results can be derived from alternative (accounting) models of regional adjustment, such as a multi-region Marxian scheme (Sherman, 1971, 1976) or a model of interregional monetary relations (Goodhart, 1975, Ch. 14).

Two main shortcomings characterize these formulations. First, they can explain regional adjustments about some trend value but cannot account endogenously for the latter (Pasinetti, 1974, Ch. 3). Thus in the interregional multiplier-accelerator model, changes over time in the cyclical course of different regions must be ascribed to changes in the time-lags of consumption and investment decisions, changes in the impact of automous expenditures and changes in the population and industrial structure of regions, all factors exogenous to the model. Second, these regionalizations of aggregate models of cyclical fluctuation do not incorporate explicit consideration of spatial structure, but simply treat each of the regions as individual 'point' economies.

A second frame of reference can be identified with the 'location-theoretic' approach, in which the spatial incidence and transmission of economic impulses is inferred from the premises of general location or central place theory (Jutilla, 1971, 1972, 1973; Sant, 1973; Curry, 1976). More specifically, the spatial pattern of cyclical fluctuation and the propagation of impulses amongst regions is derived from the hierarchical structure of commodity production and trade within the urban system, with changes in demand being transmitted downward from higher-order to lower-order centres, and outward from major areas of final demand to regions specializing in the production of intermediate and primary goods. Unfortunately, the economic landscapes depicted in these models are generally too abstract to be of much empirical interest. The geography of the space-economy is collapsed into some *a priori* geometrically-constituted Loschian or Christallerian structure (e.g. Sant, 1973) or in some cases into a simple linear transect (e.g. Curry, 1976), and the spatial pattern of relative fluctuation is deduced more from the arrangement of the central place network or from the mathematics of diffusion equations than from consideration of how *actual* regional and urban economies function. If the major shortcoming of statistical analyses of regional fluctuation is that they have been conducted in a state of what Hay (1973) has termed 'theoretical agnosticism', then it is also the case that the greater part of regional theory has in its turn been characterized by 'empirical atheism' and has failed to generate plausible propositions with which to confront empirical results. While we would agree with Curry (1976, p. 353) that theory construction should not be conditioned by the constraints of data availability, we cannot support his advocacy of a formal mathematical approach to regional analysis, since this view leads all too easily to 'theory without measurement', to models founded on wholly unrealistic assumptions and hence logically isolated from empirical assessment. The fundamental problem is that 'statistical methods are not general in the sense in which our logic is and that, outside of the range of probability schemata, they must grow out of the theory of the patterns to which they are to apply' (Schumpeter, 1939, p. 199). We would argue, therefore, that the thrust of future work in this area of economic geography should be to develop an empirically relevant theory of regional economies and their adjustment that is capable of yielding explicit hypotheses amenable to statistical analysis.

The analysis of spatial labour markets

The explanation of persistent differences among regional unemployment rates, observed in many countries, provides one of the most challenging problems to regional analysts, since conventional economic theory predicts that labour and capital flows should adjust to and thereby eliminate such differences. However, the observed long-run stability (or even narrowing) of the regional and urban wage structure would suggest that an equilibrating price mechanism is impeded in some way and that the market price of labour in high unemployment regions lies somewhat above its shadow price (Martin, 1979a, b). Hence the theoretical and empirical analysis of the observed regional and urban wage structure should provide important insights into the regional distribution of unemployment. Moreover, the question of how the relative adjustment of wages and unemployment in response to national, regional and local forces varies across the space-economy may help to identify the nature and causes of the national inflation-unemployment problem that has emerged in the 1970s, and may reveal the potentialities and limitations of alternative aggregate and regional policies.

Economic geographers have only just begun to analyse the money wage dynamics of spatial labour markets. Drawing upon the economic literature, the central dispute has concerned the empirical validity of the local 'Phillips curve', which in conventional form posits an inverse relationship between local wage inflation and the excess demand for labour as proxied by the unemployment rate, i.e.

$$(\dot{\omega}/\omega)_i = f_i'(U_i);\ f_i' < 0 \qquad (10)$$

where $(\dot{\omega}/\omega)_i$ is the proportionate rate of change of wages in region i. Empirical estimates of this relationship for different regional and urban markets in the UK and USA show that the slope, position and statistical significance of local Phillips curves are far from identical across the space economy (Metcalf, 1971; Smith and Patton, 1971; Smith and Smith, 1972; King and Forster, 1973; Marcus and Reed, 1974; Webb, 1974; Martin, 1979a, 1981; Hanham and Chang, 1981); quite often the curves are not significantly different from zero, in some cases they are positively sloped, and frequently they differ according to the period over which they are estimated. There would thus appear to be many local wage-employment relations in the economy, a feature which may have implications for the observed instability of the aggregate Phillips curve (Albrecht, 1970).

One particularly interesting result is that $d(\dot{\omega}/\omega)_i/U_i$ is consistently lower (the curve is flatter) in high-unemployment regions. A possible explanation for this is that high-unemployment regions owe their higher unemployment not so much to deficiency of demand but to the greater imperfection of their labour markets. This implies that if U_i is replaced by $(U_i - U_i^s)$ where U_i^s is the rate of structural unemployment, then $f_i^!(U_i - U_i^s)$ should be similar across regions with the same 'corrected' unemployment rates. However, neither Webb (1974) nor Martin (1979b) found convincing support for this hypothesis, and a satisfactory explanation for the systematic variation in $f_i^!(.)$ between regions is still required. The issue is of some importance, since if the slope of the curve does tend to be steeper in low-unemployment than in high-unemployment areas, the policy implication is that some shifting of unemployment from the high- to the low-unemployment regions would lower the aggregate rate of wage inflation (Lipsey, 1960; Archibald, 1969). In fact, (10) is not a correct representation of the competitive hypothesis that underpins the Phillips relation, for neoclassical theory postulates that it is the *expected real* wage that is bid up or down depending on the excess demand for or supply of labour in the market, yielding the 'expectations-augmented' curve

$$(\dot{\omega}/\omega)_i = f_i^!(U_i) + \alpha_i(\dot{p}/p)^e;\ f_i^! < 0 \quad \alpha_i \epsilon [0, 1] \quad (11)$$

where $(\dot{p}/p)^e$ is the expected rate of price inflation. The 'monetarist' version of this model assumes that workers and firms have rational expectations, implying that in the long run inflation is perfectly anticipated, $(\dot{p}/p)_t^e \rightarrow (\dot{p}/p)_t$ as $t \rightarrow \infty$, and that the wage adjustment mechanism responds fully to such perfectly anticipated inflation, $\alpha = 1$ (no 'money illusion'). Thus, while there is a whole *family* of short-run Phillips curves, in the long run $(\dot{\omega}/\omega)$ is independent of U. Now this hypothesis has implications for the mechanisms that hold the interregional wages structure together over the long run. If each spatial labour market is faced by similar price changes, and if the latter are aggregated over a similar vector of commodity weights, then given the 'no money illusion – perfect expectations' condition, regional wages may be kept in line by similar reactions to expected changes in the national cost-of-living index. Few studies have examined the impact of expected inflation on local wage movements, but such evidence as is available suggests that full long-run adjustment to expected inflation is not typical of all spatial labour markets (e.g. Martin, 1979a, 1981; Hanham and Chang, 1981). Moreover, using time-varying parameter regression (Kalman filter) methods, Martin found local estimates of the augmented model (11) to be far from stable over time, and that, in particular, α varies directly with $(\dot{p}/p)^e$. To the extent that adjustment to expected inflation varies across the space economy, the hypothesis that the observed rank-order stability of the regional (urban) wage structure is the result of offsetting changes in local wages to increases in the national consumer price index is inadequate, and some other explanation is required.

According to the 'institutional' hypothesis of wage determination, the interdependence of regional wage changes is the result not of competitive forces but of 'spillover effects' and 'earnings spread' arising from the operation of equitable and coercive comparisons in wage bargaining. In spatial terms, this hypothesis asserts that wage settlements achieved in region i reflect not only traditional market forces that may operate in that region, but also wage settlements in some other 'reference' sectors, and hence other regions, j. Several alternative formulations of this model are possible. Probably the most popular is that in which the spatial transfer of wage movements is assumed to emanate from one or more 'leading' regions, where the latter are typically defined in terms of low unemployment, or high average earnings, or some combination of the two (Thomas and Stoney, 1971; Brechling, 1974; MacKay and Hart, 1975). However, results from models of the form

$$(\dot{\omega}/\omega)_i = f_i^! \,[U_i\ (\dot{\omega}/\omega)_L] + \alpha_i\,(\dot{p}/p)^e;$$
$$f_{i_1}^! < 0,\ f_{i_2}^! > 0,\ \alpha_i \ \epsilon\ [0, 1] \quad (12)$$

where $(\dot{\omega}/\omega)_L$ is the rate of wage change in some appropriate leading region(s), have not on the whole been particularly decisive. Weissbrod (1976) has attempted to cast this approach into more explicitly geographical terms by using central place theory to derive the *a priori* diffusion of wage inflation down the urban hierarchy, from high-order (high relative wage) centres to low-order (low relative wage) centres. Although he found some support for his model from a cross-spectral analysis of urban wage changes in a small six-city system in South East Pennsylvania, the generalization of his findings is rendered doubtful by two main shortcomings shared by all 'leading-region' type models. First, the assumption that wage changes diffuse asymmetrically through the space-economy is too rigid and unrealistic; it ignores the impact of wage feedbacks and that a given region is likely to be a member of two or more overlapping orbits of comparison, implying that spillovers may occur in different directions at different times. Second, it is unreasonable to assume that there is a leading region set which consistently provides the reference sector for wage changes in other areas. The regions comprising this set could change from one time period to another, in a manner dependent upon the relative timing and incidence of wage settlements.

In reality, wage bargains are likely to be determined on an industry basis, and hence beyond the local market context. It would seem essential, therefore, to take account of the industrial mix of areas when analysing the spatial aspects of wage inflation spillovers. In a study of twenty-three metropolitan markets in north-east USA, Martin (1979b) found that cities contiguous in both geography and industry exhibited similar wage inflation patterns. The approach adopted was to remove the estimated effects of market forces (unemployment and inflationary expectations – equation (11)) from each urban wage change series, and then to treat the resultant residuals as a multiple time series, the covariance structure of which was used to identify sub-groups of cities having comparable behaviour profiles by applying S-mode factor analysis (cf. previous section). Three factors (group types) were obtained, corresponding to the three dominant types of manufacturing industry in the region. Because of the spatial specialization of production within the area, these groups also formed sub-regional systems of cities. Cross-correlation analysis revealed that cities within a given regional-industrial complex were more responsive to wage changes within that complex than to wage movements in the other groups.

Clearly, considerably more work needs to be done on the nature of the local wage determination process, and on the transmission of wage changes within and between spatial labour markets. However, the results obtained to date do indicate that further research using a spatially disaggregated framework will provide additional insights into the observed behaviour of aggregate inflation–unemployment relationships. As yet there is just no good theory of wage determination to draw upon to explain wage movements across time and space, but it is evident that such a theory will have to identify the external forces impinging upon spatial labour markets and which are transmitted by the different networks of economic, social and political linkages between them, and also the internal forces generated within the labour market itself (King, 1976, p. 300).

Forecasting urban and regional systems

A wide range of models have become available to researchers interested in using accounting frameworks to describe, monitor and predict urban and regional systems from both demographic and economic viewpoints. Consider first the variety of spatial demographic accounting methods which have developed out of traditional studies of population at the scale of the nation. Put simply these models describe the characteristics of the population of a spatial system at one point in time and take account of the changes that transform such a state into the characteristics of the population of the same system at the next point in time. The more detail that can be incorporated into the accounts on births and deaths, on age and sex structures, and on migration, the greater is the degree of understanding and control of how one state is transformed into another. Clearly of course the complexity of the accounts are also much increased and in order to cope with this the models are usually specified in matrix form. The accounting framework having been constructed, the next phase is to calibrate the accounts from Census or survey information so that forecasting may be undertaken under a variety of assumptions about the stability of the calibration parameters. The use of such matrix accounting methods has a long history (Leslie, 1945), although it is only relatively recently that a spatial dimension has been added explicitly. The work of Rogers in the late 1960s was particularly important in delimiting the field and indicating potential, but perhaps the most detailed review has been provided by Rees and Wilson (Rogers, 1968, 1975; and Rees and Wilson, 1977). The techniques available range from simple single region cohort-survival models which as the name implies survive age-specific cohorts (groups of people born in a given time period) from one time period to the next. Calibration of birth, death and net in-migration rates result from such models and it is clear that they omit much detail. This is the basic form of model used by Gilje and Campbell (1973) for their population forecasting at the scale of Greater London boroughs. The obvious refinement of such an approach is to extend it to many regions and this was first operationalized by Rogers (1966, and 1967). These multi-region models can more explicitly handle migration flows between regions using migration transition matrices. However even at this degree of accounting detail it is still not possible to describe individuals who are born in the calibration period and who subsequently migrate, as well as some other difficult-to-describe defects. Several of these deficiencies, including the migrating infant problem, were eliminated by the subsequent use of multiregional recurrence equations using life table concepts (Rogers, 1973). Another researcher undertaking pioneering work in the field is Stone who was responsible for a number of specifically accounting models as opposed to the components-of-change type models specified by Rogers (Stone, 1971a, b and 1972). In his models Stone is able to develop an accounting matrix which connects the opening and closing stocks, for a single region plus rest of the world system, for a time period with the flows between the system occurring during the time period. Out of this literature Wilson and Rees have developed a generalized spatial demographic accounts framework incorporating elements from a variety of sources but improving on them by eliminating deficiencies and making them truly spatial (Rees and Wilson, 1973, 1974, 1975). The next phase in the development of this field must be to capitalize on the empirical application of the methodologies that have been developed. The data requirements of the techniques are vast but no less constraining than the computational problems of massive matrices and both are increased by the multiregion case. As a result the applications to date have been limited to treatment at a large region disaggregation of national space or single small region and the rest of the nation examples. When the methods can be applied at the city or local authority scale their full and important potential will have been realized. The best empirical example of the use of these sorts of models to date is based in the West Yorkshire area of some fifty-one local authorities roughly approximating the conurbation (Rees, Smith and King, 1977). A vast amount of empirical research underlies this work and is fully referenced therein.

The second theme to be emphasized here concerns the use of input–output frameworks for describing and predicting interindustry interregional flows. These concepts have been around for many years (Leontief, 1936) especially at the level of the nation, but again it would appear that relatively recent developments are indicating the full potential of such approaches. Essentially input–output analysis seeks to record the flow of trade which links one part of the economy, either an industrial sector or a household/institution sector with every other part. The more parts of the economy that can be so described the greater is the degree of understanding and control that is gained. But as in spatial demographic accounts detail promotes complexity requiring matrix formulations for the empirical discipline and computational facility that they possess. The development of spatial or regional input–output analysis began in the 1950s often using the national accounts in a regional setting (Isard, 1951). Put simply these models describe the trade flows between sectors and between regions at one point in time. These raw flows are transformed into technical and areal coefficients of levels of inputs required to produce unit levels of output. Once calibrated such models can be used to determine the impact of altering certain sectors of the economy, either production or consumption, assuming that the coefficients reflecting the interindustry interregional relationships remain stable (Isard and Kuenne, 1953).

Recently there have been many developments in the methodology and empirical application of it. The studies, of course, have become more detailed as is evidenced by the Philadelphia input–output table of 496 production sectors plus 86 final demand sector disaggregation (Isard and Langford, 1971). The studies have also become more explicitly spatial, often incorporating gravity model concepts into the interregional commodity flow framework (Leontief and Strout, 1963; Edwards and Gordon, 1970; and O'Sullivan, 1970). Often a much more explicit forecasting component has been included such as in the Washington model (Tiebout, 1969). However, perhaps the most significant recent development has been in the literature specifically attempting to facilitate and call for practical planning applications of the methodology (Smith and Morrison, 1974; and Morrison and Smith, 1977). In addition to forming a suitable base for local planning information systems, the method facilitates impact assessment by estimating sectoral multipliers, thus allowing the specification of those sectors likely to have the greatest direct and indirect ramifications on the economy. Perhaps the most valuable empirical contribution in this area is Morrison's work on the Peterborough economy in which a whole series of sectoral multipliers are evaluated (Morrison, 1973a, b, c). The main lessons to be learned from these local studies are the intricate data manipulation techniques which facilitate the construction of the tables themselves. There has developed in recent years a series of valuable papers elaborating on non-survey techniques for constructing input–output tables (Hewings, 1971; Morrison and Smith, 1974). Clearly the survey method, although costly, still has overwhelming advantages but this is not to say that the non-survey methods do not have a role. Interestingly it appears that a combination of the two may be the most cost-effective way of constructing input–output tables. The technique known as the biproportional or RAS method has been used to norm the national input–output tables to regional situations by imposing row and column total regional constraints which are usually estimated through local survey methods. As was the case in the spatial demographic accounting models, input–output models have developed sufficiently in methodological terms to make possible a phase of practical application to real-world issues.

Conclusion

The review began by considering King's perceptive paper *Alternatives to a Positive Economic Geography* and its implications for quantitative studies. King called for the development of a policy paradigm for the subject, one which considers values and goals alongside facts. Research into regional development processes is probably the best example of a field of economic geography in which the variety of societal goals must be related to the possible set of public interventions. Quantitative and other research in the subject should move away from the formal analysis of not too relevant but often convenient data sets or abstract theorizing to analysis of issues that contribute to public debate. Such issues are of course difficult, often dirty, usually value-laden and certainly not ideal for pure quantified analysis. In the next phase of research, analysis has to be subservient to problem relevance – quantitative economic geography should attempt to supply useful findings to meet contemporary societal demands and help ameliorate its problems. The three areas of economic geography reviewed in detail here were chosen with just this potential in mind.

References

Airov, J. (1963) 'The construction of interregional business cycle models', *Journal of Regional Science*, 5, 1-20.

Albrecht, W.P. (1970) 'Intermarket and intertemporal differences in the relationship between the rate of change of wages and unemployment', *Mississippi Valley Journal of Business and Economics*, 6, 51-8.

Archibald, G.C. (1969) 'The Phillips curve and the distribution of unemployment', *American Economic Revue*, 59, 124-34.

Bartels, C.P.A. (1977a) *Economic Aspects of Regional Welfare, Income Redistribution and Unemployment*, Martinus Nijhoff: Leiden.

Bartels, C.P.A. (1977b) 'The structure of regional unemployment in the Netherlands: an exploratory statistical analysis', *Regional Science and Urban Economics*, 1, 88-102.

Bartlett, M.S. (1935) 'Some aspects of the time-correlation problem in regard to tests of significance', *Journal of the Royal Statistical Society*, 98, 536-43.

Bassett, K., and Haggett, P. (1971) 'Towards short-term forecasting for cyclic behavior in a regional system of cities' in M. Chisholm, A.E. Frey and P. Haggett (eds), *Regional Forecasting*, 389-414, Butterworth: London.

Beavon, K.S.O., and Mabin, A.S. (1975) 'The Losch system of market areas: derivation and extension', *Geographical Analysis*, 7, 131-51.

Berry, B.J.L. (1965) 'Identification of declining regions: an empirical study of the dimensions of rural poverty' in R.S. Thoman and W.D. Wood (eds), *Areas of Economic Stress in Canada*, Queens University Press, Kingston.

Berry, B.J.L. (1966) *Essays on Commodity Flows and the Spatial Structure of the Indian Economy*, Research Paper No. 111, University of Chicago, Department of Geography.

Box, G.E.P., and Jenkins, G.M. (1970) *Time Series Analysis, Forecasting and Control*, Holden-Day, San Francisco.

Box, G.E.P., and Newbold, P. (1971) 'Some comments on a paper of Coen, Gomme and Kendall', *Journal of the Royal Statistical Society, A*, 134, 229-40.

Brechling, F.P.R. (1967) 'Trends and cycles in British regional unemployment', *Oxford Economic Papers*, NS, 19, 1-21.

Brechling, F.P.R. (1974) 'Wage inflation and the structure of regional unemployment' in D.E.W. Laidler and D. Purdy (eds), *Inflation and Labour Markets*, Ch. 7, Manchester University Press.

Brillinger, D.R. (1975) *Time Series: Data Analysis and Theory*, Holt, Rinehart & Winston, New York.

Cant, R.G. (1971) 'Changes in the location of manufacturing in New Zealand, 1957-1968. An application of three mode factor analysis', *New Zealand Geographer*, 27, 38-55.

Casetti, E., King, L., and Jeffrey, D. (1971) 'Structural imbalance in the US urban-economic systems: 1960-1965', *Geographical Analysis*, 3, 239-55.

Cho, D.W., and McDougall, G.S. (1978) 'Regional cyclical patterns and structure, 1954-75', *Economic Geography*, 54, 66-74.

Clark, D. (1973a) 'Urban linkage and regional structure in Wales: and analysis of change, 1958-1968', *Transactions of the Institute of British Geographers*, 58, 41-58.

Clark, D. (1973b) 'The formal and functional structure of Wales', *Annals of the Association of American Geographers*, 62, 71-84.

Cliff, A.D., Haggett, P., Ord, J.K., Bassett, K., and Davies, R. (1975) *Elements of Spatial Structure: A Quantitative Approach*, Cambridge University Press.

Conroy, M.E. (1975) *Regional Economic Diversification*, Praeger: New York.

Curry, L. (1976) 'Fluctuations in the random economy and the control of inflation', *Geographical Analysis*, 8, 339-53.

Denike, K.G., and Parr, J.B. (1970) 'Production in space: spatial competition and restricted entry', *Journal of Regional Science*, 10, 49-63.

Duncan, S. (1979), 'Qualitative change in human geography – an introduction', *Geoforum*, 10, 1-4.

Edwards, S.L., and Gordon, I.R. (1970) 'The application of input–output methods to regional forecasting: the British experience' in M. Chisholm, A.E. Frey and P. Haggett (eds), *Regional Forecasting*, Butterworth: London.

Engerman, S. (1965) 'Regional aspects of stabilisation policy' in R.A. Musgrave (ed), *Essays in Fiscal Federalism*, Brookings Institute: Washington.

Fothergill, S., and Gudgin, G. (1979) 'Regional employment change: a sub-regional explanation', *Progress in Planning*, 12, 3, 155-219.

Frost, M., and Spence, N. (1978a) 'Changes in unemployment rates in Great Britain, 1963-1976', *Working Report 4*, Regional Unemployment Variations and Changing Economic Structure in Great Britain, London School of Economics and King's College, London.

Frost, M., and Spence, N. (1978b) 'The timing of unemployment response', *Working Report 5*,

Regional Unemployment Variations and Changing Economic Structure in Great Britain, London School of Economics and King's College, London.

Gandolfo, G. (1971) *Mathematical Methods and Models in Economic Dynamics*, North-Holland: Amsterdam.

Geweke, J. (1976) 'The dynamic factor analysis of economic time series models', *Report 7602*, Social Systems Research Institute, University of Wisconsin, Madison.

Gilje, E.K., and Campbell, R.A. (1973) 'A new model for projecting the population of the Greater London Boroughs', Department of Planning and Transportation, *Research Memorandum 408*, Greater London Council.

Goodhart, C.A.E. (1975) *Money, Information and Uncertainty*, Macmillan: London.

Granger, C.W.J. (1977) 'Comment on a paper by Pierce: relationships – and the lack thereof – between economic time series', *Journal of the American Statistical Association*, 72, 22–3.

Granger, C.W.J., and Hatanaka, M. (1964) *Spectral Analysis of Economic Time Series*, Princeton University Press, NJ.

Haggett, P. (1971) 'Leads and lags in inter-regional systems: a study of cyclic fluctuations in the South West economy' in M. Chisholm and G. Manners (eds), *Spatial Policy Problems of the British Economy*, 69–95, Cambridge University Press.

Hanham, R.Q., and Chang, H.-Y. (1981) 'Wage inflation in a growth region: the American sun belt' in R.L. Martin (ed), *Regional Wage Inflation and unemployment,* Ch. 7, Pion: London.

Haugh, L.D. (1976) 'Checking the independence of two covariance-stationary time series: a univariate residual cross-correlation approach', *Journal of the American Statistical Association*, 71, 378–85.

Haugh, L.D., and Box, G.E.P. (1977) 'Identification of dynamic regression (distributed lag) models connecting two time series', *Journal of the American Statistical Association*, 72, 121–30.

Hay, A. (1973) *Transport for the Space Economy*, Macmillan: London.

Hepple, L. (1975) 'Spectral techniques and a study of interregional economic cycles' in R. Peel, M. Chisholm and P. Haggett (eds), *Processes in Physical and Human Geography*, Ch. 19, Heinemann: London.

Hewings, G.J.D. (1971) 'Regional input–output models in the UK – some problems and prospects for the use of non-survey techniques', *Regional Studies*, 5, 11–22.

Isard, W. (1951) 'Interregional and regional input–output analysis: a model of a space economy', *Review of Economics and Statistics*, 33, 318–28.

Isard, W., and Kuenne, R.E. (1953) 'The impact of steel upon the Greater New York–Philadelphia urban industrial region', *Review of Economics and Statistics*, 35, 289–301.

Isard, W., and Langford, T.W. (1971) *Regional Input Output Study: Recollections Reflections and Diverse Notes on the Philadelphia Experience*, MIT Press: Cambridge, Mass.

Jeffrey, D. (1974) 'Regional fluctuations in unemployment within the US urban economic system', *Economic Geography*, 50, 111–23.

Jeffrey, D., and Webb, D.J. (1972) 'Economic fluctuations in the Australian regional system', *Australian Geographical Studies,* 10, 141-160.

Jeffrey, D., Cassett, E., and King, L. (1969) 'Economic fluctuations in a multiregional setting', *Journal of Regional Science*, 9, 397–404.

Jenkins, G.M., and Watts, D.G. (1968) *Spectral Analysis and its Applications*, Holden-Day, San Francisco.

Johnston, R.J. (1979) 'On the relationships between regional and national unemployment trends', *Regional Studies*, 13, 453–64.

Joreskog, K.G. (1969) 'A general approach to confirmatory maximum likelihood factor analysis', *Psychometrika*, 34, 183–202.

Jutilla, S.T. (1971) 'A linear model for agglomeration, diffusion and growth of regional economic activity', *Regional Science Perspectives*, 1, 83–108.

Jutilla, S.T. (1972) 'Dynamics of regional economic development', *Regional Science Perspectives*, 2, 95–110.

Jutilla, S.T. (1973) 'Spatial macroeconomic development', *Papers, Regional Science Association*, 30, 39–57.

Kalecki, M. (1971) *The Dynamics of the Capitalist Economy*, Cambridge University Press.

King, L.J. (1976) 'Alternatives to a positive economic geography', *Annals, Association of American Geographers*, 66, 293–308.

King, L.J., and Clark, G.L. (1978) 'Regional unemployment patterns and the spatial dimensions of macroeconomic policy: the Canadian experience, 1966–1975', *Regional Studies*, 12, 283–96.

King, L.J., and Forster, J. (1973) 'Wage rate change in urban labour markets and inter-market linkages', *Papers, Regional Science Association*, 34, 183–96.

King, L., and Jeffrey, D. (1972) 'City classification by oblique-factor analysis of time-series data' in B.J.L. Berry (ed), *City Classification Handbook; Methods and Applications*, 211–24, John Wiley: New York.

King, L., Casetti, E., and Jeffrey, D. (1969) 'Economic impulses in a regional system of cities', *Regional Studies*, 3, 213–18.

King, L., Casetti, E., Jeffrey, D., and Odland, J. (1972) 'Spatial temporal patterns in employment growth', *Growth and Change*, 3, 37–42.

Leontief, W.W. (1936) 'Quantitative input and

output relations in the economic system of the United States', *Review of Economic Statistics*, 18, 105–25.

Leontief, W.W., and Strout, A. (1963) 'Multi-regional input-output analysis' in T. Barna (ed), *Structural Interdependence and Economic Development*, Macmillan: London.

Leslie, P.H. (1945) 'On the use of matrices in certain population mathematics', *Biometrika*, 35, 183–212.

Lipsey, R.G. (1960) 'The relationship between unemployment and the rate of change of money wage rates in the United Kingdom, 1862–1957. A further analysis', *Econometrica*, 27, 1–31.

MacKay, D.I., and Hart, R.A. (1975) 'Wage inflation and the regional wage structure' in M. Parkin and A.R. Nobay (eds), *Contemporary Issues in Economics*, 88–119, Manchester University Press.

Marcus, R.G., and Reed, J.D. (1974) 'Joint estimation of the determinants of wages in sub-regional labour markets in the United States, 1961-1972', *Journal of Regional Science*, 14, 259–67.

Martin, R.L. (1978) 'Kalman filter modelling of time-varying processes in urban and regional analysis' in R.L. Martin *et al.* (eds), *Towards the Dynamic Analysis of Spatial Systems*, 104–26, Pion: London.

Martin, R.L. (1979a) 'Subregional Phillips curves, inflationary expectations, and the intermarket relative wage structure: substance and methodology', Ch. 3 in N. Wrigley (ed), *Statistical Applications in the Spatial Sciences*, Pion-Methuen: London.

Martin, R.L. (1979b) *The dynamics of wage inflation in a system of urban labour markets*, unpublished PhD thesis, University of Cambridge.

Martin, R.L. (1981) (ed.) *Regional Wage Inflation and Unemployment*, Pion: London.

Martin, R.L., and Oeppen, J.E. (1975) 'The identification of regional forecasting models using space-time correlation functions', *Transactions of the Institute of British Geographers*, 66, 95–118.

Metcalf, D. (1971) 'The determinants of earnings changes: a regional analysis for the UK, 1960–1968', *International Economics Review*, 12, 273–82.

Morrison, W.I. (1973a) 'The development of an urban interindustry model, I. Building input-output accounts', *Environment and Planning*, 5, 369–83.

Morrison, W.I. (1973b) 'The development of an urban interindustry model, II. The structure of the Peterborough economy in 1968', *Environment and Planning*, 5, 433–60.

Morrison, W.I. (1973c) 'The development of an urban interindustry model, III. Input-output multipliers for Peterborough', *Environment and Planning*, 5, 545–54.

Morrison, W.I., and Smith, P. (1974) 'Non-survey input-output techniques at the small area level: an evaluation', *Journal of Regional Science*, 14, 1–14.

Morrison, W.I., and Smith, P. (1977) 'Input-output methods in urban and regional planning: a practical guide', *Progress in Planning*, 7, 2, 59–151.

Northern Region Strategy Team (1975) 'Cyclical fluctuations in economic activity in the Northern Region, 1958–1973', *Technical Report 1*, Newcastle-upon-Tyne.

O'Sullivan, P. (1970) 'Forecasting interregional freight flows in Great Britain' in M. Chisholm, A.E. Frey and P. Haggett (eds), *Regional Forecasting*, Butterworth: London.

Parr, J.B., and Denike, K.G. (1970) 'Theoretical problems in central place analysis', *Economic Geography*, 46, 568–86.

Pasinetti, L.L. (1974) *Growth and Income Distribution: Essays in Economic Theory*, Cambridge University Press.

Pedersen, P.O. (1978) 'Interactions between short- and long-run development in regions – the case of Denmark', *Regional Studies*, 12, 683–700.

Pierce, D.A. (1977) 'Relationships – and the lack thereof – between economic time series, with special reference to money and interest rates', *Journal of the American Statistical Association*, 72, 11–22.

Pigozzi, B.W. (1975) 'The spatial-temporal structure of inter-urban economic impulses', *Tijdschrift voor Economische en Sociale Geografie*, 66, 272–6.

Ray, D.M. (1971) 'From factorial to canonical ecology: the spatial interrelationships of economic and cultural differences in Canada', *Economic Geography*, 47, 344–55.

Rees, P.H., and Wilson, A.G. (1973) 'Accounts and models for spatial demographic analysis, I. Aggregate population', *Environment and Planning*, 5, 61–90.

Rees, P.H., and Wilson, A.G. (1974) 'Accounts and models for spatial demographic analysis, II. Age-sex disaggregated populations', *Environment and Planning*, 6, 101–16.

Rees, P.H., and Wilson, A.G. (1975) 'A comparison of available models of population change', *Regional Studies*, 9, 39–61.

Rees, P.H., Smith A.P., and King, J.R. (1977) 'Population models' in A.G. Wilson, P.H. Rees and C.M. Leigh (eds), *Models of Cities and Regions: Theoretical and Empirical Developments*, Wiley: Chichester.

Rees, P.H., and Wilson, A.G. (1977) *Spatial Population Analysis*, Arnold: London.

Rogers, A. (1966) 'Matrix methods of population analysis', *Journal of the American Institute of Planners*, 32, 40–4.

Rogers, A. (1967) 'Matrix analysis of interregional population growth and distribution', *Papers, Regional Science Association*, 18, 177–96.

Rogers, A. (1968) *Matrix Analysis of Interregional Population Growth and Distribution*, University of California Press: Berkeley.

Rogers, A. (1973) 'The mathematics of multiregional demographic growth', *Environment and Planning*, 5, 3–29.

Rogers, A. (1975) *Introduction to Multiregional Mathematical Demography*, Wiley: New York.

Sant, M. (1973) *The Geography of Business Cycles*, London School of Economics Geographical Papers 5.

Sargent, T., and Sims, C.A. (1977) 'Business Cycle modelling without pretending to have too much *a priori* economic theory' in C.A. Sims (ed), *New Methods in Business Cycle Research*, 45–109, Federal Reserve Bank of Minneapolis.

Sayer, R.A. (1978) 'Mathematical modelling in regional science and political economy, some comments', *Antipode*, 10, 79–86.

Schumpeter, J.A. (1939) *Business Cycles: A Theoretical, Historical and Statistical Analysis of the Capitalist Process, Vol. I.* McGraw-Hill: New York.

Sherman, H. (1971) 'Marxist models of cyclical growth', *History of Political Economy*, 3, 28–55.

Sherman, H. (1976) 'Comparison of Keynesian and Marxist dynamic models', *UCR Working Paper 13*, Department of Economics, University of California at Riverside.

Smith, D.M. (1968) 'Identifying the grey areas – a multivariate approach', *Regional Studies*, 2, 183–93.

Smith, P., and Morrison, W.I. (1974) *Simulating the Urban Economy*, Pion: London.

Smith, V., and Patton, R. (1971) 'Sub-market labour adjustment and economic impulses: a note on the Ohio experience', *Regional Studies*, 5, 91–3.

Smith, V. and Smith, S. (1972) 'A note on municipal Phillips' curves', *Annals of Regional Science*, 6, 79–83.

Stillwell, F.J.B. (1969) 'Regional growth and structural adaptation', *Urban Studies*, 6, 162-78.

Stone, R. (1971a) *Demographic Accounting and Model Building*, OECD: Paris.

Stone, R. (1971b) 'An integrated system of demographic manpower and social statistics and its link with the system of national economic accounts', *Sankhya*, 33, 1–184.

Stone, R. (1972) 'A Markovian education model and other examples linking social behaviour to the economy', *Journal of the Royal Statistical Society*, A, 135, 511–43.

Thomas, R.L., and Stoney, P.J.M. (1971) 'Unemployment dispersion as a determinant of wage inflation in the United Kingdom, 1925–66', *The Manchester School*, 39 (2), 83–116.

Thompson, W.R. (1967) *A Preface to Urban Economics*, Johns Hopkins University Press, Baltimore.

Tiebout, C.M. (1969) 'An empirical regional input-output projection model: the state of Washington 1980', *Review of Economics and Statistics*, 51, 334–40.

Todd, D. (1974) 'An appraisal of the development pole concept in regional analysis, *Environment and Planning*, A, 6, 291–306.

Todd, D. (1977) *Polarisation and the Regional Problem, Manufacturing in Nova Scotia, 1960–1973*, Manitoba Geographical Studies, 6.

Trott, C.E. (1969) *A Cross-Spectral Model of an Urban System*, PhD thesis, Ohio State University, Columbus.

Van Duijn, J.J. (1972) *Interregional Model of Economic Fluctuation*, D.C. Heath: Lexington, Mass.

Webb, A.E. (1974) 'Unemployment, vacancies and the rate of change of earnings: a regional analysis', *Regional Papers III, National Institute of Economic and Social Research*, 1–49, Cambridge University Press.

Weissbrod, R.S.P. (1976) 'Diffusion of relative wage inflation in southeast Pennsylvania', *Studies in Geography 23*, Northwestern University. Evanston, Ill.

Willis, K.G. (1972) 'The influence of spatial structure and socio-economic factors on migration rates: a case study, Tyneside, 1961-66', *Regional Studies*, 6, 69–82.

Chapter 31

Urban geography

R.J. Johnston and N. Wrigley

Introduction

Urban geography came to prominence within the discipline, in Britain at least, slightly before the widespread adoption of 'quantitative geography'; since the latter occurred, however, the two have progressed in tandem, with quantitative urban studies often dominating the 'new' areas of human geography. Thus by the mid 1960s there was already a considerable volume of literature to be reviewed which had a quantitative base (Garner, 1967) and this has been built on since. As with many of the other early developments of quantitative study in geography, much of the initial stimulus came from outside both the British Isles (notably from North America) and the geographical profession (economics, sociology, regional science, and, later, psychology, provided the main sources of ideas and techniques).

The material encompassed by urban geography covers a wide range, and in highly urbanized countries such as the United Kingdom it sometimes appears that virtually the whole content of human geography is claimed by urban specialists. (For a general history and review of the content of urban geography, see Herbert and Johnston, 1978.) Nevertheless the major concerns can be reduced to two general subject areas: the description and modelling of patterns (both static and dynamic) of and in urban places; and attempts to model the choices and spatial behaviour of individuals and groups within the urban system. The first of these, in particular, involves investigations at several spatial scales, whereas much of the work in the other has focused on the micro-scale only. (In the study of migration, for example, although intra-urban flows are generally allocated to urban geography, the study of inter-urban flows is more frequently associated with population geography.)

At the macro-scale, urban geographers have focused much of their work on the search for regularities in the size and spatial arrangement of settlements, with some attention also being paid to settlement functions and the links between places. At the micro-scale, involving the study of patterns within individual settlements, there has been a somewhat similar focus on regularities in spatial distributions, both of different types of land use (commercial, industrial, residential, etc.) and of different types of user within each of those categories (shops selling different commodities, households of different composition and status, etc.). There has also been work on the links between the component parts of the urban mosaic, involving the study of movement patterns and reasons for them: this has involved considerable interaction between urban and behavioural geography.

Even if it were feasible in a chapter of this length, it is not the purpose of the present review to provide a comprehensive coverage of the work undertaken by urban geographers in recent years. (This is done in several textbooks, of which the most successful British example is Carter, 1976: for American texts see Yeates and Garner, 1976; King and Golledge, 1978.) Our focus is on quantitative urban geography, but even within this we have been selective. We have not chosen to produce an annotated bibliography of what techniques have been used by which authors, and on what towns; instead we have elected to highlight those topics in which not only has the use of quantitative approaches illuminated the study of urban geography but also the problems posed by urban studies have led to innovations in technical matters. Some of the topics with which we might have dealt if this book were divided only on a substantive basis are covered in more detail in other chapters, and so receive only brief mention by us.

City systems

The search for regularities in the organization of a system of urban settlements has its origin in two distinct research traditions. The first of these is deductive theory, produced by German geographers having a strong economic background: the prime examples are Christaller's (1933) central place theory, designed as a model of the size, location and functions of settlements as service centres, and Losch's (1943) more general theory of location. Both became available as translations from the German in the 1950s, and stimulated much interest, especially among American geographers in the first instance. The second research tradition reflects the inductively-based regionalism of Anglo-Saxon geography during the period 1920–50. For urban geographers this involved the study of spheres of influence of settlements as examples of functional regions: the classic studies were those of Dickinson on particular areas (1929, 1933), of Smailes (1946) on the urban hierarchy of England and Wales, and of Green (1950) on the identification of hinterlands through the mapping of bus services.

These two strands were brought together in some of the earliest 'quantitative urban geography' conducted by Garrison and Berry at the University of Washington, Seattle. They were intrigued by the concept of a hierarchical spatial structuring of settlement systems, and set out to test for the existence of such a spatial pattern (Berry and Garrison, 1958a), using techniques adopted from spatial studies in the biological sciences (notably the nearest neighbour technique developed by botanists). They were also attracted by the work of other social scientists (such as Beckmann, Simon, Stewart and Zipf) on linear regularities in the frequency distributions of a wide range of phenomena, which included settlements of various sizes, and in an important paper (Berry and Garrison, 1958b) introduced geographers to the mathematical and statistical analysis of such distributions. (Such analyses omit consideration of the spatial aspects of the distributions, which were central to the works of Christaller and Losch as well as, of course, to the inductive studies of British urban geographers.)

Interest in a location theory for settlements has continued unabated for more than twenty years since the appearance of Berry and Garrison's papers. Much of it has been American, however, and has given more attention to the frequency distribution of settlements of varying sizes than to the spatial arrangement of those settlements. The recent British literature includes contributions by Parr (1978) and by Beavon (1976); see also Hay and Beavon (1978, 1979). A related concern is with periodic market systems, which might be considered as nascent urban systems. A great deal of empirical work, much of it on West Africa, has been reported in recent years (for a review see Bromley, 1980) but there has also been some attempt to provide and test models of this form of spatial organization which involves the integration of spatial and temporal elements of the systems to provide an efficient marketing arrangement: a recent important contribution to this has been made by Hay and Smith (1980). On the non-spatial aspects of urban size distributions, the main British contribution has been a review by Richardson (1973) of the major competing statistical and mathematical models, first introduced to geographers by Berry and Garrison.

The main thrust of British work on central places has followed the inductive tradition of Dickinson, Smailes and Green (see also Bracey, 1952), although this has been informed by the theoretical developments achieved in North America. Bracey's work illustrates the continued use of the traditional approach, but more common has been the application of statistical procedures, notably classification methods, in some cases preceded by factor or principal components analyses (Tarrant, 1967; Caroe, 1968), to define hierarchies within spatial systems. Less use has been made of the nearest neighbour methods pioneered by Berry and Garrison, used by them to test for the existence of hierarchical classes and by King (1962) to test for regularity in spatial patterns (though see the paper by Witherick and Pinder, 1972). The main volume of work on central place patterns has been conducted in Wales, led by the example of Harold Carter (whose own studies, e.g. Carter, 1965, are in the non-quantitative Smailes/Bracey tradition). Thus Davies and Lewis (1970), for example, have applied factorial and graph theory methods to the delimitation of urban hinterlands and hierarchical systems, and Clark (1973a, 1973b) has used a range of multivariate techniques in his investigations of the links between centres in the Welsh urban system.

The most innovative piece of work in this field by a British geographer is by Robson (1973). A frequent criticism of central place theory relates to its inherent static nature, which makes it impossible to incorporate aspects of growth and decline. Robson's work was empirically grounded, in an attempt to model the changing distribution of settlement sizes in England and Wales during the nineteenth century. The basis of his explanation was a process whereby technological innovations are diffused through the urban system and he used simulation procedures to identify the likely consequences of such a process. This was followed by a more technical discussion of a statistical modelling of urban-size distributions, based on random processes (Cliff and Robson, 1978).

The concentration on central place aspects of urban settlement systems has meant that British geographers have largely eschewed studies of other urban functions and they have not, unlike their North American counterparts, undertaken many exercises involved in the classification of places according to functions. (Data problems are crucial here: one of the few studies, using a range of multivariate methods, employed county data: Spence, 1968.) One major classification was published early by non-geographers, however, and Moser and Scott's (1963) book has become something of a classic, in part because it was one of the first published examples of the use of factor analysis in urban studies.

There has, however, been some work on economic functions other than those associated with the central place model. Two main types can be identified: both have policy issues related to them. The first concerns the definition of the urban system as a set of nested regions of linked places, mainly for the reporting of census data and the description of changes in population distributions at various scales. The stimulus was Berry's work in North America, where he was responsible for the definition of what are known as 'daily urban systems', using census data on commuting. The initial work in Britain, first by Hall (1972) and then by a group at the London School of Economics (Drewett, Goddard and Spence, 1976), involved little statistical analysis and was largely concerned with the manipulation of large data files, but more recent work at the University of Newcastle-upon-Tyne (Coombes *et al.*, 1979), and commented upon elsewhere in this volume (Openshaw and Taylor), has attempted to set such definitional procedures into the wider context of regionalization and the possibility of ecological problems involved in use of the data sets created.

The other type of research is concerned, like some of that related to central place theory with the links between settlements. One aspect of it draws its stimulus from what is known as growth pole theory, taking its models from the work of economists such as Hirschman, Myrdal, and Perroux. They have argued that changes in the level and type of economic activity in one place will have consequences (positive and negative) for economic activity in neighbouring places, and geographers have translated this into a distance-decay effect (the closer a place is to a growth pole, the greater the consequences should be). One of the most detailed empirical studies in this context, using a range of multivariate techniques, was by Moseley (1973). Associated work has been concerned with the timing of the consequences. A group at Bristol, for example, investigated the degree to which the unemployment trends in neighbouring places run in parallel, using a variety of techniques relevant to the study of autocorrelation in space and time (Cliff *et al.*, 1975). Detailed analysis of their pioneering work is provided elsewhere in this volume (by Hepple): a recent evaluation of certain problems of interpretation in the context of urban systems is given by Johnston (1979a, b).

The internal structure of cities

As with the study of urban systems, the quantitative investigation of spatial patterns and regularities within cities by geographers has been stimulated by the earlier work of other social scientists, mainly economists and sociologists working in North America. Investigation of patterns of land use intensity, for example, was stimulated by the works of Hurd, Clark, and, later, Alonso and Wingo, whereas the study of land use patterns, and especially of residential area characteristics, owes much to sociologists such as Wendell Bell, Ernest Burgess, Amos Hawley, Robert Park, and Eshref Shevky. Some of this work, especially that of the latter group, was introduced to American urban geographers in the early 1940s (Harris and Ullman, 1945), but it was not until the 1960s, associated with the burgeoning quantitative tradition, that geographers undertook much detailed study of the complex mosaic that forms the internal structure of the city. Earlier work in Britain had been concerned with analysis of the townscape, in the regionalism tradition (Smailes, 1955; Conzen, 1960), but little of this was followed up in the later application of quantitative methods (Johnston, 1969c). Instead, most attention was focused on shopping centres as central places (where there had been a little work in the inductive tradition: Smailes and Hartley, 1961), and on the 'new' subject-matter of the residential mosaic.

The residential pattern

It is not clear why geographers came so late to studying the enormous detail which makes up the social geography of the city, unless it was a feeling that such work 'wasn't geography', because it dealt more with people than with their artifacts. But the publication in the United States of Mayer and Kohn's (1959) *Readings in Urban Geography* increased links with the expanding discipline of sociology (several of whose leaders in Britain have geography degrees) and the availability of census data on small areas within cities eventually led to an awakening of interest in this topic.

Within Britain, several researchers were investigating the potential of census small-area data during the late 1950s as raw material for studying urban social geography. By far the most notable of these among geographers was Emrys Jones, whose *A Social Geography of Belfast* (1960) was rapidly established as a classic, providing a model which others sought to emulate. In it his work on segregation of religious groups involved little statistical sophistication beyond the computation of simple indices, but there can be little doubt that he set the pace for the development of an area of study in which, to a considerable extent, British geographers led their

North American counterparts. Thus, for example, Herbert (1967) adapted the Shevky-Bell procedure for classifying census districts according to their population characteristics and was involved in more detailed analysis of urban social patterns (Herbert, 1973a; Herbert and Williams, 1964).

These early analyses of the residential mosaic involved little quantitative manipulation of the available data beyond the construction of some indices (later considered to be somewhat dubious in their composition: Johnston, 1976c). The major quantitative development after the mid 1960s, with the adoption of factor analysis and, more frequently, principal components analysis methods for pattern description, employed the large data matrices made available from the 1961, 1966 and 1971 censuses. Use of these procedures to identify common patterns of covariation in neighbourhood population and dwelling characteristics was pioneered by American sociologists in the 1950s, and their use by geographers was heralded by Berry (1964) and by a British sociologist (Gittus, 1964): the first major publications by geographers were produced only at the end of the 1960s, however (see the reviews by Rees, 1971, 1972), contemporaneously in Britain and the United States. In Britain the pioneering studies were undertaken at Cambridge, by Timms (1971) and by Robson (1969). Their efforts at identifying neighbourhood characteristics, and providing bases for classifications of districts within towns, were soon followed by others, producing both a plethora of individual studies and some attempts at comparative work (see, for example, the volume edited by Clark and Gleave, 1973). Their approach was given the collective term factorial ecology (coined by Sweetser and popularized by Berry, 1971). The technical base rapidly became a focus of criticism, however, (for a more general review of the use of this family of techniques in geography, see the chapter by Mather in this volume). Nevertheless, the approach rapidly achieved popularity, and was also adopted by historical geographers in their analyses of manuscript census data for nineteenth-century cities.

The models of the urban mosaic developed by, in particular, the Chicago sociologists during the 1920s contained two main components. The first was concerned with the nature and degree of residential segregation of various societal groups; the second involved its locational context (where did the different groups live within the city's morphology?). With regard to the former, critics have pointed out that the factorial ecology method, because of its basis in dimensionless numbers (the parameters of the normal curve), ignores the degree of segregation (Johnston, 1973a, 1979c); as a consequence, various alternatives have been experimented with (Johnston, 1979d; Newton and Johnston, 1976) in an attempt to integrate indices of the absolute level of segregation with their covariation with other aspects of the city's social geography. With regard to the spatial pattern of segregation, relatively little quantitative investigation has been undertaken by British geographers (Murdie, 1976), in part because the importance of the public sector in urban housing markets in Britain makes any search for the sorts of model developed in the United States largely irrelevant. There have been several attempts to develop methodologies for automated mapping of material relating to urban social geography, however, of which the work on London has probably been the most successful (Shepherd *et al.*, 1976): discussion of the statistical and technical problems of such mapping is undertaken elsewhere in this volume by Evans and Rhind.

Much of the basic work in the factorial ecology tradition has been involved with the manipulation of census data to provide a descriptive base for analyses of a city's social morphology. A few workers, however, have sought to relate the patterns identified in these analyses of census data with those described by other data sets, such as the distribution of certain types of illness and of criminal activity (this was indeed the intention of the pioneer studies by Robson and Timms, but their goal was largely overlooked in the later rush to apply the techniques to easily available data sets). Attempts to link the various data sets have involved the adaptation of a variety of multivariate techniques, including the potentially valuable canonical correlation analysis (see the recent reviews of some of this material in Herbert and Smith, 1979), and the results of factorial ecologies have also been used to provide sampling frameworks for the study of various other social phenomena for which aggregate data are not available. (In some cases, the matching of the data sets involves considerable inferential problems because of the small number of observations of, for example, telephone callers to a counselling service: Johnston, 1979e.)

Some of the work which involves relating aspects of the general social geography of the city, as revealed by factorial ecologies, to particular topics contains an assumption (often implicit only) that the nature of the social environment of a neighbourhood in some way influences the way individuals behave. This influence is often termed the neighbourhood effect, which represents a particular form of socialization. Unravelling its existence with aggregate data, and even with individual data, is not easy, however, and the available models (Johnston, 1976a, b) have not been subject to exhaustive tests. Thus although mention of the neighbourhood effect is common in the literature, as with studies of voting patterns (Taylor and Johnston, 1979), its evaluation awaits detailed methodological and technical developments. In the meantime, statements inferring such an effect are likely to involve problems of the ecological fallacy, of inferring characteristics of individuals from data referring to populations; most people recognize the existence of such problems, but have not followed the work of other social scientists which seeks to circumvent them by a variety of technical procedures (Dogan and Rokkan, 1969).

The retail pattern

The development in the late 1950s and early 1960s of a quantitative approach in (human) geography as a whole, and in urban geography as one of its major branches, was synonymous with statistical investigations of the retailing system. There were two aspects to this research. The first, and most important, was the attempt to identify hierarchies of market centres and to elaborate and test empirically the basic concepts of central place theory. Following the work of Berry and Garrison (1958a, b, c) on the identification of market centres in rural areas of the United States, it was soon shown (Berry 1959a, b, 1963) that a hierarchy of retail centres *within* cities could also be identified and could be consistently related to the hierarchy of market centres in rural areas (Berry and Barnum, 1962; Berry, 1967). In a series of detailed empirical studies of the retail structure of Chicago and the changing patterns of retail location in the city through time, Berry (1963) and his associates (Simmons, 1964; Garner, 1966) developed a classification and descriptive statistical model of the retail structure of the city and the internal structure of its retail nucleations, and investigated the spatial and non-spatial processes operating to change the retail structure. The second aspect of the quantitative research on the retail system conducted at this time was the attempt to describe the spatial arrangement of retail outlets in urban areas using stochastic point process models. Using nearest neighbour distance measures or quadrat frequency counts, Getis (1964), Rogers (1965) and Dacey (1966) compared observed location patterns of retail outlets in urban areas to various theoretical distributions (Poisson, Neyman type A, negative binomial, Double Poisson) generated by contagious or non-contagious spatial processes.

British studies of the retail system which have adopted a descriptive or inferential statistical approach have to a large extent built upon the pioneer American work in these two areas of research. Many attempts (e.g. Davies 1972a, b, 1974) have been made to assess the validity and utility of the Chicago-derived classification as a description of the retail structure of British cities; the changing pattern of retail location in British cities has been investigated (Sibley, 1971; Shaw, 1978) and there have been occasional uses of point process models (Sherwood, 1970) and publications in British journals of American research on the topic (Rogers, 1969; Dacey, 1972; Lee, 1979). However, British research has by no means been confined to such a role, and some previously neglected features of the retail environment have also been investigated or considered in greater detail. Four examples illustrate this point.

(i) Attention has been directed to particular types of retail outlet. At one extreme the development of ever larger retail outlets has been monitored, and using various social survey techniques attempts have been made to assess the impact of these developments on the established retail system (Dawson and Kirby, 1975, 1980; Rogers, 1974; Thorpe and Kivell, 1971; Thorpe and McGoldrick, 1974). At the other extreme the position and role of the small shop in the retail structure has been considered, either because the rapidly changing organization of retailing has resulted in a decline in this sector of retailing (Dawson and Kirby, 1979), or because small shops being located by population are useful indicators of urban spatial structure and can be used to develop general statements about the way the spatial form of cities changes through time and varies according to city size. In this latter context Sibley (1973, 1975) has represented the spatial pattern of small shops in two British cities as density surfaces which he then describes by trend surface equations or distance decay functions of the quadratic exponential type

$$D_S = D_O \exp(bs + cs^2)$$

where D_O is the central density and D_S is the density at a distance s from the centre. He then relates the parameters of these equations to the stage of development of the city and its rank in the city size distribution.

(ii) Attention has been paid to particular retail subsystems, for example the Asian immigrant subsystem (Cater, 1979) within the overall retail structure.

(iii) Investigations of spatial variations in retail prices within urban areas have been conducted (Campbell and Chisholm, 1970; Parker, 1974; Hudson, 1974c; Parker and Jerwood, 1974) and multiple regression methods used to identify the determinants of the observed price variations. Recently the stability of the geography of urban retail prices has been questioned (Johnston and Hay, 1979; Wrigley, 1979; Hay and Johnston, 1979).

(iv) Spatial associations between particular shop types in retail centres have been investigated (Davies, 1973; McShepherd and Rowley, 1978) using or modifying the method of sequence analysis suggested by Getis (1967, 1968), and this work has been extended using multivariate statistical analysis, surveys of the movement of shoppers, and linkage analysis methods (Johnston and Kissling, 1971; Davies and Bennison, 1978) to provide classifications of shopping streets within retail centres.

In general, most British research of this type has been content to adopt quantitative techniques used in other areas of geographical enquiry rather than pioneer new methods. For the development of new quantitative approaches it is necessary to turn to those studies which have adopted a mathematical optimization approach to the study of the retail structure of cities.

The origins of the mathematical optimization approach are to be found in the retail gravitation models of Reilly (1931) and Converse (1949) and the major improvements to these models suggested by Huff (1963, 1964) and Lakshmanan and Hansen (1965) who independently produced equivalent singly-constrained spatial interaction shopping models which could be used to predict the probability that a resident of zone i of the city would shop in zone j, or to predict the flow of cash from residents in zone i to shops in zone j. Using Wilson's notation (1974, 1976, 1978) the Huff-Lakshmanan-Hansen model can be written as

$$S_{ij} = A_i e_i P_i W_j^{\alpha} \exp(-\beta c_{ij})$$

where
S_{ij} is the flow of money from residences in zone i to shops in zone j
e_i is the *per capita* expenditure on retail goods by residents of zone i
P_i is the population of zone i
W_j the size of the shopping centre in zone j (which is taken as a measure of attractiveness)
c_{ij} is the cost of travel from i to j
α and β are parameters
and

$$A_i = 1/\sum_j W_j^{\alpha} \exp(-\beta c_{ij})$$

is a balancing factor which ensures that

$$\sum_j S_{ij} = e_i P_i$$

This model has been widely used in Britain (Manchester University, 1966; Cordey-Hayes, 1968; NEDO, 1970; Gibson and Pullen, 1972; Wade, 1973; Guy, 1977) for forecasting purposes and to help evaluate a range of alternative planning policies. The traditional use of the model has been to predict the flows of money $\{S_{ij}\}$ between zones, or retail sales $\sum_i S_{ij}$ in a particular shopping centre j, given $\{W_j\}$ the sizes of the shopping centres. By exploring a range of alternative sites and alternative sizes of development, in a trial and error manner, the model can be used (Wilson, 1974, p. 46) to provide optimum site and size solutions for individual retail developments, but it is not cast in an explicit and general optimization framework. Recently, however, Wilson (1976, 1978) and Coelho and Wilson (1976) have succeeded in extending the model in this direction, and have provided a model which estimates the optimum location and size of shopping centres in an urban system in which consumers attempt to maximize a particular welfare criterion. Unlike the traditional model in which the W_js are given, in the new model (Coelho and Wilson, 1976) the W_js are predicted along with the S_{ij} flows of money. In addition, Wilson (1978) has shown that the model can be extended to allow for competition between the objectives of different groups; consumers, retailers, and planners in the urban system, with consumers attempting to maximize welfare, retailers attempting to maximize profit, and planners attempting to influence development by giving different priority to suburban and central shopping centres. In this case, the optimum location and size of shopping centres can be determined in the following way:

$$\underset{\{S_{ij}\ W_j\}}{\text{maximize}}\ \lambda\left(\sum_{j \in C} W_j + \rho \sum_{j \in S} W_j\right)$$

$$+\mu\left(-\frac{1}{\beta}\sum_{ij} S_{ij} \ln \frac{S_{ij}}{W_j^{\alpha}} - \sum_{ij} S_{ij} c_{ij}\right)$$

$$+\nu\left(\gamma \sum_{ij} S_{ij} - \sum_j W_j \rho_j\right)$$

such that

$$\sum_{j \in C} W_j \geqslant W^c \qquad \text{(planners)}$$

$$\sum_j S_{ij} = e_i P_i \qquad \text{(consumers)}$$

$$W_j \geqslant W_j^{\min} \qquad \text{(retailers)}$$

where λ, μ and ν are weights representing the relative influences of planners, consumers and retailers respectively, ρ is the priority given by planners to suburban shopping centres relative to central shopping centres; and C and S are sets defining central and suburban. Wilson (1978; see also Poston and Wilson, 1977, and this volume) has also considered how this model can be made dynamic, and how smooth changes in control variables such as λ, μ and ν can result in discontinuous (catastrophic) changes in the retail structure. In this way quantitative studies of the urban retail pattern are now linked to the developments being made in catastrophe theory, and are at the forefront of attempts by geographers to model the evolution and genesis of urban structure.

Linking the parts

Much of the work on aspects of the internal structure of cities has been disaggregative in nature, looking at one particular aspect of the city system (residential, retail etc.) in isolation from the others. In addition, in terms of quantitative work, some elements in the system have received much more attention than others.

One major area of recent quantitative development, however, has been concerned with building models of the operation of the entire urban complex. Although used in part for descriptive purposes, the main aim of such modelling exercises (reflecting their origins in planning activities, especially those concerned with traffic flows) has been to provide planning tools with which future flow patterns (journeys to work, to shop etc., by modal type, and so on) can be predicted and the most efficient (in terms of traffic flows) layout for future development of housing, industrial and other areas can be assessed. Such representation of how the city operates its daily routines, and how it grows, would appear to be central to the concerns of urban geographers, but in fact little work has been done in this area which is widely accepted as urban geography (as would appear to be the case, for example, with the efforts of Wilson *et al.*, 1977, to calibrate such models for the West Yorkshire conurbation). Thus the development of these models, and the many associated technical issues, are discussed elsewhere in this volume by Batty.

Movement patterns within cities

As quantitative investigations of the residential and retail patterns of cities increased in the early 1960s, so urban geographers began to turn their attention to the spatial flows which serve to bind together and to permit evolution in the urban structure. Initially, attention focused on the description of movement patterns such as journeys to shop and to work and intra-urban migration flows, but soon attempts were made to model and explain these patterns using two contrasting approaches; a macro-scale approach which employed aggregate zonal data and often involved the use of spatial interaction models; and a micro-scale approach which attempted to explain movement patterns using models which focused upon individual choices and behaviour. In this section, two examples of the many types of movement patterns studied by urban geographers will be considered. Although this will involve some consideration of choice and behaviour, the general discussion of these topics will be deferred until the next section.

Intra-urban migration

Much geographical investigation of patterns of migration within cities has been based on aggregate data, notably that obtained from censuses. Spatial variations in mobility rates have been mapped (Cave, 1969) and associated with other aspects of the urban environment (Moore, 1969) and patterns of both net and gross flows between areas have been investigated in the light of existing models of urban structure (Johnston, 1969a, b). Relatively little work has been done in Britain, however, and the spatial aspects of migration patterns (such as distance and directional biases: Adams, 1969; Donaldson, 1973; Donaldson and Johnston, 1973; Poulsen, 1976) have largely been ignored. Quantitative studies have in general employed basic linear regression techniques, despite the widespread application in other social sciences of the more powerful method of path analysis for disentangling the determinants of the observed spatial variations (Pickvance, 1974; Taylor, 1969). Only the study of the movement of minority (usually coloured) groups has occasioned much innovative, quantitative work. Both Lee (1973) and Woods (1976) have investigated aspects of the segregation of these minority communities and Woods (1975, 1977) has used stochastic process models to chart the changing distribution of Birmingham's immigrant population, with particular reference to tipping-points, those thresholds above which one community rapidly yields a neighbourhood to another.

A second, behaviourist approach to intra-urban migration uses data collected from individuals relating to their reasons for moving and their choice of a new home. A framework for such investigations (Brown and Moore, 1970) identifies several stages including the build-up of stress which forces consideration of a move, the establishment of criteria for the selection of a new home, the search for an acceptable dwelling and the decision to move. Few studies have investigated the entire sequence, focusing instead on certain elements of it, such as the evaluation of potential destination choices (Johnston, 1971, 1972, 1973b; Rowley and Tipple, 1974; Rowley and Wilson, 1975); the interpretation of the resulting patterns of moves (e.g. Herbert, 1973b) as indicative of choice has been criticized by those who focus their work on the constraints within the urban system (Gray, 1975). The main theoretical contribution to this line of work by a British worker concerns the temporal aspects of searching for a new home (Flowerdew, 1976, 1978); what rules do people apply in deciding when to stop their search and how do temporal constraints influence the search process? As yet, the theoretical discussion has not been followed by empirical testing.

Shopping patterns

Attempts to describe and model the pattern of shopping trips within cities have been dominated by work in two major traditions, the central place theory tradition and the spatial interaction tradition.

In the late 1950s and early 1960s when the major concern of geographical studies of the retail system centred on the identification of a hierarchy of market centres and the elaboration and empirical testing of the basic concepts of central place theory, parallel studies (Nystuen, 1959; Berry, Barnum and Tennant, 1962; Barnum, 1966) of shopping trip patterns were conducted to illustrate that consumers used the various levels of a hierarchy of market centres in a systematic way and that consequently a set of ordered, nested, trade areas existed. Though generally confirming an orderly hierarchical structuring of shopping trips and trade areas in rural areas, within cities these important descriptive studies often revealed complex patterns of shopping trips producing overlapping, non-nested trade areas. Taken together with other observed deviations from central place model expectations, these findings led to considerable criticism of the restrictive assumptions concerning consumer behaviour found within existing location theory, resulting in many attempts to assess the validity of such criticism and to replace those assumptions of traditional location theory found wanting with more 'realistic' assumptions of consumer behaviour. As such criticisms concentrated on the behaviour of the individual consumer, the result was a trend towards micro-scale studies and an associated attempt to develop a 'cognitive-behavioural' framework for research which focused on the preferences and tastes of individual consumers, their images of the retail environment, and the manner in which their images and behaviour develop over time as they learn about the retail environment and adapt to changes in it (Cox and Golledge, 1969; Garner, 1970; Golledge, 1970; Rushton, 1969, 1971; Burnett, 1973).

British studies of the late 1960s and early 1970s mirrored these trends. Many (Hudson, 1974a; Bradford, 1975; Pacione, 1975; Parker, 1976) can be classified as 'cognitive-behavioural' in approach and a number of these were significant in pioneering or extending the application of new quantitative techniques for measuring individuals' preferences and cognitive images, including the semantic differential (Downs, 1970), the Repertory Grid method (Harrison and Sarre, 1971, 1975; Hudson, 1974b), paired comparison techniques, and various scaling procedures. Others considered the 'nearest centre' assumption of the formal central place models (Fingleton, 1975) and attempted to relax the assumption and develop shopping models which allowed for multiple purpose trips (Bacon, 1971; Evans, 1972). In addition a number of studies which were concerned solely with the quantitative description of the structure of shopping patterns within particular cities (Davies, 1973; Daws and Bruce, 1971; Daws and McCulloch, 1974) used 'diary' data collection techniques to investigate micro-scale features of urban shopping patterns. In the late 1970s, although studies of a 'cognitive-behavioural' type (Potter, 1977, 1978; Smith, 1976; Spenser, 1978; Williams, 1979) have continued to dominate, there are some signs of a reawakening of interest in 'rational' utility maximizing approaches, drawing on micro-economic approaches to the study of choice under uncertainty, to model the structure of shopping patterns in what is clearly a price-uncertain retail environment (Hay and Johnston, 1979; Wrigley, 1976).

Although the majority of geographical studies of shopping patterns in the 1960s and early 1970s developed out of the central place theory tradition, a significant number built upon the spatial interaction shopping models developed by Huff (1963, 1964) and Lakshmanan and Hansen (1965). In this context, British geographers have made significant contributions to the linked problems of calibration and computer implementation (Batty, 1971; Batty and Mackie, 1972; Batty and Saether, 1972; Mountcastle, Stillwell and Rees, 1974; Openshaw, 1973, 1975, 1977; Openshaw and Connolly, 1977), disaggregation of the models to handle different types of goods, types of households, modes of travel etc., and empirical application of the models at varying geographical scales (Mackett, 1973; Smith, 1973; Smith, Whitehead and Mackett, 1977). In addition there has been some consideration of how to fuse these macro-scale or aggregate spatial interaction models with micro-scale shopping models (Wilson, 1972), how to incorporate the hierarchical concepts of central place theory into their structure (Wade, 1973; Batty, 1978), and how to reformulate the model in an optimization framework in which total consumer welfare within an urban system can be maximized subject to constraints on total shopping travel, retailers' profits, zonal population and retail floorspace etc. (Wilson, 1976). As described above, this latter approach is linked to questions of optimum size and location of shopping centres (Coelho and Wilson, 1976), has been extended (Wilson, 1978) to allow for competition between the objectives of different groups (consumers, retailers and planners) in the urban systems, and has been placed into a dynamic modelling framework.

Although it is difficult to forecast radical changes in the direction of geographical studies of shopping behaviour in the 1980s, it is unlikely that 'cognitive-behavioural' or spatial interaction models will dominate to the same extent as in the 1970s. Two possible alternative approaches are worthy of consideration:

(i) *Stochastic shop-choice models*: considerable success has been achieved by British statisticians, notably Chatfield, Ehrenberg and Goodhardt (1966), in developing a two-dimensional stochastic model (known as the stationary purchasing behaviour model) which describes the major regularities of purchasing behaviour observed in consumer panel data. The model has been widely used in empirical research for over fifteen years (Ehrenberg 1968, 1969, 1972) and its predictions have proved so accurate and valuable that it has been adopted as a standard research tool in studies of brand choice. A logical development of previous work with the model, and one which appears to hold considerable potential for the geographer, is to use the model in the context of the choice of retail outlets, i.e. to consider shop choice rather than brand choice, and an exploratory application of the model in the context of retail outlet choice in a single British city (Wrigley, 1980) has provided encouraging results.

(ii) *Discrete choice models*: these will be discussed in greater detail below (see also Chapter 11). Suffice it to say at this point that significant progress has been made over the past ten years in developing such models, and that they have already been used extensively in transportation science to consider the consumer's choice of shopping destination.

Behaviour and choice modelling

At one level of explanation, urban structure can be seen as a product of the interaction between the past and present locational decisions of the individuals and groups operating in the urban system, and of the policies and decisions, both past and present, spatial and non-spatial, of the urban institutions and managers (planners, local government officials, developers, building society and bank managers, estate agents, etc.); these policies differentially constrain the range of locational choice available to various individuals and groups. In the late 1960s, urban geographers became increasingly intrigued by such decision-making, specifically that of individuals and households, and the dominant position occupied by the description and modelling of pattern was increasingly challenged by a concern to model and explain spatial choices ranging from short-term destination choices on shopping and recreation trips to long-term residential location and residential mobility decisions.

Two possible approaches to the study of spatial choice and behaviour were considered. The first was a utility maximizing approach. This was the standard approach in urban economics and most urban geographers were familiar with work such as that of Alonso (1964) who used a utility maximizing approach to study residential location and develop a general theory of urban land use. For many urban geographers, however, such an approach based on the assumption of rational economic man was felt to be too restrictive (and was perhaps also too mathematical in orientation). As a result, a much more diffuse 'cognitive-behavioural' approach emerged (see Thrift's discussion in chapter 22 of the 'active decision-making tradition') which focused attention on the attitudes and preferences of individuals their images of the urban environment, the way in which they search and learn about their environment and adapt to changes in it, and the links between image, preference and decision. Often this involved explicit rejection of the assumption of rational economic man and the utility maximizing framework.

In the early 1970s British urban geographers began to adopt and build upon 'cognitive-behavioural' research coming out of North America, and the impact of this approach in the study of shopping and residential mobility patterns has been discussed above. Some of the British studies of this period involved innovative technical work in the measurement of individuals' preferences and cognitive images (using semantic differential, repertory grid, cloze procedure, and paired comparison methods, etc.). Others made useful quantitative attempts to investigate the manner in which individuals search and learn about new urban environments (Hudson, 1975). In general, however, there were deficiencies in the majority of these studies when it came to establishing the linkages between images, preferences, and decisions.

During this same period, the utility maximizing approach was rarely adopted by British urban geographers; exceptions include optimal stopping rule models in the context of residential location (Flowerdew, 1976) and models of the impact of price uncertainty on consumer shopping behaviour (Wrigley, 1976). Work in this tradition in Britain, particularly in the context of residential location, tenure, housing services, and residential mobility decisions, was continued by researchers in related disciplines (often, however, with close links to geography, e.g. Evans, 1973; Apps, 1973, 1974; Davies, 1974) and some of the studies addressed the important question of the quantitative modelling of decision structures (i.e. sequential or simultaneous choice structures using various forms of separable utility functions).

Since the early 1970s, several developments have occurred which may change the manner in which urban geographers attempt to model and explain spatial choice and behaviour. The first of these concerns the progress which has been made in developing the traditional utility maximizing economic theory of consumer behaviour to encompass choice among discrete alternatives (see Chapter 11 of this volume for details). In the work of McFadden (1974, 1976, 1979), Domencich and McFadden (1975), Ben-Akiva (1973), Richards and Ben-Akiva (1975), Manski (1973, 1977), Hensher (1978a, 1979) and others, it has been shown that a logically consistent theory of consumer choice among discrete alternatives can be developed which, moreover, is intrinsically connected to computationally tractable statistical models (primarily logistic/logit models, but also probit, dogit, and generalized extreme value models). Many of the spatial choices which urban geographers have traditionally been concerned with (e.g. choice of shopping and recreation trip destinations, choice of residential location and tenure type, residential mobility decisions, etc) involve selection from a finite set of discrete alternatives, and have recently been analysed using this approach by workers in transportation science and economics (Quigley, 1976; Lerman, 1977; Hensher, 1978b; Domencich and McFadden, 1975; Richards and Ben-Akiva, 1975; Adler and Ben-Akiva, 1976; Koppelman and Hauser, 1977). The random utility models which underlie discrete choice analysis offer urban geographers the opportunity to reappraise the neglected utility maximizing approach to the quantitative modelling of spatial choice and behaviour, but despite their considerable potential the methods of discrete choice analysis are not yet widely known or used by British urban geographers.

The second development concerns the increased attention which is now being paid by British urban geographers to the study of constraints and allocation rather than preferences and demand (e.g. Robson, 1975; Bassett and Short, 1980). One root of this development lies in Pahl's essays on urban managerialism (Pahl, 1970, 1975) and the manner in which they and related work in urban sociology served to articulate a growing feeling that urban geography had overemphasized the study of demand rather than supply and had neglected the constraining policies of the urban institutions. Another root lies in what Thrift (Chapter 22, this volume) has termed the 'reactive decision-making tradition' in behavioural geography. This has origins in Swedish research rather than in the American research which had dominated the 'cognitive-behavioural' approach of the early 1970s, and it stresses the importance of the socio-spatial constraints surrounding human choice. One of the major challenges of the late 1970s and early 1980s in the quantitative modelling of spatial choice and behaviour is to fuse the developments which have been made in discrete choice analysis, and the family of statistical models which underpins these developments, with the deeper understanding which urban geographers now have of the importance of constraints and allocation in the urban system. (Another is to make such models dynamic.) Some early and partial suggestions along these lines have come from the Transport Studies Unit at the University of Oxford (Heggie and Jones, 1978) and from the American geographers Burnett and Hanson (1979; see also Burnett, 1980) but considerable research is yet required, particularly in the context of residential location and mobility decisions. The neglect of discrete choice models by British urban geographers suggests, however, that these developments, when and if they appear, are unlikely to find a ready audience amongst British urban geographers.

Retrospect and prospect

The expansion of the arsenal of quantitative techniques available to geographers from the late 1950s onwards was a considerable spur to work in urban geography; without the ability to handle large data matrices and to summarize their salient features, it is unlikely either that urban geographers would have risen to their numerical prominence within the discipline or that they would have extended their research interests as widely as they did. Thus quantification was a necessary condition for the growth of urban geography in the 1960s.

The interaction between quantification and urban geography was not a two-way process, however, in that relatively little work in the latter occasioned developments in the former. By and large, urban geographers were content to adopt, and if necessary slightly adapt, methods developed in other contexts. In particular, their work on pattern description became extremely dependent on the family of techniques associated with the general linear model, especially those members of the family (regression and principal components analysis) which are most readily understood. Not surprisingly, urban geographers were able to contribute virtually nothing to the development of these standard techniques, and their technical debates have very largely been concerned with the aptness of a particular procedure for the validation of a, usually weak, theory plus its relevance to data suffering certain limitations (such as the popular percentage: see Chapter 12 by Evans and Jones).

Occasionally, of course, some pioneering quantitative work has been undertaken within the general rubric of urban geography. Frequently, however, the critical breakthroughs are made by 'immigrants', as noted by Tocalis (1978, p. 124):

> In general, geographers' contributions to the theoretical evolution of the gravity concept were minor . . . This hesitancy . . . to explore other avenues of research is only interrupted on occasion when such men as Isard and Wilson inject themselves into the mainstream of research being done in the physical and behavioural sciences . . .

and yet, as noted previously, it is not clear whether Wilson's work on the spatial organization of cities and regions is widely accepted as 'urban geography'. (Whether this is important to Alan Wilson may be irrelevant; what is relevant is that non-recognition illustrates the generally conservative nature and technical immaturity of urban geography, even after more than two decades of work in which urban geographers are supposed to have been in the vanguard.)

The future role of quantification in urban geography is currently problematic. During the 1970s, philosophical and methodological developments within human geography as a whole have swung strongly against the earlier quantification and its associated positivist philosophy. A variety of alternative approaches has been advocated, several of which argue for greater attention to the processes of decision-making and the constraints within which these are set. Although, as emphasized in this chapter, the study of decision-making forms one of the few areas where the fusion of a substantive interest in urban phenomena with a behaviourist approach suggests the possibility of considerable technical advance in modelling the relevant processes, the philosophical base to such work is not entirely accepted. The sorts of decisions modelled, such as the choice of shopping centre, are criticized as trivial by those not enamoured with the positivist-quantitative tradition; such critics argue that the major decisions, which determine the morphology of towns and which set the constraints for the more trivial decisions, are few and particular: their investigation is not amenable to quantitative analysis, most of which is dependent on large samples and populations.

The history of academic geography over recent decades does not indicate that scholars socialized into one academic mould easily break out and colonize new areas. Thus the large number of urban geographers trained during the 1960s will undoubtedly continue to work within the paradigm to which they are accustomed. The volume of quantitative urban research is unlikely to reduce drastically overnight, therefore, but it may slowly decline if the species does not reproduce. Whether those undertaking that research will largely remain as relatively passive consumers of standard statistical procedures, or whether diminished volume will be associated with an upsurge of creative technical work, remains to be seen.

References

Adams, J.S. (1969) 'Directional bias in intra-urban migration', *Economic Geography*, 45, 303-23.

Adler, T., and Ben-Akiva, M. (1976) 'Joint choice model for frequency, destination and travel mode for shopping trips', *Transportation Research Record*, 569, Transportation Research Board, Washington, DC.

Alonso, W. (1964) *Location and land use*, Harvard University Press: Cambridge, Mass.

Apps, P.F. (1973) 'An approach to urban modelling and evaluation. A residential model. I: Theory 2: Implicit prices for housing services', *Environment and Planning*, 5, 619–32, 705–17.

Apps, P.F. (1974) 'An approach to urban modelling and evaluation. A residential model: 3. Demand equations for housing services', *Environment and Planning* A, 6, 11–31.

Bacon, R. (1971) 'An approach to the theory of consumer shopping behaviour', *Urban Studies*, 8, 55–64.

Barnum, H. G. (1966) *Market centres and hinterlands in Baden-Württemberg*, University of Chicago, Department of Geography Research Paper 103.

Bassett, K.A., and Short, J. (1980) *Housing and residential structure: alternative approaches*, Routledge & Kegan Paul: London.

Batty, M. (1971) 'Exploratory calibration of retail location models using search by golden section', *Environment and Planning*, 3, 411–32.

Batty, M. (1978) 'Reilly's challenge: new laws of retail gravitation which define systems of central places', *Environment and Planning A*, 10, 185–219.

Batty, M., and Mackie, S. (1972) 'The calibration of gravity, entropy and related models of spatial interaction', *Environment and Planning*, 4, 205–33.

Batty, M., and Saether, A. (1972) 'A note on the design of shopping models', *Journal of the Royal Town Planning Institute*, 58, 303-6.

Beavon, K.S.O. (1976) *Central place theory*, Longman, London.

Ben-Akiva, M.E. (1973) 'Structure of passenger travel demand models', PhD dissertation, Department of Civil Engineering, MIT, Cambridge, Mass.

Berry, B.J.L. (1959a) 'Empirical verification of concepts of spatial structure' in W.L. Garrison, B.J.L. Berry, D.F. Marble, J.D. Nystuen and R.L. Morrill, *Studies of highway development and geographic change*, University of Washington Press: Seattle.

Berry, B.J.L. (1959b) 'Ribbon developments in the urban business pattern', *Annals of the Association of American Geographers*, 49, 145–55.

Berry, B.J.L. (1963) *Commercial structure and commercial blight*, University of Chicago, Department of Geography, Research Paper 85.

Berry, B.J.L. (1964) 'Cities as systems within systems of cities', *Papers and Proceedings, Regional Science Association*, 13, 147–63.

Berry, B.J.L. (1967) *The geography of market centres and retail distribution*, Prentice-Hall: Englewood Cliffs, NJ.

Berry, B.J.L. (ed.) (1971) *Comparative factorial ecology, Economic Geography*, 47.

Berry, B.J.L., and Barnum, H.G. (1962) 'Aggregate relations and elemental components of central place systems', *Journal of Regional Science*, 4, 35-68.

Berry, B.J.L., and Garrison, W.L. (1958a) 'Functional bases of the central place hierarchy', *Economic Geography*, 34, 145–54.

Berry, B.J.L., and Garrison, W.L. (1958b) 'Recent developments of central place theory', *Papers and Proceedings of the Regional Science Association*, 4, 107-20.

Berry, B.J.L., and Garrison, W.L. (1958c) 'A note on central place theory and the range of a good',

Economic Geography, 34, 304–11.
Berry, B.J.L., and Garrison, W.L. (1958d) 'Alternative explanations of urban rank-size relationships', *Annals, Association of American Geographers*, 48, 83–91.
Berry, B.J.L., Barnum, H.G., and Tennant, R.J. (1962) 'Retail location and consumer behaviour', *Papers and Proceedings of the Regional Science Association*, 9, 65–106.
Bracey, H.E. (1952) *Social provision in rural Wiltshire*, Oxford University Press.
Bradford, M.G. (1975) 'Spatial aspects of urban consumer behaviour', PhD dissertation, University of Cambridge.
Bromley, R.J. (1980) 'Trader mobility in systems of periodic and daily markets'. In D.T. Herbert and R.J. Johnston (eds.) *Geography and the urban environment, volume 3*, 33-74, John Wiley: Chichester.
Brown, L.A., and Moore, E.G. (1970) 'The intra-urban migration process: a perspective', *Geografiska Annaler*, 52B, 1–13.
Burnett, K.P. (1973) 'The dimensions of alternatives in spatial choice processes', *Geographical Analysis*, 5, 181–204.
Burnett, K.P. (1980) 'Spatial constraints-oriented approaches to movement, microeconomic theory, and urban policy', *Urban Geography*, 1, 53-67.
Burnett, K.P., and Hanson, S. (1979) 'A rationale for an alternative mathematical paradigm for movement as complex human behaviour', *Transportation Research Record* no. 723.
Campbell, W.J., and Chisholm, M. (1970) 'Local variation in retail grocery prices', *Urban Studies*, 7, 77–81.
Caroe, L. (1968) 'A multivariate grouping scheme: association analysis of East Anglian towns' in E.G. Bowen *et al.* (eds), *Geography at Aberystwyth*, 253–69, University of Wales Press, Cardiff.
Carter, H. (1965) *The towns of Wales*, University of Wales Press: Cardiff.
Carter, H. (1976) *The study of urban geography* 2nd Ed., Edward Arnold: London.
Cater, J. (1979) 'Asian and non-Asian retailing in inner city areas: a preliminary report on a survey of Bradford, Leicester and Ealing', Paper presented at Institute of British Geographers' Annual Conference, University of Manchester.
Cave, P.W. (1969) 'Occupancy duration and the analysis of residential change', *Urban Studies*, 6, 58–69.
Chatfield, C., Ehrenberg, A.S.C., and Goodhardt, G.J. (1966) 'Progress on a simplified model of stationary purchasing behaviour', *Journal of the Royal Statistical Society* A, 129, 317–67.
Christaller, W. (1933) (translation 1966) *Central places in Southern Germany*, Prentice-Hall, Englewood Cliffs, NJ.
Clark, B.D., and Gleave, M.B. (eds) (1973) *Social patterns in cities*, Institute of British Geographers: London.
Clark, D. (1973a) 'The formal and functional structure of Wales', *Annals, Association of American Geographers*, 63, 71–84.
Clark, D. (1973b) 'Urban linkage and regional structure in Wales: an analysis of change 1958–1968', *Transactions, Institute of British Geographers*, 58, 41–58.
Cliff, A.D., and Robson, B.T. (1978) 'Changes in the size distribution of settlements in England and Wales, 1801–1960', *Environment and Planning A*, 10, 163–72.
Cliff, A.D., Haggett, H.P., Ord, J.K., Bassett, K. and Davies, R.B. (1975) *Elements of spatial structure*, Cambridge University Press.
Coelho, J.D., and Wilson, A.G. (1976) 'The optimum location and size of shopping centres', *Regional Studies*, 10, 413–21.
Converse, P. (1949) 'New laws of retail gravitation', *Journal of Marketing*, 14, 279–304.
Conzen, M.R.G. (1960) *Alnwick, Northumberland: A study in Town Plan Analysis*, Institute of British Geographers, London.
Coombes, M.G., Dixon, J.S., Goddard, J.B., Openshaw, S. and Taylor, P.J. (1979) 'Daily urban systems in Britain: from theory to practice', *Environment and Planning A*, 11, 565–74.
Cordey-Hayes, M. (1968) *Retail location models*, Centre for Environmental Studies, London, Working Paper 16.
Cox, K.R., and Golledge, R.G. (eds) (1969) *Behavioral problems in geography: a symposium*, Studies in Geography No. 17, Department of Geography, Northwestern University: Evanston, Ill.
Dacey, M.F. (1966) *A model for the areal pattern of retail and service establishment in an urban area*, Research Report 30, Department of Geography, Northwestern University: Evanston, Ill.
Dacey, M.F. (1972) 'An explanation for the observed dispersion of retail establishments in urban areas', *Environment and Planning*, 4, 323–30.
Davies, G. (1974) 'An econometric analysis of residential amenity', *Urban Studies*, 11, 217–25.
Davies, R.L. (1972a) 'Structural models of retail distribution: analogies with settlement and urban land-use theories', *Transactions of the Institute of British Geographers*, 57, 59–82.
Davies, R.L. (1972b) 'The retail pattern of the central area of Coventry', *Institute of British Geographers Occasional Publication No. 1, Urban Study Group*, 1–32.
Davies, R.L. (1973) *Patterns and profiles of consumer behaviour*, University of Newcastle, Department of Geography, Research Series No. 10.

Davies, R.L. (1974) 'Nucleated and ribbon components of the urban retail system in Britain', *Town Planning Review*, 45, 91–111.

Davies, R.L., and Bennison, D.J. (1978) 'Retailing in the city centre: the characteristics of shopping streets', *Tijdshrift voor Economische en Sociale Geografie*, 69, 270–85.

Davies, W.K.D., and Lewis, C.R. (1970) 'Regional structures in Wales: two studies in connectivity' in H. Carter and W.K.D. Davies (eds), *Urban essays: studies in the geography of Wales* 22–48, Longman: London.

Daws, L.F., and Bruce, A.J. (1971) *Shopping in Watford*, Building Research Establishment, Garston.

Daws, L.F., and McCulloch, M. (1974) *Shopping activity patterns: a travel diary study of Watford*, Current Paper 31/74, Building Research Establishment: Garston.

Dawson, J.A., and Kirby, D.A. (1975) 'Cwbran's coming superstore and how it may affect shopping patterns', *Retail and Distribution Management*, 3, 12–18.

Dawson, J.A., and Kirby, D.A. (1979) *Small scale retailing in the United Kingdom*, Teakfield: Farnborough.

Dawson, J.A., and Kirby, D.A. (1980) 'Urban retail provision and consumer behaviour' in D.T. Herbert and R.J. Johnston (eds),*Geography and the urban environment, Vol. 3*, 87-132, John Wiley: London.

Dickinson, R.E. (1929) 'The markets and market areas of Bury St Edmunds', *Sociological Review*, 22, 292–308.

Dickinson, R.E. (1933) 'The distributions and functions of the smaller urban settlements of East Anglia', *Geography*, 18, 19–31.

Dogan, M., and Rokkan, S.E. (eds) (1969) *Quantitative ecological analysis in the social sciences*, MIT Press: Cambridge, Mass.

Domencich, T.A., and McFadden, D. (1975) *Urban travel demand: a behavioural analysis*, North-Holland: Amsterdam.

Donaldson, B. (1973) 'An empirical investigation into the concept of sectoral bias in the mental maps, search spaces and migration patterns of intra-urban migrants', *Geografiska Annaler*, 55B, 13–33.

Donaldson, B., and Johnston, R.J. (1973) 'Sectoral mental maps: further evidence from an extended methodology', *Geographical Analysis*, 5, 45–54.

Downs, R.M. (1970) 'The cognitive structure of an urban shopping centre', *Environment and Behaviour*, 2, 13–39.

Drewett, R., Goddard, J.B., and Spence, N.A. (1976) *British cities: urban population and employment trends 1951-1971*, Research Report 10, Department of the Environment: London.

Ehrenberg, A.S.C. (1968) 'The practical meaning and usefulness of the NBD/LSD theory of repeat-buying', *Applied Statistics*, 17, 17–32.

Ehrenberg, A.S.C. (1969) 'Towards an integrated theory of consumer behaviour', *Journal of the Market Research Society*, 11, 305–37. Reprinted in A.S.C. Ehrenberg and G.F. Pyatt (eds) (1971), *Consumer behaviour*, Penguin: Harmondsworth.

Ehrenberg, A.S.C. (1972) *Repeat buying: theory and applications*, North Holland: Amsterdam.

Evans, A.W. (1972) 'A linear programming solution to the shopping problem posed by R.W. Bacon', *Urban Studies*, 9, 221-2.

Evans, A.W. (1973) *The economics of residential location*, Macmillan: London.

Fingleton, B. (1975) 'A factorial approach to the nearest centre hypothesis', *Transactions of the Institute of British Geographers*, 65, 131–9.

Flowerdew, R. (1976) 'Search strategies and stopping rules in residential mobility', *Transactions, Institute of British Geographers*, NS 1, 47–57.

Flowerdew R. (1978) 'The role of time in residential choice models' in T. Carlstein *et al.* (eds), *Human activity and time geography*, 39-48, Edward Arnold: London.

Garner, B.J. (1966) *The internal structure of shopping centres*, Studies in Geography 12, Northwestern University: Evanston, Ill.

Garner, B.J. (1967) 'Models of urban geography and settlement location' in R.J. Chorley and P. Haggett (eds), *Models in geography*, Methuen: London.

Garner, B.J. (1970) 'Towards a better understanding of shopping patterns' in R.H. Osborne, F.A. Barnes and J.C. Doornkamp (eds), *Geographical essays in honour of K.C. Edwards*, Nottingham University Press.

Getis, A. (1964) 'Temporal land use pattern analyses with the use of the nearest neighbour and quadrat methods', *Annals of the Association of American Geographers*, 54, 391–8.

Getis, A. (1967) 'A method for the study of sequences in geography', *Transactions of the Institute of British Geographers*, 42, 87–92.

Getis, A, and Getis, J.M. (1968) 'Retail store spatial affinities', *Urban Studies*, 5, 317–32.

Gibson, M., and Pullen, M. (1972) 'Retail turnover in the East Midlands: a regional application of a gravity model', *Regional Studies*, 6, 183–96.

Gittus, E. (1964) 'The structure of urban areas', *Town Planning Review*, 35, 5–20.

Golledge, R. G. (1970) 'Some equilibrium models of consumer behaviour', *Economic Geography*, 46, 417–24.

Gray, F. (1975) 'Non-explanation in urban geography', *Area*, 7, 228–35.

Green, F.H.W. (1950) 'Urban hinterlands in England and Wales: an analysis of bus services', *Geographical Journal*, 116, 64–88.

Guy, C.M. (1977) 'A method of examining and evaluating the impact of major retail developments upon existing shops and their users', *Environment and Planning A*, 9, 491–504.

Hall, P. (1972) 'Spatial structure of metropolitan England and Wales' in M. Chisholm and G. Manners (eds), *Spatial policy problems of the British economy*, 96-125, Cambridge University Press.

Harris, C.D., and Ullman, E.L. (1945) 'The nature of cities', *Annals, American Academy of Political and Social Science*, 242, 7–17.

Harrison, J.A., and Sarre, P. (1971) 'Personal construct theory in the measurement of environmental images: problems and methods', *Environment and Behavior*, 3, 351-74.

Harrison, J.A., and Sarre, P. (1975) 'Personal construct theory in the measurement of environmental images', *Environment and Behaviour*, 7, 3-58.

Hay, A.M., and Beavon, K.S.O. (1978) 'Long-run average costs and Christaller's concept of range: a note', *Geography*, 63, 98–100.

Hay, A.M., and Beavon, K.S.O. (1979) 'Periodic marketing: a preliminary graphical analysis of the conditions for part-time and mobile marketing', *Tijdschrift voor Economische en Sociale Geografie*, 70, 27–34.

Hay, A.M., and Johnston, R.J. (1979) 'Search and the choice of shopping centre: two models of variability in destination selection', *Environment and Planning A*, 11, 791–804.

Hay, A.M., and Smith, R.H.T. (1980) 'Consumer welfare in periodic market systems', *Transactions, Institute of British Geographers*, NS 5, 29–44.

Heggie, I.G., and Jones, P.M. (1978) 'Defining domains for models of travel demand', *Transportation*, 7, 119–35.

Hensher, D.A. (1978a) *A review of individual choice modelling*, Report prepared for Australian Department of Environment, Housing and Community Development, Housing and Research Grant AHR–Project 73.

Hensher, D.A. (1978b) *The demand for location and accommodation: a qualitative choice approach in a policy formulating environment*, School of Economic and Financial Studies, Macquarie University: Australia.

Hensher, D.A. (1979) 'Individual choice modelling with discrete commodities: theory and application to the Tasman Bridge re-opening', *Economic Record*, 50, 243–61.

Herbert, D.T. (1967) 'Social area analysis: a British study', *Urban Studies*, 4, 41–60.

Herbert, D.T. (1973a) *Urban geography: a social perspective*, David and Charles: Newton Abbot.

Herbert, D.T. (1973b) 'The residential mobility process: some empirical observations', *Area*, 5, 44–8.

Herbert, D.T., and Johnston, R.J. (1978) 'Geography and the urban environment' in D.T. Herbert and R.J. Johnston (eds), *Geography and the urban environment, Vol. 1*, 1–33, John Wiley: London.

Herbert, D.T., and Smith, D.M. (eds) (1979) *Social Problems and the city*, Oxford University Press.

Herbert, D.T., and Williams, W.M. (1964) 'Some new techniques for studying urban subdivisions', *Geographica Polonica*, 4, 93–118.

Hudson, R. (1974a) 'Consumer spatial behaviour: a conceptual model and empirical investigation', PhD dissertation, University of Bristol.

Hudson, R. (1974b) 'Images of the retailing environment: an example of the use of the Repertory Grid methodology', *Environment and Behavior*, 6, 470-94.

Hudson, R. (1974c) *Price variations in the urban retail environment*, University of Durham, North East Area Study Working Paper 3.

Hudson, R. (1975) 'Patterns of spatial search', *Transactions of the Institute of British Geographers*, 65, 141–54.

Huff, D.L. (1963) 'A probabilistic analysis of shopping centre trade areas', *Land Economics*, 39, 81–9.

Huff, D.L. (1964) 'Defining and estimating a trade area', *Journal of Marketing*, 28, 34–8.

Johnston, R.J. (1969a) 'Some tests of a model of intra-urban population mobility', *Urban Studies*, 6, 34–57.

Johnston, R.J. (1969b) 'Population movement and metropolitan expansion: London 1951–1961', *Transactions, Institute of British Geographers*, 46, 71–91.

Johnston, R.J. (1969c) 'Towards an analytical study of the townscape: the residential building fabric', *Geografiska Annaler*, 50B, 20–32.

Johnston, R.J. (1971) 'Mental maps of the city: suburban preference patterns', *Environment and Planning*, 3, 63–72.

Johnston, R.J. (1972) 'Activity spaces and residential preferences: some tests of the hypotheses of sectoral mental maps', *Economic Geography*, 48, 192–211.

Johnston, R.J. (1973a) 'Possible extensions to the factorial ecology method: a note', *Environment and Planning*, 5, 719–34.

Johnston, R.J. (1973b) 'Spatial patterns in suburban evaluations', *Environment and Planning*, 5, 385–96.

Johnston, R.J. (1976a) 'Areal studies, ecological studies, and social patterns in cities', *Transactions, Institute of British Geographers*, NS 1, 118–22.

Johnston, R.J. (1976b) 'Contagion in neighbourhoods: a note on problems of modelling and analysis', *Environment and Planning A*, 8, 581–6.

Johnston, R.J. (1976c) 'Residential area

characteristics' in D.T. Herbert and R.J. Johnston (eds), *Social areas in cities: spatial processes and form,* 193-236, John Wiley: London.

Johnston, R.J. (1979a) 'On urban and regional systems in lagged correlation analyses', *Environment and Planning A*, 11, 705–14.

Johnston, R.J. (1979b) 'On the relationships between regional and national unemployment trends', *Regional Studies*, 13, 453–64.

Johnston, R.J. (1979c) 'On procrustean rotations and the monitoring of policies for residential patterns', *Environment and Planning A*, 11, 463–72.

Johnston, R.J. (1979d) 'On the characterization of urban social areas', *Tijdschrift voor Economische en Sociale Geografie*, 70, 232–8.

Johnston, R.J. (1979e) 'The homes of callers to a telephone counselling service: towards a mapping of community in the city', *New Zealand Geographer*, 35, 34–40.

Johnston, R.J., and Hay, A.M. (1979) 'Variability in grocery prices', *Area*, 11, 160–1.

Johnston, R.J., and Kissling, C.C. (1971) 'Establishment use patterns within central places', *Australian Geographical Studies*, 9, 116–32.

Jones, E. (1960) *A social geography of Belfast,* Oxford University Press.

King, L.J. (1962) 'A quantitative expression of the pattern of urban settlements in selected areas of the United States', *Tijdschrift voor Economische en Sociale Geografie*, 53, 1–7.

King, L.J., and Golledge, R.G. (1978) *City, space and behavior,* Prentice-Hall: Englewood Cliffs, NJ.

Koppelman, F.S., and Hauser, J.R. (1977) *Consumer travel choice behaviour: an empirical analysis of destination choice for non-grocery shopping trips*, Northwestern University Transportation Centre, W.P. 414–09: Evanston, Ill.

Lakshmanan, T.R., and Hansen, W.G. (1965) 'A retail market potential model', *Journal of the American Institute of Planners*, 31, 134–43.

Lee, B.J.L. (1979) 'A nearest-neighbour spatial association measure for the analysis of firm interdependence', *Environment and Planning A*, 11, 169–76.

Lee, T.R. (1973) 'Ethnic and social class factors in residential segregation: some implications for dispersal', *Environment and Planning*, 5, 477–90.

Lerman, S.R. (1977) 'Location, housing, automobile ownership, and mode to work: a joint choice model', *Transportation Research Record* 610, Transportation Research Board, Washington, DC.

Losch, A. (1943) (translation 1954) *The economics of location*, Yale University Press, New Haven.

McFadden, D. (1974) 'Conditional logit analysis of qualitative choice behaviour' in P. Zarembka (ed.), *Frontiers in econometrics*, Academic Press: New York.

McFadden, D. (1976) 'Quantal choice analysis: a survey', *Annals of Economic and Social Measurement*, 5, 363–90.

McFadden, D. (1979) 'Quantitative methods for analyzing travel behaviour: some recent developments' in D.A. Hensher and P.R. Stopher (eds), *Behavioural travel modelling*, Croom Helm: London.

McShepherd, P., and Rowley, G. (1978) 'The association of retail functions within the city centre', *Tijdschrift voor Economische en Sociale Geografie*, 69, 233–7.

Mackett, R.L. (1973) *Shopping in the city: the application of an intra-urban shopping model to Leeds*, Leeds University, Department of Geography, Working Paper 30.

Manchester University, Department of Town Planning (1966) *Regional shopping centres in North West England, Part II*, Manchester.

Manski, C.F. (1973) 'Qualitative choice analysis', PhD dissertation, Department of Economics, MIT, Cambridge, Mass.

Manski, C.F. (1977) 'The structure of random utility models', *Theory and Decision*, 8, 229–54.

Mayer, H.M., and Kohn, C.F. (eds) (1959) *Readings in urban geography,* University of Chicago Press.

Moore, E.G. (1969) 'The structure of intra-urban movement rates: an ecological model', *Urban Studies*, 6, 17–33.

Moseley, M.J. (1973) 'The impact of growth centres in rural regions', *Regional Studies*, 7, 57–94.

Moser, C.A., and Scott, W. (1963) *British towns,* Oliver & Boyd: Edinburgh.

Mountcastle, G.D., Stillwell, J.C.H., and Rees, P.H. (1974) *A users guide to a program for calibrating and testing spatial interaction models*, Leeds University, Department of Geography, Working Paper 79.

Murdie, R.A. (1976) 'Spatial form in the residential mosaic' in D.T. Herbert and R.J. Johnston (eds), *Social areas in cities: spatial processes and form,* 237-72, John Wiley: London.

National Economic Development Office (1970) *Urban models in shopping studies*, London.

Newton, P.W., and Johnston, R.J. (1976) 'Residential area characteristics and residential area homogeneity', *Environment and Planning A*, 8, 543–52.

Nystuen, J.D. (1959) 'A description of consumer trips to retail business' in W.L. Garrison, B.J.L. Berry, R.L. Marble, J.D. Nystuen and R.L. Morrill *Studies of highway development and geographic change*, University of Washington Press: Seattle.

Openshaw, S. (1973) 'Insoluble problems in shopping model calibration when the trip pattern is not known', *Regional Studies*, 7, 367–71.

Openshaw, S. (1975) *Some theoretical and applied*

aspects of spatial interaction shopping models, Concepts and techniques in modern geography 4, Geo Abstracts: Norwich.

Openshaw, S. (1977) 'Optimal zoning systems for spatial interaction models', *Environment and Planning A*, 9, 169–84.

Openshaw, S., and Connolly, C.J. (1977) 'Empirically derived deterrence functions for maximum performance spatial interaction models', *Environment and Planning A*, 9, 1067–79.

Pacione, M. (1975) 'Preference and perception: an analysis of consumer behaviour', *Tijdschrift voor Economische en Sociale Geografie*, 66, 84–92.

Pahl, R.E. (1970) *Whose city?*, Longman: London.

Pahl, R.E. (1975) *Whose city?* (2nd edn), Penguin: Harmondsworth.

Parker, A.J. (1974) 'An analysis of retail grocery price variations', *Area*, 6, 117–20.

Parker, A.J. (1976) *Consumer behaviour, motivation and perception: a study of Dublin*, University College Dublin, Department of Geography, Research Report.

Parker, C.S., and Jerwood, D. (1974) 'Retailing and prediction of price patterns in urban form using curvilinear regression analysis', *Environment and Planning*, 6, 339–54.

Parr, J.B. (1978) 'Models of the central place system: a more general approach', *Urban Studies*, 15, 35–49.

Pickvance, C.G. (1974) 'Life cycle, housing tenure and residential mobility: a path analytic approach', *Urban Studies*, 11, 171–88.

Poston, T., and Wilson, A.G. (1977) 'Facility size versus distance travelled: urban services and the fold catastrophe', *Environment and Planning A*, 9, 681–6.

Potter, R.B. (1977) 'The nature of consumer usage fields in an urban environment: theoretical and empirical perspectives', *Tijdschrift voor Economische en Sociale Geografie*, 68, 168–76.

Potter, R.B. (1978) 'Aggregate consumer behaviour and perception in relation to urban retailing structure: a preliminary investigation', *Tijdschrift voor Economische en Sociale Geografie*, 69, 345–52.

Poulsen, M.G. (1976) 'Restricted lateral movement: an initial test of the locational attachments hypothesis', *Environment and Planning A*, 8, 289–98.

Quigley, J.M. (1976) 'Housing demand in the short run: an analysis of polychotomous choice', *Explorations in Economic Research*, 2, 76–102.

Rees, P.H. (1971) 'Factorial ecology: an extended definition, survey and critique', *Economic Geography*, 47, 220–33.

Rees, P.H. (1972) 'Problems of classifying sub-areas within cities' in B.J.L. Berry (ed.) *City classification handbook*, 265-330, John Wiley: New York.

Reilly, W.J. (1931) *The law of retail gravitation*, Putman: New York.

Richards, M.G., and Ben-Akiva, M.E. (1975) *A disaggregate travel demand model*, Saxon House: Farnborough.

Richardson, H.W. (1973) 'Theory of the distribution of city sizes: review and prospects', *Regional Studies*, 7, 239–51.

Robson, B.T. (1969) *Urban analysis*, Cambridge University Press.

Robson, B.T. (1973) *Urban growth: an approach*, Methuen, London.

Robson, B.T. (1975) *Urban social areas*, Oxford University Press.

Rogers, A. (1965) 'A stochastic analysis of the spatial clustering of retail establishments', *Journal of the American Statistical Association*, 60, 1094–103.

Rogers, A. (1969) 'Quadrat analysis of urban dispersion: 2. Case studies of urban retail systems', *Environment and Planning*, 1, 155–71.

Rogers, D.S. (1974) *Bretton, Peterborough: the impact of a large edge-of-town supermarket*, Manchester University Business School, Retail Outlets Unit, Report No. 9.

Rowley, G., and Tipple, G. (1974) 'Coloured immigrants within the city: an analysis of housing and travel preferences', *Urban Studies*, 11, 81–90.

Rowley, G., and Wilson, S. (1975) 'The analysis of housing and travel preferences: an approach', *Environment and Planning A*, 7, 171–8.

Rushton, G. (1969) 'Analysis of spatial behaviour by revealed space preference', *Annals of the Association of American Geographers*, 59, 391–400.

Rushton, G. (1971) 'Behavioural correlates of urban spatial structure', *Economic Geography*, 47, 49–58.

Shaw, G. (1978) *Processes and patterns in the geography of retail change, with special reference to Kingston upon Hull, 1880–1950*, University of Hull, Occasional Papers in Geography 24.

Shepherd, J.W., Westaway, J., and Lee, T. (1976) *Greater London: a social atlas*, Oxford University Press.

Sherwood, K.B. (1970) 'Some applications of the nearest neighbour technique to the study of intra-urban functions', *Tijdschrift voor Economische en Sociale Geografie*, 61, 41–8.

Sibley, D. (1971) 'A temporal analysis of the distribution of shops in British cities', PhD dissertation, University of Cambridge.

Sibley, D. (1973) 'The density gradients of small shops in cities', *Environment and Planning*, 5, 223–30.

Sibley, D. (1975) *The small shop in the city*, University of Hull, Occasional Papers in Geography 22.

Simmons, J.W. (1964) *The changing pattern of*

retail location, University of Chicago, Department of Geography, Research Paper 92.
Smailes, A.E. (1946) 'The urban mesh of England and Wales', *Transactions, Institute of British Geographers*, 11, 85–101.
Smailes, A.E. (1955) 'Some reflections on the geographical description and analysis of townscapes', *Transactions, Institute of British Geographers*, 21, 99–115.
Smailes, A.E., and Hartley, G. (1961) 'Shopping centres in the Greater London area', *Transactions, Institute of British Geographers*, 29, 201–13.
Smith, A.P. (1973) *Retail allocation models: an investigation into problems of application to the Rotherham sub-region*, Leeds University, Department of Geography, Working Paper 50.
Smith, A.P., Whitehead, P.J., and Mackett, R.L. (1977) 'The utilisation of services' in A.G. Wilson, P.H. Rees and C.M. Leigh (eds), *Models of cities and regions*, John Wiley: Chichester.
Smith, G.C. (1976) 'The spatial information fields of urban consumers', *Transactions of the Institute of British Geographers*, NS 1, 175–89.
Spence, N.A. (1968) 'A multifactor regionalization of British counties on the basis of employment data for 1961', *Regional Studies*, 2, 87–104.
Spenser, A.H. (1978) 'Deriving measures of attractiveness for shopping centres', *Regional Studies*, 12, 713–26.
Tarrant, J.R. (1967) *Retail distribution in East Yorkshire in relation to central place theory*, Occasional Paper in Geography 8, University of Hull.
Taylor, P.J. (1969) 'Causal models in geographical research', *Annals, Association of American Geographers*, 59, 402–5.
Taylor, P.J., and Johnston, R.J. (1979) *Geography of Elections*, Penguin: Harmondsworth.
Thorpe, D., and Kivell, P.T. (1971) *Woolco, Thornaby: A study of an out of town shopping centre*, Manchester University Business School, Retail Outlets Unit, Report No. 3.
Thorpe, D., and McGoldrick, P.J. (1974) *Carrefour, Caerphilly: consumer reaction*, Manchester University Business School, Retail Outlets Unit, Report No. 12.
Timms, D.W.G. (1971) *The urban mosaic*, Cambridge University Press.
Tocalis, T.R. (1978) 'Changing theoretical foundations of the gravity concept of human interaction' in B.J.L. Berry (ed.), *The Nature of change in geographical Ideas*, 65-124, Northern Illinois University Press, DeKalb,, Ill.
Wade, B.F. (1973) *Greater Peterborough shopping study: technical report*, Planning Research Applications Group (PRAG) London.
Williams, N.J. (1979) 'The definition of shopper types as an aid in the analysis of spatial consumer behaviour', *Tijdschrift voor Economische en Sociale Geografie*, 70, 157–63.
Wilson, A.G. (1972) 'Behavioural inputs to aggregative urban systems models' in A.G. Wilson (ed.), *Papers in urban and regional analysis*, Pion: London.
Wilson, A.G. (1974) *Urban and regional models in geography and planning*, John Wiley: Chichester.
Wilson, A.G. (1976) 'Retailers' profits and consumers' welfare in a spatial interaction shopping model' in I. Masser (ed.), *Theory and practice in regional science*, Pion: London.
Wilson, A.G. (1978) 'Towards models of the evolution and genesis of urban structure' in R.L. Martin, N.J. Thrift and R.J. Bennett (eds), *Towards the dynamic analysis of spatial systems*, Pion: London.
Wilson, A.G., Rees, P.H., and Leigh, C.M. (1977) *Models of cities and regions*, John Wiley: Chichester.
Witherick, M.E., and Pinder, D.A. (1972) 'The principles, practice and pitfalls of nearest-neighbour analysis', *Geography*, 57, 277–88.
Woods, R.I. (1975) *The stochastic analysis of immigrant distributions*, School of Geography, University of Oxford.
Woods, R.I. (1976) 'Aspects of the scale problem in the calculation of segregation indices', *Tijdschrift voor Economische en Sociale Geografie*, 69, 169–74.
Woods, R.I. (1977) 'Population turnover, tipping points and Markov chains', *Transactions, Institute of British Geographers*, NS 2, 473–89.
Wrigley, N. (1976) 'Uncertainty and consumer shopping behaviour', PhD dissertation, University of Cambridge.
Wrigley, N. (1979) 'Variability in grocery prices', *Area*, 11, 161–4.
Wrigley, N. (1980) 'An approach to the modelling of shop-choice patterns: an exploratory analysis of purchasing patterns in a British city' in D.T. Herbert and R.J. Johnston (eds), *Geography and the urban environment, Vol. 3*, 45–85, John Wiley: Chichester.
Yeates, M.H., and Garner, B.J. (1976) *The North American City*, Harper & Row, New York.

351

'r 32

Behavioural geography

N. Thrift

Paradigm in search of a paradigm

It is probably true to say that, although established in embryo by the mid 1960s (Gould, 1963, 1966; Wolpert, 1964, 1965), the emergence of behavioural geography as a distinct and identifiable subject area within human geography can be most easily traced to two events. One was the special session on behavioural models in geography held at the sixty-fourth annual meeting of the Association of American Geographers in Washington, DC, in 1968 (Cox and Golledge, 1969), whilst the second was Torsten Hägerstrand's presidential address to the Regional Science Association in Copenhagen in 1969 (Hägerstrand, 1970). These two events both defined the newly emergent field and also underlined the presence of two quite distinct approaches within the subject to the study of human behaviour in space and time. In the first approach, that of the Washington meeting, the individual predominantly appears as an *active* decision-maker, as *homo psychologicus*, processing what information is available through a perceptual mesh and choosing between the alternatives. In the second approach, that of Hägerstrand's address, the individual appears – to varying degrees – as a *reactive* decision-maker subject to a number of socio-spatial constraints. In the first approach the concern is with the processes that underlie individual behaviour in the belief that man is the explanation of his own actions. In the second approach the concern is with tracking individual behaviour as an object of study in itself, in the belief that society is the explanation of man's actions. This distinction between the two approaches is by no means a new one in social science (cf. Dawe, 1970; Hollis, 1977) and it is, to some extent, a caricature. In reality, of course, the neat and tidy views ascribed to each approach are blurred and inchoate, forming a continuum rather than a dichotomy.

The business of identifying the intellectual antecedents of a subject area is always hazardous, not least because it may appear that equal weight is being ascribed to each influence or that a comprehensiveness is being claimed. However, with these caveats in mind, a preliminary attempt can be made. In the first active decision-making approach, as exemplified by the Washington meeting, at least five roots can be discovered. One root was in environmental perception, in how the environment appears to individuals (e.g. Kirk, 1951; Boulding, 1956; Lynch, 1960; Appleyard, Lynch and Myer, 1964; Lowenthal, 1967). A second root was in natural hazards research, in the strategies adopted by individuals in the face of natural hazards (e.g. Kates, 1962; Gould, 1963, 1965; Saarinen, 1969). Another root was in psychology, to begin with particularly in experimental psychology but with the emphasis gradually moving from the work of Tolman on rats to the work of Piaget and Bruner on cognitive development in human beings. In addition, advances in psychometrics and measurement theory were of course crucial to the quantitative expression of these problems. A fourth root can be found in early work on choice and decision-making, both in mathematical economics (e.g. Luce, 1959) and in human geography (e.g. Huff, 1960; Wolpert, 1964, 1965; Pred, 1967). Finally there is the simple sociological fact that quantitative geographers, as the vanguard of the 'new' geography, were looking for fresh pastures, partly as a result of the constant process of search and plunder of other subject areas commonly indulged in by human geographers and partly as a result of some degree of disillusionment with the results of earlier work leading to a desire to better conceptualize the theoretical base of extant and new models. This list of five antecedents is by no means exhaustive. Consider, for instance, the influence of architectural psychology (e.g. Lee, 1964) or what might now be called early humanistic geography (e.g. Lowenthal,

1961). But it does suffice in showing the eclecticism of the first approach.

By contrast the second reactive decision-making approach, exemplified by Hägerstrand's address, had a very different and more narrowly-defined set of intellectual antecedents. The first of these antecedents was in time-budget analysis, and essentially involved the addition of locational information to time budget information on human activity, usually collected by the use of some kind of diary. Already in the 1960s this had become a well-ploughed field (e.g. Marble, 1967; Chapin and Brail, 1969). A second antecedent was in micro-demography, in work on movement into and out of small areas (e.g. Hägerstrand, 1963; Norborg, 1968). Finally, there was an interest in simulation, partly the result of Hägerstrand's now mythic work on diffusion using Monte-Carlo techniques (e.g. Hägerstrand, 1967b) and partly the result of the general interest in simulation engendered by the first enthusiastic usage of the computer (Hägerstrand, 1967a).

The two approaches, active and reactive, can now therefore be seen to have had very different ancestries which, in turn, have led to a different emphasis in terms of direction of investigation. In the first approach the emphasis has largely been inductive and has concentrated on the collection of data and its interrogation through a suitable technique. If the right technique can be found the data will talk. By contrast the second approach has placed more emphasis on deduction. It follows from these differences of emphasis that in the first approach, in line with a belief in man as explanation of his own actions, the main source of data is questioning of the individual. Attitudes or preferences are presumed to be connected, in some way, to behaviour. By contrast the second approach may rely on environmental information, usually concerning constraints, or at the very least augment the study of individual behaviour with such information. Of course within the two approaches there is a continuum of emphasis. For instance, in the second reactive approach the research strategy of Chapin has included a marked emphasis on individual predisposition and the direct questioning of individuals to obtain such information whereas the time-geographic approach of Hägerstrand is rarely concerned with the views of individuals *per se*. Then again, in recent years, the first approach has become markedly less inductive in outlook.

Finally it must be pointed out that, in this chapter, no attempt has been made to separate out the contribution of quantitative geography to behavioural geography and vice versa. In the main this is for historical reasons. Whereas with many other subject areas it is possible to establish a theoretical bloodline that stretches back beyond the 'quantitative revolution', it is hard to convincingly reconstitute the scattered set of papers that exist from before this point in time that can be labelled as 'behavioural' as in any way representing a coherent behavioural geography. In effect behavioural geography and quantitative geography grew up together, both sociologically and theoretically. As a result, in the case of the active tradition in particular, the questions that behavioural geography asked were framed quantitatively and answered quantitatively. Theory and method were often all but isomorphic. Now, with the increasing emergence of the reactive tradition as a full co-partner in the project of behavioural geography and the possibilities for synthesis between the active and reactive traditions that have become apparent therefrom, behavioural geography is gaining its own impetus. In this new, more independent behavioural geography quantitative specification and analysis has become only one facet in an emergent whole. But this is still a relatively recent phenomenon in historical terms.

Behavioural geography in Britain

In Britain, apart from a brief flowering in the Department of Geography at the University of Bristol in the late 1960s (Harvey, 1967; White, 1967; Downs, 1967, 1968; Hudson, 1970, 1971), behavioural geography has never really, to borrow a Rostowian metaphor, reached the stage of academic take-off (Cullen, 1976). Instead it has taxied along, occasionally revving up, but never gaining sufficient momentum to enable it to reach the higher plane of the institutionalized subject areas. Behavioural geography has never, for instance, had its own Institute of British Geographers Study Group. Whereas by 1973 both the active and reactive tradition had gained academic respectability in Britain, the active tradition with the publication of Downs and Stea's (1973) *Image and Environment*, and the reactive with a more diffuse set of publications (e.g. Hägerstrand, 1973; Chapin, 1974; Cullen, 1972, Cullen, Godson and Major, 1972), this respectability has never been converted into the currency of an accepted subject area, attracting large grants and with major books regularly appearing, as has happened in the United States. The difference is well illustrated in the contrast between the almost tentative tone and partially different content of a recent book on environmental images by two British authors *Images of the Urban Environment* (Pocock and Hudson, 1978) and the more assured tone of its American counterparts (e.g. Gould and White, 1975; Downs and Stea, 1977). In part this reticence may be due to the generally received nature of the subject in Britain. Thus both the active and reactive approaches identified in the preceding section were essentially academic imports into Britain from other countries, in the case of the active tradition from the United States (even if some of the leading proponents were British-born) and in the case of the reactive tradition from Sweden and the United States. In part this reticence may also be due to the fact that most British behavioural geographers were also quantitative geographers who have since moved on to other interests or who have retained only a partial interest in behavioural geography, a state of affairs that, when coupled with a virtual non-replacement of interest in behavioural geography by the younger members of the academic geography community, who have usually become involved with later paradigm shifts (e.g. radical or humanistic geography), helps to explain Cullen's call in 1976 for an alternative theoretical base for behavioural studies.

In this same paper Cullen, discussing the dilemmas of behavioural geography, points out that 'it is always a bad sign when an activity which appears superficially coherent proves impossible to define' (Cullen, 1976, p. 398). Behavioural geography *is* a fragmented subject area, although it has been possible to isolate two traditions in the preceding section, and this makes it both more appropriate and more honest to review the advances in British quantitative behavioural geography in a fragmented way according to predominant themes. That old saw 'geography is what geographers do' cannot be so lightly dismissed when the case of behavioural geography comes to court.

The active decision-making tradition

Cognitive mapping

Although Boulding's (1956) *The Image* or Lynch's (1960) *The Image of the City* are often cited as the touchstones of interest in cognitive mapping from the standpoint of British behavioural geography, Gould's classic paper of 1966 (Gould, 1973) on mapping space preferences may be a more realistic beginning. From this paper have followed over the years an enormous number of papers on 'mental maps'. Britain has been no exception. Such studies have been, in the main, concerned with the *appraisive* aspect of images, with measuring and representing spatial preferences, familiarity with places or information about places (Wood, 1970; Downs and Stea, 1973; Pocock and Hudson, 1978) and have ranged over subjects as diverse as, for instance, urban images (Goodey *et al.*, 1971; Harrison and Sarre, 1975; Pocock, 1973, 1975, 1976), residential preference (Bateman *et al.*, 1974) and place preference (Gould and White, 1968, 1975). Following Harrison and Sarre (1971) the measurement process involved in the evaluation of spatial preference can be said to have three stages – specification, scaling and generalization and inference. Obviously the first task, specification, involves obtaining environmental images in a measurable form. Any questionnaire must be designed as a compromise between specification of a minimal set of basic elements of the environment and the terms relevant to description of such an environment and the pressure of time upon the respondent. Specification of relevant elements may be obtained by direct testing out of a particular theory of mental structure or by reference to existing methodologies, normally from psychology, with their underlying theories. Various rating questionnaire approaches from psychology have been used, from semantic differential (Downs, 1970; Jackson and Johnston, 1972; Palmer *et al.*, 1977) to repertory grid (Hudson, 1974; Harrison and Sarre, 1975; Townsend, 1976, 1977; Palmer, 1978) and from thematic apperception to the method of paired comparisons (Bradford, 1975; Bather, 1976), as well as various prespecified questionnaire inputs for techniques like multidimensional scaling (e.g. Spencer, 1978). There are also a number of graphical methods including, for instance, cloze procedures (Porter *et al.*, 1975; Robinson, 1974; Dicken and Robinson, 1979). Once the analyst moves to representation of the data the most common techniques used have been factor (especially principal components) analysis and, lately, multidimensional scaling. Whereas in the case of the first technique the aim has usually been to investigate the content of the data and plot the scores as representing a preference surface, in the second case the aim has usually been to understand and represent the (assumed) underlying spatial structure of cognitive configurations of place preferences. Finally in the stage of generalization and inference there is the difficult problem of ascertaining how representative and reliable the outcome is, for instance by a secondary factor analysis (Palmer *et al.*, 1977) or a repertory grid check (Palmer, 1978).

A second aspect of cognitive mapping is the investigation of the actual morphology of cognitive maps – the *designative* aspects. This work, which does derive from Lynch's (1960) book, aims at expressing the structure of cognitive maps in terms of mental judgments on spatial concepts like proximity, dispersion, location, clustering, shape, orientation, similarity, dissimilarity and so on which, supposedly, do not involve an individual's value system. Perceived distance has been particularly widely investigated at many scales (e.g. Lee, 1964, 1970; Canter and Tagg, 1975; Madden, 1978; Ferguson, 1979) whilst, in terms of cognitive mapping, mental maps have been shown to have systematic variations in terms of error with regard to orientation and shape (Porteous, 1971, Pocock, 1972; Pacione, 1976, 1978) and to vary according to age, social class and so on (e.g. Pocock, 1975). In the United States a considerable amount of work has taken place on the examination of the cognitive configurations recovered from scaling analysis using, for instance, centrographic analysis (cf. Golledge, 1978a). This remains to be attempted by British behavioural geographers.

Finally cognitive maps have been used in the context of learning about environments. Gould's examination of the preference and information space of Swedish children (Gould, 1973) and Swedish children and adults (Gould, 1975) is the classic work here but it has, as yet, received little follow-up in Britain (Burchill, 1975) compared to the situation in the United States (cf. Moore and Golledge, 1976; Golledge, 1978a, 1978b).

For some time it has been realized that cognitive map studies need a conceptual base. But it has been hard to find one that fits. Gould (1975) has attempted to paint cognitive maps as indexes of relative location within a social information space. Others have seen them as expressions of a learned and acculturated cognitive organization created by constant interaction between internal and external demands (e.g. Moore and Golledge, 1976; Golledge, 1978a). To yet others cognitive maps represent the spatial part of more general environmental images (Pocock, 1976). The fact that reviews can be written addressed to the problem of what mental maps are (Boyle and Robinson, 1979) shows a potentially bewildering diversity of emphasis that, when combined with the pessimism of most British commentators regarding the future of cognitive mapping (e.g. Johnston, 1973; Graham, 1976; Maclennan, 1977; Boyle and Robinson, 1979) seems to imply that an academic cul-de-sac has been reached. As a social psychologist and a social anthropologist point out, in a recent review of *Environmental Knowing* (Moore and Golledge, 1976):

> Maps drawn by an individual demonstrate his knowledge of his environment, reflecting his experience of it. One overestimates the distance to a neighbouring town because one has not been there; and because one thinks it far, one is less likely to go. But where does one go from here? The deficiencies . . . lie in the failure to link image to future action and to relate experimental studies to real problems (Lloyd and Lloyd, 1980).

Diverse strategies have appeared in answer to the problems of this hiatus. Some workers have turned away from explicit spatial representation to consider the language of place (Burgess, 1974, 1978) or place and community (Irving, 1975, 1978). Others have moved to a more humanistic emphasis (e.g. Pocock, 1979; Downs and Meyer, 1978) or to policy implications (Bather, 1976; Hudson, 1976b). Yet others see the answer to the hiatus in 'more rigid and systematic testing or methods and . . . the search for different ways of eliciting the kind of information which indisputably forms the basis of our behavioural dispositions' (Boyle and Robinson, 1979, p. 79). Whatever the answer it seems certain that the understanding of *what* cognitive maps are (and if they are) requires not only new directions of search and better experiments and techniques. It also requires that the emphasis on the *individual* and the individual mind, whether drawn from psychology or humanism (Tuan, 1975), be balanced by a more than cursory consideration of the influence of *social* consciousness on individual's perceptions (cf. Pocock and Hudson, 1978, Chapter 9).

Decision-making behaviour

A link can be made between the image and future action and therefore between this and the preceding section in the fact that cognitive representation of the environment has been tied to decision-making in at least two ways. One tie between representation and decision is in the identification and scaling of preferences (Hudson, 1976b). This is still a formidable problem, practically and theoretically, and has yet to be solved (Maclennan, 1977; Benwell, 1979). The second tie between representation and decision is reflected in the amount of information an individual has at his or her disposal at any one time which must be a function of the process of search and learning and also the perception an individual has of the strength of the constraints that make a decision feasible.

Decision-making depends on a behavioural process that, for convenience, can be broken down into at least three components, deciding to decide, the search for a set of alternatives and the actual decision between these alternatives. These three components are often modelled as a simultaneous decision. However, in reality, they are not. The first component, the decision to make a decision, is an under-researched area and it is true to say that, as yet, little is known about this process. In the area of residential location a number of studies in Britain have shown that most long-distance moves are related to job changes and most short-distance moves to passage through the life-cycle (Johnson, Salt and Wood, 1974: Office of Population Censuses and Surveys, 1978) but in other subject areas, apart from work on resistance to making a decision to change to another mode because of force of habit in transport studies (Wilson, 1976; Goodwin, 1977, Banister, 1978; Blase, 1979), the decision has too often been assumed.

In the second component of the decision process, a search of the alternatives must be initiated. In the most elementary sense this means an acquisition of information, presumably involving a learning process (Golledge and Brown, 1967; Silk, 1971; Guy, 1975). In the British context Hudson (1975, 1976a) has considered shop search patterns of in-migrating university students at both shopping centre and individual shop scale levels, using multiple regression analysis. Pacione (1975), Smith (1976) and Potter (1977a, b) have also investigated information fields of consumers and found them to be closely related to social class and length of residence. In transport geography Banister (1977, 1978) has considered the problem of modal choice in terms of car availability. In the field of industrial location Townroe (1972, 1973) has considered the entrepreneur's search process while Green (1977) has considered individual firm's information levels concerning regional incentives. Similarly, in the field of residential location, a considerable body of work has been built up on the house choice process, particularly as a function of the cumulative acquisition of housing market information (e.g. Herbert, 1973; Maclennan, 1977, 1979a, 1979b; Office of Population Census and Surveys, 1978). At some point a halt must be called to any search. In the context of residential choice Flowerdew (1976) has considered this decision within the framework of optimal stopping rule models.

If it can be assumed that a choice set has been uncovered the individual still has to evaluate the alternatives in order to be able to choose between them. This evaluation has been taken to mean ranking the alternatives, a suspect assumption. Following from the papers on space preference by Rushton (1969a) and the migration decision by Brown and Moore (1970) this evaluative aspect has become a popular area of behavioural geography. However, as Beavon and Hay (1977) point out, there is a distinction between *behavioural* modelling, which attempts to model the actual decision process and *behaviouristic* modelling which, in true positivist fashion, attempts only to predict accurately the outcome or the consequences of a decision. The arguments against the rational economic man embodied in the utility theory of most of these models have, by now, been rehearsed *ad nauseam* (e.g. Hollis and Nell, 1975) and do not bear repeating here. Suffice it to say that most choice models, until recently, have been behaviouristic and if there has been any concern at all about the missing behavioural link it has usually been displayed as a *post hoc* justification of particular techniques by investing them with mystical properties like reflecting the structure of 'mind' (e.g. Golledge and Rushton, 1976).

Beavon and Hay (1977) suggest that most behaviouristic models of the decision process follow a six-stage procedure. In the first stage data on actual behaviour are collected, whether on shopping alternatives or observed trips between residence, on another highly suspect assumption that overt behaviour reveals preferences. In the third stage the perception of alternatives is recovered by a suitable scaling technique or by other methods, for instance paired comparisons data as a direct input into a contingency table format rather than as simply the initial stages of multidimensional scaling (Wrigley, 1980). In the fourth stage the alternatives are ranked relative to a consumer utility scale and in the sixth, and final stage, the expected behaviour is output. In the intervening stage two, a preference recovery calculus may be adopted whilst in the intervening stage five, a choice model, whether deterministic (e.g. Rushton, 1969b) or probabilistic based, for instance, on the work of Luce (1959) and Suppes (1961) is inserted. Typically this kind of work has been overwhelmingly a United States enterprise particularly as applied to travel demand modelling (e.g. Domencich and McFadden, 1975) and joint choice models of housing and travel demand (e.g. Anas, 1975, 1979; Quigley, 1976) involving the use of logit models (Wrigley, 1976) or as used in parts of large-scale residential location modelling (e.g. Robinson *et al.*, 1965; Ingram *et al.*, 1972; for a review cf. Senior, 1974). But there are British examples. For instance Beavon and Hay (1977), in their consideration of consumer choice of shopping centre, have applied a probabilistic model having some similarities with Tversky's (1972) well-known elimination-by-aspects model while Wrigley (1980) has used logit models to analyse categorical data on shopping preferences. Disaggregation into shopper types has also been attempted (Williams, 1979). Residential mobility decision-making has also been considered, in terms of tenure choice, by Doling (1973) and Jones *et al.*, (1978). However this style of choice modelling has predominantly been directed in Britain towards the development of disaggregate travel demand models, particularly as found in the already seminal work of Williams (1977). These models attempt to integrate multimodal choice, particularly stemming from the classic work of Warner (1962), frequency and destination choice (whether seen as sequential or simultaneous) to arrive at a behaviouristic model which is able to function at the same scale as, for instance, the large-scale entropy-maximizing transportation models but which also has a solid grounding in a proper theory of rational choice based, in Williams's case, on random utility theory.

There are many problems associated with the active decision-making approach. But perhaps the main one relates to the role of preferences. As Maclennan (1977) has pointed out, in economic usage at least, preferences reflect the idea of underlying tastes. These tastes exist independently of constraints and are therefore theoretical entities. Choices, by contrast, are preferences that are revealed with respect to a given set of constraints. Perhaps, therefore, preferences need to be more clearly distinguished from choices than has usually been the case so far. This requires a clearer perception of the role of constraints and this is where the reactive decision-making approach is particularly useful.

The reactive decision-making tradition

The second strand of behavioural geography, the reactive approach, considers spatial behaviour *per se*. Like the active approach it too has branched to form a number of specialities. Perhaps the most well-known of these is the work on space-time budgets of human activities, that is on time budgets of activities with added locational information on where each activity took place. In Britain general awareness of this work in human geography probably dates from the review paper by Anderson in 1971, a late start compared with the United States where, particularly at the University of North Carolina under the direction of F. Stuart Chapin Jr, a large amount of research (much of it of a quantitative nature) had already taken place by this date (e.g. Chapin and Hightower, 1965; Chapin and Logan, 1968; Chapin and Brail, 1969) into activity systems as the realization of propensities, emphasizing choices reflecting values. It is also late compared with Sweden where work on time-geography as an investigation, at a pragmatic level, of environmental opportunity and, at a more general theoretical level, as an investigation into the basic physical limits to human action has formally taken place at the University of Lund since 1966 (Thrift, 1977a, b; Carlstein *et al.*, 1978).

The tardy arrival of these two traditions in Britain has, however, had its benefits in producing empirical and modelling work which is an effective synthesis of them both. The synthesis is best represented by the work of Cullen (Cullen, 1978; Cullen, Godson and Major, 1972; Cullen and Godson, 1975; Cullen and Phelps, 1975). Using diary information Cullen has investigated activity patterns of various population subgroups, their degree of routinization and fixity of activities, and the amount of stress felt during the day using a variety of techniques including factor analysis, discriminant analysis and harmonic regression. Another outcome of this synthesis of the two traditions is represented by work carried out at the Martin Centre in Cambridge into the description and modelling of activity systems which, again using diary information, has included an attempt to build an entropy-maximizing model of an activity system (Tomlinson *et al.*, 1973; Shapcott and Wilson, 1978; Shapcott and Steadman, 1978). Yet another outcome of this synthesis is to be found in work by Tivers (1977) on gender role-typing of activity patterns.

However, the major recent thrust in this field in Britain has been in the area of travel behaviour. Transport studies have, without doubt, taken the initiative in Britain under the general heading of 'the activity approach' (Hay, 1978; Burnett and Thrift, 1979). The reasons for this are not hard to see. For some time there has been dissatisfaction with large-scale transportation models in terms of results, degree of explanation and expense. In particular these models proved unable to satisfactorily handle transport as a derived demand. The first reaction to this dissatisfaction was the generation of disaggregate behavioural travel demand models represented by the work of Domencich and McFadden (1975) or Williams (1977). However, these models in particular did not take into account the constraints surrounding human choice, whether in terms of coupling constraints like household interaction or capability and authority constraints like mode choice or job, which often mean that no choice exists at all. Further, by their use of the circular reasoning of revealed preference, which insists that preferences exist separately from constraints and that consumers are rational and seek optimum solutions, they revealed themselves as highly ideological. The general thrust of the criticism and the defects of these models are best revealed by the title of the major critique in this area, Heggie's (1978) 'Putting behaviour into behavioural models of travel choice'. These models were still behaviouristic. This early stage of criticism (Jones, 1979a) has now been replaced by a number of projects mainly centred around the Transport Studies Unit at the University of Oxford. Perhaps the most successful of these has proved to be the production of the 'HATS' (Household Activity-Travel Simulator) game (Jones, 1979b) which, although in a gaming tradition stretching back to Michelson (1966) and Chapin (1974) (for a review cf. Hanson and Burnett, 1980), has significantly improved on these studies. 'HATS' has proved to be a powerful and democratic survey tool for eliciting household reactions to new transport plans in a constrained situation, although its possibilities do not end there. Other approaches in transport geography in the reactive tradition have included work on the Lund time-geographic simulation model, PESASP (Program Evaluating the Set of Alternative Sample Paths) (Lenntorp, 1976), by Pickup (1978).

It does not seem unlikely that the benefits of this constraints-oriented approach to transport studies will spill over into other subject areas, for instance in to residential location modelling. There gaming techniques have already been used by Rowley (Rowley and Tipple, 1974; Rowley and Wilson, 1975; Tipple and Rowley, 1977) to gauge housing preferences of various population subgroups and there seems no immediate reason why the gaming approach could not be successfully used to consider the housing choice process in detail. For instance, a recent paper (Lyon and Wood, 1977), shows that house choice, in the case of the sample of households questionnaired in that study, could not by any means be regarded as rational, at least in the normal sense of the term. Search and evaluation were unsystematic; many respondents became bored with looking. Housing satisfaction seemed to depend on need, *not* preference, once again underlying the dangers, ideological and analytical, of assuming choice and ignoring constraints.

The prospects for behavioural geography

As the preceding section demonstrates a synthesis of the approaches to reactive decision-making has already taken place, although quantitative specification is still at a rudimentary stage. But the question remains: is a synthesis between the active and reactive approach to be deemed possible, or indeed desirable, and can such a synthesis be the subject of quantitative investigation and modelling? The answer is undoubtedly in the affirmative – to a point.

Human decision-making is a complex subject whose complexity has yet to be fully realized. Characterizing this complexity requires the recognition of at least two separate but related points. One is that scattered through everyday life and at other scales, for instance job-related house moves, are non-choice or minimal choice situations where decisions are not so much made by people as for people (we need a word for these routine and habituated, acquiescent or simply no-choice decisions). The second is that it is unlikely that human decisions are made in any rational sense as a rationating choice between competing alternatives (or even as a search towards a local optimum). This latter idea is the invention of scientific rationality, itself a specific historical and social product (Hollis and Nell, 1975; Sohn-Rethel, 1978). But human decisions are not made in this elevated realm of *theory*. They are made in specific *practical* situations and there it seems likely that they are constituted in terms of a minimal logic which is fuzzy, contradictory and ambiguous, and whose equivalences are cavalier (Bourdieu, 1977). Yet this logic works because it is played out in practice. Such a decision-making process is relatively complex since it will be full of echoes of past (again specific and practical) situations and it seems unlikely that, even if we were able to investigate such decisions, the results would give us much insight.

All behavioural models must therefore make some simplifying assumptions. That point is not at issue. What is at issue is how far these assumptions can be simplified before they become so unreal that they are also unreliable, inaccurate and ideological. It was doubts about the efficacy of rational economic man that led to behavioural (or rather, behaviouristic) geography. Now doubts about behaviouristic models are leading to a new behavioural geography. Perhaps what is needed now, *if* there is to be a quantitative specification of certain areas of human action, is a guide to what assumptions certain methods and models make and where these models can therefore be appropriately applied. As Heggie and Jones (1978) have pointed out, what we need is to define domains in which particular methods and models are valid. Such domains are specified according to the criteria of, firstly, the degree of complexity and spatial scale and, secondly, the degree of choice that needs to be assumed in the study of a particular problem. At least three simple domains suggest themselves (Heggie and Jones use a more complex four-way classification). The first domain would be the domain of large-scale studies in which the emphasis is generally pragmatic, rather than explanatory, relatively inexact estimates of numbers are all that is needed and choices can be assumed to be independent of the majority of constraints. An example of an application here might be one of the range of behavioural travel demand or consumer preference logit models of the active decision-making tradition applied at the scale of an urban area. An additional justification that can be made for this type of model at this level is that individual choices are not necessarily important in terms of what actually happens at these larger scales. The second domain consists of applications at scales where space-time and social constraints have become more important. This may involve a smaller spatial scale but, equally, it may involve selected aspects of a larger scale where more realism and model-sensitivity needs to be sought after. An application at this level might include, in activity analysis, the Lund PESASP model (Lenntorp, 1976) perhaps couched in probabilistic form so as to take advantage of the fragmented data available in most research and planning situations outside Sweden. Other applications are legion, for instance the various Markovian or semi-Markovian simulation models evolved in decision-to-move and house choice modelling (e.g. Ginsberg, 1973) and employment modelling suitably adapted to include more realistic constraint structures, and the type of travel models currently being constructed in the United States (Burnett, 1979; Burnett and Hanson, 1979) and West Germany (Brög *et al.*, 1977; Becker *et al.*, 1980) which rely on specification of stopping as attribute-seeking or on probabilistic reductions of action space, respectively. All these types of models are able to consider particular variables in detail – for instance, the complex interactions engendered by a new road, or a bus timetable, manifested as new activity patterns.

Finally, at the very detailed micro-scale (for instance, change in mode caused by village bus-service reductions, Heggie and Jones, 1978), quantitative models may be inappropriate or only partially appropriate and in need of supplementing with more qualitative information. Gaming techniques like HATS (Jones, 1979b), Brög's multi-instrument game (Brög *et al.*, 1977) or Chapin's trading stamp game (Chapin, 1974) may come into their own in this domain (see also Burnett, 1980). As the recession continues to bite, policies in most sectors of the economy, whether housing or transport, have become smaller in scale and the building of smaller-scale models to reflect this situation and the more complex interactions that are important at this scale becomes more and more appropriate. Then again, research aimed at a fuller behavioural explanation must start at this point (and also, of course, at the points where constraints are generated, in political economy and the analysis of institutions).

Of course, methods should not be confused with theory – it is certainly true to say that behavioural geography remains undertheorized. Quantitative approaches, by themselves, are no substitute for theory. Indeed they are actively dangerous when used as a surrogate for explanation since they still reflect underlying theories and practices (Lewis and Melville, 1978). But some of the new modelling work and gaming investigation, reviewed in preceding sections, does actually seem to be elaborating on the subtle and complex interaction between choice and constraint that lies at the heart of the new behavioural geography in a way which seems to be producing an *ad hoc* theory which is, at the very least, immediately satisfactory. This work should give a breathing space during which longer-term projects can be initiated which will examine the exact connection of the area of study covered by behavioural geography with other areas like political economy (Eyles, 1978) or the promotion of the process of hegemony in the activity patterns of everyday life (Tivers, 1977).

So now it is possible to see a general and admittedly *ad hoc* model of human decision-making emerging in behavioural geography that will fit most situations. This will involve a synthesis of the active and reactive approaches to decision-making. Choices (*not* preferences) can be partitioned into four subsets – perceived but non-feasible, perceived and feasible, feasible but not perceived and neither perceived nor feasible (Maclennan, 1977). Given a chosen domain, an appropriate reactive model is first applied to specify the interacting pattern of constraints that circumscribe the choice set of each and every individual in order to arrive at the feasible choice sets. Then, within each of these bounded choice sets, a conventional active decision model can be used which will be able to take into account the role of perception in choice (Thrift and Williams, 1981). Such a synthesis is already possible with the existing range of reactive and active models. However, a more satisfactory melding of the two approaches may be achieved by new methods. Programming models, suitably modified, suggest themselves as one possible avenue of research with, however, distinct domain limitations (Clarke, 1978). A more powerful methodology, which suggests itself as able to cope with a range of domains and as able to deal with *both* the circumscription of choice sets by constraints *and* the decision process, is micro-simulation (Orcutt *et al.*, 1961, 1976). Whilst this methodology still has problems in terms of deducing the response of different individuals or groups to changed situations it has an obvious potential (cf. Clarke *et al.*, 1979; this volume, Chapter 24). Another general methodology that suggests itself is Q-analysis (Atkins, 1974). However, this methodology may well prove more amenable to modelling of the process of constraint circumscription (Johnson, 1976) than the actual decision process since its insistence on the Russell Theory of types may prove inappropriate to the essential ambiguity of human decision-making (Olsson, 1975). This ambiguity in the decision process may be better served by the methodology of fuzzy sets (e.g. Pipkin, 1978). However, out of the number of papers that have now appeared in human geography on fuzzy sets not one has yet gone beyond the exegetical to the application. Fuzzy sets may well still prove to be yet another methodology easier to write about than apply. But, even from this brief review, it can be seen that the possibilities of a new and more coherent quantitative behavioural geography abound. The next ten years should see these possibilities translated into fact.

The halcyon days of behavioural geography are long gone. With them have passed the days when behavioural geographers made inflated claims for the explanatory power of their subject area. But the subject area still has its place in human geography – and not just as a straw man. Recently, it has come under fire from both humanistic and radical geography. The one criticizes behavioural geography for ignoring the contours of experience and reducing the individual to a crude automaton (e.g. Seamon, 1979), the other for ignoring social process and reducing society to a collection of individuals making decisions (e.g. Sayer and Duncan, 1977; Harvey, 1973). Both criticisms have some force. Both, to some extent, are also misplaced (Cullen, 1976, 1977). Behavioural geography occupies a distinct niche in terms of explanation of human behaviour and there is no reason to think that it cannot be amalgamated with or supplement either the radical or the humanistic approaches (see e.g. Howell, 1973). To apply the active-reactive model distinction again the kind of theoretical explanation behavioural geography can make about human behaviour is heavily circumscribed by constraints on what it can know given its emphasis on the individual and on the individual as a decision-maker. But to say that behavioural geography is therefore half-blind is not to say that it can see nothing at all. Its explanations may be limited. That does not mean that they are therefore non-existent.

Acknowledgments

I would like to thank Alan Hay, Ron Johnston and Peter Jones for their valuable comments on a draft of this paper. Of course, no blame attaches to them for the outcome.

References

Anas, A. (1975) 'The empirical calibration and testing of a simulation model of residential location', *Environment and Planning A*, 7, 899–920.

Anas, A. (1979) 'The impact of transit investment on housing values: a simulation experiment', *Environment and Planning A*, 11, 239–55.

Anderson, J. (1971) 'Space-time budgets and activity studies in urban geography and planning', *Environment and Planning*, 3, 353–69.

Appleyard, D., Lynch, K., and Myer, J. (1964), *The View from the Road*, MIT Press: Cambridge, Mass.

Atkin, R.H. (1974) *Mathematical Structure in Human Affairs*, Heinemann: London.

Banister, D.J. (1977) 'Car availability and usage: a modal split model based on these concepts', University of Reading, Department of Geography, Geographical Paper 58.

Banister, D.J. (1978) 'Decision-making, habit formation and a heuristic modal split model based on these concepts', *Transportation*, 7, 5–18.

Bateman, M., Burtenshaw, D., and Duffett, A. (1974) 'Environmental perception of residential areas in South Hampshire' in D. Canter and T. Lee (eds), *Psychology and the Built Environment*, 148–55, Architectural Press: London.

Bather, N.J. (1976) 'The speculative residential developer and urban growth', University of Reading, Department of Geography, Geographical Paper 47.

Beavon, K.A.O., and Hay, A.M. (1977) 'Consumer choice of shopping centre – a hypergeometric model', *Environment and Planning A*, 9, 1375–93.

Becker, H., Holzapfel, H., Kutter, E., and Volkmar, H. (1980) 'The analysis of activity patterns for transportation planning purposes' in P.R. Stopher, A.H. Meyburg and W. Brög (eds), *New Horizons in Behavioural Travel Research*, D.C. Heath: Lexington, Mass.

Benwell, M. (1979) 'Attitude and social meaning – a proper study for Regional Scientists' in I.G. Cullen (ed.), *Analysis and Decision in Regional Policy*, London Papers in Regional Science 9, 120–36, Pion: London.

Blase, J.H. (1979) 'Hysteresis and catastrophe theory: empirical identification in transportation modelling', *Environment and Planning A*, 11, 675–88.

Boulding, K.E. (1956) *The Image*, University of Michigan Press: Ann Arbor.

Bourdieu, P. (1977) *Outline of a Theory of Practice*, Cambridge University Press.

Boyle, M.J., and Robinson, M.E. (1979) 'Cognitive mapping and understanding' in D.T. Herbert and R.J. Johnston (eds), *Geography and the Urban Environment, Vol. 2*, 59–82, John Wiley: Chichester.

Bradford, M.G. (1975) 'Spatial aspects of urban consumer behaviour', unpublished PhD Thesis, University of Cambridge.

Brög, W., Heuwinkel, D., and Neumann, K. (1977) 'Psychological determinants of user behaviour', *European Conference of Ministers of Transport Round Table 34*, OECD: Paris.

Brown, A., and Moore, E.G. (1970) 'The intra-urban migration process: a perspective', *Geografiska Annaler*, Series B, 52, 1–13.

Burchill, P.R. (1975) 'Cognitive maps: conceptual and empirical explorations', unpublished PhD Thesis, University of Bristol.

Burgess, J.A. (1974) 'Stereotypes and urban images', *Area*, 6, 167–71.

Burgess, J.A. (1978) 'Image and identity: a study of urban and regional perception with particular reference to Kingston-upon-Hull', University of Hull, Department of Geography, Occasional Paper 23.

Burnett, K.P. (1980) 'Spatial constraints-oriented approaches to movement, microeconomic theory and urban policy', *Urban Geography*, 1, 53-67.

Burnett, K.P., and Hanson, S. (1979) 'A rationale for an alternative mathematical paradigm for movement as complex human behaviour', *Transportation Research Record*, no. 723.

Burnett, K.P., and Thrift, N.J. (1979) 'New approaches to understanding travel behaviour', in D.A. Hensher and P.R. Stopher (eds), *Behavioural Travel Modelling*, Croom Helm: London.

Canter, D., and Tagg, S. (1975) 'Distance estimation in cities', *Environment and Behaviour*, 7, 59–80.

Carlstein, T., Parkes, D.N., and Thrift, N.J. (eds) (1978) *Timing Space and Spacing Time, Vol. 2. Human Activity and Time Geography*, Edward Arnold: London.

Chapin Jr, F.S. (1974) *Human Activity Patterns in the City: Things People Do in Time and Space*, Wiley Interscience: New York.

Chapin Jr, F.S., and Brail, R.K. (1969) 'Human activity systems in the United States', *Environment and Behavior*, 1, 107-30.

Chapin Jr, F.S., and Hightower, H.C. (1965) 'Household activity patterns and land use', *Journal of the American Institute of Planners*, 31, 222–31.

Chapin, Jr, F.S., and Logan, T.H. (1968) 'Patterns of time and space use' in H.S. Perloff (ed.), *The Quality of the Urban Environment*, Johns Hopkins University Press: Baltimore.

Clarke, M. (1978) 'Towards a mathematical formulation of travel behaviour based on activity patterns', University of Oxford, Transport Studies Unit, Working Paper 39.

Clarke, M., Keys, P. and Williams, H.C.W.L. (1979) 'Household dynamics and economic forecasting: a microsimulation approach', Paper prepared for the European Congress of the Regional Science Association, London, August 1979.

Cox, K.R., and Golledge, R.G. (eds) (1969) *Behavioural Problems in Geography: A Symposium*, Northwestern University Studies in Geography 17, Department of Geography, Northwestern University: Evanston, Ill.

Cullen, I. (1972) 'Space, time and the disruption of behaviour in cities', *Environment and Planning* 4, 459–70.

Cullen, I.G. (1976) 'Human geography, regional science and the study of individual behaviour', *Environment and Planning A*, 8, 397–409.

Cullen, I.G. (1977) 'The "new" behavioural geography - some comments upon the letter by Sayer and Duncan', *Environment and Planning A*, 9, 233–4.

Cullen, I.G. (1978) 'The treatment of time in the explanation of spatial behaviour' in T. Carlstein, D.N. Parkes and N.J. Thrift (eds), op. cit., 27–38.

Cullen, I.G., and Godson, V. (1975) 'Urban networks: the structure of activity patterns', *Progress in Planning*, 4, 1–96.

Cullen, I.G., and Phelps, E. (1975) 'Diary techniques and the problems of urban life', Final report to the Social Science Research Council, London.

Cullen, I.G., Godson, V., and Major, S. (1972) 'The structure of activity patterns' in A.G. Wilson (ed.), *Patterns and Processes in Urban and Regional Systems*, 281-96, (London Papers in Regional Science 3) Pion: London.

Dawe, A. (1970) 'The two sociologies', *British Journal of Sociology*, 21, 42–63.

Dicken, P., and Robinson, M.E. (1979) 'Cloze procedures and cognitive mapping', *Environment and Behavior*, 9, 11, 351-73.

Doling, J.F. (1973) 'A two-stage model of tenure choice in the housing market', *Urban Studies*, 10, 199–211.

Domencich, T., and McFadden, D. (1975) *Urban Travel Demand: A Behavioural Analysis*, North Holland: Amsterdam.

Downs, R.M. (1967) 'Approaches to, and problems in, the measurement of geographical space perception', University of Bristol, Department of Geography, Seminar Paper 9.

Downs, R.M. (1968) 'The role of perception in modern geography', University of Bristol, Department of Geography, Seminar Paper 11.

Downs, R.M. (1970) 'The cognitive structure of an urban shopping centre', *Environment and Behavior*, 2, 13-39.

Downs, R.M., and Meyer, J.T. (1978) 'Geography and the mind', *American Behavioral Scientist*, 22, 59–77.

Downs, R.M., and Stea, D. (eds) (1973) *Image and Environment: Cognitive Mapping and Spatial Behaviour*, Edward Arnold: London; Aldine Press: Chicago.

Downs, R.M., and Stea, D. (1977) *Maps in Minds: Reflections on Cognitive Mapping*, Harper & Row: New York.

Eyles, J.D. (1968) 'Inhabitants' images of Highgate village', London School of Economics, Graduate School of Geography, Discussion Paper 15.

Eyles, J.D. (1978) 'Social geography and the study of the capitalist city: a review', *Tijdschrift voor Economische en Sociale Geografie*, 69, 296–322.

Ferguson, A.G. (1979) 'Some aspects of urban spatial cognition in an African student community', *Transactions, Institute of British Geographers*, NS 4, 77–93.

Flowerdew, R. (1976) 'Search strategies and stopping rules in residential mobility', *Transactions, Institute of British Geographers*, NS 1, 47–57.

Ginsberg, R.B. (1973) 'Stochastic models of residential and geographic mobility for heterogeneous populations', *Environment and Planning*, 5, 113–24.

Golledge, R.G. (1978a) 'Representing, interpreting and using cognized environments', *Papers of the Regional Science Association*, 41, 169–204.

Golledge, R.G. (1978b) 'Learning about urban environments' in T. Carlstein, D.N. Parkes and N.J. Thrift (eds) , op. cit., 76–98.

Golledge, R.G., and Brown, L.A. (1967) 'Search, learning and the market decision process', *Geografiska Annaler*, Series B, 49, 116–24.

Golledge, R.G., Brown, L.A., and Williamson, F. (1972) 'Behavioural approaches in geography: an overview', *Australian Geography*, 12, 59–79.

Golledge, R.G., and Rushton, G. (1976) *Spatial Choice and Spatial Behaviour: Geographic Essays on the Analysis of Preferences and Perceptions*, Ohio State University Press: Columbus, Ohio.

Goodey, B., Duffet, A.W., Gold, J.R., and Spencer, D. (1971) 'City scene: an exploration into the image of central Birmingham', University of Birmingham, Centre for Urban and Regional Studies, Research Memorandum 10.

Goodwin, P.B. (1977) 'Habit and hysteresis in mode choice', *Urban Studies*, 14, 95–8.

Gould, P.R. (1963) 'Man against his environment: a game-theoretic framework', *Annals of the Association of American Geographers*, 53, 290–7.

Gould, P.R. (1965) 'Wheat on Kilimanjaro: the perception of choice within game and learning model frameworks', *General Systems Yearbook*, 10, 157–66.

Gould, P.R. (1966) 'On mental maps', Michigan Inter-University Community of Mathematical

Geographers Discussion Paper 9, reprinted with postscriptum in R.M. Downs and D. Stea (eds), 182–220.

Gould, P.R. (1973) 'The black boxes of Jönköping: spatial information and preference', in R.M. Downs and D. Stea (eds), 235-45.

Gould, P.R. (1975) *People in Information Space: The Mental Maps and Information Surfaces of Sweden*, Lund Studies in Geography, Series B, Human Geography, 42, CWK Gleerup: Lund.

Gould, P.R. (1976) 'Cultivating the garden: a commentary and critique on some multi-dimensional speculations' in R. Golledge and G. Rushton, (eds), op. cit., 83–91.

Gould, P.R., and White, R. (1968) 'The mental maps of British school leavers', *Regional Studies*, 2, 161–82.

Gould, P.R., and White, R. (1975) *Mental Maps*, Penguin: Harmondsworth.

Graham, E. (1976) 'What is a mental map?', *Area*, 8, 259–63.

Green, P.H. (1977) 'Industrialists information levels of regional incentives', *Regional Studies*, 11, 7–18.

Guy, C.M. (1975) 'Consumer behaviour and its geographical impact', University of Reading, Department of Geography, Geographical Paper 34.

Hägerstrand, T. (1963) 'Geographic measurement of migration: Swedish data' in J. Sutter (ed.), *Les Déplacements Humains*, 61–83, Entretiens de Monaco en Sciences Humains, Première Session: Monaco.

Hägerstrand, T. (1967a) 'The computer and the geographer', *Transactions, Institute of British Geographers*, 42, 1–19.

Hägerstrand, T. (1967b) 'On Monte Carlo simulation of diffusion' in W.L. Garrison and D.F. Marble (eds), *Quantitative Geography Part 1: Economic and Cultural Topics*, 1–32, Northwestern University Studies in Geography 13, Department of Geography, Northwestern University: Evanston, Ill.

Hägerstrand, T. (1970) 'What about people in regional science?', *Papers of the Regional Science Association*, 24, 7-21.

Hägerstrand, T. (1973) 'The domain of human geography' in R.J. Chorley (ed.), *Directions in Geography*, 67–87, Methuen: London.

Hanson, S., and Burnett, P. (1980) 'The analysis of travel as an example of complex human behaviour in spatially constrained situations: measurement issues' in P.R. Stopher, A.M. Meyburg and W. Brög (eds), *New Horizons in Behavioural Travel Research*, D.C. Heath: Lexington, Mass.

Harrison, J.A., and Sarre, P. (1971) 'Personal construct theory in the measurement of environmental images: problems and methods', *Environment and Behavior*, 3, 351-74.

Harrison, J.A., and Sarre, P. (1975) 'Personal construct theory in the measurement of environmental images', *Environment and Behavior*, 7, 3-58.

Harvey, D. (1967) 'Behavioural postulates and the construction of theory in human geography', University of Bristol, Department of Geography, Seminar Paper 6; also in *Geographica Polonica* (1969).

Harvey, D. (1973) *Social Justice and the City*, Edward Arnold: London; John Hopkins University Press: Baltimore.

Hay, A. (1978) 'Transport geography', *Progress in Human Geography*, 2, 324–9.

Heggie, I.G. (1978) 'Putting behaviour into behavioural models of travel choice', *Journal of the Operational Research Society*, 29, 541–50.

Heggie, I.G., and Jones, P.M. (1978) 'Defining domains for models of travel demand', *Transportation*, 7, 119–35.

Herbert, D.T. (1973) 'The residential mobility process: some empirical observations', *Area*, 5, 44–8.

Hollis, M. (1977) *Models of Man: Philosophical Thoughts on Social Action*, Cambridge University Press.

Hollis, M., and Nell, E.J. (1975) *Rational Economic Man: A Philosophical Critique of Neo-classical Economics*, Oxford University Press: New York.

Howell, J.T. (1973) *Hard Living on Clay Street: Portraits of Blue Collar Families*, Anchor Books: New York.

Hudson, R. (1970) 'Personal construct theory, learning theories and consumer behaviour', University of Bristol, Department of Geography, Seminar Paper 21.

Hudson, R. (1971) 'Towards a theory of consumer spatial behaviour', University of Bristol, Department of Geography, Seminar Paper 22.

Hudson, R. (1974) 'Images of the retailing environment: an example of the use of the repertory grid methodology', *Environment and Behavior*, 6, 470-94.

Hudson, R. (1975) 'Patterns of spatial search', *Transactions, Institute of British Geographers*, 65, 141–54.

Hudson, R. (1976a) 'Linking studies of the individual with models of aggregate behaviour: an empirical example', *Transactions, Institute of British Geographers*, NS1, 159–75.

Hudson, R. (1976b) 'Environmental images, spatial choice and consumer behaviour', University of Durham, Department of Geography, Occasional Paper 9.

Huff, D.L. (1960) 'A topographical model of consumer space preferences', *Papers of the Regional Science Association*, 7, 159–73.

Huff, D.L., and Clark, W.A.V. (1978) 'Cumulative stress and cumulative inertia: a behavioural model of the decision to move', *Environment and*

Planning A, 10, 1011–20.

Ingram, G.K., Kain, J.F., and Ginn, R.J. (1972) *The Detroit Prototype of the NBER Urban Simulation Model*, National Bureau of Economic Research: New York.

Irving, H.W. (1975) 'A geographer looks at personality', *Area*, 7, 207–12.

Irving, H.W. (1978) 'Space and environment in interpersonal relations' in D.T. Herbert and R.J. Johnston (eds), *Geography and the Urban Environment, Vol. One*, 249–84, John Wiley: Chichester.

Jackson, L.E., and Johnston, R.J. (1972) 'Structuring the image: an investigation of the elements of mental maps', *Environment and Planning*, 4, 415–27.

Johnson, J., Salt, J., and Wood, P.A. (1974) *Housing and the Migration of Labour in England and Wales*, Saxon House: Farnborough, Hants.

Johnson, J.H. (1976) 'The methodology of Q-analysis in planning and transportation' in R.E. Matzner and G. Rusch (eds), *Transport as an Instrument for Allocating Space and Time – A Social Science Approach*, 160–80, Technical University of Vienna: Vienna.

Johnston, R.J. (1973) 'Mental maps: an assessment' in J. Rees and P. Newby (eds), *Behavioural Perspectives in Geography*, Middlesex Polytechnic Monographs in Geography 1, 6–34.

Jones, C., Gudjonsson, S., and Parry Lewis, J. (1978) 'A two-stage model of tenure mobility', *Environment and Planning A*, 10, 81–92.

Jones, P.M. (1979a) 'New approaches to understanding travel behaviour: the human activity approach' in D.A. Hensher and P.R. Stopher (eds), *Behavioural Travel Modelling*, Croom Helm: London.

Jones, P.M. (1979b) 'HATS: a technique for investigating household decisions', *Environment and PlanningA*, 11, 59–70.

Kates, R.W. (1962) *Hazard and Choice Perception in Flood Plain Management*, Research Paper 78, University of Chicago, Department of Geography: Chicago.

Kirk, W. (1951) 'Historical geography and the behavioural environment', *Indian Geographical Journal*, Silver Jubilee Volume, 52–60.

Lee, T.R. (1964) 'Psychology and living space', *Transactions of the Bartlett Society*, 2, 9–36.

Lee, T.R. (1970) 'Perceived distance as a function of direction in the city', *Environment and Behavior,* 2, 40-51.

Lenntorp, B. (1976) *Paths in Space-Time Environments: A Time-Geographic Study of Movement Possibilities of Individuals*, Lund Studies in Geography, Series B, 44, CWK Gleerup: Lund.

Lewis, J., and Melville, B. (1978) 'The politics of epistemology in regional science' in P.W.J. Batey (ed.), *Theory and Method in Urban and Regional Analysis*, London Papers in Regional Science 8, 82–100, Pion: London.

Lloyd, B.L., and Lloyd, P.C. (1980) 'Review of "Environmental Knowing" ', *Environment and Planning A*, 12, 235-7.

Lowenthal, D. (1961) 'Geography, experience and the imagination: towards a geographical epistemology', *Annals of the Association of American Geographers*, 51, 241–60.

Lowenthal, D. (ed.) (1967) *Environmental Perception and Behaviour*, Research Paper 109, University of Chicago, Department of Geography: Chicago.

Luce, R.D. (1959) *Individual Choice Behaviour: A Theoretical Analysis*, John Wiley: New York.

Lynch, K. (1960) *The Image of the City*, MIT Press: Cambridge, Mass.

Lyon, S., and Wood, M.E. (1977) 'Choosing a house', *Environment and Planning A*, 9, 1169–76.

Maclennan, D. (1977) 'Information, space and the measurement of housing preferences and demand', *Scottish Journal of Political Economy*, 24, 97–115.

Maclennan, D. (1979a) 'Information networks in a local housing market', *Scottish Journal of Political Economy*, 26, 73–88.

Maclennan, D. (1979b) 'Search in a model of housing choice' in A. Evans and M. Harloe (eds), *Urban Economics Papers Vol. 1*, Centre for Environmental Studies, London.

Madden, M. (1978) 'The perception of distance in migration in Merseyside', *Area*, 10, 167–73.

Marble, D.F. (1967) 'A theoretical exploration of individual travel behaviour' in W.L. Garrison and D.F. Marble (eds), *Quantitative Geography, Part 1, Economic and Cultural Topics*, Studies in Geography 13, 33–53, Department of Geography, Northwestern University: Evanston, Ill.

Michelson, W. (1966) 'An empirical analysis of urban environmental preferences', *Journal of the American Institute of Planners*, 32, 355–60.

Moore, G., and Golledge, R.G. (eds) (1976) *Environmental Knowing: Theories Perspectives and Methods,* Dowden, Hutchinson & Ross: Stroudsburg, Pa.

Norborg, K. (1968) *Jordbruksbefolkningen I Sverige: Regional Struktur Och Förändring Under 1900 – Talet* (The Agricultural Population in Sweden: Regional Structure and Change During the Twentieth Century), CWK Gleerup: Lund.

Office of Population Censuses and Surveys (1978) *The General Household Survey, 1976*, HMSO: London.

Olsson, G. (1975) 'Birds in Egg', University of Michigan, Department of Geography, Michigan Geographical Publications 15, reprinted 1980 by Pion: London.

Orcutt, G.H., Greenberger, M., Korbel, J., and

Rivlin, A.M. (1961) *Microanalysis of Socio-Economic Systems: A Simulation Study*, Harper & Row: New York.
Orcutt, G.H., Caldwell, S., and Wertheimer, R. (1976) *Policy Explorations Through Micro-analytic Simulation*, The Urban Institute: Washington.
Pacione, M. (1975) 'Preference and perception: an analysis of consumer behaviour', *Tijdschrift voor Economische en Sociale Geografie*, 66, 84–92.
Pacione, M. (1976) 'Shape and structure in cognitive maps of Great Britain', *Regional Studies*, 10, 275–83.
Pacione, M. (1978) 'Information and morphology in cognitive maps', *Transactions, Institute of British Geographers*, NS 3, 548–68.
Palmer, C.J. (1978) 'Understanding unbiased dimensions: the use of repertory-grid methodology', *Environment and Planning A*, 10, 1137–50.
Palmer, C.J., Robinson, M.E., and Thomas, R.W. (1977) 'The countryside image: an investigation of structure and meaning', *Environment and Planning A*, 9, 739–50.
Pickup, L. (1978) 'Space-time budgets, coupling constraints and women's career opportunities', unpublished PhD Thesis, University of Reading.
Pipkin, J.S. (1978) 'Fuzzy sets and spatial choice', *Annals of the Association of American Geographers*, 68, 196–204.
Pocock, D.C.D. (1972) 'City of the mind: a review of mental maps of urban areas', *Scottish Geographical Magazine*, 88, 115–24.
Pocock, D.C.D. (1973) 'Environmental perception: process and project', *Tijdschrift voor Economische en Sociale Geografie*, 64, 251-7.
Pocock, D.C.D. (1975) 'Durham: images of a cathedral city', University of Durham, Department of Geography, Occasional Paper 6.
Pocock, D.C.D. (1976) 'Some characteristics of mental maps: an empirical study', *Transactions, Institute of British Geographers*, NS 1, 493–512.
Pocock, D.C.D. (1979) 'The novelists' image of the north', *Transactions, Institute of British Geographers*, NS 4, 62–76.
Pocock, D.C.D., and Hudson, R. (1978) *Images of the Urban Environment*, Macmillan: London.
Porteous, J.D. (1971) 'Design with people: the quality of the urban environment', *Environment and Behaviour*, 3, 155–78.
Porter, J., Hart, T., and Machin, T. (1975) 'Cloze procedure tested in Hampshire', *Area*, 7, 196–8.
Potter, R.B. (1977a) 'The nature of consumer usage fields in an urban environment: theoretical and empirical perspectives', *Tijdschrift voor Economische en Sociale Geografie*, 68, 168–76.
Potter, R.B. (1977b) 'Spatial patterns of consumer behaviour and perception in relation to the social class variable', *Area*, 9, 153–6.
Pred, A. (1967) *Behaviour and Location, Part 1*, CWK Gleerup: Lund.
Quigley, J. (1976) 'Housing demand in the short-run: an analysis of polytomous choice', *Explorations in Economic Research*, 2, 76–102.
Robinson, I.M., Wolfe, H.B., and Barringer, R.L. (1965) 'A simulation model for renewal programming', *Journal of the American Institute of Planners*, 31, 126–34.
Robinson, M.E. (1974) 'Cloze procedures and spatial comprehension tests', *Area*, 6, 137–42.
Rowley, G., and Tipple, G. (1974) 'Coloured immigrants within the city: an analysis of housing and travel preferences', *Urban Studies*, 11, 81–9.
Rowley, G., and Wilson, S. (1975) 'The analysis of housing and travel preferences: a gaming approach', *Environment and Planning A*, 7, 171–7.
Rushton, G. (1969a) 'Analysis of spatial behavior by revealed space preference', *Annals of the Association of American Geographers*, 59, 391–400.
Rushton, G. (1969b) 'The scaling of locational preference' in K.R. Cox and R.G. Golledge (eds), op. cit., 197–227.
Saarinen, T. (1969) 'Perception of environment', Commission on College Geography Resource Paper 5, Association of American Geographers: Washington, DC.
Sayer, A., and Duncan, S. (1977) 'The "new" behavioural geography – a reply to Cullen', *Environment and Planning A*, 9, 230–2.
Seamon, D. (1979) *A Geography of the Lifeworld: Movement, Rest and Encounter*, Croom Helm: London.
Senior, M.L. (1974) 'Approaches to residential location modelling 2: urban economic models and some recent developments (a review)', *Environment and Planning*, 6, 369–410.
Shapcott, M., and Steadman, P. (1978) 'Rhythms of urban activity' in T.Carlstein, D.N. Parkes and N.J. Thrift (eds), op. cit., 49–74.
Shapcott, M., and Wilson, C. (1978) 'Correlations amongst time uses', *Transactions of the Martin Centre*, 1, 73–94.
Silk, J. (1971) 'Search behaviour: general characterisation and review of literature in the behavioural sciences', University of Reading, Department of Geography, Geographical Paper 7.
Smith, G.C. (1976) 'The spatial information fields of urban consumers', *Transactions, Institute of British Geographers*, NS 1, 175–89.
Sohn-Rethel, A. (1978) *Intellectual and Manual Labour: A Critique of Epistemology*, Macmillan: London.
Spencer, A.H. (1978) 'Deriving measures of attractiveness for shopping centres', *Regional Studies*, 12, 713–26.
Suppes, P. (1961) 'Behaviouristic foundations of utility', *Econometrica*, 29, 186–202.

Thrift, N.J. (1977a) 'An introduction to time-geography', *Concepts and Techniques in Modern Geography* 13, Geo Abstracts: Norwich.

Thrift, N.J. (1977b) 'Time and theory in human geography, part 2', *Progress in Human Geography*, 1, 413–57.

Thrift, N.J., and Williams, H.C.W.L. (1981) 'On the development of behavioural location models', *Economic Geography*.

Tipple, G., and Rowley, G. (1977) 'Sequential fixed cost evaluators: a probabilities approach', *Environment and Planning A*, 1395–1400.

Tivers, J. (1977) 'Constraints on spatial activity patterns: women with young children', Kings College, London, Department of Geography, Occasional Paper 6.

Tomlinson, J., Bullock, N., Dickens, P., Steadman, P. and Taylor, E. (1973) 'A model of students' daily activity patterns', *Environment and Planning*, 5, 231–66.

Townroe, P.M. (1972) 'Some behavioural considerations in the industrial location decision', *Regional Studies*, 6, 261–72.

Townroe, P.M. (1973) 'Industrial location search behaviour and regional development' in J. Rees and P. Newby (eds), *Behavioural Perspectives in Geography*, Middlesex Polytechnic Monographs in Geography 1, 44–58.

Townsend, J.G. (1976) 'Farm "failures": the application of personal constructs in the tropical rain forest', *Area*, 8, 219–22.

Townsend, J.G. (1977) 'Perceived worlds of the colonists of tropical rainforest, Colombia', *Transactions, Institute of British Geographers*, NS 2, 430–58.

Tuan, Yi-Fu. (1975) 'Images and mental maps', *Annals of the Association of American Geographers*, 65, 205–13.

Tversky, A. (1972) 'Elimination by aspects: a theory of choice', *Psychological Review*, 79, 281–99.

Warner, S.L. (1962) *Stochastic Choice of Mode in Urban Travel: A Study in Binary Choice*, Northwestern University Press: Evanston Ill.

White, R. (1967) 'The measurement of spatial perception', University of Bristol, Department of Geography, Seminar Paper 8.

Williams, H.C.W.L. (1977) 'On the formation of travel demand models and economic evaluation measures of user benefit', *Environment and Planning A*, 9, 285–344.

Williams, N.J. (1979) 'The definition of shopper types as an aid in the analysis of consumer spatial behaviour', *Tijdschrift voor Economische en Sociale Geografie*, 70, 157–63.

Wilson, A.G. (1976), 'Catastrophe theory and urban modelling: an application to modal choice', *Environment and Planning A*, 8, 351–6.

Wolpert, J. (1964) 'The decision process in a spatial context', *Annals of the Association of American Geographers*, 54, 537–58.

Wolpert, J. (1965) 'Behavioural aspects of the decision to migrate', *Papers of the Regional Science Association*, 15, 159–72.

Wood, W. (1970) 'Perception studies in geography', *Transactions, Institute of British Geographers*, 50, 129–41.

Wrigley, N. (1976) 'An introduction to the use of logit models in geography', *Concepts and techniques in modern geography*, 10, Geo Abstracts: Norwich.

Wrigley, N. (1980) 'Paired comparison experiments and logit models: a review and illustration of some recent developments', *Environment and Planning A*, 12, 21–40.

Chapter 33

Transport geography

A. Hay

Transport geography had a very special relationship with early quantitative geography in North America, largely due to the influence of a few individuals, notably E.L. Ullman, W.L. Garrison and E.J. Taaffe. In the United Kingdom the relationship has been less close because the leading British transport geographers of the 1950s never fully accepted the quantitative approach (this is probably a fair description of the stance adopted by O'Dell, Appleton, Sealy and Bird). The impact of quantitative methods on transport geography in the United Kingdom therefore leaned heavily on the North American model, either directly as was evident in the early work of O'Sullivan (1968) and Starkie (1967), or indirectly from the influential books published by Haggett and Chorley (Haggett, 1965; Haggett and Chorley, 1969). One consequence of this indirect learning which is clearly evident in Haggett and Chorley's work was for the content of transport geography to be subsumed under wider and more abstract categories of 'locational analysis' (Haggett, 1965), 'network analysis' (Haggett and Chorley, 1969) and 'graph theory' (Tinkler, 1979). In consequence there was an absence of any statement about the nature and content of transport geography, and a lack of detailed study of actual transport systems.

There have been several recent statements about what is, or should be, the content of transport geography. Hay (1973), Taaffe and Gauthier (1973), and Eliot Hurst (1974) have offered the fullest statements and Rimmer (1978) has made a strong appeal which links with the position of Eliot Hurst. These authors' differing viewpoints leave the reader sufficiently confused to make it worth while to identify some problems of transport geography which constitute recurring motifs within the literature, and which will be used to define the main sections of this review.

First there is the theme of *accessibility*, a concept which is widely used by geographers but which has proved extremely difficult to define and measure. It is generally recognized that accessibility is a product not only of location but also of the transport system including networks, access points and vehicle supply. Second there is an interest in *relating network form* to flow patterns, whether this is attacked by classificatory or explanatory approaches. Third, there is a recurrent interest in the *explanation of commodity flow*.

It should be stressed that this selection of three themes ignores other, important, themes, especially intra-urban passenger movement, inter-urban passenger flow and international passenger flow. This is not an attempt to dismiss these topics from the field of transport geography but a response to the space limits on this contribution and the pattern of other contributions to this volume.

Accessibility

As a naïve concept accessibility has a long history in geographic thought, not only as a descriptive variable to be measured and explained, but also as a causal variable in studies of population, retail trade, industrial growth etc. (Ingram, 1971; Pirie, 1979). It is generally recognized that accessibility is partly a function of relative location in the sense that *ceteris paribus* a central location is more accessible to (and from) an area than is a peripheral location. But accessibility is also determined by the form of permanent transport networks and the pattern of scheduled services, especially scheduled passenger services.

For some authors accessibility seems to be defined solely in terms of access to mechanized transport systems, an approach exemplified by Jefferson's classic (1929) study of the 'civilising rails'. Jefferson demonstrated the existence of a cartographic relationship between network density and the proportion of land area lying more than a specified distance from the network. In detail Jefferson's approach might be faulted because access to rail systems is confined to specific points (Neft, 1959). The contributions of Hay (1973, pp. 35–8) and Melut and O'Sullivan (1974) make the mathematical relationships behind these points explicit for geographical readers.

The idea that accessibility is conferred by both location and network form becomes very clear in studies of route factors (the ratios between route distances and geodetic distances). Timbers (1968), Volk (1968) and Hay (1971) all identify this aspect because the method of calculating route factors necessarily excludes the effects of location (centrality etc.) and therefore concentrates attention upon the effects of network form. Hay relates this to the concept of orientation and suggests that in West African networks orientation is less marked than is usually supposed.

Early attempts to define accessibility more broadly usually failed because the authors were forced to consider access to a single point (often a central point), but such exercises have continued with various forms of isochrone mapping (Forbes, 1964; Hardwick, 1974), aided in some cases by computer mapping, but without any great conceptual advance. A second alternative has been to map accessibility by private or public transport to a particular type and level of service (whether retail, medical, educational, recreational, or social). An example of this work is given by Robertson (1976). Once again computer mapping has proved a useful tool. The third alternative is to introduce some concept of choice by mapping the cumulative opportunities within a given time or distance range. The greatest problem has been the selection of a critical distance range (essentially an arbitrary choice). A point which becomes clear, especially in studies of passenger accessibility, is the multidimensional aspect of the concept not only in terms of destinations but also in terms of the various travel cost metrics. Some authors have attempted to reduce these to a single abstract measure (e.g. Johnston, 1966), while others have sought to reduce them to a single metric of generalized cost.

A second main route to the study of accessibility has been to use potential models in which the accessibility of every part of a region is considered and is weighted by the importance of that part (in terms of population, markets, opportunities etc.). The classic geographic papers on this topic (e.g. Harris, 1954) were used as the basis for studies by Clark (1966) and Clark, Wilson and Bradley (1969). All these studies use a potential model of the form:

$$V_j = \sum_{i=1}^{i=n} \frac{P_i}{d_{ij}^{\alpha}}$$

where V_j is the potential (accessibility) of the jth region,
P_i is the weight (population etc.) of the ith region,
d_{ij} is the distance (however measured) between the ith and j^{th} regions, and
α is an exponent (most commonly 1 or 2).

The degree of assurance with which such potential measures may be used has been steadily reduced over the last ten years. It was recognized from an early stage that the resulting accessibility maps can be extremely sensitive to the boundaries and extent of the study region (especially if the value of α is less than 1). It has since become clear that the results are extremely sensitive to the shapes and sizes of the areal units which are used as origin and destination zones.

A third route to establishing measures of accessibility is to use the so-called topological measures. In these a transport network (whether the routes themselves or the scheduled service network) is conceived as a graph which can be represented as a binary matrix (Tinkler, 1977). The different conventions by which such a simplification may be achieved has been the subject of an exhaustive discussion by Brook (1976); it is sufficient to note that no standardized method has emerged, and that differences between the results of substantive studies may in part reflect different practices.

The binary matrix – often called the connectivity or connection matrix and so given the symbol C – is subject to a number of analyses.

$$\sum_{j=1}^{j=n} c_{ij}$$

gives the number of links incident at a node, sometimes called the degree of the node. It can be seen as a version of the opportunities method because it measures the number of opportunities which become available by traversing one link. It is seldom a very useful measure for large and complex networks.

The attention has therefore focused upon the power series

$$S_n = C + C^2 + C^3 + C^4 + \ldots\ldots C^n$$

or, alternatively, the power series based on aC where a is a scalar $0 < a < 1$):

$$T_n = aC + a^2C^2 + a^3C^3 \ldots a^nC^n$$

A number of points about these series have become clear since their use was introduced in transport geography. Firstly, the rows of the matrices which constitute the terms of both these series in general become convergent upon the principal eigenvector of the matrix C^T (as the matrix is usually symmetric this is equivalent to the eigenvector of C). The point was made forcibly by

Tinkler (1972) and the fact that it applies also to the aC series is demonstrated by Hay (1975). As the convergence is quite rapid, and as the relative magnitudes of the cell values increase rapidly in the powering of C, the sum of the series

$$C + C^2 + C^3 + C^4 \ldots + C^n$$

will yield a matrix whose row values approximate to the principal eigenvector, except where n is small. The result is therefore equivalent to the direct principle components analysis of the original matrix (Tinkler, 1975; Hay, 1975).

The choice of n was often defined as being either the *diameter* of the network or the *solution matrix*. In the first case it is the lowest power of the matrix for which all cells in the sum are positive. In the second case it is the lowest power of the matrix for which all cells in the power are positive. If the original C matrix has ones on the principal diagonal the solution power and the diameter are equal. The physical interpretation of these powers is that at the power C^n, the cell $C_{ij} = x$ implies that there are exactly x routes between i and j containing exactly n links to be traversed. These points have become clear in papers published by Alao (1970), Tinkler (1974) and Garner and Street (1978).

The choice of the scalar a was widely subject to the condition $o < a < 1$. This condition is sufficient if the only function of the scalar is to discount the value of highly indirect routes, but it is insufficient for another purpose. Many authors have wished to avoid tedious computation of the sum of the series by considering the infinite series:

$$aC + a^2C^2 + a^3C^3 \ldots\ldots$$

which, under certain conditions, can be summed by the well known matrix expression:

$$[I - aC]^{-1} = I + aC + a^2C^2 + a^3C^3 \ldots\ldots$$

This latter expression is generally true if and only if the series converges to zero, and this can only be true if the principal eigenvalue of aC is less than unity. It was therefore necessary to choose a such that the eigenvalue of aC is less than one. This was clearly recognized by Nystuen and Dacey (1961) but was overlooked by others (e.g. Stutz, 1973) until the point was made by Tinkler (1974).

These mathematical-computational points do little to clarify the physical interpretation of the cells T_n and S_n and some geographers still seek to interpret them as accessibility indices: the most important step foward (by Tinkler, 1972) takes the discussion away from accessibility towards the relation between network form and flow which will be examined below.

The fourth possibility is to consider the D matrix, or shortest path matrix, in which the ij^{th} cell records the number of links in the shortest path between i and j. This can be used to give a number of accessibility measures including:

$$d_i = \sum_{j=1}^{j=n} d_{ij}$$

$$\bar{d}_i = \sum_{j=1}^{j=n} d_{ij}/v$$

where v is the number of vertices, giving d_i and $\bar{d}_i$ which measure gross vertex accessibility and mean vertex accessibility (the higher the value the lower the accessibility). The distribution of all the cells in D was used to great effect by James, Haggett, Cliff and Ord (1970) to characterize whole networks using not only the mean but also the second, third and fourth moments of the distribution. They did not however take this a step further to look at the distribution of the dij for individual nodes (i.e. single rows or columns of Dj). The idea that these too should be described in more detail (mean, standard deviation and skewness) appears in Hay (1977), who points out that such accessibility histograms can be constructed not only for topological distance (the D matrix) but also for cost distances (using as example data from Kissling, 1969).

The results from this research effort are rather disappointing. The original intention was to seek simple methods of measuring accessibility which would reduce the need for huge data sets of mileages, costs and times. The mathematical developments outlined above have proved extremely demanding in terms of computer store space and time, and at the same time hardware and software developments (the latter often developed for commercial application in transportation planning and traffic management) have made it possible to store, manipulate, recall and plot complex accessibility indices based on the real-life properties of transport networks. On the other hand it may be noted that the methods have provided tools which geographers (Williams, 1977) and planners (e.g. Armstrong, 1972) find useful in their studies of location questions.

The relationship between network form and flow

The desire to demonstrate an interrelationship between network form and network flow has always been present in geographical studies of transport. In some work the intent has been simply descriptive, but other authors have attempted a classification of facilities. For example Turton (1969) adapted the categorization devised by Wallace (1958) and attempted to show that it could be applied to the UK railway systems of the 1920s. Bird adapted the well-established location quotient techniques in an attempt to establish the distinctive functional roles of British seaports (Bird, 1969).

More recently Chu (1978) has attempted a Markov analysis of flows. All these studies tend to suggest a causal loop: is the network determining flow, do the flows determine the network, or are both these causal links present in a feedback system?

A major contribution to this theme resulted from the convergence of three lines of work. The first line has already been noticed: Tinkler (1972) sought to interpret the powered connection matrix of a network as a Markov type equilibrium distribution which resulted from repeated operation by the network on the flows. In his paper Tinkler speaks of rumours and of diseases but the transport interpretation was taken up by his colleague J.A. Smith (1974). She argued that the principal eigenvector of the connectivity matrix could be used to predict the pattern of flows on the Ugandan road network. It proved a weak predictor when used in isolation ($r^2 = 0.35$) but extremely effective when used in conjuction with a population based gravity model (multiple $r^2 = 0.93$).

This interpretation of the eigenfunctions of the connection matrix links directly with a paper by Morley and Thornes (1972). Like Tinkler and Smith they assume that traffic arising at a node along one link is equally likely to leave that node by any other link incident at that node. The network can therefore be described as an ergodic Markov transition matrix for which the equilibrium vector describes the distribution of those travellers on the network who allow their movements to be channelled by the network. The method was applied to data for tourist movements in a recreation area on Dartmoor. Outside geography a similar line of research was initiated by Garbrecht (1973) although he was concerned especially with regular grid networks of the type found in the United States and their effect on channelling pedestrian flows within an urban area. Once again the decision rules for behaviour at junctions was necessarily arbitrary.

A more sophisticated approach emerged from the work of economists and planners who argued that the distribution of traffic (on a given origin-destination path) between alternative routes will tend to equalize journey times by both routes (the shorter distance route attracting more traffic and thus lower average speeds).

The establishment of a relationship between network form and the channelling of flows obviously raises questions about changes in the flow pattern which will result from changes (by addition or subtraction) in the network. O'Sullivan (1978) returns to the topological measures developed in the 1960s to demonstrate the geographical extent and magnitude of such effects. It must be stressed however that this work of O'Sullivan's, like all the references in this section is concerned with the geographical allocation of flows between routes, not with the causes of the flows themselves.

The explanation of commodity flow

The study of commodity flow has always been inhibited by the problems of data (Thompson, 1974). The data sets which are available are often unsatisfactory in terms of the size and shape of the zones, the number of transport modes covered, the accuracy of commodity classification and total sample size. Certainly the United Kingdom data sets are still markedly inferior to those available in the United States. There have been however some major surveys to which geographers have had access, notably the Ministry of Transport (1964) survey of road goods and the Martech survey of port traffic (Bird 1969).

It is for these reasons that several of the early studies of commodity flow depended upon data collected by individual research workers (Starkie, 1967) and there were necessarily severe doubts about the generality of the conclusions. The first major effort was the Portbury exercise carried out within the Ministry of Transport in an attempt to predict the likely traffic through the port of Bristol. The study was important in its introduction of gravity model techniques into the published reasoning of central government. (This technique was also used by the (Roskill) Commission on the Third London Airport, 1970.) It is tempting to claim that events since the Portbury exercise have abundantly justified the predictions of the model that Bristol was unlikely to attract the vast tonnages which the city fathers needed to justify their investment.

In academic work the first major study of commodity flow in the UK was offered by Chisholm and O'Sullivan (1973) using the Ministry of Transport's data set. They adapted many of the methods used in the US (e.g. Perle, 1964, and Helvig, 1964) in an attempt to model interregional commodity flow between 78 zones in the United Kingdom. Two basic models were tried: the first was a Wilson (1970) type gravity model of the form

$$T_{ij} = A_i B_j O_i D_j d_{ij}^{-\alpha}$$

where T_{ij} is the flow between zone i and zone j,
O_i is the flow out of zone i,
D_j is the flow into zone j,
d_{ij} is the distance between i and j and A_i, B_j and α are empirically derived.

The second model used was the linear programming transportation problem.

Before publishing their main findings (Chisholm and O'Sullivan, 1973) the authors presented a number of papers, including Chisholm (1970) and O'Sullivan (1970). The book and papers together raise a number of issues which have attracted attention in the following years.

First of all it must be stressed that the models adopted seek to explain the transport flows in terms of the proximate variables of surplus, deficit, and transport cost. They do not seek the underlying causes in terms of production costs, production possibilities (Brook, 1974) or demand curves, a course of action which was impossible given the data available. The major British contribution to this problem has been Wilson's work on an entropy derivation of the gravity model (Wilson, 1970; Senior, 1979). But despite his attempts to integrate them with regional input–output methods the fact remains that his models are essentially constrained equiprobability models and the constraints can scarcely be considered as true exogenous causal variables. For example in most forms of his gravity model the total transport cost (or effort) expended in the flow system is introduced as a constraint (and thus a quasi-cause) when in reality it is itself a consequence of the system. But the other constraints too which basically ensure that no region exports in excess of supply or imports in excess of demand are also (as Hay, 1979, argues) results of the system. However useful the model and its constraints may be for representing real world systems its ability to explain such systems is severely limited. When this is understood (especially when the role of the constraints is understood) it is unsurprising to learn (Evans, 1973; Wilson and Senior, 1974) that gravity models and linear programming models are mathematically convergent. It is also unsurprising that the gravity model continues to come under attack from other disciplines (Heggie, 1969), from positivistic geographers (Pred, 1967) and radical geographers (Sayer, 1977).

Another issue was raised by O'Sullivan's (1970) paper in which he sought to interpret differences in the gravity model exponent for different regions of Great Britain as differences in accessibility. This paper was published almost simultaneously with one by Olssòn (1970) in which he stressed the incompleteness of the gravity model specification, the presence of both autocorrelation and multicollinearity, and hence the likely instability in coefficients which were determined by regression methods.

The issue was taken up in a series of papers which suggested two reasons for doubting the interpretation of different coefficients. Johnston (1973, 1975 and 1976) is mainly concerned with the effect of map pattern, thus picking up ideas suggested by Porter (1956), Olsson (1970) and Curry (1972). He argues that the range of distances from i^{th} region and the nearest neighbour distance from the ith region both affect the regression estimates of the exponent (α) in a gravity model for the ith region. Cliff, Martin and Ord (1974 and 1975) dispute Johnston's conclusion arguing that α is inversely related to the dispersion of log distances.

Curry (1972) and his associates (Curry, Griffith and Sheppard, 1975 and 1976), are more concerned with autocorrelation and collinearity in the independent variables. This effect too is disputed by Cliff, Martin and Ord (1974, 1975 and 1976): they agree that the variances are increased by the confounding of distance-friction and spatial structure, but insist that the regression method provides the best linear unbiased estimators. To some readers these problems may appear to be merely technical puzzles but not only do they continue to attract interest (Sheppard, 1979) but also if these models are used for planning purposes the problems of parameter stability and interpretation may have great practical importance (Gordon, 1979; and Pitfield, 1979).

A third issue which the original study raised (albeit obliquely) is the question of comparing model solutions with reality. Although Cox (1965) and others have felt that the correlation coefficient and significance tests are legitimate measures of fit, it is quite clear that this is wrong. The data are not sample data, the individual cell values are not independent and it is possible to have a high r^2 even when the model grossly misestimates the flows. Although various proposals have been made (Lankford, 1974, suggests using canonical correlation) the problem is not satisfactorily solved.

Backwards and forwards

The review above is not of course exhaustive of all the transport-related research published by British geographers in the last two decades. In particular it omits to mention the introduction of disaggregate passenger demand modelling which has been followed in Britain by Jones (1974). The disaggregate models have a special interest for geographers because they link with the work of Rushton (1969) through the work of authors like Girt (1977) and Beavon and Hay (1977) on destination choice. They have additional interest for quantitative geographers because they utilize the increasingly widely adopted logit models for purposes of calibration. Finally as Girt (1977) and Stutz (1976) have argued these disaggregate models may pave the way for a more 'social' transport geography (Wheeler, 1973; Rimmer, 1978). In Britain this social transport geography has been focused, once again, upon accessibility and equality of access by people to essential services (Ambrose, 1977). In urban areas the most important lead has been given by an architect (Hillman, 1973) while in rural areas there have been major studies by Moseley and his associates (1977) and by Farrington and Stanley (1978). All these authors have tended to abandon the tools of the quantitative revolution for much simpler presentations of empirical material in graph and table form. At the same time a number of authors have accepted the arguments that the use of positivist models in transport planning leads to the perpetuation of the *status quo* and that the family of gravity models in particular has implicit pareto optimality, with all that that implies for equity considerations. Sayer (1977) has also argued that the use of gravity models tends to obscure the real bases of passenger and commodity movement.

Faced with these challenges it is reasonable to inquire whether the theoretical quantitative transport geography of the 1960s was merely an episode of 'alienating mental exercise' (Eliot Hurst, 1974, p. 287) or whether it can contribute to the different concerns which are becoming the focus for transport geographers in the late 1970s and the 1980s. First, it should be noted that the quantitative transport geography of the 1960s largely lacked the elements which are essential in the new paradigm: equity measures and normative models. Although many of the methods demonstrated the existence of geographic variation in access there was no attempt to distinguish between those variations which are inescapable (for example given a spatially dispersed population and punctiform distribution of services), those which are random, and those which represent systematic and avoidable variations in the levels of welfare provided to individuals or communities. Second, although there was some interest in linear programming and other optimizing techniques, the transport geographers focused more upon the positivistic application than on normative application to real world problems.

I believe that despite these clear inadequacies in the 1970s the transport geography of the 1980s should not ignore the developments of the preceding decades. Firstly, there has been a sharpening of the concepts of transport geography (of which accessibility is only one instance) concepts which will inevitably retain a role in the viewpoints of the 1980s even if the measures attached to those concepts are viewed with scepticism or are discarded. Secondly, the studies of the 1960s and 1970s have revealed clearly the highly interconnected nature of transport systems, so that every intervention will have widespread repercussions; the positivistic geography of the 1970s however imperfect may be able to predict the unintended consequences of intentional actions. Finally, even the most radical restructuring of our society will still need to exist in a geographic space with all the consequences that has for the inequality of access and the need for transport provision.

Acknowledgment

I am grateful to Ron Johnston for comments on a first draft of this paper.

References

Alao, N. (1970) 'A note on the solution matrix of a network', *Geographical Analysis* 2, 83–8.

Ambrose, P.J. (1977) *Access and spatial inequality*, Open University Press: Milton Keynes.

Armstrong, H.W. (1972) 'A network analysis of airport accessibility in South Hampshire', *Journal of Transport Economics and Policy*, 6, 294–307.

Beavon, K.S.O., and Hay, A.M. (1977) 'Consumer choice of shopping centre, a hypergeometric model', *Environment and Planning A*, 9, 1375–93.

Bird, J. (1969) 'Traffic flows to and from British seaports', *Geography*, 54, 284–302.

Brook, C.R.P. (1974) 'The movement of goods between regions' in C. Brook and A.M. Hay, *Regional Analysis and Development – the Macro approach – the analysis of regional change,* 7–49, Open University Press: Milton Keynes.

Brook, C.R.P. (1976) 'Measures of transport network geometry, the case of the Republic of Ireland', unpublished PhD thesis, University of Exeter.

Chisholm, M. (1970) 'Forecasting the generation of freight flows in Great Britain' in M. Chisholm, A.E. Frey and P. Haggett, *Regional Forecasting*, 431–42, Butterworth: London.

Chisholm, M., and O'Sullivan, P.M. (1973) *Freight flows and spatial aspects of the British Economy*, Cambridge University Press.

Chu, D.K. (1978) 'The growth and decline of the major British seaports, an analysis of the pattern of non-fuel cargo traffic', PhD Thesis, University of London.

Clark, C. (1966) 'Industrial location and economic potential', *Lloyds Bank Review*, 82, 1–17.

Clark, C., Wilson, F., and Bradley, J. (1969) 'Industrial location and economic potential in Western Europe', *Regional Studies*, 3, 197–212.

Cliff, A.D., Martin, R.L., and Ord, J.K. (1974) 'Evaluating the friction of distance parameters in gravity models', *Regional Studies*, 8, 281–6.

Cliff, A.D., Martin, R.L., and Ord, J.K. (1975) 'Map pattern and friction of distance parameters: reply to comments by R.J. Johnston and by L. Curry, D.A. Griffith and E.S. Sheppard', *Regional Studies*, 9, 285–8.

Cliff, A.D., Martin, R.L., and Ord, J.K. (1976) 'A reply to the final comment', *Regional Studies*, 10, 341–2.

Commission on the Third London Airport (Roskill Commission) (1970) *Papers and Proceedings*, London.

Cox, K.R. (1965) 'The application of linear programming to geographic problems', *Tijdschrift voor Economische en Sociale Geografie*, 56, 228–35.

Curry, L. (1972) 'A spatial analysis of gravity flows',

Regional Studies, 6, 131–47.
Curry, L., Griffith, D.A., and Sheppard, E.S. (1975) 'Those gravity parameters again', *Regional Studies*, 9, 289–91.
Curry, L., Griffith, D.A., and Sheppard, E.S. (1976) 'A final comment on mis-specification and autocorrelation in those gravity parameters', *Regional Studies*, 10, 337–9.
Eliot Hurst, M.E. (ed.) (1974) *Transportation Geography*, McGraw Hill: New York.
Evans, S.P. (1973) 'A relationship between the gravity model for trip distribution and the transportation problem in linear programming', *Transportation Research*, 7, 39–61.
Farrington, J.H., and Stanley, P.A. (1978) *Public Transport in Skye and Lochalish*, Highlands and Islands Development Board: Inverness.
Forbes, J. (1964) 'Mapping accessibility', *Scottish Geographical Magazine*, 80, 12–21.
Garbrecht, D. (1973) 'Describing pedestrian and car trips by transition matrices', *Traffic Quarterly*, 27, 89–109.
Garner, B.J., and Street, W.A. (1978) 'The solution matrix, alternative interpretations', *Geographical Analysis*, 10, 185–9.
Girt, J.L. (1977) 'The statistical derivation of revealed spatial preference and spatial equity functions', *Environment and Planning A*, 9, 521–8.
Gordon, I.R. (1979) 'Freight distribution model predictions compared', *Environment and Planning A*, 11, 219–21.
Haggett, P. (1965) *Locational analysis in Human Geography*, Arnold: London.
Haggett, P., and Chorley, R.J. (1969) *Network analysis in Geography*, Arnold: London.
Hardwick, P.A. (1974) 'Journey to work patterns in Salisbury, Rhodesia', *Journal of Transport Economics and Policy*, 8, 1801–191.
Harris, C.D. (1954) 'The market as a factor in the localisation of industry in the United States', *Annals of the Association of American Geographers*, 44, 315–48.
Hay, A.M. (1971) 'Connection and orientation in three West African road networks', *Regional Studies*, 5, 315–19.
Hay, A.M. (1973) *Transport for the space economy: a geographical study*, Macmillan: London.
Hay, A.M. (1975) 'On the choice of methods in the factor analysis of connectivity matrices: a comment', *Transactions, Institute of British Geographers*, 66, 163–7.
Hay, A.M. (1977) 'Transport networks' in Sarre, P., Pryce, R. and Hodgkiss, A. (eds) *Fundamentals of Human Geography, Spatial Analysis, Line Patterns*, 8–39, Open University Press: Milton Keynes.
Hay, A.M. (1979) 'The geographical explanation of commodity flow', *Progress in Human Geography*, 3, 1–12.
Heggie, I.G. (1969) 'Are gravity and interactance models a valid technique for planning regional transport facilities?', *Operational Research Quarterly*, 20, 93–110.
Helvig, M. (1964) *Chicago's external truck movements*, Research Paper No. 90, Department of Geography, University of Chicago.
Hillman, M. (with I. Henderson and A. Whalley) (1973) *Personal mobility and transport policy*, Political and Economic Planning: London.
Ingram, D.R. (1971) 'The concept of accessibility: the search for an operational form', *Regional Studies*, 5, 61–7.
James, G.A., Haggett, P., Cliff, A.D., and Ord, J.K. (1970) 'Some discrete distributions for graphs with application to regional transport networks', *Geografiska Annaler B*, 52, 14–21.
Jefferson, M. (1929) 'The civilising rails', *Economic Geography*, 4, 217–31.
Johnston, R.J. (1966) 'An index of accessibility and its use in the study of bus services and settlement patterns', *Tijdschrift voor Economische en Sociale Geografie*, 57, 33–8.
Johnston, R.J. (1973) 'On frictions of distance and regression coefficients', *Area*, 5, 187–91.
Johnston, R.J. (1975) 'Map pattern and friction of distance parameters, a comment', *Regional Studies*, 9, 281–3.
Johnston, R.J. (1976) 'On regression coefficients in comparative studies of the frictions of distance', *Tijdschrift voor Economische en Social Geografie*, 67, 15–28.
Jones, P.M. (1974) 'An alternative approach to person-trip modelling', Proceedings PTRC Conference.
Kissling, C.C. (1969) 'Linkage importance in a regional highway network', *Canadian Geographer*, 13, 113–27.
Lankford, P.M. (1974) 'Testing simulation models', *Geographical Analysis*, 6, 294–302.
Melut, P., and O'Sullivan, P.M. (1974) 'A comparison of simple lattice networks for a uniform plan', *Geographical Analysis*, 6, 163–73.
Ministry of Transport (1964) *Survey of road goods Transport 1962*, published as Statistical papers 2–6, HMSO: London.
Morley, C.D., and Thornes, J.B. (1972) 'A Markov decision model for network flows', *Geographical Analysis*, 4, 180–93.
Moseley, M.J., Harman, R.G., Coles, O.B., and Spencer, M.B. (1977) *Rural Transport and accessibility* (2 vols), Centre for East Anglian Studies, University of East Anglia.
Neft, D.S. (1959) 'Some aspects of rail commuting: London, New York, Paris', *Geographical Review*, 49, 151-63.
Nystuen, J.D., and Dacey, M.F. (1961) 'A graph theory interpretation of nodal regions', *Papers and Proceedings, Regional Science Association*, 7, 25-42.

Olsson, G. (1970) 'Explanation, prediction and meaning variance: an assessment of distance interaction models', *Economic Geography*, 46, 223-31.

O'Sullivan, P.M. (1968) 'Accessibility and the spatial structure of the economy', *Regional Studies*, 2, 195–206.

O'Sullivan, P.M. (1970) 'Variations in distance friction in Great Britain', *Area*, 2, 36-9.

O'Sullivan, P.M. (1978) 'Regions of a transport network', *Annals, Association of American Geographers*, 68, 196–204.

Perle, E.D. (1964) *The demand for transportation*, Research Paper 95, Department of Geography, University of Chicago.

Pirie, G.H. (1979) 'Measuring accessibility: a review and proposal', *Environment and Planning A*, 11, 299-312.

Pitfield, D.E. (1979) 'Freight distribution model predictions compared: some further evidence', *Environment and Planning A*, 11, 223–6.

Porter, R. (1956) 'Approach to migration through its mechanism', *Geografiska Annaler*, 38, 13–45.

Pred, A. (1967) *Behavior and location, Part 1*, Lund Studies in Geography, Series B, Human Geography, No. 27, 111-12.

Rimmer, P.J. (1978) 'Redirections in transport geography', *Progress in Human Geography*, 2, 76–100.

Robertson, I.M.L. (1976) 'Accessibility to services in the Argyll District of Strathclyde: a locational model', *Regional Studies*, 10, 89–95.

Rushton, G. (1969) 'Analysis of spatial behavior by revealed space preference', *Annals, Association of American Geographers*, 59, 391-400.

Sayer, R.A. (1977) 'Gravity models and spatial autocorrelation, or atrophy in urban and regional modelling', *Area*, 9, 183–9.

Senior, M.L. (1979) 'From gravity modelling to entropy maximising: a pedagogic guide', *Progress in Human Geography*, 3, 179–210.

Sheppard, E.S. (1979) 'Gravity parameter estimation', *Geographical Analysis*, 11, 120–32.

Smith, J.A. (1974) 'Regional inequalities in internal communications: the case of Uganda', in B.S. Hoyle (ed.), *Spatial aspects of development*, 307-22, Wiley: London.

Starkie, D.N.M. (1967) *Traffic and Industry*, Geographical Papers No. 3, London School of Economics and Political Science.

Stutz, F.P. (1973) 'Accessibility and the effect of scalar variations on the powered transportation connection matrix', *Geographical Analysis*, 5, 61–6.

Stutz, F.P. (1976) *Social aspects of interaction*, Association of American Geographers, Resource Paper No. 76, Washington, DC.

Taaffe, E.J., and Gauthier, H.L. (1973) *Geography of Transportation*. Prentice-Hall: Englewood Cliffs, NJ.

Thompson, D. (1974) 'Spatial interaction data', *Annals, Association of American Geographers*, 64, 560–75.

Timbers, J.A. (1968) 'Route factors in road networks', *Traffic Engineering and Control*, 9, 392–4, 401.

Tinkler, K.J. (1972) 'Physical interpretation of eigenfunctions of dichotomous matrices', *Transactions of the Institute of British Geographers*, 55, 17–46.

Tinkler, K.J. (1974) 'On summing power series expansions of accessibility matrices by the indirect method', *Geographical Analysis*, 6, 175–8.

Tinkler, K.J. (1975) 'On the choice of methods in the factor analysis of connectivity matrices: a reply', *Transactions of the Institute of British Geographers*, 66, 168–71.

Tinkler, K.J. (1977) *An introduction to graph theoretical methods in Geography*, Catmog No. 14, Norwich: Geo Abstracts.

Tinkler, K.J. (1979) 'Graph theory', *Progress in Human Geography*, 3, 85–116.

Turton, B.J. (1969) 'British railway traffic in 1921', *Transactions of the Institute of British Geographers*, 48, 155–71.

Volk, A.R. (1968) 'A method of estimating road mileages', *O. and M. Bulletin*, 23, 204–6.

Wallace, W.H. (1958) 'Railroad traffic densities and patterns', *Annals, Association of American Geographers*, 53, 312–31.

Wheeler, J.O. (ed.) (1973) 'Transportation geography: societal and policy perspectives', *Economic Geography*, 49, 95–184.

Williams, A.F. (1977) 'Crossroads: the new accessibility of the West Midlands' in F.E. Joyce (ed.) *Metropolitan development and change in the West Midlands: a policy review*, 367–92, University of Aston and Teakfield: Birmingham.

Wilson, A.G. (1970) 'Interregional commodity flows, entropy maximising approaches', *Geographical Analysis*, 2, 255–82.

Wilson, A.G., and Senior, M.L. (1974) 'Some relationships between entropy maximising models, mathematical programming models and their duals', *Journal of Regional Science*, 14, 207–15.

Chapter 34

Political geography

R.J. Johnston

Political geography was dubbed a 'moribund backwater' of the discipline by Brian Berry (1969), because its practitioners had failed to accommodate to the 'quantitative revolution' which had swept through many other areas of geographical endeavour during the previous decade. Needless to say, he was castigated by political geographers for this remark and was criticized for his ignorance of the field. Whilst making such a criticism, however, Prescott (1972, p. 41) admitted that 'political geographers should experiment with mathematical techniques'. Relatively little experimentation has been reported, however; that undertaken has very largely been by 'intruders', whose main expertise is in other areas of geography. Recently, there has been something of a revival in political geography but much of this, as Taylor (1977) notes, has been stimulated by workers not generally characterized as political geographers. Again, therefore, study of the research application of quantitative procedures in the 'new' political geography must focus very largely on the 'imports'.

This movement into the field of political geography by those trained, and with their major expertise, in other areas of the subject leads this chapter to focus on two major themes. The first concerns the nature of the imports – what methods were applied to political topics, and with what effect? The second concerns the influence of the applications on the development of quantitative methodology – what questions were raised about particular techniques and what novel developments were occasioned?

Since there was an absence of substantial indigenous development of quantitative work within political geography, the types of work undertaken by the 'outsiders' are best understood in the context of what was happening within quantitative geography as a whole at particular periods. Two major relevant themes are identified here (Johnston, 1979a). The first is the spatial science approach which was dominant in the 1960s. Its principal focus was the concept of spatial efficiency, which was translated into operational terms as the minimization of transport costs. Models were constructed which portrayed idealized economic and social landscapes in situations where such minimization was the overriding concern of decision-makers, and empirical research compared the real world with those ideals. Associated with this work was a normative approach to spatial planning, in which the transport-cost minimization models were presented as the goals for efficient decision-making. The second theme, which has increased in popularity during the 1970s, emphasized the distribution of power in society – who has it?; how is it exercised?; what spatial elements are involved in the wielding of power?; and what spatial patterns result from its use? For the politically-inclined geographer (though not for the political geographer!) this focuses attention on the state as the nexus of power in society, rather than on the spatial arrangements in the landscape *per se*.

Spatial science and political geography

The spatial science approach to human geography in the 1960s contained a number of separate components. Only two of these are relevant to a discussion of applications in the field of political geography.

The efficient organization of space

Much of the new work of the 1960s was marked by changes in techniques and procedures, with the substantive interests reflecting a continuity of research concerns. This was in part the case with the quantitative incursions of political geography. In Britain, for example, work had been done since the 1920s on the definition of local government areas (e.g. Fawcett, 1919; Gilbert, 1939). A major argument was that such territories should reflect the interdependence of town and country, of market centre and hinterland. Such interdependence had been identified in Dickinson's (1947) classic studies and in Green's (1950) extensive investigations of bus services. Studies of the latter in Sweden had led to a redesign of the country's administrative system; in Britain town–country interdependence was recognized belatedly by the Maud Commission (Johnston, 1979b), but its recommendations were not implemented.

One of the major stimuli to the so-called quantitative revolution – Christaller's central place theory – was partly concerned with administrative areas. He produced several models of the most efficient organization for a settlement pattern, one of which, using his administrative principle, suggested the ideal pattern of local government territories based on the concept of town–country interdependence. In this, and in his other models, each settlement was centrally located within the area that it served; the area had an hexagonal shape and the whole settlement pattern comprised a system of nesting hexagons of various sizes.

Fusion of these two approaches with the statistical orientation of a new generation of researchers led geographers to ask whether existing administrative systems conformed to Christaller's model; does society organize its territory efficiently? This required measures of shape, with which to quantify the existing patterns and compare with the ideals: the relevant statistical literature was scoured for apparently useful techniques. Some of the initial work employed relatively simple shape indices. An early example was Haggett's (1965, pp. 50–3) evaluation of a sample of administrative counties in Brazil: his first index suggested that most were elongated rather than regularly hexagonal; his second found that most shared boundaries with approximately six others, which was what the model predicted. Later work (notably by Boots, 1973, 1975; see also Getis and Boots, 1978) used probabilistic models, most of them based on the Poisson, to generate expected patterns against which the real world could be compared.

As a counterpoint to those studies which compared actual patterns with ideal landscapes was work which suggested how administrative areas should be organized, given certain assumptions about the distribution of population and settlement in an area. An early example asked what the catchment areas of schools in Grant County, Wisconsin, should be so as to minimize transport costs within the constraints of school capacities. The method chosen (Yeates, 1963) was the transportation problem in linear programming, introduced to geographers a few years previously (Garrison, 1959). Comparison of actual boundaries with those predicted to be the most efficient allowed the derivation of an index of shape efficiency for various types of administrative area (Massam, 1975). Linear programming methods have also been used to suggest efficient patterns for a wide variety of tasks related to service provision by local governments – such as the location of health care facilities (for a general review, see Hodgart, 1978) – and also for politically sensitive issues such as school integration in American cities (Hall, 1973).

One particular problem which has attracted considerable attention concerns the definition of electoral constituencies. The stimulus for this was a series of decisions by the United States Supreme Court in the 1960s which required that all Congressional and State electoral districts be equal in population (Johnston, 1979b, p. 167). Virtually the whole country had to be redistricted (at three levels in most cases) and this generated much interest in the development of computer algorithms for the task, because of their speed, their precision, and their apparent neutrality. Most of this work was done by non-geographers; some of it included work on shape indices to assess district compactness and thereby avoid the associated electoral abuse of gerrymandering. Geographers became aware of this issue (see Haggett and Chorley, 1969, pp. 228–33) and one contributed to the debate among political scientists on the efficacy of various shape measures (Taylor, 1973). This latter work stimulated a related interest in the distribution of distances between randomly-selected points in districts of varying shapes (Taylor, 1971), a topic which has been very largely ignored in the literature on spatial efficiency.

The search was for optimal spatial organization within certain constraints, and was, therefore, a topic to which some geographers have devoted considerable attention. Thus Sammons (1976, 1978), for example, has developed an algorithm to produce optimal districts and Openshaw's (1978) work has shown how almost any set of constraints can be accommodated in the design of optimal divisions of a territory (which includes constituencies). Some of the algorithms have proved difficult to implement, however, especially under severe time constraints (Morrill, 1973).

A major criticism of such work, advanced in particular by political scientists such as Dixon (1968), is that the objectivity of the computer algorithms does not avoid the political implications of the optimal solutions produced. Compact districts may be spatially efficient and thus 'better' in the view of a disinterested spatial scientist (Morrill, 1976), but their effects on representation may make them unacceptable politically (Taylor and Johnston, 1979, pp. 387–91). Electoral systems based on single-member constituencies almost invariably produce biased results – in terms of the percentage distributions of votes and of seats among parties (Johnston, 1979b, pp. 56–71). Thus all districting is potentially gerrymandering (albeit perhaps unintentionally), and recognition of this has led geographers to try to identify the range of possible electoral results that different districting solutions could produce with the same underlying spatial distribution of votes. The initial work in this area was done by Jenkins and Shepherd (1972) on the possibilities for black control of school board districts in Detroit. Their procedures were adopted and set on a sounder statistical basis by Taylor and Gudgin (1976), Gudgin and Taylor (1979), and applied in a variety of contexts (e.g. Johnston and Hughes, 1978). Development of the algorithm is proving difficult, however, because with large problems the number of feasible solutions is considerable and quasi-random procedures are currently being used by the author to identify the full range; political scientists are undertaking similar work (Engstrom and Wildgen, 1977).

Interaction and political organization

The spatial efficiency theme has been pursued by geographers with studies of movement as well as of form. The widely-used gravity model analogue has been used to model the flows of people, ideas, and commodities within the movement-minimization theme and several studies in political geography have been based on this.

A major stimulus to such work was produced by the political scientist Karl Deutsch (1953, 1964) who related the intensity of interaction within a territory to the growth of national cohesion there. This idea was taken up by Soja (1968), for example, who used interaction data to index the consolidation of the British East African territories into a functional political unit. He used transaction flow analysis (Brams, 1966), a method of indexing interaction density initiated by Deutsch; it is similar to the gravity model except that it excludes the negative constraint of distance. Wittkopf (1974) also used it in a study of international aid flows. One pioneering geographical study employed the full gravity model to investigate telephonic interaction between pairs of towns in the Canadian provinces of Quebec and Ontario; it showed the role of the provincial boundary as a constraint on communications, for the volume of traffic between two towns on opposite sides of the boundary was much less than that between comparable towns (same populations and same distance apart) both in the same province (Mackay, 1958).

Flow data matrices have been used in studies of the spatial organization of the international system. One of the pioneers of this work was Russett (1967) – whose book stimulated the comment by Berry mentioned earlier. He factor analysed matrices of trade flows (some of them transformed by techniques such as transaction flow analysis) and of pairwise voting agreement at the United Nations: among geographers, similar work has been reported by Freeman (1972) on international flows of capital, commodities, and labour. Other geographers have fitted the gravity model to the rows and columns of such matrices, incorporating additional variables to index political affinities (e.g. membership of the British Commonwealth). The largest of such studies was Johnston's (1976a) expansion of Yeates's (1969) analysis of trade patterns. The former raised questions about problems of interpreting the regression coefficient for the distance variable in the gravity model. If the model is fitted for each country separately as

$$\log (I_{ij}/O_i D_j) = \log a - b \log d_{ij}$$

where I_{ij} is the trade from country i to country j;
O_i is the total export trade of country i;
D_j is the total import trade of country j; and
d_{ij} is the distance separating countries i and j:
then it is commonly observed that the value of b is greater for countries on the edge of the system than for those at its centre. This finding is counter-intuitive. The larger the value of b, the more closely constrained is the country's trade to places nearby; but the more peripheral the country in the system, the further its goods should have to travel. Johnston (1973, 1975, 1976b) argued that this paradoxical finding reflects the underlying geography of the system; a similar point was made by Curry (1972) in terms of autocorrelation among the independent variables in the gravity model. The nature of this problem, its causes and possible solutions, has been the focus of a small debate (Sheppard, 1979).

Political geography and the study of power

The works just discussed can be summarized as 'political geography without politics', which although something of a caricature does very largely capture the apolitical nature of the spatial science approach. Political conflict underlying location-type decisions was almost entirely ignored; even when the concept of political power was introduced (as in Muir, 1975) it was in terms of simplistic indices only, such as steel production and number of missiles.

The spatial science approach was based on an implicit consensus model of society which held that everybody would agree to the validity of adopting the spatially most efficient solution to any problem. Increasingly, geographers became aware that any such solution benefited some people more than others (as was the case with constituency-boundary drawing) and that those with power in society would seek to ensure that they gained the benefits. Study of the unequal struggles over resources (including space) led to moves to replace the consensus model of society by one based on conflict, especially inter-class conflict (Eyles, 1974). Some of the earliest work in this context was led by Wolpert, who introduced the concept of political compromise to locational decision-making (Wolpert, 1970): this work has been continued by his students with the more formal analyses using game theory and learning models (Seley and Wolpert, 1974).

Much of the work undertaken in the 1970s by geographers who employ the conflict model of society has shunned the use of quantitative procedures – in part on ideological grounds. Instead, they have been concerned to write general, verbal theories of the distribution and use of power, making little reference to particular realizations of that power and their representation in the landscape; where empirical work is undertaken, detailed statistical analysis is almost always deemed unnecessary. Others argue, however, that to study the distribution of power in society is to identify only the degrees of freedom which constrain individual and group action. Their realist approach leads them to investigate those degrees of freedom and what takes place within them: two main themes using a quantitative approach have been developed.

The distribution of power and influence

Most political systems comprise many vested interest groups seeking to direct the system towards particular ends. In many cases two or more such groups are in conflict over the same resources and so are engaged in political conflict aimed at ensuring that they obtain as large a portion of the scarce resources as possible. Who wins, and how? How do the various groups interact with each other in the bargaining situation? What are the spatial components and consequences of such bargaining?

In many situations answers to questions such as these can be obtained only by detailed case studies of particular groups, even individuals, in which quantification is of little use. But in other circumstances, especially where the bargaining is overt and the final decision is made by voting, quantitative analyses do provide insights. Two approaches have been canvassed. In the first, Batty (1976) has worked on a political theory of planning, aiming to provide a procedure which will unravel the influence of different actors in certain political situations. His stimulus is a book by Coleman (1973) on *The Mathematics of Collective Action*, which is based on Markovian theory. (One of Batty's papers is entitled 'A theory of Markovian design machines' (1974a); see also Batty 1974b, 1976; Batty and Tinkler, 1978.) Matrices are established which show the relative power of different actors over certain policy areas and their degree of interest in the different areas plus the policies related to them. Manipulation of the matrices identifies which actors are likely to interact with and influence which others, which problem areas are likely to be grouped together because they attract the same interest, and so on.

The second thrust draws its stimulus from the application of game theory to political issues (see Brams, 1975). The basic problem posed is: given a committee or similar body comprising groups of voters of varying sizes none of which has a majority of votes, what are the relative bargaining strengths of the different groups? One approach to this has been adopted by Johnston (1977a, b; Johnston and Hunt, 1977) who has sought to identify the relative power of different countries within the institutions of the European Communities. His analyses suggest that power is not distributed equitably, which may have clear consequences in the policy decisions of the Communities and their spatial impact; associated work suggests that design of an equitable power structure is virtually impossible (Taylor and Johnston, 1978). More recently, Johnston (1979c) has attempted to merge his approach with Batty's, to indicate which countries will bargain most with each other. The challenges for such work are considerable in their potential for analysing certain topics in political geography, but the measurement of power and influence is difficult, as an exchange between Laver (1978) and Johnston (1978a) indicates.

The exercise of power

If some people and groups have more power and influence in a political system than do others, then one would expect them to manipulate the allocation of goods and rewards to their own ends. If in addition those people and groups can be characterized either by where they live or by the place(s) they represent, then their exercise of power and influence should be reflected in the spatial allocation of public goods and rewards.

Most geographical studies of the political factor in economic and social geography have focused on particular policies only, notably those concerned with regional economic development. Recently, however, Johnston (1979b, e) has illustrated the wide range of political impacts on spatial variations in economic and social phenomena and has argued that some of these reflect the operation of the spatial components of electoral systems (Johnston, 1980a). A major stimulus to the work has been the American phenomenon known as the pork barrel, which refers to the activities of politicians seeking benefits (notably public expenditure) for their constituencies so as to advance their re-election prospects (Mayhew, 1975; Fiorina, 1977). A theory to account for these activities has been propounded by Barry (1965), Ferejohn (1975), and others. It has been tested by Johnston (1977c, d, 1978b, c, d, 1979g, 1980b) in a series of analyses relating power within Congress to the spatial distribution of federal expenditures in the United States. Most of these analyses use multiple regression and correlation: the latter raises certain problems regarding the interpretation of equations including several dummy variables each of which contains only a small number of non-zero entries (Johnston, 1980b). A similar approach has been used to analyse the closure of teacher-training colleges in England, this time employing the hypergeometric distribution for the test statistic (Johnston, 1976c).

Conclusions

A reasonable criticism of much political geography research is that it has been politically naïve; the real stuff of politics – power and influence, conflict and compromise – has very largely been ignored. This was illustrated by many of the 'political' applications of the quantitative techniques developed during the 1950s and 1960s, when normative solutions to problems of spatial organization were proposed in apparent ignorance of the fact that however 'scientifically neutral and objective' the proposal it was almost certain that its implementation would benefit some groups and disadvantage others – conflict between those pairs of groups would probably ensure non-implementation, certainly in the original form of the proposal (as happened with British local government reform in the 1970s: Johnston, 1979b). Occasionally, the social scientist's plan may be adopted as a compromise between the desires of two conflicting groups (as in Morrill, 1973), but social science evidence is rarely unambiguously interpretable (on this, see Horowitz, 1975, and Rosen, 1972). Thus in the era of spatial science dominance in geography the political field provided situations for the application of the new-found arsenal of techniques, but these exercises provided little illumination of the realities of political geography: indeed, their main contribution was probably to an appreciation of the limitations to the techniques.

This naïve approach has been replaced recently by attempts to analyse the spatial components and impacts of political activity. For some of the adherents to this new orientation, quantitative analysis has no role to play, because each realization of a process (or theory) is unique in that it occurs at a particular spatial and temporal conjuncture whose characteristics are not repeated (Gregory, 1978). Clearly there are many cases where positivist analysis, with its emphasis on generality, is irrelevant, as in the excellent analysis by Young and Kramer (1979) of the conflict between the Greater London Council and London Borough of Bromley over housing policy. But specificity and generality are two poles of a continuum rather than alternatives, and there are many aspects of political geography for which quantitative analysis is viable. Some conjunctures are repeated, at least in the main if not completely, providing large numbers of realizations of a particular set of influences whose interacting impacts can perhaps best be unravelled numerically; can one, for example, learn about the geography of federal spending in the United States only by studying each allocation decision as a unique event? To date, quantitative analysis has not been especcially successful in this field (Johnston, 1980b), except in the negative sense of apparently falsifying the pork barrel hypothesis. (See also Archer, 1980, who found that success in the pork barrel had no influence on subsequent electoral contests and Johnston, 1979e, f, on the apparent irrelevance of campaign spending for vote accumulation in Britain.) Similarly, if politicians are able to operate within certain degrees of freedom only, then quantitative analysis has a clear role to play in charting those degrees (as in the work on constituency-building) and in outlining the optimum allocation given certain constraints (Bennett and Tan, 1979). The state is now a major actor within capitalist societies, for example (see Bennett, 1980), and there are strong arguments that it should be equitable in its use of public finance for the creation of opportunities; analysis of its activities requires both normative and positive studies and makes considerable demands on quantitatively-inclined political geographers.

During the last few decades, therefore, the few geographers who have placed the adjective political before the noun have not been attracted to the quantitatively-inclined methodology introduced to the discipline through economic, urban, and physical geography. What work has been done in political geography using quantification has largely been by 'intruders', seeking to apply their tools in new ways. Some of these applications have been naïve with regard to the political substance of the topics studied, and they have not illuminated the subject-matter of political geography; some have, however, been able to sharpen their tools with the application. More recent work is less naïve in its approach and is set in a realistic theoretical framework. The importance of study in political geography is increasingly being recognized, and the value of certain types of quantitative analysis appreciated. There is a clear need for the development of relevant techniques for the particular topics, advancing the political-quantitative dialogue which has developed in recent years and removing the earlier naïve colonialism of political geography by those best characterized by the phrase 'have technique, will travel'.

References

Archer, J.C. (1980) 'Congressional incumbent re-election success and federal outlays distribution: a test of the electoral-connection hypothesis', *Environment and Planning A*, 12, 263–77.

Barry, B. (1965) *Political Argument*, Oxford

University Press: New York.

Batty, M. (1974a) 'A theory of Markovian design machines', *Environment and Planning B*, 1, 125–46.

Batty, M. (1974b) 'Social power in plan generation', *Town Planning Review*, 45, 291–310.

Batty, M. (1976) *A Political Theory of Planning and Design*, Geographical Papers 45, Department of Geography, University of Reading.

Batty, M., and Tinkler, K.J. (1978) 'Symmetric structure in spatial and social process', unpublished paper.

Bennett, R.J. (1980) *Geography of Public Finance*, Methuen: London.

Bennett, R.J., and Tan, K.C. (1979) 'Allocation of the UK Rate Support Grant using the methods of optimal control', *Environment and Planning A*, 11, 1011–27.

Berry, B.J.L. (1969) Book review, *Geographical Review*, 59, 450–1.

Boots, B.N. (1973) 'Some models of the random subdivision of space', *Geografiska Annaler*, 55B, 34–48.

Boots, B.N. (1975) 'Patterns of urban settlements revisited', *Canadian Geographer*, 19, 107–20.

Brams, S.J. (1966) 'Transaction flows in the international system', *American Political Science Review*, 60, 880–98.

Brams, S.J. (1975) *Game Theory and Politics*, The Free Press: New York.

Coleman, J.S. (1973) *The Mathematics of Collective Action*, Heinemann: London.

Curry, L. (1972) 'A spatial analysis of gravity flows', *Regional Studies*, 6, 131–47.

Deutsch, K.W. (1953) *Nationalism and Social Communications*, MIT Press: Cambridge, Mass.

Deutsch, K.W. (1964) 'Transaction flows as indicators of political cohesion' in P.E. Jacob and J.V. Toscano (eds), *The Integration of Political Communities*, 73–97, Lippincott: New York.

Dickinson, R.E. (1947) *City, Region and Regionalism*, Routledge & Kegan Paul: London.

Dixon, R.G. (1968) *Democratic Representation: Reapportionment in Law and Practice*, Oxford University Press: New York.

Engstrom, R.L., and Wildgen, J.K. (1977) 'Pruning thorns from the thicket: an empirical test of the existence of racial gerrymandering', *Legislative Studies Quarterly*, 2, 465–79.

Eyles, J.D. (1974) 'Social theory and social geography' in C. Board *et al.* (eds), *Progress in Geography*, 6, 27–88, Edward Arnold: London.

Fawcett, C.B. (1919) *The Provinces of England*, Hutchinson: London.

Ferejohn, J.A. (1975) *Pork-Barrel Politics*, Stanford University Press, Cal.

Fiorina, M.P. (1977) *Congress; Keystone of the Washington Establishment*, Yale University Press, New Haven, Conn.

Freeman, D.B. (1972) *International Trade, Migration and Capital Flows*, Research Paper 146, Department of Geography, University of Chicago.

Garrison, W.L. (1959) 'The spatial structure of the economy', *Annals, Association of American Geographers*, 49, 471–82.

Getis, A., and Boots, B.N. (1978) *Models of Spatial Processes*, Cambridge University Press.

Gilbert, E.W. (1939) 'Practical regionalism in England and Wales', *Geographical Journal*, 94, 29–44.

Green, F.H.W. (1950) 'Urban hinterlands in England and Wales: an analysis of bus services', *Geographical Journal*, 116, 64–88.

Gregory D. (1978) *Ideology, Science and Human Geography*, Hutchinson: London.

Gudgin, G., and Taylor, P.J. (1979) *Seats, Votes and the Spatial Organisation of Elections*, Pion: London.

Haggett, P. (1965) *Locational Analysis in Human Geography*, Edward Arnold: London.

Haggett, P., and Chorley, R.J. (1969) *Network Analysis in Geography*, Edward Arnold: London.

Hall, F. (1973) *Locational Criteria for High Schools*, Research Paper 150, Department of Geography, University of Chicago.

Hodgart, R.L. (1978) 'Optimizing access to public services: a review of problems, models and methods of locating central services', *Progress in Human Geography*, 2, 17–48.

Horowitz, D.L. (1975) *The Courts and Social Policy*, Brookings Institute, Washington, DC.

Jenkins, M.A., and Shepherd, J.W. (1972) 'Decentralizing high school administration in Detroit: an evaluation of alternative strategies of political control', *Economic Geography*, 48, 95–106.

Johnston, R.J. (1973) 'On frictions of distance and regression coefficients', *Area*, 5, 187–91.

Johnston, R.J. (1975) 'Map pattern and friction of distance: a comment', *Regional Studies*, 9, 281–3.

Johnston, R.J. (1976a) *The World Trade System*, G. Bell: London.

Johnston, R.J. (1976b) 'On regression coefficients in comparative studies of the friction of distance', *Tijdschrift voor Economische en Sociale Geografie*, 67, 15–28.

Johnston, R.J. (1976c) 'Resource allocation and political campaigns: notes towards a methodology', *Policy and Politics*, 5, 181–99.

Johnston, R.J. (1977a) 'National sovereignty and national power in European institutions', *Environment and Planning A*, 9, 569–77.

Johnston, R.J. (1977b) 'National power in the European Parliament as mediated by the party system', *Environment and Planning A*, 9, 1055–66.

Johnston, R.J. (1977c) 'Environment, elections and expenditure: analyses of where governments

spend', *Regional Studies*, 11, 383–94.
Johnston, R.J. (1977d) 'The geography of federal allocations in the United States: preliminary tests of some hypotheses for political geography', *Geoforum*, 8, 319–26.
Johnston, R.J. (1978a) 'On the measurement of power: some reactions to Laver', *Environment and Planning A*, 10, 907–14.
Johnston, R.J. (1978b) 'The allocation of Federal money in the United States; aggregate analysis by correlation', *Policy and Politics*, 6, 279–97.
Johnston, R.J. (1978c) 'Political spending in the United States: analyses of political influences on the allocation of Federal money to local environments', *Environment and Planning A*, 10, 691–704.
Johnston, R.J. (1978d) 'Congressional committees and the geography of federal spending in the USA: the examples of NASA and the AEC', *Area*, 10, 272–8.
Johnston, R.J. (1979a) *Geography and Geographers*, Edward Arnold: London.
Johnston, R.J. (1979b) *Political, Electoral and Spatial Systems*, Oxford University Press.
Johnston, R.J. (1979c) 'Political geography and political power', *Munich Social Science Review*, 1 (3), 5–31.
Johnston, R.J. (1979d) 'Governmental influence in the human geography of developed countries', *Geography*, 64, 1–11.
Johnston, R.J. (1979e) 'Campaign spending and votes: a reconsideration', *Public Choice*, 33, 97–106.
Johnston, R.J. (1979f) 'Campaign expenditure and the efficacy of advertising at the 1974 general elections in England', *Political Studies*, 27, 114–19.
Johnston, R.J. (1979g) 'Congressional committees and department spending: the political influence on the geography of Federal expenditure in the United States', *Transactions, Institute of British Geographers*, NS 4, 373–84.
Johnston, R.J. (1980a) 'Political geography and electoral geography', *Australian Geographical Studies*, 18, 37–50.
Johnston, R.J. (1980b) *The Geography of Federal Spending in the United States,* Research Studies Press, Eugene, Oregon and John Wiley: London.
Johnston, R.J., and Hughes, C.A. (1978) 'Constituency delimitation and the unintentional gerrymander in Brisbane', *Australian Geographical Studies*, 16, 99–110.
Johnston, R.J., and Hunt, A.H. (1977) 'Voting power in the EEC's Council of Ministers: an essay on method in political geography', *Geoforum*, 8, 1–9.
Laver, M. (1978) 'The problems of measuring power in Europe', *Environment and Planning A*, 10, 901–6.
Mackay, J.R. (1958) 'The interactance hypothesis and boundaries in Canada', *Canadian Geographer*, 11, 1–8.
Massam, B.H. (1975) *Location and Space in Social Administration*, Edward Arnold: London.
Mayhew, D.R. (1975) *Congress: The Electoral Connection*, Yale University Press, New Haven, Conn.
Morrill, R.L. (1973) 'Ideal and reality in reapportionment', *Annals, Association of American Geographers*, 63, 463–77.
Morrill, R.L. (1976) 'Redistricting revisited', *Annals, Association of American Geographers*, 66, 548–56.
Muir, R. (1975) *Modern Political Geography*, Macmillan: London.
Openshaw, S. (1978) 'An empirical study of some zone-design criteria', *Environment and Planning A*, 10, 781–94.
Prescott, J.R.V. (1972) *Political Geography*, Methuen: London.
Rosen, P.L. (1972) *The Supreme Court and Social Science,* Illinois University Press: Urbana, Ill.
Russett, B.M. (1967) *International Regions and the International System*, Rand McNally, Chicago.
Sammons, R. (1976) *Zoning Systems for Spatial Models*, Geographical Paper 52, Department of Geography, University of Reading.
Sammons, R. (1978) 'A simplistic approach to the redistricting problem' in I. Masser and P. Brown (eds), *Spatial Representation and Spatial Interaction*, 71–94, Martinus Nijhoff: Leiden.
Seley, J.R., and Wolpert, J. (1974) 'A strategy of ambiguity in locational conflicts' in K.R. Cox, D.R. Reynolds and S.E. Rokkan (eds), *Locational Approaches in Power and Conflict*, 275–300, Halsted Press: New York.
Sheppard, E.S. (1979) 'Gravity parameter estimation', *Geographical Analysis*, 11, 120–32.
Soja, E.W. (1968) 'Communications and territorial integration in East Africa: an introduction to transaction flow analysis', *East Lakes Geographer*, 4, 39–57.
Taylor, P.J. (1971) 'Distributions within shapes: an introduction to a new family of finite frequency distributions', *Geografiska Annaler*, 53B, 40–53.
Taylor, P.J. (1973) 'A new shape measure for evaluating electoral district patterns', *American Political Science Review*, 67, 947–50.
Taylor, P.J. (1977) 'Political geography', *Progress in Human Geography*, 1, 130–5.
Taylor, P.J., and Gudgin, G. (1976) 'The statistical basis of decision-making in electoral districting', *Environment and Planning A*, 8, 43–58.
Taylor, P.J., and Johnston, R.J. (1978) 'Population distributions and political power in the European Parliament', *Regional Studies*, 12, 61–8.
Taylor, P.J., and Johnston, R.J. (1979) *Geography of Elections*, Penguin: Harmondsworth.
Wittkopf, E.R. (1974) 'The concentration and

concordance of foreign aid allocation: a transaction-flow analysis' in K.R. Cox, D.R. Reynolds and S.E. Rokkan (eds), *Locational Approaches in Power and Conflict*, 301–42, Halsted Press: New York.

Wolpert, J. (1970) 'Departures from the usual environment in locational analysis', *Annals, Association of American Geographers*, 60, 220–9.

Yeates, M.H. (1963) 'Hinterland delimitation: a distance minimizing approach', *The Professional Geographer*, 15 (6), 7–10.

Yeates, M.H. (1969) 'A note concerning the development of a geographical model of international trade', *Geographical Analysis*, 1, 399–404.

Young, K., and Kramer, J. (1979) *Strategy and Conflict in Metropolitan Housing*, Heinemann: London.

Chapter 35

Geography of elections

P.J. Taylor and G. Gudgin

It has recently been suggested that elections are 'a positivist's dream' (Taylor, 1978, p. 153). From a quantitative-geographical viewpoint this is particularly the case as thousands, and often millions, of votes are counted, classified by candidates (party) and then published usually into sets of approximately equal-population areal units (constituencies). In short we are presented at frequent intervals with large sets of data that are ready made for various forms of geographical analysis. It is not surprising therefore that electoral geography has become the exception within the field of political geography and has thrived upon the quantitative revolution (Taylor, 1978).

Quantitative research on the geography of elections

It is usual to divide up the subject-matter of electoral geography into three distinct sections (Taylor and Johnston, 1979). The first area is simply the study of the pattern of votes among sets of areal units. This 'geography of voting' is amenable to analyses using standard techniques of multivariate analysis. In fact the pioneer uses of factor analysis outside their experimental psychology origins were to investigate patterns of voting and its correlates in Chicago (Gosnell and Gill, 1935; Gosnell and Schmidt, 1936) and these examples were developed as teaching aids in the first edition of Holzinger and Harman (1941). In modern electoral geography a typical example of factor analysing voting returns along with socio-economic variables can be found in McPhail's (1971) study of a mayoral election in Los Angeles. The use of factor analysis to produce independent predictor variables for regression on voting returns is illustrated in Kirby and Taylor's (1976) study of the EEC referendum. Roberts and Rumage (1965) provide an early example of the use of various multiple regression equations and their residuals to explain English inter-urban voting differences. The most interesting studies of geographies of voting from a technical viewpoint are the work of Crewe and Payne (1976) relating ecological findings to individual survey findings and the work of Miller (1978; Miller, Raab and Britto, 1974) on data-matching and drawing inferences between different geographical scales of analysis.

The second area of study is the search for 'geographical influences in elections'. This involves several topics which are reviewed in Taylor and Johnston (1979, Part Three). Very often such studies consist of careful analysis of survey data (Cox, 1969) but they also may involve multivariate ecological analysis of voting returns (Johnston 1973, 1974) and use of residuals to indicate further geographical influences (Reynolds, 1974). Cox (1968) has used a similar strategy in a pioneer use of causal modelling in geography (see also Taylor, 1969; Biel, 1972).

The third substantive area of study involves a shift of attention away from voting *per se* to the representation that is the *raison d'être* of an election. This geography of representation has included interest in districting algorithms to produce constituencies (Cope, 1971; Taylor, 1974) and the development of equations to disaggregate bias found in election results (Taylor and Gudgin, 1976a; Johnston, 1976). In the remainder of this essay we will concentrate on two particular aspects of the geography of representation which seem to offer interesting approaches to spatial analyses beyond the narrow confines of elections. Any election consists of two interacting spatial distributions (Taylor, 1973): the pattern of party voters and the pattern of constituency boundaries. The spatial meshing of these two distributions produces the election result of winners and losers. Some recent modelling exercises for both distributions will be described below.

Constituencies as arbitrarily placed quadrats

Quadrat analysis in geography is usually associated with the placing of a square grid over a limited number of points (e.g. representing towns) on a map. In the geography of elections the grid is the pattern of constituency boundaries and the 'points' become the large numbers of party voters distributed over the constituencies. In this way plurality elections can be interpreted as large probability experiments whereby individual voter preferences are aggregated to give constituency results. Nearly a hundred years ago, these results in terms of proportion voting for a particular party were found to approximate normal distributions (Edgeworth, 1898). This provided a stimulus to modelling elections as probability processes (Kendall and Stuart, 1950; Gudgin and Taylor, 1974).

The simplest model of voting is to assume purely random decision-making on the part of voters choosing between two parties. This binomial process treats each constituency as a Bernoulli trial so that the distribution of all constituency results is an approximately normal distribution (binomial) as found in reality. Unfortunately the variance produced by this particular process

$$\text{Var}(x_i) = \frac{p(1-p)}{n} \qquad (1)$$

(where x_i is the party vote proportions for the ith constituency of fixed size n and p is the overall party vote proportion) is very small compared to reality. With constituencies of 60,000 voters (approximately British-size), the standard deviation for an even party division of the vote ($p = 0.5$) is 0.002 whereas actual elections have invariably produced results with a standard deviation of between 0.1 and 0.2. The problem therefore is to develop a plausible voting process which leads to such 'large' standard deviations.

Kendall and Stuart (1950) were the first to address this problem. One of their solutions was to employ a simple Markov process in order to add some degree of dependence between voters which is missing in the binomial model. Voters are allocated to constituencies as follows: let α be the probability that if a party A supporter has been allocated to a constituency then the next voter also supports party A. Conversely β is the probability that the next voter supports party B. Contagion in voting is assured by letting $\alpha > \beta$. The degree of contagion is given by $E = \alpha - \beta$. Now for a series of trials of n allocations to produce constituency results it is known that as n increases the distribution tends to normality and that

$$\text{Var}(x_i) = \frac{p(1-p)}{n} \cdot \frac{(1+E)}{(1-E)} \qquad (2)$$

(Uspensky, 1937, p. 225).

Given this formulation we can vary E to produce any required level of variance among constituency results. Of particular interest is the standard deviation value of 0.14 which is known to underlie the operation of the cube law (Kendall and Stuart, 1950) that seat 'odds' are the cube of vote 'odds', i.e. a 2:1 vote advantage for a party will be translated into an 8:1 seat ratio. Substituting $\text{Var}(x_i) = 0.14^2$ into equation (2) produces a value for E of 0.9996 which means $\alpha = 0.9998$. This is interpreted as voter intentions changing only every 5000 allocations on average. This extremely high level of contagion is deemed unrealistic by Kendall and Stuart. Hence we are left with a simple model which may give the correct outcome but is based upon an unsatisfactory process.

Instead of allocating voters to constituencies consider allocating clusters of voters. These may be small blocks of similar housing with households of similar voting dispositions. These clusters of voters can be given some fixed levels of support for a party. We can think of working-class clusters voting 70 per cent Labour and middle-class clusters voting 70 per cent Conservative for example. Let this level of support for one type of cluster be r for party A and $(1 - r)$ for party B with a second type of cluster where r proportion of voters support B and $(1 - r)$ support A. The simple Markov scheme can now be employed to allocate these two types of clusters to constituencies with there now being dependence between clusters instead of individual voters. In this case the variance of the frequency distribution of results is given by

$$\text{Var}(x_i) = (2r-1)^2 \quad \frac{p(1-p)}{n} \cdot \frac{(1-E)}{(1-E)} \qquad (3)$$

(Gudgin and Taylor, 1979, p. 42). The value of n now refers to the number of clusters in a constituency.

We now have a simple spatial model of a plurality election. Let us consider the three elements in this expression for variance (equation (3)). The first element $(2r - 1)^2$ relates to the cluster homogeneity of the voting mosaic upon which the constituency grid is being placed. The higher r the greater the internal homogeneity of clusters. The second element $p(1-p)/n$ relates to the overall balance between parties and the size of clusters relative to constituencies. It is therefore a relative scale component. Finally $(1+E)/(1-E)$ is a measure of spatial autocorrelation in the clusters of voters. If like-clusters are found together in large zones of one party voting then α and hence E will be relatively large whereas zero autocorrelation is found where $\alpha = E = 0.5$.

Election results as described by a variance of cube law dimensions can be produced by any number of combinations of homogeneity within clusters, relative scales of clusters/constituencies and spatial autocorrelation between clusters: for instance with internal homogeneity of $r = 0.7$, clusters of 500 voters in constituencies of 250,000 voters (USA scale) so that $n = 500$ and spatial autocorrelation of $\alpha = 0.864$. Tables showing other combinations for producing the cube law variance are given in Gudgin and Taylor (1979, pp. 44–5). Which combination of spatial components actually produces a particular frequency distribution of election results can be derived by empirical investigation of actual voting mosaics. In a study of Newcastle upon Tyne using individual survey data for the February 1974 election aggregated by housing areas, Gudgin and Taylor (1979, pp. 47–52) found that the empirically derived parameters $r = 0.7$, $n = 45$ and $\alpha = 0.94$ give a standard deviation of 0.12 which is only slightly below the level of deviation of results found in British elections.

For a more detailed discussion of the assumptions underlying this modelling and effects of relaxing those assumptions reference should be made to Gudgin and Taylor (1979).

Constituencies as combinatorial structures

The term 'gerrymander' reminds us that very often we cannot expect that constituencies are arbitrarily-placed as our previous modelling has assumed. Gerrymandering is possible in part because of the large number of different ways a state can be divided up into electoral districts or constituencies. All districting agencies whether they are neutral Boundary Commissioners or flagrant partisan gerrymanders have to choose from among the very many districting solutions possible. In statistical terms there is a population of feasible solutions from which one has to be chosen. Districting is best understood, therefore, within a comprehensive statistical framework incorporating all solutions (Pulsipher, 1973; Taylor and Gudgin, 1976b) with constituencies as combinatorial structures.

Electoral districting consists of combining some set of base units (wards, county districts, etc.) into a prescribed smaller number of constituencies. The total number of solutions to such 'region-building' problems are derived in Cliff *et al.*, (1975). In the case of constituencies there are constraints (notably equality of population (voters) and spatial contiguity) which lessen the number of acceptable solutions but still large numbers of feasible solutions can be found for quite small problems.

Algorithms for listing all possible solutions have been independently developed by Garfinkel and Taylor (1969) and Shepherd and Jenkins (1970; Jenkins and Shepherd, 1972) Both suffer size limitations, however, since as the number of base units increases the number of feasible solutions increases explosively. This can be overcome by sampling all feasible solutions using a simple region-building algorithm developed by Taylor (1973, Gudgin and Taylor 1979). These techniques are being developed in an SSRC project at Sheffield University, *Regionalization and the Geography of Constituency Boundaries*, under the direction of R.J. Johnston.

Early work on the range of results to be found among populations (or large samples) of feasible solutions has suggested a definte majority-party bias. In Sunderland, for instance, of the 87 feasible solutions only 14 allow Conservatives to share one of the town's two seats with their usual 42 per cent of the town's overall vote (Taylor and Gudgin, 1976b). Hence most solutions favour Labour, the town's majority party. This empirical finding can be more formally derived (Gudgin and Taylor, 1979).

Each combination of base units into two constituencies gives only one independent result (the first constituency) since when that is built the second constituency is merely the combination of remaining base units. In statistical terms there is just one degree of freedom so that when plotted against one another on a graph such pairs of results will define a one-dimensional pattern at 45° to the axis (Figure 35.1a). This space may be divided into different regions corresponding to election results. In the top right quarter the majority party wins over 50 per cent of the vote in both constituencies and hence this is the 2–0 region. The top left quarter includes solutions where the majority party loses the first constituency and wins the second. The bottom right quarter includes solutions where the majority party wins the first but loses the second constituency. These two quarters represent 1–1 regions. Since the majority party cannot lose both constituencies and remain a majority the bottom left region will be empty.

If we can define the distribution of results along the 45° line in Figure 35.1 we find the proportion of 2–0 victories and hence majority-party bias can be derived. Since the constituency results are (for equal size base units) the mean proportion of their base units, we can use the central limit theorem and infer that these mean values along the 45° line will tend to be normally distributed about the overall mean value. Such normality quickly converges as sample size (number of base units) increases and hence a normal distribution of constituency results may not be unrealistic in the present situation. This means that there will be far more 2–0 results than 1–1 results (Figure 35.1b). The latter only occur in the tail of the normal distribution and their numbers will depend on the variance of the base-area party vote proportions (Gudgin and Taylor, 1979).

The above derivation remains to be extended to more than two constituencies and requires further empirical testing beyond Sunderland, Newcastle upon Tyne and Iowa (Taylor and Gudgin, 1976b; Gudgin and Taylor, 1979), Brisbane (Johnston and Hughes, 1978), and London (Johnston, 1978). The

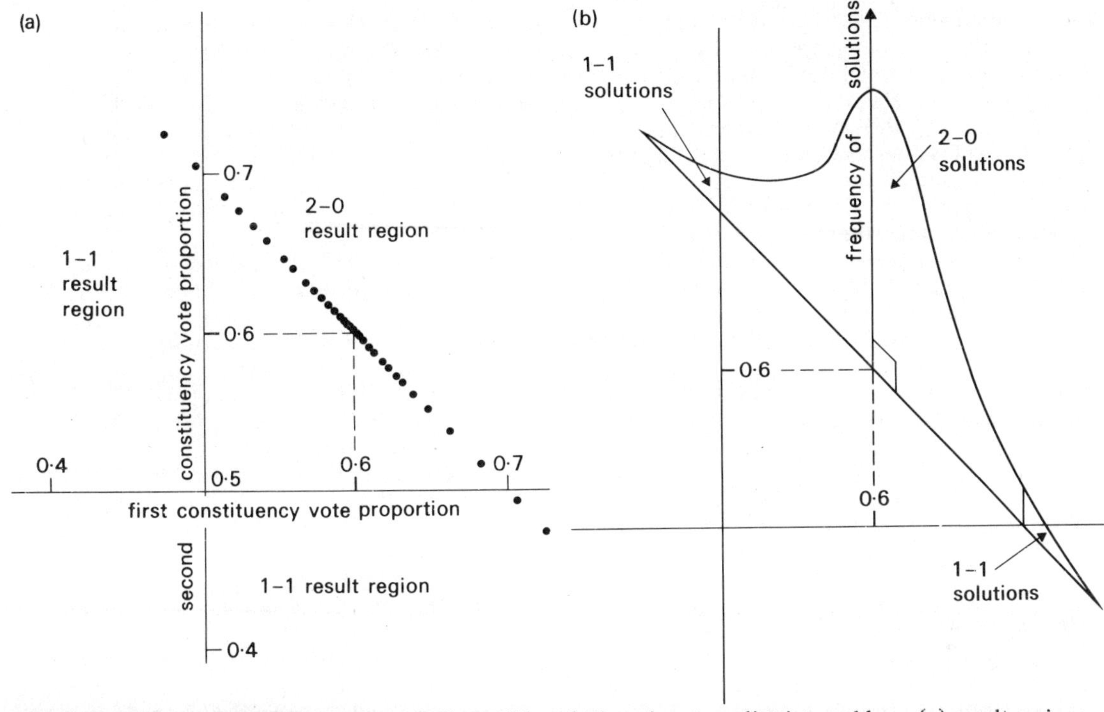

Figure 35.1 Hypothetical distribution of all possible solutions in a two-district problem: (a) result regions; (b) normal distribution of solutions

Sheffield project is working on Sheffield, Leicester, Hull and Brighton as well as Sunderland and London. This work is particularly interesting because a majority-party bias in results implies that neutral non-partisan districting agencies will most likely sample a pro-majority solution which is, of course, what gerrymanderers are normally expected to produce.

Conclusions

We have concentrated in this essay upon two examples within the geography of representation because they seem to us to have potential for being of more general utility within geographical analysis. Most of the studies briefly referred to at the beginning of this essay are of substantive interest within electoral studies but merely represent applications of well-known techniques. The modelling examples, however, do suggest that despite the surplus of data available in this field, theoretical studies can still produce insights that simple empirical analysis, however sophisticated, cannot provide. The large number of case studies of districting abuses and regression analyses of the relationship between proportions of seats and votes in an election (Tufte, 1973) do not begin to even scratch at the surface of the understanding derived from the two modelling exercises presented above (Gudgin and Taylor, 1979, Chapter 8). Elections do represent *real* applications of the modifiable areal unit problem and do consist of *real* exercises in spatial data aggregation. This is clearly a highly relevant topic where we can expect some growth in applied spatial analysis.

References

Biel, H.S. (1972) 'Suburbia and voting behaviour in the London metropolitan area: an alternative perspective', *Tijdschrift voor Economische en Sociale Geografie*, 63, 39–43.

Cliff, A.D., Haggett, P., Ord, J.K., Bassett, K., and Davies, R. (1975) *Elements of Spatial Structure*, Cambridge University Press.

Cope, C.R. (1971) 'Regionalization and the electoral districting problem', *Area*, 3, 190–5.

Cox, K.R. (1968) 'Suburbia and voting behaviour in the London metropolitan area', *Annals, Association of American Geographers*, 58, 111–27.

Cox, K.R. (1969) 'The spatial structuring of information flows and partisan attitudes' in M. Dogan and S. Rokkan (eds), *Quantitative Ecological Analysis in the Social Sciences*, 157-85, MIT Press: Cambridge, Mass.

Crewe, I., and Payne, C. (1976) 'Another game with nature: an ecological regression model of the British two-party vote ratio in 1970', *British Journal of Political Science*, 6, 43–81.

Edgeworth, F.Y. (1898) 'Miscellaneous applications of the calculus of probabilities', *Journal of the Royal Statistical Society*, 61, 534–44.

Garfinkel, R.S., and Taylor, H. (1969) 'Optimum political districting by implicit enumeration technique', *Management Science*, 16, 495–512.

Gosnell, H.F., and Gill, N.N. (1935) 'A factor analysis of the 1932 presidential vote in Chicago', *American Political Science Review*, 29, 967–84.

Gosnell, H.F., and Schmidt, M.J. (1936) 'Factorial and correlational analysis of the 1934 vote in Chicago', *American Statistical Association Journal*, 31, 507–18.

Gudgin, G., and Taylor, P.J. (1974) 'Electoral bias and the distribution of party voters', *Transactions, Institute of British Geographers*, 63, 53–73.

Gudgin, G., and Taylor, P.J. (1979) *Seats, Votes and the Spatial Organization of Elections*, Pion: London.

Holzinger, K.J., and Harman, H.H. (1941) *Factor Analysis*, Chicago University Press.

Jenkins, M.A., and Shepherd, J.W. (1972) 'Decentralizing high school administration in Detroit: an evaluation of alternative strategies of political control', *Economic Geography*, 48, 95–106.

Johnston, R.J. (1973) 'Spatial patterns and influences on voting in multi-candidate elections: the Christchurch city council election, 1968', *Urban Studies*, 10, 69–82.

Johnston, R.J. (1974) 'Local effects in voting at a local election', *Annals, Association of American Geographers*, 64, 418–29.

Johnston, R.J. (1976) 'Spatial structure, plurality systems and electoral bias', *Canadian Geographer*, 20, 310–28.

Johnston, R.J. (1978) 'Single-member European constituencies for London', *Representation* 18(72), 23–5.

Johnston, R.J., and Hughes, C.A. (1978) 'Constituency delimitation and the unintentional gerrymander in Brisbane', *Australian Geographical Studies*, 16, 99–110.

Kendall, M.G., and Stuart, A. (1950) 'The law of cubic proportions in election results', *British Journal of Sociology*, 1, 183–97.

Kirby, A.M., and Taylor, P.J. (1976) 'A geographical analysis of the voting pattern in the EEC referendum, 5 June 1975', *Regional Studies*, 10, 183–92.

McPhail, I.R. (1971) 'The vote for mayor of Los Angeles in 1969', *Annals, Association of American Geographers*, 61, 744–58.

Miller, W.L. (1978) *Electoral Dynamics in Britain since 1918*, St Martins Press: New York.

Miller, W.L., Raab, G., and Britto, K. (1974) 'Voting research and the population census 1918–71: surrogate data for constituency analyses', *Journal, Royal Statistical Society*, A 137, 384–411.

Pulsipher, A.G. (1973) 'Empirical and normative theories of apportionment', *Annals, New York Academy of Science*, 219, 234–41.

Reynolds, D.R. (1974) 'Spatial contagion in political influence processes' in K.R. Cox, D.R. Reynolds and S. Rokkan (eds), *Locational Approaches to Power and Conflict*, 203–22, Halstead Press: New York.

Roberts, M.C., and Rumage, K.W. (1965) 'The spatial variations in urban left-wing voting in England and Wales in 1951', *Annals, Association of American Geographers*, 55, 161–78.

Shepherd, J.W., and Jenkins, M.A. (1970) 'Decentralizing high school administration in Detroit: a computer evaluation of alternative strategies of political control', *Proceedings, Conference on Inter-Disciplinary Research in Computer Science*, University of Manitoba: Winnipeg.

Taylor, P.J. (1969) 'Causal models in geographic research', *Annals, Association of American Geographers*, 59, 402–4.

Taylor, P.J. (1973) 'Some implication of the spatial organization of elections', *Transactions, Institute of British Geographers*, 60, 121–36.

Taylor, P.J. (1974) 'Electoral districting algorithms and their applications', Institute of British Geographers, Quantitative Methods Study Group Working Paper Set 2.

Taylor, P.J. (1978) 'Progress Report: Political Geography', *Progress in Human Geography*, 2 (1), 153–62.

Taylor, P.J., and Gudgin, G. (1976a) 'The myth of non-partisan cartography: a study of electoral biases in the English Boundary Commission's Redistribution for 1955–1970', *Urban Studies*, 13, 13–25.

Taylor, P.J., and Gudgin, G. (1976b) 'The statistical basis of decision-making in electoral districting', *Environment and Planning*, A 8, 43–58.

Taylor, P.J., and Johnston, R.J. (1979) *Geography of Elections*, Penguin: Harmondworth.

Tufte, E.R. (1973) 'The relationship between seats and votes in two-party systems', *American Political Science Review*, 67, 540–54.

Uspensky, J.R. (1937) *Introduction to Mathematical Probability*, McGraw-Hill: New York.

Chapter 36

Quantitative geography and public policy

R.J. Bennett

Introduction

If one could write in the 1960s of a 'new' geography, it was one concerned with a revolution in technique, approach and methodology deriving from the application of quantitative methods, statistical techniques and mathematical models to the discipline. In the 1970s it has been possible to write of a new 'new' geography concerned with radicalism, relevance, or what Smith (1971, p. 153) terms a 'revolution of social responsibility'. Like the first 'quantitative revolution', this second 'revolution' has also been derived largely from North America where its inception was marked by, perhaps more than anything, the Boston meeting of the Association of American Geographers (see Prince, 1971; Smith, 1971; Berry, 1972). Since this new revolution concerned public policy at its very core, it is all the more surprising that Britain should have lagged behind. With Keltie, Patrick Geddes, Dudley Stamp and Caesar, geography in Britain had developed, from the early 1900s, a strong policy involvement and concern with the impacts of government decisions. In a very real sense the British geographers of the 1930s, 1940s, and 1950s had acted out what Berry (1972, p. 78) termed 'effective policy-relevant geography [which] involves neither the blubbering of the bleeding hearts nor the machinations of the Marxists. It involves the working with – and on – the sources of power and becoming part of society's decision-making apparatus.'

Why then did British geographical research on public policy receive such a jolt of surprise in the 1970s? The major reasons can perhaps be adduced as twofold. First, British research on public policy up to the late 1970s had been primarily concerned with economic and physical planning. Britain had implemented a distribution of industrial policy in the 1930s, reformulated as a continuing regional policy in the 1960s and institutionalized into Regional Planning Boards and Regional Economic Planning Councils (until their abolition in 1979). In addition, sweeping reforms of physical land use planning were implemented in the 1947 *Town and Country Planning Act* and subsequent legislation. No counterpart of these policies is to be found in other countries at this early date, and the influence of Geddes and Stamp can be seen in much of the thinking of this period (see Stamp, 1966). Moreover, the emphasis on economic and physical planning can be seen pervading the 1971 volume edited by Chisholm and Manners, *Spatial Policy Problems of the British Economy*, written as a tribute to Caesar. Indeed this book contains the clear statement on page 3 that the problem of 'two nations' has 'shifted from being one of class to being to an important degree a spatial problem'; and indeed was seen as the same spatial problem as the 1930s, of imbalance between the lagging and growing regions. As such, Chisholm and Manners adduce geographical concern with public policy to be one mainly of transport and land use, with the location of railways and bridges, with physical and land use planning, with the distribution of employment and the nationalized industries, with local authority administration, and with environmental impacts: an emphasis, on the one hand, on capital investment and single one-off exercises rather than continuing policy appraisal, and on the other hand, with economic as opposed to social policies. There is, for example, mention of local authority services in only one short paragraph, which seems surprising since they account for 30 per cent of expenditure on public policy, and are very significantly variable in their spatial incidence.

A second major reason for the relative neglect of wider issues of public policy in geography in Britain has been the governmental and constitutional structure. Britain is a small, centralized state, and the British constitution has attempted to suppress spatial variation. The local government structure, even prior to its reorganization in 1974, was highly consolidated; central government policies were relatively uniformly applied in different areas; and there has never been a significant degree of deliberate spatial variation in public expenditure and benefit levels. The British centralized party systems, in particular, have served to suppress rather than emphasize such variations. In the US, in contrast, the 'littleness' and fragmentation of local government, on the one hand, and the constitutional separation of powers within the federal government and between federal and state governments on the other hand, has proved to be a living laboratory of social and political conflict in which constitutional and governmental structure have been important determinants of the distribution of social benefits and economic prosperity.

The relative shock in Britain in the early 1970s, which greeted the call for relevance and radicalism from the US, should not therefore be surprising to us now. Those elements of geography in Britain up to about 1974 which would term themselves applied had neglected some of the most important aspects of policy: the social and political effects. As such, the early emphasis by Keltie (1908, p. 4) on geography as an 'applicable' discipline concerned with 'the intimate bearings of geographical conditions on collective humanity, on man in his striving after political, social and industrial development', 'bearing on human interests, history, industry and commerce' (Keltie, 1908, p. 3) had been lost in Britain to an emphasis on centralized planning of the state at aggregate level through regional policy, or to the overbearing concern with the physical structure and layout of settlements through land use planning. Although intimately affecting both social conditions and social distribution, the social and political elements in Keltie's applicable discipline had been forgotten.

A new 'applicable' quantitative geography

What, then, do we deduce from the reactions to the social challenge of the 1970s on the likely prospects for development of applied quantitative geography in the 1980s? Certainly there is no shortage of guidance in the contributions of Hägerstrand (1970), Thompson (1964), Coppock and Sewell (1976), Eyles (1974), Massam (1974), Smith (1971, 1973), Chisholm (1971), Berry (1972), Jansen (1976), as well as Chisholm and Manners (1971), which are the only major contributions. However, none of these contributions seems to offer a very satisfactory solution to the problem of specifying what a new applied geography of public policy should be, still less what the quantitative aspects of such a subdiscipline might be. We indeed have to wait in the UK for the developments of the later 1970s by Harvey (1974), D.M. Smith (1974, 1977, 1979), Johnston (1979, 1981), and Bennett (1980). These led to a fairly natural set of concerns for a new 'applicable' quantitative geography, and in this chapter such an emphasis is demarked by possession of the following attributes:

(1) construction of theories of how policy should be formulated: normative and ontological study;
(2) a direct analysis of the workings of policy as such in terms of economic, social and political impacts;
(3) assessment of the direct and indirect impacts of policies, against criteria of economic efficiency, social equity, and political effect, i.e. comparison of policy with normative goals.

As such, an applicable quantitative geography of public policy concerns analysis of the actions of the *state*, specifying its goals, measuring its performance, and assessing its social, economic and political consequences. Hence its central concern is with *public goods*: economic functions which cannot, ought not, or are not provided by private action.

Using this definition for a geography of public policy, a major component of any ensuing analysis is *necessarily* quantitative: in particular the construction of normative goals and the intricate assessment of the consequence of given policies for social distribution and economic policy. These quantitative aspects can be disaggregated into three main interrelated components, and these form the subsequent focus of this chapter. First, and most important, there is the quantification of social distribution. Any policy decision has differential impacts on different social classes and income groups. Recently Smith (1977, 1979) has thrown emphasis on one component of these impacts: geographical and social benefit distribution which he terms 'who gets what where'. However, policy may also be attacked from the revenue side of public policy: who pays how much where. The combination of the revenue and benefit terms quantifies the *fiscal incidence* of public policy: who gets what where, at what cost.

A second concern of quantitative policy assessment is with the economic impacts of policy: what are the areal consequences for growth and economic stability of a given policy decision, and what is the differential effect on the geography of price. Whilst much geographical research on regional policy has concerned the effects on economic growth, the consideration of the quantitative variation in the policy effects on the geography of price and economic stability has been largely neglected. Finally, quantitative analysis of public policy concerns the political aspects of who decides on a given policy, and what level of government is chiefly involved. Especially important is the effect of overlapping of governments and competition between governments on the level of benefits and the level of local burdens of the state.

Social distribution

The distributional effects of public policy raise the major social, political and ideological questions of who benefits, and who pays, for a given form of organization of the state. There are three main components to this aspect of who gets what where at what cost: interpersonal, intersectoral, and interregional.

(1) Interpersonal distribution: between class, income, racial, religious and client groups. Class groups have been the natural focus of radical geography, but there is a much wider range of groups in different client-need categories such as the old, the young, the sick, the unemployed, or those seeking access to particular facilities; this raises three subsidiary questions of the different forms of public goods, different categories of need, and the ability of different individuals to support the state.

(2) Intersectoral distribution: between industries, between industry and personal consumption, or between public and private consumption. This raises questions of economic efficiency, the size of the state, and a trade-off between such goals as social equity and economic efficiency.

(3) Interregional distribution: between locations, local governments, and jurisdictions. This has been a natural focus for political science and a new focus for political geography, as discussed in the previous chapter. Interregional distribution raises aggregate questions of the average, median or intergroup benefits of public policy available in different locations. This in turn leads to consideration of local public goods, and the needs and resources of different local government units, rather than of individuals directly.

Within each of these categories, the quantification of social distribution concerns the manner in which benefits and tax burdens are distributed, i.e. the extent of progression or regression of public policy with variable 'need' or 'tax ability'.

On the benefit, or expenditure, side of public policy it is possible to distinguish three broad categories of benefits:

(1) Public non-redistributive goods and benefits: available to individuals on an equal basis, e.g. police and fire protection, refuse disposal, public utilities, etc.

(2) Public redistributive goods and benefits: available to individuals or locations on the basis of measured client needs, e.g. education, hospitals, flood control, social security, industrial tax incentives, etc.

(3) Geographical goods and benefits: available to all individuals within a given jurisdiction on the basis of aggregate need, but not meaningfully divisible between individuals, e.g. urban renewal, water, sewerage, utilities, transport infrastructure, local employment policies, etc.

The first of these categories, non-redistributive goods, should be allocated on, as far as possible, a basis of equality to all, whereas the second and third categories of benefits should be allocated on the basis of need. This will require some normative construct, which can be clearly quantified, of what constitutes need, and how the extent of inequality of need can be translated into quantitatively different levels of benefits. Probably more geographical research has been directed towards assessing the resulting forms of benefit incidence than other aspects of public policy. For example, Smith (1974, 1977, 1979), *Economic Geography* (1976), Dear (1977) and Dear *et al.*, (1977) have examined benefit levels for health services. Frequently measures of aggregate benefit incidence based on regional or jurisdictional indices of 'social well-being' have been the focus for greatest attention, especially in the work of D.M. Smith. Others have emphasized the role of social class groups, or the effect on economic development (see e.g. Taylor, 1968; Davies, 1968; Weicher, 1971; Harvey, 1973; Knox, 1975; Brookfield, 1976; Wohlenberg, 1976a, 1976b; Coates, Johnston and Knox, 1977; Johnston, 1979, 1981; King and Clark, 1978).

On the revenue side of public policy the question of distribution involves measurement of differential levels of ability to pay for public goods. Various definitions of ability can be used, e.g. income, property, level of consumption, or wealth. As with benefits and needs, so with ability, a normative construct is required of what constitutes ability, and how differences in ability can be translated into quantitatively different levels of tax burdens. Very little geographical research in Britain has been concerned with the resulting distribution of equity of tax burdens in relation to the ability to pay, although Ford and Brown (1978) and Bennett (1980, 1982) are recent exceptions. In the US, work by the US Advisory Commission on Intergovernmental Relations (1962, 1971, 1978), Phares (1973), and Break (1980) provide more outstanding examples which many geographers could follow; whilst in Canada and Australia the work, respectively, of the Canadian Tax Foundation (1976) and Mathews and Jay (1972) provide other important examples by economists.

The combination of the two components of public policy, benefits and revenues, allows assessment of *final fiscal incidence*: a budget-balance measurement of who gets what, and who pays what, as a function of who they are (need or ability group) and as a function of where they live. This aggregate assessment requires quantification of a complex seven-step procedure outlined by Bennett (1980): definition of client group, assessment of expenditure need, costing of unit need provision, assessment of benefit incidence at unit levels of needs and costs, assessment of revenue ability, calculation of revenue burdens relative to ability, and final incidence assessment. Few studies have attempted such a complex task over a wide range of need, ability or geographical groups, and no large-scale study has yet emerged for Britain. Smaller-scale or more specialized studies have however been given by Short (1978) and Bennett (1982). In the US the studies by Hirsch, Segelhorst and Marcus (1964) and Greene, Neenan and Scott (1974) are two seminal studies for relatively small spatial areas (a school district in Missouri and the District of Columbia, respectively). At a more aggregate level studies by Bron, Fréville, Biehl, and Forte (in EEC, 1977), Brown (1972), and Zimmerman (1980) give regional budget incidence for various European countries. It should be hoped that other studies giving a finer mesh of ability, need or geographical groups will follow these examples.

Interrelated with the distributional assessment of the incidence of public goods are issues of the value system underlying the allocation of benefits to needs and revenue burdens to ability. Despite a flurry of recent concern with 'relevant', 'critical', and 'political' geography (e.g. Harvey, 1973; Gregory, 1978; Smith, 1977), there has been insufficient assessment of measureable, quantifiable normative goals. Smith (1977) and Lea (1979) indicate that such an issue can be resolved only by recourse to a more explicit 'welfare' theory of public policy, but this theory has been almost completely lacking in a spatial context. Such a normative welfare theory would provide an answer to Taylor's (1948, p. 137) demand that 'there can . . . be no applied geography unless there has first of all been an adequate pure geography'. Early work by Buchanan (1950, 1952) and Tiebout (1956, 1961) has posed the question of interspatial and interpersonal equity in terms of the *fiscal residuum*: that is the difference between contributions made and the value of public service provision received under public policy, should be equalized for similar individuals and should be independent of location. However, development of an explicit normative welfare theory is complicated in the geographical context by three factors, which are additional to all the complexities of the non-spatial context. First, *tapering* affects the demand and supply of public goods inducing a demand cone and price funnel for any good (public or private) and this prevents equal access to all unless they live within equal distances from the point of facility location. Attempts to incorporate this characteristic into normative theory have usually followed an algorithmic path (Scott, 1971) or have been concerned with time-space budgets, as in the work of the Swedish school (see Hägerstrand, 1970; Thrift, 1977; Pred, 1977; Carlstein *et al.*, 1978).

A second source of complexity arises from *jurisdictional partitioning* of space between different State, local, community or regional governments. Although highly desirable in matching different levels of need or senses of community identity, this creates enormous problems for achieving equity in public policy: different local governments may choose to provide a different range of public goods, tax from different revenue sources, and may allocate goods to needs and burdens to abilities under quite different criteria. Finally, *spillovers and spillins* occur between jurisdictions such that some places or people provide services to others at low or zero cost, the so-called free rider problem.

Each of these three factors creates enormous complications for a normative welfare theory of public policy in a spatial context, not only because they undermine all the conditions for the existence of a pure public good (jointness, nonexcludability, and nonrejectability; see Samuelson, 1954; Musgrave and Musgrave, 1976), but also because they set up spatial and intergroup disparities in the state. This in turn leads to resource-need disparities between people and between places which are crucial to the understanding of the fiscal component to residential segregation and of jurisdictional fiscal crises. Major steps in the direction of constructing a spatial welfare theory encompassing these complexities have been made by Teitz (1968), Sandler (1975), Sandler and Cauley (1976), *Regional Science and Urban Economics* (1976), Smith (1977), Papageorgiou (1979), Bigman and Revelle (1978, 1979), and Lea (1978, 1979), but much still requires to be undertaken. Moreover, most of this work is of American origin; the field is virtually untouched by British geographers.

Economic policy

The economic aspects of public policy concern especially three factors: the influence of public policy on price, on economic stability, and on economic growth. Each area is reviewed briefly below.

Price

The effect of public policy on price is intricately interrelated with distributional aspects: what goods cost how much as a function of where they are provided. In the economic theory of public goods (due mainly to Lindahl, 1919, Wicksell, 1898; and Ellickson, 1973), price is determined by neoclassical analysis: the price per unit of a good consumed is equal to an individual's marginal evaluation (or utility), which in turn gives the marginal rate of substitution between public goods and a so-called 'numeraire private good' representing the alternative consumption possibilities. Hence price is determined in the same way for both public and private goods; they compete with one another, and the equilibrium price solution is both Pareto- and Lindahl-optimal.

This neoclassical theory allows the effect of different revenue-raising and benefit allocation policies to be appraised on the two sides of public policy: revenue and expenditure effects.

On the revenue side, taxes act as surcharges giving a direct price effect, but of extent varying with the degree of tax shifting and tax exporting that occurs. A major problem in the geography of policy is appraising the extent of interregional and interpersonal burden shifting. The relatively few studies that have been made (Musgrave and Daicoff, 1958; Brownlee, 1960; McClure, 1967) are all by economists and all for the US. They do show, however, that a considerable proportion of tax burdens is shifted from one location to another, and from one income group to another. Shifting is especially marked for property tax on commercial activities, and severance tax on primary industry.

On the expenditure side, publicly provided benefits reduce the costs of goods to the consumer and hence increase demand, with demand often being regulated more by congestion rather than price. Whilst much attention in price analysis is centred on neoclassical comparison with the *numeraire private good*, an additional aspect of price analysis relates to geographical differences in the real costs of providing public goods. These have various components, but those of most significance are environmental factors, technical factors (such as density, shape, productivity and layout of settlements and the influence of economies and diseconomies of scale), locational variables (such as building costs, wage rates, land costs, costs of living, and distances from points of distribution), the influence of 'sunk' costs deriving from existing infrastructure and capacity, and finally the costs deriving from differences in social capital (in-place human attributes of labour training, etc.). Considerable research has been directed into each of these areas, but again the predominant body of work has been undertaken outside of geography (most notably by Stone, 1970, 1973; Newton and Sharpe, 1976; Nicholson and Topham, 1972), and considerably more attention is required in this area by geographers.

Economic stability

The effect of public policy on economic stability concerns the interrelation of the diversity of the space economy with national economic management. At a national level, both fiscal and monetary policy are employed to smooth out the economic cycle, and this economic stabilization policy is the centrepiece of Keynesian demand management. However, many geographical studies have evidenced strong differences in the structural and cyclical response of different locations during different phases of the economic cycle. The early economic studies by Vining (1945), Thompson (1965) and Brechling (1967) have been extended in both North America and Britain by King *et al.* (1972), Haggett (1971), Sant (1973), Bennett (1975), Hepple (1975), Weissbrod (1976), Bartels (1977) and Clark (1978), and demonstrate that the economic cycle leads and lags at different periods of the cycle in different regions. Although Lever (1980) has recently questioned the consistency of lead-lag relations over different cycles, it is clear that economic instability is never uniform in its geographic impacts. This in turn suggests the need for modifications of national macro-economic policy to take account of spatial differences in the constraints on demand and supply. To substantiate this view, the economists' traditional objection to such policies (as expressed for example by Hansen and Perloff, 1944), that they encourage pro-cyclical expansion of local spending (so-called fiscal perversity), can now be considered inapplicable since in modern times most public policy at local level has been markedly stabilizing rather than the reverse (see Snyder, 1973; Pommerehne, 1977).

Economic growth

Shorter-term economic stability is intricately interrelated with the longer-term economic aspects of growth and it is perhaps here that the major centre of British geographical concerns with public policy has been directed. On the one hand, the issue of regional policy has been a central concern of British geographical research on public policy (see e.g. Chisholm and Manners, 1971; Keeble, 1976; Manners, 1972). On the other hand, geographers have also directed emphasis on the issue of interregional and international development. Where the regional economist has sought recourse to neoclassical theory (see e.g. Richardson, 1969) geographers have been more willing to accept the implications of the cumulative imbalance models propounded by Myrdal (1958), Hirschman (1958) and Holland (1976): see e.g. Moseley (1974), Brookfield (1976). However, the more precise interplay of public policy with imbalances in economic growth has not often been fully developed. This requires the reconstruction of regional budgets of total incomes, and the translation of regional income into distributional attributes. Total incomes involve the local gross income or gross regional product (giving a measure of regional imbalance) and the balance of trade between regions (giving a measure of the rate of divergence of regional imbalance). The major British studies concerned with constructing interregional total income and balance of payments are those of the NIESR (Brown, 1972; Woodward, 1970) and the Northern Regional Strategy (Short, 1978). These have not been replicated to any great extent at the local level which is the one where there is a significant level of political control and policy. An important exception is the Peterborough input–output study of Morrison (1973), but the total income concept has not been pursued as steadfastly in Britain as it has in the US.

Political aspects

The political aspects of public policy draw us towards two new foci of concern: first, who decides on policy; and second, the effect of competition between local governments. The functioning of the state has become a major preoccupation of much current work (see Anderson, 1978; Taylor, 1977; Dear and Clark, 1978). However, the major focus of geographical interest should be the division of power *within* the state (see Taylor, 1979). Hence attention will be concentrated here on intergovernmental issues within the state.

No state is organized without some division of powers between different levels of government. In most countries a two- or three-level system is employed, with many other bodies overlapping each level. There are many practical, economic and social motivations for such a division of powers, but more important for the concern here, mixed levels of government have profound influences on the manner in which we must analyse public policy. In federal or mixed government structures, each government may perform differently with respect to distributional and economic effects. Each jurisdiction may treat equals equally and unequals unequally in accordance with redistribution policies seeking equity in the provision of public goods, but there is no likelihood that similar firms or people in different jurisdictions will be treated equally, nor that unequals will be treated in any similar way. Each local government, in pursuing its own programmes, will disrupt the general pattern of equity unless each local tax is equally progressive and each expenditure directed equally at the same needs in each area.

The effect of the territorial division of powers between different levels of government has led to a vigorous new focus of concern in economics and political science under the umbrella title of 'fiscal federalism', and major contributions have been made by Musgrave (1959), Musgrave and Musgrave (1976) and Oates (1972, 1977). However, despite its explicitly spatial implications, this vigorous new concern has been almost completely overlooked by geographers. There are two main foci for geographical involvement.

First, the concept of local government is the political or constitutional manifestation of a local 'sense of community': it is an expression of individualism and democratic variety. The public choice school of economics (see e.g. Tiebout, 1956; Buchanan, 1960; Buchanan and Tullock, 1969; Bish and Ostrom, 1973) suggests that the result is the formation of local service clubs, each providing a bundle of public goods in accordance with local preference patterns and needs. Whilst there has been a large body of concern in geography with the way in which local preferences and communities can be fostered, little of this has been framed in such a way as to lead to practical proposals for government organization; for example the devolution issue in the UK has been almost completely neglected by geographers. On the revenue side, practical proposals require quantitative assessment of means of interfacing different tax sources at different levels of government, having regard to the implications of tax competition and overlapping between different tax sources, and a concern to mesh multi-level taxation by credits, deductibility, and revenue-sharing. Moreover, the revenue component highlights the differences between geographical ability to pay, and hence leads to the need to develop measures of local fiscal capacity, and of means of translating local government burden and ability into individual burden and ability. On the expenditure side, multi-level government has the profound implication that it crystallizes differential need into the government structure requiring the development of intergovernmental grant programmes aimed at balancing different need and cost impacts in different locations. When combined with revenue capacity differences, variation in expenditure need is a stimulus to complex grant programmes aimed at equalizing fiscal capacity and fiscal need. The UK rate support grant is the major British programme with such a structure and this has recently attracted the attention of geographers (Bennett, 1982). Similarly housing policies, General Improvement Areas, and the Community Land Act have seen a concentration of geographical research (Harvey and Chatterjee, 1974; Bassett and Hauser, 1975; Duncan, 1974, 1977; Boddy, 1976; Bassett and Short, 1978).

A second geographical focus for concern generated by mixed levels of government is its effect on social segregation. Migration replaces choice in the market for public goods: individuals 'vote with their feet' by moving to the jurisdiction which best accords with their preference patterns (see Tiebout, 1956). Whilst this may encourage a sense of local community and introduce the market mechanism into the supply and demand for public goods, it is often linked with conscious exclusionary policies in the supply of public goods (so-called fiscal or 'snob' zoning: see Mills and Oates, 1975), and to exclusion in the housing location process. There has been a useful recent concern in geography with the activities of estate agents, building societies and public authority housing management (see e.g. Ambrose and Colenutt, 1975; Pahl, 1976; Williams, 1978). Moreover, the effect of jurisdictional competition on tax and benefit spillovers is also increasingly recognized (Smith, 1977; Bennett, 1980). But it is inevitable that research on 'snob' zoning and segregation in the school system should have developed to a much greater extent in the more diverse government system of the US, and much British work could usefully follow similar lines.

Conclusion

Whilst British geography has shown an increasing concern for 'relevance' and hence for analysis of public policy, it is only in recent years that it has really become 'applicable' to many of the more crucial social, economic and political questions. However the last five years have seen a refreshing number of thrusts into applicable quantitative geography. Central emphases of this new impetus have been a concern with access to public goods, the influence of political power on benefit distribution, and the influence of tax burdens, benefit distribution and intergovernmental behaviour on overall fiscal incidence. It is interesting that much of this work in the new school of public policy has sought study areas outside of the UK, and it would certainly seem true that policy-related research in Britain has been limited both by the form of past policies implemented and by the form of its government. Moreover, research on other countries is perhaps inevitable in the policy sphere. Since any one country gives only limited scope for experimentation in policy, it is possible to gain a wider view of possible public policy effects only by international comparisons. As a prospect for the 1980s, therefore, it can be expected that geographical research on public policy will gain increasing momentum, that international comparative studies will contribute to an increasingly important extent, and that the role of quantitative geography within that drive will become crucial in its contribution to policy impact assessment, appraisal and analysis.

References

Ambrose, P., and Colenutt, R. (1975) *The Property Machine*, Penguin: Harmondsworth.

Anderson, J. (1978) 'Geography, political economy and the state', *Antipode*, 10, 87–93.

Bartels, C.P.A. (1977) *Economic aspects of regional welfare, income distribution and unemployment*, Nijhoff: Leiden.

Bassett, K., and Hauser, D. (1975) 'Public policy and spatial structure' in R. Peel, M. Chisholm and P. Haggett (eds), *Processes in Physical and Human Geography: Bristol Essays*, Heinemann: London.

Bassett, K., and Short, J. (1978) 'Housing improvement in the inner city: a case study of changes before and after the 1974 Housing Act', *Urban Studies*, 15, 33–42.

Bennett, R.J. (1975) 'Dynamic systems modelling of the north west region', *Environment and Planning* A, 7, 525–38, 539–66, 617–36, 887–98.

Bennett, R.J. (1980) *The geography of public finance*, Methuen: London.

Bennett, R.J. (1982) *Central grants to local governments: political and economic impacts of the Rate Support Grant in England and Wales*, Methuen: London.

Berry, B.J.L. (1972) 'More of relevance and policy analysis', *Area*, 4, 77–80.

Bigman, D., and Revelle, C. (1978) 'The theory of welfare considerations in public facility location', *Geographical Analysis*, 10, 229–40.

Bigman, D., and Revelle, C. (1979) 'An operational approach to welfare considerations in applied public facility location models', *Environment and Planning*, 11, 83–95.

Bish, R.L., and Ostrom, V. (1973) *Understanding urban government*, American Enterprise Institute: Washington, DC.

Boddy, M. (1976) 'The structure of mortgage finance: Building Societies and the British social formation', *Transactions, Institute of British Geographers* NS 1, 58–71.

Break, G.F. (1980) *Intergovernmental fiscal relations in the United States*, 2nd edn, Brookings Institution: Washington DC.

Brechling, F. (1967) 'Trends and cycles in British regional unemployment', *Oxford Economic Papers*, 19, 1–21.

Brookfield, H.C. (1976) *Interdependent development*, Methuen: London.

Brown, A.J. (1972) *A framework for regional economics in the United Kingdom*, Cambridge University Press.

Brownlee, O.H. (1960) *Estimated distribution of Minnesota taxes and public expenditure benefits*, University of Minnesota Press: Minneapolis.

Buchanan, J.M. (1950) 'Federalism and fiscal equity', *American Economic Review*, 40, 583–97.

Buchanan, J.M. (1952) 'Federal grants and resource allocation', *Journal of Political Economy*, 60, 208–17.

Buchanan, J.M. (1960) *Fiscal theory and political economy*, North Carolina University Press: Chapel Hill.

Buchanan, J.M. and Tullock, G. (1969) *The Calculus of Consent*, University of Michigan Press: Ann Arbor.

Canadian Tax Foundation (1976) *The National Finance 1975–76: An Analysis of the Revenues and Expenditures of the Government of Canada*, Canadian Tax Foundation: Toronto.

Carlstein, T., Parkes, D.N., and Thrift, N. (1978) *Timing space and spacing time*, 3 vols, Arnold: London.

Chisholm, M. (1971) 'Geography and the question of relevance', *Area*, 3, 65–8.

Chisholm, M.D.I., and Manners, G. (1971) *Spatial policy problems of the British Economy*, Cambridge University Press.
Clark, G. (1978) 'The political business cycle and the distribution of regional unemployment', *Tijdschrift voor Economische en Sociale Geografie*, 69, 154–64.
Coates, B.E., Johnston, R.J., and Knox, P.L. (1977) *Geography and Inequality*, Oxford University Press.
Coppock, J.T., and Sewell, W.R.D. (eds) (1976) *Spatial dimensions of Public Policy*, Pergamon: Oxford.
Davies, B. (1968) *Social Needs and Resources in Local Services*, Michael Joseph: London.
Dear, M. (1977) 'Locational factors in the demand for mental health care', *Economic Geography*, 53, 223–40.
Dear, M., and Clark, G. (1978) 'The state and geographic process: a critical review', *Environment and Planning A*, 10, 173–83.
Dear, M., Fincher, R., and Currie, L. (1977) 'Measuring the external effects of public programs', *Environment and Planning A*, 9, 137–47.
Duncan, S.S. (1974) 'Cosmetic planning or social engineering? Improvement, Grants and Improvement Areas in Huddersfield', *Area*, 6, 259–71.
Duncan, S.S. (1977) 'The housing question and the structure of the housing market', *Journal of Social Policy*, 6, 385–412.
Economic Geography (1976) Special issue on medical geography, Vol. 52, No. 2.
EEC (1977) *Report of the Study Group on the Role of Public Finance in European Integration, 2 vols*, Commission of the European Communities, Directorate-General for Economic and Financial Affairs: Brussels.
Ellickson, B. (1973) 'A generalisation of the pure theory of public goods', *American Economic Review*, 63, 417–32.
Eyles, J.D. (1974) 'Social theory and social geography', *Progress in Geography*, 6, 27–87.
Ford, R.G., and Brown, C.J. (1978) 'Rating reform and urban structure', *Area*, 10, 8–14.
Greene, K.V., Neenan, W.B., and Scott, C.D., (1974) *Fiscal interactions in a Metropolitan Area*, D.C. Heath: Lexington, Mass.
Gregory, D.E. (1978) *Ideology, Science and Human Geography*, Hutchinson: London.
Hägerstrand, T. (1970) 'What about people in regional science?', *Papers; Regional Science Association*, 24, 7–21.
Haggett, P. (1971) 'Leads and lags in intra-regional systems: a study of cyclic fluctuations in the South West Economy' in M.D.I. Chisholm and G. Manners (eds), *op. cit.*
Hansen, A.H., and Perloff, H.S. (1944) *State and local finance in the national economy*, Norton: New York.
Harvey, D. (1973) *Social Justice in the City*, Arnold: London.
Harvey, D. (1974) 'What kind of geography for what kind of public policy?', *Transactions, Institute of British Geographers*, 63, 18–24.
Harvey, D., and Chatterjee, L. (1974) 'Absolute rent and the structuring of space by governmental and financial institutions', *Antipode*, 6, 22–36.
Hepple, L.W. (1975) 'Spectral techniques and the study of interregional economic cycles' in P. Peel, M. Chisholm and P. Haggett (eds), *Processes in Physical and Human Geography: Bristol Essays*, Heinemann: London.
Hirsch, W.Z., Segelhorst, E.W., and Marcus, M.J. (1964) *Spillover of Public Education Costs and Benefits*, UCLA Institute of Public Affairs: Los Angeles.
Hirschman, A.O. (1958) *The strategy of economic development*, Yale University Press: New Haven, Conn.
Holland, S. (1976) *Capital versus the regions*, Macmillan: London.
Jansen, A.C.M. (1976) 'On theoretical foundations of policy-oriented geography', *Tijdscrift voor Economische en Sociale Geografie*, 67, 342–51.
Johnston, R.J. (1979) *Political, Electoral and Spatial Systems*, Oxford University Press.
Johnston, R.J. (1981) *The geography of the state*, Macmillan: London.
Keeble, D. (1976) *Industrial location and planning in the United Kingdom*, Methuen: London.
Keltie, J.S. (1908) *Applied Geography: A preliminary sketch*, 2nd edn, G. Philip: London.
King, L.J., and Clark, G.L. (1978) 'Government policy and regional development', *Progress in Human Geography*, 2, 2–17.
King, L.J., Casetti, E., and Jeffrey, D. (1972) 'Cyclical fluctuations in unemployment levels in US metropolitan areas', *Tijdschrift voor Economische en Sociale Geografie*, 63, 345–52.
Knox, P.L. (1975) *Social well-being: a spatial perspective*, Oxford University Press.
Lea, A.C. (1978) 'Interjurisdictional spillovers and efficient public goods provision', presented at Association of American Geographers Conference, New Orleans.
Lea, A.C. (1979) 'Welfare theory, public goods, and public facility location', *Geographical Analysis*, 11, 217–39.
Lever, W.F. (1980) 'The operation of local labour markets in Great Britain', *Papers, Regional Science Association*, 44, 57-8.
Lindahl, E. (1919) 'Just taxation: a positive solution' in R.A. Musgrave and A.T. Peacock (eds), *Classics in the Theory of Public Finance*, Macmillan: London (1958).
McClure, C.E. (1967) 'The interstate exporting of State and local taxes: Estimates for 1962',

National Tax Journal, 20, 49–77.

Manners, G. (1972) 'National perspectives' in G. Manners, D. Keeble, B. Rodgers and K. Warren, (eds), *Regional Development in Britain*, Wiley: London.

Massam, B. (1974) 'Political geography and the provision of public services', *Progress in Geography*, 6, 179–210.

Mathews, R., and Jay, W.R.C. (1972) *Federal Finance: Intergovernmental relations in Australia since federation*, Nelson: Melborne.

Mills, E.S., and Oates, W.E. (1975) *Fiscal Zoning and Land Use Controls*, Saxon House: Farnborough.

Morrison, W.I. (1973) 'The development of an urban interindustry model', *Environment and Planning A*, 5, 369–83, 433–60, 545–54.

Moseley, M.J. (1974) *Growth Centres in Spatial Planning*, Pergamon: Oxford.

Musgrave, R.A. (1959) *The theory of public finance*, McGraw-Hill: New York.

Musgrave, R.A., and Daicoff, D.W. (1958) 'Who pays the Michigan Taxes?' in H.E. Brazer (ed.), *Michigan Tax Study: Staff Papers*, Michigan Joint Research Team: Lansing, Michigan.

Musgrave, R.A., and Musgrave, P.B. (1976) *Public Finance in Theory and Practice*, McGraw-Hill: New York.

Myrdal, G. (1958) *Economic theory and under-developed regions*, Methuen: London.

Newton, K., and Sharpe, L.J. (1976) 'Service outputs in local government: some reflections and proposals', Nuffield College: Mimeo, Oxford.

Nicholson, R.J., and Topham, N. (1972) 'Investment decisions and the size of local authorities', *Policy and Politics*, 1, 23–44.

Oates, W.E. (1972) *Fiscal Federalism*, Harcourt, Brace Jovanovich: New York.

Oates, W.E. (ed.) (1977) *The Political Economy of Fiscal Federalism*, D.C. Heath: Lexington, Mass.

Pahl, R.E. (1976) *Whose City?*, Penguin: Harmondsworth.

Papageorgiou, G.J. (1979) 'Agglomeration', *Regional Science and Urban Economics*, 9, 41–59.

Phares, D. (1973) *State-local Tax Equity*, D.C. Heath: Lexington, Mass.

Pommerehne, W.W. (1977) 'Quantitative aspects of federalism: a study of six countries' in W.E. Oates, *op. cit.*

Pred, A. (1977) 'The choreography of existence: comments on Hägerstrand's time geography and its usefulness', *Economic Geography*, 53, 207–21

Prince, H.C. (1971) 'Questions of social relevance', *Area*, 3, 150–3.

Regional Science and Urban Economics (1976) Special issue on Public Economics and Planning in Space, Vol. 6.

Richardson, H.W. (1969) *Regional Economics*, Weidenfeld and Nicolson: London.

Samuelson, P.A. (1954) 'The pure theory of public expenditure', *Review of Economics and Statistics*, 36, 387–9.

Sandler, T. (1975) 'Pareto optimality, pure public goods, impure public goods, and multiregional spillovers', *Scottish Journal of Political Economy*, 22, 25–38.

Sandler, T., and Cauley, J. (1976) 'Multiregional public goods, spillovers and the new theory of consumption', *Public Finance*, 31, 376–95.

Sant, M. (1973) *The Geography of business cycles: A case study of Economic Fluctuations in East Anglia 1951–1968*, London School of Economics, Geographical Papers No. 3.

Scott, A.J. (1971) *Combinatorial programming; spatial analysis and Planning*, Methuen: London.

Short, J. (1978) 'The regional distribution of public expenditure in Great Britain 1969/70–1973/4', *Regional Studies*, 12, 499–510.

Smith, D.M. (1971) 'Radical geography – the next revolution', *Area*, 3, 153–7.

Smith, D.M. (1973) 'Alternative "relevant" professional roles', *Area*, 5, 1–4.

Smith, D.M. (1974) 'Who gets what where and how: a welfare focus for human geography', *Geography*, 59, 289–97.

Smith, D.M. (1977) *Human Geography: a welfare approach*, Arnold: London.

Smith, D.M. (1979) *Where the grass is greener*, Croom Helm: London.

Snyder, W.W. (1973) 'Are the budgets of State and Local Governments destabilising? a six country comparison', *European Economic Review*, 4, 197–213.

Stamp, L.D. (1966) 'Ten years on', *Transactions, Institute of British Geographers*, 40, 11–20.

Stone, P.A. (1970) *Urban Development in Britain – Standards Costs and Resources 1964–2004. Vol. 1: Population Trends and Housing*, Cambridge University Press.

Stone, P.A. (1973) *The Structure, Size and Cost of Urban Settlements*, Cambridge University Press.

Taylor, C.L. (ed.) (1968) *Aggregate data analysis: political and social indicators in cross-national research*, Mouton: The Hague.

Taylor, E.G.R. (1948) 'Geography in war and peace', *Geographical Review*, 38, 132–41.

Taylor, P.J. (1977) 'Progress report: political geography', *Progress in Human Geography*, 1, 130–5.

Taylor, P.J. (1979) 'Political geography', *Progress in Human Geography*, 3, 139–42.

Teitz, M.B. (1968) 'Toward a theory of public facility location', *Papers, Regional Science Association*, 21, 35–51.

Thompson, J.H. (1964) 'What about a geography of poverty?', *Economic Geography*, 40, 283.

Thompson, W.R. (1965) *A preface to urban economics*, Johns Hopkins University Press; Baltimore

Thrift, N. (1977) 'Time and theory in human geography', 2 parts, *Progress in Human Geography*, 1, 65–101, 413–57.

Tiebout, C.M. (1956) 'A pure theory of local expenditures', *Journal of Political Economy*, 64, 416–24.

Tiebout, C.M. (1961) 'An economic theory of fiscal decentralisation' in J. Margolis (ed.), *Public Finances: Needs, sources and utilisation,* Princeton University Press, for National Bureau of Economic Research.

US ACIR (1962) *Measures of State and Local fiscal Capacity and Tax Effort*, US Advisory Commission on Intergovernmental Relations: Washington, DC.

US ACIR (1971) *Measuring the Fiscal Capacity and Effort of State and Local Areas*, US Advisory Commission on Intergovernmental Relations: Washington, DC.

US ACIR (1978) *Measuring the 'Fiscal Blood Pressure' of the States – 1964 – 1975*, US Advisory Commission on Intergovernmental Relations: Washington DC.

Vining, R. (1945) 'Regional variations in cyclical fluctuations viewed as a frequency distribution', *Econometrica*, 13, 183–213.

Weicher, J.C. (1971) 'The allocation of police protection by income class', *Urban Studies*, 8, 207–20.

Weissbrod, R. (1976) *Diffusion of relative wage inflation in Southeast Pennsylvania*, Studies in Geography No. 23, Northwestern University Press: Evanston, Ill.

Wicksell, K. (1896) 'Finanztheoretische Untersuchungen und das Steuerwesen Schwedens' in R.A. Musgrave and A.T. Peacock (eds), *Classics in the Theory of Public Finance*, Macmillan: London (1958).

Williams, P. (1978) 'Building Societies and the inner city', *Transactions, Institute of British Geographers*, NS 3, 23–34.

Wohlenberg, E.H. (1976a) 'An index of eligibility standards for welfare benefits', *Professional Geographer*, 28, 381–4.

Wohlenberg, E.H. (1976b) 'Public assistance effectiveness by States', *Annals, Association of American Geographers*, 66, 440–50.

Woodward, V.H. (1970) *Regional Social Accounts for the United Kingdom*, Cambridge University Press.

Zimmerman, H. (1980) 'The regional impact of interregional fiscal flows', *Papers, Regional Science Association*, 44, 137-48.

Part 6

Quantitative geography teaching

In this final section, we turn our attention to the teaching of quantitative geography in British universities, polytechnics and schools. Since its formation, the QMSG as a body, and many of its members as individuals, have played a central role in introducing quantitative teaching into, first, higher education curricula, and then school curricula.

As a body the QMSG has contributed to teaching in a number of ways. First, it has provided a forum for the teachers of quantitative geography in higher education. Its meetings serve a pedagogic role for these teachers, and have been the avenue by which many of the new trends in quantitative teaching, which Kirby discusses in Chapter 37, have been introduced into the subject. Second, QMSG publications include the teaching orientated series of booklets known as CATMOG (Concepts and Techniques in Modern Geography). Although several of these monographs are oriented towards a fairly high level undergraduate/graduate audience, the aim of the series has been, in part, to provide teaching material at a price students can afford. It is also noteworthy, as Kirby points out, that no other study group of the IBG possesses such a collection of pedagogic material. Moreover, it is significant to note how widely several recent text books in quantitative geography have drawn on these monographs. Third, many QMSG meetings have been concerned with topics of interest to teaching-oriented journals, and papers from these meetings have subsequently been published in these journals. For example, the *Journal of Geography in Higher Education* has recently published papers from QMSG meetings relating to exploratory data analysis (Cox and Anderson, 1978, 1980; Ehrenberg, 1979), mathematics teaching for geographers (Wilson, 1978; and also JGHE special issue 1978), and computer-assisted learning in the teaching of statistical methods (Silk, 1979); and a relatively new journal, *Teaching Statistics*, intends to publish similar articles. Fourth, as noted in Chapter 1, the QMSG has been involved in Social Science Research Council sponsored research training courses for graduate students (Hepple and Wrigley, 1979) and these have provided a useful and important training ground in the methods of the subject. Fifth, the QMSG, in a number of surveys (Robson, 1970; Unwin, 1974; Kirby, this volume), has provided the only systematic information which exists on the resources and hours devoted to the teaching of quantitative methods to geography students in higher education. In addition, individual members of the QMSG have supplemented the Group's role by their work on committees, as authors, as editors, by participating in computer program exchange schemes and as advisors.

What then has been the result of these many efforts of the QMSG in the teaching sphere? In the late 1960s and early 1970s, Robson (1970) was able to point to a dramatic increase in both the time devoted to, and number of staff involved with, training in quantitative techniques. Moreover, Unwin (1974) was able to demonstrate that geographers had become major users of computers and that British Geography departments often had a relatively high hardware investment. In a follow-up survey conducted in 1979 for this volume, however, Kirby found quantitative teaching had reached some sort of watershed. In terms of both the staff numbers involved and the contact hours with students, quantitative teaching had ceased to grow in importance and, indeed, its relative importance within the geography degree structure was probably declining a little. This change reflects the growth in the 1970s of new concerns such as those with radicalism, social relevance, and social responsibility. It may also arise in part from the critique of positivism, and the associated attack on quantitative geography which has emerged. By itself a lack of expansion or slight contraction in quantitative teaching in undergraduate geography courses is in no sense an unhealthy trend for geography. More worrying is the fact that this has been accompanied in some quarters by the emergence of an undesirable gap between quantitative geographical research and quantitative teaching, and between other areas of geographical research and quantitative teaching. As can be deduced from our comments in Chapter 1, we do not accept this division. We believe that many concerns of the 'new' geography of social relevance, with its applied and normative emphasis, require approaches which are, to a significant extent, necessarily quantitative. Hence, it is in the best interests of geography that an organic and creative link between geographical research and quantitative teaching can be re-established in those quarters in which it has lapsed. This may involve restructuring courses in quantitative geography and increasing graduate research training in quantitative methods. It certainly will involve sensitive consideration of which staff members teach such courses (for example the practice that many junior staff are obliged to teach first-year courses in statistical methods, often reluctantly, is to be deplored), and consideration is also required of the way in which quantitative methods courses relate to other courses in the undergraduate curriculum. It will also necessitate an awareness and sensitive use of the new generation of low-cost micro-computers now available.

References

Cox, N.J., and Anderson, E.W. (1978) 'Teaching geographical data analysis: problems and possible solutions', *Journal of Geography in Higher Education*, 2, 29–37.

Cox, N.J., and Anderson E.W. (1980) 'In defence of exploratory data analysis', *Journal of Geography in Higher Education* 4, 85–9.

Ehrenberg, A.S.C. (1979) 'A note of dissent on data analysis', *Journal of Geography in Higher Education* 3, 113–16.

Hepple, L.W., and Wrigley, N. (1979) 'SSRC research training course', *Area*, 11, 261–2.

Journal of Geography in Higher Education (1978) Special issue on teaching mathematics to geographers: includes papers by Gregory, Haining, Bennett.

Robson, B.T. (1970) 'The teaching of quantitative techniques', *Area*, 2, 58–9.

Silk, J.A. (1979) 'The use of classroom experiments and the computer to illustrate statistical concepts', *Journal of Geography in Higher Education*, 3, 13–25.

Unwin, D.J. (1974) 'Hardware provision for quantitative geography in the United Kingdom', *Area*, 6, 200–4.

Wilson, A.G. (1978) 'Mathematical education for geographers', *Journal of Geography in Higher Education*, 2, 17–28.

Chapter 37

The universities

A. Kirby

Introduction

Unwin has recently suggested that 'virtually all levels of geographical education now involve some quantitative work' (Unwin, 1978, p. 342). However it has been noted by Gregory that despite this ubiquity of effort, very little discussion takes place of the ultimate aims of this quantitative input: 'the limited attention given to what we teach (let alone how we teach it) would suggest that we do not believe that this matter is particularly important' (Gregory, 1976, p. 399). As one example of this relative lack of concern for quantitative teaching within higher education, it is perhaps significant that there exists no published information on the contemporary situation. No data on course content or hours taught have been collected since the QMSG survey presented at the Annual IBG Conference of 1969 (Robson, 1970); indeed an attempt to update Robson's study in 1978 received a 'miserable response' (Shepherd, 1979).

In consequence, when we try to appraise the current state of quantitative teaching in British geography departments, it is necessary to fall back upon a mixture of sources. This section assesses some of the literature recently published in and around the quantitative-education field, and compares the trends indicated within the literature with information collected by questionnaire. Four main developments are indicated within the field, namely the growth of computer-assisted learning; the development of mathematics and mathematical modelling; the introduction of exploratory data analysis, and the expression of these (and established topics) in a new generation of quantitative texts.

Computed-assisted learning

Within the teaching of statistical and technical material, the emphasis has shifted markedly from the use of small, illustrative data sets to the processing of large, realistic research problems. In consequence 'the ability to understand and use a computer is as important as the ability to understand and use quantitative methods' (Mather, 1976, vii).

Some geography departments have long traditions of encouraging students to use interactive terminals for the processing of data and the application of statistical methods: such practice was routine a decade ago at the London School of Economics for example. More recently, however, the revolution in computer technology that has made the mainframe computer potentially obsolete has fuelled computer use. Many departments can now afford to purchase mini-computers, although the description belies the storage, speed and peripheral capability of such machines.

In step with hardware development, it is normal to find the improvement and proliferation of software. In the research field, this has already occurred, with programs being disseminated by journal (e.g. *Environment and Planning*) and organizations such as the Geography Algorithm Group (Campbell, 1978). There is as yet no system which complements the GAPE program exchange for schools, although many users of GAPE operate within higher education (Shepherd, 1976). As Silk has however argued, computer-usage during quantitative methods practicals can be particularly useful. His example is placed within the context of illustrating the mechanics of sampling experiments, but his remarks could obviously be applied to a larger series of phenomena, such as fitting various trend-surfaces (Silk, 1979a).

Although functioning programs may not be readily available, several texts are now on hand to aid both lecturer and student (Baxter, 1976; Dawson and Unwin, 1976; Mather, 1976). All offer advice on programming from the standpoint of the needs of a spatial analyst, including for example material on mapping and graphical displays (for a review, see Shepherd, 1977).

Mathematics

Although there is not as yet a large body of research material that involves the use of mathematical modelling, it is clear that vigorous developments are occurring (see Part 4). In the pedagogic context, several writers have emphasized, either explicitly or implicitly, the benefits to be gained from a mathematical input to the curriculum. Wilson, for example, has stressed the development of deductive thought, expressed in mathematical notation, as a counterbalance to much empirical geography (Wilson, 1972). Haggett has pointed to the interdisciplinary advantages of a universal language, whilst Haining, in a practical vein, has reminded us that students require mathematical training in order to cope with the fruits of mathematized research (Haggett, 1975; Haining, 1978). An interesting argument is mounted by Bennett, who observes that mathematics is a desirable language that provides 'meaning without colouring' (Bennett, 1978).

Although these and other writers propose mathematical education, they generally agree that there exists a good deal of student resistance to such courses. Haining points to obvious issues (such as a general lack of numeracy), but also to the important point that whilst mathematics must be taught in an applied context, the types of situation in which it may be useful are however complex systems. He hints that interdisciplinary mathematical social science has probably more potential than a little mathematical education within geography (Haining, 1978).

A fundamentally different view is taken by Wilson, who outlines the mathematical input to the degree courses offered within the Leeds Geography Department (Wilson, 1978). His remarks are essentially optimistic, as befits a Head of Department responsible for one of only two mathematics texts for specific geographic use (Wilson and Kirkby, 1975; see also Sumner, 1978).

To conclude this section, it may be noted that the emphasis given by the above authors to mathematics does not relate to their use in the teaching of statistics. An examination of the CATMOG (Concepts and Techniques in Modern Geography) series of introductions to particular quantitative topics reveals that matrix notation *is* favoured, although it is perhaps worthy of note that not all statistics teachers favour mathematics: 'the inferential use of techniques cannot be understood unless the student grasps the statistical concepts upon which they are based . . . there seems little point to me in asking students to learn matrix algebra' (Silk, 1976, p. 56; see also Cox and Anderson, 1978).

Exploratory data analysis

A radical departure from the mainstream of statistics is the field of 'exploratory data analysis', which is normally associated with the name of Tukey, although many members of the Quantitative Methods Study Group were introduced to the basic concepts by Besag (1975). An outline of the 'method' is provided by Cox (1978), and Cox and Anderson (1978). They reiterate worries concerning the applicability of statistical analysis to spatial series, given problems of autocorrelation and non-normality, and emphasize a three-stage process of graphical analysis, numerical description and some relevant statistical testing. Such an approach emphasizes the importance of handling data and extracting as much of its information as possible, using simple but informative measures such as medians, and 'stem and leaf' displays. The pedagogic value of such an intimate involvement with data cannot be questioned, but it may be noted in passing that not all researchers are convinced of the problems of using 'classical' tests with spatial data (see Gudgin and Thornes, 1974).

Quantitative literature

Within the last four years, a flurry of texts has appeared that attempt to marry statistical teaching with geographical requirements (Ebdon, 1977; Hammond and McCullagh, 1974, 1978; Lewis, 1977; Smith, 1975). In the main, however, it cannot be said that these texts have come to terms with the issues that face geographical researchers; instead, they have essentially applied 'standard' tests and statistical manipulations to spatial data sets. In this sense, they are directly in line with a tradition dating back to Gregory (1963) and Haggett (1965).

It is only very recently that a number of books have appeared which specifically present the problems of spatial analysis. As reviews have pointed out, Taylor is, for example, the first to have dealt with the central issue of spatial autocorrelation in an undergraduate text (Taylor, 1977a). Johnston has also outlined the specific problems of applying 'the linear model' in the spatial domain (1978), whilst Silk has explored the implications of using a geographical sampling frame within the process of significance-testing (1979b).

Within this discussion, mention should again be made of the CATMOG series. Although reservations may be voiced concerning the high level of sophistication of some of the twenty-one titles that exist, it is noteworthy that no other study group of the IBG possesses such a collection of pedagogic material: for a favourable review, see Massam (1979). In a more practical context, the Quantitative Methods group has also pioneered research training for postgraduates; the first residential meeting, lasting two weeks, was held at the University of Bristol early in 1979. It is likely that as quantitative teaching becomes relatively less important within the undergraduate curriculum, such courses will become increasingly important for those entering research (Hepple and Wrigley, 1979; Hyman, Mortimer and Peake, 1979).

Questionnaire results

Introduction

As suggested above, little up-to-date information exists on the state of quantitative teaching; consequently a brief questionnaire was dispatched to all those institutions in the UK possessing a geography department. Forty-five replies were received. Details of the questionnaire and those institutions replying are available in Kirby (1979), along with a fuller discussion of the results. A précis of the findings is given below.

Quantitative teaching

In his study, undertaken in 1969, Robson noted that the mean annual number of hours given to teaching quantitative methods was 86.3 (ranging from 22 to 230); furthermore, almost one-third of staff were involved in such teaching (Robson, 1970, p. 58). Although only twenty-eight departments (all within universities) were involved in that study, it can be seen that these results are very different from those summarized in Tables 37.1–3.

Table 37.1 Statistics teaching, universities and polytechnics

Year of Course	*Modal contact hours, for each academic year*
First	20–25 hrs. (mean: 30 hrs)
Second	20–35 hrs (mean: 38 hrs)
Third*	0 hrs (mean: 8 hrs)

*or Scottish Honours year.

Statistics The 1979 study asked for contact hours broken down into the three main areas of statistics teaching, mathematics and computing. With respect to the first, a student attending a 'typical' university or polytechnic can expect to receive between 20–25 hours of statistics instruction in the first year, and 25–35 hours the following year. Several institutions offer no statistics at all at year One: few offer courses in year Three (or Honours year). (With respect to the differences between universities and polytechnics, it is of interest that whilst some of the latter offer no quantitative teaching, and draw upon other departments' expertise, the only two *integrated* quantitative courses, covering statistics, mathematics and computing, are outside the universities.)

Table 37.2 Mathematics teaching, universities and polytechnics

Year of Course	*Modal contact hours, for each academic year*
First	0 hrs (mean: 6 hrs)
Second	0 hrs (mean: 6 hrs)
Third*	0 hrs (mean: 4 hrs)

*or Scottish Honours year.

Mathematics If we continue to consider the 'typical' geography department, we find that little specialized mathematics training is usual at any level: indeed, only eighteen institutions offer any competence in this field.

Table 37.3 Computing teaching, universities and polytechnics

Year of Course	*Modal contact hours, for each academic year*
First	0 hrs (mean: 9 hrs)
Second	0 hrs (mean: 11 hrs)
Third*	0 hrs (mean 5 hrs)

*or Scottish Honours year.

Computing Similarly, an undergraduate can expect to receive very little programming instruction *vis-à-vis* the use of the computer, although thirty-three departments offer some computing taught either internally or externally.

Summary Despite the attention given to computing and mathematics within the literature, it appears from these results that few undergraduates receive any real instruction as to their applications. Whilst this does not deter many students from using programs of one sort or another in their dissertation work, it seems likely that a gap is building up between those researching in the field of mathematical modelling, and those able to fully digest the results.

In relation to statistics, the findings hint at a reduction in contact hours for most students since 1969, and most departments stated that quantitative teaching was holding steady or had declined in amount in recent years.

Quantitative teachers

Although the Quantitative Methods Study Group has always boasted a large membership, this is not reflected in the numbers of departmental staff who teach quantitative courses. Typically, two or three individuals are responsible for the range of such courses, although most departments involve demonstrators, and many draw upon the staff of computing departments. In very few departments is there the degree of involvement noted by Robson a decade ago, although Durham boasts eight staff teaching statistics, aided by six demonstrators!

Clearly, quantity does not automatically suggest quality, and neither dozens of contact hours nor large numbers of staff members guarantee qualitative success in terms of student understanding. The figures reported here may thus be interpreted in different ways. Similarly, the fact that a teacher does not carry out research, using the techniques (s)he teaches, or does not publish material of a pedagogic nature, need not indicate that such an individual is a poor quantitative teacher. None the less, a divorce between teaching and research has been identified with misgivings in other parts of the discipline (see Taylor on political geography, 1977b).

The questionnaire indicated that relatively few of the departments responding had members who utilize the results of their statistical or mathematical analyses in their teaching, whilst even fewer (18) institutions had staff who published any pedagogic material (such as for example a CATMOG).

Reading material

One obvious result of the situation with respect to publications is that most departments do not have a text of any description pertaining specifically to their courses. Many recommend a variety of sources to their students, although a great deal of overlap exists between the titles favoured. Hammond and McCullagh (1974) is a popular source, particularly at introductory level, whilst the modernity of many of the other recommended texts is interesting: Johnston (1978), Taylor (1977a) and Silk (1979b) are all well represented. CATMOGs are recommended in a sizeable number (but, it must be said, a minority) of departments.

Computer hardware

The questionnaire did not attempt to obtain information on the *consumers* of this teaching; none the less, it is interesting to note that two out of five arts graduates from Reading Geography Department in 1978 entered fields such as computer programming and systems analysis. Statistics such as these illustrate (besides the vagaries of the job market) that geography is a subject that is unusually quantitative in its approach, something borne out by the provision of computer hardware within the average department. Unwin noted five years ago that geography is well provided for, and the recent questionnaire indicates a wealth of computing power (Unwin, 1974). Thirty-two institutions reported sole use of an on-line terminal, whilst twenty-five now possess a departmental minicomputer, and several more are in the process of purchasing such a machine, or the more recent generation of micro-computer: (only three 'minis' were reported in 1974: Unwin, p. 202). Although these may have been purchased initially for research, twenty institutions also use them for teaching purposes.

Trends

As one might expect, the computer revolution is seen by most departments as being the source of most of their course changes in the foreseeable future. In answer to an open-ended question, most respondents indicated two things. First, that in terms of content, a 'steady state' had now been reached, and little was likely to be added to quantitative courses, beyond a little mathematics in some departments. Second, the possession of mini- or micro-computers was likely to involve students far more in their statistics practicals, and this will have major benefits with regard to project and dissertation work.

Conclusion

It is difficult to summarize the themes outlined here, not least because the sources of information are so open to misinterpretation. None the less, some general remarks may be made. First, it is clear that quantitative teaching has reached some sort of watershed. In terms of staff numbers and contact hours, it has ceased to grow in importance: indeed, its relative importance within the degree structure is probably falling away. Conversely, the profession as a whole is on the brink of a major change in educational practice, due to the advent of near-ubiquitous micro-computers (the same applies, to a lesser degree, to the school system). The implications of this are enormous, as practitioners have noted with respect to other disciplines (Batty,

1979). This need not, nor should not, imply bouts of programming by rote: rather it should provide an opportunity for students to get to grips with the subject, generating data, choosing and applying tests, fitting curves, all with the immediacy offered by the computer.

All this will require a good deal of commitment from teachers, and it remains to be seen just how much commitment there exists in teaching, as opposed to research, in this difficult field. Here, more than anywhere else, sympathetic presentation is vital if student attention and comprehension is to be maintained. Nor should it be forgotten that the student body itself is changing. It is already moving towards numeracy by the time it arrives within higher eduction. As Gregory has observed, 'if the recent quantitative, model-based, process-orientated development is to proceed, then we must ensure that our students can cope with it' (1976, p. 399). It is as important though for us to ask ourselves: 'can *we* cope with *them*'?

References

Batty, J.M. (1979) 'Computers and design', *The Planner.*

Baxter, R.S. (1976) *Computer and statistical techniques for planners*, Methuen: London.

Bennett, R. (1978) 'Teaching mathematics in geography degrees', *Journal of Geography in Higher Education*, 2 (1), 38–47.

Besag, J. (1975) 'On the use of exploratory data analysis in human geography', Paper presented to IBG Quantitative Methods Study Group, Applied Stochastic Processes Study Group, University of Swansea.

Campbell, W.J. (1978) 'Computer algorithms for spatial data', *Area*, 9, 106–8.

Cox, N. (1978) 'Exploratory data analysis for geographers', *Journal of Geography in Higher Education*, 2 (2), 51–4.

Cox, N., and Anderson, E. (1978) 'Teaching geographical data analysis: problems and possible solutions', *Journal of Geography in Higher Education*, 2 (2), 29–37.

Dawson, J., and Unwin, D. (1976) *Computing for geographers*, David & Charles, Newton Abbot.

Ebdon, D. (1977) *Statistics in geography: a practical approach*, Blackwell, Oxford.

Gregory, S. (1963) *Statistical methods and the geographer*, Longman, London.

Gregory, S. (1976) 'On geographical myths and statistical fables', *Transactions of the Institute of British Geographers*, NS 1 (4), 385–400.

Gudgin, G., and Thornes, J.B. (1974) 'Probability in geographic research: applications and problems', *Statistician*, 23 (3, 4), 157–78.

Haggett, P. (1965) *Locational analysis in human geography*, Arnold: London.

Haggett, P. (1975) 'Mathematics in human geography: a personal view of recent liaisons' in J.I. Clarke and P. Pinchemel (eds), *Human Geography in Britain and France*, Social Science Research Council, London.

Haining, R. (1978) 'Mathematics in the geography curriculum', *Journal of Geography in Higher Education*, 2 (1), 29–37.

Hammond, R., and McCullagh, P.S. (1974) *Quantitative techniques in geography*, Oxford University Press (2nd edn, 1978).

Hepple, L.W., and Wrigley, N. (1979) 'SSRC research training course: the organisers' view', *Area*, 11 (3), 261–2.

Hyman, H., Mortimer, J., and Peake, L. (1979) 'SSRC research training course: the view of three research students', *Area*, 11 (3), 262–3.

Johnston, R.J. (1978) *Multivariate statistical analysis in geography*, Longman, London.

Kirby, A.M., (1979) 'The state of the science: quantitative education in higher education', *Papers on Education in Geography*, 3, University of Reading.

Lewis, P. (1977) *Maps and statistics*, Methuen: London.

Massam, B. (1979) 'Dear Diary: Comments on CATMOG', *Journal of Geography in Higher Education*, 3, 54–63.

Mather, P. (1976) *Computers in Geography*, Blackwell: Oxford.

Robson, B.T. (1970) 'The teaching of quantitative techniques', *Area*, 2 (1), 58–9.

Shepherd, I.D.H. (1976) 'Bridge that gap with GAPE', *Area*, 8 (3), 173–4.

Shepherd, I.D.H. (1977) 'Computerteach: printed resources for the computer-assisted geography curriculum', *Journal of Geography in Higher Education*, 1 (2), 52–9.

Shepherd, I.D.H. (1979) Personal Communication.

Silk, J.A. (1976) 'Some problems of the teaching of statistics to undergraduates', *Reading Geographer*, 6, 50–62.

Silk, J.A. (1979a) 'The use of classroom experiments and the computer to illustrate statistical concepts', *Journal of Geography in Higher Education*, 3 (1), 13–25.

Silk, J.A. (1979b) *Statistical concepts in geography*, Allen & Unwin: London.

Smith, D.M. (1975) *Patterns in human geography*, David & Charles, Newton Abbot.

Sumner, G.N. (1978) *Mathematics for physical geographers*, Arnold: London.

Taylor, P.J. (1977a) 'Political geography', *Progress in Human Geography*, 1 (1), 130–5.

Taylor, P.J. (1977b) *Quantitative methods in*

geography, Houghton-Mifflin: Boston, Mass.
Unwin, D.J. (1974) 'Hardware provision for quantitative geography in the United Kingdom', *Area*, 6 (3), 200–4.
Unwin, D.J. (1978) 'Quantitative and theoretical geography in the United Kingdom', *Area*, 10 (5), 337–43.
Wilson, A.G. (1972) 'Theoretical geography: some speculations', *Transactions of the Institute of British Geographers*, 57, 31–44.
Wilson, A.G. (1978) 'Mathematical education for geographers', *Journal of Geography in Higher Education*, 2 (2), 17–28.
Wilson, A.G., and Kirkby, M.J. (1975) *Mathematics for geographers and planners*, Oxford University Press.

Chapter 38

The schools

M.G. Bradford

In the last fifteen years there has been a revolution in geography teaching in British secondary schools. It has not occurred everywhere at the same time or pace, or to the same extent, but there are now few schools left untouched by it. Unfortunately to many this revolution is called 'the quantitative revolution' and the changes it has brought the 'new geography'. Both have 'come down' from the universities along with graduates, supposedly, but not always enthusiastic converts. The main agents of change, however, have been a set of experienced teachers, many of them now promoted to other educational or administrative posts, who were influenced in the mid 1960s by university promulgators of a different geographical methodology (Chorley and Haggett, 1965). Their pressure and that of the universities has led to new 'A' level syllabuses which lay more emphasis on thinking than memory, on concepts than facts, and on analysis and explanation rather than description. At the same time these syllabuses, and the new 'O' level and CSE syllabuses, demand a much greater degree of numeracy than in the past. But this 'revolution' does not just reflect a handing down of ideas that characterized geographical research in the later 1950s and early 1960s in order to close the gap between universities and schools, it reflects major changes in educational thinking and organization, technological progress and demands of society which have in many ways increased the impact in schools of the methodological changes in the subject. In order to understand the revolution in geography teaching and to pinpoint its problems, it is necessary to review the methodological changes and their direct impact in the light of these other influences. The character and impact of this revolution can be best seen in the classroom, but it is most easily described and measured via the syllabuses, exams, texts, journals and resource materials for schools, and by the kind of fieldwork carried out, and sometimes submitted as part of exams.

Syllabus change began with the Oxford and Cambridge Board in 1969. Cambridge followed suit, but it was not until the mid seventies that the two largest boards, JMB and London, introduced their new syllabuses. A few boards have still not made major changes but for those which have Jones (1978) identifies four major emphases of content – 'contemporary processes in the natural environment; location and spatial organization of man's social and economic activity; a regional geography which stresses concepts and principles in preference to descriptive detail; and use of statistical techniques'. At worst these syllabuses encapsulate university geography of the early 1960s with an emphasis on spatial patterns rather than the processes producing them, the use of techniques as an end rather than a means (especially true of nearest neighbour analysis (JMB syllabus B)), and a tendency to ignore problems that do not lend themselves to a scientific approach and measurement. At best they require more analytical and conceptual thinking, more rigour, less memory and description, and more interpretation and explanation. The JMB actually specifies its weighting as 60 per cent understanding principles and concepts, 40 per cent knowledge of information.

The emphasis on concepts and principles has largely meant an importation of basic models in human geography like central place theory and von Thunen's agricultural land use model. Too often just the patterns, hexagons or rings, are remembered and the way that they are produced, given the simplifying assumptions, not understood (Bradford, 1977). So the emphasis on concepts in the syllabuses is not always carried through into the teaching. At worst these models are taught, or at least written about in examinations, either in isolation from or to the exclusion of other relevant material, so that, for **example, von Thunen's model becomes agricultural** geography. At best the models offer one approach to explaining location and land use in a given context, nineteenth century capitalist countries, and are useful teaching devices on which to build an understanding of the present world.

Examinations of these theories seem to have followed a similar path in most boards. When the theories are first introduced, questions are specifically on them, whether the question is in an essay or a data response form. This too easily permits the poorer candidates to regurgitate from memory, often incorrectly, as with the regional detail of past exams. K = 3, 4 and 7 abound. Later papers require much more application of the ideas or simply ask questions on subject matter for which the models may be just part of an appropriate answer.

The examination of statistical techniques has been by a number of forms, but it too has followed a trend (Bowler, 1978). At first most questions involved an element of calculation, usually with the interpretation of a test statistic. Increasingly questions have become more geographically significant with less trivial manipulation for the technique's sake. Very few, as yet, ask the student to choose an appropriate test for a problem. Bowler analyses the main content of papers to be 'data collection, frequency curves, measures of central tendency and dispersion, non-parametric correlation, network and nearest neighbour analysis'. With many syllabuses requiring, for example, the study of sampling, but few the standard error of the mean, there is a grave danger of cook-book learning with little understanding. Although a survey (Department of Education and Science, 1974) in 1971–2 revealed that less than a third of schools were teaching quantitative methods, the proportion is undoubtedly higher now, but the problem of the depth to which the quantitative methods should be understood is no closer to being solved.

However, there are signs that the introduction of statistical techniques into school geography is beginning to find its proper place. To many teachers, techniques were, perhaps still are, the 'new geography'. Too often their importance has been overemphasized. At worst, by some advocates, they have been confused as an end rather than a means, which has led to many trivial problems being studied in order to produce some numbers; by some opponents (Brayne, 1974) they have been viewed as numeracy replacing literacy, as the 'new' replaced the 'traditional' geography. Fortunately, it seems a more balanced view is emerging, with statistical techniques being one of a set of tools which would be used by numerate and literate young people, who have been trained to think geographically.

The greater emphasis on statistical techniques has itself given a boost to fieldwork projects, which is sometimes the way application of techniques is assessed. The methodological changes in university geography in combination with educational trends of learning by experience have been reflected in school fieldwork to change some of its character from look-see and coach tour, to setting up experiments for hypothesis testing or slightly less rigorously 'searching reality' (Daugherty, 1974). Much of this work has been of a high standard, perhaps with a bias towards hydrological and urban geography. The dangers, in addition to trivial problems, have been over-zealous data collection, with little idea of what to do with masses of data; saturation of some areas, especially new towns and coastal resorts, with questionnaire surveys; if some examples from *Teaching Geography* are anything to go by, some poor formation of hypotheses, e.g. 'that soil character has undergone direct influence by related factors'; and by totally ignoring the more traditional field techniques, the wasting of opportunities to see, at first hand, many interesting geographical features that a field area might offer. Overall, though, there have been major advances in school fieldwork recently, with the best types combining hypothesis testing and look-see approaches.

With this revitalization of fieldwork, it is perhaps surprising that new syllabuses and resource material tend to play down physical geography. The JMB 'A' level syllabus C gives it a much reduced role. Geography for the Young School Leaver does not have a physical geography package. The Oxford Geography Project (Rolfe *et al.*, 1974), a well-received and widely used set of texts pre-'O' level, has little or no physical geography.

An analysis of four journals used in schools, *Geography* (1958, 1963, 1968, 1973, 1978), *Geographical Magazine* (1968, 1973, 1978), *Classroom Geographer* (1973–4, 1977–9), and *Teaching Geography* (1975–9) bears out this relative neglect of physical geography (Table 38.1). At best the ratio is 2 to 1, human to physical, in *Teaching Geography*. In many of the others there is hardly any physical geography at all. This is perhaps surprising when at research level much of the push for measurement, so central to the 'new geography', came on the physical side. In school geography the impact of change has undoubtedly been greatest on the human side.

Beginning around 1964–5 in *Geography* and in 1969 to a much lesser extent in the *Geographical Magazine*, articles appeared which may be said to reflect the 'new geography', being theoretical, locational analysis, systems, hypotheses testing, and quantitative in orientation. But in neither journal do these articles dominate yet, with at most *Geography* in 1973 having just over a quarter; and very rarely do these articles themselves use analytical as against descriptive techniques. Even in *Classroom Geographer*, which includes by far the greatest number of 'new geography' articles, only a sixth employ techniques which would not have been used before the 'quantitative revolution'. Although so much is heard of this revolution, there are relatively

Table 38.1 Contents analysis of some school periodicals

	Classroom Geographer		*Teaching Geography*	*Geography*					*Geographical Magazine*	
	1973/4	*1977/9*	*1975/9*	*1958*	*1963*	*1968*	*1973*	*1978*	*1973*	*1978*
Articles specifically on quantitative techniques, computing	5	3	13	0	0	0	0	1	0	2
Games	10	15	5	0	0	0	1	0	0	0
Articles with 'new geography' approach/ content										
Human (a)	9	6	7	0	0	0	2	1	1	8
(b)	6	12	11	0	0	1	2	2	0	6
(c)	4	3	1	0	0	1	0	0	0	0
Physical (a)	1	2	1	0	0	1	3	2	0	6
(b)	2	5	7	0	0	0	3	1	0	5
(c)	0	2	2	0	0	0	1	0	0	0
Review articles (syllabus)	32 (5)	23 (4)	101 (30)	5 (0)	7 (1)	5 (3)	7 (1)	7 (3)	5 (0)	5 (2)
Other more traditional and miscellaneous articles.	15	6	75	29	40	33	27	26	100	113

(a) no techniques used
(b) only simple descriptive techniques
(c) analytical techniques

few articles specifically on quantitative techniques. In these journals at least, the revolution in school geography is shown to be much more than just a 'quantitative revolution'. From these journals, of at least equal or greater importance has been the contribution of games, most of them being non-quantitative, decision-making exercises. This trend reflects the late 1960s focus on decision-making and is an addition to, rather than a replacement of, existing teaching methods. The revolution then in schools has been one of ideas rather than just techniques.

Its impact has been increased by the movement of educational thinking away from chalk-talk to child-centred learning. Educational reorganization and mixed ability teaching has further necessitated such a change of emphasis. The 'new geography' has lent itself to worksheets, games and practicals, which permit Piaget's ideas of self-discovery and experienced centred learning. It has fitted well into Bruner's (1961) ideas of an accumulative rather than additive learning of concepts rather than facts. Its principles and even its techniques are best learned in a spiral curriculum where concepts are returned to, refined and extended during a child's education. Thus for those already introduced to the ideas of friction of distance, etc., in the lower school, the new 'A' level syllabuses do not come as an abrupt change as they have to those trained on a traditional 'O' level.

Changes in assessment techniques have also aided the adoption of 'new geography'. It lends itself much more than 'traditional geography' to questions requiring data handling, interpretation, response and analysis. The introduction of these types of questions, to complement essay questions, highlights the problem of teaching, not just assessing geography at this level: to obtain data on significant problems that may reflect complex situations without simplifying them so much as to make them meaningless. Increased demands for internal assessment has given more weight to assessed fieldwork, where much innovation has and can be introduced. However, along with society's desires for 'relevance', educationists' ideas of self-discovery, and the need for accessibility to data, this has often led to an overemphasis on the local area, to the neglect of the study of foreign countries. More recent trends in geographical research with a structuralist approach and a political economy bias would perhaps suggest that this is an unfortunate direction to follow at school level.

In addition to these trends in educational thinking and organization which have reinforced the impact of 'new geography', there has been the effect of new technology, most notably the computer. This has been used so that real data can be stored and analysed (Computer Assisted Learning in Upper School Geography, Robinson, 1979). More directly the Geographical Association Package Exchange (GAPE), run from Loughborough University, offers about twenty programs, already used by over 400 teachers. These programs permit time-saving by doing the bookkeeping during games, perform simulations showing the effect on output variables of changed input values and make simple the task of standard statistical routines. Although not yet widely used, computers can relieve the boredom of excessive data manipulation, can allow real problems with large data sets to be analysed, and particularly when used interactively can facilitate the comprehension of difficult concepts and interdependencies via simulation. Here, as in the past with its committee on Models and Quantitative Techniques (Chorley, 1969) the Geographical Association via a working group is playing its part in overseeing this innovation. Hopefully teachers will resist the danger of computers, that is the fascination and frustration that so many research students feel with them, and use the machines that PTAs may be purchasing in the future with an appropriate perspective.

All these changes have produced a somewhat different 'A' level geographer. Using figures for the JMB, which previous investigations have shown to reflect national trends and which accounts for one quarter to one third of those entering 'A' level, an analysis was made covering the period 1974 (10,053 candidates) to 1978 (10,461 candidates), during which the new syllabus B was introduced. The 'A' level geographer is now more likely to be male (56.6 to 57.6 per cent), more likely to take Maths (19.3 to 19.6 per cent), much less likely to have taken history (31.5 to 26.5 per cent) and English (35 to 31.3 per cent); and much more likely to have taken Economics (20.6 to 23.4 per cent); in short, less an artist and more a social scientist.

Geography in schools has changed remarkably in the last decade or so. It has been one of the most dynamic departments which has undoubtedly increased its status at a time when integrated courses, like Humanities and Environmental Studies, have threatened its base in the lower school. But it must not stop changing just because it is called the 'new geography', and because so many teachers have devoted so much energy to change already. The subject has to change to reflect new ideas and a new world. It cannot afford to become a petrified forest of 1960s geography with an overemphasis on pattern, local rather than foreign, technique as end rather than means, and a false objectivity that ignores underlying values. It must add divergent and creative thought to the convergent thinking that hypothesis testing tends to produce. Finally, in a continued use of theories, models and quantitative techniques, it should ensure their appropriateness, geographically, by the applicability of their underlying values and assumptions, and educationally in the time taken learning them, in their contribution to concept formation and not least in their capacity to motivate pupils to think geographically.

Acknowledgment

The author would like to thank Frances Hewitt for her assistance in the preparation of this chapter.

References

Bowler, I. (1978) 'Quantitative methods in the 'A' level geography syllabuses', *Teaching Geography*, 3 (3), 113–15.

Bradford, M.G. (1977) 'Some problems of varying overlap between secondary and higher education', *Journal of Geography in Higher Education*, 1 (1), 80–6.

Brayne, M. (1974) 'Neophilia, literacy and advanced level geography', *Classroom Geographer*, February, 23–6.

Bruner, J.S. (1961) *The process of education*, Harvard University Press, Cambridge, Mass.

Chorley, R.J., and Haggett, P. (eds) (1965) *Frontiers in Geographical Teaching*, Methuen: London.

Chorley, R.J. (1969) 'The standing committee on the role of models and quantitative techniques in geography teaching', *Geography*, 52 (2), 1–10.

Daugherty, R. (1974) 'Data Collection', Science in Geography No. 2, Oxford University Press.

Department of Education and Science (1974) *School Geography in the Changing Curriculum*, Education Survey 19, HMSO: London.

Jones, P. (1978) 'The new geography at 'A' level: a look at question papers and syllabuses', *Teaching Geography*, 3 (4), 161–3.

Robinson, R., and Boardman, D. (1979) 'How data can help in the sixth form', *Teaching Geography*, 4 (3), 117–21.

Rolfe, J., Dearden, R., Kent, A., Grenyer, N., and Rowe, C. (1974) *Oxford Geography Project*, Oxford University Press.

Contributors

M.G. Anderson, Department of Geography, University of Bristol, University Road, Bristol BS8 1SS.

M. Batty, Department of Town Planning, University of Wales, Institute of Science and Technology, King Edward VII Avenue, Cardiff CF1 3NU.

R.J. Bennett, Department of Geography, University of Cambridge, Downing Place, Cambridge CB2 3EN.

M.G. Bradford, Department of Geography, University of Manchester, Manchester M13 9PL.

G.P. Chapman, Department of Geography, University of Cambridge, Downing Place, Cambridge CB2 3EN.

R.J. Chorley, Department of Geography, University of Cambridge, Downing Place, Cambridge CB2 3EN.

M. Clarke, Department of Geography, University of Leeds, Leeds LS2 9JT.

A.D. Cliff, Department of Geography, University of Cambridge, Downing Place, Cambridge CB2 3EN.

N.J. Cox, Department of Geography, University of Durham, Science Laboratories, South Road, Durham DH1 3LE.

E. Culling, Department of Geography, London School of Economics, Houghton Street, London WC2 2AE.

I.S. Evans, Department of Geography, University of Durham, Science Laboratories, South Road, Durham DH1 3LE.

R.I. Ferguson, Department of Earth and Environmental Science, University of Stirling FK9 4LA.

A.C. Gatrell, Department of Geography, University of Salford, Salford M5 4WT.

G. Gudgin, Department of Applied Economics, University of Cambridge, Sidgewick Avenue, Cambridge.

P. Haggett, Department of Geography, University of Bristol, University Road, Bristol BS8 1SS.

R.P. Haining, Department of Geography, University of Sheffield, Sheffield S10 2TN.

R. Harris, Department of Geography, University of Durham, Science Laboratories, South Road, Durham DH1 3LE.

A. Hay, Department of Geography, University of Sheffield, Sheffield S10 2TN.

L.W. Hepple, Department of Geography, University of Bristol, University Road, Bristol BS8 1SS.

R.J. Johnston, Department of Geography, University of Sheffield, Sheffield S10, 2TN.

K. Jones, Department of Geography, University of Reading, Reading RG6 2AH.

P. Keys, Department of Geography, University of Hull.

A. Kirby, Department of Geography, University of Reading, Whiteknights, Reading RG6 2AH.

R.L. Martin, Department of Geography, University of Cambridge, Downing Place, Cambridge CB2 3EN.

P.M. Mather, Department of Geography, University of Nottingham, University Park, Nottingham NG7 2RD.

J.A. Matthews, Geography Section, Department of Geology, University College, P.O. Box 78, Cardiff CF1 1XL.

S. Openshaw, Department of Town and Country Planning, University of Newcastle upon Tyne, Newcastle upon Tyne NE1 7RU.

J.K. Ord, Department of Statistics, University of Warwick, Coventry, Warwickshire CV4 7AL.

P. Rees, Department of Geography, University of Leeds, Leeds LS2 9JT.

D.W. Rhind, Department of Geography, University of Durham, Science Laboratories, South Road, Durham DH1 3RL.

K.S. Richards, Department of Geography, University of Hull, Hull HU6 7RX.

M.L. Senior, Department of Geography, University of Salford, Salford M5 4WT.

J. Silk, Department of Geography, University of Reading, Whiteknights, Reading RG6 2AH.

N.A. Spence, Department of Geography, London School of Economics, Houghton Street, London WC2A 2AE.

P.J. Taylor, Department of Geography, University of Newcastle upon Tyne, Newcastle upon Tyne NE1 7RU.

R.W. Thomas, Department of Geography, University of Manchester, Manchester M13 9PL.

J.E. Thornes, Department of Geography, London School of Economics, Houghton Street, London WC2A 2AE.

N. Thrift, Department of Human Geography, School of Pacific Studies, The Australian National University, P.O. Box 4, Canberra ACT 2600, Australia.

D.J. Unwin, Department of Geography, University of Leicester, Leicester LE1 7RH.

H.C.W.L. Williams, Department of Geography, University of Leeds, Leeds LS2 9JT.

A.G. Wilson, Department of Geography, University of Leeds, Leeds LS2 9JT.

N. Wrigley, Department of Geography, University of Bristol, University Road, Bristol BS8 1SS.

Index